Lecture Notes in Networks and Systems

Volume 133

The series "Lecture Notes in Networks and Systems" publishes the latest developments in Networks and Systems—quickly, informally and with high quality. Original research reported in proceedings and post-proceedings represents the core of LNNS.

Volumes published in LNNS embrace all aspects and subfields of, as well as new challenges in, Networks and Systems.

The series contains proceedings and edited volumes in systems and networks, spanning the areas of Cyber-Physical Systems, Autonomous Systems, Sensor Networks, Control Systems, Energy Systems, Automotive Systems, Biological Systems, Vehicular Networking and Connected Vehicles, Aerospace Systems, Automation, Manufacturing, Smart Grids, Nonlinear Systems, Power Systems, Robotics, Social Systems, Economic Systems and other. Of particular value to both the contributors and the readership are the short publication timeframe and the world-wide distribution and exposure which enable both a wide and rapid dissemination of research output.

The series covers the theory, applications, and perspectives on the state of the art and future developments relevant to systems and networks, decision making, control, complex processes and related areas, as embedded in the fields of interdisciplinary and applied sciences, engineering, computer science, physics, economics, social, and life sciences, as well as the paradigms and methodologies behind them.

**** Indexing: The books of this series are submitted to ISI Proceedings, SCOPUS, Google Scholar and Springerlink ****

More information about this series at http://www.springer.com/series/15179

Svetlana Igorevna Ashmarina ·
Valentina Vyacheslavovna Mantulenko
Editors

Current Achievements, Challenges and Digital Chances of Knowledge Based Economy

Editors
Svetlana Igorevna Ashmarina
Applied Management Department
Samara State University of Economics
Samara, Russia

Valentina Vyacheslavovna Mantulenko
Department of Applied Management
Samara State University of Economics
Samara, Russia

ISSN 2367-3370 ISSN 2367-3389 (electronic)
Lecture Notes in Networks and Systems
ISBN 978-3-030-47457-7 ISBN 978-3-030-47458-4 (eBook)
https://doi.org/10.1007/978-3-030-47458-4

This Springer imprint is published by the registered company Springer Nature Switzerland AG
The registered company address is: Gewerbestrasse 11, 6330 Cham, Switzerland

Contents

Knowledge and Information as Basic Values of a New Economic Paradigm

Knowledge and Information as an Object of Criminal Law Protection

A. V. Beliakov

Abstract Recently, the number of registered crimes in the field of information and telecommunication technologies has increased several times. The relevance of developing issues related to information security is increasing. The solution of this problem is not possible without working out the legal mechanism. In our country, this problem is generally resolved. There are a number of regulations that define the mechanism under consideration. Criminal law plays a key role in this case. In this paper, we have identified about 40 structures that provide for criminal liability for violation of the rules for handling information. The selected set determined the need for ordering them by classifying them. Thus, two groups of acts were identified, depending on the approach of the legislator in describing the objective features of the crime. In the course of studying the special literature, a natural tendency was revealed that determines the need for a theoretical rethinking of some basic concepts in criminal law. For example, the need for a broader interpretation of the concept of object of crime is justified.

Keywords Information security · Unauthorized access · The subject of crime · The tools of crime

1 Introduction

In accordance with presidential decree No. 204 of May 7, 2018 [1], one of the key goals is to achieve breakthrough scientific, technological, and socio-economic development of the Russian Federation. To achieve this goal, the same legal act provides for the implementation of tasks related to accelerating the technological development of the Russian Federation and increasing the number of organizations that implement technological innovations. The government of the Russian Federation is tasked with creating a national program (project) on the topic "Digital Economy". Specifying the tasks, the decree contains a number of targets, which in particular should include:

A. V. Beliakov (✉)
Samara State University of Economics, Samara, Russia
e-mail: Belekov2007@ya.ru

S. I. Ashmarina and V. V. Mantulenko (eds.), *Current Achievements, Challenges and Digital Chances of Knowledge Based Economy*, Lecture Notes in Networks and Systems 133, https://doi.org/10.1007/978-3-030-47458-4_1

the creation of a stable and secure information and telecommunications infrastructure for high-speed transmission, processing and storage of large amounts of data, accessible to all organizations and households; the use of mainly domestic software by state bodies, local governments and organizations. As we can see, the solution of the problem of information and telecommunications infrastructure security is a task without which it is impossible to achieve the strategic goals set for the state. It is extremely difficult to ensure security in modern society without working out its legal basis.

Before this study, the problem of information protection is put on the resolution. The relevance of this study is also evidenced by statistical data. Thus, according to the Prosecutor General's office of the Russian Federation in 2017, the number of crimes in the field of information and telecommunication technologies increased from 65,949 to 90,587. Their share of all criminal acts registered in Russia is 4.4%—this is almost every 20 crimes. The most common of them are the so-called "computer crimes" [6]. In 2019, the number of such crimes reached 294,400 [5]. These questions are also the subject of active discussion in the special literature [3, 4, 7–9, 12].

2 Methodology

To substantiate the problem, in particular to prove the existence of violations in the use of information, a statistical research method was used. In this study, using the analysis of the dispositions of articles of criminal law, the features associated with information were identified, which were used as a basis for grouping the corresponding elements of crimes.

In the course of the study of various regulations that define the mechanism of information protection in Russia, a comparative legal method was used. Based on the results of the analysis of articles of the Criminal code of the Russian Federation, a list of crimes aimed at protecting information was synthesized, which in turn was grouped according to the relevant grounds.

3 Results

Currently, the mechanism of criminal legal protection of information in Russia has developed and exists. However, for its most effective operation, it is necessary in addition to applied problems. In particular, it is necessary to revise the concept of defining the subject of crimes. In criminal and civil legislation, the concepts of property and information should be fixed. It is necessary to determine the regularities of systematization of criminal law norms that protect public relations related to the use of information.

4 Discussion

The Constitution of the Russian Federation proclaims and guarantees the right to freedom of information. In this case, the exception is information that constitutes a state secret. The legislation defines information as an object of public, civil and other legal relations. It also declares the principle of free transfer of information between individuals, as well as the possibility of limiting it by federal law [2]. Thus, there are two types of information: public and restricted access.

Information depending on the order of its presentation or distribution subdivided into: information, freely available; information provided under the agreement of the persons participating in the relevant relations; information that in accordance with Federal law shall be provided or disclosed; information which dissemination in the Russian Federation is restricted or prohibited.

The same law imposes a number of restrictions on the use of information in order to protect it. In particular, the following types of information are protected: information constituting a state secret; information constituting a commercial secret, official secret and other secret; information obtained in the performance of professional duties; personal data.

Liability for violation of the rules for using information is provided for, including criminal law. So, for example, M., acting intentionally, realizing the illegality of his actions, out of personal interest, through correspondence using an Internet messenger, asked B. to illegally replace the SIM card of the subscriber number no., the owner or actual user of which he was not, to which B. as a result of his persuasions agreed, thereby M. declined by asking and persuading B. illegal access to legally protected computer information contained in the database and modification of computer information, i.e. the Commission of a crime. Then B., M. inclined to the Commission of the offence by instigation, wanting to be of service to M. in accordance with a prior agreement with him, that is, acting out of personal interest, while at his workplace, using his official position specialist sales office, intentionally, without legal grounds (references and a written statement of the subscriber, its representative or the actual user ID) for access to personal data of the subscriber, information about the provided telecommunication services and SIM cards, acting unlawful in violation of the job requirements through regular software, using the account of an employee of the above-mentioned sales office, she carried out illegal access to the legally protected computer information contained in the database—information about the communication services provided to the actual user of the subscriber number, after which she replaced the SIM card numbers by making appropriate changes to the database, which led to modification of the computer information contained in the database. After completing the procedure for replacing the SIM card, M., who is not the owner and user of the specified subscriber number, acting intentionally, received a SIM card from B. At the same time, the registered subscriber of the number, its representative or the actual user of the specified subscriber number did not contact the sales offices about replacing the SIM card. The court also qualified B.'s actions under article 272 of the criminal code [11].

However, the legislator has different approaches to formulating rules for the misuse of information. In particular, if we consider the violations we are interested in from the point of view of the elements of crime, they can be divided into two groups. Compositions in which information acts as a means of committing a crime. These include:

1. Article 128.1 Libel.
2. Article 185.3. Market manipulation.
3. Article 185.6. Misuse of insider information.
4. Article 336. Insulting a soldier.
5. Article 207. Deliberately false report of an act of terrorism.
6. Article 298.1. Libel against a judge, juror, Prosecutor, investigator, person conducting an inquiry, bailiff.
7. Article 306. Deliberately false denunciation.
8. Article 319. Insulting a government official.
9. Article 200.6. Deliberately false expert opinion….
10. Article 237. Concealment of information about circumstances that pose a risk to life or health.
11. Article 287. Refusal to provide information to the Federal Assembly of the Russian Federation or the accounting chamber of the Russian Federation.
12. Article 307. Knowingly false testimony, expert opinion, or incorrect translation.
13. Article 308. Refusal of a witness or victim to testify.
14. Article 330.2. Failure to comply with the obligation to submit a notification that a citizen of the Russian Federation has citizenship (citizenship) of a foreign state or a residence permit or other valid document confirming the right to permanent residence in a foreign state.
15. Article 140. Refusal to provide information to a citizen.
16. Article 170.2. Entering deliberately false information in the boundary plan, technical plan, survey report, land survey project or land plots or map-plan of the territory.
17. Article 171.1. Production, acquisition, storage, transportation or sale of goods and products without marking and (or) applying information provided for by the legislation of the Russian Federation.
18. Article 172.3. Failure to include information about funds placed by individuals and individual entrepreneurs in the financial documents of accounting and reporting of a credit institution.
19. Article 185.1. Malicious evasion of disclosure or provision of information defined by the legislation of the Russian Federation on securities.
20. Article 185.5. Falsification of the decision of the General meeting of shareholders.

Structures in which information acts as a subject of criminal law protection. These include:

1. Article 137. Violation of privacy.

2. Article 138. Violation of the secrecy of correspondence, telephone conversations, postal, telegraphic or other messages.
3. Article 138.1. Illegal circulation of special technical means intended for secret obtaining of information.
4. Article 146. Violation of copyright and related rights.
5. Article 147. Infringement of inventive and patent rights.
6. Article 155. Disclosure of the secret of adoption).
7. Article 180. Illegal use of means of individualization of goods (works, services).
8. Article 183. Illegal receipt and disclosure of information constituting commercial, tax or banking secrets.
9. Article 272. Unauthorized access to computer information.
10. Article 273. Creating, using, and distributing malicious computer programs.
11. Article 274. Violation of the rules for using computer information and information and telecommunications networks.
12. Article 275. Treason.
13. Article 276. Espionage.
14. Article 283. Disclosure of state secrets.
15. Article 283.1. Illegal receipt of information constituting a state secret.
16. Article 284. Loss of documents containing state secrets.
17. Article 310. Disclosure of preliminary investigation data.
18. Article 311. Disclosure of information about security measures applied to judges and participants in criminal proceedings.
19. Article 320. Disclosure of information about security measures applied to an official of a law enforcement or Supervisory authority.
20. Article 274.1. Undue influence on the critical information infrastructure of the Russian Federation.

This method of grouping has already been used in special literature, but significant changes in the current legislation have determined the need for its correction and addition. Based on the goals of this study, the second group of compounds is of the greatest interest. At the same time, it should be noted that in this work, the object of the crime will be understood as public relations protected by criminal law, which are harmed by the crime or create a real threat of harm. Signs of the object are: public relations, the subject of crimes, the victim. Thus, information that is not a public relation cannot be the object of a crime.

Based on the content of the analyzed articles, the category under consideration is closest in content to such a feature of the composition of crimes as the subject. Currently, the subject of crimes is usually understood as the object of the material world, by affecting which a person causes harm to public relations themselves. This approach is called a "real concept". However, recently, taking into account the current trends in the informatization of public relations, there have been comments about the need to expand this approach. For example, it is proposed to recognize information as the subject of crimes against property, since it has all the characteristics of property. These include the value and ability to act as an object of civil relations [10].

However, it is impossible to fully agree with this point of view. Information can be different in content, and therefore different in meaning. As some authors rightly point out, such a property of information as copying allows its owner not just to transfer it to other people, but to share it, increasing the number of copies. In this case, for example, illegal copying of information that even has a monetary equivalent does not have the signs of theft described in criminal law.

One of the reasons for this scientific discussion is the absence of the concepts of property and information in criminal and civil legislation, as well as the uncertain status of the latter as an object of civil law relations. The above suggests that it is quite difficult to adapt the current legislation by a new interpretation of the subject of a separate crime. Rapidly developing information technologies require more radical changes in the theoretical interpretation of the subject of crimes, namely a more extensive one.

In the Russian criminal law and its branch legislation, there are structures that traditionally protect property rights. Their construction is mostly based on the concept of "theft" in the traditional (real) sense, but now there is a need to describe these actions in relation to information.

To substantiate the proposed hypothesis, it is necessary to consider in more detail some of the features of the objective side, with which the legislator describes criminal misconduct in relation to information. These include: collection, distribution, disposal of information, demonstration; access to information; obtaining information; misappropriation, illegal use, acquisition, storage, transportation; disclosure of secrets; unauthorized access; unauthorized destruction, blocking, modification, copying; violation of the rules for processing or transmitting information; issuance, transfer or loss of information; illegal impact on the critical information infrastructure of the Russian Federation.

The above indicates a much broader range of illegal actions that can be committed in relation to information as the subject of a crime in comparison, for example, with property in the traditional (real) sense. However, it is not possible to commit theft of information under the current legislation.

5 Conclusion

The formal framework does not allow you to fully produce a full-fledged study that would allow you to fully and reasonably answer the questions posed. However, the identification of the problem, and the justification for its resolution, is also an initial, but still a significant part of the study. For this reason, the conclusions in this paper will be formulated not only affirmative theses, but also suggestions for solving some problems in future studies.

The above allows us to state that at present the mechanism of criminal legal protection of information in Russia has developed and exists. However, for its most effective operation, it is necessary to solve a number of fundamental problems in addition to applied ones. In particular, the question of the conceptual apparatus remains unresolved. Also, the variety of criminal law norms devoted to the protection and lawful use of information determine the need to resolve the issue of their classification and place in the Russian criminal law system. The division proposed in this paper is only the beginning for more thorough work.

References

1. Decree of the President of the Russian Federation No. 204 of May 7, 2018. On national goals and strategic objectives for the development of the Russian Federation for the period up to 2024. http://kremlin.ru/events/president/news/57425. Accessed 26 Feb 2020
2. Federal law of July 27, 2006 N 149-FZ. On information, information technologies and information protection. https://www.wto.org/english/thewto_e/acc_e/rus_e/WTACCRUS58_LEG_369.pdf. Accessed 26 Feb 2020
3. Frolova EE, Polyakova TA, Dudin MN, Rusakova EP, Kucherenko PA (2018) Information security of Russia in the digital economy: the economic and legal aspects. J Adv Res Law Econ 9(1):89–95
4. Jara AJ, Bocchi Y (2019) GEO-trust: geo-aware security protocol for enabling cross-border trustable operations and data exchange in a global digital economy. In: de Oliviera FC (ed) Proceedings of the 1st sustainable cities Latin America conference. IEEE, New Jersey, pp 1–6
5. Petrov I (2020) The scammers went online. The number of crimes committed using IT technologies has increased dramatically. Russian Newspaper—Federal Issue, 17(8071). https://rg.ru/2020/01/28/rezko-vyroslo-chislo-prestuplenij-sovershaemyh-s-pomoshchiu-it-tehnologij.html. Accessed 20 Feb 2020. (in Russian)
6. Report on crimes committed using modern information and communication technologies. https://genproc.gov.ru/smi/news/genproc/news-1431104/. Accessed 19 Feb 2020. (in Russian)
7. Saarenpää A (2017) Information law revisited. In: Schweighofer E, Kummer F, Hötzendorfer W, Sorge C (eds) 20 years of IRIS: trends and communities of legal informatics. Weblaw AG, Bern. https://jusletter-it.weblaw.ch/en/issues/2017/IRIS/information-law-revi_239b6b89a0.html__ONCE&login=false. Accessed 02 Feb 2020
8. Sandalova VA (2019) The modern state of the institute of banking secrecy in the conditions of digitization of banking services. In: Studies in computational intelligence, vol 826, pp 85–92
9. Spindler G (2019) Digitalization and corporate law—a view from Germany. Eur Co Financ 16(1–2):106–148
10. Stepanova KV (2018) Information as the subject of a crime against property. Soc Law 2(64):68–72
11. Verdict No. 1-262/2019 of July 22, 2019 in case No. 1-262/2019 of the Anzhero-Sudzhensky city court of the Kemerovo region. https://sudact.ru/regular/doc/P3OmRGV4c11r/. Accessed 22 Feb 2020. (in Russian)
12. Wienke A, Friese K (2018) Legal aspects of digitalization in the medical sector. Laryngo-Rhino-Otologie 97(10):713–716

Information Space Concept of Interaction Between Digital and Innovative Economy

O. A. Bulavko and L. R. Tuktarova

Abstract The article investigates possibilities of effective formation of the innovation economy (knowledge economy), information technologies that have a positive impact on the development of the digital economy. Improving the digital economy development can increase the economic growth rate, the GDP share, and complete the implementation of national projects. This goal can be achieved through a close interaction between the innovative economy, development of production processes, digital and information technologies. Indicators of the innovative activity and innovativeness are also of great importance in this aspect. The authors identify innovative development tools based on digital and information technologies that allow shifting the vector of interaction between institutions and enterprises that contribute to increasing the growth of both production and knowledge-intensive potential.

Keywords Digital economy · Innovative technologies · Digital technologies · Innovative economy · Intellectual capital · Economic growth rates

1 Introduction

Currently, the transition to the development of high-tech industries and the formation of a new information society is particularly relevant for all Russian regions. At the same time, there is an increasing need for software development and implementation, and the release of new specialized equipment. In accordance with the new requirements, Russian enterprises should reorient themselves to high-tech products, while increasing innovative indicators and the number of cooperation agreements with high educational institutions and secondary professional organizations. In our opinion, digital and information technologies and information support programs are

O. A. Bulavko (✉) · L. R. Tuktarova
Samara State University of Economics, Samara, Russia
e-mail: vikigor163@mail.ru

L. R. Tuktarova
e-mail: tuktarovalr@rambler.ru

S. I. Ashmarina and V. V. Mantulenko (eds.), *Current Achievements, Challenges and Digital Chances of Knowledge Based Economy*, Lecture Notes in Networks and Systems 133, https://doi.org/10.1007/978-3-030-47458-4_2

the main tools for stimulating the growth of the innovative economy, which contribute to the development of enterprises and allow shifting the trajectory towards high-tech industries. A lot of authors consider the impact of human capital formation and the impact of the Internet on economic growth [1, 7, 9, 10].

At the moment, the development of information technologies is very rapid, and the number of users on the Internet is growing every day. Any file presented in the digital form can be copied, modified, and distributed. This is why multimedia products are vulnerable to illegal use. In connection with this problem, various methods of information protection are being developed. Digital data protection is provided by cryptography and stenography methods. The main difference between these methods is that cryptographic methods hide the contents of the file by encryption, while stenographic methods hide the fact that any information is transmitted.

Information technologies cover all areas of human and state activity and are becoming increasingly important in the development of the economy, positively affecting the national welfare and receiving so-called "digital dividends". According to this report, "digital economy" is defined as a paradigm for accelerating economic development through digital technologies. Thus, the digital economy and information technologies are a powerful tool for effective development of the innovative economy [13].

2 Methodology

In the course of this research, methods of stochastic, retrospective, prospective, and comparative analysis were applied. The authors use empirical research methods, as well as methods of analogy, interpretation, comparison, and generalization. The applied statistical methods enabled to obtain quantitative characteristics that allow avoiding minor deviations in the research process.

3 Results

When forming elements of a new economic policy, there is a strong correlation between the economic development rate and the share of investment in human and fixed capital in the GDP. Combining these factors produces approximately 85% of the economic growth. At the same time, a very important proof, according to some authors, is the fact that centralized functions, such as research and development, asset optimization, corporate planning (strategies, investment planning), are of great value for business [7].

A positive factor for the development of the innovative economy, namely the knowledge economy, is provided by ICT use in education, both at the secondary and higher levels. This topic is considered to the works of Tam et al. [12]. In Table 1, the authors present the interaction of intellectual capital and the innovative economy.

Table 1 Requirements of the innovative economy for intellectual capital

Characteristic	Content	Requirements of the innovative economy
Motivation	The combination of internal and external driving forces for innovations	Active communication and interaction of innovators in the market
Knowledge	The result of the learning process, formally confirmed by documents on education	Areas: scientific, technical, humanitarian, social Professional knowledge: engineering, IT-technologies, bioengineering, physics, medicine, chemistry, economics, entrepreneurship, etc.
Skills	Set of mastered ways to perform actions.	
Abilities	Knowledge and skills implemented in practice	
Investments	Funds for health maintenance, education and professional development	Human capital is the basis for the growth of an innovative economy

Source authors

Investments in human capital are the most important factor in shaping its image that meets the requirements of an innovative economy. Investments in education, science, and health care contribute to the formation and development of human capital from the perspective of a process approach to the development of innovations in the region. Human capital is a qualitative characteristic of the influence of the human factor on the results of the economic activity in general and innovations in particular. The analysis and improvement of the terminological apparatus allowed us to formulate a number of key conclusions for the development of innovative activities and the formation of a new technological order [8].

The intensity of innovative activity and innovative processes determine the level of economic development and the nature of economic growth. Effective deployment of innovative processes in the Russian economy is associated with the strengthening of its scientific and innovative potential, the orientation of scientific organizations to the solution of socio-economic problems.

The development of the innovative activity in Russia looks very modest compared with indicators of the European Union countries, including former Eastern European countries, as well as Japan, a number of countries in Central, South and North America, New Zealand, and Australia. Russia is inferior even to the average economic development countries, such as Mexico, Estonia, Latvia, Slovenia, Hungary, etc.

The main indicator that characterizes the effectiveness of innovative activities is the share of innovative products in the total volume of industrial products. According to the opinion of a number of Russian scientists, which the authors also adhere to, the threshold value of the considered indicative indicator is 15%.

At present, the potential for economic growth based on the export-raw material model is almost exhausted. It is necessary to switch to an innovative development way, which opens opportunities for modernizing the Russian economy, increasing the pace and efficiency of its development based on the intensive use of domestic intellectual potential and the development of advanced technologies.

It is supposed to bring a resource base under the Russian economic development in the form of scientific and technical development, high-tech goods and services and on this basis to diversify and improve the quality of the economic growth, to raise the efficiency of the primary resources usage. Accelerated development of the scientific and educational complex and high-tech industrial sectors will create an effective growth potential for the domestic economy. Increasing the number of signed cooperation agreements between institutions, universities and enterprises will allow graduates to improve their practical skills in real life, expand the number of business contracts, and put enterprises on an innovative development path. This restructuring will require major investments in the development of the scientific and educational complex, updating the material and technical base of the industry, and creating new high-tech industries.

As a tool for the development of digital technologies that can improve the efficiency both in the production and in the high-tech industries, the authors propose a computer shorthand method, the main directions of which are: digital stenography methods and methods focused on the data format. These signals include images, video data, and audio files.

Often, there are also problems with copyright infringement. One of the most popular and effective methods in the field of stenography for copyright protection and protection against unauthorized copying is to embed shorthand inserts in the protected object—labels that carry a certain identifier of the copyright holder. These labels are called digital watermarks. Currently, the problem of copyright protection focuses on the protection of intangible property. The main component of intangible property is the intellectual property. This problem has become particularly important with the development of digital technologies and the Internet. One way to solve this problem is to embed digital watermarks in multimedia files, in particular in digital images and video data.

4 Discussion

The spread of high technologies in the industrial sphere shows the need to spread and develop scientific and technological developments based on certain knowledge, which are a prerequisite for long-term economic growth and prosperity. At the same time, a synergistic effect can be obtained thanks to the breakthrough of the fourth industrial revolution, defined as "Industry 4.0", which is based on the "digitalization" of industry and production. Schulze [11] in his research proves that the integration of high technologies and traditional industry into a network of active players grouped around the "authorized" state characterizes the European model. This model is based on elements of reindustrialization, and a sufficient background in additive technologies, digital design and modeling. Technological breakthroughs in many industries are possible with implemented developments and funded research.

Research and development is possible using intellectual capital, which, according to some authors [5], consists of human capital and intellectual assets. The problem of studying intellectual capital and information technologies in the space of interaction with the digital economy has not been sufficiently studied. In order to study this interaction, the following main aspects are proposed. In the course of our research, it is necessary to consider the process of interdependence between information technologies, production, the digital economy and human capital. The problem of information literacy and the development of information and communication technologies were considered in the works of Feldvari and Varga [6]. Questions of development of neuro-network digitalization were studied by Dyatlov et al. [4]. Problems of digital economy include the implementation of multimedia products that are vulnerable to their illegal use. In connection with this problem, various methods of information protection are being developed. Digital data protection is provided by cryptography and stenography methods.

5 Conclusion

An important aspect of the interrelation between the digital and innovative economies is tools for implementing the intellectual capital in the framework of the digital economic paradigm. One of the problems (copyright protection) can be solved using a "demonstration of ownership" scenario. If the author of an intellectual product wants to prove the fact of authorship then after creating this product, he embeds a persistent watermark in it, which can uniquely identify him as the owner. Due to the specific nature of this application, it should be as resistant as possible to a wide range of distortions, such as linear and nonlinear filtering, lossy compression, cropping, and others. Thus, the innovative economy and information technologies are rightly considered as the most important factor in the development of the digital economy, the renewal of industrial production, and human resources aimed at achieving economic, social, and knowledge-intensive results.

For the Russian Federation, the transformation of economic relations and the transition to a new technological order is important, as it allows us to overcome the trends of increasing technological and intellectual lag, reduce the dominance of rent-oriented attitudes in the society and ultimately take a worthy place among the world leaders of the globalizing world. For the authors, the motivation for choosing transformational mechanisms for the formation and development of the digital economy was the multidimensional nature of the topic, covering a wide range of issues related to the formation of coalitions as a separate mechanism for accelerating modernization processes, transforming risk into a specific type of resource for the development of innovative systems, using human capital in the interests and methods of regulating the information and innovative economy. The study of these issues allowed us to form a more complete picture of the modernization mechanisms aimed at improving the development of innovation policy in the conditions of improving the digital economy development. All the above-mentioned problems of transformation of the world regulatory system into a global management mega-system require new scientific understanding, classification and identification of its structural elements, structurally and functionally subordinated by sublevels, based on the use of an interdisciplinary entropy-synergy approach, as well as the development of an organizational mechanism for resolving poly-system contradictions of the global economy and ensuring its sustainable development in conditions of increasing hyper-competition [3].

In the context of growing informatization of the economy, the quality of economic growth becomes much more important, expressed in such indicators as: the quality of life, the environmental factor, competitiveness, strengthening resource conservation; improving the quality and diversity of products; increasing the profitability of high-tech industries, and sustainable economic development.

The most important obstacle for companies' innovative activity during the crisis was the high uncertainty of the development prospects of sales markets. The uncertainty factor for the economy's prospects was also enhanced by the low predictability of government actions [2].

References

1. Aleshkova DV, Greshnova MV, Smolina ES, Popok LE (2020) Research of efficiency of tax stimulation of innovative entrepreneurship. In: Ashmarina SI, Vochozka M, Mantulenko VV (eds) Digital age: chances, challenges and future. Springer, Cham, pp 80–84
2. Bulavko OA (2013) Industrial and investment policy in the post-crisis modernization of Russian industry. Thesis of the PhD, St. Petersburg State University of Economics, Saint-Petersburg
3. Dyatlov SA, Vasiltsova VM, Vasiltsov VS, Bezrukova TL, Bezrukov BA (2015) Methodology of management innovation hypercompetition. Asian Soc Sci 11(20):166–169
4. Dyatlov SA, Lobanov OS, Gilmanov DV (2017) Digital neuro-network economy: institutions and development technologies. Publishing House of St. Petersburg State Economic University, Saint Petersburg
5. Evans N, Price J (2017) Human capital and information asset management behaviours in law firms. In: Tsui E, Cheung B (eds) Proceedings of the 14th international conference on intellectual capital knowledge management & organizational learning. Hong Kong Polytechnic University, Hong Kong, pp 63–71
6. Feldvari K, Varga M (2017) Young people as human capital—what type of information literacy education is required? In: Tonkovic AM (ed) Proceedings of the 6th international scientific symposium on economy of Eastern Croatia—vision and growth. Ekonomski Fakultet Osijeku-Fac Economics Osijek, Croatia, pp 318–326
7. Haini H (2019) Internet penetration, human capital and economic growth in the ASEAN economies: evidence from a translog production function. Appl Econ Lett 26(21):1774–1778. https://doi.org/10.1080/13504851.2019.1597250
8. Ioda EV, Bulavko OA, Khmeleva GA, Ioda YV (2013) Modernization mechanisms for the formation of a new technological order. Monograph, Samara Academy of State Municipal Administration, Samara
9. Konovalova ME, Kuzmina OY, Salomatina SY (2020) Transformation of the institution of money in the digital epoch. In: Ashmarina S, Mesquita A, Vochozka M (eds) Digital transformation of the economy: challenges, trends and new opportunities. Advances in intelligent systems and computing, vol 908. Springer, Cham, pp 315–328

10. Schekoldin VA, Bogatyreva IV, Ilyukhina LA (2020) Digitalization of labor regulation management: new forms and content. In: Ashmarina S, Vochozka M, Mantulenko V (eds) Digital age: chances, challenges and future. Lecture notes in networks and systems, vol 84. Springer, Cham, pp 137–143
11. Schulze PW (2017) Future of industry: the fourth revolution—functions of state and society. Econ Revival Russia 2(52):39–46
12. Tam VC, Chan JWW, Li SC, Pow J (2018) Developing and managing school human capital for information and communication technology integration: a case study of a school-based e-learning project in Hong Kong. Int J Leadersh Educ 21(4):447–461. https://doi.org/10.1080/13603124.2017.1318958
13. World Bank Group (2016) Development of the digital economy in Russia. World Bank report of December 20, 2016. http://documents.vsemirnyjbank.org/curated/ru/413921522436739705/pdf/EAEU-Overview-Full-RUS-Final.pdf. Accessed 07 Feb 2020

Information Technologies Significance in Higher Education in Context of Its Digitalization

T. V. Gromova

Abstract The problem is that the education system does not meet the challenges of the digital economy. The digital economy requires competent personnel, to whose training is necessary to improve education system and specialized training, adapt didactic programs in accordance with digital challenges. The goal of the paper is to promote the introduction of information technologies in the framework of digitalization in the education system and professional training of students by improving the efficiency of the educational process. This is done by analyzing and identifying the distinctive features of the teacher of distance and traditional learning systems, revealing the essence of the main models of distance education, elements of the work of the distance learning teacher. The research methods included both empirical (observation, comparison, etc.) and theoretical (analysis, synthesis, modeling, etc.). The article considers the objective factors of integration of information technologies in the higher education system, and the relationship between the functions of information technologies and their methodological and communicative potential in education.

Keywords Digital economy · Digital education · Information and communication technologies · Distance education · Distance learning · Distance learning teacher

1 Introduction

Radical socio-administrative and cultural transformations happening nowadays are the motive of innovations that penetrate all spheres of public life: from the economy to education. The education system should provide society with a confident transition to a digital age focused on productivity growth and human needs. In 2017 Ministry of education and science of Russia launched the project "Modern digital educational

T. V. Gromova (✉)
Samara State University of Economics, Samara, Russia
e-mail: gromova73@yandex.ru

S. I. Ashmarina and V. V. Mantulenko (eds.), *Current Achievements, Challenges and Digital Chances of Knowledge Based Economy*, Lecture Notes in Networks and Systems 133, https://doi.org/10.1007/978-3-030-47458-4_3

environment in the Russian Federation", which aims to create conditions for systematically improving the quality and expanding opportunities for continuing education aimed at all residents groups by the enlargement digital learning space of Russia [2].

Digital education, on the one hand, makes it possible to qualitatively change the methods and organizational forms of education; on the other hand, it provides training of personnel for the digital economy. Digital education is primarily information and communication technologies (ICT) that can be used in the traditional learning process involving a huge number of students, not limited to physical space. However, it is more appropriate to use ICT as part of distance learning (DL).

Many researchers have studied the advantages and disadvantages of using ICT in the EP. Thus, Santos et al. [9] wrote about the growing interest in accepting new communication technologies in teaching context, enhancing the digital transformation of universities.

Pollock et al. [8] as well as Ron [6] noted the increasing role of information technologies (IT) in modern campuses, especially in areas such as data collection, analytics and predictive modeling, and required campus leaders to rethink IT structures and determine the best ways to maximize IT potential.

The factors that prevent the widespread use of ICT in education around the world, according to L. Starr, include: lack of funding to support technologies, lack of training among recognized teachers-practitioners, lack of motivation and the need for teachers to use ICT as a means of learning [10].

Analyzing the role played by digital tools such as social media and mobile devices the authors of paper [7] focused on vision impairment. The influence of information technologies on the education system is mainly manifested in the globalization and internationalization of scientific and pedagogical knowledge. These trends place increased demands on both the staff of universities and the students themselves. The modern information society requires high-class professionals, regardless of their specialization. Their key competencies should be strategic thinking, the ability to navigate in dense flows of information, the ability to structure and analyze it. Certainly, education is the main tool aimed at developing these competencies.

This article discerns the relationship between the functions of information technologies and their methodological and communicative potential in the educational process (EP); examines and analyzes: DL models, differences between a DL teacher and a teacher of a traditional educational system, elements of a DL teacher's work.

2 Methodology

The study used methods of theoretical analysis of literature, observation, comparison, analysis, synthesis, modeling, etc. The work is devoted to the significance of ICT in educational process on the principle of observation, comparison, study of sources, analysis, synthesis, modeling it consistently examines the differences in the activities of the teacher of traditional and DL systems; DL models; elements of the DL teacher's work; DL models.

Striving to improve the quality of education is a key task that educational institutions around the world are working on. Definitely, in this regard the introduction of ICT in the EP plays a primary role. In the traditional sense, education carried out with the help of information technologies is a process of indirect communication among trainers and learners, directed at the systematic assimilation of knowledge, skills and abilities provided by the educational standard.

The introduction of ICT in the higher education system allows, firstly, to ensure equal access to educational resources (electronic libraries, cases, educational platforms), regardless of the student's place of residence, his state of health etc.; secondly, it forms students' ability to build independent educational strategies, and also contributes to the development of independent work skills with information, which increases the analytical potential of students [1, p. 29].

The first experience of using information technologies in higher education is observed abroad. According to statistics, the introduction of ICT in the EP has reached significant proportions in universities in the United States and Singapore. In the USA, distance education has almost a century of history and is a leading trend in the educational system. In total, more than 30% of students study in DL programs, and the distribution of ICT in education is constantly growing.

Implementation of ICT in US education system has significantly changed not only the methodological, but also the economic infrastructure of universities, since university leaders had to decide on the redevelopment and construction of a number of educational buildings adapted for the implementation of mixed educational programs [11].

In 2009, McKinsey rated Singapore's educational system as the best in the world, which is not surprising, since Singapore initially relied on the "economy". In this regard, Singapore is making global investments in education, creating a single space that is optimally prepared for ongoing learning process [11].

Simultaneously, the prospects offered by ICT encourage many Russian universities to join these processes. Currently, most Russian universities offer distance learning not only in certain areas of training, but also in particular disciplines. The introduction of ICT in the EP and the transfer of educational technologies to remote sites, in addition to pedagogical perspectives, also open the possibility of combining different areas of knowledge, which, in general, contributes to the formation of interdisciplinary trends.

3 Results

Implementation of ICT in the sphere of education makes it possible to qualitatively change the methods and organizational forms of training, making it more comfortable and accessible. As noted earlier, ICT is a major part of the process of modernizing education. They make it possible to implement remote communication between teachers and students.

The functions of ICT in the EP are entirely determined by their didactic potential. These functions are implemented both at the subject-subject level ("teacher-student", "teacher-university administration") and at the subject-object level ("student-information", "student-educational infrastructure of the university").

ICT significantly transform the standard type of relations in the "teacher-student" and "student-information" systems. The consequence of these changes is to increase the flexibility and mobility of teachers and students. On the one hand, ICT encourage teachers to stimulate the creative activity of students and on the other—expand the prospects for their cognitive activity.

The main DL models were adopted by Russian universities from foreign practice. The "distributed classroom" model is based on transferring the course content to the desired location using ICT (in this case, the localization of students becomes irrelevant). Students can even be at several points at the same time—this does not reduce the effectiveness of the model. The consequence of this model introduction is the expansion of student groups through their virtual integration.

The model of "independent training" does not provide for the binding of students to a specific time—course. Students receive extensive theoretical and methodological support for the EP, have the opportunity to familiarize themselves with the course program in order to present the logic of mastering the material and adjust it to their own educational strategy. The attribute of this model is feedback from the teacher, which is also provided through the introduction of ICT.

"Combined model" is a synthesis of DL with classical classes. On the one hand, the completeness of the methodological support of the course allows using the provided theoretical material for independent study, and on the other, the student is not isolated from the group, because interactive tools in the structure of the course provide integration with other members [5].

The "independent training" and "combined model" have proven themselves best in DL system. If the "combined model" involves a small proportion of classroom sessions, then "independent learning" is concentrated exclusively in the virtual space, and students' dependence on the time and localization of the EP is reduced to zero.

Each model is effective in its own way, and the choice of model depends solely on the training tasks. The specificity of DL is that its greatest effectiveness is achieved by introducing already established specialists into the educational environment (through professional development and retraining).

It should be noted that DL model, implemented with the use of ICT, assumes that the student audience has formed competencies: self-realization, motivation in learning, focus on a high standard of achievement, initiative. The presence of these competencies allows for the most effective use of ICT in the EP.

Informatization of the EP has changed the very concept of pedagogy. If in the traditional model students, coming to the lecture, did not have initial information about the content of the material, then the distance form assumes the presence of preliminary awareness, in which students can ask clarifying questions to the teacher. In addition, the new system is significantly changing the existing format of relations between the student and the teacher. In the new information model, the subjective factor of such relationships is erased, which creates a lot of difficulties in building interdisciplinary communication.

If in the classical model of education the role of the teacher was reduced to relay and modify knowledge, then the new educational model, focused on the digital training format, imposes new functions on the teacher. The analysis of scientific sources and practical experience allowed us to identify differences in the activities of the teacher in traditional and DL systems, presented in Table 1.

The table shows that the activities of both teachers differ greatly. The activity of a DL teacher is the activity of providing educational, methodological, psychological, pedagogical, and organizational assistance to students with wide use of ICT.

With a significant share of independent work in the DL, high-quality student learning is a well-thought-out system of teachers' support for students in the EP. Under the teachers' support, we understand the system of interrelated actions implemented in diverse forms through various elements of teaching activities (primary diagnosis, teaching planning focused on objectives setting, creating psychologically comfortable conditions and encouraging students, managing communication, self-analysis, control) [3].

Table 1 Comparative characteristics of the teacher's work in distance and traditional education

Attribute	Teacher (distance learning)	Teacher (traditional learning)
Prevailing activities	Provides assistance and support	Teaches
Approach to academic information	Corrects and instructs students' autonomous activities	Transmits training course content
Relationships with students	Provides support, being a source of academic knowledge	The main content informer
Communication with students	Communicates through distance technologies	Directly communicates with students
Position at webinars/seminars	Encourages students involvement	Frequently acts much more actively than students
Approach to information and distance learning technologies	Freely uses them in searching information, interacting with and implementing feedback from students	Applies them most often only for information search
Teacher—student relationships	Builds partnership relationships, attaching importance to the students incentive	Creates/maintains relationships of subordination ("knowing-not knowing")
The nature of the progress assessment	Monitors the performance of control tests and tasks distantly	Monitors students at the lessons

Source author

Although the use of ICT is becoming commonplace in all areas of life, their impact in education is not as significant as in other areas. Teaching is a specific activity, and first-rate education is connected with qualified teachers who have a high degree of personal contact with students. The use of ICT in education creates conditions for more student oriented learning.

4 Discussion

The paper attempts to help solve the problem of the discrepancy between the education system and the challenges of the digital economy through its modernization. Improving the effectiveness of the EP should also occur by determining the relationship between the functions of ICT and their methodological and communication potential, the ability to use the results of the analysis of differences between the DL teacher and the teacher of the traditional system, establishing the elements of the DL teacher's work and the types of DL teacher's support for students.

The popularization of ICT in education surely has not only advantages; this process is associated with a number of obvious difficulties (this includes the material costs of ensuring the smooth functioning of distant sites, and the search for the necessary experts, and the complexity of the methodological support of the EP in new conditions). Noting the advantages of introducing ICT into the EP, the article nevertheless does not ignore the evident problems associated with the informatization of education, which are manifested, in particular, in the complete or partial loss of live communication by the participants of interaction.

Methodological difficulties in providing distant EP consist in the fact that educational and methodical complexes should be developed specifically for this format of pedagogical technologies with the possibility of remote sites. In addition, they should be presented in an interactive form, so that students have access not only to the curriculum, but also to the entire set of tasks and theoretical content of the course with methodological explanations of the teacher.

The introduction of ICT in the higher education system determines a new range of responsibilities for the management of higher education institutions. Supervisors of courses, as well as deans, heads of departments, should not only be methodically prepared, but also be required to pass the necessary retraining in terms of technical competence [4]. Of course, this can significantly complicate the work of people whose activities have not previously been related to the field of ICT.

The new training format obliges University managers and higher-level officials to develop a system for assessing the quality of the EP, taking into account current trends (accreditation), as well as a system for training and retraining teachers (attestation and certification).

Currently, the introduction of ICT in education has not yet reached global proportions. In most European universities there is a combination of distance educational technologies with traditional forms of education (lectures and seminars held in the classroom).

For students, the value of education carried out on electronic platforms is due to the fact that it allows them to significantly save time resources. At the same time, the disadvantage of such training is a weak emotional connection with the course teacher, which is very important for students who are getting higher education for the first time. In this connection, "combined" strategy is supposed to be optimal at the initial stage, allowing, on the one hand, to significantly save the resources of students, and on the other hand, to maintain an emotional connection with the course coordinator (teacher, observer), which, of course, stimulates individual achievements.

As areas for further research, we can note the issues of training/retraining/professional development of DL teachers, methodological support of the EP, protection of educational electronic resources and copyright on them in the information and educational environment.

5 Conclusion

Summing up, it can be concluded that introduction of ICT in learning in the context of its digitalization significantly contributes to the efficiency of EP. The paper reveals the features of the teacher of distance and traditional learning systems, the essence of the main models of distance learning, the types of support for students in DL, as well as elements of the work of the DL teacher. New conditions require new teachers; in this regard, special training of DL teachers is of particular importance. Informatization of education in the context of its digitalization is undoubtedly changing the education system in the XXI century. It creates a basis for the transition to a new stage targeted on preparing professionals that will surely be in demand in their business, actively use mobile and internet devices and are aimed at lifelong training/professional development through ICT.

References

1. Akimova OB, Shcherbin MD (2018) Digital transformation of education: timeliness of educational and cognitive independence of students. Innov Proj Programs Educ 1:27–34
2. Decree of the government of the Russian Federation of July 28, 2017 on approval of the program "Digital economy of Russia". http://static.government.ru/media/files/9gFM4FHj4PsB79I5v7yLVuPgu4bvR7M0.pdf. Accessed 15 Feb 2020
3. Gromova TV, Belousov AI (2015) Functional components of the model of teacher's pedagogical activity in the distance learning system. Bull Samara State Aerosp Univ (Natl Res Univ) 14(2):248–260. https://doi.org/10.18287/2412-7329-2015-14-2-248-260
4. Instefjord EJ, Munthe E (2017) Educating digitally competent teachers. A study of integration of professional digital competence in teacher education. Teach Teach Educ 67:37–45. https://doi.org/10.1016/j.tate.2017.05.016
5. Nikolić V, Kaljevic J, Jović S, Petković D, Dachkinov P (2018) Survey of quality models of e-learning systems. Phys A 5111:324–330. https://doi.org/10.1016/j.physa.2018.07.058

6. Oliver R (2002) The role of ICT in higher education for the 21st century: ICT as a change agent for education. https://www.researchgate.net/publication/228920282_The_role_of_ICT_in_higher_education_for_the_21st_century_ICT_as_a_change_agent_for_education. Accessed 15 Feb 2020
7. Pacheco E, Lips M, Yoong P (2018) Transition 2.0: digital technologies, higher education, and vision impairment. Internet High Educ 37:1–10. https://doi.org/10.1016/j.iheduc.2017.11.001
8. Pollock K, Schwartz C, Buck D (2018) Information technology and its future role in student success. Educ Rev 29. https://er.educause.edu/articles/2018/1/information-technology-and-its-future-role-in-student-success. Accessed 15 Feb 2020
9. Santos H, Batista J, Marques RP (2019) Digital transformation in higher education: the use of communication technologies by students. Procedia Comput Sci 164:123–130. https://doi.org/10.1016/j.procs.2019.12.163
10. Starr L (2001) Same time this year. http://www.education-world.com/a_tech/tech075.shtml. Accessed 15 Feb 2020
11. Valeeva RZ (2014) Distance learning in the US educational system. Modern Pedagogy 4. http://pedagogika.snauka.ru/2014/04/2136. Accessed 07 Feb 2020

«Knowledge» and «Information» in the Structure of Modern Rationality and Human Activity

A. E. Makhovikov, A. V. Guryanova, and T. G. Stotskaya

Abstract The problem of definition and correlation of «information» and «knowledge» concepts relevant for the rational activity of the modern human is analyzed in the article. The authors are sure both of these concepts should be considered strictly within a certain type of scientific rationality. The modern post-non-classical type of scientific rationality connects the human activity more with the emotional than with the intellectual perception of knowledge and information. So the world around us is no longer a result of cognition, but an object of pleasure connected with experience of new feelings and emotions. Within the latest rational tradition knowledge and information are functionally performing the same tasks: they are the main tools for ensuring the vital activity of the modern human. In the modern world pragmatism and utilitarianism have become the main characters of knowledge and information. They both are turned into a form of specific pragmatic matrix of rational activity typical for the modern human. The last one values most of all the pertinence of information and the performativity of knowledge.

Keywords Information · Knowledge · Rationality · Human · Information society · Homo economicus

1 Introduction

Modern information society is a result of the development of scientific rationality. In other words, society is characterized as informational in the case of its basic conditions of reproduction depend on scientific knowledge [1]. Today we are facing rapid

A. E. Makhovikov · A. V. Guryanova (✉)
Samara State University of Economics, Samara, Russia
e-mail: annaguryanov@yandex.ru

A. E. Makhovikov
e-mail: shentala_sseu@inbox.ru

T. G. Stotskaya
Samara State Technical University, Samara, Russia
e-mail: stotskaya@yandex.ru

S. I. Ashmarina and V. V. Mantulenko (eds.), *Current Achievements, Challenges and Digital Chances of Knowledge Based Economy*, Lecture Notes in Networks and Systems 133, https://doi.org/10.1007/978-3-030-47458-4_4

growth of information, its expansion to the development of science and scientific technologies [3]. For this reason it's difficult to identify the status of «knowledge» and «information» and the nature of their relations in the structure of rational activity.

Modern writers often interpret these concepts as completely identical or interchangeable. It's especially typical for a level of everyday consciousness, but the authors of famous theoretical studies can't make the question clearer. Some of them don't see this problem at all; the others don't distinguish these concepts replacing one with another; the third ones interpret «information» as the most «perfect» state of «knowledge» typical for the modern society. In such a situation it's necessary to clarify the correlation between these concepts and to find out the real meaning of «knowledge» and «information» in the modern sense.

2 Methodology

The set of philosophical and logical methods is consistently used in the article. It includes, first of all, a method of philosophical analysis that allows realizing the main purpose the present research—finding out an sense and correct meaning of two related concepts—«knowledge» and «information»—in the framework of the modern post-non-classical rationality. Secondly, it's a dialectical method that helps to trace an evolution of different types of rationality from its classical to the post-non-classical stage. Thirdly, one of the most important methods of the present study is a comparative method with a help of which the features of classical, non-classical and post-non-classical types of rationality are compared, as well as the certain correlation between «knowledge» and «information» taking place in each of them.

3 Results

3.1 Knowledge and Information in the Context of Classical Rationality

It's necessary to mention that modern conceptual status of «knowledge» and «information» in structure of rational human activity hasn't been unchanged over time. So, these concepts can't be qualified as the new ones for the European cultural tradition. The concepts of «knowledge» and «information» have been known since the period of ancient classical rationality. For example, the concept of «information» (lat. «informacio»—acquaintance, explanation, presentation, awareness) in its original meaning means a message, an awareness of the state of affairs, a data about something transmitted by people. At the same time, it's related to the process of cognitive human activity. It isn't also endowed with such characteristics as objectivity and truth.

In accordance with the classical type of rationality, knowledge is considered as a result of the reality cognition process verified by socio-historical practice. It's an adequate reflection of reality by human consciousness fixed in the forms of representations, concepts, judgments and theories. In this sense, knowledge is in contrast to information, because it always contains the dialectic of absolute and relative truth in itself. So, the human feels sure that he's involved in the cognitive process directed from unknowing to knowledge, from superficial comprehension of reality to its essential and complex understanding.

Within the boundaries of the classical type of scientific rationality, there is no need to demarcate the concepts of «knowledge» and «information». Clear definitional and functional differences between these concepts were also absent among the authors developing the conceptions of «postindustrial information society» [2] or a «knowledge-based society» [7].

3.2 A New Status of Information in the Structure of Rational Human Activity

The first attempt of understanding the new status of information and its increasing role in the structure of rational human activity was made by Shannon and Weaver [9]. They both paid primary attention to the growth of technical components of mass and other communications, to the huge increasing of the volume of transmitted information. In these circumstances the need of measuring information in order to improve the conditions of its transmission to the consumer has become evident. And in 1948 American scientists proposed a probabilistic version of information understanding. In its frameworks all messages can't be considered as information, but only those of them that reduce the uncertainty of the recipient. The last one must also be able to overcome the uncertainty by choosing the most important information for him. Information here is interpreted in terms of structure, organization and heterogeneity, which are reflecting the syntactic aspect of its content.

The development of cybernetics has led to the emergence of information management. In its framework information is considered as a copy of an object significant for the system, created on the basis of its displaying, representing in a different form and adapted for using in the process of managing. At the same time Wiener gave his well-known definition: information is information, not matter or energy [12]. It can be measured, transmitted, stored, sold and bought. It also has a unique character to change the state of the system. Information is considered from the viewpoint of mathematical (statistical) methods of measurement. It doesn't depend on the method of transmission, forms of signals in communication channels and types of material carriers. It's also independent of the transmitted messages with their specific content.

The related approach to the new understanding of information of all the above mentioned scientists is obvious. They both describe the symbolic structure of messages and accent the syntactic aspect of information. Therefore, information is defined as a kind of ideal message that reduces or completely eliminates the uncertainty in choosing one or more possible alternatives. And this choice should be remembered.

3.3 Problem Aspects in Understanding the Essence of Information

It's well known that the syntactic aspect of understanding information isn't the only one. There is also a semantic aspect of information that considers its sense and meaning and a pragmatic aspect that takes into account its value and usefulness. The unity of these three aspects becomes fundamentally important in defining the concept of information. But it's difficult to combine them in one definition, since they often contradict and exclude each other. For example, «information» can be considered as a data organized in a certain way and becomes an object of communication. It also can be interpreted as a data placed in the context, organized and structured—and thus endowed with meaning. It's obvious that there are at least two meanings of «information»: it's a special form of knowledge and an event that loses the character of information in the process of updating [1].

In this context, information remains significant for the human only when it induces changes within the communication system, while its certainty doesn't cancel its surprising character. Thus, the most important feature of information is its ambivalence. It's an event and at the same time something different from it, it reproduces both knowledge and unknowing. Therefore, information in its essence shouldn't be correct, but only plastic, constantly transmitting an ambivalence of knowledge and unknowing.

This contradiction in understanding an essence of information in the structure of rational activity of the human leads some researchers to the conclusion that information is a reflection of objective reality [4]. This definition is a reason of much more mixing between the concepts of «information» and «knowledge». It can't explain the fact that false information, disinformation or falsified information sometimes has a necessary meaning for the human. It turns out that any of these forms of information can remain in demand, be effective and even useful for the human from the viewpoint of his personal decisions. In contrast to information, knowledge is either there or not. Unknowing doesn't reflect the need of the human for rational activity development. This situation is caused by the specific historical nature of rationality itself.

3.4 Historical Types of Scientific Rationality and Their Specific Characters

It's well known that rationality had three different types in the course of its historical evolution—classical, non-classical and post-non-classical. The emergence of these types of scientific rationality, according to academician V.S. Stepin, was due to the fact that classical, nonclassical, post-nonclassical science involve different types of reflection on activity: from exclusion of everything that doesn't belong to the object from the explanation procedures (classic), to understanding the relation between the explained characteristics of the object with characteristics of the methods and operations of activity (non-classic), to understanding the value and purposeful orientations of the subject of scientific activity and their correlation with social goals and values (post-non-classic) [11].

The post-non-classical type of rationality dominating in the modern information society has several distinctive features. The scientific knowledge production is no longer a search for fundamental natural laws, but a process connected with knowledge application in the sphere of social needs significant for potential consumers. Besides, the specificity of post-non-classical rationality is that the rational activity of the human connects more with emotional than with intellectual perception of knowledge and information [6]. So, the world around us is no longer a result of purposeful cognition, but an object of pleasure united with experience of new feelings and emotions.

In these conditions, the choice of perceived information or knowledge is determined by such essential features of post-non-classical rationality as pragmatism and utilitarianism. As a result, the modern science is changing from the interdisciplinary nature of scientific research to the transdisciplinarity. That's why the production of modern scientific knowledge is a real hybrid of fundamental research activity focused on finding the truth, and applied researches setting a useful effect as a goal.

3.5 Knowledge and Information in the Post-non-classical Rationality

Forming a new paradigm of scientific rationality has changed much the previous meaning of «knowledge». Now the cognitive process isn't an adequate reflection of an object, but a human activity aimed at the construction of this object or the reality as a whole. And there is no other reality for the human. Therefore, «knowledge» is no longer a reflection of some objective reality, but a substrate with a help of which the reality is forming. There isn't a human creating his own image of reality,—exactly the opposite: created by subjective representations and knowledge reality is forming the human. Cognition here has the task of regularizing the inner world of the human, rather than explaining the essence of its objective being. In such conditions it seems improper to raise the question of its truth, since the criteria aren't proof, verification and justification, but trust, suitability, acceptability, etc.

Within the new post-non-classical type of rationality knowledge is no longer a foliation of ideas, nor an individual experience, but a complex, self-organizing system phenomenon having special ways of socialization and institutionalization. It covers all spheres of social life and has its own specific forms of being. As we can see, «information» and «knowledge» have fundamentally changed their nature, sense and meaning, in contrast to classical rationality.

Within the frameworks of post-non-classical rationality, information is included in the context of cognitive activity. Now the main goal of cognition is to provide information control of the environment and the inner cognitive states of the human, leading to his adaptation and survival. There are no other requirements for information in this case. So, information doesn't need to be true, but only useful and plastic. Its most significant features are efficiency and pragmatism for achieving a certain goal.

Information is first of all evaluated not for its truth (as information that had a status of «knowledge»), but only for completeness, correctness and other parameters helping to choose the right reaction from number of real or potential opportunities. Even false information or misinformation remains significant in this case. However there are many doubts about finding a correct measure between the truth and the benefit. Immediate benefit and personal interest seem increasingly important today.

Knowledge and information are functionally performing the same tasks within the post-non-classical rational tradition. Therefore, they can be considered as instruments for providing the life and the rational activity of the human. The last one always means a choice of a certain event in a horizon of possibilities. In this regard, it's difficult to say whether information or knowledge has a greater influence on the reality forming and constructing by the person. This is fully reflected in an actual problem of correlation between logico-methodological and cognitively-structural (information) forms of understanding.

4 Discussion

Modern researchers often describe modern society as a «network society» [5] characterizing it as a whole system. The activity of all its elements is interconnected and coordinated by functional programs. The network society is always a combination of information flows, technologies and capital. It's based on a «message», but not on a person. So, it doesn't matter who is the carrier of messages—it's important what kind of messages form the communication system. In this context, «information» has a meaning of «knowledge», and this leads to the constructive system changes. In fact, information is knowledge representing the limits of available capabilities [8].

Such interpretation is typical for the modern information society. Any system needs information for effective management of its activity. And any change in its field leads to transformations of the whole system. But the change itself is also the information. Information in the modern society reflects its ability of transmission or communication. So, the person can define «information» like «knowledge». But this «knowledge» is fundamentally different from its interpretation within the framework of classical rationality. In the post-non-classical rationality «knowledge» can be defined through the concept of «pertinent information». Therefore, the main criterion of knowledge is the principle of performativity [8].

That's why knowledge and information have another specific character. They turn into a kind of pragmatic matrix of rational activity of the modern human, formulating the most effective and adequate rules and regulations in the certain existential situation. In another way, knowledge and information can't play a role of the most important resource transforming into market goods and services [10]. On the one hand, knowledge is already digitalized in the modern conditions, but on the other—it's the best method of digitalizing the knowledge itself.

In this situation, the binary relation of the concepts «knowledge» and «unknowing» is irrelevant, because the truth of the certain knowledge can be verified or guaranteed with great difficulty. So, it's possible to reach the point of absurdity by refusing of all the previous successfully applied strategies of scientific and technical development under the pretext of calling them «untruthful».

5 Conclusion

In conclusion, it's necessary to accent that the situation described above with the definition of «information» and «knowledge» in the rational activity of the human is actual for the information society only. Here the post-non-classical rationality is playing a leading role. As for the classical type of rationality, it doesn't face such difficulties at all. It separates essentially «information» and «knowledge», correlates these concepts through the criteria of reflection, objectivity and truth. The specificity of the mental tradition of post-non-classical rationality has turned information and knowledge primarily into a product. The last one can be exchanged or sold as an intellectual private property. Another specific character of knowledge and information in the post-non-classical rationality is the same: they both are turned into a kind of pragmatic matrix for the modern human. So, they formulate the most effective personal rules and regulations in accordance with the modern economic situation.

In these circumstances such a standard model of the person as «homo economicus» can be considered as a cognitive one. The human makes a rational maximizing choice in accordance with it. So, the main condition for using information or knowledge is realization of the human's own needs. The rational value of information and knowledge becomes neutral for the human. They both are evaluated only by the effectiveness of their practical use. Thus, their basic characteristics are the pertinence of information and the performativity of knowledge. They are undoubtedly the most important for the modern human.

References

1. Behman G (2010) Society of knowledge—a brief overview of theoretical searches. Quest Philos 2:113–126 (in Russian)
2. Betelin VB (2019) Challenges and prospects of education in the postindustrial information society. Herald Rus Acad Sci 89(3):231–239
3. Dynkin AA (2019) Response to grand challenges: the dimension of social sciences and humanities. Bull Rus Acad Sci 89(2):120–124
4. Elyakov AD (2012) Information and society (problems of modern society). Samara State University of Economics, Samara (in Russian)
5. Guryanova A, Khafiyatullina E, Petinova M, Frolov V, Makhovikov A (2020) Technological prerequisites and humanitarian consequences of ubiquitous computing and networking. In: Popkova EG, Sergi BS (eds) Digital economy: complexity and variety vs. rationality, vol 87. Lecture notes in networks and systems. Springer, Cham, pp 1040–1047
6. Guryanova A, Khafiyatullina E, Petinova M, Astafeva N, Guryanov N (2020) Social, psychological and worldview problems of human being in digital society and economy. In: Popkova EG, Sergi BS (eds) Digital economy: complexity and variety vs. rationality, vol 87. Lecture notes in networks and systems. Springer, Cham, pp 244–250
7. Kulikov SB (2016) Russian way to the knowledge-based society. Foresight 18(4):379–390
8. Makhovikov LA, Makhovikov AE (2019) Information and knowledge in the structure of rational activity of the modern person. Sci XXI Century Curr Dir Dev 2(2):152–159 (in Russian)
9. Shannon C, Weaver W (1964) The mathematical theory of communication. The University of Illinois Press, Illinois
10. Shestakov A, Noskov E, Tikhonov V, Astafeva N (2017) Economic behavior and the issue of rationality. In: Popkova EG (ed) Russia and the European union: development and perspectives. Contributions to economics. Springer, Cham, pp 327–332
11. Stepin VS (2003) Self-developing systems and post-non-classical rationality. Quest Philos 8:5–17 (in Russian)
12. Wiener N (1953) Ex-prodigy: my childhood and youth. Simon and Schuster, New York

Information as Basic Value of Revenue Maximization in Evolving Public Procurement

D. Khvalynskiy

Abstract An information about competitor's value of the contract became the basic term of revenue maximization in contemporary public procurement. The standard revenue-maximizing auction reduces information rents of the bidders. Such reduction is no longer optimal when the auction's winner may resell the contract to another bidder, and the auctioneer has informational uncertainty about possibility of such resale. If bidders' values become publicly known after the auction losing bidders could compete to buy the object from the winner. This form of resale no longer reduces the information rents of the low-value bidder, as he could still secure the same rents by selling the object in resale. The paper overviews research on auction theory & auctions with resale. The author attempted to find a better design of the auction in resale conditions, providing telling the truth about bidders' value. Implementation of the proposed auction may increase the revenue of standard procurement auctions for the auctioneer.

Keywords Auctions · Bid · Auction formats · Vickrey auction · Second-price auction · Optimal auction

1 Introduction

Auctions are on the rise and have become very useful tool of procurement of goods and services in public and private sectors. Especially over the past few years, their development has been connected with the mass use of the internet, which made the auctions accessible to a wider range of bidders. The number of bidders is then an important component for revenue maximization. The higher is the number of bidders, the lower are the prices in the procurement auction. A wide range of commodities are procured by different kinds of auctions. More than 40% of large firms (over $100 million earning), surveyed in North America, are using auctions for procurement.

D. Khvalynskiy (✉)
Altai State University, Barnaul, Russia
e-mail: hdms@email.ru

S. I. Ashmarina and V. V. Mantulenko (eds.), *Current Achievements, Challenges and Digital Chances of Knowledge Based Economy*, Lecture Notes in Networks and Systems 133, https://doi.org/10.1007/978-3-030-47458-4_5

In connection with the public sector auctions are used in award procedure of public procurements.

Procurement auctions for goods, works and services are often followed by resale in secondary markets, because not all sellers could participate in the primary market. For instance, a company may be registered after the time the government decides to buy goods, works and services, or it may reject to participate in the auction because it has a high participation cost, it may also has tax duties or bad reputation that can prevent a seller from participating in auction, the buyer may be constrained to buy only from domestic firms, and so on. In future the company may decide to sell goods to the winner with lower price than in the procurement auction.

That is why, procurement auctions are often followed by a resale of the contract to another seller, whose valuation is lower. If bidders' values become publicly known after the auction losing bidders could compete to buy the object from the winner. This form of resale no longer reduces the information rents of the low-value bidder, as he could still secure the same rents by selling the object in resale.

In that paper the author attempted to find a better design of the auction in resale conditions. To develop such an auction it should be provided telling the truth about strongest bidders' value. The paper overviews the basic auction theory, surveys the existing literature research on auction design, proposes new auction format, discusses the design process and overall success of the new auction, and provides necessary arguments on its revenue maximization. None of the writing is technical, the author lets alone equations for the further works.

2 Methodology

In the paper the author used methodology of the Auction theory, which is related to Nobel laureate William Vickrey, who is considered its founder and was the first to examine auction bidding under a game theoretic perspective [12]. The research on auction theory began with the independent private values model (IPV). In that model, a single indivisible object is to be procured from one of risk-neutral bidders. Each bidder has an information about the value of the contract to himself, but not to the other bidders. All values are independently drawn from continuous distribution. Bidders are assumed to compete; that's why, the auction is analyzed as a non-cooperative game.

There are four standard types of auctions:

- English auction,
- Dutch auction,
- sealed-bid auction (first-price auction),
- Vickrey auction (second-price sealed-bid auction).

Each auction mechanism has predetermined rules. These rules include:

- the way in which the bids are made,

- the way in which the auction's winner is chosen,
- the way in which the price is paid.

In a Dutch procurement auction the auctioneer begins with a very low price and then rises it continuously until the first bid. That bidder stops the auction and wins the contract for the price named.

A seller in a sealed-bid procurement auction chooses his bid by trading off the probability of winning (by placing the lowest bid) against the profitability if he does win (lower bid means lower price and lower profit). An English procurement (reverse) auction begins with a high price, and then the auctioneer lowers the price until only one bidder remains. In the Vickrey procurement auction a bidder should submit a sealed bid equal to the value of the contract to himself. Bidding lower than one's value can never give a positive profit, and bidding higher means passing up profitable opportunities. As a result, the lowest value seller wins the procurement auction and gets the price set by the next lowest value seller.

Vickrey described, that in Dutch & sealed-bid procurement auctions sellers submit their bids in the same way, based on the number of bidders, individual valuations, and the distribution of valuations among sellers. That is why sealed-bid and Dutch auctions types are strategically equivalent. An English auction yields the same outcome as the Vickrey auction in the IPV environment, regardless of risk attitudes: the item is sold by the lowest value seller at a price equal to the second lowest seller value. For both auctions bidding up to the individual valuations of bidders is their dominant strategy. In both auction types, it is optimal for every seller to fully reveal his value, no matter what other sellers do. This means staying in the English auction until the bid lowers below his value, and bidding his actual value in the Vickrey auction. That is why both auctions are strategically equivalent too.

The main result in auction theory is the Revenue Equivalence Theorem, according to which risk-neutral sellers in the independent private values model lead to identical expected revenues for the buyer in Dutch, English, sealed-bid, and Vickrey auctions. Nobel laureate Roger Myerson showed how to design an optimal auction, maximizing the buyer's expected revenue. That auction reduces information rents of the bidders. Such reduction is no longer optimal when the auction's winner may resell the contract to another bidder, and the auctioneer has informational uncertainty about possibility of such resale. Myerson's theory only applies when the resale factor is not considered. But real procurement auctions do not usually meet this assumption.

The importance of the resale factor has been known since works of P.R. Milgrom [6]. In the last 5 years there were not a lot of papers that cover the major issues in the auctions with resale [5, 7, 11, 13]. The latest works belongs to M. Pagnozzi and K.J. Saral. They show that the presence of a resale after auction significantly rise strong sellers' bids compared to the no resale treatment [8]. The experimental results of M. Pagnozzi and K.J. Saral illustrate how an increase in the probability of resale market can result in less efficient auction outcomes [9].

3 Results

In procurement practice sealed-bid auctions are popular if the number of sellers is small. If the number of sellers is high the uncertainty about the final price is high. Usually sellers could not observe the number of sellers participating in the sealed-bid auction. That is why sellers are not sensitive to the actual number of competitors in sealed-bid formats and bidding is mainly driven by their beliefs about competitors. This implies that buyers get more revenue in sealed-bid auctions if there is high uncertainty about the final price and sellers overestimate competition, which is more likely if the actual number of sellers is small. That is why sealed-bid and Dutch auctions become superior in terms of revenue for the auctioneer when bidders are risk-averse or when there are specific asymmetries in bidders' valuations (there is one significantly stronger bidder which is expected to outbid all weaker bidders).

The research that followed Vickrey paper showed that sealed-bid and Dutch auctions are not superior in terms of revenue when bidders' valuations are depended (affiliated) on the valuations of other bidders. If this is the case, English auction become significantly superior in terms of revenue because only during the English auction bidders receive information about other bidders' valuations to update their personal valuations.

Nevertheless participation of only one real bidder in the auction (formally, up to a dozen affiliated sellers) will lead to awarding the contract at the reserve price in English and Vickrey auctions. Accordingly, if we rise an uncertainty of the bidder about the final price in these auctions, we would significantly decrease the winner's bid in the auction and rise auctioneer's revenue. However, in the moment when auction have just begun the number of bidders become to be essentially static (fixed). Consequently, the bidder is able to find out other bidders and to call them to collude, to use the "winner's curse" information to frighten of other bidders.

Thus, the author found that it is necessary to rise an uncertainty of the bidder about the final price until the end of English auction. It could be done easily by allowing the possibility to contract after the end of the auction with the lowest bid made. Based on the analysis of constrains and problems of standard auctions, the author proposes a new auction design, that has advantages compared to standard auctions. It is a hybrid format of Vickrey & English auctions. It implies the contracting based on competitors' bids.

The auction has several stages:

1. Sellers make their sealed bids on the contract.
2. Sellers review the lowest bid made in the auction and should take it or reject it. If the lowest price wasn't accepted by any bidder, the seller proposed that price become the winner of the auction. If the lowest price was accepted by other bidders, in the auction remain the bidder made the lowest bid & only two bidders accepted it first. All other bidders exit the auction.
3. If there are several bidders remained, sellers make their sealed bids on the contract again. If the lowest price wasn't accepted by any bidder, the seller proposed that

price become the winner of the auction. If the lowest price was accepted by other bidders, only the bidder accepted it first becomes the winner of the auction.

4 Discussion

There are several papers that consider the revenue increasing in the auctions with second-chance offer, where the seller makes a take-it-or-leave-it offer to the second highest bidder from the auction [1, 3, 4]. C.L. Elisa, G. Lewis, M. Mobius, and H. Nazerzadeh also showed that the randomized allocation incentivizes high-valuation bidders to buy-it-now [2]. That is also approved by raising revenue in auctions with sniping, which is the practice of waiting until the last minute before the English procurement auction ends to submit a bid that just barely beats the lower bid and gives rival bidders no time to respond. That also makes a randomized allocation, and the literature suggests that sniping is an effective and prevalent strategy within an eBay auction [10].

The new auction the author proposed combines second-chance offer (the seller may accept other seller's bid) with randomized allocation (only the bidder accepted first the lowest bid becomes the winner of the auction). It could provide a higher expected revenue for the auctioneer because of following arguments:

1. The advantage in revenue of the proposed auction model over the English auction is as follows: the object is sold to the bidder who accepted the lowest bid before other sellers, so there is no strategy to bid higher to the value of the contract to bidder. A seller bids only once, without seeing others' bids, so the bidder with the lowest value of the contract cannot rise his bid to the second lowest value among the bidders that guarantees the win and contracting like in English auctions. Bidding higher than one's value means passing up profitable opportunities, because the seller exits the auction in case of accepting his bid by other sellers. That is why there is a dominant seller's strategy to bid his value like in the Vickrey auction. The opportunity to bid only once eliminates high vulnerability of the new auction to collusion (in English auctions every bidder may submit several bids that ensure that any deviation from a collusive agreement is severely punished by overbidding the «renegade's bid». So if any bidder attempts to obtain more than its agreed share, the strongest bidder will overbid him and will leave that bidder with no agreed shares).
2. The advantage in profitability of the proposed auction model over the Vickrey auction is as follows: the auction price is equal to the bid accepted by anyone or unaccepted by everyone. If the bid is not accepted by competitors, the auctioneer gets more revenue than in Vickrey auction, because the winning price is equal to the bid. The seller exits the auction in case of accepting his bid by other sellers.
3. The advantage in profitability of the proposed auction model over sealed bid auction and Dutch auction is as follows: the ability to accept other sellers' bids uses the linkage principle, which lowers the price depends on all other bidders'

information in an auction. The seller is able to change his evaluation of the contract depending on bids of other sellers. So the lower is the winner's information rent and hence her expected surplus, and so the lower is the final price in the auction.

5 Conclusion

In standard auctions the contract price tends to be the second-lowest valuation among bidders. Resale possibility in procurement auctions no longer reduces the information rents of the low-value bidder. In the paper, the author attempted to develop a new auction model based on the analysis of constrains and problems of standard auctions. The new auction maximizes the revenue of procurement when resale is allowed, providing the truthful revelation to be dominant strategy (players have tell the truth about their value of the contract and this is their dominant strategy). It is a hybrid format of Vickrey & English auctions. It implies the contracting based on competitors' bids. The model provides the dominant strategy of bidder to declare a real value of the contract for himself. Some advantages in a revenue in comparison with standard auctions are described.

References

1. Bagchi A, Katzman B, Mathews T (2014) Second chance offers in auctions. J Econ 112(1):1–29
2. Elisa CL, Lewis G, Mobius M, Nazerzadeh H (2014) Buy-it-now or take-a-chance: price discrimination through randomized auctions. Manag Sci 60(12):2927–2948
3. Kaplan TR, Zamir S (2014) Advances in auctions. https://ssrn.com/abstract=2412429. Accessed 02 Feb 2020
4. Joshi S, Sun YA, Vora PL (2005) The privacy cost of the second-chance offer. In Atluri V (ed) Proceedings of the 2005 ACM workshop on privacy in the electronic society. Association for Computing Machinery, New York, pp 97–106
5. Loertscher S, Leslie MM (2017) Auctions with bid credits and resale. Int J Ind Organ 55:58–90
6. Milgrom PR (1987) Auction theory. In: Bewley T (ed) Advances in economic theory: fifth world congress. Cambridge University Press, Cambridge, pp 1–32
7. Pagnozzi M, Saral KJ (2017) Demand reduction in multi-object auctions with resale: an experimental analysis. Econ J 127(607):2702–2729
8. Pagnozzi M, Saral KJ (2019) Efficiency in auctions with (failed) resale. J Econ Behav Organ 159(C):254–273
9. Pagnozzi M, Saral KJ (2019) Entry by successful speculators in auctions with resale. Exp Econ 22:477–505
10. Roth AE, Ockenfels A (2003) Last-minute bidding and the rules for ending second-price auctions: evidence from eBay and Amazon auctions on the Internet. Am Econ Rev 92(4):1093–1103
11. Susin I (2017) Auctions with resale: a survey. HSE Econ J 21(2):333–350

12. Vickrey W (1961) Counterspeculation, auctions, and competitive sealed tenders. J Financ 16:8–37
13. Virág G (2016) Auctions with resale: reserve prices and revenues. Games Econ Behav 99:239–249

Public Diplomacy in the Context of Cross-Border Knowledge-Based Cooperation Between Universities

A. Zotova

Abstract At present, Russian universities are increasingly involved in the processes of globalization of education and science, which, in our opinion, are components of public diplomacy. The education system has become the most important institution of modernity in the state. The main forms of contacts between Russian teachers and foreign colleagues are participation in joint conferences, and the implementation of joint research projects. Such forms of scientific cooperation are typical for teachers of almost all specialties. A prerequisite for participation in the cross-boarder university cooperation and support of the process is the full awareness of all stakeholders on the entire spectrum of issues (principles, goals, objectives, expected results). Questions of forecasting and exclusion, or at least compensation of the possible negative consequences of choosing specific options for the implementation of cross-border university cooperation, also require an answer. The paper analyzes the existing strategies of cross-border cooperation at universities in terms of implementing issues of public diplomacy.

Keywords Public diplomacy · Cross-border cooperation · University · Knowledge · Education

1 Introduction

World economic cooperation and economic relations in the trade, financial and transport sectors are being successfully built taking into account the knowledge of the economic, cultural and institutional features of all participants in global processes. The modern globalized world requires more and more businesses to operate at international level. This reality has been widely acknowledged by governments throughout the world, both in developed and developing countries [1, 4, 6]. The practical implication of this understanding is the widespread creation of governmental and non-governmental agencies, supporting the international business activities of

A. Zotova (✉)
Samara State University of Economics, Samara, Russia
e-mail: azotova@mail.ru

S. I. Ashmarina and V. V. Mantulenko (eds.), *Current Achievements, Challenges and Digital Chances of Knowledge Based Economy*, Lecture Notes in Networks and Systems 133, https://doi.org/10.1007/978-3-030-47458-4_6

national enterprises through information, consultancy and sometimes even financial resources.

Public diplomacy differs from traditional diplomacy, which is carried out through people with a special profession (diplomats, politicians, intelligence officers). Public diplomacy is the means by which the government of one country tries to influence the society of another so that it, in turn, influences its government. This instrument was widely used in various countries to promote the brand of the country [9, 13]. Issues of public diplomacy were especially sharply discussed during the moments of crises in Russian-American relations [15] and in Russia-EU relations after 2014 [10].

In Russia these instruments were historically implemented with the help of universities. Russian higher educational institutions and state organizations seek to develop contacts with foreign graduates of educational institutions in order to use their potential to strengthen relations between Russia and foreign countries, as well as develop business cooperation and partnership with foreign graduates. Much work is being done among national alumni associations, in which regional alumni meetings are organized in Europe, Asia, North Africa and Latin America. For the period from the beginning of the 50s of the XX century to the 1991, more than 1 million foreign specialists from 172 countries of the world were trained at USSR universities. Foreign students who come to study in Russia are convinced by their fellow countrymen who previously studied in our country as a convincing source of information about Russian universities. This is what makes working with foreign graduates and their national associations (associations) particularly important.

In addition, graduates have significant scientific and professional potential for cooperation in various fields (political, humanitarian, cultural, scientific, technical and economic). Nowadays the main areas of cross-border cooperation between universities are the development of academic incoming and outgoing student mobility, franchising, validation, pre-university training, teaching Russian as a foreign language, etc. Sources of interaction can be both budgetary and extrabudgetary funds (grants). By attracting young foreign students to study in the Russian Federation today and creating comfort conditions for them, the Russian government and educational bodies shape future interaction with other countries, positively affect Russia's international ratings and raise its credibility. In Russia, as in the USA, the educational system does not divide students into "yours" and "ours". All students study on the same conditions. In most universities, visitors do not have to pay more for tuition than local ones. More than 15,000 visiting young people can come to Russia and study at its higher educational institutions for free, living in hostels on the same conditions as the Russians. The competition among applicants for allocated quotas is 4.5–5%. To increase the country's credibility, it is planned to continue providing preferential places in universities for foreigners and increasing their number.

2 Methodology

The study is based on general organizational and economic mechanisms and forms of cross-border cooperation:

- meetings of regional leaders, forums with the participation of business circles, scientists, and the public,
- institutions of representation of enterprises and businesses in the regions,
- foreign economic infrastructure in the regions,
- trade, logistics and transport zones in the border area,
- implementation of joint projects.

Planning and analysis of activities for the implementation of cross-border cooperation is based on the following fundamental principles: (1) increasing the availability of education quality for the interested subject; (2) an increase in the financial resources of cross-border cooperation both at the expense of the budget and by attracting extra-budgetary funds, which guarantees stable development of the process.

Many cross-border cooperation events are currently in the experimental stage and, naturally, according to the results of a scientific assessment of their effectiveness, they will undergo correction. Main effects of cross-border cooperation were studied such as: growth of direct regional budget revenues from training and residence of foreign students, same increase in indirect income affecting on the economy of the region in the long run.

Indirect economic benefits include: job creation (employment, wages, taxes); increase in number of innovative high-tech companies and, accordingly, an increase in volumes of high technological production; foreign growth investment in the region, the creation of new joint enterprises and more [7, 14]. In separate studies [14] some attempts were made to calculate the economic and demographic effects in the regional cuts that get the largest Russian cities from study migration and also develop typology of Russian territories on contribution educational migration to the regional economy. To assess the economic effect can be used different methods, for example: accounting for the direct costs of the presence of foreign students in the region (tuition fees, housing, transportation, food, leisure and so on); an estimation of expenses of foreign students, going from the cost of living in the region; an estimation of expenses of foreign students, going from the amounts recommended by the Ministry education and science of the Russian Federation or living in Russia.

3 Results

State policy of internationalization of higher education in most of the countries is based on an understanding of the importance of joining universities into a single world educational space, respectively the impact of domestic education on international standards honors [3, 5, 11]. The internationalization of higher education for universities today implies, in addition to student and teaching synergies, program and curriculum reform, collaboration in research through networks and associations, open and distance learning without borders, regional and foreign-cooperation between institutions, international division of labor and other activities. Recently, the main task for Russian universities is to get international recognition—due to the entry of Russian universities in international rankings. This is performed through the development of scientific follow-ups and educational programs—through integrated international participation in various forms. It is completely justified with a formalized result of this approach through the indicators of international activities of the university. Among them are the number of international students, international educational programs, including with teaching in foreign languages and program of double diplomas, the number of foreign teaching staff and other indicators.

Foreign youth is quite easily adapted and mastered in Russian cities. It was revealed that the following factors influence adaptation: climate, personality-psychological relations, pedagogical system, life, interpersonal communication, living in international hostels. Important is not only the creation of a sufficient number of benefits for students from other countries, but also their assistance in adapting to the territory of a complex and distinctive Russia. People come to Russia to study mainly from the CIS countries, China, and African states. The number of students from developed countries, including European ones, does not exceed 2000 people. In 2016–2017, there was a large increase in students from countries such as: India, China, Vietnam, African states, Kazakhstan (Table 1).

The forecast for the future is that by 2025, the number of foreign students in the Russian Federation will grow three times. But still there are regions where the number of foreign students is not so numerous, and the universities try to expand their cross-border cooperation strategies. Taking Samara region as an example, the number of foreign students was about 4439 in 2019 and is planning now to increase it by 20%.

Table 1 The number of foreign students in the regions of the Russian Federation (top-10 regions) in 2020

Region	The number of students, percentage
Total number in Russia	237 538 (100%)
Moscow	59 289 (25%)
St. Petersburg	23 030 (9.7%)
Omsk region	11 286 (4.8%)
Tomsk region	10 033 (4.2%)
Tatarstan	7 972 (3.4%)
Novosibirsk region	7 087 (3.0%)
Rostov region	5 806 (2.4%)
Chelyabinsk region	5 358 (2.2%)
Astrakhan region	5 137 (2.2%)
Altai region	4 806 (2.0%)

Source author

Traditionally, there are six components of internationalization:

- international mobility of students, teachers and administrative staff of the university,
- mutual recognition of foreign qualifications through the ECTS credit system,
- revision/adjustment of curricula and programs (educational content), development of joint programs and double degree programs,
- international program mobility,
- the work of countries, regions and universities in attracting foreign students and creating partnerships,
- reaching a strategic level in participation in international partnerships and cross-border networks.

Most of the universities in Samara region include them in their university strategies. Thus, the strategy of Samara University focuses on academic mobility programs; double degree programs; joint participation in scientific conferences; writing and publishing joint scientific articles; joint research; participation of visiting professors at Samara University. But the formalized internationalization strategy is not still developed at the university. Samara Polytech presents internationalization strategy at their website and it focuses on ensuring its integration into the international university community, obtaining additional opportunities for accelerated development within the framework of: academic student and teacher exchange programs; joint educational programs; exchange of educational products and participation in joint research grants. So, Samara Polytech international strategy makes special accent on grant activities but doesn't give any strategic indicators. Samara State Medical University only starts the formalized process of internationalization strategic planning although 340 foreign students studied at the faculty of training even in 2015 (about 70 students more from Kazakhstan). Primarily, Samara State Medical University attracts students from neighboring countries—from Kazakhstan, Tajikistan, Belarus, Armenia, Azerbaijan, Kyrgyzstan. But the organizational development of internationalization process started only in 2019 with new international cooperation department set up at the university.

Samara State University of Economics indicates in its internationalization strategy that gradual integration of the university into the international educational space requires the compliance with international standards of academic programs quality and a high level of internationalization of the educational process. Within the framework of the concept of international activities development in SSUE, there are two important trends for university international activity. These are obtaining international accreditations, double master's degree programs development and obtaining a PhD degree in cooperation with foreign academic partners.

An important long-term project within the framework of international activity development is obtaining international accreditations of educational programs and membership in leading international professional associations. International accreditation is quality assessment of the program in relation to established standards. This is a continuous process of consulting, as well as the improvement of educational technologies that ensure program uniqueness, strengthen the prestige and market position of the university. International accreditations, as well as the membership in professional organizations is an essential condition into ensure the opportunity to participate in prestigious international rankings. Thus, the further development of international cooperation is one of the key activities and main competitive advantage of Samara State University of Economics.

The examples of the universities from Samara region give us the general understanding of the problem common for other regions with low numbers of international students too: regional universities are focused on domestic activities mainly. These universities interact with people who already speak Russian or organize training language courses for teaching Russian language to future foreign students. They have a few foreign lecturers or professors. In turn, Russian teachers very rarely come to foreign universities.

The universities are trying to intensify cross-border cooperation through a number of projects. Samara State University of Economics is now finishing the project "Enhancement of higher education and corporate sectors integration in accordance with new social environment" (ENINEDU) which is performed in the consortium together with two leading Kazakhstan universities (Kazakhstan National University "Al-Farabi" and Eurasian National University. L.N. Gumileva). The overall objective of the project is to expand cooperation and align interests between business and the higher education system in all countries of the consortium.

Another project which is now being developed at SSEU is new educational program aimed at raising the effectiveness of enterprises in CIS area through raising the specialists able to perform the support activities in the field of commercial diplomacy. The uniqueness of this program is that the region of Russia involved in the project is a highly multinational region of the Russian Federation, on the territory of which representatives of different cultures are historically interacted. The Volga region is traditionally a contact zone of Indo-European Slavic, Finno-Ugric and Turkic-speaking peoples, which represents miniature model of Eurasia. The region is multi-confessional character, which allows to consider its experience as an example of successful integration of various religious practices. The significant feature of the Samara region in the Volga region is the role of transfer—the territory of economic, legal, cultural, linguistic communications between the West and the East. So, the joint program, aimed at studying Eurasian cultural and economic integration, will allow accumulating and broadcasting the experience of intercultural interaction in the region in the educational space of the countries of Eurasian continent.

4 Discussion

The university became the initiator, unifier and distributor of international experience, as well as the center of its improvement. But there is no universal model of globalization and internationalization of education: each university should implement its own approach to this process, based on clearly formulated tasks, regional specifics and expected results [2, 8, 12]. Another disputable issue is that financial and economic factors have a serious impact on the development of cross-border cooperation between universities. Established quotas and imbalances in demand for educational programs sometimes do not allow meeting planned targets. Also in this area should take into account reputational risks.

5 Conclusion

Over the years, Russian education has become more accessible to foreigners and adapted to various ethnic groups, since studies are being carried out on the specifics of the national psychological characteristics of various peoples. Given these, it is possible to create conditions under which students coming to Russia from other countries will take home not only good impressions, but also a large amount of knowledge. For Russian universities cross-border cooperation allows to increase the flow of academic mobility (incoming and outgoing), stimulate participation in international scientific and educational projects. The process is structured through the following stages: the use of new educational technologies in the educational process, expanding the range of educational programs, improving the quality of education, expanding international cooperation, additional characterization of the reputation of the university, the effective use of material and technical (classrooms, libraries, technical teaching aids, hostels and other) and human resources (teachers, teaching support and administrative staff). There is an urgent need to study, generalize and discuss the processes occurring in the field of higher education, forms of international cooperation, best practices and development prospects. In the context of globalization, higher education implements the function of a generator of knowledge and the means of its transmission, is the basis for the national development of regions.

Acknowledgments The publication was inspired by the project 574060-EPP-1-2016-KZ-EPPKA2-CBHE-SP "Enhancement of Higher Education and Corporate Sectors Integration in Accordance with New Social Environment" (ENINEDU) co-funded by the Erasmus+ Program.

References

1. Apanovich M, Teteryuk A (2019) Business diplomacy as a tool for realization of economic interests of domestic companies at the present stage: problems and prospects. Theor Prob Polit Stud 8(4A):40–54. https://doi.org/10.34670/AR.2019.45.4.047
2. Chiocca E (2020) Hearts and minds: goal-orientation and intercultural communicative competence of ROTC cadets learning critical languages. Intercult Educ 31(1):102–132. https://doi.org/10.1080/14675986.2019.1666247
3. Coffey B (2019) Teaching economics across cultures: challenges and lessons learned in Tajikistan and Kyrgyzstan. Intercult Educ 30(6):415–428. https://doi.org/10.1080/14675986.2018.1540108
4. Collins N, Bekenova K (2019) Digital diplomacy: success at your fingertips. Place Brand Public Dipl 15:1–11. https://doi.org/10.1057/s41254-017-0087-1

5. Flores S, Park T, Viano S, Coca V (2019) State policy and the educational outcomes of English learner and immigrant students: three administrative data stories. Am Behav Sci 61(14):1824–1844. https://doi.org/10.1177/0002764217744836
6. Golan G, Manor I, Arceneaux P (2019) Mediated public diplomacy redefined: foreign stakeholder engagement via paid, earned, shared, and owned media. Am Behav Sci 63(12):1665–1683. https://doi.org/10.1177/0002764219835279
7. Golubinov I (2017) Strategies of modern university internationalization. Izvestiya Samara Sci Centre Rus Acad Sci Soc Humanit Medicobiol Sci 19(3):16–20
8. Jæger K, Gram M (2020) Enduring not enjoying? Emotional responses to studying abroad among Danish and Chinese students. Intercult Educ 31(1):1–15. https://doi.org/10.1080/14675986.2019.1673651
9. Kim HS (2018) When public diplomacy faces trade barriers and diplomatic frictions: the case of the Korean wave. Place Brand Public Dipl 14:234–244. https://doi.org/10.1057/s41254-017-0076-4
10. Ociepka B (2019) (Un)successful years: EU countries' cultural diplomacy with Russian Federation. Place Brand Public Dipl 15:50–59. https://doi.org/10.1057/s41254-018-00113-3
11. Osler A, Pandur I (2019) The right to intercultural education: students' perspectives on schooling and opportunities for reconciliation through multicultural engagement in Bosnia and Herzegovina. Intercult Educ 30(6):658–679. https://doi.org/10.1080/14675986.2019.1626576
12. Saffer A, Yang A, Morehouse J, Qu Y (2019) It takes a village: a social network approach to NGOs' international public engagement. Am Behav Sci 63(12):1708–1727. https://doi.org/10.1177/0002764219835265
13. Samuel-Azran T, Ilovici B, Zari I (2019) Practicing citizen diplomacy 2.0: "The hot dudes and hummus—Israel's yummiest" campaign for Israel's branding. Place Brand Public Dipl 15:38–49. https://doi.org/10.1057/s41254-018-00111-5
14. Vashurina E, Evdokimova Y (2017) Developing foreign student recruitment system: regional model. Univ Manag Pract Anal 21(1):41–51. https://doi.org/10.15826/umpa.2017.01.004
15. Velikaya A (2019) Russian—U.S. public diplomacy dialogue: a view from Moscow. Place Brand Public Dipl 15:60–63. https://doi.org/10.1057/s41254-018-0102-1

Information and Knowledge: From a Resource to a Key Factor in Progress

S. A. Sevastyanova and A. L. Sevastianova

Abstract In the age of social transformations associated with changes in the environment, technical and communication capabilities, as well as moral attitudes, the problem of determining a value system, reference points for the transition to a new format of the information society arises. This is necessary for comprehending the new realities, understanding the laws governing the functioning of the socio-economic system, determining development goals, principles and management mechanisms. The purpose of this article is to offer the scientific community arguments for the formation of a value system in the context of economic and social changes associated with the widespread adoption of digital technologies. Based on these arguments, reasoning is built on the importance of knowledge in a new format of social relations, forming the conceptual basis for building a new economic paradigm. The article provides a conceptual and philosophical analysis of the concepts of "information" and "knowledge", the prerequisites and prospects for the transition to digital economic models, using social and hierarchical approaches to the definition of the value concept. Information is considered as an intellectual resource for the formation of knowledge. Human intelligence transforms information into knowledge as a result of comprehension, systematization, structuring, interpretation.

Keywords Knowledge based economy · Values · Digital economy

1 Introduction

Russia's entry into the status of a society with a digital economy is a priority for the development of the state. Since the mid-2010s, Russia has adopted a number of documents declaring the state support for measures to introduce digital technologies

S. A. Sevastyanova (✉)
Samara State University of Economics, Samara, Russia
e-mail: s_sevastyanova@mail.ru

A. L. Sevastianova
Saint-Petersburg State University, Saint-Petersburg, Russia
e-mail: alex-sandrine@mail.ru

S. I. Ashmarina and V. V. Mantulenko (eds.), *Current Achievements, Challenges and Digital Chances of Knowledge Based Economy*, Lecture Notes in Networks and Systems 133, https://doi.org/10.1007/978-3-030-47458-4_7

in various areas of production, business, finance, education, public services, and law. The key documents are the Development Strategy of the Information Society of the Russian Federation for 2017—30 years and the national program "Digital Economy of the Russian Federation". Priority digital technologies were identified: big data, quantum technologies, components of robotics, sensors, neurotechnologies, artificial intelligence, new production technologies, industrial Internet, distributed registry systems, wireless technologies, virtual and augmented reality technologies.

The special significance of the task has been reflected in scientific research of recent years. Their main result was the formation of an understanding of the special role of informatization and digitalization in the concept of a new economic paradigm. In this latest scientific theory, an active process of comprehending new realities, understanding the patterns of functioning of the socio-economic system, determining goals, principles and management mechanisms is taking place. Standing on the verge of radical changes associated with changes in the environment, technical and communication capabilities of mankind, moral attitudes, we are wondering about the definition of a system of values, reference points for the transition to a new format of post-industrial society. It is necessary to direct the complex mechanism of socio-economic relations both to satisfy the personal needs of each person, to ensure the high quality of life of each individual, and the sustainable development of society as a whole.

2 Methodology

The analysis of scientific publications on the research topic, as well as a conceptual and philosophical analysis of the basic concepts of "information" and "knowledge" provide a basis for understanding the place and role of these concepts of the value system in the context of economic and social changes associated with the widespread adoption of digital technologies. Social and hierarchical approaches to the essence of the value concept determine our point of view regarding this analysis. The causal analysis of the observed social transformations leads to the conclusion about the regularity and timeliness of the change in the economic paradigm.

3 Results

The objective nature and pace of development and implementation of digital technologies leave no doubt that in the nearest future many aspects of public relations will be reviewed. To discuss problems of a legal, ethical, political nature, material relations, and many others, it is necessary to define the system of value concepts on the basis of which the position of the individual and social groups and the state will be formed. The analysis allows us to conclude that the importance of knowledge as a tool for the production of new judgments, and therefore the progress of mankind, is unchanged. Knowledge cannot be replaced by the results of using digital technologies without losing the fundamental dominant of a person over his creations. The basis for the knowledge production is the possession of information. The redundancy of information gives rise to a number of problems that need to be addressed, including the legal field. Moreover, the availability of information is a fundamental value of civil society.

4 Discussion

Under the influence of objective changes in the environment, the level of technology development and their implementation in various spheres of human activity, changes in social relations and their management mechanisms are inevitable. There is a need to transform the traditional approach to the concept of shared values [2]. A conceptual understanding of the changes, expressed in the formation of a new economic paradigm, is a factor determining similar processes in other areas, for example, in education [5]. Following the economy, we should expect a change in the educational paradigm as a reaction to the new demands of society for training personnel for the labor market, for the education of a new generation to be ready to interact in changing conditions [11]. Over the next decade, the main participants in public relations will be representatives of generation-Z, young people born in 2000–2010. Understanding the distinctive features of the generation, based on data from sociological and psychological studies and forecasts, makes it possible to smooth out acute angles in the economy, education, and the social sphere, and brings us closer to understanding the psychological and sociological conditions and prospects of the educational system.

The main features of the new generation, according to various sources, include:

1. The use of electronic devices for communication, training, information, self-organization, entertainment.
2. Multitasking. The quality that allows performing several types of activities at the same time cannot be unambiguously assessed as negative or positive. On the one hand, the quality of operations may be reduced, and on the other, this feature allows you to initially process a significantly larger amount of information, highlighting the more important and filtering out the insignificant.
3. Clip thinking, fragmentation and inconsistency of information perception due to its redundancy. Reducing the depth of the new information development, the superficiality of the knowledge assimilation [3].
4. Focus on personal development and self-realization, including when choosing a profession, life priorities [10].
5. The desire for the priority use of resource-saving technologies and materials; awareness of environmental responsibility.
6. Intellectual, emotional, and psychological overloads, and as a result—a decrease in attention and concentration ability, a negative impact on health (lack of sleep, walking in the fresh air, physical activity, etc.).

7. A high degree of involvement in social communities through communication in social networks.
8. Greater use of alternative forms of training (online courses, video lectures, etc.)
9. Self-education, moving of focus from a teacher to a student.

Thus, in the new realities, man is changing. Accordingly, changes cannot but affect literally all aspects of the functioning of society [6]. One of the most important signs of a change in the concepts of society under the influence of digitalization is a change in the labor market [1]. Already, we are witnessing a change in its structure, manifested in the redistribution of labor resources in the direction of increasing the number of workers employed remotely, people in creative professions, and reducing a number of "routine" professions. Demand is constantly high for specialists whose responsibilities cannot be formalized to the level of equipment usage, technologies, artificial intelligence. Russia also has its own specifics of demand, related to the level of wages. It reflects the "pain points" of labor relations and allows you to determine positions that require increased attention from the state, the adoption of special support measures in the form, for example, of national projects, programs, etc. First of all, the education system should respond to changes in the labor market. Despite the complexity and formalization of the system, there are signs of changes: the spread of distance and on-line training on various platforms, the introduction of an automated control system for the interaction of all participants in the educational process, the availability of information resources and communication tools, the new areas of training emergence, the possibility of continuous self-education and development abilities, regardless of location, age, physical and other abilities. Improving information literacy in the personal space of a person simplifies the introduction of digital technologies in domestic and financial issues, issues of interaction with authorities, health care institutions. Ordering a taxi, booking a hotel, paying bills, filling out documents through the public services service, chatting with members of the virtual community has become commonplace for many citizens, including those of mature age.

Of course, all these factors are systemically associated with changes in the economy. The subjects of economic relations, their goals, principles, methods of implementation and management are changing [4]. All this we call the change of economic paradigm. Digitalization has become a catalyst for the emergence of the latest models of relations between a seller and a consumer of goods and services [12]. On the basis of data analysis and consumption forecasting, an optimal level of production supply, maintenance and demand satisfaction is achieved. The use of technologies for processing big data, artificial intelligence, and the analysis of the digital footprint allows you to choose the optimal parameters for these relationships for the client. Organization of business contacts on digital platforms allows business partners to optimize the costs of time, labor, finance, resources. Expanding the services provided is the most humane form of struggle for the customer. Today, no one is surprised at the reminder in a mobile device not to forget to take an umbrella, order water in the office or pay utility bills. Suppliers of goods and services want to be close to the client, create a friendly image and a good reputation to continue a profitable relationship. Already, artificial intelligence and machine learning technologies are ready to offer a person a choice of ways to solve his problems, taking into account personal preferences. The production Internet takes to a new level the processes of management, innovation and production optimization. Tracking product quality, monitoring production waste and energy consumption, simulation to identify weaknesses, digital design of new products are signs of a reliable modern manufacturer.

All the transformation processes described above and their results entail the idea of revising economic priorities. There is clearly a need for the formation of a new techno-economic paradigm, and with it, the definition of new goals for economic development. A promising scientific theory is born that reflects new objective realities, statements of actual problems and approaches to their solution, principles and laws of the functioning of the socio-economic system, methods of managing the system and methods for assessing effectiveness [8]. To set the vector of the direction of theoretical research, the primary task is to determine the basic values of the new system, to rethink the place and role of current guidelines. Is increased consumption good for future generations? Is it worth it to increase production and in what ways? What should competitiveness be based on? Many of these questions are based on the study of the value system of future generations.

Speaking about the category of value, it should be noted that society cannot function outside the system of values, it is a reflection of the culture, history and level of development of society. The development of philosophical thought laid the foundation for numerous scientific theories on the understanding of values, which were subsequently combined in the scientific field of axiology. Like many authors, we tend to consider the social rather than the natural aspect of the value concept as a priority, without supporting the objective-naturalistic approach. To a greater extent, we are close to the Marxist concept, which closely connects values with social being and the practical activities of people, expressing the objective socio-historical nature of the concept. In a general sense, value is a kind of material or intangible entity that has a certain value for members of a given society. The value category is variable, subject to change depending on the context and various conditions. The heterogeneity of being implies the heterogeneity of the significance of phenomena, their hierarchy, the possibility of modulation, the transition from one level to another, between an individually personal level, social and universal or basic. The value picture of the world is binary: on the one hand, it is a system of moral standards inherent in this community, and on the other hand, it also acts as an assessment tool through which the subject determines his attitude to the phenomena of surrounding reality. The value category is the result of the refraction of the world image concept in the human mind and its interpretation, and the totality of values can be considered as a separate value aspect of the general world image [9].

In our understanding, information is data about various facts, presented in a symbolic, visual or verbal form for the purpose of subsequent transmission. The basis of information is a combination of data, observation results. Primarily processed data constitute the information space. For example, instrument readings entered in the table; video clips viewed after extracting from a video surveillance device, news agency messages, etc. Information is nothing but an intellectual resource. Why can information be called a resource? Because virtual data is not a value "on their own", as it does not exist outside the human consciousness. The value of information is realized in the possibility of processing to meet the various needs of the individual and society. Just as, for example, natural resources become valuable only when they are extracted, processed and sold for the benefit of man (or have such a perspective). The effectiveness and direction of its use depends on the method of presentation and interpretation of information. From a legal point of view, information resources mean documents and their arrays in information systems. However, in a broader sense, any significant information to be transmitted and further used can be called an information resource. Information resources in their value to society are comparable with other strategic resources (natural, energy, raw materials, etc.), since they serve as "raw materials" for the production of intangible products.

Information, being an intellectual resource, must go through the stages of receipt, processing, storage, and implementation. Each of these stages is associated with a number of objective problems. For example, obtaining information is a process whose dynamics show incredible growth rates in connection with informatization and globalization. Moreover, the development of electronic communication tools gives rise to talk about a new phenomenon—the transformation of the information society into a network. The degree of information accessibility, the speed of its creation and dissemination raises the problem of redundancy of information. The volume of new information exceeds the ability of a person to process it, but again digital technology comes to the rescue of a person. Cloud storage and computing are increasingly being used to store and process large amounts of data. Under these conditions, the problem of the reliability of the information arises. No Internet resources today can guarantee the objectivity of published information, and the information itself can be interpreted subjectively or intentionally distorted. Hence the problem of the security of personal information from undesirable publicity. The relevance of information is also a question: in some areas, the situation can change quite quickly, data can become outdated before any actions are performed on the basis of it.

Since we consider information as a resource, it is reasonable to think that the processing of information will lead to getting some valuable substance, and this substance is knowledge, which may be defined as a complex of information and all kinds of connections between its items, used to develop new judgments or make decisions. Knowledge is information passed through the prism of understanding, systematization, structuring, and interpretation. Information is the result of observation, some evidence. In order for this data to become a stimulator of any action, a factor of progress, the participation of intelligence is necessary. At the moment, this is not the result of the work of artificial intelligence, but the result of the mental efforts of man. It is at this stage that the intellectual potential of the information consumer is revealed: a member of the society, researcher, scientist, businessman, politician, etc. In scientific discussions today, an opinion is formed that artificial intelligence can serve as a better moderator of economic development than knowledge [7]. In our opinion, this position is premature, although it is promising and worthy of attention. Knowledge is not an antagonist of a creative idea. On the contrary, creative thought only has value when it is a superstructure of knowledge. If we talk about scientific research, it is the increment of knowledge that has arisen as a result of information processing that is the goal of all previous research activities. This increment of knowledge serves as the basis for improving the living conditions of a person, developing new technologies, creating management methods, understanding the laws of functioning and dynamics of systems development at various levels—from everyday to absolute. Therefore, knowledge is valuable in a material sense, as evidenced by the rapid development of the legal field in the direction copyright and intellectual property. The struggle for the best minds is the struggle for the possibility of increasing knowledge. Knowledge allows you to expand the horizons of the situation, set a new task and look for methods of solution.

Are information and knowledge values in the modern world? Of course yes. Social values have a double meaning. Firstly, as part of consciousness, they are the basis for the formation and preservation in minds of attitudes that help the individual to take a certain position, express a point of view, give an assessment. Secondly, values act in a transformed form as motives for activity and behavior, since the orientation of a person in the world and the desire to achieve certain goals inevitably correlate with the values included in the personality structure. This fully applies to the categories discussed in this article. Knowledge is the basis of a person's understanding of phenomena and processes in the world around him, in the consciousness of man himself at the household and professional levels. "Elite" knowledge is the basis of the latest scientific developments, progress and security. In our time of high technology, the struggle for the best minds is becoming part of the state policy of advanced countries. The ability to receive and transmit knowledge is the most important factor in the development of the individual and society as a whole.

Note that there is a difference between the "weight" of the value concepts of "knowledge" and "information". Above, we discussed the view that information is primary, and knowledge is a consequence of the "processing" of information by consciousness. And if the processing of information is entrusted, for example, to artificial intelligence, will the result of this processing be knowledge? I don't think so. This will be a new level of information, the basis for the "next" knowledge. Thus, we come to the conclusion about the hierarchical structure of these values: the concept of "knowledge" dominates the concept of "information", it is a concept of a higher level.

5 Conclusion

In conclusion, we add that freedom of information is one of the basic values of a democratic society. A person deprived of the opportunity to receive reliable information is deprived of the opportunity to be in the know about events, objectively perceive the situation, analyze, predict processes in various fields: social, economic, political, environmental, technical, etc. In restricting information, there is a fine line between security, common sense, ethical restrictions, health care on the one hand and authoritarianism, restriction of personal freedoms of citizens, unfair competition on the other. The ability to transform information into knowledge is the most valuable quality of human intelligence, which is a key factor in progress.

References

1. Abdrakhmanova K, Vishnevsky G, Gokhberg L, Dranev Yu, Zinina T, Kovaleva G, Shmatko N et al (2019) What is the digital economy? Trends, competencies, measurement. HSE, Mosow
2. Glauner F (2019) Redefining economics: why shared value is not enough? Compet Rev Int Bus J Inc J Global Compet 29(5):497–514
3. Glukhov A (2019) Generation Z's digital literacy: a social network view. Tomsk State Univ J Philos Sociol Polit Sci 52:126–137
4. Goldfarb A, Tucker C (2019) Digital economics. J Econ Lit 57(13):3–43. https://doi.org/10.1257/jel.20171452
5. Martyakova E, Gorchakova E (2019) Quality education and digitalization of the economy. In: Lecture Notes in Mechanical Engineering, pp 212–218
6. Nazarov D, Fitina E, Juraeva A (2019) Digital economy as a result of the genesis of the information revolution of society. In: Nazarov A (ed) Proceedings of the 1st international scientific conference modern management trends and the digital economy: from regional development to global economic growth. Advances in economics, business and management research. Atlantis Press, Paris, pp 351–356
7. O'Donovan N (2020) From knowledge economy to automation anxiety: a growth regime in crisis? New Polit Econ 25(2):248–266
8. Romanova O, Akberdina V, Bukhvalov N (2016) Common values in the formation of the modern techno-economic paradigm. Econ Soc Changes Facts Trends Forecast 3:173–190
9. Sevastianova A (2018) Pragmalinguistic analysis of value-event discourse (on the material of speeches of the USA political leaders). PhD dissertation, St. Petersburg State University, St. Petersburg
10. Skrbiš Z, Laughland-Booÿ J (2019) Technology, change, and uncertainty: maintaining career confidence in the early 21st century. New Technol Work Empl 34(3):191–207
11. Yavorskiy MA, Milova IE, Bolgova VV (2020) Legal education in conditions of digital economy development: modern challenges. In: Ashmarina SI, Mesquita A, Vochozka M (eds) Digital transformation of the economy: challenges, trends and new opportunities. Advances in intelligent systems and computing, vol 908. Springer, Cham, pp 455–462
12. Zaki M (2019) Digital transformation: harnessing digital technologies for the next generation of services. J Serv Mark 33(4):429–435

Digital Forum as a Challenge to Jurisdiction of a Modern State

S. K. Stepanov and A. V. Nektov

Abstract The article investigates the regulatory impact of the state in the Internet. The authors consider a problem of applying law in the digital environment. Since the law application is based on the state coercion, there is a limitation of the state jurisdiction in the Internet. The reason for the emergence of a problem with the law application is the lack of a key territoriality principle in the digital environment for determining the state jurisdiction. Trends for the formation of an independent digital forum and its own enforcement mechanism are identified. Possible types of regulation of the digital forum are highlighted. The application of the conclusions formulated in this article is necessary to determine a further vector of legal regulation of the digital environment and its methodological foundations.

Keywords Jurisprudence · Internet · Jurisdiction · Law enforcement · Digital law

1 Introduction

Born in the forties of the last century in the United States, the Internet was released to wide public use in the 1980s. In 1997, the United States proclaimed basic principles of the state regulation of the Internet space in the form of the Framework for Global Electronic Commerce. Published by the Presidential administration, they became a model for building regulatory systems of the Internet space for many other countries [4]. Since then, the Internet, as well as the attitude of society to it, has changed significantly. There has been a clear transformation of the legal regulation of public

S. K. Stepanov (✉)
Moscow State Institution of International Relations (University) of the Ministry of Foreign Affairs of the Russian Federation, Moscow, Russia
e-mail: s.stepanov@inno.mgimo.ru

A. V. Nektov
University of Birmingham, Birmingham, UK
e-mail: artem_nektov@outlook.com

S. I. Ashmarina and V. V. Mantulenko (eds.), *Current Achievements, Challenges and Digital Chances of Knowledge Based Economy*, Lecture Notes in Networks and Systems 133, https://doi.org/10.1007/978-3-030-47458-4_8

relations related to its use. In 2018, in connection with the case of the famous messenger Telegram [5], a new stage in the history of the Internet and its legal regulation has begun in the Russian Federation, which seems to be far from unambiguous.

It is known that the jurisdiction of a state is determined primarily based on certain territorial borders. Even in the United States and England, where the possibility of extraterritorial legal regulation is allowed, there is a presumption against extraterritorial application of state law [19]. It is also believed that the right is based on compulsion. But what if the state cannot enforce the law in the Internet space? Should the loss of the territoriality principle in the digital environment be followed by the loss of the state control over the Internet? Can this serve as a basis for the formation of a new legal regulation that is still unknown to society?

In an attempt to answer these questions, researchers predicted either an absolute separation of the Internet space from off-line reality in the context of legal regulation [8], or they took a more moderate view, allowing for a cautious extension of the state jurisdiction to the digital environment [17]. However, before the Telegram case, there was no situation of a total failure in the law enforcement mechanism, and there were no clear restriction indicators of the state regulation. This enables an attempt to analyze the impact of the Internet on the state jurisdiction, as well as to formulate a forecast for the future, based on trends in legal regulation observed in the past and present.

2 Methodology

The research methodology is aimed at identifying new trends in the law development in the era of digitalization. The research was based on both general scientific methods of cognition and special methods. We used methods of analysis, synthesis, system and structural approaches to identify possible ways of legal regulation in the digital environment. The formal legal method was used to study court decisions on the subject of our research. Applying a comparative approach, we were able to identify a certain analogy between the comity doctrine in the private international law and the current position of the state in the Internet space. To achieve a full understanding of the text of scientific works, linguistic, systematic, logical and historical methods of interpretation were combined. The categories of the general theory of law were widely used in the work. Thus, in order to justify a gradual transition from the Westphalian model of regulation to the Post-Westphalian model, we discussed the problems of jurisdiction and state sovereignty in the era of digitalization.

3 Results

The results obtained in the research work allow us to formulate the main trends in the development of law in the era of digitalization. The state's monopoly on rule-making is being lost in the digital environment. Many subjects claim to regulate behavior in the Internet. The main way to regulate behavior is social (corporate) norms, the source of which is a contract. At the same time, it should be understood that the jurisdiction and sovereignty of a state currently need to be rethought. We have proved that the territoriality principle of a legal norm is not reflected in the Internet environment. In the Telegram case, we also demonstrated that the state does not have enforcement in the Internet. In this regard, we note that the regulation of behavior in the Internet forms a new digital forum, which has its own mechanism for enforcing the behavior rules and has a contractual nature.

4 Discussion

On April 13, 2018, the Tagansky court of Moscow ruled in favor of the Federal service for supervision of communications, information technology and mass communications, thereby allowing the well-known Telegram messenger to be blocked [5]. At the same time, the court's decision explicitly stated: "… to oblige the Federal service for supervision of communications, information technologies and mass communications and other persons to stop creating technical conditions for receiving, transmitting, delivering and (or) processing electronic messages of Internet users" [5, p. 7], distributed by means of information systems and (or) programs for electronic computers, the operation of which is provided by Telegram Messenger Limited Liability Partnership, including "by restricting access to the specified information systems and (or) programs for electronic computers, until the specified organizer of information distribution in the Internet fulfills the obligation established by law" [5, p. 7]. However, Telegram continues to work until now, and statistics are provided about the sharp growth in popularity of Telegram channels.

It was from this point in the development of law, in our opinion, that a new round began. What happened? What circumstances allow us to say that the traditional theory of law is subject to revision?

The traditional theory of law is based on coercion. It is permissible to use physical coercion against someone who commits an offense [11]. At the same time, supporters of this view refer to I. Kant. Since I. Kant considers positive-legal compulsory order compatible with the ideal of freedom and tries to justify this order. I. Kant is forced to interpret any offense as an "obstacle to freedom" and thus make the coercion used against the violator compatible with freedom [9].

Telegram's actions are illegal from the point of view of traditional theory, since Telegram failed to fulfill its obligation to provide the Federal security service of the Russian Federation with the information necessary for decoding received, transmitted, delivered and (or) processed electronic messages. In accordance with the theory of H. Kelsen [10], the act of coercion (blocking a Telegram) acts as a reaction to illegal behavior determined by the law. However, the state, represented by the authorized bodies, cannot ensure complete blocking of the Telegram. This raises two fundamental questions. Is it needed to regulate the interaction of individuals in the Internet? And if so, how can such regulation be theoretically justified?

R. Posner gave a very convincing answer to the first question [15]. Regulation in the Internet is necessary for four reasons. First, the Internet makes it easier to anonymously distribute and receive obscene material (such as child pornography). Second, it is believed that the Internet does not control the accuracy of information that can be misleading. Third, because the Internet provides access to a potentially large audience, it increases the potential harm. Fourth, the Internet often promotes anti-social behavior. However, the word "regulation" does not mean coercion. For example, if false information is disseminated, it is assumed that private demand for fact-checking will eventually lead to the Internet gaining the same effective control over the accuracy of information as traditional media [15]. It seems that in the first and fourth cases a direct state intervention is necessary, for example, blocking such Telegram channels. However, such actions of the state may have side effects, as potentially socially significant but unconventional statements are censored. And since Telegram channels are open to everyone, including government agents, society can protect itself by monitoring them rather than prohibiting them.

The answer to the second question about the way of theoretical justification of regulation in the Internet is more complex. One possible explanation is the theory of forums, the most interesting way proposed by P. Prodi. At the same time, the forum should not be understood only as a physical place of justice administration. There are both an external forum where positive law prevails with its inherent compulsion, and an internal forum where the main focus is on a sense of justice, conscience. Among other things, P. Prodi points to the existence of other forums that overlap [16]. At the same time, borders have movable borders and a constantly shifting dividing point.

The state is a monopoly of coercion in the external forum. As a result, positive law, largely under the influence of the Internet and globalization, has acquired two completely abnormal characteristics: ubiquity and autoreference. Thus, omnipresence means that the state has penetrated vast areas of everyday life that were not regulated by positive law, but were regulated by ethical norms or customs. Autoreference means a very dangerous misconception that any conflict or problem can be solved with the help of positive law: society is caged, entangled in an increasingly dense network, which was not the least reason for the failure of welfare state [16].

Geographical boundaries of a state have traditionally defined limits of the exercise of a state's sovereignty and, consequently, its *jurisdiction*. That is, historically, they were considered, first, as a limit of the possibility to implement a mechanism of the state coercion. Secondly, these are usually limits of geographical borders that outlined the *legitimacy* of the existing legal regulation, and therefore of the state. The primary recognition of a state is carried out primarily by persons living in a certain territory and within that territory. Recognition of the people as a sole source of power determines the legitimacy of a state and legal regulation. Third, geographical boundaries as indicators of changes in legal systems are ingrained in the legal consciousness. It is the intersection of geographical borders of a state that is most often associated with the termination of the law of one state and the beginning of the law of another state. Thus, if a person crosses the border of a state, even common knowledge of the law is sufficient to understand that the fact of entry obliges to follow the rules established in the country of entry [8].

Since the second half of the XX century, a new forum has been formed, which has recently been actively asserting itself. We are referring to a digital forum. The Similarity of processes of forming a new digital forum with the formation of forums during the Middle Ages and the Enlightenment (mainly the forum of merchants or commercial law, lex mercatoria), has given rise to some modern researchers to call a new system of legal regulation "lex informatica" [18]. Of course, the forum should be described through enforcement mechanisms existing in it. We can easily find such mechanisms in the digital environment, for example, blocking and deleting an account in a social network. Is this compulsion comparable to compulsion in an external forum?

Similar discussions have already been observed in the literature, for example, H. Kelsen said that morality is also a normative device that prescribes sanctions; prescribing a form of behavior, indicates that one should react in a certain way to the opposite behavior that goes against morality. This behavior is discussed by the society members. They should respond to it with acts of disapproval, namely, censure, expressions of contempt, and so on. Morality differs from law in that reactions prescribed by it, that is, its sanctions, are not of a coercive nature, in contrast to the sanctions provided by law, that is, in contrast to legal sanctions, physical coercion cannot be used to assert them if they are resisted [10]. This issue is debatable. For example, is it a physical compulsion to block the account of a well-known blogger who uses it to extract income?

However, it is most interesting to find grounds for coercion in the digital environment. If the state does not have a monopoly on physical coercion in the digital environment, then the state itself regulates the digital environment only in part. L. Lessig points out that for the state regulation it is necessary: (1) to know who the regulation is addressed to; (2) where the subject to which the regulation applies is located; (3) the regulation subject is behavior. Here, the scientist notes that it is difficult to meet these conditions for successful regulation in the digital network [13].

Of course, the state, represented by authorized bodies, develops a number of technologies and ways to regulate the Internet, but there are other entities that have the ability to regulate the behavior of people in the Internet. Of particular interest is the conflict between controlling claims. For example, the messenger Telegram officially recognized as blocked in Russia continues to work successfully: Russian citizens actively use the messenger and create Telegram channels. Consequently, the territorial norm of positive law loses its value for the Telegram user, since the user agreement becomes the main source of behavior rules for such a person. The principle of the territoriality norm, so fundamental for the legal structure of the last centuries, is disappearing, since a new digital forum is being created, where the role of regulator is equally claimed by both the ordinary programmer who wrote the corresponding code, and the state.

It is worth noting that the state in this "competition" for the position of a digital forum regulator in some sense loses traditional advantages of a sovereignty carrier, since the Telegram case demonstrates in a new way the effect of blurring geographical boundaries in the digital space. The change in the usual form of legal regulation space indicates the loss of a decisive role for the state's sovereignty. The traditional form of jurisdiction and the possibility of applying a coercive mechanism are being lost. Previously it was considered as possible to implement state enforcement, which fully applies to subjects whose relations are connected with the digital forum, that is, they originated and mainly took place in the Internet environment. It was believed that a single mechanism for managing an authoritative system of Internet root DNS servers was sufficient to compel entities using the global network [3]. The inaccuracy of forecasts can be explained by the intensity of this technology development, and therefore social relations associated with it: what seemed unthinkable twenty years ago, today seems quite possible. All this demonstrates the incompatibility of the current system of law based on the territorial theory of sovereignty with the digital forum. According to some researchers, just as it is necessary to take into account the historical and cultural context of a separate legal system for the legal regulation construction, so in the case of a digital forum, it is necessary to take into account special conditions that differ significantly from the usual form of building public relations.

The failure of the state enforcement mechanism in the case considered by the Tagansky district court of Moscow is a consequence of the lack of borders in the Internet, the ability to use "proxy" encryption and other mechanisms to circumvent the court's decision to ban activities on a certain territory. According to some researchers, sovereignty as the consolidation of the state's jurisdiction in the conditions of outlined borders, and therefore within the potential exhaustion of resources, was formed in order to ensure the need for distributive dissemination of rights to the goods available on the territory [8]. In a situation where there are no borders in the Internet and resources are inexhaustible, the function of distributive dissemination of the state's benefits loses its significance in relation to the digital environment. The state loses the basis to be a legitimate regulator in the digital forum.

The formation of a special enforcement mechanism in the digital environment, which the state does not have a monopoly on, resembles a complex process of forming a legal norm. Thus, the moral norm and the legal norm were separated only when the concept of the legal norm was formed within the normative universe as a norm formulated and associated with the imperial power, and therefore opposed to the moral norm. Now a new regulation is deduced from the legal norm. In the sphere of physical coercion of the state, there are still issues of intellectual property, compensation for harm caused in the Internet (cyber delicts), and some others. However, control over the implementation of such regulation is not assigned to the state (as it is the case with legal norms), but to many entities. Thus, N. Fraser sees the principle of "universal interest" ('*all-affected principle*') as the basis of a possible post-Westphalian model of regulation. According to this principle, all persons affected by the social institution (in this case, the digital forum) have a moral basis to determine issues of justice (i.e. the ability to exercise coercion) in relation to subjects participating in this institute [6]. The basis of the legitimizing factor for the implementation of coercion is their belonging to the institution, which, accordingly, contributes to the emergence of rules that determine forms of their social interaction as legitimate and illegal. Thus, their legitimacy will be given not by the territory outlined by state borders, as it is defined in the Westphalian sovereignty model, but by belonging to a public institution. The authors also note that the need to create an institutional system of public enforcement is caused by the focus of such groups on the most effective dispute resolution, rather than on determining who should resolve this dispute [6].

Previously, the principle of "universal interest" was allowed only in conjunction with the principle of state territory. Although this principle is still applied in certain areas, such as in the electoral law of most countries, where there is a national citizenship qualification for certain positions, this principle is significantly weakened in other areas, including in the digital forum. Globalization strengthens the distinction between state territoriality and social efficiency, which is achieved by a proper degree of ordering actions. However, the inevitable, in our opinion, appearance of a digital forum is not instantaneous. Isolation, coexistence and absorption of one forum by another can be quite long. Hypothetically, the construction of a new post-Westphalian model of the state can take place in two forms. The first form consists of ineffective resistance to urgent changes in the state. The second form is expressed in the assistance of the state gradually delegating the management functions for those public relations that are within the scope of the digital forum. Researchers allow the possibility of fixing the primacy of the second form by establishing the comity principle, which is inherent in the doctrine of private international law in England and the United States. In the interpretation of the American doctrine, the comity principle means the sovereign's refusal to regulate relations, which should be followed by the court of this state, deciding on jurisdiction in a particular dispute, in favor of the court whose jurisdiction is justified by greater awareness of the case, the applicable law, supported by other incentives [8].

This approach is based on an assessment of the state's interests in resolving a dispute. Researchers who belong to the English legal tradition interpret the concept of politeness in a much more modest way, namely as a method of interpreting existing rules and policies of law, a method used in creating legal regulation. The meaning of this concept for the English law was reduced by A. Briggs to two principles: (1) "placing and demonstrating mutual trust and confidence in the foreign judicial institutions, not interfering with them, and determining the precise conditions by which this is to be done" [1, p. 91]; and (2) "giving full faith and credit to, or respecting the conclusiveness of, the acts of foreign institutions, and working out exactly what this mean" [1, p. 91]. European continental law does not apply the concept of comity. When determining jurisdiction, a continental judge does not see his task as weighing interests of individual sovereign states. Rather on the contrary the true European ideal for determining jurisdiction would be to determine the right on *ex post* criteria that are absolutely neutral in relation to the state interests and are enshrined in the legislation. This approach is based, first, on the theory of the settlement of the legal norm [19], and, second, on the desire for codification inherent in the continental tradition of law. This tendency of the continental legislator is determined by the prerequisites of the French revolution of 1789, one of the goals of which was to limit the absolute discretion of the Imperial courts. Then the view that it is impossible to assess the interests of another state was legitimized in the fact that the codification enshrines the interest of a separate nation, which is limited to a separate territory [19]. If we start from the destruction of the territoriality principle in the digital environment, we can allow the use of the comity concept as a political and legal mechanism for differentiating the traditional forum of the Westphalian model and the digital forum. The application of this doctrine will not eliminate the desire of the state to retain its sphere of influence and regulation, but will contribute to a significant degree of restraint in reserving the right to implement this regulation.

The need to identify certain borders of the state in the regulation of the digital environment is convincingly demonstrated by C. Reed and A. Murray when formulating the principle of the form equivalence [17]. Researchers are convinced that if the state retains the ability to regulate the digital environment, it should limit itself only to what is covered by its legitimation. State regulation can be legitimized by the equivalence of forms of public relations that exist offline to those public relations that arise in the online environment [17]. Compliance with justice principles is the second principle of determining the state legitimacy, which is formulated by researchers. However, the latter principle is rather vague, because, as the authors themselves note, the requirement of honesty and fairness usually comes from a utilitarian concept of the common good, which does not always correspond to the interests of the majority [17]. The use of this principle as the sole and guiding one for determining legitimacy seems quite ineffective also because the digital environment is conditioned by a greater degree of individualism, and it is difficult to outline even an approximate majority in terms of which it would be fair and just to justify a particular state intervention. In an attempt to make such a calculation, we can reach a depressing number equal to all users of the global network. It is obvious that the number of minority groups whose interests the generalized concept of honesty will contradict will increase accordingly.

Thus, we can assume that all regulation will be reduced to four types: (1) regulation by law (law in the naive positivist's view); (2) regulation by social norms; (3) regulation by market and competition laws (markets); (4) regulation by technical norms (something we might call "architecture") [12]. Let us look at each type in more detail.

As we have already indicated, positive law regulates issues of intellectual property and harm. This means that the state has a possibility of coercion in cases where the norms of objective law are violated, under the condition that the form of social relations is equivalent. Social (corporate) norms are accepted by subjects of Internet relations themselves. These rules do not originate from the state, but they also have a mechanism of enforcement in violation cases. To such rules in the first place, it is necessary to include the user agreement.

The peculiarity of regulation in the Internet is that subjects of Internet relations actively use the technology achievements to regulate their behavior and the behavior of other subjects, sometimes even unconsciously. L. Lessig called this property of the regulator "plasticity", that is, "the ease with which this regulator can be changed or modified" [12, pp. 511–512]. This can also include the ability of participants to perceive it in Internet relationships. How many people think when they agree to the terms of a user agreement? This provision is another difference from positive law, in which the rule-making process is highly bureaucratic. In the digital environment, the thesis about the spontaneous formation of social regulators receives support. The theory of "spontaneous formation" appeared in the XIX century in the works of the economist C. Menger. The scientist characterizes a social phenomenon by the fact that it occurs as an unintended result of individual (pursuing individual interests) human aspirations, as an unforeseen result of individual ideological factors [14]. Subsequently, this view of social institutions was supported by F.A. Hayek, who criticized the concept of H. Kelsen, since the latter was unable to comprehend either the essence of the idea of development or the existence of a spontaneous social order [7]. The theory put forward by C. Menger and supported by F.A. Hayek makes it easy to justify the nature of regulation in the digital environment [7, 14]. Due to the above mentioned feature of the subjects' behavior in the digital environment, which cannot be expected, but which can be predicted (if sufficient information is available), therefore it should be regulated more by the *ex post* method than by the *ex ante* method, and rather by principles than by developing specific norms [2]. The main form of interaction between subjects in this case is a contract. It is recognized that it is best adapted to regulating the digital environment. Contractual regulation of relations is less dependent on a specific national jurisdiction based on the territoriality principle. Enforcement is thought possible both in the Westphalian model of regulation and in the post-Westphalian model. However, the enforcement mechanism may differ.

Of course, the market and prices themselves have the ability to regulate, since they set boundaries for both collective and individual behavior. For example, prices and the cost of accessing certain information posted on the Internet have a significant impact on consumer choices. If we present the Internet as a market of ideas and opinions, then as R. Posner has convincingly proved, any direct coercion of the state in such a market will have a detrimental effect [15]. From this point of view, it is worth considering indirect methods of the state influence in the Internet environment. For example, in an attempt to bring the Russian multinational company Yandex, which owns the Internet search system of the same name, as well as Internet portals and services in several countries, under its direct control, the possibility of introducing representatives of several companies with a significant state participation to the Board of Directors was discussed. Yandex takes the second place in the Russian market after Google [20] and is of considerable interest for the state from the point of view of the influence opportunity on the digital sphere. Subsequently, the possibility of the state control was mitigated by the creation of a special "public interest fund", which will include the company's executives, major higher education institutions and three independent directors. "Golden share" of Yandex will be transferred to this fund. Of course, this method of influence, although legitimate, is anomalous for the model under consideration. However, this quasi-governmental impact should be taken into account in order to see the breadth of potential indirect influence on the digital environment.

The fourth type of regulation sets boundaries for behavior of Internet relations subjects through network protocols, software used in this environment, encoding, passwords and locks. All this affects the dynamics of the user's activity in the virtual space. In this regard, L. Lessig believes that technical regulation can replace the laws of the real world to some extent, allowing more effectively discipline network users [12]. This type of regulation is justified at least by the fact that initially even domain name regulation was created by engineers [8]. Thus, applying the theory of forums, we observe a plurality of regulation. On the one hand, the state claims to regulate the digital environment, having a monopoly on physical coercion in an external forum. On the other hand, the digital environment resembles a layer cake, in which only one layer—objectified to the outside world—can be subject to state regulation. Other subjects, for whom the territory principle is not important, claim to regulate the rest of these relations.

5 Conclusion

Regulation of the behavior of individuals in the Internet has certain features. First, the state does not have a monopoly on norm-making. Second, in the Telegram case, we have demonstrated that the state is not capable of carrying out enforcement in the Internet environment. The concept of state sovereignty and jurisdiction has traditionally been associated with geographical borders of a state. When forming a digital forum, the state geographical borders lose their traditional meaning. Consequently, the traditional form of jurisdiction and the possibility of applying a coercive mechanism are lost. The principle of the territoriality norm, so fundamental for the legal structure of the last centuries, is disappearing. All this makes us agree with the view of building a new post-Westphalian model of a state, an integral part of which is the gradual delegation of public relations management in the digital forum. All regulation in the digital forum is reduced to four types: (1) regulation by law; (2) regulation by social norms; (3) regulation by market and competition laws; (4) regulation by technical norms. In our opinion, the main type of regulation will be social norms that have a contractual nature.

References

1. Briggs A (2011) The principle of comity in private international law. Academie de Driot Recueil des Cours 354:65–182
2. Cave J (2013) Policy and regulatory requirements for a future internet. In: Brown I (ed) Research handbook on governance of the internet. Edward Elgar, Cheltenham, pp 143–167
3. Charlesworth A (2000) The governance of the Internet in Europe. In: Akdeniz Y, Walker C, Wall D (eds) The internet, law and society. Pearson Education, Harlow, pp 47–78
4. Clinton WJ, Gore Jr AA (1997) Framework for global electronic commerce. https://clintonwhitehouse4.archives.gov/WH/New/Commerce/read.html. Accessed 26 Feb 2020
5. Decision of the Tagansky district court of the Russian Federation of April 13, 2018 N 2-1779/2018. https://www.mos-gorsud.ru/rs/taganskij/cases/docs/content/9c02a662-9f68-44cf-bbd7-7c8b10a3d814. Accessed 26 Feb 2020
6. Fraser N (2005) Reframing justice in a globalizing world. New Left Rev 36:69–88

7. Hayek FA (1967) Kinds of rationalism. Studies in philosophy, politics and economics. University of Chicago Press, Chicago
8. Jonson DR, Post D (1996) Law and borders—the rise of law in cyberspace. Stanford Law Rev 48(5):1367–1402
9. Kant I (2014) Dreams of a spirit-seer. Forgotten Books, London
10. Kelsen H (1979) Allgemeine Theorie der Normen. Manz, Wien
11. Kelsen H (2014) Pure theory of law, 2nd edn. Alef Press Publishing House, St. Petersburg
12. Lessig L (1999) The law of the horse: what cyberlaw might teach? Harvard Law Rev 113:501–549
13. Lessig L (2006) Code: Version 2.0. Basic Books, New York
14. Menger C (1883) Untersuchungen über die Methode der Sozialwissenschaften und der politischichen Ökonomie insbesondere. Duncker Humblot, Leipzig
15. Posner R (2001) Frontiers of legal theory. Harvard University Press, Cambridge
16. Prodi P (2017) History of justice: from pluralism of forums to modern dualism of conscience and law. Gaidar Institute Press, Moscow
17. Reed C, Murray A (2018) Rethinking the jurisprudence of cyberspace. Edward Elgar, Cheltenham
18. Reidenberg JR (1997) Lex informatica: the formulation of information policy rules through technology. Texas Law 76(3):553–594
19. Ryngaert C (2008) Jurisdiction in international law. Oxford University Press, New York
20. SEO Auditor (2019). Search engine rating. https://gs.seo-auditor.com.ru/sep/. Accessed 26 Feb 2020

Towards a New Format of Regional Integration: Co-creation and Application of Technologies

G. A. Khmeleva and T. Czegledy

Abstract The articles studies the process of transition towards new format of regional integration, based on economy digitalization. The authors contemplate on the problem of technological development acceleration. To increase the regions' digital maturity, a wider deployment of technology co-creation possibilities is necessary. The fact of the matter is that the use of modern tools to receive digital dividends is insufficient. An additional complication is the complexity and interdisciplinary nature of the process of digital technologies co-creation for the territories development. Countries, regions, municipalities equally face threats of global challenges. It is becoming apparent that only jointly they can find answers to global challenges. Following corporations, regional and local authorities must consider new formats for integrating regions into world economic relations. To do this, the authors of the article analyzed the process of technologies co-creation using hackathons as an example.

Keywords Sharing economy · Regional integration · Innovation · New technologies · Hackathon

1 Introduction

The sharing economy provides a fresh look at the process of technologies co-creation when analyzing the countries integration. Information and communication technologies provide ample opportunities for collaboration [4]. Researchers argue that values co-creation by representatives of different regions is a complex process of interaction between participants, and in the modern world it is possible only if information and communication technologies are available and accessible. It is important that participants understand and clearly differentiate their roles in this process [1].

G. A. Khmeleva (✉)
Samara State University of Economics, Samara, Russia
e-mail: galina.a.khmeleva@yandex.ru

T. Czegledy
University of Sopron, Sopron, Hungary
e-mail: czegledy.tamas@uni-sopron.hu

S. I. Ashmarina and V. V. Mantulenko (eds.), *Current Achievements, Challenges and Digital Chances of Knowledge Based Economy*, Lecture Notes in Networks and Systems 133, https://doi.org/10.1007/978-3-030-47458-4_9

Of major importance the ability of subjects to cooperate with foreign competitors to solve technological problems and capture opportunities of penetrating foreign markets, protecting their positions and promoting technological innovations. Research has indicated higher quality and increased share of cross-border knowledge. Moreover, modern cross-border studies have a shorter technological cycle, which is no less important given the requirements for a high technological development rate to support competitiveness [13, 14].

Researchers point out that traditional approaches to regional economic integration in the future may prove to be ineffective, since digital solutions provide people and enterprises with the opportunity to communicate with each other directly, regardless of state borders presence between them [5].

The digital era expands the understanding of economic integration. The regional integration success directly depends not only on the regional enterprises ability to compete in foreign markets and integrate into global value chains, but also on the local community involvement, including business, in the creation of smart cities [9]. It is no coincidence that there are increasingly more articles in the field of artificial intelligence, big data, and mobile technologies, since these are the basis of urban infrastructure [11]. Freudendal-Pedersen and Kesselring [7] point to overestimation of the effective technologies issue, while in fact a much bigger problem is the underestimation of social cohesion, integration and coherence of territories with citizens' mobility trend growing.

However, there are risks in joint efforts to create technology. Urry [16] warns about them, since smart cities are still in their infancy, large IT companies are striving to use digital technology to expand market power. Efforts must be made to avoid the "cost of digitalization" and explore alternative digital options [10]. All the more so because the digital technologies "safe" practices expansion in economic activities increases the digital maturity of the region [3].

Digital technology is driving the uprise of such new regional integration format as hackathons. Researchers note that it is hackathons that enable the mobilization of business, people, city and regional authorities around issues related to urban space [10].

2 Methodology

The authors found their study on the Chesbrough' theory of open innovation [2] and the theory of sharing economy [8], taking into account three important aspects: the hackathon as a process of technology creation is both global, regional and local. As a global phenomenon, hackathons can potentially bring together representatives of different countries, ideas and have a variety of formats. In a regional aspect, hackathons are held in the capitals of countries or large regional centers with developed information and communication infrastructure. In a local aspect, the quantity, organization specifics and the issues of hackathons are determined by specific requests from the local community, including business, and, rarely so far, the authorities. For these purposes, three data collection strategies have been used. First, the authors referred to the content analysis of digital platforms (over 50) on the study topic. The world's largest digital platform, Hackathon.com, allowed collecting information by geographic and industry breakdown of hackathons from 2015 to January–April 2020. The European Commission official website (https://ec.europa.eu/) enables the study of technologies co-creation with the direct interest of the state. Besides, information of documents and blogs freely distributed on the Internet was studied in order to understand the differences in the main organization strategies and, more importantly, alternative hackathons motivated by social innovation.

3 Results

It is believed that the sharing economy differs from the traditional model of the corporation in that digital platforms are used here for the sale, rental or transfer of goods and services free of charge. With collaborative consumption, special attention is paid to costs at all stages of creation and sales. The theory of open innovation has shown an excellent model of cost optimization in technology creation such as a joint forum of developers, including programmers, managers, designers. For a limited time, they create software at the request of the organizers.

Hackathons are a good tool for "creative destruction" according to Schumpeter [12], since shared interests of participants in the creative search for new products and services solutions come together here [10]. Hackathons are the quintessence of a situation in which social, technological and organizational mechanisms are applied in such a way as to reorient the knowledge, skills and technologies possessed by corporations to form the desired image of the territory's future.

The analysis showed that in the world all hackathons are used to co-create technologies, since according to the largest digital platform Hackathon.com, in 2018 5636 hackathons were organized. This is 40% more than in 2016. The hackathons geography is expanding and has now reached global coverage, since the fact of hackathons organizations are registered on all continents.

If initially hackathons emerged as a mechanism for attracting talents and reducing costs exclusively for the IT sector, now the idea of hackathons is expanding and is not limited to the software development. In recent years, the interest in events in industrial and non-profit spheres has grown at a faster pace; these topics are currently the most popular (Fig. 1).

More than a thousand hackathons were dedicated to the sharing economy and the Internet of things. A focus on topics of technical and organizational solutions creation for a smart city, including the Internet of things, artificial intelligence, big data, transport, energy, can be noted.

Not only corporations, but educational institutions and municipalities are increasingly organizing hackathons. Hackathons in educational institutions often do not have a specific topic, but are designed to enable students to bring their concerns and

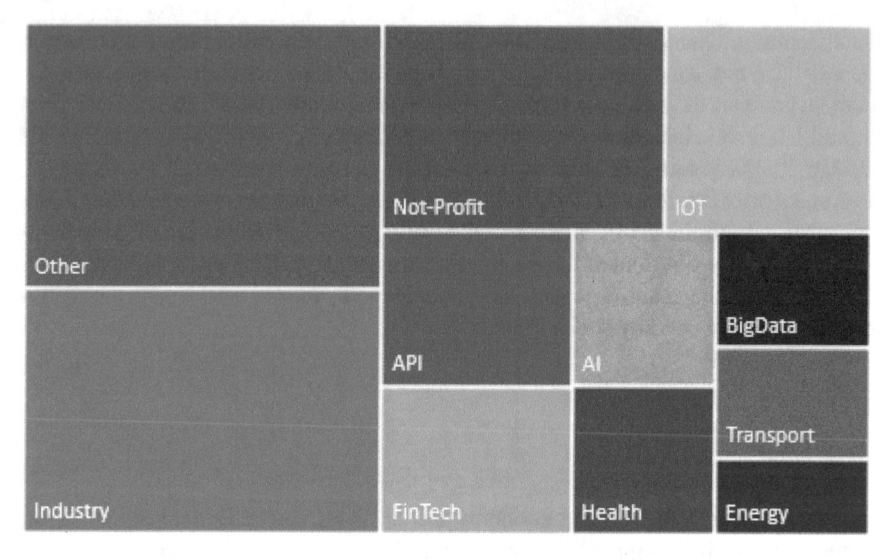

Fig. 1 Hachathons structure in 2015–2020 (January–April) (*Source* authors)

develop their idea into a holistic product within a limited time. The best projects and products are evaluated by a jury with the employers' participation.

Hackathons are increasingly gaining international caliber. Thus, an ALL DIGITAL week is held annually in Europe, and it includes events not only on the basis of European centers of digital competencies, libraries, schools, universities. In 2019, the third global transport hackathon took place in Stockholm, with request for the software and hardware prototypes development for the transport sector, including passenger transportation and cargo freight, innovation and entrepreneurship, intermodalism, sustainable development and technology. It was attended by teams from Sweden, France, Germany, India, Brazil, Chile, etc.

Another form of international hackathon is the network format. Thus, the global legal hackathon is the world's largest competition for the creation and promotion of legal IT projects. In 2019, more than 6000 people from 46 cities of 24 countries participated in it. This hackathon is based on a hosting form of organization. During one weekend the teams based on a network of hosts (host organizations) in different countries offer ideas and develop technological solutions for personal gain (business and legal practice) and the public good (good government, equal access to legal systems). The semi-finals are held virtually, and the finals are hosted by one of the participating countries.

The second Space Apps Global Challenge for programmers, scientists, designers, writers, engineers, builders, technologists and other professions will be run in cities around the world in 2020, where teams will interact with free and open data from the National Aeronautics and Space Administration to solve real problems on Earth and in space. In 2019, Space Apps brought together more than 29,000 participants in 71 countries. 225 events were held.

The first global hackathon, Datathon 2020, is planned to take place in more than 30 countries around the world in an online format. Teams will work on real business cases from different areas of machine learning, artificial intelligence, and data science. All developed solutions will be open source and will be publicly available for use and further improvement.

The Odyssey 2020 global Blockchain hackathon is dedicated to building an interconnected, multi-stakeholder ecosystem to meet the global challenges of humanity: the fast and energy efficient digital power market commons, smart meter data autonomy for prosumers, Conscious Cities, Nature 2.0, Internet of Logistics, Public Safety and Security, Energy Singularity, etc. Teams in 48 h develop specific solutions for practical cases, which the organizers then commit to bring to pilot stage.

The issues of using the created prototypes are usually determined at the stage of hackathons organization. There are several options. Thus, if the hackathon is organized in the interests of corporations, then the corporation gains the results. At best, the participants are left with a practical case experience. But increasingly, hackathons create products for sharing, such as at Datathon 2020, which was mentioned above.

Another example is the BigData 2019 European hackathon, which terms stipulated that prototypes created during it could be published as an article or study. The open sources analysis revealed the following key motives of the hackathon participants and hackathons results:

- actualization of participants' internal need for communication with colleagues when creating innovations,
- participants view hackathons as an opportunity to gain practical experience, rewards (if prizes are provided), as well as a kind of holiday where they can chat with advanced colleagues, establish contacts and, possibly, get to know the future employer. An analysis of published interviews with developers provides insight into their attitude to product creation as an art, when they strive to show all their skills,
- the presence of company representatives helps to understand the specifics of the task.

The causes of failure can be described as follows:

- inability of the developed technologies interesants to clearly set tasks,
- due to low admission barriers to participation, there is a mismatch of the participants qualifications with the set tasks requirements.

This is especially noticeable in multidisciplinary teams solving the problems of smart cities development. Global hackathons data analysis provides argument that among the organizers there is not only business, but also educational institutions, state and municipal structures. There is a view of hackathons as driver of new research and innovation for society. Dedicated to global challenges, hackathons allow academic and research institutions to collect invaluable material for possible solutions to problems and further use it for scientific purposes, ensuring knowledge generation in modern life laboratories and "experimental cities" [6]. For state and municipal structures, hackathons can provide good opportunity to better understand the needs of society. But they are not using it very actively yet [13].

4 Discussion

In the digitalization era, the successful integration of regions into the global economic space is determined by the participants ability to use Internet technologies and new formats for shared activities in regional economic activities. Organization and participation in international hackathons opens up a new opportunity to integrate the region's economy and increase its digital maturity [4]. Joint development and use of technology reduces the time and cost of their creation. Internationalization of developments is instrumental in keeping abreast of global trends and achievements. Due to cloud technology and the Internet, collaboration and expertise can be executed online [15]. However, there are certain hackathon problems that need to be considered, since digital technologies designed for sharing can affect the urban economy sustainability. Researchers point to the facts of digital platforms such as Uber and Airbnb changing the transport and housing sectors in cities, creating problems for urban governance structures seeking to balance the disruption of market relations [13]. Perng [10] uses hackathons as an example to demonstrate that such cooperation fragments are crucial for the digital future, as they strengthen the momentum of joint search for new technological solutions.

5 Conclusion

Thanks to hackathons, innovative entrepreneurs reduce resources needed for product development and testing, creating innovative business solutions using public and online hackathons. It is important to note that the hackathons movement embraces more and more new areas. In 2019, less than a third of technology companies organized hackathons, while there was a gradual growth in industry and services. Having emerged as a means of finding talent and solving narrow tasks for which companies lacked their own resources, hackathons become an important part of the organizational culture of companies seeking innovation. Moreover, hackathons can be powerful tool for developing smart cities and integrating territories into the global economy. Therefore, it is important to continue hackathons study; city and regional authorities should pay attention to new integration formats, which, when used skillfully, would make it possible to attract local residents more extensively and provide digital dividends for specific territories. At the same time, it is important that local authorities and corporations remain open throughout the process of creating and using technologies, respecting the interests of residents.

Acknowledgments The reported study was funded by RFBR and FRLC according to the research project № 19-510-23001.

References

1. Breidbach CF, Maglio PP (2016) Technology-enabled value co-creation: an empirical analysis of actors, resources, and practices. Ind Mark Manag 56:73–85. https://doi.org/10.1016/j.indmarman.2016.03.011
2. Chesbrough HW (2006) Open innovation: the new imperative for creating and profiting from technology. Harvard Business Press, Boston
3. Chirkunova EK, Khmeleva GA, Koroleva EN, Kurnikova MV (2020) Regional digital maturity: design and strategies. In: Ashmarina S, Vochozka M, Mantulenko V (eds) Digital age: chances, challenges and future, vol 84. Lecture notes in networks and systems. Springer, Cham, pp 205–213. https://doi.org/10.1007/978-3-030-27015-5_26
4. Czegledy T, Fedorenko RV, Zaichikova NA (2020) Modelling of software producer and customer interaction: nash equilibrium. In: Ashmarina S, Vochozka M, Mantulenko V (eds) Digital age: chances, challenges and future. ISCDTE 2019, vol 84. Lecture notes in networks and systems. Springer, Cham, pp 298–307
5. Erokhin D (2019) Connectivity in the digital age. Digital futures of trade and economic cooperation in Eurasia. Challenges and opportunities of economic integration within a wider European and Eurasian Space. Background paper. IIASA, Laxenburg
6. Evans J, Karvonen A, Raven R (2016) The experimental city. Routledge, London
7. Freudendal-Pedersen M, Kesselring S (2016) Mobilities, futures & the city: repositioning discourses—changing perspectives—rethinking policies. Mobilities 11(4):575–586. https://doi.org/10.1080/17450101.2016.1211825
8. Hamari J, Sjöklint M, Ukkonen A (2016) The sharing economy: why people participate in collaborative consumption. J Assoc Inf Sci Technol 67(9):2047–2059. https://doi.org/10.1002/asi.23552
9. Horelli L, Saad-Sulonen J, Wallin S, Andrea A (2015) When self-organization intersects with urban planning: two cases from Helsinki. Plann Pract Res 30(3):286–302. https://doi.org/10.1080/02697459.2015.1052941
10. Perng SY (2019) Anticipating digital futures: ruins, entanglements and the possibilities of shared technology making. Mobilities 14(4):418–434. https://doi.org/10.1080/17450101.2019.1594867
11. Pink S, Fors V, Mareike G (2018) The contingent futures of the mobile present: automation as possibility. Mobilities 13(5):615–631. https://doi.org/10.1080/17450101.2018.1436672
12. Schumpeter J (1942) Capitalism, socialism, and democracy. Harper & Bros, New York
13. Trefler D (2004) The long and short of the Canada—U.S. free trade agreement. Am Econ Rev 94:870–895
14. Sharp D (2018) Sharing cities for urban transformation: narrative, policy and practice. Urban Policy Res 36(4):513–526. https://doi.org/10.1080/08111146.2017.1421533
15. Su ChY, Lin BW, Chen ChJ (2016) Knowledge co-creation across national boundaries: trends and firms' strategies. Knowl Manag Res Pract 14(4):1–13. https://doi.org/10.1057/kmrp.2015.14
16. Urry J (2016) What is the future? Polity, Cambridge

The Concept of “Information” in the Law of the Digital Society

T. Cherevichenko and A. G. Galkin

Abstract The research purpose is to analyze problematic aspects of regulating information as an object of legal relations in the era of digital revolution. The author considers the problem of correlation between the concepts of “information” and “data” in the Russian and international law, as well as doctrinal views on this issue. Classification of information in legal acts, the essence of the legal regime of information will be discussed. Information in the modern world is one of the strategic resources of social and economic development of any state. It is the development of new information and communication technologies that determines the effective development of the economy and society as a whole. And this is not possible without proper legal regulation of information, which should be based on a clearly formulated concept of “information” and delineated boundaries between it and such category as “data”. The problem is rooted in the identification of the concepts “information” and “data” in Russian legislation. This, in turn, makes it difficult to define the legal regime and distribute rights to specific information. In order to eliminate the existing problems in the legal regulation of information, the ways of reforming the Russian legislation are proposed.

Keywords Information · Data · Privacy · Access · Distribution

1 Introduction

The term “information” was known in ancient times and was used by such outstanding minds as Cicero and Augustine. The middle ages introduced this concept into everyday use and since the XV century it has been understood as research, education, communications, and knowledge [8]. In the famous Russian dictionary, there is such

T. Cherevichenko (✉)
Samara State University of Economics, Samara, Russia
e-mail: Cherev_777@mail.ru

A. G. Galkin
Kuban State Agrarian University named after I.T. Trubilin, Krasnodar, Russia
e-mail: alex-amway@inbox.ru

S. I. Ashmarina and V. V. Mantulenko (eds.), *Current Achievements, Challenges and Digital Chances of Knowledge Based Economy*, Lecture Notes in Networks and Systems 133, https://doi.org/10.1007/978-3-030-47458-4_10

a definition of information: it is (1) information about the world around us and the processes occurring in it, perceived by a person or a special device; (2) messages informing about the state of affairs, about the state of something; scientific, technical and newspaper; mass media (print, radio, television, cinema). In this case, the word "information", according to the same dictionary, means: (1) knowledge in some area; (2) news, message; (3) in some combinations: knowledge, idea of something [13]. The concept of information is located in the Federal law of 27.07.2006 N 149-FZ "On information, information technologies and information protection" [5] according to the provisions of which information is made up of data that has various forms of provision. And information includes messages and data. The value of information is in the fact that humanity from one generation to another passes the "baggage" of accumulated knowledge and skills. Here it is appropriate to give an example of the Pythagorean Theorem, which was developed by him in the VI century BC and on which modern knowledge and developments of modern mathematicians are based. Information has become so abundant that it has generated and continues to generate new technologies for collecting, storing, transmitting, processing, analyzing, and using it [20].

2 Methodology

The methodological basis of this research includes general scientific and special legal methods of cognition. During the study, the following general scientific methods were used: general logical, dialectical, system-structural, formal-logical methods. The author also used special methods: formal-legal and historical-legal. All these methods were used together, which enabled a comprehensive study of the claimed topic.

3 Results

As we can see in the Russian legislation, the definition of information is not ideal. Its drawback lies in the wording itself—one phenomenon is characterized through a list of others that are ambiguous and indisputable too. Therefore, this definition can be considered incorrect. It is proposed to define information at the legislative level as data, regardless of the form of its presentation (messages, knowledge or a set of perceptions), which is the object of legal relations within the framework of the above mentioned federal law [21]. It is difficult to agree with this formulation, since the concept of "set of perceptions" is also not specific and indefinable. This will generate controversies both in theory and in the field of law enforcement. It is necessary to agree with the proposal to fix the features of information that characterize it as an object of legal relations at the legislative level. Information has a certain set of properties, the totality of which is different for different authors. In our opinion, a more correct enumeration of such properties was formed by V.A. Kopylov. He refers to them: physical inseparability, separateness, information content of the object, distribution, organizational form, and instantiation [11].

However, not all features inherent in information are fixed by the legislator. In the Russian legislation, the following features are fixed: diversity, expressiveness in the material world, the presence of links between information and its consumer, value, reliability, sufficiency. Information is divided into known and unknown. Unknown information can be divided into conditionally available and unavailable information. The boundary of their separateness lies within their scientific validity: the first group is scientifically justified, but the second group is not. Based on the subject structure, information is divided into: socially valuable, valuable for a specific group or for a single person (individually valuable). Any type of information is subject to protection—preventing it from being influenced by the environment. Such protection means, first, protection against distortion, and second, unauthorized distribution. Based on the information value, the access mode to it is also determined. Based on this, the Russian legislator divided the information into publicly available and with limited access.

Information that is restricted differs from public information by the presence of a specific property—confidentiality. The legislator defined it in Art. 2 of the Federal law of 27.07.2006 N 149-FZ "On information, information technologies and information protection": the mandatory for a person (who has access to certain information) requirement not to transfer such information to third parties without the consent of its owner [5]. This is the reason for the concept of secrecy in the Russian law. In court decisions, secrecy is defined as "unknown to a wide range of people" [3, 19]. So, Federal arbitration court of Volgo-Vyatka District in the Decision № A39-1941/2010 [4] stated that in accordance with article 53 of the Law on communications, data about subscribers (users of communication services) and communication services provided to them which became known to the operators (legal persons or individual entrepreneurs that provide telecommunications services on the basis of a license) in the performance of the contract on rendering of communication services, are confidential and are to be protected in accordance with the legislation of the Russian Federation [14].

As we can see, it is the presence of the privacy property in information that causes the need for protective measures, which are implemented in restricting access to it. It should be noted that this state of affairs was formed after the adoption in 2006 of the Federal law N 149-FZ "On information, information technologies and information protection" [5]. Up to this point, information was divided into two types. The first is a state secret. The second is information whose distribution was restricted. Some scientists identified a third type of such information—interstate secrets [22]. An attempt was made to create a different classification based on the turnover of information: completely circulating information; information that is restricted in circulation, including information that is restricted by law to its owner and various groups of persons are required to ensure its non-disclosure; information that is withdrawn from circulation (state secret). But the turnover of information does not reveal its essential characteristics and does not allow determining its value for the owner.

This does not make it possible to determine the necessary mode of information security—access restrictions. In the Soviet law, information with restricted access was divided into: state secrets, official secrets, and for official use. Each of these categories had a security label: "very important" and "top secret", "secret", the latter did not have a label. Only state secrets were subject to legal regulation. Only this type of secret had a sign of confidentiality. Today, it is the presence of the above attribute in information that causes the application of a special legal regime to it. The sign of which is the volume (size) of access to this kind of information. Each action should have specific goals. This is also true for providing access to confidential information. The purpose of "access to information"—the ability to obtain information and use it—is enshrined in the Russian legislation [16]. However, such a concept as "access" is not used in the Constitution of the Russian Federation, and other terms are used: search, receive [1].

4 Discussion

Information should be defined as data that is transformed and interpreted by a person. Since the concept of "information" is inextricably linked to a personality and consciousness, it is advisable to fix the concept of "information" in the legislation by listing the features inherent in it: inalienable; separability; information content of the object; dissemination; organizational forms; instantiation. Information may be subject to different legal regimes. The legal regime is the procedure and nature of regulating the legal status of specific information and the distribution of rights to it. State protection of information is determined by the presence of a specific feature—confidentiality.

A.V. Susloparov defines information as data that does not have physical characteristics, is stored on a material carrier, and transmitted using coded signals [17]. In the Russian legislation, the concepts of "information" and "data" are identified. In the Federal law of 27.07.2006 N 149-FZ (ed. of 02.12.2019) "On information, information technologies and information protection", information is defined as messages and data. In Federal law No. 152-FZ of 27.07.2006 (ed. from 31.12.2017) "On personal data" is fixed that personal data is any information relating to a directly or indirectly specific or identifiable individual (subject of personal data) [5, 6]. In the international law, these concepts are separated. The International Organization for Standardization (ISO) defines data as a multiple-interpretable representation of information in a formalized form suitable for transmission, communication or processing [9], and information as the result of interpretation (meaning) of such a representation [10]. Based on this, scientists conclude that a single volume of data can generate different information [15]. This view is supported by the OECD, stating that information that can be extracted from data depends not only on the data itself, but also on the existing analytical capabilities for linking data and extracting new knowledge from it. These capabilities depend not only on the available metadata, analytical technologies and methodologies, but also on skills and knowledge already available to the data processor [12]. The identification of concepts creates a problem of definiteness of the legal regime and distribution of rights to data. As we pointed out above the same amount of data generates different types of information with different legal regime which is the order and nature of regulation of the legal status of information. It should be noted that the Russian legislation uses different concepts: "information" and "data". Foreign scientists also do not tend to identify data and information, believing that data is transformed into information after entering the human consciousness [7].

Based on this, information is mediated as a process occurring in the human mind and the result of which is the interpretation of what it perceives as data between the human mind and a certain stimulus, in contrast to which data is a "simple representation of information or some idea" [14]. There is also an understanding of information as a set of data in contrast to knowledge, which is understood as thinking and reasoning based on data when comparing and classifying them [2]. It should be said that in the scientific literature it is sometimes proposed to consider information and data as identical concepts, and information and messages—as synonyms [18]. However, it does not find broad support. In our opinion, the most correct way to understand data is to understand it as a component of information that is transformed into it after realizing them and making them meaningful by a person. "Materials" is also synonymous with "information".

5 Conclusion

The modern world is impossible to imagine without the information cycle. Today, information is understood as the engine of our world. Every year its value only grows. With its help, mankind passes on accumulated knowledge from generation to generation, which becomes the basis for new research and development in all sciences. Its research is a topic of scientific works in various fields. In this case, the information is considered and studied from various aspects. But without specifying its concept, features, access modes and protection measures in legislation, a breakthrough in all sciences, technologies and spheres of the social life is simply impossible. In the Russian legislation, the concepts of "information" and "data" are identical concepts. In the international law, they are separate. The concept of information is inextricably linked to human (Homo Sapiens). It is data (information, messages) that, being a meaningful by a person, becomes information.

References

1. Afanasyeva OV (2012) Access to information. Global trend in reforming a state. Issues State Municipal Adm 3:159–172
2. Brillouin L (1960) Information science and theory. Fizmatgiz, Moscow

3. Decision of the Federal Antimonopoly Service of the Volga-Vyatka District dated December 1, 2010 in case No. A39-2405/2010. http://www.consultant.ru/cons/cgi/online.cgi?req=doc&base=AVV&n=41379#05508108446133866. Accessed 06 Dec 2019
4. Federal arbitration court of Volgo-Vyatka District (2010) Decision № A39-1941/2010 in the case of N A39-1941/2010. http://www.consultant.ru/cons/cgi/online.cgi?req=doc&ts=12457541560938418823791661 7&cacheid=BFEDF58CB91AB3D9C9903DB464DF41B1&mode=splus&base=AVV&n=41569&rnd=A298A8BA6C7EB1E9005708DF02EBE6EC#koi0iiyiey. Accessed 06 Dec 2019
5. Federal Law of July 27, 2006 No. 149-FZ (as amended on December 2, 2019) "On Information, Information Technologies and Information Protection". http://www.consultant.ru/document/cons_doc_LAW_61798/. Accessed 06 Dec 2019
6. Federal Law of July 27, 2006 No. 152-FZ (as amended on December 31, 2017) "On Personal Data". http://www.consultant.ru/document/cons_doc_LAW_61801/. Accessed 06 Dec 2019
7. Floridi L (2013) The ethics of information. Oxford University Press, Oxford
8. Gabbay DM, Woods J, Thagard P (2008) Philosophy of information. Elsevier, Oxford
9. ISO/IEC 2382:2015 Information technology – Vocabulary. http://docs.cntd.ru/document/1200139532. Accessed 06 Dec 2019
10. ISO/IEC/IEEE 24765:2010 Systems and software engineering – Vocabulary. https://www.cse.msu.edu/~cse435/Handouts/Standards/IEEE24765.pdf. Accessed 06 Dec 2019
11. Kopylov VA (2002) Information law. Yurist, Moscow
12. OECD (2014) Data-driven innovation for growth and well-being. Interim synthesis report. https://www.oecd.org/sti/inno/data-driven-innovation-interim-synthesis.pdf. Accessed 06 Dec 2019
13. Ozhegov SI, Shvedova NYu (2008) Explanatory dictionary. Onyx, Moscow
14. Piragoff KD (1993) National report: Canada. Int Rev Penal Law 64(1–2):201
15. Savelyeva AI (2019) Towards the concept of data regulation in the digital economy. Law 4:174–195
16. Serova OA (2019) Problems of development of the methodology of civil law research in the digital age. Methodol Probl Civil Law Res 1(1):351–362. https://doi.org/10.33397/2619-0559-2019-1-1-351-362
17. Susloparov AV (2010) Computer crimes as a type of information crimes. Ph.D. thesis, Krasnoyarsk: Siberian Law Institute of the Ministry of Internal Affairs of Russia
18. Susloparov AV (2017) Conformity of the Criminal Code of the Russian Federation to the principle of unity and definiteness of terminology by the example of the terms "information", "materials" and "data". Russ Laws: Exp Anal Pract 11:94–98
19. The determination of the Supreme Arbitration Court of the Russian Federation of November 28, 2012 No. BAC-15662/12. http://www.consultant.ru/cons/cgi/online.cgi?req=doc&base=ARB&n=306406#08507605707331156. Accessed 06 Dec 2019
20. Zakharov AL, Sidorova AV (2016) Information inequality: history and modernity. Econ Legal Issues 100:17–21
21. Zeynalov ZZ (2010) Problems of defining information as an object of informational legal relations. Inf Law 1(10):6–9
22. Zvereva EA (2007) Legal regulation of information support of business activity in the Russian Federation. Abstract of the Dr. thesis, Moscow State University, Moscow

Gender Extremism in the Internet Age

M. A. Yavorsky and S. V. Mikheeva

Abstract The article presents the author's view on the interpretation of illegal, violent acts motivated by hatred or enmity, based on the biological characteristics of an individual or social group, as well as promoting the superiority or inferiority of a person on the basis of sexual (gender) identification. The authors of the article, based on the analysis of positions, scientific approaches of Russian and some foreign scientists, the practice of applying the norms of criminal and administrative-tort legislation of Russia and foreign countries, the content of sites and pages of social groups in the Internet network, consider the essence and content of gender-oriented (gender or sexual) extremism, formulate its definition, and reveal its features. The article is of theoretical and practical interest for further research and improvement of the mechanism of countering extremism. The results of the research could be used by teachers and students of secondary and higher educational institutions in preparation for classes of anti-extremist orientation.

Keywords Gender · Internet · Counteraction · Extremism · Radicalism · Feminism

1 Introduction

Extremism in all its manifestations is a serious threat to modern society. The effectiveness of the system of measures to counteract extremism currently depends on the objectivity of the study of its causes and conditions, the depth of analysis of its essence and content, understanding of its forms, types and specific manifestations. The works of foreign scientists are devoted to countering the spread of various types of extremism on the Internet: Aly et al. [1], Bertram [3], Conway [6], Costello et al. [7],

M. A. Yavorsky (✉)
Samara State University of Economics, Samara, Russia
e-mail: yavorm@mail.ru

S. V. Mikheeva
Samara Law Institute of the Federal Penitentiary Service of Russia, Samara, Russia
e-mail: rio-mikheeva@mail.ru

S. I. Ashmarina and V. V. Mantulenko (eds.), *Current Achievements, Challenges and Digital Chances of Knowledge Based Economy*, Lecture Notes in Networks and Systems 133, https://doi.org/10.1007/978-3-030-47458-4_11

Oksanen et al. [12], and others. In Russia, these problems have also become the subject of close study by many researchers. However, upon closer examination of the content of scientific papers it turns out that the spread of gender-oriented radical ideology in cyberspace, and in fact, gender (sexual) extremism, is almost not paid attention. In this situation, it is necessary to talk about the demand for scientific research on the spread of gender-oriented extremism on the Internet. The goal of the research is associated with the need to systematize existing normative approaches, with Simultaneous scientific generalization of theoretical views on the essence and content of gender extremism.

2 Methodology

The authors present their own vision of the phenomenon of gender-oriented extremism. Dialectical and metaphysical methods are among the universal methods of scientific research. The main method in this work is the method of analyzing the special literature, views, and scientific approaches of Russian and foreign scientists on the problem of research, and the practice of applying the norms of anti-extremist criminal and administrative-tort legislation in Russia and foreign countries. The system-structural method made it possible to generalize the main directions of legal thought and research in the subject under consideration and determine the vector of their further development. This study was conducted in two phases: At the first stage, problem, purpose and methods of the study were determined, the scientific literature on the subject of the study was analyzed, as well as the legislation regulating the issues of countering extremism. At the second stage, the main conclusions obtained during the analysis of scientific literature and legislation were formulated, and the publication of this study was prepared. The materials for applying anti-extremist legislation are fairly representative. This allowed us to consider a number of significant issues related to the gender aspects of extremism.

3 Results

Today, there is a steady increase in the use of the global information and telecommunications network "Internet" as a platform for spreading extremist ideology, propaganda of hatred and enmity. It is the modern mass media that often contribute to the growth of not just criminal, but extremely aggressive and extremist tendencies in society. The fact that crimes and administrative offenses committed by both men and women of different age categories who are representatives of one of the two biological sexes, it would be at least incorrect to dispute. However, this circumstance does not exclude the possibility of manifestations of extremism based on gender (sexual) hostility. Fortunately, Russia has not yet developed large-scale prerequisites for the development and subsequent spread of hatred or enmity towards citizens, depending on their belonging to a particular gender.

4 Discussion

We will consistently present arguments about the possibility of gender (sexual) extremism, as well as specific facts.

1. It is well known that gender (sex) affiliation does not always correspond to gender-role behavioral attitudes of the individual associated with the biological sex of the individual. Actually genders (social sexes) in contrast to biological sexes, which, as we know, are usually divided into two (male and female), can be much more. An example is transgenderism, which has become widespread in some countries in South-East Asia. For example, in Thailand, there are five genders: men, women, and the so-called "third gender", or "kathoey", which includes three gender categories (men who identify themselves as women in different degrees, as well as women with varying degrees of masculinity and femininity, and with corresponding social and gender-role functions).
2. In a number of countries, incitement to hostility and affront to human decency by gender or affiliation to any social group is criminalized. Thus, criminal liability for committing crimes based on social motives is provided for in the criminal laws of Vietnam, Kazakhstan, Lithuania, Turkmenistan, Turkey and other countries. Gender (gender) is reflected in the legislation of, for example, Andorra, Spain, Luxembourg and the Netherlands, etc. In a number of countries, the legislator left the list of social motives for committing crimes open, not excluding the gender motive.

   The anti-extremist legislation of Russia also contains an indication of the social affiliation of citizens. Extremism in Russia includes acts aimed at promoting the exclusivity, superiority or inferiority of citizens due to their social affiliation. Criminal law refers to such characteristics, including the gender of a person, as well as belonging to a social group. Social groups, although they have a sign of stability, can nevertheless disappear and re-emerge. Moreover, over time, new socially-structured entities appear. An example is the existing informal youth movements, historical and Patriotic clubs, associations of environmental orientation of it. All of them together form the social structures of society, i.e., social groups. At the same time, members of such social groups may well become the object of extremist actions based only on belonging to a particular group. Thus, social identification can be carried out, including through social functions, roles, models and forms of behavior of the individual. Therefore, gender, as well as the gender of an individual, must be considered as belonging to a particular social group.
3. Modern researchers of the problems of extremism, although they do not directly indicate the presence of "sexual extremism" or "gender extremism", do not exclude the possibility of committing crimes and other illegal acts for these reasons. So S.A. Markaryan, who, while investigating the motives of crimes against the individual, comes to the conclusion that the subject of the extremist motive is stable social groups formed, including on the basis of demographic characteristics (women and men) [10].

According to L.N. Nadolinskaya, radical masculinist and feminist trends that call for an open gender war with irreconcilable confrontation between the two sexes are prominent representatives of sexual extremism [11].

Some researchers, for example, N.O. Avtaeva, speaking about extremist manifestations of gender intolerance, identify them only with discrimination against women and sexism [2]. At the same time, extremist feminism or extremist feminism originated in the second half of the twentieth century in Europe and America and acquired a sharp, sharp and deeply irrational protest against the background of the "production" and "kitchen" revolutions [9]. In this regard, we note the position of M.A. Butayeva, who not only allows the existence of sexual extremism, but also actively uses this construct in relation to radical feminist groups [4]. Thus, it can be argued that in many ways the interests of the sexes are evaluated as opposite by feminist ideology, while the methods that radical feminist movements use to achieve their goals are directed directly against men. In many works of feminists, you can find not only a denial of traditional family values, historically formed in society ideas about the family, but also propaganda or agitation that excite sexual hatred and hostility, as well as calls for violence against men, up to their physical destruction. As an example, we can cite the "works" of the ideologist of radical feminist theory, Andrea Rita Dworkin.

Male chauvinism or sexism are also forms of ideological justification for violence between the sexes. However, today direct misogynistic maxims are relatively rare, and feminism, despite its modest history, currently dominates and tends to different levels of extremist worldviews. Moreover, male chauvinism and radicalism appear more as a sharply negative response of men to radical feminism. At the same time, both male chauvinism and radical feminism are characterized by conflict, a high level of which may have a pronounced attitude to the use of violent means to achieve their goals. In recent years, in Europe and North America, violent forms of misogyny have been widespread in far-right organizations such as the Men's rights movement (MRM). MRM cultivates hatred of women and resentment towards them. This misogynistic rhetoric ranges from disrespecting women to promoting violence against them.

Extremely radical accents and blatant propaganda of gender extremism are contained in the "SCUM Manifesto", published by V. Solanas—an American radical feminist, calling on the pages of this "opus" (just think!) to the complete destruction of men as biologically defective individuals [13]. It is not necessary to conduct comprehensive psychological and linguistic studies of these texts in order to make sure that they contain open calls for extremist activities. Moreover, the above-mentioned "SCUM Manifesto" is available to the General public: it is freely available on the global Internet, distributed through Internet marketing and electronic libraries. In other words, it is already possible to speak with confidence about the dissemination of information containing a negative emotional assessment and forming a negative attitude, inciting the Commission of violent acts against a specific social group, which, in our opinion, is consistent with the signs of a crime under Art. 282 of the Criminal Code of the Russian Federation [8].

The activities of the feminist movement in Russia is also at times extremely aggressive and hostile in nature. So in the period from 2013 to 2016, L. Kalugina, an activist of the feminist movement, posted statements and materials on her Vkontakte page and on the LiveJournal blog platform that humiliate the male sex and incite hatred towards men. A psychological and linguistic examination revealed signs of inciting hatred in two of Kalugina's publications. A criminal case was opened against the radical feminist under part 1 of art. 282 (Criminal Code)—incitement to hatred and enmity on the basis of sex. Subsequently, due to the partial decriminalization of article 282 (Criminal Code) [8], the case was dismissed. At the same time, the presence of aggressive rhetoric in feminist posts, as well as statements that contain signs of humiliation and calls for violence against men, was also recognized by experts of the non-governmental organization dealing with issues of xenophobia and nationalism—"SOVA Center for Information and Analysis" [15].

Another example of gender extremism is the case of Vladislav Pozdnyakov, the Creator of the "Male State" community in The "Vkontakte" social network, who, along with Patriarchal and nationalistic ideas, promoted misogyny. On November 20, 2018, the man was found guilty under part 1 of article 282 (Criminal Code) and sentenced to 2 years of probation. The court of Nizhny Novgorod also overturned Pozdnyakov's sentence in connection with the partial decriminalization of the criminal law provision providing for liability for such acts [16].

In August 2018, a 27-year-old resident of the Kaliningrad region, Artur Smirnov, was charged under part 1 of article 282 (Criminal Code) [8], who published texts and images directed against women in Vkontakte, in particular, an image of a man striking a woman, and a text in which the author speaks in harsh terms about women, their inherent qualities and behavior [8, 14].

4. Current Russian legislation provides for liability, including for an administratively punishable form of extremism: article 20.3.1. (incitement to hostility or hatred as well as affront to human decency) Code of Administrative Offences of the Russian Federation [5]. Criminal liability in accordance with the new version of article 282 occurs in respect of a person who was previously brought to administrative responsibility for a Similar act within one year. Practice of bringing to administrative responsibility under article 20.3.1. (Code of Administrative Offences) in the Russian Federation has already developed [5].
5. Analysis of the content of more than 30 feministic and muscular sites, pages of social groups and blogs in the Internet network obviously indicates a growing trend of gender conflicts. This process, as a biosocial phenomenon, is described in detail by sociologists E.A. Turintseva, E.V.Reshetnikova and V.V. Popova in their work "Gender conflict through the eyes of youth: Biosociological aspect" [17]. The reasons for this conflict, according to researchers, are biosocial dysfunction of the sexes, cognitive dissonance, eclecticism (mixing) of immanent (natural) hierarchies of the sexes.

   More than half of the Internet sources analyzed by us revealed materials (texts, photos, images, memes, audio recordings) containing a negative assessment (or directly aimed at forming such an assessment) in relation to both men and women.

Some materials contained calls for illegal actions, including violent ones. The communities of supporters of radical feminism in the social network "Vkontakte", whose content contains separate man-hating posts, can include such groups as: "Overheard feminism", "Stupid prostate", "Misandria" and others. Among the communities of radical masculinism can be identified groups: "anti-Feminist", "Anti-Feminism and-LGBT", "Anti-Feminist left front", "Forum of the Men's Movement", etc.

5 Conclusion

We can conclude that all of the above is quite correlated with our thesis that extremism in any of its manifestations is possible where the interests of subjects (individuals) that are opposed to each other in ideological, or philosophical terms collide. A diametrically opposite perception of the world around us, of social processes in it, is possible from the point of view of different approaches, including from the position of gender or other social self-identification. At the same time, this perception can be radical, including violent. Therefore, we believe that the danger of gender extremism is currently properly underestimated, and in some cases it is simply ignored. Based on the above, it is possible to formulate a definition of gender (sexual) extremism, which should be understood as the commission of any illegal criminal or administratively punishable acts motivated by hatred or enmity on the grounds of gender (sex) identity (self-identification) of an individual or social group, as well as public calls for the commission of such acts. We also believe that any justification for radical feminist or male chauvinistic (sexist) statements or appeals motivated by the gender (sex) of a person or social group should be immediately suppressed, and those who allow them should carry administrative or criminal responsibility in accordance with current legislation.

In modern society, unfortunately, there is a change and mixing of traditional, biologically predetermined and historically determined social roles of men and women, assuming specific (approved and shared by the majority of people!) ways of behavior, fixed by social norms, customs and traditions. This in turn leads not only to the loss of distinguishable images of men and women, but also to radical ideas and actions. We believe that any denial of the significance of the only possible, genetically determined, binary gender structure of society, and even more lenient attitude to the facts of gender extremism, motivated by hatred of a person on the basis of gender, is unacceptable.

References

1. Aly A, Macdonald S, Jarvis L, Chen TM (2016) Introduction to the special issue: terrorist online propaganda and radicalization. Stud Terror Confl 40(1):1–9

2. Avtaeva NO (2010) Hate speech in modern mass media: the gender aspect. Vestnik Lobachevsky Univ Nizhni Novgorod 4–2:811–813 (in Russian)
3. Bertram L (2016) Terrorism, the internet and the social media advantage: exploring how terrorist organizations exploit aspects of the Internet, social media and how these same platforms could be used to counter-violent extremism. J Deradicalization 7:225–252
4. Butaeva MA (2010) Discourse of gender extremism. Sociosphere 2:13–14 (in Russian)
5. Code of Administrative Offences of the Russian Federation, Law No. 195-FZ, 30 December 2001. https://legalacts.ru/kodeks/KOAP-RF/. Accessed 11 Jan 2020 (in Russian)
6. Conway M (2016) Determining the role of the internet in violent extremism and terrorism: six suggestions for progressing research. Stud Confl Terror 40(1):77–98. https://doi.org/10.1080/1057610x.2016.1157408
7. Costello M, Hawdon J, Ratliff TN (2017) Confronting online extremism: the effect of self-help, collective efficacy, and guardianship on being a target for hate speech. Soc Sci Comput Rev 35(5):587–605. https://doi.org/10.1177/0894439316666272
8. Criminal Code of the Russian Federation, Law No. 63-FZ, 13 June 1996. https://legalacts.ru/kodeks/UK-RF/. Accessed 11 Jan 2020 (in Russian)
9. Gevorkova KV (2008) Feminism as extremism. Philos Law 2:55–58 (in Russian)
10. Markarjan SA (2012) Motives as means of differentiation of criminal liability for crimes against the personality. Abstract of thesis. All-Russian Research Institute of the Ministry of Internal Affairs of Russia, Mahachkala (in Russian)
11. Nadolinskaya LN (2010) Masculinism and feminism as varieties of gender extremism. Ekstremizm.ru. http://www.ekstremizm.ru/publikacii/item/635-maskulinizm-i-feminizm-kak-raznovidnosti-gendernogo-ekstremizma. Accessed 09 Jan 2020 (in Russian)
12. Oksanen A, Hawdon J, Holkeri E, Näsi M, Räsänen P (2014) Exposure to online hate among young social media users. Soc Stud Child Youth 18:253–273. https://doi.org/10.1108/S1537-466120140000018021
13. Solanas V (1996) SCUM manifesto. AK Press, San Francisco
14. Sova (2018) A resident of the Kaliningrad region is accused of inciting hatred towards women. https://www.sova-center.ru/misuse/news/persecution/2018/08/d39886. Accessed 11 Jan 2020 (in Russian)
15. Sova (2019) Criminal prosecution of Omsk feminist Kalugina stopped. https://www.sova-center.ru/misuse/news/persecution/2019/02/d40611. Accessed 11 Jan 2020 (in Russian)
16. Sova (2019) Nizhny Novgorod: the sentence of the creator of the community "Men's state" was canceled. https://www.sova-center.ru/racism-xenophobia/news/counteraction/2018/10/d40116. Accessed 11 Jan 2020 (in Russian)
17. Turintseva EA, Reshetnikova EV, Popova VV (2016) Gender conflict through the eyes of youth: biosociological aspect. Knowl Underst Skill 4:171–184. https://doi.org/10.17805/zpu.2016.4.30 (in Russian)

Semantic Metalanguage for Digital Knowledge Representation

V. Dobrova, N. Ageenko, and S. Menshenina

Abstract The article addresses the problem of knowledge representation in digital systems in terms of cognitive science. The authors consider formalized semantic metalanguage as a tool to Simulate the semantic structure of sentences and thoughts. The created on the basis of frame-scenario model SESAME metalanguage is described and discussed, that allows to represent the meaning structure of standard events and situation in an artificial formalized language and combine linguistic and extralinguistic information. SESAME is a "matrix of meanings" for making natural language concepts suitable for "semantic calculations" using computer technology.

Keywords Metalanguage · Knowledge presentation · Meaning structure · Modeling · Frame

1 Introduction

The world is immersed in digitalization. The state program "Digital economics of the Russian Federation" [3] designed up to 2024 and "The Strategy of scientific and technological development of the Russian Federation until 2035" [19] determine knowledge as one of the basic directions for digitalization. Communications and data move to the "clouds", information and knowledge are transformed into algorithms, and the world becomes "algorithmic". This situation poses a number of unsolvable problems, one of which is the mathematization of thinking, and as a result, the development of a new algorithmic language for the digital world and artificial intelligence. This should be a mathematically rigorous language—a metalanguage—of human-machine communications, equally understood by people in

V. Dobrova (✉) · N. Ageenko · S. Menshenina
Samara State Technical University, Samara, Russia
e-mail: victoria_dob@mail.ru

N. Ageenko
e-mail: l-2402@yandex.ru

S. Menshenina
e-mail: menshenina.mail@mail.ru

S. I. Ashmarina and V. V. Mantulenko (eds.), *Current Achievements, Challenges and Digital Chances of Knowledge Based Economy*, Lecture Notes in Networks and Systems 133, https://doi.org/10.1007/978-3-030-47458-4_12

ordinary communication and their "digital counterparts" in computer communications. Since any knowledge should be expressed verbally or with the help of other signs, the task of world representation will be feasible only if the processes of natural language communication are provided with a sufficiently developed metalanguage.

A metalanguage can be defined as a language by which some other language properties are described and studied [7]. The "metalanguage" is one of the key concepts in modern philosophy, logic, IT, linguistics, etc. as it is used to research the verbal languages, to study logical and mathematical calculi, to display the whole spectrum of languages connections and to characterize the languages relations and the subject areas described with their help.

However, the term metalanguage is used ambiguously. It is used in relation to languages to describe the verbal formulations of meanings that corresponds to the traditional philosophical and logical definition. The variety of interpretations can be explained as follows.

The first approach considers any natural language as a combination of an object language and a metalanguage which we use to speak of an object language [2]. This kind of natural language understanding explains the existence of such expressions as meaning, denotation, affirmation, signification, presupposition, synonymy, etc. in a metalanguage.

The second interpretation contrasts the semantic metalanguage to the object language as another sign system that enables the direct reflection of the object language expression structure, revealing and objectifying it. So, the above-mentioned expressions according to the second understanding are considered as the object language expressions, requiring explication in terms of metalanguage [6, 20].

The task of knowledge representation is, first of all, the task of creating semantic metalanguages [9]. Despite the fact that in our time there are strong metalanguages of the phonetic and morphological levels, the language semantics still has a poor arsenal of tools that prevents an explicit representation of the meaningful side of speech. Moreover, the semantic metalanguage is an integral tool for studying the object language in the communicative and cognitive aspects. It is a necessary component of language modeling. The purpose of the metalanguage is to model and compactly describe the content plane of the object language.

The metalanguage of the semantic description of a natural language is also called the semantic language that is used to represent the meanings of various units of the language [10]. Traditional linguistics describes mainly an expression plane—the directly observable patterns. The semantic metalanguage explicates what is not directly observable, describes a content plane of a natural language that is an unobservable meaning structure of the language units. This is, so to speak, a "grammar of semantics", the identification of the content plane schemes by modeling them, i.e. by describing on the basis of indirect data of the model. The specificity of such a metalanguage is that it is not the only possible way of describing an object language and allows alternative methods of description.

The metalanguage classifications used to describe semantics vary significantly. First of all, non-verbal and verbal metalanguages are distinguished. The first include languages that use not only the words and other language units, but also drawings, schemes, mathematical constructs (dependency trees, graphs, etc.). Phrases in such a metalanguage have no verbal equivalents. An example of non-verbal metalanguage is the semantic representation of Melchuk and Polguère [13, 14] "Meaning-Text" model. The difficulty of classifying non-verbal languages is explained by their great differences; however, they must be accompanied by any verbal comments or interpretations, that is, the introduction of a third-order language.

Verbal metalanguages are divided into natural and artificial. Speaking of natural metalanguages (when one or another natural language acts as a metalanguage), it is necessary to take into account the coincidence and mismatch of the metalanguage and the object language. An example of a coincidence is the language of interpretations which is used to define words in explanatory dictionaries. Examples of mismatch are presented in bilingual explanatory dictionaries, in which a word of one language is associated with its translation into another language—a word or an expanded formula. The possibility of coincidence of metalanguage and object language, according to Krongauz [10] is a rather non-trivial fact. Indeed, it is the natural language that can serve as a metalanguage for describing itself. In this case, as a rule, we are talking about two of its functions, namely the functions of the object language and metafunction. A special status was given to this phenomenon by Jacobson [8], who considered the metalanguage function to be the main one.

It should be noted that the coincidence of the object language and metalanguage can lead to difficulties and problems associated with the interpretation of words; the simplest of these is a tautology. Analytical philosophers present the metalanguage as an ideal and insist on the elimination of tautologies, that implies the number of restrictions for the natural language in metafunction.

According to Solovyev and Ivanov [18], there are several requirements to the semantic metalanguage. First of all, it must be universal to express a variety of semantic meanings accurately enough. Secondly, it should include procedural elements in order to build models of cognitive processes of natural language processing on its basis and implement them in appropriate computer programs. Thirdly, such a language should be sufficiently organic and intuitively understandable. The experience prove that these goals can be effectively achieved with the help of artificially created formalized languages that are also called formal calculi, formalized systems, abstract calculus, formalisms or logistics systems, formal logic, etc.

2 Methodology

To solve the problem of language meaning modeling it is possible to use an artificial metalanguage. Its preference for natural language is caused by the fact that the latter tends to collapse information and is characterized by a variety of isosemantic means of transmitting thought and the presence of polysemy. The listed properties make it difficult to be used as a tool for scientific modeling.

Formalization provides models with advantages over object descriptions in natural languages such as uniformity and unambiguity.

Metalanguages are divided into verbal and non-verbal and have common features in the structure, which includes an information component (glossary) and a procedural component (a set of rules for operating characters). In our work, the metalanguage apparatus is based on the principles of propositional logic, predicate logic, and practical reasoning logic.

As noted above, models created on the basis of a natural language can be conveniently described by applying theoretical models that take the form of artificial metalanguages. For this reason, we chose an artificial language as the tool for the analysis. The advantage of the formal description over the natural language is that it (a) is unified, (b) is reduced, (c) is generalized. It is common for all situations of the class in question and reveals their invariant structure, which may not be visible when comparing various descriptions in natural language.

Having analyzed many definitions, it is necessary to highlight the main characteristics of formalized systems. This is an uninterpreted calculus in which the class of expressions is usually given inductively [11], which involves a complete abstraction from the meaning of the words of the language used and attention to the conditions governing the use of these words in theory.

Accordingly, a formalized language is used both for the purpose of representing logical forms of real contexts of a natural language, reproducing logical laws, and for expressing methods of correct reasoning on logical theories of a given language [1].

The construction of such an artificial language of logic begins when its alphabet is specified—a set of initial, elementary symbols, where it is customary to include logical symbols (signs of logical relations and operations, for example, quantifiers and propositional connectives), illogical symbols (i.e., parameters of descriptive components from natural language), as well as technical characters (for example, brackets). Further, from the simple language signs the rules for the construction of complex signs are formulated—different types of correctly constructed expressions are set, the most important forms of which are the analogues of natural language statements, i.e. formulas. The main advantage of a formalized language is the effective definitions of any of its morphosyntactic categories: the problem of belonging of one arbitrary symbol or chain of alphabet symbols to any class of language expressions can be solved in a finite number of steps. Sometimes formalized languages, in addition to the alphabet and the construction rules, include a number of transformation rules—deduction procedures, rules of transition. It is significant that in such cases, a formalized metalanguage acquires the status of logical calculus.

Thus, a formal language is the code on which a formal theory (formal system) is created, that is a text on a given code. However, this is an abstract theory (a scheme of the symbols relationship), which is not applied to a specific area of reality (subject area). To apply it to a specific subject area, it is necessary to subject it to semantic interpretation, i.e. to give specific meanings to the abstract symbols from the given subject area. An interpreted (i.e., filled with subject content) formal system is called calculus.

Calculus can be represented in the form of a formal apparatus for operating knowledge of some kind, which is based on clear rules, and which allows you to provide an accurate description of a particular class of problems [16], or an algorithm for solving subclasses of this class. A logical calculus is built on the basis of a formalized language; for this, the composition of the source symbols is specified, and then, according to well-defined rules, formulas of the calculus under consideration are created from these symbols. After the interpretation is added to the calculus, which attaches meaning to its original symbols and formulas, the calculus turns into a formalized language for describing a certain subject area.

The use of formalized languages in research make it possible to apply the principle of reduction, which reduces the whole variety of surface phenomena to a limited number of typical (invariant) schemes and in this way to reveal the underlying patterns in the field under study.

The feasibility of developing a formal description in an artificial metalanguage is determined by a number of reasons: it reduces or eliminates the interference of the metalanguage and the object language; it significantly increases the uniqueness and accuracy of the description of the object; it makes possible to determine the exact degree of homomorphism of frames to be compared; it makes such descriptions suitable for applied research. It has been correctly noted that descriptions of this kind are programmable and can be set when modeling intelligent systems by various means: in the language of frames, scripts, etc. [5].

In linguistics and cognitive sciences, several iconic systems have been developed, which can be considered formalized (i.e., reduced/unified/generalized), and which are based on different methodological grounds: situational semantics, graph theory, symbolic logic. This includes semantic networks, declarative and procedural semantic representations, as well as cognitive structures, which are called memory organization configurations, scripts, frames.

So, the function of semantic metalanguage is a compact representation of the content of an object language. Unlike those metalanguages, with the help of which the expression plane of an object language is described, a semantic metalanguage clearly expresses what is hidden, i.e. content. The specificity of the semantic metalanguage is that it's not the only possible means of modeling the object language and does not exclude alternative modeling tools.

The essence of the author's concept of metalanguage is that texts in this language allow you to describe the structure of a natural language utterance, verb, verb derivatives and phraseologisms in a single way—in the form of a poly-predicate frame in which the connections between individual "acts" are explicated. "Little drama" is hidden in the utterance.

As the denotative area of the semantic metalanguage, the primary concepts of cognitive science were used—"event", "state of affairs", "situation". A significant role here is played by the participants (actants) of the events. Any actant has a set of attributes called functional and characterizing him as one of the participants in the event in question. One participant in the event is able to play several roles, and several actors can have the same role in the event. The role of the participant is determined by his place in the structure of the event. An event is also characterized by properties such as its position in space and duration in time.

At a new stage of cognitive science development researchers recognize the fact that the most effective way of representing knowledge is to schematize it, for example, in the form of frames. A frame is a technique for studying the structure of knowledge and the principles of their organization in the language system [17]. A person, trying to learn a new situation for himself or take a fresh look at already familiar things, selects from his memory a certain data structure (image), which we called a frame so that, by changing individual details in it, make it suitable for understanding a wider class of phenomena and processes. A frame is a data structure for representing a stereotypical situation [15].

The frame method makes it possible to effectively describe not only linguistic objects (meanings of natural language units), but also cultural objects (typical situations reflected in lexical meanings) using a uniform system of methodological categories. This is consistent with the idea of Minsky that linguistic and non-linguistic ideas do not have a significant difference: a significant part of understanding and reasoning based on common sense is much like the transformation of linguistic structures and their manipulation [15].

The created metalanguage of the meaning structure description based on frames got the conventional name SESAME, formed from the enumeration of the metalanguage characteristics: Standard Events and Situations Artificial MEtalanguage. This metalanguage, created in the course of our research, is proposed as a methodological tool for semantic research.

The SESAME artificial metalanguage glossary contains three classes of units: (1) terms, (2) operators, (3) propositional variables. From the metalanguage units according to its rules, formulas are created. An example is the expression "WANTS x, A" that is translated into a verbal language like this: "A certain subject wants (wants, aspires, needs) a certain state of affairs". For example: *Captain Grant's children yearn for their father. Bill is thirsty. Soames Forsyth wants a new home.*

The frame serves as a schematic representation of the situation and models it. In our model, a frame is a multicomponent object, a semantic representation that embodies a certain amount of data, representing a certain stereotypical situation or fragment of reality.

The frame-scenario creation begins with a multi-step analysis of the natural language units definitions. Later with its help actants and circonstants of an event/situation are found, presented in a word, phrase, text. Next, the resulting detailed dictionary definition is translated into an artificial metalanguage, and then a script is created with a sequence of scenes included in it. The final phase is the semantic interpretation of the frame script on the terminals by giving the script elements specific values and meanings. The scenario can be presented in the form of a network consisting of nodes; the nodes are filled with so-called "tasks", which contain typical features of the described situation.

As a resulting example, let us consider the semantic representation of a subject-practical event. The action here is initiated by an agent that is guided by a specific motive, has a goal, has internal and external potentials to perform this action, at least in general terms, has an idea about the course of the planned action, has the ability

Table 1 Terminal interpretation

№	x (agent)	m (material)	i (instrument)	v (value)	verbs
1	Weaver	Thread	A loom	Cloth	To make
2	Carpenter	Logs	An axe	A cottage	To build
3	Baker	Dough	An oven	A loaf	To bake
4	Tailor	Fabric	A needle	A dress	To sew
5	Smith	Cast-iron	A hammer	A grate	To forge

Source authors

to predict the action, adjust its course and evaluate the result. All this is reflected in the frame "Production".

(1) INITIAL SITUATION:
IF ((HAS x, m) AND (USES x, i)) THEN CAN x (CREATES x, v).
((HAS x, i) and (HAS x, m)) IN PERIOD t^0.
(2) WISH x: (WANTS x (HAS x, v)) IN PERIOD t^1.
(3) POTENTIAL x: CAN x (USES x, i).
(4) INTENTION x: (INTENDED x (CREATES x, v)) IN PERIOD t^2.
(5) ACTION x: (USES x, i) IN PERIOD t^3.
(6) CONSEQUENCE: (HAS x, v) IN PERIOD t^4.
(7) EVALUATION: PURPOSEFUL (ACTION x).

On the terminals of the frame, the variables get the interpretation that is presented in Table 1.

Thus, the script model allows to represent the meaning not just as a collection of semes [4], but in the form of an expanded text written in a metalanguage, where the lexical meaning is presented as a narrative in which the relations of the actors are shown as a sequence of states and events.

3 Results

In the course of the study, we obtained the following main results. The starting point to study the sentence meaning is a semantic predicate, which, due to its specificity, determines the quantity and quality of the actants dependent on it. Thus, the structure of the verb meaning is a kind of "model" of the semantic structure of the sentence. The semantic structure of a sentence forms its meaning, while the surface structure of a sentence determines its lexical content, the sequence of words and the nature of syntactic connections. Our cognitive and logical approaches allowed us to establish the absence of isomorphism between the semantic and surface structures of the sentence.

We created a formalized metalanguage that makes it possible to simulate the semantic structure of sentences. It meets a number of requirements: (1) an accurate description of heterogeneous values; (2) the possibility of improvement and expansion; (3) intelligence. The created frame-scenario model allows to describe the meaning of the verb in the form of text composed in an artificial formalized language and combining linguistic and extralinguistic information. Since the verb can model not one but a number of elementary events connected with each other, the situation acts as the content of the sentence. This means that every actant can fulfill a number of roles in a situation, depending on its specifics.

The semantic role is a predicate meaning fragment that reflects the actant-predicate features. Along with semantic roles in linguistic cognitive science, situational roles are postulated in the framework of situational semantics. The content of the sentence is characterized by the presence of a number of characteristics (morphological, syntactic, lexical and semantic) and contains information as a unit that combines all of these components.

The authors' research is focused on the functioning of natural intelligence, the focus of knowledge of which is the classical philosophical tradition, linguistics and mathematics. Only philosophy gives a generalized idea of a holistic picture of the world and processes of thinking, and mathematics is "the art of calling different things by the same name", and linguistics allows us to convey meanings through language.

On this scientific and theoretical basis, an algorithmically constructed metalanguage was developed—a network language for modeling collective thinking with the logic of "common sense". Its semantics is formed by an ordered set of universal concepts, and syntax is formed by the rules for their ordinal recording as mathematical and logical objects. SESAME is a "matrix of meanings" for making natural language concepts suitable for "semantic calculations" using computer technology. Its application is organized by ordering ordinary discrete texts with a linear sequence of words using a metalanguage, which is an integral network structure of universal meanings.

4 Discussion

The analysis of everyday language in analytical philosophy showed what facts of the objective world are hidden behind complex and diverse language constructions that express these facts in indirect and intricate ways, and sometimes distort them. According to this methodological approach the surrounding world and its fragments (state of affairs) have a certain, objective structure independent of human consciousness and language that needs to be identified in the course of the world cognition [12]. Cognitive linguistics also shares this statement in its believe that the situation, which is denoted by a linguistic utterance, has a unified objective structure independent of the utterance's structure and reflected in the utterance.

It should be noted that, firstly, the surrounding reality has an infinitely complex multi-level structure (from micro- to macro-level); human capabilities to identify this structure are limited. Secondly, the reflection of reality is not a "mirror-dead act", but a "living contemplation"; the human being reflects the world based on his own pragmatic attitudes. All this leads to the fact that linguistic statements do not reflect the absolutely objective and the only situation structure, but their "cut", moreover, pragmatically oriented.

For example, considering the situation from Dostoevsky's novel in which Katerina Ivanovna dies of consumption, the role of a heroine's death causer can be attributed to Koch's wand, or a disease as a whole, or the miserable living conditions of an unhappy woman, or Marmeladov who brought her to such a state, or cruel and indifferent society. From the cause-and-effect chain, any of the links can be brought into the focus of attention, depending on the context in which this situation is considered. The observer's position isolates a pragmatically determined "cut" from the infinitely complex structure of the situation and reflects it in the structure of the utterance. Thus, it is difficult to talk about some unique and absolutely objective situation structure, independent of its perception and understanding by the subject. The structure of the situation, which is reflected in the statement, is partly determined by the subject's view of the object.

A frame script is an initial, literal, nonlinear thought expressed on a subject code. The frame does not depend on the language, although it depends on the pragmatic attitude of the speech producer. Further, the verbal case frame is an intermediate stage from thought to word. The composition of the case frame is determined by the characteristics of the verb. This is "the thought speech." Further, the surface structure is already a verbal speech, i.e. the thought fully reflected in words.

5 Conclusion

In the framework of cognitive science, objects of analysis are described in such a way that the speaker (a person or a speech processor) can use language units adequately, in any contexts and communicative situations.

In the 20th century, linguistics was developed under the influence of exact sciences and almost became a precise science itself, adopting a number of methods from cybernetics, information theory, and other similar disciplines. It became associated with formalized sign systems. But now the humanitarian approach to language prevails. Linguists have largely returned to operating with vague categories such as "linguocultural concept", "linguistic gestalt", "linguistic mythology", etc., which are not applicable or can hardly be applied to logical mathematical tools. In our opinion, these approaches do not exclude, but complement each other, together creating a three-dimensional picture of the object. As a result of their complex application, scientists begin to see the inextricable unity of the structural "skeleton" of the language and the "flesh" of its semantic content. Language appears not as a mechanism, but as an organism, alive and flexible.

The theory of the structure of language semantics is in demand today by a number of sciences and, in particular, the science of the representation of knowledge in electronic systems—part of the program of work in the field of creating artificial intelligence. The structure of the human language, and through it the image of the world structure, can be transmitted to the electronic brain using formalized languages that are accessible to its understanding.

Clarity of thinking and clarity of presentation of the meanings of the surrounding reality takes on special significance in the era of the growing influence of digital technology on all aspects of our lives. The ability to think holistically and constructively becomes necessary for information communications in the digital world—the world of signs and symbols with mathematical abstractions and formal logics.

The success of the digital transformation of the society is the ability to understand, display and reproduce reality in circuits and logics that are equally applicable in human-human, human-machine, machine-machine communications. The need for this increased during the transition to digitalization of data.

The need for changes in the organization of thinking and bringing it into line with the requirements of mutual understanding in all types of information communications determines the meaning of semantic metalanguages. Their application can contribute to imparting the necessary rigor to the language of communication in information systems, the development of skills to express thoughts in a form that allows you to translate them into algorithms.

References

1. Bocharov V, Voishvillo E, Dragalin A, Smirnov V (1980) On problems of the evolution of logic. Soviet Stud Philos 18(4):31–52. https://doi.org/10.2753/RSP1061-1967180431
2. Curry H (1980) Some philosophical aspects of combinatory logic. Stud Logic Found Math 101:85–101. https://doi.org/10.1016/S0049-237X(08)71254-0
3. Digital economics of the Russian Federation – National program. http://government.ru/rugovclassifier/614/events/. Accessed 10 Feb 2020
4. Dobrova V, Kistanova O (2017) Meta-languages: nature and characteristic features. Voprosy Kognitivnoy Lingvistiki 1(50):114–117. https://doi.org/10.20916/1812-3228-2017-1-114-117
5. Faber P, Cabezas-García M (2019) Specialized knowledge representation: from terms to frames. Res Lang 17:197–211. https://doi.org/10.2478/rela-2019-0012
6. Goddard C, Wierzbicka A (2013) Words and meanings. Oxford University Press, Oxford. https://doi.org/10.1093/acprof:oso/9780199668434.003.0001
7. Gvishiani N (2017) The metaconcept word-combination and its transformation in English computer-corpus discourse. Voprosy Kognitivnoy Lingvistiki 2:15–25. https://doi.org/10.20916/1812-3228-2017-2-15-25
8. Jakobson R (1965) Quest for the essence of language. Diogenes 13(51):21–37. https://doi.org/10.1177/039219216501305103
9. Jakus G, Milutinović V, Omerović S, Tomažič S (2013) Concepts, ontologies, and knowledge representation. Springer, New York. https://doi.org/10.1007/978-1-4614-7822-5_4
10. Krongauz M (1994) Word formation and linguistics. Russ Linguist 18:379–387. https://doi.org/10.1007/BF01650153
11. Laurini R (2017) Geographic knowledge infrastructure. ISTE Press, Elsevier, London. https://doi.org/10.1016/b978-1-78548-243-4.50002-5

12. Lebedev M (2007) Prospects of modern concepts of reliability of knowledge. Voprosy Filosofii 11:119–132
13. Melchuk I (2015) Dependency in language. In: Wright JD (ed) International encyclopedia of the social & behavioral sciences. Elsevier, London, pp 182–195. https://doi.org/10.1016/b978-0-08-097086-8.53005-0
14. Melchuk I, Polguère A (2002) A formal lexicon in the meaning – Text theory (or how to do lexica with words). Comput Linguist 13(3–4):261–275
15. Minsky M (1975) A framework for representing knowledge. In: Winston PH (ed) The psychology of computer vision. McGraw-Hill Book, New York, pp 211–281
16. Musen M (2014) Knowledge representation. In: Sarkar IN (ed) Methods in biomedical informatics. Academic Press, Cambridge, pp 49–79. https://doi.org/10.1016/b978-0-12-401678-1.00003-8
17. Nazaruks V, Osis J (2017) A survey on domain knowledge representation with frames. In: Damiani E, Spanoudakis G, Maciaszek L (eds) Proceedings of the 12th international conference on evaluation of novel approaches to software engineering, MDI4SE, vol 1. SciTePress, Porto, pp 346–354. https://doi.org/10.5220/0006388303460354
18. Solovyev V, Ivanov V (2015) Knowledge-driven event extraction in Russian: corpus-based linguistic resources. Comput Intell Neurosci 2016:4183760. https://doi.org/10.1155/2016/4183760
19. The Strategy of scientific and technological development of the Russian Federation until 2035. http://kremlin.ru/acts/bank/41449. Accessed 10 Feb 2020
20. Wierzbicka A (2009) Language and metalanguage: key issues in emotion research. Emot Rev 1:3–14. https://doi.org/10.1177/1754073908097175

Entrepreneurship in Digital Era: Prospects and Features of Development

E. V. Rybakova and M. A. Nazarov

Abstract The purpose of the study is to identify the main issues of doing business in the digital economy era and determine the directions for further improvement of state regulation of electronic commerce in Russia. The authors discuss basic concepts of the digital economy, analyze its role in the development of entrepreneurship, including small and medium companies, and identify trends that indicate the digitalization of various spheres of public life. The work reflects the main components of the digital economy, considers the impact of digitalization on business structures using Russian and foreign companies as examples. The study is based on the papers of Russian and foreign scientists dedicated to the development of entrepreneurship in digital economy.

Keywords Digitalization · Entrepreneurship · Business development · E-commerce · Innovations

1 Introduction

The problems of business sector development are becoming very relevant, especially in the era of the transformation of traditional ideas about doing business. This new era is mainly related to implementation of digitalization processes in all spheres of human life. The formation of a new paradigm of economic development necessitates a comprehensive rethinking of both legal and economic aspects of entrepreneurial activity.

The introduction of digital technology entails increased competition, creating threats for existing leaders from new waves of innovation. In the digital era, economies of scale were achieved through the construction of large production complexes. The deployment of such industries required a significant investment of time

E. V. Rybakova · M. A. Nazarov (✉)
Samara State University of Economics, Samara, Russia
e-mail: good_mn@mail.ru

E. V. Rybakova
e-mail: e_ryb@bk.ru

S. I. Ashmarina and V. V. Mantulenko (eds.), *Current Achievements, Challenges and Digital Chances of Knowledge Based Economy*, Lecture Notes in Networks and Systems 133, https://doi.org/10.1007/978-3-030-47458-4_13

and resources and brought significant costs. In digital business structures, there is a combination of low costs with the easy scalability of IT platforms. This allows the most successful of them to reach significant proportions in a short time.

Nowadays, the process of transformation of classical industries is becoming irreversible. The digital economy is changing the face of entire industries: telecommunications, printing, tourism, passenger transportation, etc. Thanks to digital technology, the specialization of business structures is increasing. This occurs in industries where manufacturers develop customized offers for small markets within the system. For example, new services are emerging in the retail sector, such as food delivery from restaurants that do not deliver on their own.

2 Methodology

The following methods were used in this study: theoretical (dialectical logic, rational knowledge and others); diagnostic (the diagnostic analysis of the state and reasons, questioning and testing); empirical (facts description, measuring and generalization of the research results and others).

The information base of the research is the data of the analytical research institutes, centres of the World Bank (the International Bank of Reconstruction and Development, the International Association of Development, the Multilateral Agency of Investments' Guarantees), international consulting companies and analytical agencies (McKinsey, Bloomberg), publications on the problems of the introduction and usage of innovations and digital technologies in business activities.

3 Results

The development of the digital economy contributes to the growth of the purchasing power of the population, as digital platforms and trading platforms create intense price competition. Trading platforms, such as Yandex.Market, not only allow you to buy goods at the best price, but also know its characteristics by reviews of real users and by comparing your choice with other options.

This forces competing sellers and manufacturers to provide high quality products and services and lower prices. Some digital companies have begun to offer free additional services, such as GPS navigation.

The development of digitalization significantly saves entrepreneurs time on such necessary procedures as registration of an enterprise, opening a current account, registration with the tax service, etc. The introduction of information services, such as, for example, "Online appointment for inspection", the possibility of electronic submission of documents through the Multi-functional centers, significantly reduce the transaction costs of business entities. The possibility of electronic signature of

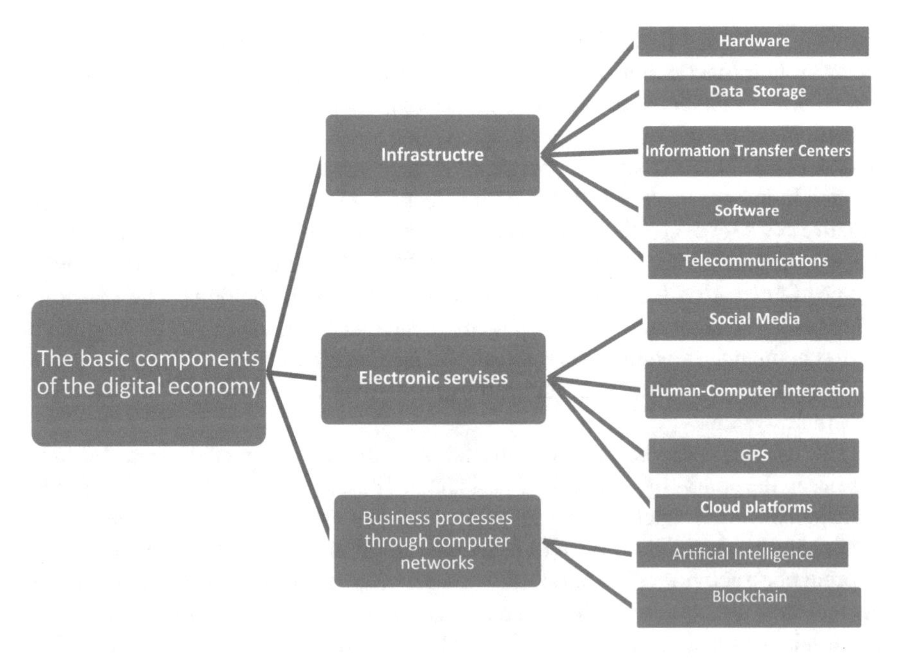

Fig. 1 The basic components of the digital economy (*Source* authors)

all constituent documents undoubtedly accelerates the process of registration and the start of work of a small business.

Among the basic components of the digital economy there are the following (Fig. 1):

As we see it, e-commerce can also be considered as a new type of economic activity in the context of the digital economy. The positive dynamics of the turnover of various products with the help of such well-known online stores as Amazon, Apple-iTunes, Ozon indicates the growth of attractiveness of entrepreneurship in this area [2].

Especially interesting is the use of artificial intelligence (AI) in transactions through online stores, which allows such transactions to exist on behalf of the economic agent creating this online platform.

Recent advances in the construction of supercomputers make it possible to bring their computational capabilities closer to the capabilities of the human brain, whose computing power is estimated at about 1020 FLOPS (FLOPS—the number of floating point operations per second) [5].

We should not confuse AI with conventional automation. AI has a very important distinguishing feature—it can change the algorithm of its work based on the data it collects. Moreover, these changes are far from always programmed by logical algorithms but are the conclusions that AI systems make themselves.

The use of digital technologies, namely cryptography in the financial markets, contributes to the creation of another fundamentally new type of entrepreneurial activity—mining. Mining is the cornerstone process in the field of the cryptocurrency system functioning. This is a computational cryptographic process of creating new blocks, a chain of records, in other words, confirmation of transactions. Now mining is controlled by large pools with which small miners share their resources. The network depends on the operators of large data centers, many of which are located in the provinces of China, where electricity is cheap. Miners in China, having invested a lot of capital in mining capacities, now have computing "superpowers" for monopolizing the cryptocurrency market. In order for mining to be recognized as a type of enterprise, its legal consolidation is necessary. It should be said that some attempts at such consolidation are already being made. So, the bill on the regulation of digital financial technologies is already being considered, which clearly defines all the basic concepts related to the development of financial technologies in the modern economic system.

The increased use of digitalization tools facilitates the interaction of business entities in the course of economic transactions. Thus, the use of electronic document management accelerates the process of procurement of work, goods, services, which significantly reduces the time of transactions, and as a result, increases the efficiency of production activities. The possibility of electronic payments for goods and services also leads to an increase in the speed of ongoing processes. In addition, all modern electronic payment services are quite reliable and provide a high level of security.

Another important thing of the implementation of entrepreneurial activity is to ensure the rights and legitimate interests of market participants. In our opinion, the creation and implementation of an electronic justice system is necessary, which will be able to consider economic disputes in efficient and quick way, draw up the necessary acts, and give competent conclusions.

It should be noted that experiments are already underway to introduce certain elements of digitalization in judicial practice. For instance, during the process of court hearings conversations are recorded using a special program that recognizes speech and captures dialogs in an electronic document. Previously, as you know, this work was performed by the court secretary. There are already programs that can consider standard cases and make a preliminary decision. Such facts suggest that in the near future court hearings will be virtual in nature, where electronic documents, audio and video recordings, digital images, etc. are used as evidence. However, large enterprise structures based on traditional technologies, digitalization can pose a serious threat in the form of loss of competitiveness in case they do not introduce new ideas. Even in traditional industries, methods of analyzing large amounts of data are increasingly being applied to gain new knowledge and make effective management decisions.

The Internet of things can improve the quality of equipment operation, increase labor productivity, and increase energy efficiency. Additional opportunities open up with the development of innovations such as additional reality, unmanned aerial vehicles (drones), robotics, and artificial intelligence.

Innovative entrepreneurial structures are introducing digital tools in a wide variety of sectors. For example, in the oil and gas industry, this allows you to develop a comprehensive virtual field model, increase the efficiency of the drilling process, monitor fields remotely, etc.

In modern conditions, companies are striving to increase labor productivity not by increasing output, but by optimizing the structure of the enterprise and the number of personnel. Automation of production has become an ongoing process. Digital technology leads to a reduction in the number of middle-skilled workers. Robots replace workers on conveyors, information systems perform operations for which office workers were previously responsible.

Digital platforms that allow users to search and compare different offers have entered a fragmented market, previously owned by traditional companies. This has found wide application in particular in the field of tourism. For example, Aviasales.ru—ticket sales, Booking.com, Ostrovok.ru—hotel room reservation.

Thus, strong market positions today are moving from physical intermediaries and asset owners to consumers and digital platforms, which are platforms through which suppliers and consumers find each other and conduct transactions directly without intermediaries (for example, Booking.com, Ostrovok.ru). This allows you to increase the utilization of assets (housing, taxis, etc.) to suppliers connected to these platforms. At the same time, the opportunities for operating assets for owners who are not represented on such digital platforms are significantly reduced.

Digital business structures are entering new markets. As digital companies take leading positions in one market, they seek to more actively develop related areas, which then become the main ones. For example, Amazon.com went beyond selling books, offering consumers a wide range of consumer goods, later created its own publishing platform, began providing cloud and logistics services to other companies, and producing electronic consumer goods (e-books).

In Russia, the OZON online store, like Amazon.com, provides logistics services to other companies. Yandex entered the taxi market and began to analyze large amounts of data. Thus, in many sectors, entrepreneurial structures have to constantly monitor the market, while analyzing whether there are threats of the emergence of actively developing competitors, counteract these threats, and also increase their competitiveness through digitalization.

4 Discussion

In scientific sources, there are several approaches to the definition of "digital economy": digital economy as the way of doing business on the Internet; digital economy as a system of relations based on the use of digital technologies; digital economy as a specific production organization [7, 11].

The spheres of digital economy implementation are growing every year, and now the most developed are the following: e-business; Internet banking; social sphere; education; telecommunications; Information Systems; industry.

Four main trends in the technological aspect in the formation of the digital economy can be distinguished: development and practical application of mobile technologies, business analytics, the use of cloud computing, social media; global social networks such as Facebook, YouTube, Twitter, LinkedIn, Instagram, etc. [3, 4, 6, 8].

Under current conditions, serious investments are needed in the future competitiveness of the national economy, industries, and entrepreneurial structures [1, 10]. The spread of digital technology at a natural pace will not give the expected result and will aggravate the backlog of Russia from the most developed countries. Therefore, advancing strategies should be developed, quickly adapting and implementing technological achievements of leading companies, developing active cooperation with educational and research organizations, as well as tracking the needs, preferences and habits of consumers.

The use of digital technologies improves the business and investment climate due to the availability of state services for registration of legal entities, certification and accreditation, obtaining permits, declaring and paying taxes, customs escort, developing an entire ecosystem of business services (logistics services, mobile banking), and increasing transparency in conducting business (electronic sites for tenders and procurements, feedback portals).

Significant effect can be achieved by state participation in promising entrepreneurial projects. It can be implemented either through the creation of joint ventures, or in the form of a state order in the most popular areas of digital development. State participation will help in financing promising digital projects, especially at the stages of scaling up, introduction into production and the transition to the use of developments on an industrial scale.

Startups, innovative companies and small enterprises also benefit from the use of digital technologies, as they optimize costs, increase the profitability of assets and increase the profitability of new investments. In addition, in the digital era, you can quickly enter the global market, become accessible to consumers anywhere in the world.

In the context of global digitalization of business, basic ideas about the possibilities and ways of making profit are changing. The law "On the Strategy for the Development of the Information Society in the Russian Federation for 2017–2030" adopted by the Government of the Russian Federation forces the scientific community to take a fresh look at many established mechanisms and models for coordinating business entities. The vertical type of interaction of economic factors is being replaced by a horizontal one, which practically erases the territorial boundaries of markets, due to the introduction of modern investment tools such as crowdfunding, ICO, etc.

A significant tendency of recent years in the field of investment activity and investment decision making is the introduction of cloud technologies, Blockchain technology, Big Data, which significantly reduce the level of transactional and transformational costs. These technologies allow you to create decentralized online services based on "smart contracts". Blockchain accelerates the process of making investment decisions in terms of interaction with a large number of counterparties and allows you to increase transaction security. Experts note a significant increase in the use of digital communication channels in the adoption and implementation of investment decisions by various market participants [9].

5 Conclusion

In conclusion, it can be noted that the transformation of the paradigm of socio-economic development, due to digitalization, inevitably leads to a modification of entrepreneurial activity, the emergence of its new forms, tools and methods of coordination. The expansion of the spectrum of entrepreneurship leads to the emergence of a qualitatively new imperative of development, which is based on increasing the well-being of society.

Entrepreneurship structures should constantly search for innovative solutions and business models based on digital technologies. Product development should be more flexible. You need to experiment with new business models, products, ideas, and technologies. A passive position can lead to a loss of competitiveness.

Mastering the technologies of Industry 4.0, such as Internet of things, 3D printing, virtual reality, touch interfaces and advanced robotics, will allow entrepreneurial structures to take advantage of the development of these areas and become at the forefront of this group of technologies.

Large companies should efficiently and quickly adapt and deploy ready-made platform solutions and services, but also actively participate in the formation of the market, creating partnerships with other industry companies and solution developers.

References

1. Bondarekno V (2019) Digital economy: a vision from the future. J Econ Sci Res 3(1):16–23. https://doi.org/10.30564/jesr.v3i1.1402
2. Belousov Y, Timofeeva O (2019) Methodology for defining the digital economy. World New Econ 13(4):79–89. https://doi.org/10.26794/2220-6469-2019-13-4-79-89
3. Bunz M, Meikle G (2018) The internet of things. Polity Press, Cambridge
4. Denisov I, Karakhanyan G (2013) Technology as a moving force of economic processes. Theory Pract Soc Dev 8:324–326 (in Russian)
5. Fomin E, Konovalova M (2018) Features of business sector development in the conditions of formation of the digital model of economy. Econ Law Issues 125:51–53. https://doi.org/10.14451/2.125.51 (in Russian)
6. Kuntsman A, Miyake E (2016) The paradox and continuum of digital disengagement: denaturalising digital sociality and technological connectivity. Media Cult Soc 41(6):901–913. https://doi.org/10.1177/0163443719853732
7. Nazarov M, Mikhaleva O, Fomin E (2019) Digital economy: Russian taxation issues. In: Mantulenko V (ed) Proceedings of the international scientific conference "global challenges and prospects of the modern economic development". The European proceedings of social and behavioural sciences, vol 57. Future Academy, London, pp 1269–1276. https://doi.org/10.15405/epsbs.2019.03.129
8. Plantin J-C, Punathambekar A (2018) Digital media infrastructures: pipes, platforms, and politics. Media Cult Soc 41(2):163–174. https://doi.org/10.1177/0163443718818376
9. Ravi C, Manimaran P (2020) Introduction of blockchain and usage of blockchain in internet of things. In: Rajput D, Thakur R, Basha S (eds) Transforming businesses with bitcoin mining and blockchain applications. IGI Global, Hershey, pp 1–15. https://doi.org/10.4018/978-1-7998-0186-3.ch001

10. Savchenko T, Rogov V, Shadov G, Verhozina V (2015) Metodological approaches to the management of innovative development of an enterprise. Asian Soc Sci 11(8):243–252. https://doi.org/10.5539/ass.v11n8p243
11. Smirnov E (2018) Evolution of innovative development and prerequisite of digitalization and digital transformations of the world economy. Russ J Innov Econ 8(4):553–564. https://doi.org/10.18334/vinec.8.4.39696 (in Russian)

Technological Paradigms of Digital Competences Development

M. V. Kurnikova, I. V. Dodorina, and V. B. Litovchenko

Abstract Since the technological paradigms are the theoretical basis for digitization, the purpose of the study is to develop a unified methodological approach to the definition of the content and processes of digital competencies formation on the basis of these paradigms. The unity of the approach is achieved by clarifying the concept, defining the inventory, composition, structure of digital competencies within the boundaries of technological paradigms and on the basis of the European Qualification Framework (EQF). The authors have used the methods of comparative and semantic analysis, contextual analysis of policy-based documents at the supranational level to justify the methodological provisions of the development of digital competencies in the Russian policy as part of key objectives and priorities of training personnel possessing competence for the development and implementation of digital technologies, increasing the digital literacy among the population. The practical significance of the results of the study is connected with the effectiveness of the strategies for the development of industries and subjects of the Russian Federation, the validity of the choice of educational programs.

Keywords Digitalization · Paradigms · Digital competences · Digital literacy

1 Introduction

The scale and depth of digital power is rapidly transforming the traditional industrial economy. As part of digital technologies, augmented reality, artificial intelligence, Big Data are called "cross-cutting technologies" as they transform not only the way

M. V. Kurnikova (✉)
Samara State University of Economics, Samara, Russia
e-mail: mvkurnikova@gmail.com

I. V. Dodorina · V. B. Litovchenko
Samara State Transport University, Samara, Russia
e-mail: dodorina@mail.ru

V. B. Litovchenko
e-mail: vip.vereneya@mail.ru

S. I. Ashmarina and V. V. Mantulenko (eds.), *Current Achievements, Challenges and Digital Chances of Knowledge Based Economy*, Lecture Notes in Networks and Systems 133, https://doi.org/10.1007/978-3-030-47458-4_14

of production and business models, but also employment, into new forms and states. In the absence of consensus on the details of their impact, most researchers are in clear concurrence about understanding the consequences. Digital technologies will create new jobs and forms of employment, based on (1) robotic product and service industries, (2) automated management, based on digital platforms, crowdsourcing, and the "sharing" economy.

These prospects need competences of the human of a digital society, the digital production worker and manager, set by the labor market, the state and the society—the need for digital competences enabling to effectively meet the challenges of professional activity and everyday life.

In scientific publications and special literature, the concept of a "*digital competence*" is represented in the context framework [3, 7, 10, 16], covered in the simplified sense of "new", "modern" competencies, thus, the development of a unified methodological approach to the definition of the content and processes of digital competencies formation seem to be highly relevant. The integrity of the approach is achieved by: (1) analysis of technological paradigms leading to digital transformation, clarification of the concept of a "digital competence"; (2) definition of the composition and structure of digital competencies within the boundaries of technological paradigms and based on the European Framework of Qualifications (EQF); (3) developing methodological provisions for the development of digital competencies within Russian policy as part of key goals and priorities aimed at training personnel with competence for the development and implementation of digital technologies, for the growth of digital literacy among the population, which is the objective of the research.

2 Methodology

The research is aimed at developing a unified methodological approach to the creation of the digital competencies of the human being of the digital society, a worker in the digital economy on the basis of technological paradigms. The following methods have been involved to achieve this objective:

1. The method of comparative analysis to research into the technological paradigms, which set the system of requirements to universal competence of a person of the digital society, to professional competence of the employee engaged in the digital economy.
2. The method of semantic analysis of the composition and structure of digital competencies to identify the inventory and structure of knowledge, competences and skills in working with digital technologies.
3. The contextual analysis of supranational policy-based documents that develop a system of measures aimed at digital competencies.

The implementation of these methods allows to clarify the concept of digital competencies that meet the requirements of the digital society and economy to a person and a professional, to develop methodological provisions for the development of digital literacy among the population, digital competencies of workers based on the research of foreign best practices.

3 Results

The type of production based on digital technologies has been called the digital economy, in which the widespread and rapidly accelerating dissemination of information creates radically new realities in the organization of work and learning. The digitalization-generated shifts in occupations, labor markets, and forms of employment have led to deficient demand for digital skills, which require an analysis of the technological paradigms that have "triggered" digital transformation and explain the nature of societal and economic change as a result of digital transformation.

3.1 Technological Paradigms in Researching into Digital Competences

The concept of a "paradigm" was first used in the works of the American historian of science T.S. Kuhn as a "model of the production and solution of research problems" [9]. In addition to generally accepted paradigms (divided by most specialists of a certain area of knowledge), there are private paradigms (reflecting research models from the author's point of view, serving to solve a private problem). Thus, to solve the applied problem of changing employment landscape in the digital era, Canadian researchers of the University of Prince Edward Island engaged in a Social Sciences and Humanities Research Council (SSHRC) proposed two paradigms [1]:

(1) The first paradigm—"that technology is getting work through automation"—is the digital development of the ideas of Taylorism based on a technological innovation. According to those who share these views, computers have been able to achieve human to super-human levels of performance in a range of tasks, and the companies linked to the Internet of Things can use big data and analytics to develop algorithms that speed efficiency, increase productivity and dramatically lower the marginal cost of producing and distributing goods and services [13]. The implications of the changes, within this paradigm, are presented by the declining need for workers and even technological deployment of industries, and its use sets priorities for developing professional competencies in India and China investing in ICT and engineering education;
(2) The second paradigm—"Technology is changing Communication, Collaboration and Knowledge creation"—uses knowledge economics concepts to describe

Table 1 Technological paradigms and requirements for digital competencies

	Technology is replacing work through automation (*professional competencies*)	Technology is changing communication, collaboration and knowledge creation (*universal competencies*)
Categories of competencies	Digital competencies for ICT professionals	Digital competencies for everyone
Levels of competencies	Creation of new technologies Creative use and adaptation of existing technologies	Basic use of technologies
Composition of competencies	Complex programming skills Knowledge of complex algorithms	Understanding the cultural context of the Internet environment Ability to communicate in online communities Ability to create and distribute content Skills to use digital technology for self-development

Source authors

the role of modern ICT in communication. This paradigm implies two implications for the labor market: (i) the use of ICT for solving professional problems leads to the need for digital competencies for almost all jobs; (ii) the change of means of collaboration based upon digital platforms ("platformization") highlights the importance of developing universal competencies—soft skills (compromise, problem solving, cooperation, creativity).

Explaining the content of digital transformations, the presented technological paradigms determine the unity of the methodological approach to the content of universal and professional digital competencies (Table 1).

3.2 The Composition and Structure of Digital Competences

The technological paradigms clearly define the digital component of the concept of a "digital competence", while the competence component is defined by the European Framework for Qualification Definition (EQF) [5, 15]. As a level of learning within the boundaries of "knowledge—skills—attitudes", the concept of competencies allows to form the composition and structure of professional digital knowledge and skills, to justify their development within the boundaries of professional qualification, to adapt to the needs of a digital worker, to effectively use in a variety of management tasks [6]. In defining digital competence, the European Commission emphasizes the conscious and responsible use of digital technologies, the ability to digital cooperation aimed at solving problems within established security boundaries [14]. In the personality of a person of the digital society, digital competencies are specified as certain types of literacy, and information, computer, communicative, media literacies are used to clarify the composition of universal competencies necessary to achieve the individual's own goals (Table 2).

3.3 A Policy-Based Approach to Digital Competencies

The shift to an active digital agenda shapes a supranational policy that in most countries aims public administration at digital transformation of the economy and the development of digital competencies. As part of this policy, projects have been developed in developed countries, as well as programs of creating high-tech jobs and adequate digital competencies embedded in the broader scientific, technological, and

Table 2 The composition and structure of universal and professional digital competences

Source	Types of competencies
World Development Report by the World Bank: Digital Dividends [14]	*Cognitive*—literacy, numeracy, and higher-order cognitive skills, verbal ability, problem solving, memory, and mental speed (*universal*) *Social and behavioral*—Openness to experience, conscientiousness, extraversion, agreeability, and emotional stability, self-regulation, grit, mind-set, decision making, and interpersonal skills (*universal*) *Technical*—Technical skills developed through postsecondary schooling or training or acquired on the job, skills related to specific occupations (for example, engineer, economist, IT specialist) (*universal*)
Analysts of the World Economic Forum [12]	*The 10 skills you need to thrive in the Fourth Industrial Revolution*: complex problem solving, critical thinking, creativity, people management, coordinating with others, emotional intelligence, judgment and decision making, service orientation, negotiation, cognitive flexibility (*universal*)
DigComp 2.0 (The Digital Competence Framework for Citizens by the European Commission) [4]	*21 competencies in 5 competence categories* (*professional*): 1. Information and data literacy 2. Communication and collaboration 3. Digital content creation 4. Safety 5. Problem solving

(continued)

Table 2 (continued)

Source	Types of competencies
The New Foundational Skills of the Digital Economy, developed by Burning Glass Company [13]	*Human Skills*: critical thinking, creativity, communication, analytical skills, collaboration, and relationship building *(universal)* *Domain Knowledge*: strategy, economics, marketing, communications/PR, talent development/HR, R&D/product development (*professional*) *Digital Building Block Skills*: AI/Data analyst, big data and data management, software development, information security (*professional*) *Business Enabler Skills*: project management, decision making, visualization, data communication (*professional*)
Target Competency Model 2025, developed by BCG within the study "Russia 2025. Resetting the talent balance" [11]	*Cognitive skills* (self-development, self-discipline, management skills, achieving results, solution of non-routine tasks, adaptability) (*universal*) *Socio-behavioral skills* (communication, interpersonal skills, intercultural interaction) (*universal*) *Digital skills* (creating systems, information management) (*professional*)

Source authors

innovation agenda. In supranational policy, digital competencies are a priority and a basic right that implements citizens' expectations from the information society [8].

The supranational policy, capable of forming a methodological basis for the development of digital competencies in the policy of the Russian Federation, includes two strategic initiatives (Table 3).

The analysis shows that the content of these initiatives is aimed at (1) improving professional skills in the field of digital competencies for the labor force and consumer literacy among the population; (2) upgrading educational programs capable of providing "digital literacy" by means of personalized learning paths; (3) analyzing and forecasting the needs for digital competencies. It should be noted that the increased interest in digital competence on the part of supranational structures is caused by the complexity of the relationships between citizens, corporations, territorial socio-economic systems, which on a global scale leads to the network, horizontal, synchronized integration of "all with all". In such conditions, the supranational digital policy has to be implemented by means of expert platforms, networks of competence centers, pools of initiatives and projects, partnerships at the international level. For the Russian Federation, a unified coordinated approach within the framework of the supranational policy of developing digital competencies will achieve a synergistic effect, realize the goals of the national digital agenda.

Table 3 The structure of the supranational policy aimed at digital competences

	Digital education action plan [5]	Building digital competencies to benefit from frontier technologies [9]
Source	The European Commission	The United Nations
Key objective	Support of technology use and the development of digital competences in education	Ensuring effective participation in the current and future world as well as to benefit from existing and emerging technologies
Priorities	Making better use of digital technology for teaching and learning; Developing digital competences and skills; Improving education through better data analysis and foresight	Forecast of digital skills and competencies for the 21st century; Existing and emerging technologies for education; Building digital competencies through educational frameworks; Creating an enabling environment: Investment in infrastructure and institutional development; Establishing initiatives that promote entrepreneurship in the digital economy; Support collaboration among all stakeholders, including at the international level

Source authors

4 Discussion

Information as a basic value sets the challenge for the Russian Federation, which in its importance is comparable to the universal elimination of illiteracy at the beginning of the 20th century. The strategic foundations for the formation of digital competencies are laid down by the National Program "Digital Economy", the road map "Personnel and education", which justifies the indispensability of the digital transformation of the economy, the need to stimulate the development of digital technologies [11].

The methods of developing digital competences are detailed in the above documents, however, not sufficiently clear, which is due both to the complexity of tasks and to the lack of methodology for the development of digital competencies in Russian policy. Technology paradigms "Technology is replacing work through automation" and "Technology is changing Communication, Collaboration and Knowledge creation", a supranational policy for the stimulation of digital competencies allowed us to work out such a methodology of developing digital competencies in the structure of universal competencies (literacy of the population), professional competencies (employed in the economy) and to present them in the form of a scheme (Fig. 1).

According to the authors, the formation of digital competencies should be accompanied by the development of the infrastructure of the digital ecosystem designed to reduce the digital divide of the first level (unequal access to technology). And if the infrastructure part of the "digital divide" is gradually going down—even in

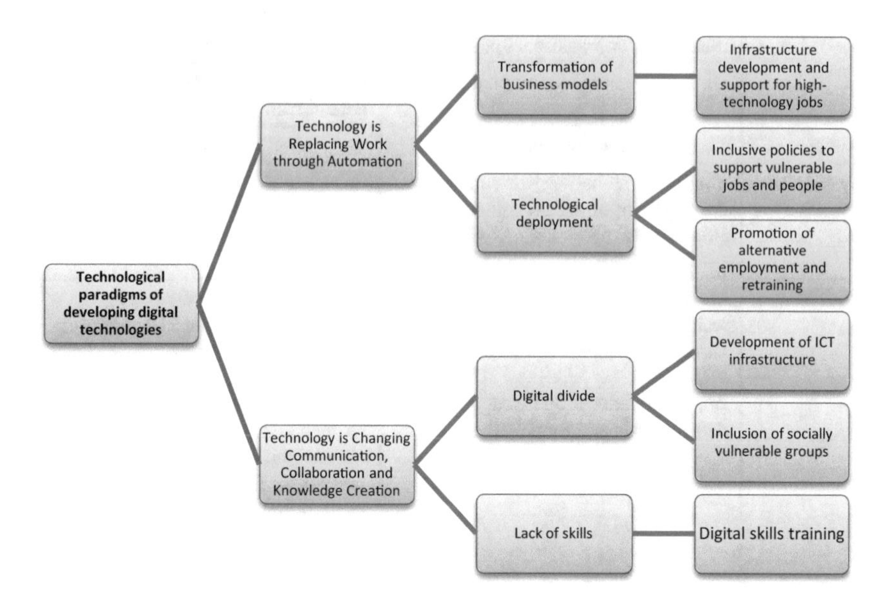

Fig. 1 The methodology for developing digital competences based upon technological paradigms (*Source* authors)

the countryside, the level of Internet penetration in households exceeds 70%—the digital divide of the second level, associated with the differences between people in the level of digital competencies, still remains high [4]. The second-level divide is due to both personal factors (age, education, income) and economic conditions (level of socio-economic development of the territory, urbanization). For any reason, it impedes the expected effects of digitization. The digital divide in the transition to a new technological order requires not just a set of measures, but a deep theoretical and methodological basis for the development of digital competencies [2].

Technological paradigms and their implementation in the supranational policy aimed at developing digital competencies can change the essential foundations of the educational process in the Russian Federation, the forms of its organization, transform universities into "digital" and "digitalizing", set new strategic objectives.

5 Conclusion

In summary, the need for digital competences is the result of the irreversibility of digital economic transformation, but the process of its formation reflects a well-founded approach to governance. The authors assume that the comparative analysis of technological paradigms allows to define a system of requirements to digital competences within professional (for those engaged in the digital economy) and universal (for citizens of the digital society) competencies, and the semantic analysis of the composition and structure of digital competencies—a list of knowledge and skills in each group of competencies. The contextual analysis of the supranational policy-based documents provides a basis for the effectiveness of the digital competency factor, achieved by professional development in the field of digital competencies for the labor force and consumer literacy among the population. The means of developing digital competence is the modernization of educational programs, the creation of programs of "digital literacy" among the population, personalized learning trajectories, as well as the analysis and prediction of the need for digital competencies. The result of the development of digital competencies is the overcoming of digital divide of the second level (the problem whose solution in the Russian Federation is far from complete), the final transition of the country's economy to a new technological order. The proposed methodology can be used while working out documents for strategic planning of industries and subjects of the Russian Federation. The support mechanisms based on the composition and structure of digital competencies will help to increase the validity of the choice of training directions, the effectiveness of vocational education programs, additional education and training programs.

References

1. Altass P, Wiebe S (2017) Re-imagining education policy and practice in the digital era. J Can Assoc Curric Stud 15(2):48–63
2. Bolgova EV, Grodskaya GN, Kurnikova MV (2020) The model for meeting digital economy needs for higher education programs. In: Ashmarina S, Mesquita A, Vochozka M (eds) Digital transformation of the economy: challenges, trends and new opportunities, vol 908. Advances in intelligent systems and computing. Springer, Cham, pp 542–556. https://doi.org/10.1007/978-3-030-11367-4_54
3. Boutenko V, Polunin K, Kotov I, Sycheva E, Stepanenko A, Zanina E, Lomb S, Topolskaya E (2017) Russia 2025: resetting the talent balance. The Boston Consulting Group. http://image-src.bcg.com/Images/russia-2015-eng_tcm26-187991_tcm9-192725.pdf. Accessed 01 Feb 2020
4. Chirkunova EK, Khmeleva GA, Koroleva EN, Kurnikova MV (2020) Regional digital maturity: design and strategies. In: Ashmarina S, Vochozka M, Mantulenko V (eds) Digital age: challenges, challenges and future. Lecture notes in networks and systems, vol 84, pp 205–213. https://doi.org/10.1007/978-3-030-27015-5_26
5. European Commission (2018) Digital education action plan. https://eur-lex.europa.eu/legal-content/EN/TXT/?uri=COM:2018:22:FIN. Accessed 01 Feb 2020
6. Fraile MN, Peñalva-Vélez A, Mendióroz Lacambra AM (2018) Development of digital competence in secondary education teachers' training. Educ Sci 8(3):104. https://doi.org/10.3390/educsci8030104
7. Grey A (2016) The 10 skills you need to thrive in the fourth industrial revolution. World Economic Forum. https://www.weforum.org/agenda/2016/01/the-10-skills-you-need-to-thrive-in-the-fourth-industrial-revolution/. Accessed 01 Feb 2020
8. Ilomäki L, Paavola S, Lakkala M, Kantosalo A (2016) Digital competence – an emergent boundary concept for policy and educational research. Educ Inf Technol 21:655–679. https://doi.org/10.1007/s10639-014-9346-4
9. Kuhn TS (2012) The structure of scientific revolutions: 50th anniversary edition (with an introduction by Ian Hacking). University of Chicago Press, Chicago and London

10. Markow W, Liu P-C, Hughes D (2019) The new foundational skills of the digital economy: developing the professionals of the future. https://www.burning-glass.com/wp-content/uploads/New_Foundational_Skills.pdf. Accessed 01 Feb 2020
11. National program Digital economy of the Russian Federation. http://tadviser.com/index.php/Article:National_program_Digital_economy_of_the_Russian_Federation. Accessed 01 Feb 2020
12. Navas-Sabater J, Petrov O (2018) The EAEU 2025 digital agenda: prospects and recommendations – overview report (English). World Bank Group, Washington, DC
13. Rifken J (2016) How the third industrial revolution will create a green economy. New Perspect Q 33(1):6–10. https://doi.org/10.1111/npqu.12017
14. The World Bank (2016) World development report 2016: digital dividends. https://www.worldbank.org/en/publication/wdr2016. Accessed 01 Feb 2020
15. Vuorikari R, Punie Y, Carretero Gomez S, Van Den Brande G (2016) DigComp 2.0: the digital competence framework for citizens. https://ec.europa.eu/jrc/en/publication/eur-scientific-and-technical-research-reports/digcomp-20-digital-competence-framework-citizens-update-phase-1-conceptual-reference-model. Accessed 01 Feb 2020
16. Yazon A, Ang-Manaig K, Buama Ch, Tesoro J (2019) Digital literacy, digital competence and research productivity of educators. Univ J Educ Res 7:1734–1743. https://doi.org/10.13189/ujer.2019.070812

Innovative Digital Economy of Regions: Convergence of Knowledge and Information

E. Chirkunova, V. Y. Anisimova, and N. M. Tukavkin

Abstract The transformation of the modern economy is associated with the transition to digital technologies in both the industrial and social sectors. The purpose of this article is to test the hypothesis that currently knowledge and information are, on the one hand, tools of convergence (catching up development) of regions to achieve structural shifts in the innovative digital economy of the Russian Federation, on the other hand, differences in regional spending on information and communication technologies lead to divergence of innovative digital development of regions. On the basis of official statistical indicators the authors analyze the dynamics of innovation digital development in the regions, estimate the divergence range of indicators values by regions and show the innovative development rating of Russian regions. The dependence of the regional innovative development on financing for the ICT sector and effective local management is noted. It was found that Russian regions differ significantly in the development level of the innovative digital economy and the number of personnel engaged in research and development.

Keywords Digital economy · Knowledge and information · Convergence · Regions of the Russian Federation

1 Introduction

The main feature of the 4th industrial revolution (Industry 4.0) is the transformation of enterprises' activities for the widest possible use of robots, artificial intelligence and Big Data in production and other business processes. This transformation is

E. Chirkunova (✉)
Samara State University of Economics, Samara, Russia
e-mail: ekchirkunova@gmail.com

V. Y. Anisimova · N. M. Tukavkin
Samara University, Samara, Russia
e-mail: ipaniSIMova@yandex.ru

N. M. Tukavkin
e-mail: tnm-samara@mail.ru

S. I. Ashmarina and V. V. Mantulenko (eds.), *Current Achievements, Challenges and Digital Chances of Knowledge Based Economy*, Lecture Notes in Networks and Systems 133, https://doi.org/10.1007/978-3-030-47458-4_15

associated with the transition to digital technologies that use real-time data about production processes. Current data allows you to reduce production costs, improve the product quality, and ensure rapid production of goods that meet current customer needs. The changes caused by the Industry 4.0 are revolutionary in their nature, since they affect various social relations [9]. It is estimated that dozens of professions will disappear from the labor market in the next decade, but the digital economy will need more than 2 million IT specialists in the next 10 years.

The national digital economy employs 2.3 million people. The state order for training IT-specialists has increased by 70% over the past two years—from 25,000 to 42.5 thousand budget places [22]. However, only 15% of graduates are ready for immediate employment and effective work in the industry. The average period of adaptation of a young specialist in the workplace is from six months to a year.

The Russian mobile economy has created 1.2 million jobs, and another 430,000 will appear in the next five years. At the same time, 55.5% of jobs were provided by mobile operators and Internet service providers. On average, the number of mobile app developers will grow by 38.4% per year [23].

Digital transformation leads to a reduction of jobs that require average qualifications and an increase in demand for highly qualified creative specialists [20]. As a result of technological progress, up to 6.7 million working places may disappear, and another 20 million will have completely different qualification requirements. At the same time, new jobs will be generated, including those related to cognitive technologies and algorithmizable processes (IT-specialists, machine learning, Big Data, and robotics) [7].

The penetration level of online technologies in the Russian education reached 1.1%. The volume of the online learning market is projected to grow from 20.7 to 53.3 billion rubles within five years [6]. Developments in the IT sphere and communication technologies are the main focus of R&D investments—from 1.3 to 3.7% of the total investment volume. There are 150 business incubators and technology parks and 112 technology transfer centers nowadays in the country [21].

Despite the traditionally high level of specialists' training and the penetration of the Internet into the economy, small businesses are not sufficiently developed in the Russian economy. There are only about 700 companies in Russia that can be called technology startups. At the same time, the lion's share of them is located in Moscow and St. Petersburg. Although almost every regional center has a technopark, there is no development there. 40% of startups use b2b services in various fields (data analysis, advertising platforms, payment aggregators, financial services, logistics solutions, etc.). This segment is growing significantly faster than the user segment. Technological sectors are beginning to develop actively: artificial intelligence, robotics, and the Internet of things. However, many market participants note a very low level of entrepreneurial culture, a shortage of IT-personnel and positive practices in the traditional activity areas [16].

2 Methodology

In the 21st century, "knowledge, information, technology" have become an integral resource of the innovation infrastructure and the framework of the innovation economy for the development of all spheres of our society [2, 10, 11, 19]. Even P. Drucker noted that the knowledge transformation is associated with the growth of its codification, the gradual convergence of scientific (explicit) knowledge with skills, abilities (implicit knowledge) [5]. At the present development stage, there is no universally recognized methodological system for measuring the innovative digital economy. Nevertheless, in the last decade, due to the urgency of the problem of innovative development of regions, a lot of applied research is being carried out, devoted to the study of these aspect, including methodological issues of evaluating the innovative activity. At the same time, the methodological assessment of innovative processes in the region is complicated by the fact that this approach, first of all, is still quite new in the practice of the regional innovation management. A large number of different approaches have been proposed to test the presence or absence of convergence in the academic literature, ranging from simple statistical methods (estimating the dynamics of standard deviation) to the use of complex econometric models. The works of Pinkovsky and Sala-i-Martin [13] are the most cited studies in this field. The article by Borsi and Metiu [3] investigates the actual convergence in the context of institutional changes, and the authors conclude that there are convergence clusters. In the works by Quah [15], Pontines and You [14], Rughoo and You [17] the problems of evaluating general and club convergence are considered.

It is possible to distinguish approaches of national statistical bodies and professional associations (organizations) to the assessment of the innovative activity in regions. The article aims to prove or refute the established opinion that the convergence of knowledge and information contributes to the development of an innovative digital economy and eliminates the socio-economic inequality in the regions. The research database was Rosstat statistics of the Russian Federation (Federal State Statistic Service), the official database of departmental statistics EMISS, as well as the results of a study of the information society of the Ministry of Communications of the Russian Federation, Higher School of Economics.

3 Results

At the present stage, the most relevant are two directions of the economic transformation that ensure its transition to a new economic paradigm: (1) technical and economic re-equipment of enterprises aimed at using digital technologies; (2) personnel policy related to investments in human capital and the reproduction of social relations that ensure the development of human capital and the susceptibility of the economic system to innovations. At different stages of the innovation process, the results of innovative activity are embodied in different forms (Table 1) [3].

Table 1 Differentiation of innovative activities at different stages of innovation process

Stages	Business	HEIs, Russian Academy of Sciences, research institutes	State regulation
Access to information	Ideas	Scientific work; inventions	Formation of human capital
The generation of knowledge	Utility models; prototypes; industrial designs	Selection achievement; topology of integrated circuits	Creation and support of high-tech industries
The creation of technologies and tools	Know-how; trademarks; commercial designations	Information models of business processes; e-campus model; IT-service	E-government
Production of innovative products	Finished innovative products	The creation of technology transfer centres	Increasing the share of high-tech products in GDP; increasing the volume of innovative products

Source authors based on [3]

The convergence of knowledge and information is manifested in strengthening of the dominant position of some large regions, strengthening of the influence of megacities and regional capitals on the development of other regions in several directions:

- number of employees engaged in research and development, as a percentage of the employed population (A1),
- use of Internet platforms by organizations, as a percentage of the total number of organizations surveyed (A2),
- use of information and telecommunication networks by the population (A3),
- organizations that had a website (A4),
- number of organizations that used special software tools (A5),
- number of personal computers per 100 employees, units (A6), including those with the Internet access (A6*),
- ICT costs (A7).

The discrepancy between the values of indicators that reflect the level of ICT development in the regions of the Russian Federation is shown in Table 2.

It should be noted that the infrastructure part of the "digital divide" is gradually eroding—even in the villages the penetration rate of Internet households is above 70%, and the use of modern digital technology becomes more relevant, as Russia still has a great digital divide of the second level associated with differences between people's level of digital competence.

Table 2 Range of values of indicators of the level of ICT development by regions of the Russian Federation in 2017

Indicators	Maximum		Minimum	
	Value	Region	Value	Region
A1, %	2,51	Moscow Region	0,13	Ingush Republic
A2	100	Ingush Republic	68,4	Republic of Dagestan
A3	95,5	Yamalo-Nenets Autonomous District	63,0	Republic of Chechnya
A4	68,4	Saint Petersburg	30,6	Republic of Buryatia
A5	98	Ingush Republic	52,3	Republic of Dagestan
A6	78	Moscow	34	Republic of Dagestan
A6*	59	Moscow	21	Republic of Dagestan
A7, mln rub	116400,7	Moscow Region	359,8	Republic of Kalmykia

Source authors based on [8]

The widespread penetration of digital technologies and their rapid development creates new requirements for the qualification of the workforce: the possession of digital and related competencies becomes a necessary condition for any professional activity, each employee should be able to work in a high-tech digital environment, quickly retrain and think creatively.

As a result, the assessment of the digital economy in regions is carried out using the Index "Digital Russia" (IR) based on seven sub-indices (B1–B7), which reflect either some institutional conditions for the formation of the digital economy, or the effects of its implementation [18]:

B1—regulatory and administrative indicators of digitalization,
B2—specialized personnel and training programs,
B3—availability and formation of research competencies and technological reserves, including the level of research and development work,
B4—information infrastructure,
B5—information security,
B6—economic indicators of digitalization,
B7—social effect of the application of digitalization.

Table 3 shows the values of the sub-indices and the final index obtained at the end of 2018. From 85 regions, there are 3 leading ones, the average level regions, and the last two regions in the list.

As can be seen from Table 3, the leading regions have a higher index value, not only because of the level of the infrastructure component, but also because these subjects were transforming to a new way of digital economy long before the appearance and implementation of the special state program. The implementation of national projects in the Russian Federation, in our opinion, will allow us to move to a convergence of indicators of the development level of the digital economy due to the convergence of knowledge and information [1].

Table 3 Rating of Russian regions by the level of the innovative digital economy, 2018

Rank	RF subject	B1	B2	B3	B4	B5	B6	B7	IR
1	Moscow	79,45	77,63	72,93	82,49	66,09	67,67	69,74	75,14
2	Republic of Tatarstan	76,47	69,36	87	79,51	68	76,53	63,49	74,74
3	Saint Petersburg	67,13	75,46	74,64	71,53	79,11	74,8	80,79	74,55
…									
7	Moscow Region	64,51	74,21	71,49	80,09	63,1	65,78	72,24	71,86
8	Bashkortostan Republic	73,17	62,57	73,09	75,46	80	68,09	73,75	71,29
9	Leningrad Region	71,93	75,49	69,65	75,08	65	66	66,43	71,25
….									
15	Voronezh Region	65,47	71,99	65,38	69,04	71,7	64,05	68,76	68,51
16	Tula Region	70,6	67,17	68,23	65,49	68,4	69,97	69,98	68,02
17	Samara Region	65,44	67,5	75,19	61,5	69,68	65,6	73,48	67,87
…									
62	Republic of Dagestan	47,6	43,84	47,57	49,18	49,5	47,4	50,37	47,42
63	Republic of Crimea	46,15	50,01	48,77	41,49	47,98	46,89	48,55	47,97
…									
84	Ingush Republic	38,42	38,22	38	38,34	38,8	37,75	37,25	38,15
85	Jewish Autonomous Region	39,51	37,56	37,1	37,22	33,59	38,2	36,69	37,2

Source authors based on [18]

4 Discussion

Innovative digitalization in regions cannot be effective if there is a gap between the digitalization of activity areas and the low level of users' knowledge and skills. No matter what "smart" equipment is purchased, and no matter what modern digital technologies are applied into the production sphere, their use will not bring a significant effect if people are not prepared for these transformations and appropriate digital security is not provided [4].

The further development of the innovative digital economy should be ensured by the security of information systems, networks and software applications against digital attacks, the creation of common IT-platforms, the "digitalization" of public services, and the creation of B2B online platforms [12].

The formation of new knowledge to support the digital economy depends largely on the application of an integrated approach to the development of the country's human capital, which includes three areas: increasing the population number in each region; changing the quality of the education system; and the country's ability to retain, attract and use talented and qualified personnel. The development of innovative digitalization in the regions will improve the quality of life of the population through the creation of comfortable smart cities, smart jobs and public safety.

5 Conclusion

One of the key problems of the transition to a new economic paradigm in Russia is the significant gap between the values of the production digitalization and the digital training level of the population in 85 regions of the country.

To solve the backlog of regional enterprises in the economy digitization in 2017–2020, the following allocations from the Federal budget were planned for the implementation of national technology initiatives (NTI): in 2017—2.0 billion rubles; 2018—2.4 billion rubles; 2019—1.8 billion rubles; 2020—1.6 billion. But effective innovation is impossible if the current management system does not have the appropriate processes, technologies, and tools to enable forming to take effective and timely management decisions towards achieving innovative results.

In Russia, individual scientists and organizations are engaged in convergent development of the future, but the state is not systematically and institutionally active in these issues. The role of science, education, modern technologies and the transfer of information through digital platforms should be a priority in creating a modern society, not only potentially, but also really equal opportunities in it. The process of convergence of knowledge and information contributes to the acceleration, development, and improvement of promotion along the chain: creativity-innovation-products. To implement this chain, it is necessary to change the paradigm of training specialists and move from the disciplinary to the interdisciplinary principle of training.

Acknowledgements The research was carried out with the financial support of the Russian foundation for basic research, the project "Development of financial support mechanisms for the strategic development of the industrial complex of the Samara region". Contract no. 18-410-630001/18.

References

1. Bezborodov, JuS (2018) International legal methods and forms of legal convergence. Monograph. Prospekt, Moscow (in Russian)
2. Bogatyrev VD, Kononova EN, Martyshkin SA, Chirkunova EK, Khmeleva GA (2016) Innovative system of the regional industrial complex. Monograph. Samara University, Samara (in Russian)
3. Borsi MT, Metiu N (2015) The evolution of economic convergence in the European Union. Empirical Econ 48:657–681
4. Chirkunova EK, Khmeleva GA, Koroleva EN, Kurnikova MV (2020) Regional digital maturity: design and strategies. In: Ashmarina SI, Vochozka M, Mantulenko VV (eds) Digital age: chances, challenges and future. Lecture Notes in Networks and Systems, vol 84. Springer, Cham, pp 205–213
5. Drucker P (1994) Post-capitalist society. Harper Business, New York

6. EdTech Think Tank (2017) Research on the Russian market of online education and educational technologies. https://assets.website-files.com/58c30a8e570c9ea96dae660b/59af909f8211b90001e59ad8_edumarket_cut_rus.pdf. Accessed 11 Feb 2020
7. European Commission (2018) Digital single market. https://ec.europa.eu/digital-single-market/en. Accessed 11 Feb 2020
8. Federal Service of State Statistics (2019) Regions of Russia. Socio-economic indicators. Rosstat, Moscow (in Russian)
9. Gokhberg L (ed) (2019) What is the digital economy? Trends, competencies, and measurement. HSE, Moscow (in Russian)
10. Kononov OA, Kononova OV (2015) Knowledge management based on information and communication technologies. Curr Probl Econ Manag 3:79–85 (in Russian)
11. Krichevsky GE (2017) Introduction to the NBICS-technologies. NBICS-Sci Technol 1(1):27–54 (in Russian)
12. Naumkin M (2018) Five trends of Russia's digital economy in 2018. Rusbase. https://rb.ru/opinion/ekonomika-rossii/. Accessed 30 Jan 2020 (in Russian)
13. Pinkovsky M, Sala-i-Martin X (2016) Newer need not be better: evaluating the pen world tables and the world development indicators using nighttime lights. Federal Reserve Bank of New York. https://www.newyorkfed.org/medialibrary/media/research/staff_reports/sr778.pdf?la=en. Accessed 11 Feb 2020
14. Pontines V, You K (2015) Asian currency unit (ACU), deviation indicators and exchange rate coordination in East Asia: a panel-based convergence approach. Jpn World Econ 36:42–55
15. Quah DT (1997) Empirics for growth and distribution: stratification, polarization, and convergence clubs. J Econ Growth 2:27–59
16. RAEC, Google, & OC&C (2018) Digital horizons: ecosystem of the IT-entrepreneurship and start-ups in Russia. https://raec.ru/upload/files/oc-c-raec.pdf. Accessed 20 Feb 2020 (in Russian)
17. Rughoo A, You K (2016) Asian financial integration: global or regional? Evidence from money and bond markets. Int Rev Financ Anal 48:419–434
18. Skolkovo (2018) Index "Digital Russia": ratings. http://www.skolkovo.ru/public/ru/press/news/96-news-research/4749-2018-10-18-digitalrussia/. Accessed 11 Feb 2020 (in Russian)
19. Spires HA, Paul CM, Kerkhoff SN, Kerkhoff SN (2019) Digital literacy for the 21st century. In: Mehdi Khosrow-Pour DBA (ed) Advanced methodologies and technologies in library science, information management, and scholarly inquiry. IGI Global, Hershey, pp 12–21. https://doi.org/10.4018/978-1-5225-7659-4.ch002
20. Stepanenko A, Sycheva E, Zanina E, Lomp S, Topolskaya E (2017) Russia 2025: from cadres to talents. Boston Consulting Group. https://www.bcg.com/ru-ru/default.aspx. Accessed 11 Feb 2020
21. Sukharev OS (2018) Development financing: the solution of the structural-distributive problem. Finance: Theory Pract 22(3):64–83. https://doi.org/10.26794/2587-5671-2018-22-3-64-83
22. The program "Digital Economy of the Russian Federation" (approved by the order of the Government of the Russian Federation of July 28, 2017 N 1632-p). http://government.ru/docs/28653. Accessed 11 Feb 2020 (in Russian)
23. Yudina MA (2018) Social prospects of the digital economy of the Russian Federation 2017-2030 project. Stand Living Russ Reg 1(207):60–65. https://doi.org/10.24411/1999-9836-2018-10007 (in Russian)

State's Legal Policy in the Era of Digitalization

A. P. Korobova

Abstract The article analyzes the essence of state's legal policy, proposes that it is directly related to the development and implementation of strategic legal ideas. The nature of these ideas is investigated, their characteristic features are revealed: connection with law, fundamental nature, strategic nature, consolidation in official sources. There is a direct relationship between legal policy and the legal life of society, legal practice, their mutual influence and dependence on each other. The question of changes in the legal policy of the state caused by the widespread of information technology, the digitalization of processes in all spheres of public life, including, of course, the legal sphere, is considered. It is emphasized that in recent years legal policy has acquired additional means and channels for its implementation and, therefore, should become clearer and more understandable to its addressees. At the same time rapidly changing living conditions dictate the need to revise the key ideas of the legal development of society, expressed in the concept of legal policy, at least in terms of securing the means of implementing a legal strategy. Herein general theory of law, within the framework of which the theory of legal policy is developed, could be of great help.

Keywords Legal policy · Legal strategy · Law · Digitalization · Strategic legal ideas

1 Introduction

The legal policy of the state directs the legal life of society in a certain direction, gives it a solid character and meaning. It has a decisive strategic importance for legal practice, is implemented in each of its varieties, actualizes, highlights one or another of its socially significant functions, affects the choice of legal and organizational tools and forms used in everyday legal practice. The types of legal practice itself, for example, the practice of law-making, the practice of law enforcement, the practice

A. P. Korobova (✉)
Samara State University of Economics, Samara, Russia
e-mail: a.p.korobova@inbox.ru

S. I. Ashmarina and V. V. Mantulenko (eds.), *Current Achievements, Challenges and Digital Chances of Knowledge Based Economy*, Lecture Notes in Networks and Systems 133, https://doi.org/10.1007/978-3-030-47458-4_16

of interpreting legal norms, etc. Although it is formed objectively, in many ways, under the influence of the legal policy pursued in the state, the circle of its subjects is determined, as well as objects, content, actions, operations, stages and results.

At the same time, legal practice acts as the sphere of implementation of the strategy and tactics of legal regulation, of those ideas, attitudes, principles and priorities that the state develops in relation to the legal sphere [4]. In the process of lawmaking, application, interpretation and systematization of legal acts, as well as legal education of the population, one can fully assess their relevance, realism, and ability to exert an organizing impact on legal life. Legal practice makes it possible to improve the tools of legal policy, to adjust the main lines of its implementation. This is the main indicator of its effectiveness.

Thus, legal policy and legal practice are closely related, are in constant interaction. The study of legal policy problems is of particular relevance in the modern period of significant systemic transformations in all spheres of public life associated with the spread of information technology [2, 8, 12]. Thanks to digitalization, the legal policy has received fundamentally new opportunities both from the point of view of the sources of its formation, and from the standpoint of those forms and means by which it can be implemented. The purpose of this study is to identify the essence of legal policy, the nature and basic features of legal ideas of a strategic nature, designed to form the basis of legal policy in the digitalization era.

2 Methodology

In order to understand what the legal policy of the state is, we turn, first of all, to the formal legal research method, which, as you know, helps to form concepts, determine the essential characteristics and nature of phenomena, and to distinguish one phenomenon from another. The methods of analysis and synthesis make it possible to identify most characteristic features of strategic legal ideas, the observation method allows to assess the perception of key legal attitudes by citizens, and the systematic approach allows to present a set of strategic legal ideas as an internally consistent, integral basis of state legal policy. Using induction and deduction, as well as the logical method, it becomes possible to draw significant conclusions in the study. The historical method and the method of comparative legal research help to track the changes in the legal life of society and, accordingly, in legal policy that we have observed in recent decades.

3 Results

Legal policy is always connected with the activities of the state in the field of legal regulation, with the formation of a specific strategy for legal development. The essence of legal policy is to develop and implement strategic legal ideas, which express such a strategy. Strategic legal ideas, as their name implies, relate to law, its scope or, in other words, the legal field. Moreover, on the one hand, they accumulate in themselves the attitude of people to existing law, and on the other, they express ideas about the law as an ideal, about how it should be.

The widespread use of Internet resources in recent years allows for sufficiently high-quality and comprehensive monitoring of such "legal" sentiments and to respond quickly in law-making activities. This is of relevance in the light of the forthcoming amendments to the current Constitution of the Russian Federation.

Strategic legal ideas have a direct organizing effect on relations in the legal sphere and are aimed at its transformation in the direction of improvement, and, whenever possible, elimination of negative phenomena. At the same time, they are the mirror in which reality, everyday legal practice are reflected. For example, at the moment, the urgent issue is the formation of the legal framework for regulating the digital economy, the improvement of the regulatory framework for the regulation of the information sphere, etc.

Ideas are the starting point of any activity. Moreover, only such a policy has practical meaning and justifies its social purpose, which bases on a system of basic principles that allow a legal policy to acquire such important qualities as stability and internal coherence. But these are not simple ideas, these are strategic ideas. And this means that they should be as just as possible and should anticipate future socio-economic and political changes in society, and, accordingly, in legal reality.

From the foregoing, one more important sign of strategic legal ideas follows—their official nature, their need to be expressed in some official source of the state. And here the rapidly evolving reality provides completely unlimited opportunities for the publication of the legal strategy, for familiarizing the widest possible range of recipients. Thus are the presence of official websites of government bodies and officials, and pages on social networks, and information and legal databases.

4 Discussion

The peculiarity of legal policy, as emphasized by V.A. Rudkovsky is predetermined by those ideas, ideals, legal views, priorities that make up its value basis, its connection with legal awareness. And though the purpose of legal policy is objective, it always remains its integral feature, without it, legal policy cannot exist as an independent political and legal phenomenon, even if the specific content of the principles in the process of historical development of legal policy changes [9].

As for the legal nature of strategic legal ideas, it can be interpreted in two ways, but it is similar to the way the legal ideology is interpreted. Firstly, strategic legal idea relates to existing law and represents its assessment, approval, positive attitude or, on the contrary, rejection, negative, skeptical attitude. Take, for example, the attitude of the population to the problem of the death penalty: public opinion polls irrefutably show that the vast majority of the population of our country support its preservation as an exceptional measure of punishment for committing serious crimes and crimes of special gravity. And this situation is understandable, given the specifics of the Russian mentality: the question of a possible mitigation of punitive policies cannot be taken positively by people who are at times close to the idea of a "hard hand" and who are still sometimes nostalgic for the period of the personality cult, iron "order" and discipline in the state. Secondly, strategic legal ideas are not only about real-life, but also about the ideal, desired law, about how it should be in order to fully fulfill its social mission—the regulation of public relations. These are ideas about the legal life of society, about such values of the legal sphere as the rule of law and civil society, equality and justice, rights and freedoms.

Strategic legal ideas are designed to improve the quality of legal life, eliminate or at least weaken the manifestation of its negative aspects, for example, such phenomena as deformation of legal consciousness (legal nihilism, legal idealism, legal infantilism, legal cynicism [1]), criminal manifestations, corruption, gaps, conflicts and other imperfections of legislation, judicial errors, etc. As noted by N.N. Voplenko and Y.Y. Vetyutnev, the legal idea in its essence is the idea of raising certain interests and conscious human needs to the status of a legal norm [11]. The formation and development of legal ideas go through several different levels, each of which can cover different sections of society.

The first step is the level of individual citizens, namely scientists, specialists, and professionals. Here a legal idea arises and for the first time acquires the outlines of future clear legal formula. The second step is the level of social groups and individual segments of the population. At this level, so to speak, "ideas take hold of the masses," gaining numerous followers and opponents. The first include those social groups and individual segments of the population whose interests this idea meets. Opponents—on the contrary, those whose interests it contradicts. And here the task is to work out a reasonable compromise, to take into account the maximum possible interests of each and everyone. The third, final stage in the development of legal ideas is the legislative and related legal implementation level. At this stage, ideas are embodied in legal regulation, acquire the ability to regulate social relations and go into the stage of practical implementation.

Relatively speaking, a legal idea undergoes a certain evolution. First of all, it moves from its inception to awareness, i.e. "mastery of the masses," and then it falls to the legislator. In turn, from the legislator it goes all the way to the law enforcer, people's perception of the laws adopted and subject to implementation. In the end, the idea realized in law is again subjected to the final professional assessment and again returns to where it originated. The data obtained on the effectiveness of the law make it possible to generate new ideas, improve, supplement and change the law, repeal it, etc. Thus, the idea enters a new phase of its development and will go through all those stages again, but it will already have to be filled with qualitatively different content. It is obvious therefore that any progressive social idea, including a legal one should develop ahead of the social consciousness and the level of social ties. This is precisely its creative power, its driving potential [5]. The emergence of legal policy, therefore, begins with an idea. Let's consider it in more detail.

The most important characteristics of legal ideas of a strategic nature include their fundamental nature, global nature, and fundamental nature. This is manifested in the fact that their implementation affects the life of the whole society, leads to significant consequences on the scale of the whole society, and not just a specific person or group of persons. They predetermine the legal life of society, the direction of its course, and generally affect the process of formation of statehood, the choice of the main path of development by modern society. Therefore, the importance of strategic legal ideas cannot be underestimated. In addition, the legal ideas under consideration are strategic in nature, indicate the long-term prospects for the development of society, those priorities that society chooses as the main ones. Such strategic goals include the formation of the foundations of the digital economy, improvement of the legal regulation of processes associated with the use of information technology, digitalization. If the ideas that were the basis of legal policy change, therefore, legal policy itself should also change.

In relation to individual varieties of legal policy, these goals and objectives can be specified, become more private. For example, in recent years, the problem of forming the concept of judicial policy has taken on particular relevance, given the active reform of the field of justice in the country.

One of the most important sources of strategic legal ideas are such sciences as jurisprudence, political science, sociology, philosophy, therefore, the need for their scientific substantiation can also be included in the characterization of these ideas. The wide use of scientific knowledge and the rich tools available to science allows us to simulate various legal situations [10], to calculate what effect will be achieved as a result of the adoption or cancellation of one or another regulatory act, how the population will respond to the laws adopted, whether they correspond to its aspirations. This makes it possible to save significant tangible and intangible funds involved in the adoption of legislative decisions, to direct the energy of the legislator in the right direction.

A special role in the process of developing strategic legal ideas is played by the theory of state and law, because it, being a methodological science, is designed to at least develop the concept of legal policy, its goals, objectives, priorities, principles, means, forms of implementation, as well as the concept of these very strategic legal ideas, as long as we agree with the conclusion that the essence of legal policy consists in their development and implementation.

In the context of what has been said, one cannot fail to mention one more problem. In recent years, the concept of "digitalization of law" has become widespread. It seems that the general theory of law should pay close attention to it. The fact is that the term itself raises some doubts about the legitimacy of its use: after all, law is a system of norms, institutions, and how can norms be digitized? It seems that it would be more appropriate to talk not about the digitalization of law, but about the digitalization of sources, forms of law.

Unfortunately, it cannot be said that today, legal science fully fulfills its function. On the one hand, the reasons should be sought in the properties of science itself, which is not formed immediately, but for years, decades, even centuries. Legal science, being a fairly conservative branch of knowledge, often lags significantly behind the rapidly changing conditions of society, does not have time to adequately reflect them, respond quickly to changes in society, especially if they are as dramatic as we have in recent decades. It is enough to say that we have a completely, fundamentally updated system of legislation. On the other hand, there is clearly insufficient attention to legal science on the part of state authorities, a lack of desire for active cooperation, and the use of theoretical baggage accumulated by legal science in the practice of state building. The fact that reforms are being implemented is what we all see and often get the opportunity to feel for ourselves, but far from everyone understands what their ultimate goal is. Any scientific theory, as you know, is systemic, conceptual in nature. Therefore, another feature of strategic legal ideas is that they together form a monolith, each such idea is closely related to other ideas, the implementation of one of them invariably entails the implementation of the other. Given the above, one of the most important pressing tasks today is to develop a concept (or doctrine) of the state's legal policy adjusted for digitalization processes.

The experience of state policy in different countries indicates that the state and society will not be able to achieve positive political and socio-economic results without a specific strategy. When the ideological component of politics is in a state of crisis, the goals of the current policy are not understood by the population, this inevitably provokes a crisis of the will of those in power, generates insecurity in the future among the population and, as a result, a low level of trust in government, social confrontation. Ultimately, the progressive development of society is inhibited [7].

The concept of legal policy is precisely designed to make all activities in the legal sphere transparent and understandable to ordinary citizens and thereby provide support for the state-led reform course in all areas of public life. Scientists have been talking about the need for the formation of a concept, ideology of legal policy for a long time. R.S. Bainiyazov, in particular, emphasizes how dangerous it is not to take into account the legal spirit of society in the legal policy of the state. This leads only to one thing—to its inefficiency. The legal spirit of the people should be considered at all stages of legal policy [3]. It is gratifying to see that significant steps towards the preparation of the concept of legal policy have been taken for a long time (meaning the Concept of legal policy developed by the Saratov branch of the Institute of State and Law of the Russian Academy of Sciences) [6]. Hopefully the work done by scientists will be appreciated as well as demanded by the authorities.

5 Conclusion

Thus, legal policy can be interpreted as ideologically oriented activity aimed at transforming the sphere of legal regulation, arranging the legal life of society. Obviously all activities in the legal sphere should be subject to a single plan, common goals and objectives, have clear priorities supported by the people, contain such ideas that meet the interests of most people, be carried out in accordance with certain reasonable principles and in pre-established forms. Legal policy should be transparent and understandable to people, and modern digital realities provide all the possibilities for this.

References

1. Adriaenssen A, Paoli L, Visschers J, Karstedt S (2019) Taking crime seriously: conservation values and legal cynicism as predictors of public perceptions of the seriousness of crime. Int Crim Justice Rev 29(4):317–334. https://doi.org/10.1177/1057567718824391
2. Aburomman AA, Reaz MBI (2017) A survey of intrusion detection systems based on ensemble and hybrid classifiers. Comput Secur 65:135–162. https://doi.org/10.1016/j.cose.2016.11.004
3. Bainiyazov RS (2001) Legal realism and legal policy. Legal Policy Legal Life 2:27–34
4. Gatti M (2016) The log in your eye: is Europe's external promotion of religious freedom consistent with its internal practice? Euro Law J 22(2):250–267. https://doi.org/10.1111/eulj.12162
5. Livshits RZ (2001) Theory of law. Beck, Moscow
6. Malko AV (2001) On the concept of legal policy in the Russian Federation. Legal Policy Legal Life 2:174–177
7. Malko AV, Shundikov KV (2003) Goals and means in law and legal policy. SGAP Publishing House, Saratov
8. Oluwatimi O, Bertino E, Damiani ML (2018) A context-aware system to secure enterprise content: incorporating reliability specifiers. Comput Secur 77:162–178. https://doi.org/10.1016/j.cose.2018.04.001
9. Rudkovsky VA (2003) On the principles of legal policy. Legal Policy Legal Life 4:6–14
10. Schuck A, Calati R, Galynker I, Barzilay S, Bloch-Elkouby S (2019) Suicide crisis syndrome: a review of supporting evidence for a new suicide specific diagnosis. Behav Sci Law 37(3):223–239. https://doi.org/10.1002/bsl.2397
11. Voplenko NN, Vetyutnev YuYu (2004) Institutionalization as a law of state legal life. Legal Policy Legal Life 3:21–31
12. Xu G, Li H, Tan C, Dai Y, Liu D, Yang K (2017) Achieving efficient and privacy-preserving truth discovery in crowd sensing systems. Comput Secur 69:114–126. https://doi.org/10.1016/j.cose.2016.11.014

Statistical Analysis of the Population and Household Digitization

N. V. Proskurina and Y. A. Tokarev

Abstract The ease of access to information and communication technologies for the citizens is one of the major priorities, that are stated in the government program «Information Society». Information and communication technologies are implemented into everyday routine of inhabitants. This paper attempts to statistically analyze the digitalization of households in terms of information and communication technologies accessibility. The major statistical indicators set of households digitization was formed and analyzed. Cross section dynamic analysis of information and communication technologies use by the Russian population was made. The regions of the Russian Federation (RF) were classified according to the level of households and individuals digitization. Integrated indices were used in the comparative analysis of the RF regions. The differentiation distance in household digitization was measured.

Keywords Digitization · Information and communication technologies · Population · Households · Regions · Statistics

1 Introduction

In 2011, the government program «Information Society» was launched, aimed at improving the standard of living through the introduction of information networks and information and communication technologies (ICT) into daily life activities of citizens. The international indexes, which are calculated annually, represent the key indicators of the program's implementation. The RF's upturn in the international ranking indicates that the situation is improving and that the country is approaching international standards. Although for the most part, the program is designed to

N. V. Proskurina (✉) · Y. A. Tokarev
Samara State University of Economics, Samara, Russia
e-mail: nvpros@mail.ru

Y. A. Tokarev
e-mail: tokarev_ya@mail.ru

S. I. Ashmarina and V. V. Mantulenko (eds.), *Current Achievements, Challenges and Digital Chances of Knowledge Based Economy*, Lecture Notes in Networks and Systems 133, https://doi.org/10.1007/978-3-030-47458-4_17

improve a large number of indicators in various areas within the country, and in particular to upgrade the availability of information and communication technologies for the population.

The situation with the introduction of information and communication technologies and the emergence of a new development course, namely the information society, can be considered in two ways. On the one hand, the facilitation of public access to the global net and ubiquitous handling of desk computers and new software by various social strata and professional communities reflects the growing demand of Russians. On the other hand, not all citizens are ready to use and apply ICT. Internet usage in Russia has a fairly disparate level of application in the regions. Akin to all technologies, they affect the socio-economic situation of the household, as well as represent the media, which is the link between the household, its members, and the outside world. The above provides a sufficient reason to consider monitoring household access to ICT and the extent of its use by the population as one of the major constituents of the monitoring system for the modern information society development, which is meant to invigorate the application of information technologies in the household sector of the economy.

2 Methodology

The data array of the research is official information of the Federal State Statistic Service, results of the Federal statistical questionnaires and surveys on the use of ICT and information and communication networks by the citizens, data from the Unified Interdepartmental Information and Statistic System. Methodologically and theoretically this study is based on electronic resources, materials from expert publications, periodicals, as well as academic research on the subject matter that has been carried out by both domestic and foreign authors.

The following statistical analysis methods are used in this paper: tabular, graphical methods for visualizing statistical data, structural and dynamic analysis, time series analysis, and methods of multidimensional statistical analysis (cluster analysis). To determine the integrated index of ICT use by the households and by the population (the digitalization index), the methodology of calculating the information society development index was applied. The methodology was provided by the Ministry of Digital Development, Communications and Mass Media of the Russian Federation [5]:

$$P_{ICT_{hh}} = \sum_{i=1}^{6} \frac{1}{6} * P_i$$

Where, P_i—normalized values for the information society (households and population) development index in the constituent entities of the Russian Federation.

The obtained values for each indicator are normalized and translated into an estimate in the range from 0 to 1. For indicators calculated as a % of the total population (households), the normalized value is 100%; for other indicators it is the maximum value of the indicator in the aggregate.

The assessment of the distance is carried out by the ratio of 20% of regions with the highest values of the index of the households and the population ICT use, and 20% of regions with the minimum values of the integrated indicator. The data were processed with STATISTICA 13.3 и EXCEL programs.

3 Results

The exploration of the availability and use of ICT by households and population (that is, the presence of ICT at home and the use of it by one or more individuals from the household, either at home or elsewhere) and their main indicators of resulted in the following findings.

In comparison to 2014, in 2018, the percentage of households, which used personal computers grew by 3.9%. Annually, the portion of households using a personal computer increased by 0.8%. Such insignificant growth rate and the fact that after 2017 the share of households using personal computers decreased by 2.7% in 2018 is due to the fact that more and more affordable alternative mobile devices with Internet access are entering the market. Moreover mobile devices combine most of the functions performed by desk computers.

The annual growth in the proportion of households with Internet access was 2.6%. In 2018, the share of households changed by 13.9% compared to 2014.

There is a differentiation between urban and rural settlements. But at the same time, in 2014, *PJSC Rostelecom* signed an agreement on laying fiber-optic communication lines to settlements with a population of 250 to 500 people. This reduced the gap between urban and rural settlements to 12.6 percentage points, while in 2014 the difference was 23.3 percentage points (Fig. 1).

Among the most significant reasons for abandoning the Internet is the reluctance of the population to use it. Thus, in 2014, the portion of the citizens who did not have the desire to master modern technologies was 54.7%, and in 2018, it declined by 5.72 percentage points or 10.5%. The decrease by 1.03 percentage points or by 16.8% was observed among the population that does not use the Internet due to lack of technical infrastructure. The data on the population that refuses the Internet for security reasons remained unchanged.

In 2014, the Internet was mostly popular among the 25–29 aged. In 2018, the portion of the population with the Internet usage at this age decreased by 17.4%. In 2018, the largest proportion of the population was aged 30–34. Every year, the share of 55 aged and older increases. So, in 2018, compared to 2014, the share of the 55–59-year-olds increased by 35.8%, 60–64-year-olds percentage increased by 98.9%, and the portion of 65–72-year-olds increased by 57.8%. Thus, digitalization is increasingly spreading among the elderly citizens (Fig. 2).

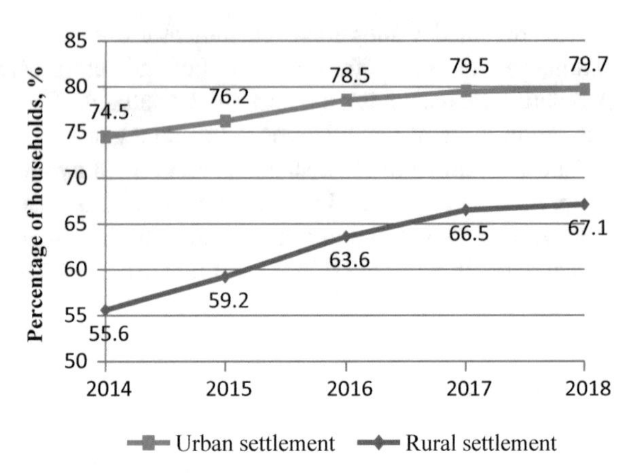

Fig. 1 Percentage of households with Internet access by settlement type for 2014–2018 (*Source* authors based on [6])

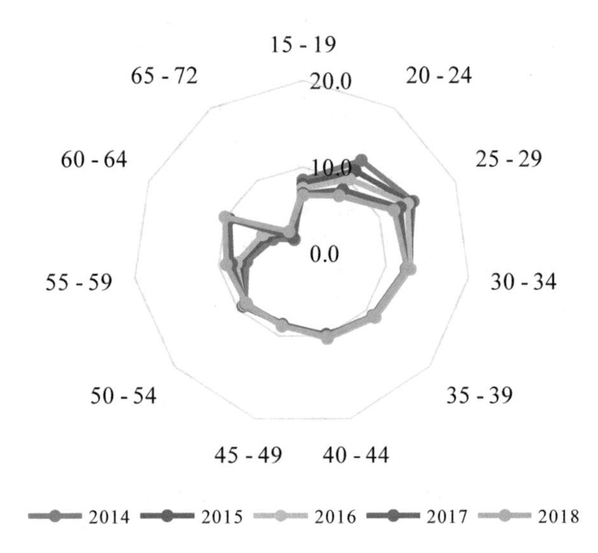

Fig. 2 Distribution of the households with Internet access by the age group in 2014–2018 (*Source* authors based on [6])

The increase in the share of the population receiving state and municipal services through the Internet from 2014 to 2018 is considered to be a very important issue in the government program. In 2018, compared to 2014, this indicator increased by 414.2%. The annual increase is 50.6%. Measures, implemented in the framework of the subprogram «e-Government» to improve the portal, i.e. adding a growing list of services, promoting the campaign by the Ministry through the publications in social networks and web portals, are enhancing the public confidence and involving more and more citizens.

One of the main objectives of the government program «Information Society» is to eliminate digital inequality throughout the Russian Federation and reduce the differentiation according to the degree of ICT use in the households.

The homogeneous regional groups according to the level of the population and households digitalization were identified with cluster analysis methods. A multidimensional classification of Russian regions (excluding *Sevastopol*, the Republic of Crimea and the Republic of *Ingushetia* due to the lack of data for 2014) was carried out for 2014 and 2018.

The main indicators of ICT accessibility and application in the households and among the population from the Information Society Development Official Monitoring in the Russian Federation were selected for the analysis [6]:

X_1—the number of mobile phones per 100 households, units;
X_2—the share of households with a personal computer in the total number of households, %;
X_3—the number of personal computers per 100 households, units;
X_4—the share of households with Internet access in the total number of households, %;
X_5—the share of the population that used the Internet to obtain state and municipal services in the total population surveyed, %;
X_6—the share of the population that used the Internet to order goods and (or) services in the total population, %.

The typology for 2014 and 2018 make sense, since the calculated values of the significance level p for each variable are less than the set significance level 0.05, and the intergroup and intragroup variances are different. Therefore, the division of the population of the Russian Federation into 3 groups (clusters) is quite reasonable.

In 2014, the cluster with a high level of ICT use included 7 regions, namely Moscow, St. Petersburg, the Republic of *Sakha*, *Khanty-Mansiysk*, Yamalo-Nenets Autonomous *Okrug*, Chukotka Autonomous *Okrug*, Murmansk *Oblast*. The average level of ICT use is observed in 22 constituent entities of the Russian Federation. These are mainly the entities of the Central, Volga, and North-Western Federal districts. The lowest level is in 53 regions, including the North Caucasus Federal district.

Since the launch of the government program, there has been an improvement in the use of ICT by the population. In 2018, two regions left the average level group and moved to the group with a high level of ICT use, while 17 regions improved their position and moved from the low level group to the group with an average level of ICT use (Table 1).

Table 1 The RF's constituent entities clustering by the level of ICT use in 2014 and 2018

The level of ICT use by the population (clusters)	Number of entities in the cluster	
	2014	2018
1 Cluster—Low level	53	36
2 Cluster—Average level	22	37
3 Cluster—High level	7	9
Total:	82	82

Source authors

Table 2 The RF's leading and outsider constituent entities in terms of household digitization for 2014 to 2018

Entity rating according to digitization index	2014	2018
1	Khanty-Mansi Autonomous *Okrug*—Ugra	Yamal-Nenets Autonomous *Okrug*
2	Moscow	Khanty-Mansi Autonomous *Okrug*—Ugra
3	Yamal-Nenets Autonomous *Okrug*	Moscow city
4	St. Petersburg	Murmansk *Oblast*
5	Murmansk *Oblast*	Moscow *Oblast*
…	…	…
78	Voronezh *Oblast*	Republic of Adygea
79	Republic of Kalmykia	Amurskaya *Oblast*
80	Bryansk *Oblast*	Mari El Republic
81	Tambov *Oblast*	Transbaikal region
82	The Republic of Dagestan	Jewish Autonomous *Oblast*

Source authors

The assessment of distance in the distribution structure of the RF's constituent entities by the level of ICT use, calculated by the Ryabtsev index, equals to 0.211. It showed a significant level of distance for 2014 and 2018.

The leading regions and outsider regions were identified in terms of the level of digitalization of households and the population for 2014 and 2018. The calculation of the integrated index (digitalization index) for all RF's constituent entities was used to perform this task (Table 2).

The highest value of the index in 2014 was in the Khanty-Mansiysk Autonomous *Okrug* (0.662), which means that the values of the indicators of ICT availability and use by the population and households in relation to the normative ones amounted to 66.2%. In 2018, the leader was Yamalo-Nenets Autonomous *Okrug* (0.854), which improved its position, rising from the third position in five years. The last place in 2014 was occupied by the Republic of Dagestan (0.365), but in 2018 it came out of the top 5 outsider regions. In 2018, the lowest index value was in the Jewish Autonomous *Oblast* (0.470).

The integrated indicator was applied to estimate the distance in the level of ICT use by the population and households of the RF's constituent entities. Over the five years under analysis, the distance of entities by the level of digitalization decreased slightly, namely from 1.4 times in 2014 to 1.3 times in 2018.

4 Discussion

Cross-country analysis shows significant differences between Russia and the EU countries in terms of access to the Internet and the digital economy, as well as their impact on GDP and social processes in the country [1]. Information and communication technologies play a paramount role in the development of the information society. The Russian Federation is not on the last place in the world rankings for the level of ICT development. The data available in the Russian Federation on the level of ICT use in the economy and households over a decade allow modeling the impact of ICT factors on the country's economic growth [12].

The impact of Internet use on the wellbeing of the population comprises both positive effects and potential risks [4]. The efficacy of technological advances on human development depends on income distribution and availability of ICT services. It should be noted that for low-income subgroups, inequality has a positive effect on ICT availability, while for high-income households, this effect is reversed [2].

For instance, the introduction of ICT has a positive and statistically significant impact on income diversification and is more beneficial for low income rural households in China [8]. The correlation of the availability and use of information technology in rural Mexican households and various socio-economic and demographic factors, namely age, level of education, type of employment, and geographical location, was traced in [10]. The common factors for innovation and ICT is also being investigated at the regional level, taking into account not only the use of ICT by households, but by firms as well [3].

It is argued, that the expansion of technological advances in households has an effect on the intensity of regional economic development and the activation of the entrepreneurial potential of households [11].

To characterize the digital component the living standard, the following attributes were selected in [9]: the availability of digital goods, digital skills, the quality of the social sphere and services under digitization, government electronic services to ensure the security of information activities of people. The Russian regional index of the digital component of people's quality of life (RRIDPQL) is determined for the entire country, as well as for its Federal districts and regions [9].

Based on the analysis of statistics, assessment of the costs for firms and households in the digital sphere is made. The share of public expenditure on digitalization in GDP is determined [7]. That allows the researchers to understand the level of digitalization in the domestic economy and the degree of its transition to the digital transformation.

The analysis of the previous academic research showed that, despite the relevance of the research topic, there is still an understudied problem of digitization (access and use of ICTs) of households and population in terms of their digital divide at the regional level.

5 Conclusion

In the transition to the information society, the eponymous government program, which was launched in 2011, has increased the share of the population using the Internet. It was achieved with the lay of fiber-optic communication lines carried out by *PJSC Rostelecom*. The company has connected new access points in separated settlements of the country, reducing the information differentiation of regions. Also, the number of users is increased by the growing number of affordable state-of-the-art devices entering the market. An open platform that combines a huge number of sites for buying and selling goods, as well as ongoing activities of *Roskomnadzor* to protect users' personal data, is gaining confidence and attracting new "inhabitants" of the Internet, which, in turn, attracts and increases the share of the population moving from offline to online service. At the same time, to reach the set goals and to fully move to the targeted information society, it is necessary to conduct more courses, teaching the ITC skills. It might be vital to attempt reducing the costs for the Internet connection equipment, as well as to cope with the inequality of regions in information supply more and more effectively.

References

1. Afonasova MA, Panfilova EE, Galichkina MA, Slusarczyk B (2019) Digitalization in economy and innovation: the effect on social and economic processes. Pol J Manag Stud 19(2):22–32. https://doi.org/10.17512/pjms.2019.19.2.02
2. Ali MA, Alam K, Taylor B, Rafiq S (2019) Do income distribution and socio-economic inequality affect ICT affordability? Econ Anal Policy 64:317–328. https://doi.org/10.1016/j.eap.2019.10.003
3. Billon M, Marco R, Lera-Lopez F (2017) Innovation and ICT use in the EU: an analysis of regional drivers. Empirical Econ 53(3):1083–1108. https://doi.org/10.1007/s00181-016-1153-x
4. Castellacci F, Tveito V (2018) Internet use and well-being: a survey and a theoretical framework. Res Policy 47(1):308–325. https://doi.org/10.1016/j.respol.2017.11.007
5. Development assessment methodology information society in the subjects of the Russian Federation. Annex No. 1 to the minutes of the meeting Regional Informatization Council Government Commission on use of information technology to improve quality living and living conditions business activities dated April 20, 2016 No. 172pr. https://digital.gov.ru/uploaded/files/metodika-otsenki-urovnya-razvitiya-informatsionnogo-obschestva-v-subektah-rf-proekt.pdf. Accessed 01 Feb 2020 (in Russian)
6. Federal State Statistic Service (2020) Monitoring the development of the information society in the Russian Federation. https://gks.ru/folder/14478. Accessed 15 Jan 2020 (in Russian)
7. Korobeynikova EV, Ermoshkina CN, Kosilova AF, Sheptuhina II, Gromova TV (2019) Digital transformation of Russian economy: challenges, threats, prospects. In: Mantulenko V (ed) Proceedings of the international scientific conference "global challenges and prospects of the modern economic development". European proceedings of social and behavioural sciences, vol 57. Future Academy, London, pp 1418–1428. https://doi.org/10.15405/epsbs.2019.03.144
8. Leng C, Ma W, Tang J, Zhu Z (2020) ICT adoption and income diversification among rural households in China. Appl Econ. https://doi.org/10.1080/00036846.2020.1715338. Accessed 01 Feb 2020

9. Litvintseva GP, Shmakov AV, Stukalenko EA, Petrov SP (2019) Digital component of people's quality of life assessment in the regions of the Russian Federation. Terra Economicus 17(3):107–127. https://doi.org/10.23683/2073-6606-2019-17-3-107-127
10. Marlen M-D, Jorge M-R (2020) Internet adoption and usage patterns in rural Mexico. Technol Soc 60:101226. https://doi.org/10.1016/j.techsoc.2019.101226
11. Podol'naya NN (2015) Spreading information and communication technologies in households and the economic development: regional perspectives. Nat Interests Prior Secur 38(323):22–32 (in Russian)
12. Tkachenko D, Duc LM (2018) Methodological approaches to assessing the impact of information and communication technologies on the national economy. Econ Qual Commun Syst 2(8). https://cyberleninka.ru/article/n/metodicheskie-podhody-k-otsenke-vliyaniya-infokommunikatsionnyh-tehnologiy-na-natsionalnuyu-ekonomiku/viewer. Accessed 05 Feb 2020 (in Russian)

The Digital Economy: Challenges and Opportunities for Economic Development in Russia's Regions

S. V. Domnina, O. A. Podkopaev, and S. U. Salynina

Abstract The relevance of the study is explained by the significant potential of the digital economy in terms of its impact on the acceleration of economic development of the city, the region and the state as a whole, as well as improving the life quality. This article is aimed at studying the process of digital economy development at the regional level and defining indicators to assess the impact of digitalization on the economic condition of the region and the population's welfare. The article deals with the issues of genesis of digital economy in Russia, identifies its strengths and justifies the impact of digitalization on population's welfare in the region. The key research methods to this problem are methods of graphical and tabular data presentation, methods of statistical analysis, which make it possible to consider comprehensively the issues of improving techniques for assessing the quality of the digital economy and their impact on population's welfare in the region. Materials of the article are of practical value for the calculation of populations' welfare index in the region by using the improved methodology based on comparative analysis by the regions of the country.

Keywords Population's welfare index · Information and communication technologies · Life quality · Digital economy · Digitalization · Economic development of the regions

S. V. Domnina (✉)
Samara State University of Economics, Samara, Russia
e-mail: swdomnina@mail.ru

S. V. Domnina · O. A. Podkopaev · S. U. Salynina
Samara State Institute of Culture, Samara, Russia
e-mail: podkopaev@smrgaki.ru

S. U. Salynina
e-mail: salyninasu@mail.ru

S. I. Ashmarina and V. V. Mantulenko (eds.), *Current Achievements, Challenges and Digital Chances of Knowledge Based Economy*, Lecture Notes in Networks and Systems 133, https://doi.org/10.1007/978-3-030-47458-4_18

1 Introduction

Digitalization of the economic system contributes to the solution of modern social and global problems, expansion and facilitation of communications between the state, legal entities and individuals, significantly improves the quality of social services, increases productivity, creates new opportunities for entrepreneurship and labor activity, education and permanent improvement of professional qualifications and competencies. "Big Data", formed in the process of digitalization development in the national economy is becoming one of the main assets of our state, civil society and business structures.

The concept of digital economy is closely linked to those market segments where value added is created through digital (information) technologies. However, everything is not so clear. The category "digital economy" has different segments and is considered in the works of scientists from different positions. In addition, there are sectoral and regional specifics of digital economy formation. Thus, there are serious differences in the development of digital economy in agriculture [21], manufacturing industry [3], in urban and rural areas [16].

Information and communication technologies will allow creating an efficient urban infrastructure [17], and the digital platform will make it possible to create smart social systems in the city [11]. Thus, issues related to the development of the digital economy are relevant today. At the state level Russia has adopted the Program "Digital economy of the Russian Federation" [18] and a number of other strategic documents in this area [4, 23, 24]. Since the Russian Federation has a strong differentiation of the regions in terms of economic development and population's welfare, the methodology issues for assessing the impact of digital economy on regional development rates and population's welfare in the region are also becoming relevant.

The article considers the impact of digital economy on regional economic development and welfare. The goal of this research is to study the process of digital economy development at the regional level and determine the assessment indicators of digitalization impact on the economic condition of the region and the population's welfare. Objectives of the study:

- to identify the strengths of the digital economy of Russia,
- to identify indicators to assess the level of digital economy development and analyze their dynamics in the districts and regions of Russia,
- to modify the population's welfare index in the regions by introducing quantitative and qualitative indicators of economy digitalization.

The object of the study is the digital economy. The subject of the study is socio-economic relations in the process of digital economy at the regional level in the Russian Federation.

2 Methods

The following research methods were used: analysis and synthesis, deduction method, graphical and tabular data presentation methods, statistical analysis methods. To assess the impact of the digital economy development on the population's welfare in the region the modified welfare index "population's welfare index" in the region is assumed to be applied, it was proposed by the authors earlier [5]. In particular, the indicators of information and communication component of the welfare were defined, its quantitative and qualitative indicators were determined, which show the level of digital economy development in the regions.

3 Results

Analysts from the Internet Technology Centre (ROCIT) annually determine the digital literacy index. Today this index is 5.99 (an increase of 5.7% per year) [20]. Russia is ranked 45th in the world (among 176 countries) by the ICT Development Index and is among the first 30% of countries [1]. This indicates a fairly high potential for the development of the digital economy in Russia.

Russia is ranked 45th in the world by the Global Innovation Index, 38th by the Global Competitiveness Index, 10th by the Global Cybersecurity Index, 35th by the E-Government Development Index [1, 2].

In 2018, the contribution of the digital economy to Russia's GDP was 2.7% [2]. In absolute terms, the value of this indicator is 3.9 trillion rubles. Analysts of the Association of Electronic Communications divide the ecosystem of the digital economy into 7 "hubs": marketing and advertising; finance and trade; infrastructure and communications; media and entertainment; government and society; cyber security; education and personnel. Hubs contributed to the Russian economy in 2018, respectively: "Finance and Trade"—1953.4 billion rubles, Marketing and Advertising—262.9 billion rubles, Infrastructure and Communications—106.2 billion rubles, Media and Entertainment—75 billion rubles [19].

The conducted research allowed us to highlight the strengths of the Russian digital economy. So, in the Russian Federation the science and innovation infrastructure are generated which is represented by the development institutions aimed at the stimulation of innovative processes. The institutional environment is formed to create new markets and make them functioning by means of digitization of economic processes; new information and communication technologies (ICT) facilitate new prospects and preconditions and increase the efficiency of domestic economy and population welfare; the high potential of human capital is saved up and generated [4, 18, 24]. Despite Russia's strengths in the development of the digital economy, 11.8% of ICT organizations lack their own funds for technological innovations; 10.6% of ICT organizations believe that the cost of technological innovations is very high; 8.3% of ICT organizations think that the economic risk of introducing technological innovations is high; 7.1% of ICT organizations believe that there is a lack of financial support from the state to introduce technological innovations; 3.6% of ICT organizations believe that there is a lack of financial support from the state [1]. In this connection, the main strategic documents and programs for the development of the digital economy in the Russian Federation define the goal of improving the life quality through information and telecommunication technologies [23]. The development of digital economy also affects the population's welfare at the regional level, to evaluate it we used the population's welfare index in the region [5]. In this study, to improve the methodology for assessing the level and quality of digital economy development and its impact on the population's welfare in the region, it is proposed to modernize population's welfare index within a comparative approach.

In particular, in addition to the "number of personal computers per 100 employees" and "cost of information and communication technology", we propose to add six additional quantitative indicators and three quality indicators of information and communication technology component of the population's welfare index (Table 1). The introduction of new indicators in the information and communication technology component of population's welfare index comes from the fact that the digital economy has evolved significantly over the past 7 years and new quantitative and qualitative determinants of these changes have emerged.

To obtain the qualitative assessment of information and communication component of the population's welfare index, it is necessary to implement a sociological study. These indicators will allow taking into account the impact of the digital economy on the development of the region and population's welfare.

Analyzing the quantitative aspect of information and communication component of population's welfare index for the Russian Federation in 2018, it is observed that the Volga Federal District takes 5th place among the 8 districts of Russia. At the same time, it has a high value of the indicator "share of the population using the Internet to receive state and municipal services in electronic form, in the number of people aged 15–72 years, who received state and municipal services, %" (3rd place). In terms of the indicator "Share of subscribers to fixed broadband Internet access" the Volga Federal District is ranked 4th among the 8 Russian districts (Fig. 1).

When analyzing the distribution of the subjects of the Russian Federation in the Volga Federal District according to the main indicators of the digital economy development in 2018, it should be noted that the Samara Region takes the third place among the five regions of the Volga Federal District with a high level of economic development and prosperity, behind the Republic of Tatarstan and Nizhny Novgorod Region (Fig. 2).

At the same time, the Samara region in the Volga Federal District has a low value of the indicator "share of the population using the Internet to receive state and municipal services in electronic form, in the population aged 15–72 years old who received state and municipal services", ranked 54th among all the constituent entities of the Russian Federation.

Table 1 Proposals for upgrading the information and communications technology component of population's welfare index

Year	Quantity indicators	Quality indicators
2011 (initial model)	Number of personal computers per 100 employees The cost on information and communication technologies	Workplace quality
2019 (proposals on increasing welfare index)	The share of subscribers to fixed broadband Internet access, % The share of subscribers of mobile broadband access to the Internet, % The share of households with broadband Internet access in the total number of households, % The share of mobile broadband Internet subscribers, % The share of the population using the Internet in the total population aged 15–74 years, % The share of households with broadband access to the Internet in the total number of households, %	Quality of Internet access
	The share of the population using the Internet to order goods and services in the total population aged 15–74 years, % The share of the population that uses the Internet to receive public and municipal services in electronic form, in the population aged 15–72 years who received public and municipal services, %	Quality of communication security for ordering goods and services over the Internet
	The share of the population that uses the Internet to receive public and municipal services in electronic form, in the population aged 15–72 years who received public and municipal services, %	Quality of communication security for obtaining state and municipal services in electronic form

Source authors

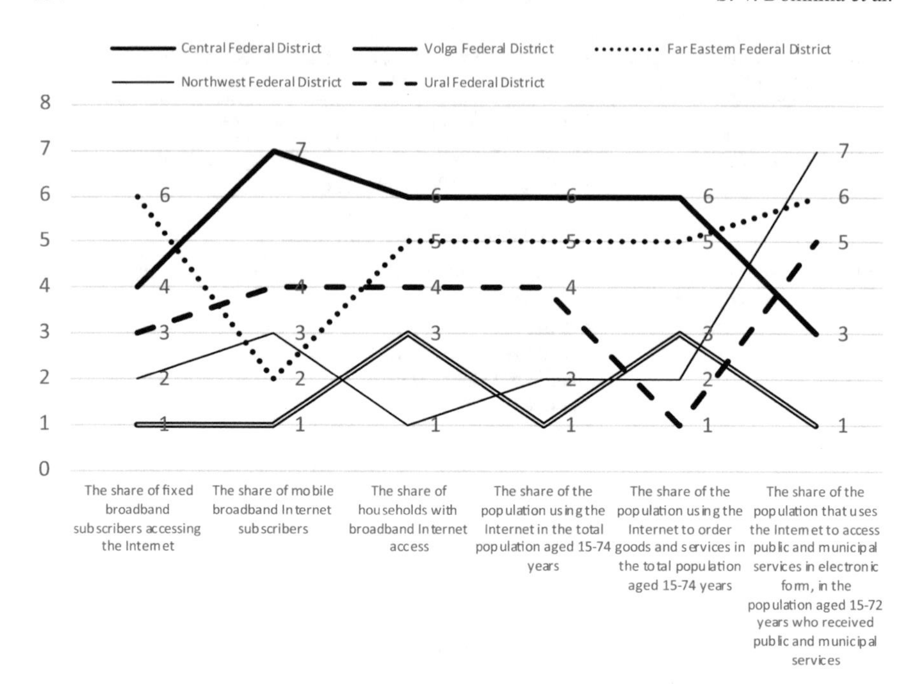

Fig. 1 Distribution of the Russian Federation Districts by key indicators of the digital economy development in 2018 (place occupied by the Russian Federation Districts by indicators) (*Source* authors based on [1])

4 Discussion

In modern scientific research the digital economy is considered as a driving force of production, the factor of its transformation into intellectual production [15], the impact of digital economy on the sustainable development of the companies and various economic activities is studied [6]:

- the tourism industry through the Internet and web technologies [10],
- on agriculture with environmental entrepreneurship through new forms of e-agriculture [21],
- manufacturing industry through the development of a digital manufacturing base [3].

There are articles which researched the role of social trade as an engine of sustainable development of the city information economy [13]. The issues related to the definition of the main drivers of the digital business models, which include key performance indicators, individualization, efficiency and communication are also relevant issues [9]. The issues related to the opportunities of digital literacy in society and its impact on the economy sustainability are researched [22]; the role of

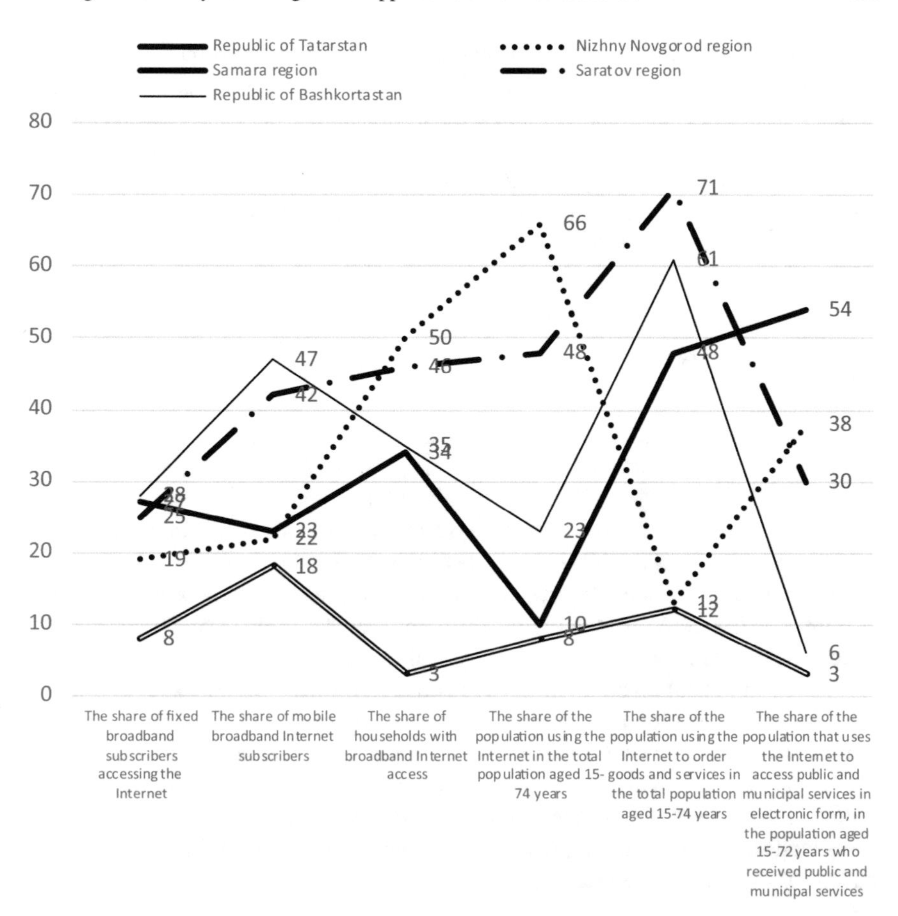

Fig. 2 Distribution of the territorial entities of the Russian Federation of the Volga Federal District by key indicators of digital economy development in 2018 (ranked by the entity of the Russian Federation by indicators) (*Source* authors based on [1])

active innovative civic participation in the creation of urban infrastructure through information and communication technologies [17] are studied.

The new digital age requires organizations to use data in such a way that it is efficient in terms of other production factors and thus provides a new quality of economic growth and information value in the economy [7]. The impact of digital finance on financial availability and stability is also relevant [14]. There are studies on the role of the urban environment in promoting digital entrepreneurship and overcoming institutional resistance to innovation [8], on the importance of the digital platform for smart social systems in the city [11], and on the differences and gaps in digital technologies in urban and rural areas [16].

Recently, there are research works on the creation of an innovation management model in a globalized digital society [12], on the role of digital education, the relationship between education, the digital environment and the labor market, and the impact of digital competences on the labor market [25]. However, these works do not consider digital economy as the most important institutional factor in the development of regional economic system and do not study the impact of digital economy on regional economic development and welfare.

5 Conclusions

In this study, the authors reviewed the genesis of Russia's digital economy, analyzed its strengths, and conducted the study on the impact of digitalization development on the population's welfare in the region. It is known that the digital economy has an impact on the population's welfare in the region, its assessment should be carried out through the population's welfare index in the region [5]. In order to improve the methodology for assessing the level and quality of digital economy development and its impact on the welfare of the region, the authors of this study have modernized this indicator by introducing new quantitative and qualitative indicators in the information and communication component of the population's welfare index. This is due to the fact that the development of the digital economy in the last 7 years has changed significantly. The proposed indicators will make it possible to study precisely the impact of the digital economy on the development of the region and its population's welfare. Further prospects of the study are connected with the calculation of population's welfare index by the improved methodology and comparative analysis of the regions of the Russian Federation.

References

1. Abdrakhmanova GI, Vishnevskiy KO, Gokhberg LM (eds) (2019) Digital economy indicators in the Russian Federation: 2018: data book. HSE, Moscow
2. Abdrakhmanova GI, Vishnevskiy KO, Gokhberg LM (eds) (2019) Digital economy: 2019: summary of statistics. HSE, Moscow
3. Beckmann B, Giani A, Carbone J, Koudal P, Salvo J, Barkley J (2016) Developing the digital manufacturing commons: a national initiative for US manufacturing innovation. Procedia Manuf 5:182–194. https://doi.org/10.1016/j.promfg.2016.08.017
4. Decree of the President of Russian Federation "Strategy of scientific and technological development of the Russian Federation" dated 1 December 2016, No. 642. http://kremlin.ru/acts/bank/41449. Accessed 19 Feb 2019
5. Domnina SV (2011) Technique to generate and analyze an integrated wellbeing index to conduct interregional comparisons. Reg Econ Sociol 3:70–77 (in Russian)
6. Garifova L (2014) The economy of the digital epoch in Russia: development tendencies and place in business. Procedia Econ Finance 15:1159–1164. https://doi.org/10.1016/s2212-5671(14)00572-3

7. Garifova LF (2015) Infonomics and the value of information in the digital economy. Procedia Econ Finance 23:738–743. https://doi.org/10.1016/S2212-5671(15)00423-2
8. Geissinger A, Laurell C, Sandström C, Eriksson K, Nykvist R (2018) Digital entrepreneurship and field conditions for institutional change – investigating the enabling role of cities. Technol Forecast Soc Chang 146:877–886. https://doi.org/10.1016/j.techfore.2018.06.019
9. Härting R-C, Reichstein C, Schad M (2018) Potentials of digital business models – empirical investigation of data driven impacts in industry. Procedia Comput Sci 126:1495–1506. https://doi.org/10.1016/j.procs.2018.08.121
10. Hojeghan SB, Esfangareh AN (2011) Digital economy and tourism impacts, influences and challenges. Procedia Soc Behav Sci 19:308–316. https://doi.org/10.1016/j.sbspro.2011.05.136
11. Jucevičius R, Patašienė I, Patašius M (2014) Digital dimension of smart city: critical analysis. Procedia Soc Behav Sci 156:146–150. https://doi.org/10.1016/j.sbspro.2014.11.137
12. Kadar V, Moise IA, Colomba C (2014) Innovation management in the globalized digital society. Procedia Soc Behav Sci 143:1083–1089. https://doi.org/10.1016/j.sbspro.2014.07.560
13. Oleynikova E, Zorkina Y (2016) Social commerce as a driver of sustainable development of the information economy of the city. Procedia Eng 165:1556–1562. https://doi.org/10.1016/j.proeng.2016.11.893
14. Ozili PK (2018) Impact of digital finance on financial inclusion and stability. Borsa Istanbul Rev 18(4):240–329. https://doi.org/10.1016/j.bir.2017.12.003
15. Paritala PK, Manchikatla S, Yarlagadda PKDV (2017) Digital manufacturing – applications past, current, and future trends. Procedia Eng 174:982–991. https://doi.org/10.1016/j.proeng.2017.01.250
16. Philip L, Cottrill C, Farrington J, Williams F, Ashmore F (2017) The digital divide: patterns, policy and scenarios for connecting the 'final few' in rural communities across Great Britain. J Rural Stud 54:386–398. https://doi.org/10.1016/j.jrurstud.2016.12.002
17. Praharaj S, Han JH, Hawken S (2017) Innovative civic engagement and digital urban infrastructure: lessons from 100 smart cities mission in India. Procedia Eng 180:1423–1432. https://doi.org/10.1016/j.proeng.2017.04.305
18. Program "Digital economy of the Russian Federation" approved by the order of the Government of Russian Federation dated 28 July 2017, No. 1632-p. http://government.ru/docs/28653/. Accessed 19 Feb 2019
19. RAEC (2018) Runet economy/digital economy of Russia 2018. https://raec.ru/activity/analytics/9884/. Accessed 19 Feb 2019
20. ROCIT (2017) Digital literacy index up 5.7% in 2017. https://rocit.ru/news/digital-literacy-2017. Accessed 19 Feb 2019
21. Şerbu RS (2014) An interdisciplinary approach to the significance of digital economy for competitiveness in Romanian rural area through e-agriculture. Procedia Econ Finance 16:13–17. https://doi.org/10.1016/S2212-5671(14)00768-0
22. Sharma R, Fantin A-R, Prabhu N, Guan C, Dattakumar A (2016) Digital literacy and knowledge societies: a grounded theory investigation of sustainable development. Telecommun Policy 40(7):628–643. https://doi.org/10.1016/j.telpol.2016.05.003
23. State program "Information Society (2011–2020)" approved by the Decree of the Government of Russian Federation dated 15 April 2014, No. 313. http://government.ru/docs/11937/. Accessed 19 Feb 2019
24. Strategy of development of the information technology industry in Russian Federation for 2014–2020 and or the perspective up to 2025 approved by the order of the Government of Russian Federation dated 1 November 2013, No. 2036-p. http://government.ru/docs/8024/. Accessed 19 Feb 2019
25. Ţiţan E, Burciu A, Manea D, Ardelean A (2014) From traditional to digital: the labour market demands and education expectations in an EU context. Procedia Econ Finance 10:269–274. https://doi.org/10.1016/S2212-5671(14)00302-5

Transformation of Public Administration in the Interests of Digital Economy Development

A. V. Pavlova and S. I. Ashmarina

Abstract The article is devoted to the research of the category "digital economy", the analysis of the transformation of the nature of socio-economic relations and changes of the society development paradigm in the digital economy conditions. The authors identify some problems of managing the digital economy and ways to solve them, as well as highlight a special role of the nature of government management in the digital economy. The article investigates functioning principles of digital public administration and formation directions of the state management of the digital economy.

Keywords Digital economy · Digital public administration · Transformation of public administration

1 Introduction

"Digital economy" is a concept as widespread as it is ambiguous and has multiple interpretations. For the first time "digital economy" as a concept appeared in 1995 in the works of several researchers, namely in the work of Negroponte "Being digital" [3] and Tapscott "Digital economy" [4].

Over the past 25 years, many concepts have appeared that replace each other, in particular, the Internet economy, digital economy, electronic economy, digital sector, e-business, digitalization, digital society, information society, etc. But observing this development convinces us that the process of digitalization, which is launched in many areas of the modern society, is not identical to the digital economy, if there is one.

Why do we doubt the existence of a "digital economy"? After all, we observe so many "traces" of its existence:

A. V. Pavlova (✉) · S. I. Ashmarina
Samara State University of Economics, Samara, Russia
e-mail: 930895@list.ru

S. I. Ashmarina
e-mail: asisamara@mail.ru

S. I. Ashmarina and V. V. Mantulenko (eds.), *Current Achievements, Challenges and Digital Chances of Knowledge Based Economy*, Lecture Notes in Networks and Systems 133, https://doi.org/10.1007/978-3-030-47458-4_19

- digital economy—economic activity;
- digital economy—a system of economic, social and cultural relations based on the use of digital information and communication technologies;
- digital economy—a global network of economic and social activities implemented through platforms such as the Internet, as well as mobile and sensor networks;
- digital economy—a business process automation;
- digital economy—a set of business models, methods of production, management, which is based on information technologies;
- digital economy—a state of the economic system in which all business processes are organized online, and efficiency is provided by information and communication processes;
- digital economy—a result of production digitalization;
- after all, digital economy is Russia's national project for 2019–2024 [5].

2 Methodology

A closer study of such a phenomenon as "digital economy" leads us to the conclusion that the digital economy is not a fundamentally different form of the economic system, it is the evolutionary development of the "traditional" economy, which is based on the use of electronic rather than analog means [1, 2].

The economy can be defined by any term: market, innovative, green, resource-saving, digital, etc., but all categories of the traditional economy—objects and subjects of economic relations, product, service, and economic operations are equally applicable to each of them, including the digital economy.

However, despite the fact that the "digital economy" is only a new form of manifestation of the economy itself, it is characterized by a transformation of the very nature of socio-economic relations and a change in the development paradigm of our society, and this determines the complexity of managing the digital economy.

First, as we have already pointed out, there is no clear, well-established concept of the digital economy. Secondly, at the present stage, there is no methodology and international standards for measuring the digital economy, as a result, economic entities and economic operations are an artificially allocated statistical totality. Third, the digital economy is an open system with hyperglobal markets. Fourth, it leads to the emergence of new trade forms, the Internet of things (IoT), and new intangible products. Fifth, there is a problem of trustworthiness and reliability of information received from business entities because of the lack of observability of individual transactions (the possibility to go into the shadow), as well as the lack of structure and fuzziness of transactions from the point of view of accounting. Sixth, the discrepancy between the development level of information technologies of companies and public administration, which creates prerequisites for the disintegrating nature of public management of socio-economic development in the country.

For the purposes of this study, we have systematized a list of indicators of the national project "Digital economy of the Russian Federation" [5] and its federal components. The analysis of the content of these components showed that they do not reflect the effectiveness of the digital economy, not all types of economic activities and not all business categories participate in the survey. The analysis of transformation directions of the public administration, which has a direct impact on the nature of the socio-economic development in the country, was also carried out. Characteristics that determine the digital nature of the development of the modern Russian economy were highlighted. This methodological approach allowed us to identify some problems of digital economy development, as well as ways to solve them.

3 Results

One of the key problems is the complexity of the observation object. In our view, the first step towards resolving this problem is to distinguish between digital economy and digitalization.

The next problem, or rather a task, is the identification of subjects and objects of the digital economy. The solution of this problem requires the definition and establishment of criteria for assigning economic entities to a statistically observable group, for example, by the share of high-tech products produced, or by the share of digital services provided, or by the level of automation and digitalization of business processes of economic entities.

A similar task is faced in relation to economic operations that form the digital economy: production and distribution of products, financial and non-financial services, consumption, and accumulation. It is necessary to develop a system of clear criteria that will allow referring business operations to the digital economy.

At the present stage, there are separate indicators that characterize only digitalization processes, so it is necessary to create a system of indicators that will characterize all aspects of the digital economy: resources, the number of organizations by type of еруШК economic activity, the gross output of digital and high-tech products and services, profit and value added received in the digital economy. In particular, the solution may be to develop satellite accounts of the digital economy, which would allow determining the contribution of the digital economy to the growth of the country's economy.

Today, there is a problem of covering all economic entities operating in the segment of the digital economy. This problem can only be solved by mandatory registration of all organizations, individual entrepreneurs, self-employed, and households on the on-line platform, which will allow economic entities to submit reports in a convenient format and participate in mandatory surveys.

Optimization of primary data collection and processing processes, which can be implemented on the basis of the previous solution, will solve another problem—insufficient trustworthiness and reliability of information at the level of economic activities and regions.

It should also be noted that there is no unified methodology for collecting, processing and analyzing information about the digital economy. The solution to this problem is obvious—it is necessary to develop a methodology for collecting, processing and research of the digital economy at the regional level and based on the type of economic activity, which will determine their contribution to the digital economy of the country. It is possible to fully use results of this methodology if it is approved and used centrally.

And for the purposes of managing the digital economy, it is necessary to allow a flexible use of statistical information by all interested users, which is solved by the free placement of operational information on the Rosstat website.

Ensuring access to reliable information is one of the tasks of the modern public administration. The key task of changing the public administration in the digital economy is a complete transition to data-based management (data-drive management). This method of developing and making decisions is based on collecting and analyzing the amount of information that will be sufficient to make the most optimal decision.

The transition to data-based or digital management, in our opinion, should be based on a number of principles. Figure 1 shows the defining principles of the state digital governance.

The principle of data primacy is that the priority in the activities of public authorities should be collection, structuring and provision of open access to high-quality consistent data by various non-governmental agents (businesses, non-profit organizations, and other government agencies).

The implementation of the principle of platform unity of data storage is provided by the organization of receipt of all data flows in a single data lake, which will ensure their harmonization with each other, connectivity and consistency.

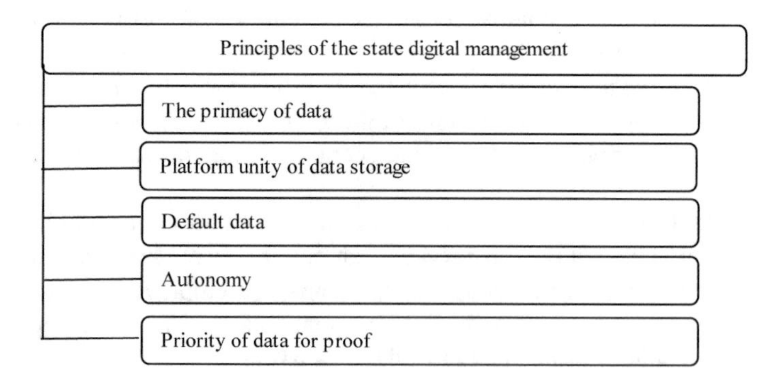

Fig. 1 Defining principles of the state digital management (*Source* authors)

The default data principle is that information arising from the activities of public authorities should be sent to the data lake automatically and in a form that provides the necessary level of data connectivity. At the same time, data received from non-state agents (otherwise the picture will not be complete) should not cause additional costs, that is, they should be collected within the framework of existing processes.

The principle of autonomy applies if the functions of the state regulator and the functions of data collection and storage do not belong to the same state body. The principle of priority of data for proof is implemented if the development and implementation of government decisions in each of the areas is based on the collected and presented data and if the amount of data for decision-making is not enough—the increase in the regulatory burden on business, the state, and society must be abandoned.

4 Discussion

It is obvious that the transition to digital public administration, which will be able to ensure socio-economic development in a new format, requires a digital transformation of public administration itself. Only after passing this transformation, is it possible to manage based on the principles given above. And in order to ensure the digital transformation of public administration, changes are needed in several directions.

First, *infrastructure and tools*. We need an access to modern digital infrastructure, including data centers, as well as to cloud solutions and software. Within a government agency, it is necessary to ensure that both stationary and mobile devices could be used.

Second, *processes*. It is necessary to analyze existing processes, reengineer them using methods of process optimization, lean manufacturing, design thinking, transfer of processes to a digital environment, embedding data-based decision-making algorithms in the processes, evaluation processes to determine what data they provide, certain indicators of processes quality, and in the subsequent monitoring and updating processes.

Third, *data*. It is necessary to analyze the available data and their quality, draw up maps of the data necessary for decision-making, and ensure their completeness, availability, quality, and relevance. Public authorities need to participate in the development of a clear and logical data architecture to create a unified data management environment on the part of the state, as well as in the creation of registers of state information systems, they have to create the ability to access data in real time, while also ensuring the data security.

Fourth, *models*. It is necessary to initiate projects related to the implementation of decision-making models, ensure the introduction of not only mathematical methods of data analysis, but also artificial intelligence into decision-making processes, and establish constant updating of models and their validity.

Fifth, *personnel*. It is necessary to create conditions for attracting specialists to the civil service who have the most advanced competencies for digital transformation and working with data, to train civil servants in new formats that develop skills for learning new materials, develop professional competencies for digital transformation, and allow them to master a lot of new things.

Sixth, *culture*. All changes will not be possible if you do not promote the implementation of changes, reduce the resistance to change, and develop a teamwork. It is important to implement the "right to error" principle and the ability to experiment when implementing projects and solutions in the framework of the digital transformation.

5 Conclusion

The digital transformation of public administration will change three types of management that are associated with different ways of doing things: processes, projects, and innovations.

"Process management" is something that all large organizations are usually engaged in, as well as our state bodies; this is all current activities.

"Projects" means activities to create something new.

"Innovation" is something that is almost absent in the public administration today. It is the application of really new management models, new products, and new processes aimed at creating an effective digital economy.

As a result, the public authorities will be at the development level of companies in the context of modern technologies usage, and they will become part of the platform that supports decision-making and the implementation of state functions. This will lead to the following target conditions.

The target state of process management—all solutions that can be algorithmized are converted to automatic/algorithmic execution. The process of making such decisions and their results are data providers for analytics and other decision-making for all economic entities.

The target state of project management—the creation and implementation of new projects, construction projects, etc. occur with access to real-time data on the implementation progress, if necessary with the presence of digital twins of the created objects, employees with a high level of competence in project management. This area includes projects for creating new processes and digital products; solutions to some of the tasks of the state policy that are implemented using flexible project management methods (agile), by employees with high teamwork skills.

The target state of innovations—a significant part of government employees (at least 10%) are fully engaged in development and innovations in the field of management and solving citizens' problems. Employees have modern competencies which they are constantly improving, they are customer-oriented, have motivation and opportunities for their work [6].

Culture, a culture of working with data, a culture of implementing change, and a culture of openness, is a prerequisite for the digital economy and the digital public administration.

References

1. Ashmarina SI, Zotova AS, Mantulenko VV (2016) New values forming among Russian entrepreneurship under the influence of globalization. In: Kliestik T (ed) Globalization and its socio-economic consequences, 16th international scientific conference proceedings: proceedings of 16th international scientific conference on globalization and its socio-economic consequences. Univ Zilina, Faculty of Operation and Economics of Transport and Communications, Rajecke Teplice, pp 58–64
2. Guryanova AV, Smotrova IV (2020) Transformation of worldview orientations in the digital era: humanism vs. anti-, post- and rans-humanism. In: Ashmarina S, Vochozka M, Mantulenko V (eds) Digital age: chances, challenges and future. ISCDTE 2019, vol 84. Lecture notes in networks and systems. Springer, Cham, pp 47–53. https://doi.org/10.1007/978-3-030-27015-5_6
3. Negroponte N (1996) Being digital. Vintage Books, New York
4. Tapscott D (1997) The digital economy: promise and peril in the age of networked, intelligence. McGraw-Hill, New York
5. The national program "Digital economy of the Russian Federation" dated 24.12.2018. http://government.ru/rugovclassifier/614/events/. Accessed 27 Feb 2020
6. Troshina EP, Mantulenko VV (2020) Influence of digitalization on motivation techniques in organizations. In: Ashmarina S, Vochozka M, Mantulenko V (eds) Digital age: chances, challenges and future. ISCDTE 2019, vol 84. Lecture notes in networks and systems. Springer, Cham, pp 317–323. https://doi.org/10.1007/978-3-030-27015-5_38

Information Technologies for Ensuring Sustainable Development of Organisations

Informatization of Labor Regulation as Basis for Ensuring Sustainable Development of Enterprises

I. V. Bogatyreva and L. A. Ilyukhina

Abstract The article investigates the IT usage in the labor regulation at enterprises in the Samara region as an important factor of ensuring their sustainable development. The labor rationing digitalization is determined by the importance of forming and implementing a digital potential in the production management, the need to implement best practices of digitalization into labor rationing, and the high complexity of calculations caused by the production complexity. The research purpose is to study modern methodological approaches, software products and automated systems, and to structure application areas of IT in the labor regulation. The study is based on the analysis of scientific literature on the informatization issues of labor regulation and its impact on the productivity and competitiveness of enterprises, analysis of the best practices of labor specialists' work automation. The authors use methods of quantitative, qualitative and logical analysis. Monitoring data collection and studying information about the use of information technologies at enterprises allowed the authors to obtain reliable research results, formulate reasonable conclusions, structure modern software products and automated systems in the field of labor regulation describing their functional capabilities and summarizing the experience of work informatization in the sphere of labor regulation for a group of Samara enterprises.

Keywords Labor regulation · Information technologies · System automation · Efficiency of automation

1 Introduction

Sustainability of the company's development in modern conditions is based on the application of information technologies and automated intelligent systems at all

I. V. Bogatyreva (✉) · L. A. Ilyukhina
Samara State University of Economics, Samara, Russia
e-mail: scorpiony70@mail.ru

L. A. Ilyukhina
e-mail: laresa@inbox.ru

S. I. Ashmarina and V. V. Mantulenko (eds.), *Current Achievements, Challenges and Digital Chances of Knowledge Based Economy*, Lecture Notes in Networks and Systems 133, https://doi.org/10.1007/978-3-030-47458-4_20

stages of the production and labor processes of manufacturing products and performing works. The use of artificial intelligence in production solves a number of problems of an economic, organizational, and socio-psychological nature. The solution of such tasks as improving the productivity, reducing the labor intensity and production costs, optimizing the staff number, the growth of employment and the qualification potential, formation of an effective system of organization, regulation and remuneration of labor creates conditions necessary for the strategic development and sustainable improvement of its competitiveness.

Labor regulation as the most important element of production management is a mandatory factor of the stability in the implementation of innovative changes inherent in the modern economy. The use of information technology in labor rationing is determined by the specific content of labor specialists' work, which includes a variety of labor functions related to the development of norms and standards, work to optimize the use of the working time, and assessment of conditions of regulation and standardization of production processes.

Currently, domestic enterprises have accumulated experience in using digital technologies in the sphere of labor regulation, but their potential is not widely used. The existing automated systems for processing labor information at Russian enterprises are mainly related to work on accounting for labor productivity and wages, while the work on automated rationing of technological processes, that is, on setting standards for products and new technologies, is not sufficiently developed and has not been developed methodically [10].

In some enterprises, informatization does not cover a whole range of tasks of labor rationing, such as accounting and reporting on labor rationing, analysis of the state of labor rationing, development of calendar plans for replacing and revising standards, reducing the labor intensity of products and monitoring their implementation. Thus, the relevance of issues on ensuring the sustainable development of enterprises based on the use of information technologies in the labor regulation is determined by the following factors:

- increasing importance of the formation and implementation of a digital potential in the organization and production management, including the labor regulation as one of its most important functions,
- the need for studying and systematizing the experience of labor standards digitalization for the purpose of its dissemination and implementation in the management practice of Russian enterprises,
- high labor intensity of labor rationing calculations caused by the constant increase and complexity of production.

2 Methodology

The following methods of IT application in the sphere of labor rationing are possible:

- setting time and output standards for production,
- development of standards and specifications for labor rationing,
- calculation of optimum norms of service and norms of number with use of economic-mathematical methods,
- results processing of studying the costs of working hours (photos of a working day, time observations),
- drawing up reporting on the conditions of labor rationing and quality of existing standards [10].

The following methodological approaches are used in informatization of labor regulation in order to ensure the sustainable development of enterprises:

1. Analytical and calculation method of rationing, based on the use of the developed technological process and relevant regulatory materials (time standards) for labor rationing. It is used for the product standardization and calculation of labor standards in the current production.
2. In order to develop normative materials for the labor regulation, the following methods are used: graphoanalytic, statistical data processing (correlation, regression, and variance analysis), experiment planning theory, and image recognition method.

The theory of experiment planning allows determining the necessary volume and specific values of duration factors for which time-keeping observations should be carried out, provides reliable data on labor costs for the development of normative materials. The correlation analysis helps to understand whether there is a relation between the measured values, to determine the closeness of the relation between the execution time of operation elements or the number of employees and the influencing quantitative factors. The regression analysis makes it possible to reasonably select and obtain a specific type of relation between the spent time (number) and the influencing factors and assess its accuracy.

The variance analysis is the basis for analyzing the impact of qualitative factors on the duration of the labor process. A comprehensive assessment of the influence of quantitative and qualitative factors is made using the correlation analysis with the method of image recognition. Its essence is that qualitative factors are assigned a quantitative attribute, all observational data are classified with a certain degree of probability according to the assigned attributes, and then the influence of these factors on the time spent or the number is determined using the correlation analysis.

Traditional methods of developing standards for time-keeping data do not allow us to fully analyze the organization of labor and determine the most rational content of the labor process, methods of its implementation, organization and maintenance of the workplace, in order to lay them in the standards.

All these requirements are met by scientifically-based microelement time standards as a measuring tool for individual movements and their universal complexes, which allow covering all types of the human labor activity and ensuring the equal intensity of labor standards.

1. Calculation of optimal service rates or staff number rates using economic and mathematical methods, in particular, the Queueing theory. This theory in labor rationing is the optimal ratio between the number of incoming service requirements and the number of service devices, by which the costs are minimal.
2. The study of working hours and labor processes is one of the main research methods used in the work of specialists in the labor organization and regulation. These methods include working time photography, duration and photo chronometry.
3. The condition of the labor regulation is assessed with a number of indicators for which the company is accounting: indicators of the quality of existing regulations; coverage of work (employees) with the labor regulation; indicators characterizing the work organization in the labor standardization; indicators characterizing the scope and effectiveness of changes and revision of standards.

3 Results

Sustainable economic development of any enterprise and organization involves the use of information technology in the labor regulation in the following sections:

Calculation of Labor Standards in the Process of Technological Preparation and Production of Goods Based on the Automation
In the digitalization age, modern organizations use systems for calculating time norms using information technologies. These are autonomous systems and systems for computer design of technological processes and their regulation.

The first class of systems for automated calculation of time standards for the designed technological processes include software and methodological complexes for automating the regulation of mechanical cutting, which are based on general machine-building standards.

The second class of systems is more promising, since the initial information provided for computer design of technological processes is also used for calculating time norms, which consequently reduces the complexity of its preparation. These systems are more efficient, since they do not require special preparation of a large amount of initial data obtained during the design of technological processes. Such systems represent a higher level of rationing.

Preparation of Standard Materials for Labor Rationing
The state of the technical regulation at enterprises depends on the regulatory framework, whether there are standards for all types of work and what quality they are. The development of time standards involves determining the dependence of the execution time of operation elements on factors that affect their duration.

The basis for the development of the staff number standards is to establish the relation between the number of employees and the labor intensity on the production maintenance and management.

The quality of standards depends on how well the volume of necessary input data is determined and the choice of factors is made, to what extent their impact on the duration of the operation or the complexity of maintenance and management is accurately determined.

The domestic practice has a fairly large complex of various automated systems for the design and development of normative materials for the labor regulation. These include varieties of systems of microelement standards, such as information system "Micro-norm", software "MOST", software "FN-prof" (Table 1).

Calculation of optimal service rates or staff number rates based on the use of economic and mathematical methods. The task of establishing optimal service standards and staff number standards is to find the most effective ratio between the number of equipment (or working places) and the number of workers. In practice, equipment and workplace maintenance systems can be divided into "with a possible expectation of service" and "without a possible expectation of service" systems.

Calculations of the staff number in systems "without a possible expectation of service" do not have any special difficulties and are made in the traditional way according to the complexity and volume of work (repair, transport, quality control in mass production, etc.). The maintenance peculiarity of equipment and working places in the system "with a possible expectation of service" is that the maintenance operations are performed irregularly, at any moment there may be downtime of equipment and working places waiting for service, and there may be downtime of maintenance workers waiting for service requirements.

In mechanical engineering, this system is used by at least 25% of the total number of auxiliary workers (adjusters, duty locksmiths and electricians). The same scheme is used by the main workers. In the textile industry, a similar scheme is used by craftsmen assistants who perform inter-repair services, and multi-machine weavers.

When rationing the labor organized according to the system "with a possible expectation of service", it is necessary to determine such service standards (the number of equipment per worker) or such staff number standards (the number of workers for a certain amount of equipment) that would ensure a minimum of the total cost of employees and possible losses from equipment downtime per work unit. Such norms can be considered economically justified, or optimal.

Study of working time costs and labor processes.

The list of works on the study of the structure of working hours and the content of labor processes performed by a specialist in the sphere of labor regulation includes the following activities:

- assessment of the actual costs of working time to perform works and operations and their structure,
- analysis of the content of labor processes and methods of work performed by the organization's personnel,
- optimization of the use of working time by the company's personnel in order to eliminate losses and unproductive costs.

Labor rationing specialists spend about 25% of their working time studying the costs of working time, which makes it necessary to automate them. As an example,

Table 1 Modern software products and automated systems in the field of labor regulation

Software for microelement and factor for regulation of the work		
IS "Micro-norm"	This program is intended for automating the calculation of time norms by the method of microelement rationing for manual labor processes that occur in various types of work in different industries. It is one of the main tools of the lean production, because it allows building a rational labor process, finding the loss of working time, standardizing the work, and as a result, reducing the time of the output cycle of a production unit and reducing its cost	The program provides work with video files for recording labor processes with the function of frame-by-frame viewing and selecting appropriate microelement movements from the database The program allows you to: – determine the actual labor costs of the staff; – identify time losses due to microelement analysis of the labor process; – increase the production through building an effective working process
Software "MOST"	This software (based on MS Excel) is designed for the microelement normalization of processes and operations. There is an intuitive interface	The work in this program is carried out on two sheets: "Processes" and "Operations". On the Operations sheet, you can normalize work operations by adding and removing sequences. Operations can be saved in the database, and if necessary, they can be downloaded for editing. On the Processes sheet, you can create processes from normalized operations. Data on them can also be saved in the database and, if necessary, uploaded from there You can also create a PDF report file that shows both the process itself and the operations that are included in it. An important feature of the software is the ability to export and import processes between users
Software "FN-prof"	The "FN-prof" program is designed for factor rationing	The software allows you to determine the number of employees by establishing a linear relation between the number of employees and a factor

Source authors

Table 2 Total balance of working hours of a labor rationing specialist according to photos of the working week

Type of work	Duration	
	Minutes	%
Working with labor standards (revision and calculation)	370	15,4
Taking photos of the working day and conducting time-keeping observations	480	20,0
Processing photos of the working day and timekeeping observations	130	5,4
Registration and verification of orders	290	12,1
Contacts with masters and technologists to discuss working issues	180	7,5
Organizational work and production meetings	120	5,0
Delivery of documents to the department of labor and salary organization	50	2,1
Breaks	200	8,3
Preparation of reports on the implementation of standards, salary structure, etc.	100	4,2
Development of a roadmap for optimizing the use of working time	480	20,0
Total	2400	100,0

Source: authors

Table 2 shows the results of the working day of a labor specialist of one Samara enterprise. The main part of the working day of the labor specialist is spent on developing measures to optimize the use of working time (20%) and conducting observations and processing their results (25%).

Analysis of the State of Labor Rationing

Now at the Russian industrial enterprises, the certain experience of information technologies application is already accumulated to solve tasks in the field of labor organization and rationing (Table 3).

Some Samara enterprises possesses the system that allows developing time rates for the labor operation elements which help to work out time standards by means of the analytical method. The system consists of the complex of software support to develop time rates and rationing of work flows (Fig. 1).

Most of the studied enterprises use “patchwork” automation. It means that an enterprise implements separate systems (on management functions and production), and then tries to unite them in the single integrated system. It also fully concerns work on labor rationing. Thus, the development of the integrated system of automated performance in the field of labor rationing is necessary, and it is an important direction in improvement of labor rationing at enterprises in modern conditions. If the saving in the production sphere doesn’t arise or it is hard to determine, then we are limited to calculation of savings in the management sphere. As calculations show, the work automation on labor rationing is rather effective (Table 4).

The carried out research shows that under the conditions of full, limited, individual production specialists in labor rationing spend up to 70% of their working

Table 3 State of work automation on labor rationing at industrial companies of Samara

Estimation criteria		
Automation object	Computer software	State of automated calculation system on labor
Samara enterprise 1		
Accounting of labor input of goods; calculation of authorized personnel; calculation of average labor grade of core and auxiliary production; handling of documents, coming from different subdivisions; maintenance of normative documentation	Package Microsoft Office; program for calculation of authorized personnel on the basis of MS Excel; program for calculation of average labor grade of core and auxiliary production. Local data base of the enterprise is created. Data bases of time standards and quotations and other normative documentation are formed. There is available, but still not functional automated system for calculation of machine time and time per unit "KOMPAS-AVTOPROEKT"	There is practically no AS of calculation of standards. The capabilities of computer technology are not used to the full extent
Samara enterprise 2		
Maintenance of normative documentation	Data bases of time standards and quotations and other normative documentation are formed in the program MS DOS	There are no special automated programs on calculation of standards
Samara enterprise 3		
Accounting of price and time wages, bonuses, additional payments and accidental benefits; calculation of average earnings; reassessment of benefits and perks; calculation of deductions from wages of workers; accounting of subsist; accounting of pension for working pensioners; preparation of record development according to all criteria of accounting of labor and wages; completing forms of statistical reporting	Program system on processing of accounting information on wages. Data on workers' salary is provided in display form and files. Local data bases are formed	AS of data processing is only related to organization of work on payroll accounting. There is no AS of labor rationing

(continued)

Table 3 (continued)

Estimation criteria		
Automation object	Computer software	State of automated calculation system on labor
Samara enterprise 4		
Accounting of piece-rate wages; recording of production runs of pieceworkers and performed operations during the base period; planning and recording of working time for all categories of workers; evaluation of labor input of goods; assessment of main workforce size; evaluation of production load; determination of planned and actual volumes of subdivisions; calculation of production capacity; product costing; performing technical-economic and production scheduling calculations	Software product of the automated workplace (AW) "Normirovshchik". Data on workers' salary is provided in display form and files. Local data bases and single data base of the enterprise are created	AS of data processing is quite powerful: it is intended both to automation of labor rationing and payroll accounting. However, available software is morally and physically outdated

Source authors

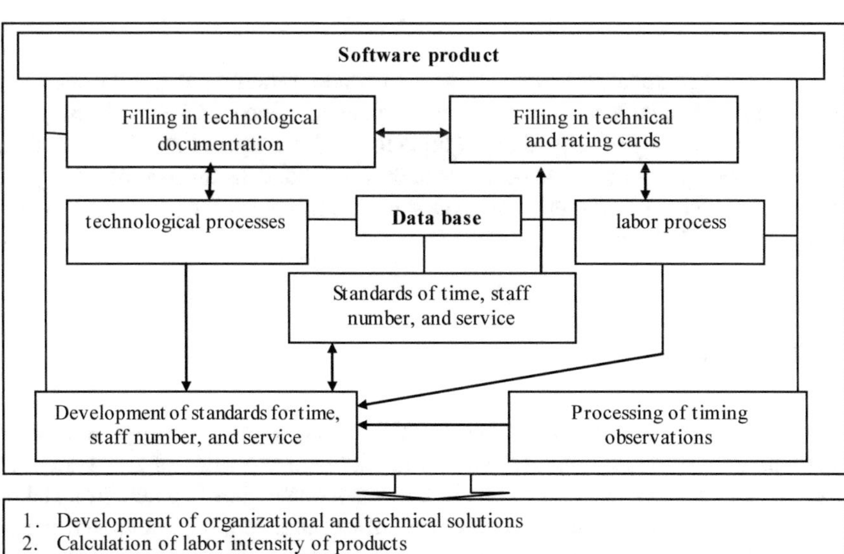

Fig. 1 Structure of the information system of organization and labor norm setting (*Source* authors)

Table 4 Labor input of work performance on labor rationing in the shopfloor

Activities	Quantity per year	Labor input of one activity, hour		Annual labor input, hour	
		Manual	In automatic mode	Manual	In automatic mode
Calculation of new standards	2060	0,5	0,05	1030	103
Changes of standards	1835	0,16	0,02	293	37
Compilation of a report on fulfillment standards	12	4,0	0,2	48	2,4
Compilation of a report on reduction of labor input	4	8,0	0,3	32	1,2
Report on financial results of activity	4	2,5	0,15	10	0,6
Logging(filing) of standards and rates	5530	0,14	0,01	774	55
Total				2187	199

Source authors

hours for setting standards. Work automation on labor rationing makes it possible to significantly facilitate the work of process engineers, reduce the labor intensity and increase the efficiency of using their working time. When organizing labor rationing, where in recent years there has been a tendency to reduce the amount of time spent training engineers, the efficiency of using these technologies can increase the labor productivity by 5–7 times.

4 Discussion

Under conditions of the digital economy and the gradual transformation of the content of the labor process, requirements to quality characteristics of the workforce increase and there is a need for highly qualified personnel, development of its labor potential, the creation of an effective system of organization, regulation and remuneration of labor, streamlining and standardizing production processes based on the lean production principles, the formation of working groups [11]. Thus, enterprises create favorable conditions for increasing the labor productivity and their further sustainable development. The results of the analysis of domestic and foreign experience on the research subject revealed a significant gap between Russia and the developed countries of the world in a number of important parameters that affect the productivity. Today, there is an objective need to systematize productivity management tools and identify those that can be applied in the short term and with minimal costs by identifying intra-production reserves [1]. The main factors of the labor productivity in these conditions are the application of innovations in the production [7]; the policy of allocating funds for information technologies at the company level [3]; the state of labor rationing at the enterprise [9]; the automation of processes that reduces the probability of errors in the production because of the human factor [6]; the work standardization [5]; the reduction of working time losses [4].

Informatization and implementation of new computer programs in the process of developing labor norms and standards, analysis of their quality, assessment of the labor regulation condition at an enterprise are issues that are studied by many domestic and foreign scholars. The use of digital technologies in the microelement labor regulation and their further development are considered in the works of Sukhanova and Pikalin [12], Malinin et al. [8]. The development of information technologies is a necessary element of improving existing and newly created software tools for designing labor standards [2]. Thus, with the development of modern information technologies, labor rationing takes on new forms and content, which allows us to consider it as a process integrated into the company's activities aimed at achieving its sustainable growth.

5 Conclusion

In the conditions of artificial intelligence use, application of automation in the production and increasing competition between enterprises, the stability of their economic position is ensured by reducing production and labor costs. This can be achieved by improving labor standards, including by regularly reviewing standards and regulations, developing a regulatory and methodological framework, and using information systems based on modern computer technologies. Thus, the need to develop an integrated system of labor regulation in cooperation with other production management systems based on the introduction of information technologies is the basis for sustainable development of enterprises.

References

1. Bogatyreva I, Simonova M, Privorotskaya E (2019) Current state of labour productivity in the economy of developed countries. In: E3S Web of Conferences, vol 91, p 08022
2. Bychin VB, Novikova EV (2018) Labor rationing as an element of effective internal management in modern conditions. Russ J Labour Econ 5(1):77–86
3. Chun H, Kim J-W, Lee J (2015) How does information technology improve aggregate productivity? A new channel of productivity dispersion and reallocation. Res Policy 44(5):999–1016

4. Collewet M, Sauermann J (2017) Working hours and productivity. Labour Econ 47(C):96–106
5. Espinosa-Garza G, Loera-Hernández I, Antonyan N (2017) Increase of productivity through the study of work activities in the construction sector. Procedia Manuf 13:1003–1010
6. Jahangiri M, Hoboubi N, Rostamabadi A, Keshavarzi S, Hosseini AA (2016) Human error analysis in a permit to work system: a case study in a chemical plant. Saf Health Work 7(1):6–11
7. Kurt S, Kurt Ü (2015) Innovation and labour productivity in BRICS countries: panel causality and co-integration. Procedia Soc Behav Sci 195(3):1295–1302
8. Malinin SV, Bakhtizina AR, Startsev GN (2016) Methods of work measurement in the coordinated system of modern production. Bull USPTU Sci Educ Econ Ser Econ 3(17):90–101
9. Schekoldin VA, Bogatyreva IV, Ilyukhina LA, Kornev VM (2018) The development of IT-technologies in labour standardization and quality assessment of standards: challenges and ways of their solving in Russia. Helix 8(5):3615–3628
10. Schekoldin VA, Bogatyreva IV, Ilyukhina LA (2020) Digitalization of labor regulation management: new forms and content. In: Ashmarina S, Vochozka M, Mantulenko V (eds) Digital age: chances, challenges and future, vol 84. Lecture notes in networks and systems. Springer, Cham, pp 137–143
11. Simonova MV, Ilyukhina LA, Bogatyreva IV, Vagin SG, Nikolaeva KS (2016) Conceptual approaches to forecast recruitment needs at the regional level. Int Rev Manag Mark 6(5):265–273
12. Sukhanova AV, Pikalin YuA (2017) Microelement work quota setting. Sci Bus: Dev Ways 12(78):50–52

The Development of the Domestic Pharmaceutical Industry in the Context of Digitalization

A. Izmaylov, A. Saraev, and Z. Barinova

Abstract In this article, the problem of the impact of digitalization is investigated in connection with its influence on the most significant elements of the pharmaceutical industry. The main aspects of the impact of digitalization are considered on such areas of the pharmaceutical industry as the development of new types of drugs and medicines; the effect of digitalization on drug development steps is reflected in the data. The main areas of influence of Big Data technology on the development of the pharmaceutical industry are highlighted. The sources of information used were analytical and statistical data from leading world consulting organizations, such as the Boston Consulting Group, Ipsos Healthcare Russia, McKinsey, as well as data from the RBC analytical portal.

Keywords Analysis · Pharmaceutical industry · Digitalization · Digital economy

1 Introduction

Economic reality is formed under the conditions of clustering at the end of the second decade of the 21st century [13] and the introduction of digital technologies [3], which affects the economic relations of market entities at all levels. Digitalization of the Russian economy will increase the country's GDP by 4.1–8.9 trillion rubles by 2025 from the point of view of McKinsey analysts [11]. Today digitalization is a megatrend that has an impact on most social and economic processes in the world. The health problem of the country's population is mainly based on the healthcare system [13], it is based on the pharmaceutical market, which performs a socially significant function

A. Izmaylov (✉)
Samara State University of Economics, Samara, Russia
e-mail: airick73@bk.ru

A. Saraev · Z. Barinova
Samara State Medical University, Samara, Russia
e-mail: saraeff10@mail.ru

Z. Barinova
e-mail: barinovazv@gmail.com

S. I. Ashmarina and V. V. Mantulenko (eds.), *Current Achievements, Challenges and Digital Chances of Knowledge Based Economy*, Lecture Notes in Networks and Systems 133, https://doi.org/10.1007/978-3-030-47458-4_21

[2, 5], and the pharmaceutical industry. Therefore, the impact of digitalization on the entire industry is of interest not only to identify its main manifestations, but also to make forecasts for the development of the studied and related industries, which allow developing tools to manage them.

Modern realities create an environment for pharmaceutical manufacturers in which it is necessary to apply quick, decisive, constructive and sometimes very costly measures, without which enterprises risk losing their competitiveness and market position [7]. It is about adapting pharmaceutical manufacturers to new conditions [9], which require a thorough analysis of key areas of economic activity. In December 2017, major players recognized at the Big Data in Pharma conference that without digital technologies, there would be no full-fledged development. Moreover, one of the market leaders, R-Pharm, believes that the time has come to revise the concept of closed research and development and create new products in the open information space. The main drivers of market growth are large companies with more than 500 employees [6]. According to the analytic data of the same organization, in 2019 such companies will spend more than 140 billion US dollars on big data. According to the forecasts of the Boston Consulting Group, by 2025 in Russia, the turnover of the big data industry will reach 100 billion US dollars [1].

2 Methodology

In research methods of strategic and innovative analysis, methods and means of economic, design and logical analysis are used. The theoretical basis of the study, the works of domestic and foreign authors in the field of pharmaceutical production management and other areas related to the production and sale of innovative products, were. The basis of the study is general scientific research methods: technique of system analysis, scientific abstraction and generalization, comparative and categorical analysis, and empirical description.

3 Results

To understand the problems, the industry is considered taking into account the influence of key trends in the field of digitalization. The pharmaceutical industry has as its main engine the creation of innovative medicines and preparations based on modern achievements of science and technology [14]. Today, pharmaceutical manufacturers can develop molecules used in the manufacture of products that for several decades were not even known in theory. One of the features that characterize this industry is conservatism [12]. However, progress is actively changing the content of intra-industry specific processes that form the basis for the development of the entire pharmaceutical industry [8]. In the pharmaceutical industry, the process of developing a new product is not just a modernization of the old, but also a significant breakthrough in certain areas of treatment [10]. It includes a set of procedures: experimental, research, experimental, clinical development over a long period is characterized by high cost (Fig. 1).

If you look at the development stages presented in the figure, you can see that the individual elements have different durations and require different financial injections. At the stage of clinical trials, phase №1 refers to the first use in humans, tolerance testing, phase №2 means dose testing, tolerability, phase №3 tolerance, effectiveness, comparison. According to various experts, the cost of the entire process can be from 1 to 11 billion US dollars. The costs of developing an innovative medicine can be divided as follows [1]: 24.8%—for research, development and preclinical studies; 57.6%—for clinical trials (57.6% are divided among the three stages as follows: 8.1%—for the first stage, 12.8%—for the second stage and 36.7% for the third stage); 6.1%—for registration; 9.1%—for post-marketing research; 1.9%—other expenses.

As can be seen from Fig. 1, the constituent elements of a single process are influenced by the digitalization. In this regard, some parameters of these elements undergo changes of substantial character, and change the financial and time costs

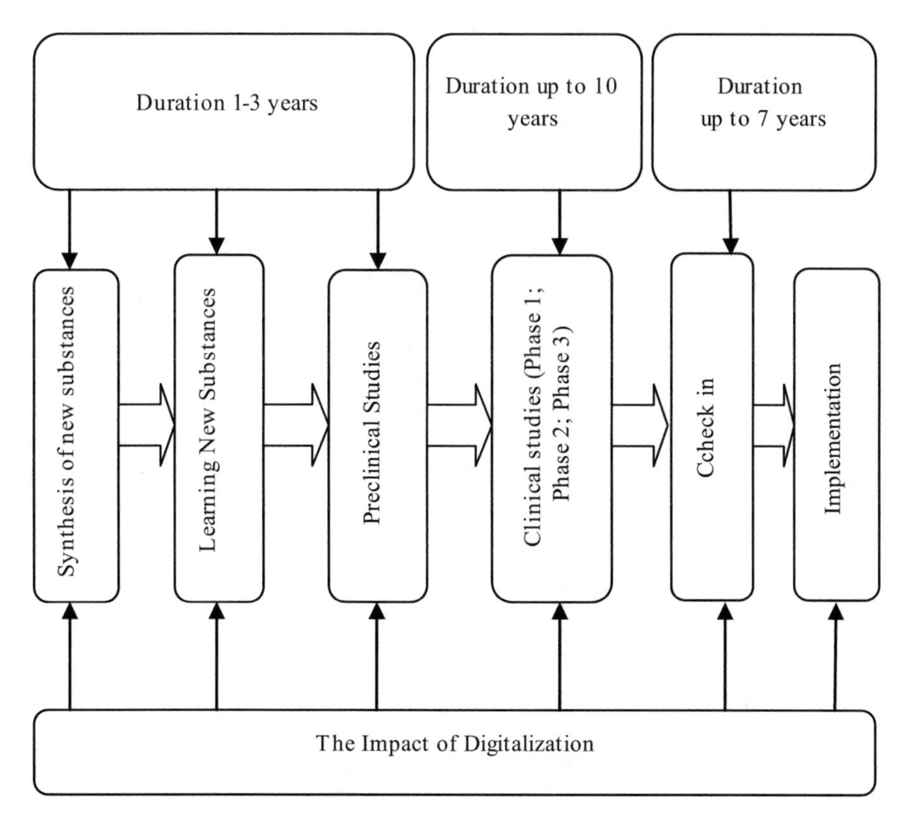

Fig. 1 The process of creating drugs (*Source* authors)

of the organization. In order to assess the impact of digitalization on the individual elements of the process of creating a new drug was made Table 1.

In the following table, we have reflected the impact of digitalization on the main elements of the period of the creation of a new drug. The use of modern technologies in the field of synthesis of new substances is a new stage in the historical development of pharmaceutical production. Earlier surveys of new substances were a long and complicated process of chemical experiments with very significant scientific and financial aspects that required a lot of time. Today computer technologies reached such a level at which all the conditions to create new substances can be modeled. It enables to reduce various costs and increase the efficiency of the implementation of this phase of the creation of original drugs. The created molecules were subjected to thorough analysis and research before using the active substance in the production. At the current stage of technology development, you can reduce the time required for this step and reduce your costs.

Preclinical testing (clinical trials) is probably the most important stage. First of all, it is connected with the question of safety of life and health involved in the testing of people and animals. The possibility of using virtual models of human physiology in the conditions of this stage of research is most effective in terms of speed and safety of implementation. People are leaving the risk zone because of the programs used at this stage, because they actually do not participate in tests, and now this is actively happening in the digital world. If the registration of a new drug was previously a stage requiring at least the remaining steps of creating the drug [4], now, with the possibility of using Blockchain and big data technologies, access is made to make databases much easier, avoiding the bureaucratic burden, it's beneficial to reduce the time required for administrative procedures.

The effects of digitalization on marketing and sales will also be considered. Today, the transition from traditional marketing to digital marketing is being actively discussed. An important issue in the context of the influence of digitization in the pharmaceutical industry is the issue of labeling and subsequent traceability of products (tracking and tracing), as well as counteracting the spread of counterfeit products. Another interesting aspect of the influence of modern trends in the pharmaceutical industry is the expansion of the scope of predictive analytics (predictive analytics) as a tool for collecting, processing and systematizing information about the set of purchases of individual consumers in order to formulate an actual marketing strategy for a particular pharmaceutical company on each site. Also important is the use of digital technology to stimulate demand for the products. For example, the company partnered with developers of various applications for devices used by patients with chronic diseases in order to prevent and reduce the negative impact of the disease on quality of life. This approach, in which the pharmaceutical manufacturer promotes its products through specialized applications allow you to stimulate the demand for pharmaceuticals certain types. According to the company "Ipsos Healthcare Russia" specializing in pharmaceutical market research, the growth rate of manufacturers are reduced, therefore, from the headquarters, owners, investors there is a demand for improving the efficiency of the activities, while the sources of efficiency are in the area of investment in digital technologies [9].

Table 1 The effect of digitalization on the individual stages of drug development

№	Stage	Impact of Digitalization	Outlook
1	Synthesis of new substances	Transferring the share of procedures related to the development, testing, testing of the latest drug molecules on the plane of virtual reality, allowing simulating physiological processes, for research	Reduce time overhead. Reduction of financial risks. Reduction of necessary time costs
2	Learning new substances	The emergence of the possibility of modeling the natural environment for researching a new substance for the possibility of further application	Reduce time overhead. Modeling environment close to natural conditions. The cost of human resources, the cost of reagents is reduced, the need for equipment is minimized
3	Preclinical animal studies	Creation of specialized technologies that allow replacing the stage of preclinical and clinical studies with a set of similar procedures, but occurring in specialized modeling programs based on the use of Big Data technology The use of programs that fully simulate the physiology of animals and humans	Minimizing the need to attract animals for research
4	Clinical researches		Reduce time overhead. Reduce financial costs. Minimization of damage to the health of animals involved in preclinical studies. The maximum reduction in the risk of causing time to the health of patients participating in clinical trials (at all phases)
5	Check in	The use of specialized systems based on the use of the Big Data and blockchain technologies for the implementation of the drug registration phase	Reduce time overhead. Simplification of bureaucratic workload. Reducing the financial costs of the procedure
6	Implementation	The emergence of the trend of the active use of Big Data and Blockchain technologies for combining databases for the purpose of deep analysis of consumer demand characteristics, business processes, and the use of mathematical algorithms to identify implicit correlations in large volumes of information	Expanding the scope of predictive analytics as a tool for collecting, processing, and systematizing information about the set of purchases of individual consumers in order to formulate an actual marketing strategy

Source authors

According to the global Digital IQ study, in 2017 the confidence of the heads of healthcare and pharmaceutical companies in the field of digital literacy of their companies is growing contrary to the general downward trend among companies in all other areas. In 2015, this indicator amounted to 62%, and in 2016, the indicator increased to 65%, while over the same period the indicator among other companies decreased from 66% to 52%. Moreover, managers see the growth of revenue as the main benefit in the digital literacy of companies. An important aspect of the development of the pharmaceutical industry is the use of the Big Data technology. In the recent past, companies were forced to destroy data older than 9 months due to the high cost of storage. This limited the analytical capabilities of the company and narrowed the horizons of strategic planning. However, the Big Data technologies today allow you to have access to massive amounts of data, followed by mathematical algorithms and obtain valuable information to improve business operations.

4 Discussion

The impact of digitalization on the development of the pharmaceutical industry is without a doubt one of the main trends in the last decade. Information technology significantly affects all stages of development, clinical research and product sales. However, despite the fact that the diffusion of information technology in the pharmaceutical industry is aimed at accelerating and simplifying, increasing safety and predicting the processes, new technologies will inevitably create new challenges. For example, the emergence of the possibility of calculating the degree of probability of the need for certain medicines in certain regions of the country, depending on a certain set of factors, cannot be fully implemented due to the fact that the productive capacity of processors today has not yet reached the level necessary for this.

5 Conclusion

Drawing conclusions about the impact of Big Data on the development of the pharmaceutical industry, it is necessary to highlight the key areas in which the domestic pharmaceutical industry is currently developing:

1. Predictive modeling of analysis to identify possible problems:
   Using modern technology, it becomes possible to use predictive modeling to develop a specific drug individually, for an individual person. This approach allows us to talk about minimizing the costs of patients on drugs and treatment and treatment that does not have a particular benefit for the patient and focusing on drugs has a guaranteed result. In this case, medications are created on the basis of the patient's genetic data, information about his chronic diseases, and lifestyle data.

2. Improving the effectiveness of clinical trials, and above all, by working with global databases, searching and selecting patients that are most suitable for testing, as well as simultaneous online monitoring of patients participating in the studies. It is considered important to predict side effects during drug testing, resorting to the Proctor method, which includes an analysis of 48 different characteristics of drugs.
3. Improving the effectiveness of communications between specialists from related industries to organize collaboration with insurance companies, database managers, as well as scientists. The combination of databases of various interacting areas can increase the potential for the use of "Big Data" by pharmaceutical companies. Scientists who are not related to a particular pharmaceutical manufacturer can submit their developments to interested colleagues from other industries, which makes the element of communication between specialists an accessible and serious engine of the latest ideas in the manufacturing sector.
4. The impact of Big Data on marketing component of the activities of pharmaceutical manufacturers is primarily in the application of Big Data analytics in the distribution of the directions of flows of finished products in areas where these products is available or projected demand.

Thus, a study of the influence of the digitalization on the development of the pharmaceutical industry, the pharmaceutical industry is among the industries actively enjoying the fruits of reprovisioning megatrend. Reviewed aspects of the impact of digitalization on the individual stages of the life cycle of medicines as one of the main catalysts for the development of the pharmaceutical industry indicate the depth of penetration "figures". The result of such influence is the emergence of intra-industry local trends. First, the increase in the productive capacities of information systems is used by manufacturers that allows to store and process large amounts of data, which gives the opportunity to obtain valuable information in the marketing aspect in the field of strategic planning. Secondly, some phases of creating innovative drugs to become faster and better be faster and better, reducing both time and financial costs. Thirdly, it is cheaper for the storage and processing of information. This reduces the financial cost to the pharmaceutical manufacturers and to extend the coverage of information layers with the aim of analytical processes. Fourth, increase the quality of logistics processes due to the technology of Blockchain. Fifth, reduce the cost of paper documents. It is especially necessary to allocate the emerging trend of unmanned production, based on maximum exclusion from production. The reason for this trend is, first of all, distort the information used in the production, which negatively affects the entire production.

Currently, the company is in the process of convergence between technologies and pharmaceutical manufacturers. One of the most significant achievements such consolidation may be forming a common vision of sustainable health in the world. Unification of common efforts, both by pharmaceutical manufacturers and on the part of companies in the sphere of information technologies made it possible to develop shared approaches to common rules to create pharmaceutical products of new generation.

References

1. European Federation of Pharmaceutical Industries and Associations (2019) The pharmaceutical industry in figures. https://www.efpia.eu/media/413006/the-pharmaceutical-industry-in-figures.pdf. Accessed 27 Feb 2020
2. Evstratov AV (2018) The main trends and prospects for the development of the pharmaceutical market in the Russian Federation: monograph. Volgograd State Technical University, Volgograd
3. Evstratov AV, Khalatyan SG (2018) Justification for the application of the logistic approach in the management of material flows in the pharmaceutical market. Bull Volgograd State Tech Univ 10(220):44–49
4. Evstratov AV, Pushkarev ON (2018) Analysis of the market structure of pharmaceutical manufacturers in the pharmaceutical market of the Russian Federation. Bull Volgograd State Tech Univ 6(216):67–73
5. Gushchina EG (2012) Mechanisms of interaction of socially significant markets: monograph. LAP LAMBERT Academic Publishing GmbH & Co., Saarbrucken
6. Ipsos (2016) Health research: Ipsos healthcare. https://www.ipsos.com/ru-ru/25-i-vypusk-daidzesta-healthindex-360deg. Accessed 07 Feb 2020
7. Kostromin PA (2015) Import substitution of drugs in Russia in terms of volume, nomenclature and quality. Theory Pract Soc Dev 9:73–77
8. Mamedyarov ZA (2017) Trends and prospects of the Russian pharmaceutical industry and the applicability of world experience. WORLD (Mod Innov Dev) 8(4(32)):772–780
9. Poverinov AI, Kunev SV (2018) Problems of implementation of the industry standard GMP as a factor in reducing the competitiveness of pharmaceutical manufacturers. Bull Mari State Univ 1(13):100–105
10. Pushkarev ON, Evstratov AV (2018) Some features of drug marketing. Bull Econ Law Sociol 1:31–33
11. RBC (2017) The digital economy will increase Russia's GDP by 8.9 trillion rubles by 2025. https://www.rbc.ru/technology_and_media/05/07/2017/595cbefa9a7947374ff375d4. Accessed 27 Feb 2020
12. Sapir EV, Karachev IA (2016) Features of the global pharmaceutical market and the problems of its development by Russian companies. Russ Foreign Econ Bull 8:97–111
13. Terelyansky PV, Soboleva SYu, Sobolev AV (2017) Public and private costs of the process of cluster formation. Management 5(4):4–7
14. Vorontsova NA (2018) Ways to overcome the recession in the development of new drugs in the pharmaceutical industry. Bull Volgograd State Tech Univ 1(211):59–66

The Development of the Russian Consumer Lending Market Under Digitalization

O. Y. Kuzmina and M. E. Konovalova

Abstract Over the past few years, the growth rate of bank consumer lending in our country has slowed down, but despite this, this segment of the financial market is one of the fastest growing areas of banking, which is largely due to the intensive use of digital technologies. The problem of the development of the consumer lending market in Russia is quite disproportionate due to the low level of competition, which leads to market failures in the form of information asymmetry. The purpose of the study is to provide a theoretical and methodological substantiation of key areas for improving the Russian institute of consumer lending in the digitalized economy. To achieve this purpose, the authors applied a systematic approach, which allowed us to consider the problem comprehensively and propose steps to solve it.

Keywords Digital technology · Banks · Consumer pricing · Competition

1 Introduction

The formation and effective functioning of the consumer lending market are the most important conditions for ensuring the continuous process of social reproduction, the stability of the financial system, achieving high rates of economic growth, and taking into account the high social significance of consumer lending. It can also be considered as a factor in the political and social stabilization of society, smoothing out increased social differentiation of citizens and increasing their level of trust in authorities [1, 11].

The special role of the consumer credit system in the socio-economic life of society makes banking institutions pay much more attention to the development process of this market segment. Improving the forms of lending, borrower solvency assessment systems, increasing the effectiveness of risk management programs, expanding

O. Y. Kuzmina · M. E. Konovalova (✉)
Samara State University of Economics, Samara, Russia
e-mail: mkonoval@mail.ru

O. Y. Kuzmina
e-mail: pisakina83@yandex.ru

S. I. Ashmarina and V. V. Mantulenko (eds.), *Current Achievements, Challenges and Digital Chances of Knowledge Based Economy*, Lecture Notes in Networks and Systems 133, https://doi.org/10.1007/978-3-030-47458-4_22

the product line of banks, considering the emergence of a new digital space, make comprehensive studies of both fundamental and applied nature relevant to identify the characteristics of these processes in the consumer lending market in Russia.

Theoretical, methodological and applied issues of consumer lending are widely covered in both domestic and foreign economic literature. At the same time, the problem of organizing bank consumer lending, considering current socio-economic conditions for the development of the digital economy, cannot be classified as solved ones both at theoretical and practical levels.

2 Methodology

As basic tools in the work, the principles of reproductive, functional, systemic and institutional approaches were used, as well as general scientific methods of cognition (analysis and synthesis, induction and deduction, abstraction) and tools of economic-statistical and mathematical analysis, which allowed us to systematize and supplement knowledge, describing the modern system of consumer lending, aimed at improving the efficiency of banking in the context of digitalization.

3 Results

The banking market is an integral part of the financial system of any country in the world. Therefore, its development should be built on principles ensuring systemic stability, which should be achieved due to the growth and quality of banking services provided. Only with the efficiently functioning banking system we can we talk about improving the welfare of citizens, the country's economic security, sustainable economic growth, taking a leading position in the domestic economy in international competitiveness and innovative development ratings [12, 13].

It should be noted that the establishment of the advanced market economy objectively provides for the development of a qualitatively new consumption system, which poses the problem of realizing economic interests of economic entities that often need constantly generating sources of financial resources [9]. One of these sources is the activity of commercial banks providing consumer loans to the population. The growing role of consumer lending contributes to the expansion of consumer demand, the acceleration of trade, creates conditions for expanded social reproduction and ensuring the sustainability of the national economy from all kinds of external influences on it. A higher standard of living provided by a qualitatively new system of consumption of economic goods can defuse the social tension arising in the general situation and make our citizens feel richer and happier.

However, like any economic instrument, consumer lending carries not only advantages, but it brings many risks, which in no case should be forgotten [6]. The solution to the problem of improving the banking management system, especially in this area of banking services, should become an integral element of the banking sector. Consumer lending should be built on uniform principles to streamline its process. For this, it is necessary to consider the current practice of Russian banks, and the foreign experience of colleagues in applying digital technologies in the financial sector of the economy.

There are a lot of financial institutions involved in direct lending to citizens in the Russian financial market. These include banks, credit organizations that do not have a banking license, pawnshops, leasing companies and consumer credit cooperatives. If we limit the scope of our research to the market of consumer lending, we get the feeling that there are only two participants in this market: in fact, banks and their customers (individuals), but this is incorrect.

In the context of the modern development of the consumer credit institution, it is necessary to emphasize once again that this institution consists of many organizations that have both horizontal and vertical coordination links. A separate banking structure cannot afford to independently deal with issues of financial and credit support, information support of creditors and borrowers, reduce problem debts or credit risks, and stimulate business entities to search for new effective forms of interaction. It requires a whole system of specialized organizations serving the consumer credit market.

The institutional infrastructure of consumer lending is aimed at creating conditions for providing and getting consumer loans, increasing the degree of accessibility of borrowed funds for the population, ensuring the unity of all stages of the lending process, that is, performing the so-called backbone function.

The infrastructure can be considered not only from an institutional point of view as a system of organizations and institutions that ensure the effective functioning of the consumer lending market, but also from a functional point of view. Then it appears as a system of economic and legal standards controlling the interaction of participants in this market segment.

The institutional infrastructure is two-tier. The first level includes financial institutions that carry out the direct lending process. The second level is formed by auxiliary financial structures that ensure the functioning of organizations of the first level. These include insurance companies, collection agencies, clearing centers, credit bureaus. Do not forget about those which are not directly part of the institutional infrastructure of consumer lending, but they create an institutional field in which organizations of the two levels indicated above are implemented by state regulatory bodies, especially by the Central Bank of the Russian Federation.

The institutional infrastructure of consumer lending is a well-structured system with both vertical and horizontal coordination chains. The same applies to the functional infrastructure, which in horizontal section represents an enlarged grouping of infrastructure elements by fields of activity, and in vertical—various levels of functioning of this system, ensuring the life of the bank as a separate business unit.

The internal sphere of the functional infrastructure of consumer lending includes, for example, analytical services of the bank, digital technologies that monitor credit transactions, analysis of the state of debt and the progress of repayment of credit obligations by customers, electronic bank document management with the use of digital signatures [8]. All these structural elements allow us to consider the bank as a single institutionalized information and analytical space.

Summing up the above, we should note that the infrastructure of consumer lending, whether it is institutional or functional, should be viewed from the perspective of a systematic approach as a certain structure of elements, some of which are participants in the consumer lending market. Their abundance, the presence of fairly complex relationships between them affects the development of the consumer lending market in our country.

The Russian system of consumer lending began the process of formation quite recently in comparison, for example, with the United States of America or the countries of Western Europe [10]. This affects both the size of the consumer lending market and its degree of segmentation.

At the end of 2013, a law on consumer credit was adopted in our country, which led to a significant reduction in the number of claims and court cases, increasing the level of borrowers' confidence in the banking system.

Currently, the consumer lending market continues to develop, showing positive dynamics after a temporary lull in the post-crisis period. According to many experts, Russians are getting used to living on credit, moving from savings to a consumer model of behavior. It should be noted that the growth in the loan portfolio was based on an increase in mortgage loans and unsecured consumer loans. The positive dynamics of consumer lending is explained by the relatively soft monetary policy of the regulator, which has repeatedly reduced the key rate in recent years, which undoubtedly affected the decrease in interest rates. So, the average interest rate on ruble loans to individuals for a period of more than 1 year in November 2018 amounted to 12.4% per annum, whereas in January 2018—13.5%. The largest increase was demonstrated by the volume of the market of mortgage loans and unsecured consumer loans—22.2% and 21.8%, respectively, for January–November 2018 [2].

In the Russian banking market, consumer lending services, in particular, express lending, POS lending in shopping centers, have been quite popular since the 2000s. The undisputed leader in this market in terms of the volume of consumer loans granted is Sberbank PJSC, which accounts for more than 45% of all provided consumer loans. Such credit organizations as JSC "Russian Standard Bank", JSC "Alfa-Bank", LLC "Home Credit and Finance Bank" and others have good dynamics of consumer lending.

The banking system of modern Russia is in permanent transformation, which is reflected primarily in the number of existing credit organizations. The number of operating credit organizations licensed to carry out banking operations as of January 1, 2019—484 (44 were non-bank credit organizations), which is 77 units less, or 13.7%, less than the same indicator by January 1, 2018 [7].

The key factor influencing the decrease in the number of credit organizations is the policy pursued by the mega-regulator aimed at enlarging credit organizations and increasing their authorized capital (the total registered capital of operating credit organizations increased by 20.3 billion rubles in 2018 to 2655.4 billion rubles as of January 1, 2019) [4]. Another factor determining the decrease in the number of credit organizations is the active introduction of information and communication technologies, the growth of economic digitalization, which directly changes traditional ways of providing banking services.

Thus, the development of electronic services by banks, the use of Internet banking have reduced the physical presence of credit organizations in regions [14]. In 2018, the total number of internal structural divisions of credit organizations (their branches) decreased by 3503 units and amounted to 29,783 as of January 1, 2019 against 33,286 as of the same date in 2018 [7].

If we talk about the structure of the consumer lending market, it cannot be called highly competitive, since leaders are clearly presented on it—as a rule, these are banks with state participation, such as Sberbank and VTB, and a significant number of smaller players that compete among themselves to attract borrowers. Sberbank's share in this market is more than 45%, which rather indicates the presence of a monopoly in the consumer lending market.

Based on the calculation by Lind and CBR-Composite, it was found out that Sberbank is included in the list of leaders in 83 out of 85 constituent entities of the Russian Federation. VTB Bank (PJSC) is the competitor which according to these indicators and the leading player in 68 regions.

Over time, the composition of leaders in the field of consumer lending remains unchanged, which is largely due to the high level of information and technological barriers that are build by banks-leaders, thereby ensuring high customer loyalty to this financial product. Provided that the number of banks engaged in lending to individuals at the end of 2018 was 197 participants.

The GAP market activity indicator has grown in most regions, indicating deterioration in the quality of the competitive banking environment. The HHI index used as an auxiliary tool only confirms the current situation; over the past 2018, the share of assets of the five largest banks in the total assets of the consumer segment increased from 42 to 55.8%, and the HHI value increased from 829 to 1108.

As for Lerner and the Boone indicators, these mathematical tools also confirm the presence of a concentration process in the consumer banking sector. The results allow us to conclude that there is a classic situation described in the model of Forheimer in the Russian market of consumer lending. There is one dominant leader in the competitive environment, which has earned its special position in the eyes of consumers due to its long history of existence in the market, the quality of products provided, advertising, business reputation, etc. In addition, the bank-leader is characterized by a low-cost structure resulting from the economies of scale, and its size allows it to have an effective management system and resources for the implementation of technological innovations. Consumers of its products are confident that the bank-leader is more than capable of predicting the dynamics of market conditions and partly determine the dynamic development of the banking market.

This model is confirmed in practice and by a system of measures implemented by Sberbank in the field of pricing policy. The banking structure determines price benchmarks in the consumer lending segment. Firms-followers only imitate its pricing model, and due to the "price umbrella" provided by the leader, they can generate additional income. According to the Bank of Russia estimates, the level of correlation between consumer lending rates of Sberbank and the bank's share in the region is so high that it is not possible to challenge the above provision [3].

The situation that has developed in the segment of Russian consumer lending, of course, does not correspond to classical notions of the level of effective competition, and clearly illustrates the discontent on the part of individual banking institutions about the dominant position of leading players. However, from a macroeconomic perspective, the leader's business model looks X-efficient. Since the key player is able to accumulate profits for new achievements of scientific and technological progress, in particular, digital technologies, and sell financial products at prices lower than its competitors. But this does not mean at all that the competitive environment should not change. Its development can largely be associated with the creation of favorable conditions for financial institutions surrounding the leader.

4 Discussion

As for the prospects for the consumer credit market, the following points can be noted. The dynamics of consumer loans will be largely determined by the overall development of the banking sector in the next period. So, experts believe that at the end of 2020, banks are expected to record financial performance (profit of credit institutions will reach 1.8-1.9 trillion rubles), and the profitability of banking institutions will return to the pre-crisis level [3].

The main reason for such a significant increase in the financial result is the sweeping of the banking industry from unreliable large players, as well as the reduction in losses of rehabilitated credit institutions. Despite the fact that there is still the problem of under-provisioning (and this is more than 15% of capital), the stability of the banking sector is increasing, which allows us to speak about fairly stable, to a greater extent, positive dynamics of credit organizations, and, in particular, the entire list of services provided by them. Under these conditions, consumer lending will occupy a significant position in the structure of banking services, but there are trends that indicate a change in the situation in this market.

According to some analysts, such factors as: saturation of consumer demand from the population will contribute to a gradual decrease in consumer lending; a slight increase in real incomes of citizens; measures introduced by the regulator to limit the growth of lending to the population; lack of prerequisites indicating a potential reduction in interest rates. The current situation in the consumer lending market poses serious challenges for the Bank of Russia, which should take into account the following circumstances in its policy:

1. The transmission mechanism of monetary policy may have a less significant impact on the consumer lending market, both from credit organizations and households. This is due to the fact that, firstly, a decrease or stabilization of the key rate by the mega-regulator may have a weak effect on lowering interest rates

on loans, since credit organizations are more oriented toward a business model focused on high-risk borrowers. To increase the efficiency of the transmission mechanism, banks will have to reorient themselves to other business strategies, which is not obvious in the face of increased competition and the absence of positive dynamics of macroeconomic indicators. Secondly, the demand for loans is characterized by a rather low elasticity, especially for the category of citizens with a relatively low income level, who are mainly consumers of this banking product, which sufficiently limits the Bank of Russia in terms of exposure.

2. The increase in lending in the face of rising inflation expectations leads to the accumulation of problems among commercial banks in the consumer lending segment and requires more precise adjustment of the prudential supervision system. Excessive inflationary expectations of borrowers may encourage them to assume unreasonable credit obligations, causing the accumulation of debt burden. Consumers with relatively low and unstable incomes that are elastic to external shocks, receiving additional funding, increase the risks for the financial system, which ultimately threatens to lose its stability and reduce the possibility of obtaining additional resources for borrowers.

Problems in the consumer lending market, caused by the impact of both endogenous and exogenous factors, the most important of which are digitalization and a low level of competition, require the Central Bank to pursue an appropriate policy aimed at protecting both credit organizations and consumers of financial services from making excessive risks, unreasonable obligations, which should help to increase their financial stability [5].

5 Conclusion

The development of the consumer lending market is currently one of the factors contributing to the formation of conditions for high-quality economic growth. An increase in the purchasing power of the population creates prerequisites for an increase in aggregate demand, which stimulates enterprises and organizations to increase the volume of output. As shown above, the consumer lending market cannot be called highly competitive, since its main players are large banks mainly with state participation.

The problem of unfair competition and closely related problem of information asymmetry, and, consequently, low transparency of banks, especially in the field of consumer lending, are one of the key ones that require special attention [15]. Marketing campaigns carried out by banks do not fully provide potential borrowers with all the necessary information about all aspects and nuances of lending in a particular bank, and, in particular, about the real cost of a loan. Information transparency should be beneficial, first of all, to the credit institution, since this affects the attitude of the borrower towards it, increasing the level of both personalized and depersonalized trust. The growth of institutional trust creates prerequisites for a more efficient functioning of the entire banking system, fundamentally changing the relationship between the borrower and the lender, forming a horizontal type of interaction between business entities. A high level of trust in financial institutions reduces transaction costs, operational risks, including loan defaults, which increases the stability and reliability of the banking system as a whole. The increase in citizens' institutional trust in banks is largely due to their use of digital technologies that can minimize the human factor, make the system more transparent, and, therefore, minimize the existing failures of the banking market.

References

1. Agarwal S, Bos M (2019) Rationality in the consumer credit market: choosing between alternative and mainstream credit. In: Haughwout A, Mandel B (eds) Handbook of US consumer economics. Elsevier, Amsterdam, pp 121–139
2. ARB (2019) Consumer lending volumes. https://arb.ru/banks/analitycs/-10262764/. Accessed 06 Oct 2019 (in Russian)
3. Bank of Russia (2019) Statistical bulletins. https://arb.ru/banks/analitycs/-10262764/. Accessed 06 Oct 2019 (in Russian)
4. Banki.ru (2019) Rating of banks. Consumer credit. https://www.banki.ru/services/responses/?date=22.05.2019&product=credits. Accessed 06 Oct 2019 (in Russian)
5. Cai S, Zhang J (2020) Exploration of credit risk of P2P platform based on data mining technology. J Comput Appl Math 372:112718
6. Casanova L, Cornelius PK, Dutta S (2018) Banks, credit constraints, and the financial technology's evolving role. In: Ikeda S (ed) Financing entrepreneurship and innovation in emerging markets. Elsevier, Amsterdam, pp 161–184
7. Central Bank of the Russian Federation (2018) Bank of Russia annual report. https://www.cbr.ru/Collection/Collection/File/19699/ar_2018.pdf. Accessed 06 Oct 2019 (in Russian)
8. Keys BJ, Wang J (2019) Minimum payments and debt paydown in consumer credit cards. J Finance Econ 131(3):528–548
9. Konovalova ME, Kuzmina OY, Salomatina SY (2020) Transformation of the institution of money in the digital epoch. In: Ashmarina S, Mesquita A, Vochozka M (eds) Digital transformation of the economy: challenges, trends and new opportunities, vol 908. Advances in intelligent systems and computing. Springer, Cham, pp 315–328
10. Kontolaimou A, Kounetas K, Mourtos I, Tsekouras K (2012) Technology gaps in European banking: put the blame on inputs or outputs? Econ Model 29(5):1798–1808
11. Kuzmina OY, Konovalova ME, Chulova ES (2020) Transformation of the banking system as a way to minimize information asymmetry. In: Ashmarina S, Vochozka M, Mantulenko V (eds) Digital age: chances, challenges and future, vol 84. Lecture notes in networks and systems. Springer, Cham, pp 12–18
12. Luzzetti MN, Neumuller S (2016) Learning and the dynamics of consumer unsecured debt and bankruptcies. J Econ Dyn Control 67:22–39
13. Mikhed V, Vogan M (2018) How data breaches affect consumer credit. J Bank Finance 88:192–207
14. Salihu A, Metin H, Hajrizi E, Ahmeti M (2019) The effect of security and ease of use on reducing the problems/deficiencies of electronic banking services. IFAC-Pap OnLine 52(25):159–163
15. Sharma P (2017) Is more information always better? A case in credit markets. J Econ Behav Organ 134:269–283

Supply Chain Resilience and Risk Analysis in Construction

I. V. Yakhneeva

Abstract The success of construction project depends on effective risk management throughout the supply chains. This article considers the problem of risk analysis in construction due to uncertainty and poor information sharing. The study has two main objectives: to identify construction project risks and determine their impact on supple chain resilience. To reduce supply chain vulnerability, the author proposes an approach to analysing construction supply chain resilience based on measuring the deviations of performance indicators. The proposed approach allows to predict disruptions and improve supply chain resilience.

Keywords Supply chain resilience · Risk analysis · Disruption risk · Construction supply chain · Information sharing

1 Introduction

Many suppliers, contractors, retailers, warehouse terminals, transport and logistics companies have complex relationships within the local and global supply systems. Uncertain environment is the primary source of supply chain risk. Uncertainty has increased recently due to interconnected trends such as changed consumer preference, competition on a global scale, more complicated supply chains and greater variety of goods with shorter product lifecycles [11]. Uncertainty affects the reliability of risk assessment and the quality of managerial decisions. As a result, total costs increase either due to risk underestimation or due to risk revaluation. In the first case it increases losses, in the second case it increases the costs of preventive measures.

The effort to cope with disruptions of supply variability, increase the service level and optimize transaction costs makes it necessary to take into account various risk factors related to supply chain partner's activities. At the same time, redesigning the business processes and interconnections between supply chain participants

I. V. Yakhneeva (✉)
Samara State University of Economics, Samara, Russia
e-mail: rinadoo@yahoo.com

S. I. Ashmarina and V. V. Mantulenko (eds.), *Current Achievements, Challenges and Digital Chances of Knowledge Based Economy*, Lecture Notes in Networks and Systems 133, https://doi.org/10.1007/978-3-030-47458-4_23

can change the risk profile and thus create greater supply chain vulnerability. Managers must cope with reducing vulnerability that means reducing the likelihood of a disruption and increasing supply chain resilience.

Most definitions relate resilience to the ability of the supply chain to withstand risks and recover from failures. Christopher and Peck defined supply chain resilience as the ability of supply system to return to its original or move to a new, more desirable state after being disrupted [3]. Sheffi and Rice Jr. point out that supply chain resilience is the firm's ability to absorb disruptions. It enables the supply network to return to standard conditions faster and thus has a positive impact on firm performance [11]. Ponomarov and Holokomb described supply chain resilience as the adaptive capability of the supply chain to get ready for unexpected events, respond to disruptions and recover from them by maintaining continuity of operations at the desired level of connectedness and control over structure and function [10]. Chopra and Sodhi distinguish supply chain resilience from supply chain efficiency. The goal of supply chain efficiency is financial performance, while supply chain resilience is directed at risk reduction [2].

The researchers in Russia have similar views, in particular the definition used in logistics is "reliability of logistic system" [9, 13]. It means an operation safety of supply chain. The level of system reliability depends on the robustness of its elements and their interconnections. Improving the system reliability contributes to the growth of its efficiency, but such growth is limited by the ratio of costs and profitability. Ivanov et al. conceptualized a review of proposed definitions and identified the drivers and elements of supply chain resilience. Key features of supply chain resilience considered by researchers are agility, visibility, flexibility, collaboration and information sharing [6].

Agility is defined as the ability of supply chain participants to respond smoothly and cost-efficiently to market changes and transform those changes to business opportunities [9]. Supply chain visibility is considered as a transparent view of upstream and down-stream inventories, demand and supply conditions, and production and purchasing schedules [3]. Flexibility is defined as the ability of enterprises to respond to market changes adapting their activity to environment with minimum effort [4]. In other words, it is the ability to adapt the supply chain configuration to changes in market. In such a case risk is a detector of bottlenecks and constraints. Collaboration means the ability of supply chain firms to work together, coordinating their activity toward common goals [3, 7]. It enables to reduce uncertainty and increase the readiness for the occurrence of unfavourable events. Moreover, collaboration is described as an important driver of supply chain resilience.

As for information sharing, the researches analyse the importance of this feature in mitigating risk due to disruptions [6, 8]. Also, they investigate the impact of information sharing on uncertainty, supply chain robustness and reducing/increasing the bullwhip effect [8, 15]. We suppose that all features of supply chain resilience rely on information sharing. In particular, it is impossible to achieve the visibility without full and accurate information. Uncertainty caused by the lack of the data, incomplete information or inaccurate reporting about potential risk affects the ability of companies to cope with the consequences of disruption. Information on risks and disruption events in the past should be accumulated in order to analyse and prevent risks in the future. The exchange of information and experience between partners makes it possible to develop an integrated database. It helps companies to predict development of the situation coordinating their strategies and action plans. Major international companies, such as Intel and General Motors created rapid response centres collecting data to avoid disruptions and losses and to coordinate actions of supply chain partners [11]. The centres find out the real facts and provide information on measures to prevent losses and ongoing activities. Providing actual and reliable data convinces companies that their partner is in control of the situation.

Construction projects have different kinds of risks due to their complexity and complicated relationships between the partners. Working with temporary actors require high coordination efforts to align the business processes. Therefore, an effective risk management throughout the construction supply chain is critical to abscond from overrun in cost and time that if not controlled properly, will finally result in project failure [12].

2 Methodology

The research method is based on estimation of the correlation between deviation of supply chain performance indicators and targets. This approach may be particularly relevant in the case of modelling losses and disruption prevention. The study is based on data provided by regional construction companies. The process is divided into three stages: the first step is to determine SC performance indicators and targets; the second step is to estimate deviation of performance indicators and its occurrence rate; and the third step is to assess the impact of deviations on SC resilience using multiple correlation and regression analysis.

3 Results

The main purpose of the developer is to complete the project on time with the required quality under conditions of the project budget. Therefore, the top SC performance indicators are: endowment of materials and equipment by the company; production, transport and warehouse capacity utilization; service level. At every stage of the construction project it is necessary to organize and coordinate the interaction of the parties involved in construction process, i.e. developers, suppliers of construction machinery, suppliers of building materials, contractors. Detailed analysis of construction supply chains identifies the deviations of performance indicators from actual values and assesses their impact on the achievement of targets. The analysis revealed three types of deviations:

- deviation in execution of the work. It is attributed to a discrepancy of actual work volume to planned indicators due to the lack of materials and technical resources, unfavourable external factors (weather conditions), coordination problems (1),
- deviation in delivery caused by a mismatch of actual deliveries due to lack of coordination, fragmented information flows, one-side optimization (2),
- deviation in the use of equipment arising from a lack of material resources and staff, due to adverse external factors (3).

The basic indicator reflecting the supply chain performance is the intensity of deviations. An assessment of intensity makes it possible to identify SC features with the highest variation. The calculation is presented in Tables 1 and 2.

High variability of deviations caused by random factors, unsustainable use of certain types of machines, and the mismatch between standard value and the actual use. Change in deviation rate does not allow assessing the impact of deviations on performance indicators . Therefore, it is necessary to consider the change in indicators

Table 1 Density analysis

(1)		(2)		(3)	
Number of deviations	Density	Interval %	Density	Interval %	Density
1	0.0197	0–10	0.0052	0–10	0.0338
2	0.0296	10–20	0.0098	10–20	0.0154
3	0.0211	20–30	0.0128	20–40	0.0089
4	0.0113	30–40	0.0144	40–60	0.0051
5	0.0056	50–60	0.0139	60–80	0.0031
6	0.0042	60–70	0.0130	80–100	0.0017
7	0.0028	70–80	0.0112	100–150	0.0017
8	0.0042	80–90	0.0088	150–200	0.0004
9	0.0014	90–100	0.0070	200–250	0.0002

Source author

Table 2 Statistical indicators of deviations

Indicators	Deviations		
	Execution of work	Delivery	Use of equipment
Standard deviation, %	1.95	24.70	109.70
Variation coefficient	0.64	0.52	0.89
Observed value	29.73	57.78	98.04
Critical value	12.59	14.07	16.92

Source author

Table 3 Calculation of correlation between deviations

	D(x)	D(y)	σ(x)	σ(y)	r	R^2
(1)–(2)	0.1433	0.2502	0.3786	0.5002	0.7860	0.6179
(1)–(3)	0.1433	0.1109	0.3786	0.3330	0.1998	0.0399
(2)–(3)	0.2502	0.1109	0.5002	0.3330	0.1051	0.0111

Source author

both separately and in relation to each other. Table 3 shows the results of evaluating the correlation between different types of deviations.

According to the estimations, the shortage of materials equal to one day leads to longer construction time equal almost three days of work with a probability of 0.9. Thus the performance indicators are closely correlated. This is partly explained by materials requirement. The delivery frequency varies from several times a day to once every few months. In addition, construction supply chain is characterized by relatively poor coordination between developers, contractors and suppliers due to lack of communication. The statement is supported by the survey of organisations from Samara region. According to our research data, only 36.7% of regional suppliers are ready to share information on delivery time [14]. 26.5% of the respondents indicated that their customers share information on inventory level, 12.2% indicated that the customers share information on sales forecast, and only 8.2% of the respondents receive stock control data from their customers. One third of organizations used automatic data exchange between internal and external information systems. Another problem is the quality of communication within a company. The study showed that participants estimate the quality of communication between supply chain partners most highly while the quality of information sharing inside the company is estimated lower.

Other values indicate a less close correlation between deviations in execution of the work and the use of equipment as well as between deviations in delivery and the use of equipment. A probability of 0.4 shows that the execution of the work doesn't depend on the use of equipment. The use of equipment depends on endowment of construction materials with a probability of 0.7%. It can be explained by the fact that there is the possibility of operative resource reallocation if the building sites are not far from each other. The next step is to determine risk management measures that contribute to achieving SC resilience. It is presented in Table 4.

There is an important proposition to be taken into account. Supply chain vulnerability is defined by reliability of the weakest supply chain partner. Thus construction supply chain resilience depends on reliability of each participant. Changes in one supply chain element may change risk levels of other elements. Systematic monitoring of supply chain elements and an estimation of their conformity to supply chain performance indicators are required.

Identification of the most significant risk elements and monitoring of deviations allow us to predict disruptions and develop risk management measures. In most cases, if companies strive for continuous and reliable activity, they use risk management

Table 4 Risk management decisions in construction supply chain

Risk factor	Impact of risk	Management decision	SC feature
Weather	Scarcity of building materials Disruption of construction deadlines	Increase in capacity Schedule float Buffer inventory	Flexibility
Machinery breakdown	Disruption of construction deadlines	Redeployment resources by types of work	Agility
Supply shortage	Shortage of material Downtime of machines Downtime costs	Redeployment materials Material substitution	Agility
Delay in delivery	Shortage of material Cost overrun	Backup supplier	Flexibility
Excess of expenses	Excess of budget	Risk sharing partnership	Collaboration

Source author

tools to reduce the probability of supply chain disruptions. However, such practice does not take aboard the ability of the supply chain to recover from the disruption and, moreover, to achieve superior performance on the basis of experience gained.

4 Discussion

Scientific papers identified reducing supply chain vulnerability and improving supply chain resilience as the key risk management objectives [7, 11]. Juttner and Maklan argue that low vulnerability does not mean high sustainability as far as there is no direct impact of risk management on system vulnerability [7]. It should be noted that the authors call identification and risk assessment "risk management". Such point of view is disputable. Firstly, risk identification and risk analysis are the key components of risk management. Secondly, risk prevention, its localization, distribution, compensation and other risk management methods improve supply chain resilience making the supply system less vulnerable, even if it is highly sensitive. Thus the purpose of supply chain risk management is to achieve optimum level of supply chain resilience, when the system remains sensitive restoring after disruption.

A new challenge to supply chain participants is posed by the fourth industrial revolution (Industry 4.0). It is changing not only the manufacturing industry but also the construction industry and its supply chains. Dallasega et al. argue that many issues in construction supply chains relate to distance between the actors. So the authors analyse synchronization between suppliers and the construction site using the concept of proximity. The main point is that Industry 4.0 technologies influence technological, organizational, geographical and cognitive proximity dimensions [5]. Brusset and Teller point out that supply chain participants need improved methods to review the complex factors that contribute to the development of resilient supply chain [1]. The managers must not only use information technology tools to integrate their internal processes but also use other supply chain management software to integrate their partners, including suppliers, customers, distributors, and logistics service providers. These efforts improve cooperation by sharing sales data and forecasts and allowing continuous inventory adjustments. Track and trace technologies using by logistic providers allow to alert them to events that affect service level.

Information sharing based on digital technologies makes it possible to enhance coordination in supply chains. Data exchange concerns the situation with demand, inventory, delivery terms, production schedule, available capacities, costs. However, despite the obvious advantages offered by information technologies, there are some disadvantages, for instance, information systems implemented by partners cannot be integrated. The research of communication between partners in supply chains has shown that the most part of the companies is focused on search for appropriate information from the various sources. In addition, suppliers are more open to information sharing while customers show the less readiness for an exchange of the data [14]. As a result, the supply chain has a low visibility and agility. Another important issue is a cost-benefit analysis. The experience of regional construction companies shows that economic motivation does not always give the desired effect due to the complexity of supply chain, geographical distance between different actors, lack of resource planning. Nevertheless, the expenses on supply chain readiness for change may be reasonable. Achieving the targeted supply chain resilience requires all parties to be involved. Moreover, it may require the supply chain firms to change internal business processes, a corporate culture and a practice of information sharing. Hence, the next stage of research might be related to developing optimization models of supply chain resilience to minimise total cost and the recovery time of a disrupted element of the supply chain.

5 Conclusion

Construction industry is characterized by high complexity of logistics, strict sequence of work, a need for coordination efforts of many suppliers and contractors. Construction is affected by a range of external factors, including climatic, economic, administrative and others. The challenge now is to make construction supply systems sufficiently sustainable. It means that supply chain must be able to thrive from disruptive events. Disruptions in construction projects are often difficult to foresee. The certain problem is how potential damage can be identified before a disruptive event occurs. This paper aims to find a way to detect supply chain risks in construction. The proposed analysis technique reveals the links between supply chain performance indicators and their impact on supply chain resilience. There are two important factors that allow achieving supply chain resilience. The first one is information sharing that enables visibility and collaboration with supply chain partners. The second one is analytics, i.e. use of intelligence and analytical tools to support supply chain and risk management functions.

References

1. Brusset X, Teller C (2016) Supply chain capabilities, risks, and resilience. Int J Prod Econ 184:59–68. https://doi.org/10.1016/j.ijpe.2016.09.008
2. Chopra S, Sodhi MS (2014) Reducing the risk of supply chain disruptions. MIT Sloan Manag Rev 3:73–81
3. Christopher M, Peck H (2004) Building the resilient supply chain. Int J Logist Manag 15(2):1–14
4. Erol O, Sauser B, Mansouri M (2010) A framework for investigation into extended enterprise resilience. Enterp Inf Syst 4(2):111–136
5. Dallasega P, Rauch E, Linder Ch (2018) Industry 4.0 as an enabler of proximity for construction supply chains: a systematic literature review. Comput Ind 99:205–225. https://doi.org/10.1016/j.compind.2018.03.039
6. Ivanov D, Hosseini S, Dolgui A (2019) Review of quantitative methods for supply chain resilience analysis. Transp Res Part E Logist Transp Rev 125:285–307. https://doi.org/10.1016/j.tre.2019.03.001
7. Juttner U, Maklan S (2011) Supply chain resilience in the global financial crisis: an empirical study. Supp Chain Manag 16(4):246–259
8. Li H, Pedrielli G, Lee LH, Chew EP (2017) Enhancement of supply chain resilience through inter-echelon information sharing. Flex Serv Manuf 29(2):260–285
9. Lukinskiy VS, Panova YN, Strimovskaya AV (2017) Integrated supply chain management: theories, models and methods. Logist Supp Chain Manag 3(80):40–56
10. Ponomarov S, Holcomb MC (2009) Understanding the concept of supply chain resilience. Int J Logist Manag 20(1):124–143
11. Sheffi Y, Rice JB Jr (2005) A supply chain view of the resilient enterprise. MIT Sloan Management Review 47(1):41–48
12. Shojaei P, Seyed HA (2018) Development of supply chain risk management approaches for construction projects: a grounded theory approach. Comput Ind Eng 128:837–850. https://doi.org/10.1016/j.cie.2018.11.045
13. Uvarov SA (2010) Problems in the theory of reliability of logistics systems and supply chains. Cargo Passeng Transp Serv 1:35–39
14. Yakhneeva IV (2013) Modeling and designing of delivery systems in conditions of risk. Biblio-Globus, Moscow
15. Yang T, Fan W (2016) Information management strategies and supply chain performance under demand disruptions. Int J Prod Res 54(1):8–27

Basic Principles for Evaluating the Enterprise Performance in the Digital Economy

E. M. Pimenova

Abstract In modern conditions, Russian business began to require a significant acceleration of processes of developing and making flexible management decisions. These tasks can be solved by taking as a basis accurate and most realistic results of assessing the profitability and business activity of enterprises (these two terms in practice are combined by the concept of "efficiency of an industrial enterprise"). The critical consideration of various methods used for evaluating the enterprises' effectiveness is relevant, theoretically and practically significant. This aspect determines goals, objectives and a compositional platform of this contribution. The article is devoted to the analysis of such concepts as "profitability" and "business activity"—the main criteria for evaluating the business performance. The system statistical information about enterprises' activities results in the Russian Federation is quite limited (closed commercial information). At the same time, using indirect indicators available in open statistical data, it is possible to assess how effective the activities of enterprises are. Therefore, the author uses methods of system and factor analysis, empirical, diagnostic, retrospective, predictive, stochastic, and other methods during the research. Special attention is paid to the relevance and necessity of using information technologies to ensure sustainable development of enterprises in the Russian market.

Keywords Efficiency · Profitability · Business activity · Method of calculating indicators · Information technologies

1 Introduction

In modern conditions of the Russian market development, the task of increasing the efficiency of domestic enterprises is solved at all levels of the national economy. Efficiency is the most important goal of financial management. Since the main goal of

E. M. Pimenova (✉)
Samara State University of Economics, Samara, Russia
e-mail: pimenova-elena@rambler.ru

S. I. Ashmarina and V. V. Mantulenko (eds.), *Current Achievements, Challenges and Digital Chances of Knowledge Based Economy*, Lecture Notes in Networks and Systems 133, https://doi.org/10.1007/978-3-030-47458-4_24

financial management is to maximize the wealth of owners, we can say that efficiency (profitability) is a very important determinant of productivity [9].

A loss-making business has no chance of surviving, and a highly profitable business is able to reward its owners with the maximum return on their investment. Thus, the ultimate goal of the company's operation is to maximize profits to ensure the sustainability and efficiency of the business in modern market conditions [12].

Sustainable development is a key issue for the modern society. In addition to corporate efforts, consumers should be responsible for protecting the environment and care of sustainable coexistence between future generations and natural ecosystems [2]. Sustainable development of the enterprise is predetermined by the efficiency of its operation. So, the main goal of any company's operation is a financial result, which does not give a lot of information about the business itself, either at a particular moment, or in the near or, especially, distant future. Business exists not only to get immediate benefits, but to ensure that the financial result will steadily increase in the future by a successful development scenario.

An enterprise can be considered as an effective one if three conditions are met simultaneously:

- the financial result achieved during the reporting period is higher than its level by competitors,
- the expected growth rate of the financial result in the near future is higher than the financial result of competitors in the reporting period,
- the company invests enough funds in the development of its activities in the longer term.

If at least one of these conditions is not met, the company cannot be considered as effective one. But this is a simplistic judgment. At the same time, a dynamically developing business is becoming more complex in modern conditions. For quick decisions management, timely and cost-effective purchase of components, effective inventory management, highly professional accounting, personnel support, and distribution of goods, industrial enterprises with their functional divisions require an increasing interdepartmental information flow. In this case, effective information systems will increase the efficiency of the entire enterprise, primarily by reducing costs and improving logistics [8]. Thus, the use of information technology to assess the business performance is extremely important, because it allows determining ways to maximize profits and ensuring the most significant increase in the operational efficiency.

2 Methodology

Digital economy is a separate segment of the economy that represents a set of investments aimed at increasing the economic efficiency of enterprises by developing new technological solutions, as well as by the further development of existing technologies. This economy type allows businesses to invest in IT in order to reduce costs of producing their own products, constantly develop their own portfolio of offers, and improve the production efficiency [4]. Innovative development of an industrial enterprise with the help of digital technologies is always aimed at increasing the efficiency of its activities and forms appropriate reserves for this growth. But in practice, different types of innovations have different effects on the economic efficiency, i.e. they determine different ways (reserves) to increase the formation effectiveness of innovative and technological potential of the enterprise. Innovations imply an increase in the production efficiency in some cases—by increasing the demand for products, improving their quality, expanding the production of goods with new properties; in other cases—by reducing costs as a result of using new, more advanced and cost-effective technologies. Nowadays, there are many types of innovations in the activities of industrial enterprises. And here it is most appropriate to identify and analyze those innovations that have a direct impact on the efficiency of an industrial enterprise through new technologies, organizational and managerial structures [7].

New approaches (methods) to assessing the enterprises' effectiveness appear regularly—these are the simplest methods of factor analysis, and methods of process-oriented analysis of efficiency. However, they should be used very carefully in practice, since not each of them can give correct, most realistic evaluation results.

An integrated approach is the most common method for evaluating the effectiveness of an enterprise. Its use is based on a certain system of indicators that must meet three main requirements: comprehensive coverage of the research object, interrelation of indicators, verifiability of indicators (clarity of the information base and algorithm for calculating indicators is required). In the economic literature, it is widely believed that there are two main criteria for evaluating the performance of an enterprise: profitability indicators, indicators of business activity.

At the same time, there are many profitability indicators, which are determined by the ratio of the economic effect (profit) and resources (costs). Business activity indicators are often identified with turnover indicators that illustrate the movement of an asset (liability) or its components in the studied period. By considering the dynamics of turnover coefficients, we can draw a conclusion: how many turnovers during the analyzed period are made by certain assets (liabilities) of the enterprise. The duration of one such turnover shows an indicator called the turnover rate (the inverse of the coefficient multiplied by the number of days in the analyzed period), which characterizes the time of conversion of the cost item into cash. When the turnover rate increases, we can conclude that the company's solvency is strengthened.

Business activity (turnover) indicators do not have universal regulatory restrictions, since these coefficients contain industry specifics. Most often, when considering the methodology for analyzing the business activity, it is proposed to calculate values of these indicators for 3–5 years, and consider the resulting dynamics: the growth of indicators indicates an increase in the intensity of the usage of enterprise resources. Sometimes it is suggested to make comparisons with the performance indicators of competitors, as well as with the average industry indicators.

A comprehensive analysis of performance indicators (profitability and business activity) is carried out with varying degrees of detail depending on the volume and reliability of the information available to the analyst. As a result of this comprehensive study, indicators that reflect all aspects of economic activity are simultaneously and consistently studied. On the basis of identifying deviations from a certain comparison base (plan, previous year, competitors' indicators), generalizing conclusions about the company's performance are made.

The enterprise is a single complex that includes various structural divisions and functions (based on the very essence of the enterprise). Therefore, in the course of evaluating the effectiveness of company's activities, we cannot limit ourselves to reviewing results of individual functions and divisions—this will not allow us to achieve a significant result. They can only be considered in relation to each other. At the same time, it is necessary to establish relations between individual structural divisions, and then determine performance indicators for each of them. Structural divisions may not have their own financial results (in contrast to the company as a whole), even if they are engaged in the sale of products, receive revenue. But this revenue is created by the work of the entire enterprise, not just by their work. Therefore, the issue of evaluating the performance of a company's structural unit separately is more complex than the same issue for the company as a whole, and it needs to be worked out separately.

3 Results

The issue of the economy digitalization in modern conditions of the Russian market (and the world economy as a whole) is one of the priorities. With the help of digitalization, the company can increase the efficiency of its activities by several percent. To achieve the maximum success in the market, the company needs to put digitalization as the basis for working out its development strategy [6].

The main incentive for the digital transformation of the Russian enterprise is the ability to withstand competition both in the domestic market and in the external one. Domestic enterprises have many problems; in particular, they lack funds to purchase assets (including digital ones) and to develop new technologies. We can note a significant increase in the volume of intellectual resources, but the level of their commercialization is very low. All this leads to an increase in innovation and technological lag behind other countries, preventing them from creating unique products (including digital ones).

The threat of losing competitive positions for Russian companies is extremely high. This is evidenced by the constant increase in costs to overcome market and administrative barriers, to establish links and contacts with partners and consumers. Obviously, in such circumstances, a digital breakthrough is difficult. But at the same time, it can be the way out of their crisis [5]. The main driver of digital transformation should be the increase in the efficiency of Russian enterprises. To do this, when working out a business development strategy, the company's management need the most detailed and reliable results of evaluating the profitability and business activity—it was previously noted that they are traditionally considered as the main efficiency criteria. The study of the economic literature and our own experience in the analytical work imply that we can achieve this maximum realism of the evaluation results only if we conduct it according to the algorithm presented in the Fig. 1.

At the first stage (defining the goal and objectives of the performance analysis), there is often a problem of correct, competent goal setting that would cover all the planned results, could justify the need for this analysis and would be its main reference point. In order to formulate the goal in the best possible way, it is necessary to clearly define: what does the company expect to achieve as a result of the performance assessment? Another stage (determination of methods and techniques for the performance evaluation) is complicated by the lack of a universal methodology that would be suitable for enterprises of all industries, all organizational and legal forms. In the market conditions, the existence of such a technique would not be particularly

Preparatory stage:

1. Defining the purpose and objectives of the performance analysis.
2. Selection of the research subject and object.
3. Development of a performance assessment program.
4. Setting deadlines and performers for the evaluation.
5. Determination of methods and techniques for the performance assessment.
6. Collection and preliminary assessment of initial information.

Main stage:

1. Analysis of the external and internal environment of the company:

a) assessment of the organizational structure of the enterprise management;

b) assessment of the impact of various external and internal factors on the company's performance.

2. Analysis of the company's results:

a) analysis of the main technical and economic indicators;

b) assessment of the main performance indicators of the enterprise that determine the effectiveness of its activities (revenue, income and expenses, profit).

3. Analysis of the efficiency of an industrial enterprise:

a) analysis of the dynamics of profitability and business activity indicators;

b) factor analysis of production profitability and sales profitability.

Final stage:

1. Integration of performance evaluation results.
2. Development of recommendations for improving the efficiency:

a) development of recommendations to address identified deficiencies;

b) justification of existing reserves for improving the efficiency;

c) control over the implementation of the developed measures.

Fig. 1 The flowchart for evaluating the effectiveness of the company's activities (*Source* author)

justified: its use would give superficial results without much specification. The solution to this problem for the company is to develop its own methodology for evaluating the performance (independently or with the involvement of third-party specialists) using digital technologies. In addition, it is possible to use methods developed by other companies in the same field of business activity.

Another difficulty is the collection and preliminary assessment of initial information. It is a time consuming process that requires analysis of all types of activities, and almost all employees are involved in its implementation. In order to avoid problems in preparing the necessary information, management should conduct orderly activities, avoid confusion in documents and information that moves between different departments (structural divisions).

The complexity of the main stage depends on the chosen methodology for evaluating the effectiveness. At this stage, additional costs may arise (payment for the work of third-party specialists who will perform calculations using methods unknown to employees of the enterprise itself, or especially labor-intensive ones). It is impossible to eliminate this drawback of the main stage. However, it should be borne in mind that these costs can be reimbursed in the future if the goals planned during the performance assessment are achieved, if the information received is perceived by employees. In any case, these costs are investments in the future, since they can improve the efficiency of the company and, as a result, strengthen the financial condition and stability of its operation in the future. The analysis itself is time-consuming and labor-intensive. It requires highly qualified employees, as well as the ability to work with raw data, formulate conclusions, and develop recommendations for increasing the efficiency. In order for the company's employees to be able to carry out the final stage of evaluation, it is necessary to send them to professional (re-)training, conduct the own training sessions. Special attention should be paid to training staff to perform analysis using digital technologies. All these measures together will help to achieve maximum results. To sum up, we note once again that there may be some difficulties in evaluating the business performance, which can only be solved by a high level of professional training and experience of employees who conduct the performance analysis.

4 Discussion

The topic of the digital economy is extremely relevant at the present time. Interest in it arose in 2017 during the rise of the crypto-currency profitability. Further, various structures, enterprises, and individuals became interested in this issue. Accordingly, they began to deal with digitalization issues at the state level: the program "Digital economy of the Russian Federation" was developed [10]. It stated that the state policy of the Russian Federation in the field of digital economy is focused on creating conditions for its full development.

However, the term "digital economy" is understood differently in different countries. Due to specific structural features of the economies of different countries, their strategic priorities in the digital economy development differ. In some countries, the service model of management is implemented, so for them the digital economy is perceived as a tool for improving the efficiency of transactional relations. Other countries (including the Russian Federation) have production and industrial specialization represented by mining and processing industries, therefore, for these countries, the digitalization process provides an increase in added value based on integration processes between various domestic and foreign enterprises. Thus, we can conclude that the digital economy is an addition to the material production, since it gives new competitive advantages to the national product and helps to generate higher added value. In addition, it is an independent sphere (it is on the basis of progressive digital solutions that a product is produced and various services are provided) [3].

The transition of Russian industrial enterprises to a new digital level (which is stipulated by the program "Digital economy of the Russian Federation" [11]) should have the character of a digital breakthrough: due to the development of its digital competitive advantages, our country is going to enter the group of leading economies in the world within the next 15–20 years. And while some "digital" indicators (e.g., the level of digitalization, the digital economy share in GDP, the degree of technology development and the degree of technological backwardness and other), our country is not included in the group of leaders, it is possible give a favorable prognosis for achieving government strategic objectives (this is indicated by the speed of digital transformations and the growth in GDP due to the digitalization over the last 3–5 years).

The financial component of digitalization plays a crucial role in an efficient resource allocation, economic growth, and job creation. Having an efficient enterprise to stimulate and maintain the growth of the state economy is a requirement for all countries [11]. In Russia, the construction of a digital model of the domestic economy is based on the industrial and production model of the economy. Therefore, the main leverage of the digital breakthrough is to obtain sufficient finance for the development and acquisition of digital assets by increasing the efficiency of industrial enterprises.

5 Conclusion

At present, the term "digital economy" has become quite often used in scientific and political circles. Many theoretical and methodological developments in the field of digital economy are reflected in the state strategic planning documents. Business performance is influenced by various factors: fiscal policy, institutional structure, social structure, environmental conditions, education and health of the country's citizens, etc. [1]. However, the quality of the regulatory system is still a key factor for the efficiency and productivity. In addition, important is the level of adaptation of the public policy to changes in the economy. Increasing the transparency of the public policy and simultaneously limiting bribery and corruption contribute to the improvement of business efficiency and productivity. Thus, the rule of law and the country's credit rating indicator (which assesses political, economic, and financial risks and emphasizes the credit quality of a particular country), as well as the high level of efficiency of domestic enterprises, are of great importance in the current development of the digital economy of the Russian Federation.

References

1. Bris A, Caballero J (2017) Business efficiency and productivity. Res Knowl. https://www.imd.org/research-knowledge/articles/february-2017-criterion/. Accessed 29 Feb 2020
2. Chang H-H, Tsai S-H, Huang C-C (2019) Sustainable development: the effects of environmental policy disclosure in advertising. Bus Strategy Environ 28(8):1497–1506. https://doi.org/10.1002/bse.2325
3. Dyachenko OV (2018) Production relations in the conditions of transition to the digital economy. Bull Chelyabinsk State Univ 12(422):7–18. https://doi.org/10.24411/1994-2796-2018-11201
4. Egorov DV (2017) Financial aspects of the digital economy. Banking 12:38–40
5. Galimova MP (2019) Readiness of Russian enterprises to digital transformation: organizational drivers and barriers. Bull Ufa State Pet Technol Univ Sci Educ Econ Econ Ser 1(27):27–37. https://doi.org/10.17122/2541-8904-2019-1-27-27-37
6. Istomina EA (2018) Methodology assessment of trends in the digital economy of industry. Bull Chelyabinsk State Univ 12(422):108–116. https://doi.org/10.24411/1994-2796-2018-11212
7. Khalabuda Y, Nikolaev M (2014) Increase of efficiency of industrial enterprises activity on the basis of innovations of various types. Procedia Econ Finan 16:299–302. https://doi.org/10.1016/S2212-5671(14)00805-3
8. Madanhire I, Mbohwa C (2016) Enterprise resource planning (ERP) in improving operational efficiency: case study. Procedia CIRP 40:225–229. https://doi.org/10.1016/j.procir.2016.01.108
9. Malik H (2011) Determinants of insurance companies profitability: an analysis of insurance sector of Pakistan. Acad Res Int 1(3):315–321
10. Program, "Digital economy of the Russian Federation" (approved by decree of the Government of the Russian Federation of 28 July 2017 no. 1632-R). http://static.government.ru/media/files/9gFM4FHj4PsB79I5v7yLVuPgu4bvR7M0.pdf. Accessed 28 Feb 2020
11. Roghaniana P, Raslia A, Gheysari H (2012) Productivity through effectiveness and efficiency in the banking industry. Procedia Soc Behav Sci 40:550–556. https://doi.org/10.1016/j.sbspro.2012.03.229
12. Sivathaasan N, Tharanika R, Sinthuja M, Hanitha V (2013) Factors determining profitability: a study of selected manufacturing companies listed on Colombo Stock exchange in Sri Lanka. Eur J Bus Manag 5(27):99–107

Digitization of JSC "RZD" Electric Networks Activity: Problems and Prospects of Development

T. B. Efimova and E. V. Chupac

Abstract Railway transport is characterized by complex technological processes and infrastructure which must be modernized through the implementation of digital technologies. Digital energetics which is the basis of infrastructure of the fourth industrial revolution plays a special role in this respect. The existing energetics must be transformed into the "smart" one. These transformations will concern customer services, the network itself, infrastructure, consumers, suppliers and other components. In the article, the authors consider issues of creating a new service allowing realization of the search of solution of analysis and discovering of "barrier locations" of electric power transmission, as well as the issue of new technological connections. For the purpose of modeling, the language UML was used allowing displaying of static structure and dynamic behavior of the developed service. The respective business-processes were studied, long-term perspectives were evaluated, and priorities of the existing industry problems which may be solved by introducing modern digital technologies were defined.

Keywords Digital technologies · Energetics · Technological connection · UML

1 Introduction

JSC "RZD" is a monopolist in railway transportation on the territory of the Russian Federation, it has an extensive network of electric power transmission lines designed for provision of transportation process and transmission of electric power to other legal and physical entities. Subject to the current legislation of the Russian Federation JSC "RZD" is considered to be a network organization, its responsibilities include ensuring reliability and quality of power supply of consumers, as well as technological connection of new consumers.

T. B. Efimova (✉) · E. V. Chupac
Samara State University of Economics, Samara, Russia
e-mail: TB_Efimova@mail.ru

E. V. Chupac
e-mail: chupack.ew@yandex.ru

S. I. Ashmarina and V. V. Mantulenko (eds.), *Current Achievements, Challenges and Digital Chances of Knowledge Based Economy*, Lecture Notes in Networks and Systems 133, https://doi.org/10.1007/978-3-030-47458-4_25

Transmission of electric power using network equipment is the JSC "RZD" additional type of business, it is realized by the specialized branch Transenergo and its regional departments of power supply. The latter fulfill functions of meeting the demand in electric power, its purchasing, transmission and distribution to JSC "RZD" divisions, as well as conduct activities in electric power transmission to consumers for the purpose of raising profits.

At present technological connection in the JSC "RZD" is a developing business demanding new approaches and revision of the existing practice. An important criterion determining the networks development trend is finding out limitations of the value of the power consumed from electric networks supply centres. This criterion is necessary for the purpose of further elimination of the mentioned limitations by means of procurement of additional power and modernization of network equipment. One of the tools of solution of the issue of analysis and discovering "barrier locations" of electric power transmission and the issue of new technological connections is creation of a new service, an information system, which will make it possible to take into account points of electric power supply and their technical characteristics. It will also enable analysis of unoccupied capacities, receiving reports about points of electric power supply and their power supply centres, including indication of capacities analysis for the purpose of technological connection, as well as that of power capacities expansion in view of their deficit.

2 Methodology

Various methods, such as SWOT, VCM (out-of-date models), BRP—a more advanced approach to development of business strategy from the point of view of Maciaszek—are used for the purpose of adequate systems planning [6]. Necessary information may be also assessed with the help of templates ISA. Using these approaches in complex allowed obtaining an effective solution; its implementation will enable an engineer of the department of electric networks operation of Administration of railway power supply to increase his labor productivity. With the help of the language BPMN the following business processes were singled out and analyzed as the ones necessary for building the system:

- analysis of activity of organizations and structural divisions of JSC "RZD" in respect to technological connection of the third persons to electric networks of JSC "RZD" and facilities of JSC "RZD" electric networks to facilities of network organizations,
- analysis of the presence of the volume of the transformer power free for technological connection of consumers by power supply centres, substations and distribution points,
- development of projects of technical terms to agreements about conducting technological connection,
- registration of subscribers concluding an agreement about power supply,

- introduction of electronic databases of agreements and reports in respect to indicators of reliability and quality of electric power,
- conducting work with applicants: compiling answers to appeals, carrying out interviews and questionnaires of consumers about the quality of services and maintenance, reception and processing of consumers' telephone appeals by the hot line, participation in development of events aimed at formation of a customer-oriented approach.

Further the basic requirements to the system from the suggested users, i.e. the department engineer, the chief and the administrator, were defined. After analysis of the obtained information and building the business model of requirements and the project of the technical specification a number of UML-diagrams were developed, the latter being precedents of usage with the respective documentation, succession of activities, conditions, components and classes. An important issue is architectural designing taking into account operation of customer and server components. Nodes and diagrams of UML expansion allowed realization of physical architectural designing.

3 Results

Administration is using a large number of various digital platforms which effectively interact with each other and constitute a unified information space [3]. As a result of analysis and modeling of the corresponding business processes and requirements of the suggested users, the project of the developed system was created (methodology UML was used). On the basis of this project a pilot project of the information system of the engineer of the department of electric networks operation was implemented. The program product being developed will allow automation of the following activities:

1. Reporting of points of electric power supply and their technical characteristics for the total of supply centres.
2. Carrying out of analysis of free capacities by the power supply centre on the whole and by each point of electric power supply separately.
3. Automatic receiving of reports by points of electric power supply and their supply centres with indication of capacities analysis for technical connection and expansion of capacities in case of their deficit.

Input data for the developed program product are: the name of the counterparty; position of the counterparty's manager; title of the point of the power supply centre; title of PS; border of balance participation; voltage level de facto, connected capacity; maximum capacity; stated capacity; normative losses; the category of reliability; title of the supply point; name of the consumer; category of the consumer; title of the consumer facility. Reports are considered to be output data. After authorization the main menu becomes accessible, here it is possible to choose one of the points—"Counterparties' reference book", "Create/change the power centre point", "Create/change the supply centre point", "Report", "Exit". The form "Counterparties' reference book" allows creation and looking through the list of existing counterparties. With the help of the screen form "The supply centre point" it is possible to create, change and look through the information about the supply centre point.

The screen form "Supply point" is designed for creation, changing and looking through information about supply points connected from the chosen point of the supply centre. When creating a supply point, first of all the point of the supply centre is chosen from the reference book, and the respective data are entered.

In order to form a report it is necessary to click the button "Reports" in the main menu and choose the type of report:

- report by the presence of free capacity which is implemented via the search of power supply centre points with the filter by the name of the supply centre point, the title of PS, the counterparty's name. For each supply centre point a report is formed according to the balance of maximum capacity by means of subtraction from the maximum capacity value by the supply centre point the sums of connected capacities by each supply point relating to the given supply centre point. The results of the report are formed in MS Excel,
- report by the supply centre points which is implemented via the search of supply centre points with the filter by the title of the supply centre point, the title of PS, the counterparty's name, the RF subject. On completion of search all data about the supply centre points with the function of choice of the data necessary for transmission are entered in MS Excel,
- report by points of supply which is implemented via the search of supply points with the filter by the title of the supply point, the category of consumer, the name of the consumer, the title of the supply centre point, the title of PS, the counterparty's name, the RF subject. On completion of search all data about the supply points with the function of choice of the data necessary for transmission are entered in MS Excel.

4 Discussion

At present serious dialogues about creating a unified platform for power supply on the whole are conducted: the publication by Kloppenburg and Boekelo [4] is devoted to digital media providing new types of interaction with energetic assets and users and a real situation on the power supply market. The group of researchers (Prodromos, Efthymiopoulos, Nikolopoulos, Pomazanskyi, Irmscher, Stefanov, Pancheva, Varvarigos) suggests their own proposals for integration of the platform S/W [8]. Methodology of creating an open data repository and the respective software for data processing developed for the purpose of consolidation of an hourly copy of the historical data set into the equivalent model of the North European network is considered in the work by Vanfretti et al. [10].

A detailed thematic research illustrating efficiency of the industrial Internet platform for industrial systems is described in the work by Wang et al. [11]. In the article by Van der Aalst et al. [9] characteristics of modern popular digital platforms are considered on the whole, valuable conclusions about their further development are made. Application of cloud technologies for joint use of resources, as well as characteristics of the platform are considered in the publication Cornetta et al. [2]. JSC "RZD" activities concerning power supply, peculiarities of creation of its own digital platform are considered in the article by Pogorelova et al. [7]. Some issues of secure digital up-to-date networks are considered in the publication by Al-Rubaye et al. [1].

Scientists of transport industry of this country underline the importance of modeling and forecasting in solution of strategic tasks [5]. It is supposed to create a digital seamless transport system on the basis of the existing services; any provided service will be accessible in any point of this system using a single click. For this purpose work in the field of big data, as well as creation of efficient digital platforms, financial and marketing business institutes for the JSC "RZD" are necessary.

5 Conclusion

As a result of the conducted research it is necessary to note that the idea of digitization of energetics consists in creating the information and telecommunication infrastructure, hardware and software ensuring a technological opportunity of application of solutions of the industrial Internet, as well as activities in improvement of regulatory and standard technical documentation and measures of personnel and informational support. The basis of digitization is automation, including introduction of intellectual reporting of electric power use [12]. The developed program product conforms to this concept allowing automated collection of the necessary data, their accumulation and making reports necessary for work and analysis by increasing efficiency of the administration engineer work.

References

1. Al-Rubaye S, Rodriguez J, Al-Dulaimi A, Mumtaz S, Rodrigues JJPC (2019) Enabling digital grid for industrial revolution: self-healing cyber resilient platform. IEEE Netw 33(5):219–225
2. Cornetta G, Mateos J, Touhafi A, Muntean GM (2019) Design, simulation and testing of a cloud platform for sharing digital fabrication resources for education. J Cloud Comput Adv Syst Appl 8:12
3. Efimov A, Chupac E (2018) The uniform corporate automated control system infrastructure of JSC "RZD" in terms of the functionality of the metering. Prob Dev Enterp Theory Pract 3:21–24 (in Russian)

4. Kloppenburg S, Boekelo M (2019) Digital platforms and the future of energy provisioning: promises and perils for the next phase of the energy transition. Ener Res Soc Sci 49:68–73
5. Lapidus B (2019) Improvements in energy efficiency and the potential use of hydrogen fuel cells in railway transport. Bull Res Inst Railw Transp 78(5):274–283 (in Russian)
6. Maciaszek L (2016) Requirements analysis and systems design. Williams Publishing House, Moscow
7. Pogorelova E, Efimov A, Efimova T (2019) Development prospects of a digital platform for Kuybyshev power supply central office. In: Mantulenko V (ed) international scientific conference global challenges and prospects of the modern economic development. The European proceedings of social and behavioural sciences, vol 57. Future Academy, London, pp 590–596
8. Prodromos M, Efthymiopoulos N, Nikolopoulos V, Pomazanskyi A, Irmscher B, Stefanov K, Pancheva K, Varvarigos E (2018) Digitization era for electric utilities: a novel business model through an inter-disciplinary s/w platform and open research challenges. IEEE Access 6:22452–22463
9. Van der Aalst W, Hinz O, Weinhardt C (2019) Big digital platforms growth, impact, and challenges. Bus Inf Syst Eng 61(6):645–648
10. Vanfretti L, Olsen S, Arava V, Laera G, Bidadfar A, Rabuzin T, Jakobsend SH, Laveniusa J, Baudettea M, Gómez-López JF (2017) An open data repository and a data processing software toolset of an equivalent Nordic grid model matched to historical electricity market data. Data Brief 11:349–357
11. Wang J, Xu C, Zhang J, Bao J, Zhang J (2020) A collaborative architecture of the industrial internet platform for manufacturing systems. Robot Comput-Integr Manuf 61:1–11
12. Zhukov O (2019) Aspects of digitization of electroenergetics and electrotechnical inspection. Eur Sci Assoc 5–2(51):121–125 (in Russian)

Analysis of Ethnic Tourism Resources and Content Placement on a Digital Platform

E. A. Solentsova, M. V. Krzyzewski, and A. A. Kapitonov

Abstract Analysis of touristic potential of territories located remotely from major megalopolises and infrastructural objects appears to have relevance in contemporary conditions. The research group from the Samara State University of Economics is working on a project oriented towards creating a digital platform providing wide access to reliable information about ethnocultural and historical objects, geographically isolated from infrastructural centers. With the support from the All-Russian public organization "Russian Geographical Society", a research project "Shentala—Beautiful Valley" is underway: an interactive ethnocultural and historical atlas of landmarks located in remote areas of northern Samara Oblast. The following article describes the potential ethno-touristic resources to be found in Shentalinsky municipal district of Samara Oblast. In the context of modern digital technology development, this will contribute to the formation of a creative economy capitalizing on the cultural heritage of the area. A basis for the concept of ethno-tourism is also established. Research materials will be integrated with the implementation of an ambitious innovative project: an informational and analytical internet resource servicing the territory of Samara Oblast, Russian Federation and the countries of the CIS.

Keywords Interactive Atlas · Digital platform · Ethno-tourism · Regional tourism · Samara region

---

E. A. Solentsova (✉) · A. A. Kapitonov
Samara State University of Economics, Samara, Russia
e-mail: solentsova2009@mail.ru

A. A. Kapitonov
e-mail: kapalex807@gmail.com

M. V. Krzyzewski
Samara Bartenev V.V. Multiprofile College, Samara, Russia
e-mail: k_mikhail_73@mail.ru

S. I. Ashmarina and V. V. Mantulenko (eds.), *Current Achievements, Challenges and Digital Chances of Knowledge Based Economy*, Lecture Notes in Networks and Systems 133, https://doi.org/10.1007/978-3-030-47458-4_26

1 Introduction

The relevance of this study stems from the necessity of promoting the tourist sector of the Samara region and integrating the investigated objects into the global information space. The authors have analyzed the existing ethnocultural resources located in Shentalinsky municipal district.

The study aims to investigate the ethnocultural potential of Shentalinsky municipal district. Research objectives set by the authors include: provide a scientific justification for existing objects of potential ethno-touristic interest located in the investigated territory; make a content analysis of materials pertaining to the expedition route; develop the structure of the core sections of the interactive platform.

The popularity of ethno-tourism in the modern world is growing steadily. Combined with the culture industry, it appears to be a relevant, well-demanded industry in many different regions of the world. Chinese scientists evaluate the potential of folklore tourism (ethno-tourism) in the country to be growing, albeit still in the beginning stages of development. The analysis of the existing digital platform for tourism in China hasn't yet reached a mature stage, argue the author [2]. International research in the area of ethno-tourism covers the most exotic regions of the world. Digital platforms are becoming a pertinent tool for popularization of architectural heritage in Brunei Darussalam [3], cultural interpretation and knowledge exchange between indigenous peoples of Northern Australia [4].

Today, ethno-tourism is becoming increasingly popular all over the world. Many different definitions of this type of tourism can be encountered in literature. This is caused by differing understandings of the nature of the phenomenon. Some authors equate ethno-tourism with nostalgic tourism, stressing the historical connections tourists may have with the visited destination. Others believe that ethno-tourism is primarily oriented towards visiting minor peoples least touched by civilization. This study provides its own definition for ethnic tourism, based on the assumption that every ethnic group possesses a unique culture, capable of attracting interest of tourists. Ethnic (ethnographic, ethnocultural) tourism is a type of cultural-cognitive tourism, the purpose of which is the direct exposure of tourists to the unique features of traditional culture and living conditions of different peoples (ethnic groups). This form of tourism entails visits to demonstrational communities that have retained cultural specificity, ethnographic museums, as well as national festivals.

2 Methodology

The leading research method of this study is the method of field research, followed by content analysis of data pertaining to the ethnocultural specifics of various communities found in Shentalinsky district. In the process of conducting this study, the authors have analyzed the various ethnocultural specifics of peoples of Shentalinsky district. Furthermore, they have identified communities which could be used as demonstrational when organizing and conducting ethno-tourist visits. The results of this study can be used when organizing ethnocultural tours in Shentalinsky District as well as other Districts of Samara Oblast.

3 Results

Projected Structure of the Interactive Atlas. The research group aims at creating a unified digital platform containing a collection of tourists visit routes and descriptions of ethnocultural objects, to be used concurrently in a single interactive space. Structural development of the atlas entails the creation of the core sections and pages for the widget. The main page includes: the logo, the title, the menu, social media buttons and a photo-and-video slideshow linked to the region's page. The widget's search bar contains zoom-in and zoom-out tools for the interactive map equipped with links to target pages with information resources. The continuous resource updating function allows the user to track upcoming events. For this purpose, a "subscribe to newsletter" button is included. Integrating the platform with the already existing informational and analytic platform operating in Russia, https://turatl.ru/, will provide users with high-quality service and constantly updated resources.

Methodological Justification for Ethno-Touristic Objects. All kinds of ethnocultural manifestations, both in the material and spiritual spheres, can serve as ethno-touristic resources. Of major interest are traditional crafts. They become the hallmark of the particular locality where they are found. Traditional crafts possess national specificity. Often, crafts go far beyond internal demand, becoming an industry. Tourists like purchasing various objects created by local craftsmen as souvenirs. That way they learn about the traditional crafts of that particular ethnic group and the specifics of their material culture.

The most important place in traditional material culture is occupied by the dwelling. The dwelling is the key part of the life-support system of any ethnic group. The traditional dwelling is very closely connected to the surrounding nature and environmental conditions. It clearly reflects the material culture and social structure of a population. The building materials, the construction techniques, the placement of household buildings, the peculiarities of interior design—all this, and more, is thoroughly infused with ethnic specificity. Even today, when construction technologies are becoming modernized and traditional building materials are being replaced with modern ones, ethnic specificity is often preserved in regards such as interior and exterior planning and design. Tourists may visit residential buildings that have been converted into museums as well as traditional dwellings that still serve in that capacity for the locals—of course, with permission from the inhabitants. Rural residents are known to be more open and hospitable than urbanites. The term "urban jungle" was coined for a reason—large cities harbor an atmosphere of alienation, atomization, universal mistrust. Because of this, ethno-tourist attractions do not only allow visiting traditional dwellings, but also offer lodgings in such places. Many tourists desire a break from the commotion of the city and wish to immerse themselves into the atmosphere of a specific ethnic group's traditional culture. They would prefer a modest dwelling built using traditional technologies to a lavish hotel.

The size and shape of the dwelling, the exterior design and interior decoration—all this piques the curiosity of an inquiring tourist. Of major interest to tourists is the national costume. The traditional costume reflects the aesthetic values of an ethnic group, serves to help in understanding its worldview. Everything from the shape and cut of clothing to its color, has significance. Various costume elements often have an ancient history going back centuries. Clothing has always served as method of visual identification of a person's age, social status, occupation. Clothing for everyday and festive occasions; men's, women's and children's costumes—all have their own unique characteristics. Furthermore, traditional costume even within a specific ethnic group contains local variations. Traditional clothing serves as an important source for distinguishing between different groups within a specific people. The women's ethnic costume is particularly persistent. As a rule of thumb, it serves better to highlight national specificity, is conserved better and lasts for longer in everyday use.

As it is known, traditional costume usually also includes, aside from clothing itself, footwear, headwear and accessories. Some of those elements are phased out of everyday use more quickly, others persist for longer. Headwear, footwear, various ornaments—each of those elements is unique in its own way. Often tourists can purchase some elements of the national costume as a souvenir, sometimes the whole costume. Such a purchase can be made in a traditional rural settlement of that ethnic group. Moreover, today, modern factory-produced clothing made in a traditional manner is sold in large urban stores in various countries. Residents of some rural locales often wear such clothing specifically to entertain visiting tourists. It is also worn by various folklore groups during their performances. Such clothing retains the national "flavor", even though it doesn't reflect the specificities of individual local groups within the ethnicity.

The material culture of any people is richly reflected in their cuisine. National dishes can tell a lot about an ethnicity's history and about the activities of its members. Staple foods used in traditional cuisine, cooking methods, meal times—all of this is particular to each ethnicity. Many tourists seek to learn more about a specific ethnicity through their national cuisine. Moreover, tourists find traditional cuisine attractive for using organic components, lacking modern chemical additives. Tourists try various national dishes and beverages with great interest. Then, upon returning home, they sometimes prepare those foods themselves, using traditional recipes.

The material culture of an ethnicity is also vividly reflected in its traditional means of transportation. Tourists enjoy not only observing such means of transportation, but also traveling using them, for both short and long distances.

Religious structures of various peoples are also used as ethno-touristic resources, what makes this type of tourism adjacent with religious tourism. By synthesizing the two types of tourism, it's possible to develop a combined form of ethno-religious tourism. Such a form of tourism aims at acquainting tourists with various peoples, their religious beliefs and houses of worship. Another name for this type of tourism is ethno-confessional tourism. When conducting this form of tourism, one should take into account the religious specifics of various ethnicities. This includes the existence of sacred places from which outsiders are barred, as well as limitations on communicating with other religious groups, which are in place in some communities.

Of interest, pertaining to planning and development of ethnographically oriented tours, are places that are linked to former religious beliefs of a particular people but have ethno-confessional lost their sacral character in modern times. No less important, when organizing ethnic tours, are the features of traditional spiritual culture which are enshrined in folklore, festivals and rituals. Some aspects of spiritual culture have more persistence compared to attributes of material culture, many of which disappear among various peoples due to modern processes of economic development. Folklore is a vivid expression of a people's spiritual culture, their worldview. Especially valuable for the touristic image of any territory are folkloric narratives dedicated to that area. In particular, tourists enjoy listening to legends describing the origins of local geographical names, historical tales, stories about fantastical creatures that allegedly inhabit or have inhabited the area. Legends about venerated animals and plants also serve well to illustrate the traditional worldview of a people, give a vivid account of the relationship between humans and nature. Some works of folklore have ancient origins and can be traced back to the oldest myths and legends. At the same time, while retelling folk narratives, storytellers often modify them by adding new elements, which are brought about by changes in modern society and, consequently, culture as well.

Particularly popular among tourists are various ethnic festivities. Such colorful events give participants a great opportunity to exercise their ethnic consciousness, demonstrate a particular ethnic group's role and place in global culture. At those festivities, one can simultaneously acquaint themselves with many different ethnographic peculiarities: national dress, cuisine, crafts and, most importantly, national music and ethnic dance. Even if a song is being sung in a completely alien language, performers' voices and inflections, sounds of ethnic musical instruments are still memorable. And how much interesting information about national culture can be learned from traditional dance! Ethnic dances reflect the soul of a people, the peculiarities of a national character. Dance movements often carry information about the traditional material culture and occupations typical for the particular people. Members of various ethnic groups often perform traditional dances specifically for visiting tourists. And the tourists do not simply observe; sometimes they join the dancers.

Analysis of Potential Ethnocultural Objects of Shentalinsky District of Samara Oblast. Shentalinsky District is located in the north of Samara Oblast, bordering the Republic of Tatarstan. In the Middle Ages these lands belonged to Volga Bulgaria; afterwards they were conquered by the Mongols and became part of the Golden Horde. Upon the fall of the Golden Horde, the area passed on to the Kazan Khanate. In the 16th century, after the Russian conquest of Kazan, the land was incorporated into the Russian state [6].

Contemporary toponymists trace the origin of the name "Shentala" from Turkic languages. "Shentala" in the Bulgar language means "new land" ("shene", "shane"—new, "tala"—steppe, land). However, there is another, more recent theory which has gained some traction in ethnographic literature. It claims that the word “Shentala” comes from German: “schönes Tal”—“beautiful valley”. Allegedly, the phrase was exclaimed by a German engineer who was working on railroad construction in the area sometime during the early 20th century. However, one finds more compelling the opinion of professional researchers, which states that the name of the area has no relation to the German language, instead emerging much earlier [1]. In the first half of the 18th century, construction of the Novo-Zakamskaya fortified line took place there. Some traces of those fortifications can still be found in the district.

One of the peculiarities of Shentalinsky District is the ratio of ethnic groups residing in its territory. In Samara Oblast overall, Russians significantly outnumber other ethnicities—according to the 2010 census, Russians make up 85.6% of the total population. They are followed by Tatars (4.1%), then Chuvash (2.7%) and Mordvins (2.1%). In Shentalinsky District, Chuvash people are the predominant ethnicity. According to the same census, their population is 5,510 or 34.2% of the total population of the District. Russians make up 28.1% of the population, or 4,525 people. Russians are followed by Tatars—3,124 (19.1%) and Mordvins—2,527 (16.1%). Therefore, it appears necessary, as a matter of priority, to investigate the ethnocultural characteristics of Chuvash people of Shentalinsky district, which must be taken into account when planning ethno-touristic routes.

The Chuvash are a Turkic people. Chuvash settlements in the area were notable for their lack of planning. People built houses to “bunch up” together with others of the same family, creating clannish clusters called “kurmysh”. This tradition has not entirely become a thing of the past: even now some Chuvash villages have no street planning, houses are unordered.

Both dwellings and other household buildings were constructed from wood, sometimes adobe, in relatively rare cases—stone. Like other peoples of the area, the Chuvash used to make thatched roofs covered with wooden poles. Today, slated or metal roofs are preferred. In the past, some Chuvash households had sheds constructed out of wattle. Bunk beds (nary) were a characteristic interior element of a Chuvash house, having many functions: a sleeping place, a shelf for storing bedding and laundry, and a table for entertaining guests. In winter months nary were also used to shelter young livestock. The house had many items of domestic make, woven or embroidered.

In older times dwellings had rather austere exteriors—the tradition to decorate window shutters only emerged in the 20th century. Today many houses are decorated with fancy ornaments, while embroidered washcloths and other domestically produced items are often preserved inside.

The ethnic cuisine prioritizes dishes made with rye flour: sour bread, meat and fish pies (huplu), various forms of flatbread. At the same time, the traditional menu also features many meat dishes made from beef, mutton, horse meat and fowl. The most famous Chuvash dish is sharttan, which serves as a traditional delicacy. It is a type of sausage made from a sheep’s stomach and filled with meat, garlic and spices. Dairy products also occupy a major place in the ethnic cuisine. There is also a traditional alcoholic drink—sara, beer made using rye or barley malt. This drink has an important role in Chuvash ritual culture.

Regarding the national costume of Chuvash people living in Samara Oblast, one must note that it possesses a number of distinctive local traits. Of particular interest is the traditional dress of Chuvash women. It consists of a white tunic (white being considered a sacred color among the Chuvash), an apron and a wide variety of accessories: necklaces, pectorals, ornaments worn on the back and the belt. Clothing was richly decorated with geometric shapes. A woman's status as well as the amount of children she had, etc., could be determined by the ornaments she wore. A diverse variety of materials was used in creating the ornaments: coins, beads, even coral and cowry shells, sewn together with lace and red calico. Fascinating is the amount of coins used to decorate the Chuvash female costume. Young girls wore masmak—a type of headgear consisting of a cloth frontlet decorated with coins. Older girls could wear a small hat called tuhya, which sometimes had a pointed top. Tuhya were decorated with beads and silver coins. Headgear worn by mature Chuvash women was called hushpu, and the various groups within the Chuvash ethnicity had their own local variations of it. Chuvashs of Shentalinsky District belong to the Transvolga group. Their hushpu are shaped somewhat like a helmet, and are richly decorated with coral and silver coins of varying sizes. The frame of a hushpu was made from canvas and rough woolen string. The top was left open, and a cylindrical shape would later be sewn onto it. A characteristic detail of this type of headgear is the tail, which was also decorated with coins or beads.

Men's traditional costume included tunic and trousers, as well as outerwear consisting of a canvas caftan and a broadcloth coat. In the winter season warm woolen coats were worn. Bast shoes or lapti were a traditional footwear, leather boots were also worn. Researchers testify that in the first half of the 20th century, the traditional Chuvash male costume was already practically phased out of everyday use. The female costume remained in use for longer. Some Chuvash people still keep old traditional costumes, which are worn alongside with modern, stylized clothing during national festivals.

Of major interest is the spiritual culture of Chuvash people. Among Chuvash people, including those living in Samara Oblast, there are practitioners of different religions. Most religious Chuvash people follow Orthodox Christianity, but Samara Oblast also has several villages where Christians coexist with Chuvash Pagans. These Chuvash Pagans are often called "unchristened Chuvash". Meanwhile, they refer to themselves as "true Chuvash".

The chief god of the Chuvash pagan pantheon is Tura (Syulti-Tura). There are also other gods: Pigambar—the patron of livestock, Hevel—the sun deity. However, even back in the 19th century researchers noted that many functions formerly attributed to other Chuvash deities were moved under the chief god's jurisdiction. This is partially confirmed by field research findings made by the authors of this study. Some Chuvash pagans they met, for example, in the Staroye Afonkino village of Shentalinsky District, affirmed that Tura is the only god.

The Chuvash pagan religion also has negative forces. The main adversarial figure is Shuytan, who lords over the dark powers. Another famous pagan character is Kiremet. He was believed to have the ability to cause harm, so prayers were held in specifically designated places in order to placate him.

Researchers note that paganism facilitated the preservation of many elements of ancient Chuvash culture. Even today pagan Chuvash people participate in various traditional rituals that were observed by their ancestors many centuries ago. Throughout generations, knowledge of traditional ritual culture is passed down from the old to the young. This fact can be vividly attested by Chuvash funerary and remembrance rites. Pagans sometimes have their own cemeteries, which can be located either nearby Christian ones or far apart from them. If Christian Chuvash cemeteries are virtually indistinguishable from those of other Orthodox peoples, pagan Chuvash cemeteries are quite distinct. A distinctive Chuvash ritual involves erecting a peculiar grave pillar called yupa. Yupa were usually made from wood: men's graves were marked with pillars of oak, while women were honored with linden ones. Sometimes those could be made from stone as well. Yupa can be found decorating the graves of Chuvash pagans to this day. In older times, those pillars were only erected in Autumn—specifically in October, for which the month is accordingly named "yupa" in the Chuvash calendar. Today yupa are erected on the fortieth day after the funeral, or on the day of the funeral proper. There is, of course, a clear connection between Orthodox Christian influence and the change in pillar erection times. However, that didn't affect the substance of the ritual, which has retained peculiar elements, such as "inviting" the deceased away from the cemetery, crafting the pillar, dressing it in the deceased's clothes and, finally, installing the pillar. The funeral was followed by a memorial vigil, which featured special songs and dances, preparation of a ritual porridge and a symbolic "send-off" for the deceased, accompanied by jumping over a bonfire. The mourning for the deceased was reflected in special "yupa yurri" songs. Chuvash pagans rather frequently have memorial gatherings for their deceased.

To this day, Chuvash pagans venerate ancestor spirits as well as guardian spirits of certain locations. Chuvash Christians celebrate Orthodox holidays and festivals. In rural areas, celebrations dedicated to a particular church or its patrol saint are popular. A part of the Chuvash population of Samara Oblast confesses Islam. This demographic was strongly influenced by Tatar culture and language. Because of this, other Chuvash people usually call Chuvash Muslims tatars, sometimes Chuvash Tatars. Like some other peoples, the Chuvash people believed that birch trees should not be planted in one's home; they used to say that "birch brings despair". There are stories about people who have violated this proscription, proceeding to die early or go bankrupt.

Chuvash folklore also contains descriptions of supernatural creatures. One of such creatures is the arsuri—the Chuvash equivalent to the Tatar shurale. Like the shurale, the arsuri likes to frighten people and can even tickle one to death. However, he tries to keep his distance from human settlements because he is extremely afraid of dogs. Today, some Chuvash people identify the arsuri with the devil. Old people describe him as a nimble creature covered in fur, sporting a tail and a pair of horns. The word "arsuri" is now also used to describe a type of person who is both cunning and arrogant, shameless. Regarding marriage, Chuvash people have never followed strict guidelines when it came to nationality, religion or age of the bride or the groom. The bride could be several years older than the groom: while boys were married off at an early age of 15–17, girls could often wait until 25–30. The traditional Chuvash wedding was distinguished by a number of unique features, even bride kidnapping used to be practiced. It should be noted that the Chuvash people living in Shentalinsky District have managed to preserve many unique features of their traditional culture. This becomes particularly apparent during national festivals, where folk groups perform.

Traditional Chuvash holidays are currently undergoing a resurgence. One of those holidays is the Surhuri, which is dedicated to welcoming the New Year. According to tradition, the New Year's Eve was when one was supposed to divine their fortune for the upcoming year. If older family members usually tried to predict future harvests, the girls were of course more interested in the topic of marriage. The most popular divination method was called surhuri, and that name later came to encompass the entire cycle of winter celebrations. The method goes as follows: a young woman enters a dark barn and grabs a sheep's leg. The sheep's color determines the future husband's hair color (black sheep means a brunette husband, a white one means he is going to be blonde).

Savarni, a holiday similar to the Russian Maslenitsa, is another festivity that is being held. The specifics of this holiday varied in the different locales inhabited by the Chuvash people of Samara Oblast. For example, in Shentalinsky and other northern Districts the holiday was based around riding troika sleighs around the settlement and chanting. An important place among the celebrations of the spring/summer cycle was occupied by the Uyav. During the Uyav, young people would participate in circle dances (khorovods), make merry and organize various games. In recent times this holiday has taken on a new character.

Shentalinsky District is home to a few very characteristic Chuvash settlements. One of them is the village of Tuarma, where the "Ullah" ("Get-togethers") folk ensemble is based. While other folklore groups tend to wear modern costumes made to resemble traditional ones, during their performances "Ullah" use authentic vintage clothing inherited from their ancestors. The "Ullah" ensemble has gained fame reaching far beyond their home district. The artists of this collective have performed in a number of cities in France, Belgium, the Netherlands, Germany, Switzerland. It is remarkable that Western audiences have had a chance to see and hear Chuvash performers. Maybe one day foreign tourists will visit "Ullah" themselves and hear beautiful Chuvash songs in the performers' natural environment. In 2003 "Ullah" was given a honorary title, which is awarded for merit in preserving and popularizing traditional culture—"artistic folk collective".

Another fairly unique village of Shentalinsky District is Staroye Afonkino, which is populated by Chuvash people of all religions: Christian, Muslim, Pagan. In the village cemetery, Orthodox crosses, Pagan yupa pillars and Muslim gravestones can be found right next to each other. This village is rather famous: according to the locals, Staroye Afonkino is frequently visited by various scientists. For the sake of the local population's welfare, there must be efforts to encourage touristic visits to this multi-religious Chuvash settlement.

The Russian population of the area used to practice a wide variety of crafts; some of them have been preserved to this day. Village industries included sled-making, chest-making, wheel-making, rim-making, etc. Houses were built mostly of wood. In the second half of the 20th century, several different kinds of front porches became popular: an open platform under a canopy, a semi-circular platform, or a veranda. Windows, pediments, cornices, porches, sometimes even gates were richly decorated with carvings. Among popular carving motifs we can name plants (flowers and leaves), birds and animals, and "pharaonki"—anthropomorphic images depicting human-fish hybrids. Solar symbols were also depicted occasionally.

At the turn of the 20th century, the characteristic dress of a Russian woman consisted of the "pair"—a skirt and a blouse. Married women wore hair-covering hats (volosniki), while during festivities the kokoshnik was a popular type of headgear. Various scarfs and shawls were also worn. Women's outer clothing universally consisted of fur coats (shubas and tulups), chapans. Various accessories were widely used in the traditional Russian costume: earrings, necklaces, hand accessories, clothing ornaments.

Researchers testify than the traditional costume of a male peasant had less variety to it. At the same time, it was more resistant to changes. The main elements of the male costume were a canvas shirt and pants of home-spun cloth. Sharovary or broader pants worn to be tucked into boots could also be encountered. Visored caps (kartuz) were the ubiquitous headgear, replaced in winter with the warm ushanka hat. The influence of neighboring Turkic peoples can be attested by the fact that some Russians frequently wore such Eastern headgear as tubeteikas or canvas hats.

Winter clothing consisted of a fur coat, a scarf and mittens.

Slant-weave bast shoes served as the traditional everyday footwear for both men and women. In colder times those were worn over onuchi—long and thin strips of thick woolen or hempen cloth, used as footwraps. Bast shoes were still worn up until the middle of the 20th century. Today they are only sometimes used in folk performances. During celebrations, men wore leather boots. In the winter months, felt valenki and bakhily (boots made out of sheep skin) were worn. Some older people still have old wooden lasts used to felt valenki.

With great enthusiasm the locals observe Christian holidays: Easter, Christmas, the Baptism of the Lord. On Christmas, young children would go caroling around the village, and were rewarded with gifts (usually various sweets) and small amounts of money. Festivals with ancient pagan pasts are also celebrated.

Maslenitsa, the eighth week before Easter, is the festival dedicated to the transition between winter and spring. The main features of this festival were thin Russian pancakes (blini) which symbolized the Sun, as well as the effigy of Lady Maslenitsa. It was made out of straw, dressed in women's clothing, paraded around the city, then burnt. This celebration still enjoys popularity in Russia. The birch tree is one of the symbols of Russian culture. It's believed to be the guardian tree of women and young girls. At the same time, in the ritual culture one could notice a connection between the birch tree and the world of the dead. According to older local residents, many birch bundles were made during the Pentecost celebrations and placed in various places.

Among the summer celebrations, the most famous and popular one is the Ivana Kupala, which falls on the 7th of July—at this time, peasants were preparing to gather harvests. On Ivana Kupala, Russians all over the country would burn bonfires, collect herbs for protection against the dark forces, use wreaths to divine futures and purify themselves with water. The bonfire-lighting part of the ritual is not practiced in Samara Oblast However, there is a tongue-in-cheek tradition of dumping water on people from a well or the window of your house, as a means of "purification". Today, Ivana Kupala is still celebrated in rural areas.

Holidays dedicated to local churches and their patron saints were also celebrated. Russian settlers in Shentalinsky District would usually settle next to Chuvash people and Mordvins, less frequently next to Tatars. Those settlers mostly came from Kazan, Penza and Kursk governorates—settlers from Kursk, for example, founded the village of Altunino. The material culture of Tatars was very multifaceted. They practiced both crop farming as well as large-scale cattle breeding. Various crafts and industries were also practiced: leathermaking, felting, even jewellery. Also, very widespread was the art of embroidery.

Usually, the primary material used in the construction of dwellings was wood. Huts and household buildings made of wattle also existed. In modern times, brick houses have become a frequent sight. Sheds were sometimes built out of stone, some of them have stood the test of time. Adobe was rarely used. In older times, roofs were usually thatched and covered with wooden poles. These days, thatched roofs have been phased out in favor of slate and metal. The interior was characterized by benches, chairs and shelves installed in a peculiar way. Like Chuvash people, Tatar houses used to have bunk beds, which simultaneously served as a sleeping place, a dining table and a place to entertain guests. Bedding would be placed into a nearby closet when the bed was used for gatherings.

Tatar houses have managed to retain plenty of vivid traditional character. The colorful exterior, the decorated windows, the many-colored fabrics—all this lets one immediately identify a Tatar house. Inside, the walls are often decorated with a variety of embroidered cloths, prayer rugs, homespun fabrics. A richly decorated quotation from the Quran would be placed in the most visible spot of the house.

Tatar traditional clothing is distinguished by combinations of vivid colors. Both the male and female ethnic costumes had the same basis: tunic, loose pants, a fitted kamzol and a shirt with a close-fitting collar (beshmet). A woman's tunic would additionally be decorated with frills and appliques or a special pectoral ornament called izu. Meanwhile men preferred to wear a comfortable and loose-fitted, shawl-collared robe over their kamzol. In winter, fur coats were worn. Tubeteika, the flat-topped headgear worn by Tatar men, has become very famous. As for women's headgear, they had many local variations depending on a particular Tatar group. Most popular were towel-shaped headgear called tastar; little hats called kalfak, which were decorated with pearls and gold-embroidery; and coverings called erpek, which are decorated using tambour embroidery. Takja, the young girl's hat, had a soft and flat top, but a rigid frame. It was usually made from velvet, usually green or maroon, and decorated with ornaments.

Traditional footwear consisted of leather boots—ichigi, as well as embroidered shoes, both soft- and hard-soled. Bast shoes were also worn. Ethnic tatar food is stunningly diverse—it contains a vast variety of flour-based, dairy and meat foods. Flour was used extensively to make bread. Different forms of dough: yeast, unleavened or sweet, were used to make pies (belesh, echpuchmak) with a variety of fillings—potato, carrot, beetroot, meat, etc. Meat was a common ingredient in soups and main courses. Beef, mutton and horse meat were consumed. The latter, for example, was salted and used in sausages. The most popular drink is tea, which is served hot with milk or sour cream. The most famous sweets are chak-chak and shelpek.

In Tatar settlements of Samara Oblast one can clearly distinguish the local peculiarities of traditional spiritual culture. Various folklore narratives are passed down from generation to generation. Tatar folklore closely resembles that of other Turkic peoples. Tatars have a story about the White Snake, who is the queen of snakes. One must put a handkerchief underneath her, and, according to belief, she will shed her skin onto it. The rook is considered an untouchable bird and the appearance of a raven is believed to bring bad luck. In the past, the birch tree was also considered an unlucky tree which was not to be planted near one's dwelling. Today, this belief is being forgotten and one can find birch trees growing nearby Tatar houses.

Tatar folklore also features the dragon Azhdaha, who comes from the mountain and kidnaps children. He was a popular bogeyman for frightening disobedient children. Another adversarial mythological character is the yuha elan ("serpent"). This snake-like creature seduces men by turning into a woman and can even have children with humans. But one day it inevitably turns back into a serpent and strangles the man to death. The backyard can become occupied by the albasty, who loves to ride horses and can exhaust one's horses to death. There are also benevolent entities, such as the iy ana ("mother of the house"). According to lore, she must be given an offering of fresh bread before she accepts a new person into the house.

There is a resurgence in the tradition of creating elaborate family trees. Today, even school-age kids are encouraged to actively participate in making them. The most colorful Tatar festival is, without a doubt, the Sabantuy—the "plough festival" dedicated to the planting of summer crops. In the old times, the festival had no set date, as the exact planting season varied from year to year based on weather and the state of the soil. Nowadays, Sabantuy is celebrated in the early summer, in June. Sabantuy involves various amusing competitions: sack races, bobbing for coins in a bowl filled from kefir, pillow fighting atop a log, etc. Of course, such a celebration cannot go without music concerts, where folk collectives perform next to pop musicians.

The village of Deniskino could serve as Shentalinsky District's demonstrational Tatar community. The village has two mosques, one build in 1909 and another founded at the turn of the 20th century. The folkloric ensemble "Chulpan" ("Morning Star"), which was awarded with the title of "artistic folk collective" in 2003, is based in Deniskino. Deniskino celebrates Sabantuy every year, and has been doing it for many decades now.

Unlike the other major ethnic groups of Shentalinsky District, Mordvins belong to the Finno-Ugric language group. Mordvins of Shentalinsky District, like most Mordvins of Samara Oblast, belong to the Erzya sub-ethnicity.

The basic dwelling of the Mordvins of Samara is the wooden hut (izba). In the past, the roofs used to be thatched, then covered with wooden poles. As the older residents of Mordvin settlements reminisce, in the summer the thatching straw was removed and fed to the cattle, the roofs were then re-thatched again. Straw was also used as kindling. In modern times, very few traditional Mordvin thatched-roof adobe huts are left. Houses used to have few windows, but with time their number increased and they began to be decorated with fancy carvings. Inside, houses often contain old objects made from wood, clay or bast. In the old times, houses used to contain bunk beds as well as the high bunk constructed above the hearth (polati)—both were slept on. In the poorest houses, people slept directly on the floor on beds of straw.

The basis of the Mordvin national costume, both male and female, was formed by a tunic called panar. Women's headdresses were rather distinctive—for example, a sort of bonnet ("soroka") distinguished by a tall headband attached to a rigid frame. Slant-woven bast shoes, made in a distinctive Mordvin style, were the characteristic footwear both for men and women. Such shoes were made from five or seven strips of bast, and had triangles of inverted bast in the front. Bast shoes were usually worn with white legwraps made from thick canvas cloth. Even today, some Mordvin villages still have people who can weave traditional bast shoes.

Mordvin boots were made with folds on the collar, and had a narrowed shoebox. Both the Erzya and Mokshya ethnic groups favored "creaky" boots, with tall heels. Wealthy people wore boots made of goatskin or morocco leather. To celebrations and special occasions, Erzya women wore half-boots—koty, decorated with brass medallions. At the turn of the last century, another version of koty shaped like deep galoshes became popular. Valenki were a popular everyday footwear for the winter months. Today, some elderly women still own certain elements of the traditional costume. During national holidays and folklore festivals, stylized pseudo-traditional costumes are worn.

The Mordvin cuisine was mostly based on the fruits of agricultural labor, with most staple foods being plant-based. Bread could be made from any kind of flour—rye, wheat, sometimes even oat or barley. To this day, pancakes remain the most popular and beloved dish (Mordvin pancakes are particularly thick and soft). Mead has an important place in the Mordvin diet. The oldest alcoholic drink consumed by Mordvins is braga. There are also plenty of non-alcoholic, ancient drinks such as kvass, diluted sour milk, birch and maple juice.

The traditional Mordvin spiritual culture also has a number of distinctive characteristics. The folklore of the Mordvins of Samara Oblast is peculiar. Mordvins are an Orthodox Christian people; however, one could notice many vestiges of ancient beliefs in their culture. Mordvins also have a belief that planting a birch tree next to one's house brings misfortune.

In Mordvin villages, one can hear stories about supernatural creatures that have allegedly inhabited that area in the past. According to local legends, any banya (sauna) is inhabited by a Banya-ava (Banya-avaka)—the guardian spirit of the building. Using the banya after midnight was prohibited, otherwise the Banya-ava could strangle the offender. However, the creature also had a more positive side to it; after all, the banya was a very beneficial place that allowed people to keep clean, warm and healthy. In the older times, women gave birth in the banya and Banya-ava was believed to help during birth. Even today, in many Mordvin settlements of Shentalinsky District the bride is bathed in the banya on the day before the wedding. This ritual signifies saying goodbye to the mistress of the banya, and purification by water. Some places still have groves containing sacred trees, not that long ago prayers were still being held there.

One of the most colorful Mordvin festivals is the Mastorava ("Mother Earth"), which may be called an equivalent of the Tatar Sabantuy. Participants wear traditional clothing, sing ethnic songs and perform local dances. The tradition of celebrating the Mastorava was revived in the Shentalinsky District in 2018.

In Mordvin settlements, traditions are still observed when it comes to celebrating various family holidays. For example, as was mentioned above, it was customary to bathe the bride in a banya before the wedding. The wood and the water for this had to be brought inconspicuously from a house where the young woman didn't live. After the wedding ceremony the now-wife would give the attendees gifts: ribbons for women, handkerchiefs for men. Then there was a solemn procession escorting the newlyweds to the husband's house, accompanied with special chants.

Mordvins, like other peoples residing in Shentalinsky District, like to hold vivid festivals, where artistic collectives perform. Mordvins have a wide variety of musical instruments. One of those instruments is the nudey—a flute made out of a type of cane which the Mordvins also call nudey. It's a small instrument that nevertheless produces a fairly melodic sound. In the past, many Mordvin settlements had craftsmen who could produce a nudey. From among modern instruments, violin, clarinet and the bagpipes are used. The accordion is also popular. Traditional melodies are alternated with modern-day tunes.

One of the most characteristic public events was the Velen ozks prayer, which was held on 12th of July, and brought children and adults together. After saying their prayers, asking for good harvests and good fortunes, young people would hold games and dances. Older people, meanwhile, ate porridge and drank beer, while women ate a special ritual chicken. On the New Year's Eve, young women told each other's fortunes. One of the most common methods of divination involved tossing a valenok out in the street—the future groom would come from the direction where the felt boot points. Another method involved fetching a bucket of water at night, then lowering a ring into the water, lightning a candle and looking into the mirror. The reflection predicted the future. During dry seasons, the residents of Mordvin settlements would pray for rain with the guidance of a priest.

Mordvin villages also celebrate Orthodox holidays. Those are celebrated almost the exact same way they are among Russians; for example, the Pentecost involves making wreaths, same way it does in Russian villages.

In Shentalinsky District, in the village of Staraya Shentala, one can find the historical and ethnographic museum named after M.I. Chuvashev. He a famous folklorist of the 20th century who dedicated his life to collecting, recording and processing verbal Mordvin folklore. Aside from exhibits telling the story of the researcher's life and work (photos, books, notes etc.), the museum features a copy of a traditional Mordvin hut, models of national clothing as well as working tools and other items of Mordvin material culture (for example, cheke—a device for spinning yarn, kardamo—an antique type of towel, tarvaz—a type of sickle, etc.).

The village of Bagana can serve as the Mordvin demonstration settlement of Shentalinsky District. The folk collective "Chilisema" ("Dawn") is based here; it sings local songs. The master craftsman S.V. Sidorov, proficient in producing the nudey musical instrument, also lives here.

4 Discussion

The interactive resource, which is currently being developed by the research group, will be integrated with the existing "Touristic and recreationary routes of lesser cities and villages of Samara Oblast" platform. Today, all kinds of digital resources for tourists are being created in the world, including cloud-based ones [5] or mobile internet platforms [7]. Different digital resources are designed to meet a common goal. The promotion of historical, cultural, ethnographic potential of agrarian regions through online platforms will help attract tourists and create new hotspots of economic growth.

Ethnic tourism is adjacent to agrarian tourism, what makes it less expensive from an economic standpoint: it doesn't require constructing hotels and restaurants. Instead, lodgings are provided by local residents, what contributes to the immersion into the atmosphere of traditional culture. Other types of tourism are also adjacent to ethno-tourism—religious tourism, event tourism, etc.

Samara Oblast has a significant resource potential, which must be harnessed for organizing ethnographic tours. Traditional settlements of the various peoples living in the Oblast have preserved their local specificity with regards both to material and spiritual culture. At the same time, one must admit that ethno-tourism in the region isn't well-developed. Including ethnocultural content dedicated to the studied region into the Interactive Atlas will allow us to begin forming a sharing economy.

5 Conclusion

The Interactive Atlas platform is based around research materials on objects of ethnic heritage in Shentalinsky District of Samara Oblast. Over 20 tourist routes were hosted on the portal for the purpose of attracting the attention of potential tourists to the studied objects, and promoting the preservation of the historical and cultural legacy of the area among the general public.

Research of the area's ethnic culture is warranted due to the multi-ethnic character of its territory, where many locally specific traditions and rituals have been preserved. The formation of each people's unique culture transpired in large part due to religious differences. The authors paid careful attention to this factor, highlighting the unique features of a number of settlements where even ancient pagan rituals have been preserved. The population of the area was distributed according to the Russian state's objectives. Because of this, a large part of settlements in the area were founded in the early 18th century, following the construction of the Novo-Zakamskaya fortified line, which marked the southeastern borderland of the Russian Empire.

Analysis, processing of field research findings and the methodological justification of studied objects have been carried out by the authors with the objective of uploading the data to an interactive online platform, providing wide access to accurate information about the historical and cultural objects of the area.

Acknowledgements This research project has been conducted with the financial support of the all-Russian social organization "Russian Geographical Society" within the framework of the scholarship project: "'Shentala—the Beautiful Valley': ethnocultural and historical atlas of landmarks in the rural north of Samara Oblast" Contract № 30/2019-P.

References

1. Galimov ShKh (2017) Brief information on the Tatar settlements of the Samara region. In: Galimov ShKh (ed) Tatars of the Samara territory – historical-ethnographic and socio-economic essays. Samara Regional Scientific Universal Library, Samara, pp 486–611
2. Gong W (2018) Innovation strategy of local folk culture tourism product design under the background of creative and cultural industries. J Adv Oxid Technol 21(2):201809075
3. Lopes RO, Malik OA, Kumpoh AA-ZA, Lee SCW, Liu Y (2019) Exploring digital architectural heritage in Brunei Darussalam: towards heritage safeguarding, smart tourism, and interactive education. In: O'Conner L (ed) Proceedings of the 5th international conference on multimedia big data, BigMM 2019. IEEE, New Jersey, pp 383–390
4. McGinnis G, Harvey M, Young T (2020) Indigenous knowledge sharing in Northern Australia: engaging digital technology for cultural interpretation. Tour Plan Dev 17(1):96–125
5. Nan R, Zhang H (2019) Multimedia learning platform development and implementation based on cloud environment. Multimed Tools Appl 78(24):35651–35664
6. Timashev VF (2001) How it was: essays on the history of the Shentalinsky district. Your Opinion, Samara
7. Xiao Y (2019) Tourism marketing platform on mobile internet: a case study of WeChat. J Electron Commer Organ 17(2):42–54

Assessment of the Digitalization Level of the Regional Investment Process

L. K. Agaeva

Abstract This paper investigates issues of investment processes digitalization in the regions of the Russian Federation in the context of modern global challenges. The development of the digital economy requires a radical change in approaches to the formation of investment attractiveness and interaction with investors. The purpose of this research is to study and assess the level of digitalization of the investment process in the region. The author used methods of quantitative and qualitative economic analysis of investment process. The study revealed that the digitization of the investment process is regulated by the regional investment standard. This standard implies the creation of an investment portal for attracting investments and interacting with a potential investor. The assessment of the digitalization level of the investment process in the Samara region showed a good organizational level, which will significantly improve the investment opportunities of the region.

Keywords Digitalization · Digital economy · Investment process in the region · Region investment attractiveness · Investment standard · Investment portal

1 Introduction

In modern development conditions, digitalization of socio-economic processes in the region provides a good incentive for their activation. The investment process in the region is moving to a new format that requires mobility and simplification of all its elements. Only this will increase the level of investment attractiveness of the region for potential investors. A significant role in attracting investment is played by the creation of a number of investment development institutions and increasing investment attractiveness [2]. It is necessary to create a unified digital system of interaction between investors, authorities and investment objects.

L. K. Agaeva (✉)
Samara State University of Economics, Samara, Russia
e-mail: Liliya.agaeva@yandex.ru

S. I. Ashmarina and V. V. Mantulenko (eds.), *Current Achievements, Challenges and Digital Chances of Knowledge Based Economy*, Lecture Notes in Networks and Systems 133, https://doi.org/10.1007/978-3-030-47458-4_27

Digitalization of investment processes in the region is mainly within the competence of the authorities, which form an investment platform that accumulates information on investment projects being implemented in the region. The digital technologies used to activate investment processes in the regions today include various investment portals that use the "one window" principle. In the Russian Federation, to increase the investment attractiveness of the region, a system of investment standards has been created that allows the creation of a single algorithm for the regional investment process based on the best investment practices. Thus, the investment processes should meet the current challenges of the digital economy in order to increase the regional investment attractiveness and further economic growth.

2 Methodology

The author used various methods of economic and statistical analysis, such as an explanatory analysis of the dynamics of a phenomenon or process, a method of qualitative assessments of the content and structure of an object or process, and a graphical method of analysis. To analyze the dynamics of investments in fixed assets of the Samara region, we used a statistical method of horizontal comparative analysis; a graphical method for representing dynamics and trends; next, a qualitative critical analysis of the structure of the investment standard of the Samara region, as the main document for improving the efficiency of the investment process in the region, with the identification of the main structural elements and their content. Using the method of qualitative analysis, we could identify the compliance of the investment portal of the Samara region with the main requirements imposed by state authorities.

3 Results

Samara region is the leading region of the Russian Federation for a long period of time included in the top two investment leaders. Nevertheless, investment processes in the region do not have sufficient efficiency to make it a leader in attracting investments. This is confirmed by the dynamics of indicators of the investment activity in the region.

Figure 1 shows that during the period from 2008 to 2018, the investment process in the region had an unstable trend. The volume of investments in fixed assets in the Samara region has tended to increase until 2014, and since 2015 there has been a decrease in the indicator. This trend is primarily related to the outflow of capital from the region because of economic sanctions imposed on Russia in 2014. During this period, there is also a decrease in the investment rating of the Samara region according to the estimates of foreign and domestic rating agencies. Thus, the deterioration of the investment process in the Samara region requires urgent approaches to improving its efficiency. The main direction in solving this task is the effective digitalization of

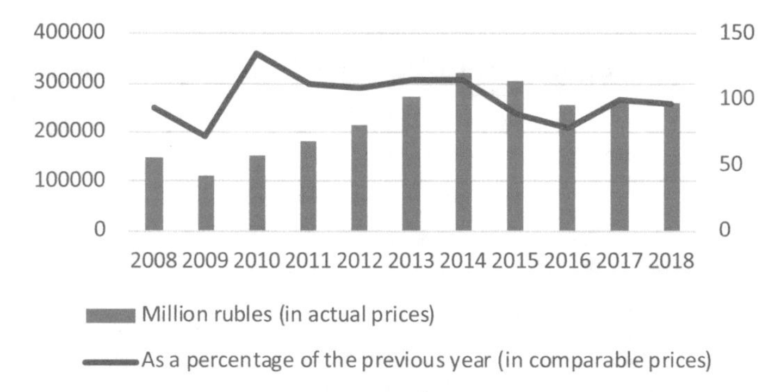

Fig. 1 Dynamics of investments into fixed capital in the Samara region in 2008 and 2018 (*Source* author based on [5])

the investment process in the region, in order to increase the investment attractiveness and attract potential investors.

To create favorable conditions for doing business in the regions, the Agency for Strategic Initiatives has developed a Regional Investment Standard. The standard includes 15 best investment practices used by economically successful regions. According to the instructions of the President, the standard is mandatory for implementation by all regions of the country. The structure of the regional investment standard is shown in Fig. 2.

The analysis of the investment standard of the Samara region showed that all the structural elements of this Standard are not fully represented in the region. Within the framework of creating conditions for business development, there is no mechanism for training personnel in the investment sphere. The greatest gaps in the application of the standard were identified in the direction of providing guarantees to investors. There are no such systems as the procedure for assessing the regulatory impact at the regional level, inclusion of energy consumers in the REC, training, improving and evaluating the competence of employees of the regional team. As part of the implementation of the investment standard of the Samara region, the investment portal «Invest in Samara» of the Fund «Agency for investment attraction of the Samara region» under the Ministry of economic development and investment of the Samara region was created to attract investments and work with investors in this direction. Table 1 presents a qualitative analysis of the investment standard of the Samara region.

The Internet portal should provide a visual representation of investment opportunities of the region, its investment strategy and infrastructure, potential investment directions, as well as collection and prompt consideration of complaints and appeals from investors. Table 2 provides an analysis of the implementation of the main requirements for investment portals in the region.

The investment portal of the Samara region, created to attract investment and work with investors, almost completely meets the requirements of the investment standard of the region. The disadvantage of this portal is the lack of video broadcasts of council

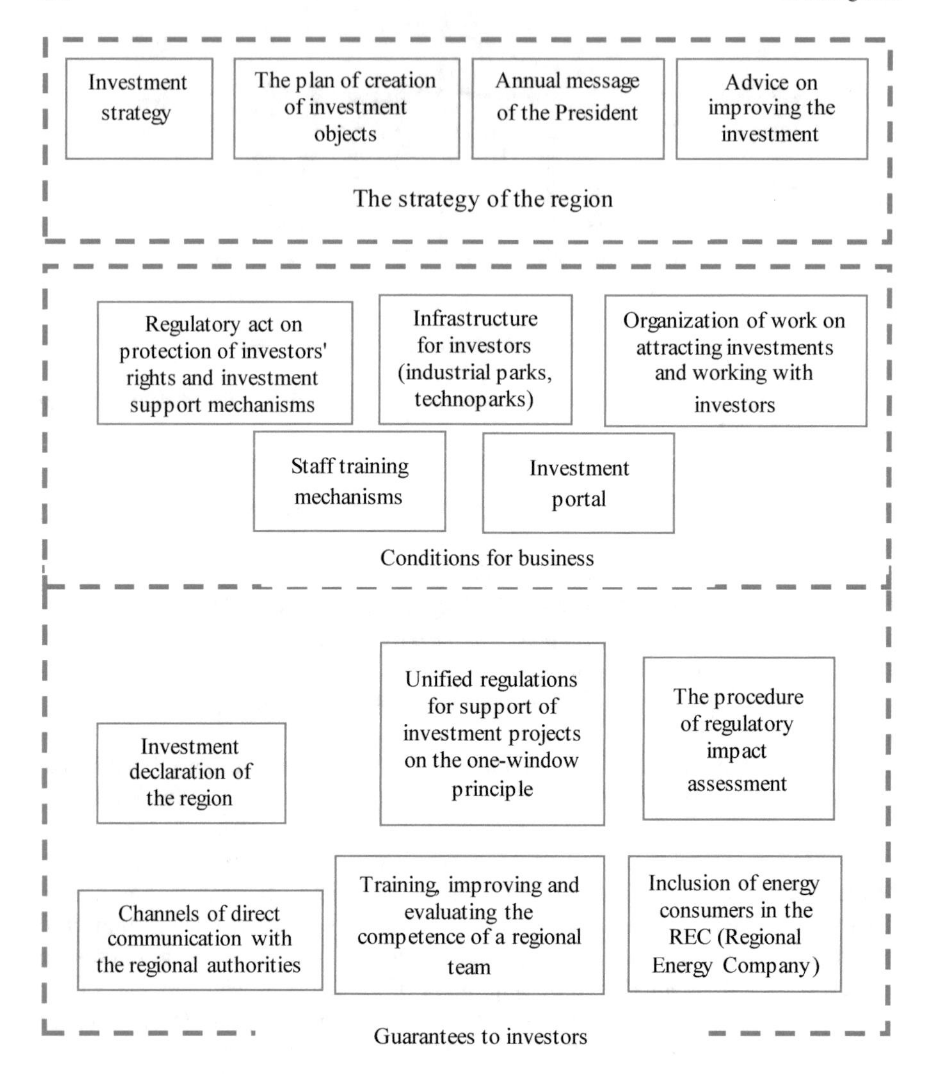

Fig. 2 The structure of the investment standard in the region (*Source* author based on [5])

meetings on improving the regional investment climate. Thus, it can be noted that the investment process in the Samara region is almost completely adapted to the conditions of the digital economy and will allow in the future, while eliminating some shortcomings, to achieve high indicators of the regional investment attractiveness.

Table 1 Analysis of the investment standard of the Samara region

Standard element	Availability	Description
The strategy of the region		
Investment strategy of the region	Available	Main directions of the investment strategy of the Samara region until 2030
The plan of creation of investment infrastructure	Available	Plan for creating investment facilities and the necessary transport, energy, social, engineering, utility and telecommunications infrastructure of the Samara region
Annual message	Available	
Advice on improving the investment climate	Available	The current collegial advisory body is the council for improving the investment climate in the Samara region
Condition for business		
Regulatory act on protection of investors' rights and investment support mechanisms	Available	Law of the Samara region of 16.03.2006 No. 19-GD "On investments and state support for investment activities in the Samara region"
Infrastructure for investors (industrial parks, technoparks)	Available	1. Special economic zones Special economic zone of industrial and production type «Togliatti» 2. Existing infrastructure: Center for innovative development and cluster initiatives Industrial Park «Preobrazhenka» Industrial Park «Togliatti synthesis» Innovative business incubator of the Samara region technopark Togliatti business incubator Neftegorsky business incubator JSC Guarantee Fund of the Samara Region (micro credit company) Agro-industrial Park «Samara» Technopark in the field of high technologies «Zhigulevskaya valley» Industrial Park «Chapaevsk» Non-profit partnership «Regional innovation center» Business incubator of Kinel-Cherkassky district of Samara region Samara business incubator Nanotechnology center of the Samara region AVTOVAZ industrial Park Stavropol industrial Park 3. Infrastructure under construction and planned: Industrial Park «Preobrazhenka» Industrial Park «Chapaevsk»

(continued)

Table 1 (continued)

Standard element	Availability	Description
Organization of work on attracting investments and working with investors	Available	Unitary non-commercial organization Fund «Agency for investment attraction of the Samara region»
Staff training mechanisms	Not available	–
Investment portal	Available	Ministry of economic development and investment of the Samara region «Invest in Samara»
Guarantees to investors		
Unified regulations for support of investment projects on the one-window principle	Available	Unified regulations for supporting investment projects based on the "one window" principle that are being implemented and (or) planned for implementation in the Samara region. Approved by the Government of the Samara region on March 13, 2014 N 126
Investment declaration of the region	Available	Investment Declaration of the Samara region. Approved by the resolution of the Governor of the Samara region from 10.04.2013 No. 85
The procedure of regulatory impact assessment	Available	–
Inclusion of energy consumers in the REC (Regional Energy Company)	Available	–
Channels of direct communication with the regional authorities	Available	Presented on the investment portal of the Samara region «Invest in Samara»
Training, improving and evaluating the competence of a regional team	Not available	–

Source author

4 Discussion

The digitization of the investment process is currently insufficiently studied, but there are some studies of domestic and foreign scientists in the field of digital economy that affect the issue in one way or another. Exploring general processes of the digital economy, the authors Y.Y. Lee and M. Falahat argue that, based on the wave of digitalization, organizations can produce and sell their products and services around the world using digital platforms with reduced costs and time savings. This may also relate to the investment process [4]. Other authors note that, using the example of social media companies such as Facebook and Twitter, the components of digital operations of companies, considered as separate assets or as an economic unit, may be qualified as "investments" in countries where such companies offer their services, even if the digital operations do not correspond to generally accepted ideas about

Table 2 Compliance of the investment portal «Invest in Samara» of the Samara region with the requirements of the Investment standard of the region

Requirements	Availability
Information about the regional investment attractiveness	+
Investment strategy of the region	+
Plan for creating investment facilities and necessary transport, energy, social and other infrastructure in the region	+
Line for direct appeals	+
Rules of interaction of investors with officials and authorities of the subject of the Russian Federation and local self-government	+
Description of measures to support investments and investment projects and how to apply for them	+
Information about plans and results of the council's meetings to improve the investment climate and video broadcasts of its meetings. It should be possible to order a guest pass for council meetings	(there is no video broadcast of meetings)
Reports and activity plans of specialized organizations for attracting investments and working with investors	+
Information about investment projects planned and implemented in the respective region	+
The Internet portal should be maintained in two or more languages (Russian and English)	+

Source author

investment activities for the purposes of international investment law [3]. Domestic researchers note that the greatest increase in investment in fixed assets is provided by such factors as the use of broadband networks and electronic document management. Strengthening positions on these factors is the most promising direction for improving the regional investment strategy [6]. Other Russian researchers argue that the development level of digital technologies is an important factor affecting the movement of global investment flows. Under the influence of this factor, there are trends of shifting global investment flows. The authors define the characteristics of modern Russia from the perspective of global investment flows in the context of the digital economy, which determines the low attractiveness of the Russian economy for international investors [1].

5 Conclusion

The development of the digital economy requires a radical transformation of all economic processes in the country. Digitalization is also important for the development of the investment process in the region, since it allows optimizing the procedure for attracting potential investors. Assessment of the digitalization level of the investment process in the Samara region showed its compliance with modern requirements of the digital economy. This process is regulated by the state and includes the development of the regional investment standard. The standard regulates the digitalization of the process of attracting investments and interacting with potential investors. Investment standards are recommended for widespread use and have an optimal structure based on the best practices of the Russian regions. The Samara region also uses the regional investment standard in the investment process. However, not all structural elements of the standard are implemented at the region level. On the basis of the investment standard, the Samara region has its own investment portal, created to attract investment in regional projects, as well as to interact with investors. The investment portal has a structure that meets almost all requirements of the investment standard. Thus, the assessment of the digitalization level of the investment process in the Samara region allows us to conclude that the region has good indicators in this direction. The digitalization process will further increase the investment attractiveness of the region.

References

1. Bogoviz AV, Lobova SV, Ragulina JV (2019) Shift of the global investment flows in the conditions of formation of digital economy. In: Popkova E (ed) The future of the global financial system: downfall or harmony. Lecture notes in networks and systems, vol 57. Springer, Cham, pp 1216–1223
2. Fadeeva AYu (2016) The role of the regional investment web portal in increasing investment attractiveness of a region. Manager 2:60
3. Horváth E, Klinkmüller S (2019) The concept of 'investment' in the digital economy: the case of social media companies. J World Invest Trade 20(4):577617
4. Lee YY, Falahat M (2019) The impact of digitalization and resources on gaining competitive advantage in international markets: the mediating role of marketing, innovation and learning capabilities. Technol Innov Manag Rev 9(11):26–38
5. Samarastat (2020) Investment. https://samarastat.gks.ru/investment. Accessed 02 Feb 2020
6. Vlasov MV (2019) Sustainability of a regional investment strategy: factors of a digital economy. Amaz Investig 8(23):140–147

Economic Efficiency Assessment of Investments in Production Digitalization

A. A. Chudaeva

Abstract Production digitalization demands investment. Evaluating the cost-effectiveness of such investments means measuring the investment against the results that production digitalization will achieve. The building of financial model has to consider the impact of digitalization on the determinants of inflows and outflows and the risks generated by this process. So it is necessary to carefully analyze the production processes and the changes that will occur in them in connection with the implementation of digital technologies. The article seeks to study the issues that arise in the process of evaluating the economic efficiency of investments in production digitalization, including those ones related to the financing of such projects.

Keywords Production digitalization · Economic efficiency of investments · Risk

1 Introduction

The development of digital technologies leads to the changes in society and business. It is a commonly accepted opinion that digitalization makes it possible to improve the efficiency of enterprises. However, making conclusions about the effectiveness of any decision at an enterprise requires grounds. To evaluate efficiency means to compare the results with the costs to achieve them, taking into account the interests of the investor. It is obvious that the introduction of digital technologies in the activity of production enterprises demands huge investments. There is no point in digitizing individual production processes or production sites, because processes and sites without digital technologies will become bottlenecks. Total digitalization is an expensive process.

One of the digitization tools is the introduction of corporate information systems, software tools that integrate data on different areas of the enterprise. The Federal State Statistics Service of the Russian Federation keeps records of the companies that use special software [6]. And the data of this company indicate that the share of

A. A. Chudaeva (✉)
Samara State University of Economics, Samara, Russia
e-mail: chudaeva@inbox.ru

S. I. Ashmarina and V. V. Mantulenko (eds.), *Current Achievements, Challenges and Digital Chances of Knowledge Based Economy*, Lecture Notes in Networks and Systems 133, https://doi.org/10.1007/978-3-030-47458-4_28

enterprises, which used special software to manage automated production and/or certain technical means and technological processes, is small, and it increases slightly every year. But the data of Rosstat do not allow drawing a conclusion about the number of enterprises using ERP-systems, which include transformation of production processes.

The results of the research conducted jointly by the company "Tsifra" and the Ministry of Industry and Trade of the Russian Federation [7] also testify to the low interest of industrial enterprises in the production digitalization: 55% of industrial enterprises of the Russian Federation spend less than 1% of their budget on the digitalization and development of information infrastructure, and only 6% of enterprises spend more than 5% of the budget.

Therefore, the production digitalization at industrial enterprises in the Russian Federation is going slowly, only a small number of domestic companies have production facilities based on digital technologies. These include the production facilities: PJSC "KamAZ", concern "Kalashnikov", "RusAl", "Petrozavodskmash" and others [1].

2 Methodology

The author used such research methods as analysis, synthesis, description and comparison. Their application is determined by the theoretical nature of the study which includes the following stages: stating the problem, analyzing information on this topic, comparing and describing different scientific views on the researched issues, synthesis of different approaches to the problem.

3 Results

The problem of production digitalization is multidimensional. The first problem, which, in fact, generates the others, is the assessment of the project aimed at production digitalization. Grounds for such project should be based on cash-flow methodology. Cash-flows aggregate the information describing the decisions put in the project and their consequences in value terms. Consequently, expenses associated with the production digitalization will be accumulated in project outflows, and cash inflows will be accumulated in inflows. Here the key issue is the identification of factors that affect project inflows and outflows and that are related to the digitalization process.

Reducing transaction costs through digitalization is the essence of digital economy built on digitalization. Other results of digitalization are less obvious—it is problematic to interpret the consequences of incorporating digital technologies into production. This is due to the huge number of factors that affect the production enterprise. To predict the factor that will influence the production system of an enterprise is a difficult task, the solution of which is multi-optimal in the conditions of uncertainty of digital technologies' functioning in the conditions of a particular enterprise.

The data's incompleteness and inaccuracy used in the process of assessment of investments' efficiency generate risks, which should be evaluated and taken into account in the calculations. Digital technologies are innovative and, therefore, the risks of the projects aimed at the production digitalization should be assessed as innovative and, consequently, high. The issues of accepting or non-accepting the risks by the investor are solved at the stage of technical-economical feasibility of the study putting the risks in calculations of investments' efficiency in three areas.

The first area is connected with the initial information on the project aimed at production digitalization. This information should take into account the expenses on reserved industrial capacities (it depends on the branch), and the expenses connected with maintenance and repairing service (TS) of introduced technologies (the concept of planned preventive maintenance). However, grounds for costs, their values and frequency in project outflows are not an easy task, as the behavior of digital technologies has a probabilistic character. All above-mentioned reasons imply to build the probabilistic model of the technical system's behavior. The authors of the work [3] give such an example of costs on the maintenance. It is supposed that the machine breakage in the period of time 1 has a probability of 20% and costs 500€. If damage occurs, the probability of the damage occurring in 2 decreases to 5%. If no damage occurs, the probability of 2 increases to 25%. Consequently, there are several variants of events during the calculation period of the project, which means that there are many scenarios with different efficiency results.

The second way to include risks in the calculations of investment efficiency is to justify the rate of return, which is crucial for an investor. The higher the risk, the higher the investor's requirements to the profitability of such project. Rate of return, which is used to determine the discount rate, affects the NPV of the project (the higher the rate of return, the lower the NPV) and is used as a guideline for assessing the acceptability of the project's profitability, expressed by the internal rate of return. All this testifies to the fact that due to the risks included in the rate of return, the value of which is defined by expert evaluation in innovative projects, projects aimed at production digitalization may be considered ineffective.

The third area of taking risks into calculations at investments' efficiency is carrying out the sensitivity analysis that defines the influence of external factors on efficiency of the investment project. Also in the third area the method of scenario analysis and of a break-even point is realized. In addition to the elaborated issues in production digitalization, another aspect should be taken into consideration, it is the enterprise's labor force and associated costs of salaries, which are cash outflows. Consequently, the reduction of personnel costs is one of the tendency for improving the efficiency of investment projects. And at first sight there is no contradiction here because it is supposed that production digitalization will lead to total robotization and automation. However, in order to determine whether the situation will be the same in each case, it is necessary to carry out calculations that give us the required number of personnel for the enterprise, the necessary level of its qualification and the average salary of such specialists in the market.

The process of production digitalization leads to attracting highly qualified specialists (programmers, maintenance and repair specialists with software, etc.). And the enterprises will introduce digital technologies at that time when it is economically feasible to buy new equipment, new technology than to pay for the work of a person capable to perform the same functions.

At present, the level of salaries in many countries is such that it is not obvious that it is profitable to replace human labor with technical means and digital technologies. Non-alternative loss of jobs due to digitalization can lead to extremely adverse social consequences. However, a more likely scenario is one in which employment patterns change, the labor market and training approaches are transformed. Thus, according to the WEF report, in the nearest future robots will destroy more than 75 million jobs worldwide and create 133 million new ones [11].

4 Discussion

Currently, research is widely discussing issues related to the production digitalization in connection with the implementation of the Industry 4.0 concept. The authors of the works put different emphasis on the description of the problems associated with the assessment of investments' efficiency in the production digitalization.

So, the authors of work [5] write that industrial production in the conditions of digitalization is connected with the complication of technological chains in more complex network structures and note that enumerated circumstances lead to the necessity of using iterative algorithms and matrix models at defining the prime cost of the products. Algorithms and models are differentiated by labor input and possibilities of modelling various economic circumstances of the enterprise [2]. And these factors necessarily should be considered at building the financial model describing cash flows of the enterprise, connected with digitalization process.

The article [12] focuses on defining the approach to collecting digital data. The authors proposed the concept of building the database and proposed recommendations for the introduction of digital twin in the production systems of small and medium-sized enterprises, which will provide a multimodal data collection and evaluation.

In the study [4] the authors proposed a business case calculator based on the data collection method for making investment decisions, developed in the framework of the research described in the article [3]. The author of the article [10] suggested that the digitalization itself does not lead to the increase in productivity at the production enterprise, the huge role is given to such intangible assets as human and organizational capital of the enterprise. This means that these parameters should also be taken into account when assessing the economic efficiency of investments aimed at production digitalization.

The study [3] describes the features of the industrial Internet of Things (IIoT) and the issues of economic effect assessment, architectural structure and infrastructure which refers to the industrial environment. The issues covered by the authors of described publications can be taken into account in the process of evaluating the investments' effectiveness in production digitalization.

5 Conclusion

In the conditions of digitalization some kinds of production (including means of work) become unified with cloud technologies without which the physical part of production cannot function, that leads to the necessity to support and improve production by its manufacturer throughout all life cycle. Therefore, it is essential to consider the costs associated with these processes as part of determining the effectiveness of digitalization investments.

In addition, the use of digital technologies allows some of the production processes to be carried out on the territory of the consumer (for example, software installation and adjustment), which leads to the changes in the costs structure of the enterprise. The opportunity to network the equipment of the enterprise on the basis of digital technologies leads to the automation of production processes and their optimization, which makes it possible to prevent emergency situations, including the simulation of the equipment's behavior through software tools.

Production digitalization based on the software enables to expand the functionality of technical systems, which can lead to such consequences as the elimination of some physical components and changes in functions of technical systems (equipment). It means that it is extremely difficult to predict how the smart technical systems are applied (these are technical systems with software) [3]. In addition, in the conditions of digitalization, industrial production can create standardized platforms that can be customized to meet specific customer requirements, and it will result in reduced stocks and in scale effect. This has an impact on the working capital and current costs of the enterprise, and should therefore be taken into account when assessing the investments' effectiveness in digitalization.

The problem of realistic assessment of inflows and outflows of the projects aimed at production digitalization entails difficulties in financing attraction. Different sectors of economy have different problems with digitalization and possibilities of its financing. The industries that show growth and make profit which can be invested in the production development and can rely on the capital inflow into their projects from outside through the generated rate of return. Industries that do not show growth and have low rates of return are not able to implement digitalization projects on their own. Consequently, it is necessary to attract credit funds, the cost of using which is not available to all enterprises because of the long payback period of investigations aimed at production digitalization. And here it is necessary to notice that the state helps the enterprises to solve the problem with the means shortage to realize the projects of production digitalization by means of various instruments.

Granting subsidies is one of ways of solving above-described problem. Thus, in 2019 the Ministry of Industry and Trade of the Russian Federation announced the competitive selection to receive subsidies from the federal budget for the development of industrial digital platforms (the rules for granting subsidies came into force on April 30, 2019 [8]. As a result of the first stage of this competition, 61 projects were selected [9]. And this is the second launched selection of the projects in terms of work instruments of the federal project "Digital Technologies". According to ANO "Digital Economy", it is planned to forward two billion rubles of subsidies from the federal budget [9] on the development of industrial digital platforms in 2019. The second way is to provide loans on co-financing terms. In Russia, the Fund for Industrial Development was created to provide loans for projects aimed at implementing digital and technological solutions to optimize production processes at the enterprise.

Both for obtaining subsidies and for co-financing from the Fund for Industrial Development, it is necessary to provide the business plan for an investment project aimed at production digitalization. As part of the feasibility study, special attention should be paid to the risks in this project. Thus, definition of economic investment's efficiency in the projects aimed at production digitalization, is connected with the whole complex of problems of data searching and processing because data's incompleteness and inaccuracy generate risks. There are many of them, and possible risk circumstances should become the subject of assessment as the risk cost can become substantial in comparison with those positive effects which the project targets.

References

1. Belser M (2018) Digitalization of industry: a fashion trend or a prerequisite for staying competitive? https://promdevelop.ru/tsifrovizatsiya-promyshlennosti-modnyj-trend-ili-neobhodimoe-uslovie-dlya-sohraneniya-konkurentosposobnosti/. Accessed 02 Feb 2020
2. Jeschke S, Brecher C, Meisen T, Özdemir D, Eschert T (2016) Industrial internet of things and cyber manufacturing systems. In: Jeschke S, Brecher C, Song, H, Rawat DB (eds) Industrial internet of things. Springer, Cham, pp 3–19. https://doi.org/10.1007/978-3-319-42559-7_1
3. Joppen R, Kühn A, Hupacha D, Dumitrescu R (2019) Collecting data in the assessment of investments within production. Procedia CIRP 79:466–471
4. Joppen R, Lipsmeier A, Tewes C, Kühn A, Dumitrescu R (2019) Evaluation of investments in the digitalization of a production. Procedia CIRP 81:411–416
5. Kozlova T, Zambrzhitskaia E, Simakov D, Balbarin Y (2019) Algorithms for calculating the cost in the conditions of digitalization of industrial production. IOP Conf Ser Mater Sci Eng 497:1–9. https://doi.org/10.1088/1757-899X/497/1/012078
6. Proportion of organizations that used personal computers by subjects of the Russian Federation (in percent of the total number of surveyed organizations of the corresponding subject of the Russian Federation) as of 27.07.2018. http://www.gks.ru/free_doc/new_site/business/it/it7.xls. Accessed 02 Feb 2020
7. Razumnii E (2018) Minpromtorg assessed readiness of the Russian enterprises for digitalization. RBC. https://www.rbc.ru/technology_and_media/03/07/2018/5b3a26a89a794785abc9f304. Accessed 02 Feb 2020
8. Resolution of the Government of the Russian Federation dated 30.04.2019 N 529 (ed. on 26.09.2019) On approval of the rules for granting subsidies to Russian organizations for reimbursement of part of the costs for development of digital platforms and software products for the purpose of creation and/or development of production of high technology industrial products. http://www.consultant.ru/document/cons_doc_LAW_324050/. Accessed 02 Feb 2020

9. RIA News (2019) The Ministry of industry and trade has selected 61 projects on digital solutions for industry. https://ria.ru/20190905/1558328608.html. Accessed 02 Feb 2020
10. Schneider M (2017) Digitalization of production, human capital, and organizational capital. Impact Digit Work 21:39–52. https://doi.org/10.1007/978-3-319-63257-5_4
11. Souter A (2019) How digital transformation will change the labor market in Russia. https://www.forbes.ru/karera-i-svoy-biznes/371537-kak-cifrovaya-transformaciya-izmenit-rynok-truda-v-rossii. Accessed 02 Feb 2020
12. Uhlemann TH-J, Lehmann C, Steinhilper R (2016) The digital twin: realizing the cyber-physical production system for industry 4.0. Procedia CIRP 61:335–340. https://doi.org/10.1016/j.procir.2016.11.152

Legal Regulation of Realtor Activity in Conditions of Using Information Technologies

F. F. Spanagel

Abstract The relevance of the issues investigated is due to the incompleteness of the legal regulation of real estate activities and the uncertainty of the legal status of a realtor. The purpose of the study is to determine the status and identify urgent problems of legal regulation of realtor activities, taking into account the normative and legal framework for the relevant business and professional activities, and to develop recommendations. The methods of research of the topic are formal legal analysis and synthesis, ensuring that the whole set of legal rules is considered. Results: In the article, the legal rules governing relations with the participation of realtors are considered on the basis of the criteria of their industry affiliation, public and private legal character of influence on participants of corresponding relations, and also on a circle of legal communications regulated by them. This article may be useful for specialists in the field of law, economics, as well as for those involved in law-making activities.

Keywords Legal regulation · Classification of legal rules · Realtor activity · Information technology

1 Introduction

Despite the theoretically limitless possibilities of digitalization, the world is becoming increasingly unpredictable. In the conditions of growing uncertainty in the trends of development of the world economy, the states use various measures to ensure the sustainability of national economy [7]. Particular importance is given to ensuring the stability of the real estate market. It was the mortgage lending crisis in the United States that caused the worst global financial, and then economic crises at the beginning of the 21st century.

For the effective use of legal means to ensure market stability, an economic analysis of real estate legislation and transactions with real estate seems interesting and

F. F. Spanagel (✉)
Samara State University of Economics, Samara, Russia
e-mail: shpanur@yandex.ru

S. I. Ashmarina and V. V. Mantulenko (eds.), *Current Achievements, Challenges and Digital Chances of Knowledge Based Economy*, Lecture Notes in Networks and Systems 133, https://doi.org/10.1007/978-3-030-47458-4_29

promising, including not only the consequences of the impact of law on contracts with real estate objects, but also a more complex problem of the impact of changes in modern legislation on real estate turnover [4].

Equally useful can be a comprehensive comparative legal and economic study of the impact of the legal environment in general on the performance of commercial organizations. A necessary segment and an important condition for the real estate market are services provided by real estate organizations, and vice versa, the real estate activity reflects all the features and general patterns of real estate turnover.

Real estate organizations and their clients, including potential ones, face numerous legal challenges. The legal status of realtors remains uncertain, which makes the issues chosen as the subject of the study relevant.

2 Methodology

The following methods were used in the course of the study: analysis, synthesis, comparison, generalization, allowing a comprehensive review of the legal framework of real estate activities.

The methodology of the study included an analysis of scientific literature on the subject of research, as well as the regulatory framework for real estate activities and the practice of implementing the law; conclusions were drawn and a publication was prepared. In this regard, it seems reasonable to conclude that legal requirements in any country are a significant determinant of profitability and overall performance of commercial organizations adjusted for risk [3]. This rule fully applies to real estate organizations whose legal regulation cannot be considered satisfactory.

At present, it is not possible to consider the legal regulation of real estate activities without taking into account the use of the results of the accelerating digital revolution by realtors and other participants in the real estate market [8].

3 Results

The introduction of information technologies into different business spheres has had a noticeable impact on the real estate market, causing significant changes due to many startups, the formation of a shared consumption economy, the use of Blockchain technologies and many other innovations, including those in proptech. The rapidly growing potential of information technologies makes it possible to expand the range of objects of property rights and other rights in rem. Thus, mass introduction and active use of three-dimensional (3D) division of real estate objects into separate 'layers' will lead to a further increase in the variety of real estate objects [5].

There are relations between the participants of the real estate market that differ by their economic and legal nature, by the composition and legal status of the subjects, by legal consequences of their actions (inaction), by types and degrees of risks of transaction participants, by types of real estate objects, by their legal regime and by other criteria. Differences in the breadth of information technology coverage of certain segments of real estate services are significant.

With the acceleration of the processes of digitalization of all aspects of society, the state of the domestic real estate market, including development activities, indicates its sustainable conservatism in comparison with other segments of the economy. Compared to other real estate turnover participants, realtors are even less active in using digital innovations, which makes this area of activity the least transparent. In many cases, realtors abuse their rights and violate the rights and legitimate interests of their clients.

In addition to the above reasons, realtors have a negative impact on the introduction of innovations by the shortcomings in the legal ordering of their activities. The state legal policy in relation to real estate activities does not seem clear and consistent, therefore, we consider the study of the system of legal regulation of relations with the participation of realtor organizations relevant.

We consider it necessary to analyze the legal rules regulating the relations with the participation of real estate organizations and realtor-entrepreneurs according to two interrelated, but independent criteria: (1) their industry affiliation and (2) the breadth of their scope. The study of the main array of legal rules that determine the legal status of realtors and regulate relations with their participation allows us to conclude that they have different sectoral legal nature, which must be taken into account when analyzing the effectiveness of legal regulation of this type of entrepreneurial activity.

The combination, or better, the system of legal rules that regulate the functioning of the real estate market and the services market adjacent to it, should provide a comprehensive impact of legal means on realtors and their counterparties, which is achievable only with an optimal combination of public law and private law means in a well-coordinated mechanism of legal regulation of relations of this kind.

As well as other spheres of business, real estate activity is in the sphere of legal regulation of both public and private law, therefore, a unified or at least coordinated approach is required when choosing legal means of influence on this segment of the real estate market and, at the same time, on the service sector. Participants in such activities when working with customers, including when drawing up contractual relations, must take into account tax, criminal, administrative and other public legal consequences of actions (inaction) of both the client and their own.

In practice, experienced realtors, when forming contractual relations, proceed not only from the provisions of enabling rules, but also necessarily (and above all) take into account the content of prescriptive and prohibitive (protective) legal rules in order to exclude or minimize the risks of numerous penalties and other sanctions of public nature. While recognizing the importance of public law regulation of real estate activities, we nevertheless conclude that the bulk of the legal regulations governing them has obvious private law nature. An analysis of civil law reveals that most of its legal provisions can be used, directly or indirectly, or at least taken into account in the process of organizing and conducting real estate activities.

The work of real estate organizations with customers—legal entities also requires taking into account the consequences of their insolvency and especially the bankruptcy of customers or organizations selling them their assets in the form of real estate. It is necessary to take into account legal, economic [9] and organizational risks when executing such transactions or invalidating them.

It is necessary to recognize the obvious direct regulatory impact on real estate activities of a huge set of legal rules on transactions, on civil law terms, including statute of limitations, on representation, on general provisions on obligations and on contractual obligations, and many provisions of some other civil law institutions.

Recently, more and more often the organization and implementation of real estate activities are carried out through the formation of a network of real estate organizations under the franchise system based on a commercial concession agreement. In such cases, it becomes necessary to include in the sphere of real estate activity a set of legal rules on exclusive rights of the right holder. In addition to the analysis of the industry affiliation of legal rules on real estate activities, it is interesting to classify the same set of legal rules according to the circle of relations regulated by them. This separation is based on criteria from general to particular and includes three groups of existing legal rules: general–special–exceptional.

The first group of legal rules regulating relations, including those with the participation of real estate organizations, are legal provisions of general nature (general rules). The main, even overwhelming, part of the legal rules regulating economic relations in our country and fixed at the federal and regional levels of legislation and by-laws of all levels has the most general character. The second group of rules of the corresponding legal array is formed from the rules of a special nature dedicated to the regulation of relations in the sphere of real estate and services, including those involving realtors (special rules).

The selection of the third group of rules is aimed at identifying all legal provisions governing relations exclusively or mainly with realtors themselves and their activities (exceptional rules). We have to admit a paradoxical situation—almost total absence of exceptional legal rules (third group) both for the realtors themselves, and for real estate activities—such an important segment of the services market.

At the same time, the absence of specific legislation on realtors and real estate activities does not give grounds to assert a complete absence of their legal regulation due to the existence of numerous legal rules for the first and second groups. Reflection of the shortcomings of the legal regulation of real estate activity, indicating its imperfection, is a high level of illegal actions in the field of real estate services. This conclusion applies not only to the characterization of the legal framework of real estate activity, but also to the same extent to the quality and efficiency of implementation of relevant legal rules. To such legal ties of realtors can be attributed, for example, such when the real estate company buys ‘on occasion’ the object of real estate at a significantly lower price than the market price and sells it to its customer as someone else’s, found and specially purchased for this particular customer.

Such options should be recognized as fairly common in the practice of real estate organizations. In these cases, it is not uncommon for a real estate organization to issue ownership of such a property to the nominal owner, with the person then selling the property to a buyer found by a real estate organization. At the same time, often the real estate organization also ‘sells’ to such a buyer its service of ‘search’ of an apartment or other real estate object he needs. The unsuspecting buyer pays for the unreceived 'realtor' service. Dutch real estate agencies use a similar tactic to circumvent the law [1].

Such facts are not revealed in all cases, but most often in the process of criminal procedure (during interrogation of witnesses, analysis of electronic and other correspondence, seizure and examination of primary and other accounting and financial documents of real estate organizations, etc.). The civil law form of protecting the rights of such consumers in court is often not effective enough. It should be noted that in such cases, in parallel with criminal liability for fraud, realtors may be held liable for tax fraud, and they may also be sanctioned for the legalization (laundering) of money or other property acquired by other persons through criminal means.

With this in mind, a high level of latency of illegal real estate activities should be noted. Unfortunately, not only statistical, but even expert estimates of the volume of such a market are not given in the sources available to us. Two types of illegal activities of real estate organizations and realtors—individual entrepreneurs should be taken into account: (1) activities without state registration as a business entity (illegal); (2) activities of registered entities, that are not reflected in their financial statements (semi-legal).

We see high rates of latent administrative offenses and even crimes (such as fraud, commercial bribery, abuse of authority and others) in this segment of the domestic services market. At the same time, the number of civil offences in the sphere of real estate activities is incomparably high. Among them are refusal of a realtor to sign a written contract on real estate services (in order to avoid responsibility and often to be able to unilaterally change the terms of an oral agreement); unreasonable overpricing of their services; receiving fees for declared but not actually rendered services (for example, for the alleged inspection of 'legal cleanliness' of the real estate object by the realtor, for verbal 'guarantee' of absence of legal and other risks on acquired object and (or) cooperation with contractor, and many other things).

Currently, there is an urgent need for government bodies to develop and approve a strategy for legal regulation of real estate activities, as well as to identify tactical steps for its implementation by adopting an appropriate 'road map' as a common way of planning innovations in modern Russian public administration.

At the same time, it seems useful to use both the existing domestic experience in legal regulation of certain types of business activities and the best foreign practices in regulating the activities of real estate organizations.

The accumulated experience in legal regulation of certain types of professional activities, such as valuation [6], legal and other activities that are related to real estate, also seems useful. Thus, the foreign literature notes the special contribution of practicing lawyers in ensuring legal certainty by monitoring the legality of transactions (primarily with real estate) and protecting the interests of their participants [2]. It also justifies the need to further regulate legal advice, for example, on mortgage transactions related to residential real estate, due to their complexity and possible conflicts of interest [11].

In addition to lawyers, realtors play an important role in the preparation and support of real estate transactions. When comparing the value of real estate, legal and other types of services in the real estate market and in law-making work it is necessary to take into account the growing trend of recent time—that is diffusion of real estate services with other types of consulting services in the real estate market. At the same time, the level and quality of legal regulation of real estate activity is significantly inferior to the legal regulation of lawyers, appraisers, financial consultants and other real estate market participants.

4 Discussion

The difficulties of optimal legal regulation of operations with real estate are caused by its high degree of dependence on the general state of the national economy and the guarantee of rights to real estate by the Russian state, which directly affects the legal attractiveness of the national environment for business, including real estate. In the international competition for such an attractiveness of business environment, Russia has recently achieved tremendous results and entered the top ten countries in the world in terms of such indicators as registration of rights to real estate and transactions with it [10]. The features of the modern real estate market are the involvement of an extremely wide range of different entities, active development of the latest digital technologies [9] in the field of investment, financial, trade, intermediary (including real estate) services.

On the basis of a systematic interpretation of civil law, the author answers positively to the question of the admissibility of introducing the term 'realtor organization' into a 'scientific circulation' as a non-profit organization engaged in income-generating activities.

The legal regulation of relations with the participation of realtors in the real estate sector reflects the complex interconnection of changes in the market economy and the legal superstructure of modern society. The variety and inconsistency of options for contractual and public legal relations regarding real estate objects is growing, the number of litigations related to the execution of so-called real estate contracts is increasing, and other negative trends can be seen.

Non-transparency of the real estate services market, large turnover of illegal real estate activities contribute to the existence of the 'black' criminal segment of this market, entail numerous cases of violations of antitrust laws, violations of the rights of participants in real estate transactions.

One of the most important factors contributing to the existence of non-transparency of this segment of the service market, the illegality and dishonesty of its participants is the complete absence of special legislation on real estate activities and proper control by the state, including supervisory and other law enforcement agencies.

5 Conclusion

In today's service market, the variety and scope of services offered by realtors to real estate transaction participants, and the amount of knowledge they need and use when providing diverse and high-quality services is enormous. These are legal, marketing, engineering and construction, expert, operational and management, analytical, evaluation, informational, psychological and other types of knowledge, that are gradually complementing the competence of realtors. With a wide variety of digital perspectives, their possibilities are not timeless and limited by the financial resources of the state and the potential of a particular national economy. To an even greater extent, the obstacles and difficulties of introducing and using digital technologies are evident on the scale of a separate commercial organization. Thus, the feature of most domestic construction organizations is their marked lag in the rate of introduction of these technologies in advanced construction corporations of developed countries.

To an even greater extent than in the construction industry, the 'digital' lag is observed in the real estate services market. Many innovations are not available to small and even medium-sized real estate organizations, not only because of the conservative nature of their management, but above all because of the extremely limited economic potential in the conditions of long-term stagnation in the domestic real estate market. The noted shortcomings in the work of domestic realtors significantly reduce their competitiveness and investment attractiveness and even threaten to partially or even completely replace them by participants in related business activities (lawyers, financial consultants and others). The early adoption of a well-prepared special law on real estate activity will contribute to its ordering and legalization.

References

1. Belloir A (2019) Navigating the rules: Legal Provisions and Dutch estate agencies: tactics to circumvent the law. In: Vols M, Schmid CU (eds) Houses, homes and the law. Eleven Publishing, The Hague. https://ssrn.com/abstract=3481644. Accessed 30 Jan 2020
2. Carosi MB (2018) Legal certainty and transaction security – an economic-comparative reflection. Politica del Diritto 3:449–506
3. Ghosh C, Petrova M (2020) The effect of legal environment and regulatory structure on performance: cross-country evidence from REITs. J R Estate Financ Econ. https://doi.org/10.1007/s11146-019-09742-8. Accessed 30 Jan 2020
4. Kowalczyk C, Nowak M, Źróbek S (2019) The concept of studying the impact of legal changes on the agricultural real estate market. Land Use Policy 86:229–237
5. Lehavi A (2020) The future of property rights: digital technology in the real world. http://dx.doi.org/10.2139/ssrn.3516096. Accessed 30 Jan 2020
6. Neznamova AA, Volkova MA, Smagina OS, Efimova OV (2019) Legal regulation of real estate appraisal services. Opcion 35(19):2337–2365
7. Sattarova NA, Shokhin SO (2018) Selected issues of state administration in the field of ensuring financial security. Perm Univ Bull Jurisprud 40:167–185. https://doi.org/10.17072/1995-4190-2018-40-167-185
8. Schwab K (2017) The forth industrial revolution. Crown Business, New York

9. Spanagel FF, Belozerova OA, Kot MK (2020) Analysis of legal and economic risks for entrepreneurs in digital economy. In: Ashmarina SI, Vochzoka M, Mantulenko VV (eds) Digital age: chances, challenges and future. Lecture notes in networks and systems, vol 84. Springer, Cham, pp 548–556
10. The World Bank (2020) Doing business 2020. https://www.doingbusiness.org. Accessed 30 Jan 2020
11. Zunzunegui F (2019) Advice on the real estate credit law. https://ssrn.com/abstract=3410936. Accessed 30 Jan 2020

Processes of Informatization in the Accounting of an Enterprise: The Methodological Aspect

S. V. Andreeva

Abstract Processes of implementing information technology in the management practice of companies have brought about the need for addressing issues of not only technical, but also of methodological nature. The relevance of the study topic is due to insufficient elaboration of issues of formation of theoretical and methodological facilities for comprehensive informatization of accounting processes in an enterprise. The objective of the study is to advance scientific knowledge in the field of methodological support of informatization of accounting activities of economic entities. In order to achieve the objective, systemic and structural methods were applied, as well as a case study method. As a result of the study, a functionally oriented structure of the organization's accounting system is presented, serving as a proposed methodological basis for a comprehensive structured informatization of the accounting of an enterprise. The proposed structure enables the coherence of substantive and technical aspects of informatization of accounting processes.

Keywords Accounting informatization · Accounting system · Functionally oriented accounting system

1 Introduction

Processes of informatization of various dimensions of human activity, the emergence of a digital society, the need of economic entities for information transparency of the business environment have caused the issues of development of accounting and information systems at the scale of individual companies and the national economy as a whole to become particularly relevant. There is a particular issue of prospects for the development of bookkeeping as the most important element of the accounting system of an economic entity.

Bookkeeping has a special place in the management system. It provides an objective reflection of the enterprise's performance, which serves the objectives of its

S. V. Andreeva (✉)
Samara State University of Economics, Samara, Russia
e-mail: apollonia2563@mail.ru

S. I. Ashmarina and V. V. Mantulenko (eds.), *Current Achievements, Challenges and Digital Chances of Knowledge Based Economy*, Lecture Notes in Networks and Systems 133, https://doi.org/10.1007/978-3-030-47458-4_30

sustainable and continuous development. The modern period of development of bookkeeping is characterized by active implementation of tools for its automation. It improves the quality of accounting information, minimizes labor costs, reduces the time of processing of accounting operations, enhances supervision and expands the analytical capabilities of accounting [9]. The use of computer systems in accounting can increase the speed of collection, transmission and processing of information, reporting and maintenance of records [1]. At the same time, IT specialists predict the extinction of bookkeeping in the near future; government officials exclude the profession of a bookkeeper from the list of modern professions. In view of this, the question arises: how sound are these views?

The range of views on the future of bookkeeping and the degree of their radicalism is quite broad. On the one hand, the development of information and digital technologies is seen as a factor in the extinction of bookkeeping: it is believed that information technology will make it possible to provide information resources for the business enterprise outside the bookkeeping system [2]. Any functions of the latter (oversight, registration, documentation, recording in ledgers, etc.) will be managed by electronic document circulation systems, various mobile applications and computer software. In addition to the point of view of the unconditional substitution of bookkeeping, opinions are expressed about its various transformations, for example, into management accounting, which is also facilitated by information technology [7].

An alternative approach is as follows: informatization processes and the development of digital technologies will make bookkeeping more efficient and flexible system for the formation of accounting and analytical resources [10]. For example, recent studies suggest the importance of "big data" for bookkeeping and auditing [4]. Based on the results of an extensive review of publications, D. Knudsen, concludes that it is possible to expand the boundaries of bookkeeping and the bookkeeper profession through the influence of digitalization [5]. The diverse range of views on the prospects of bookkeeping in the age of digital transformation of society requires further research. Scientists call for evaluating and redefining the concept of an information accounting system in the context of the emerging socio-economic changes [3]. We believe that a deeper study of the substantive side of this process is necessary in addition to the reflection of the technical aspects of accounting informatization. The information infrastructure of the accounting process is not only a combination of software and hardware of computer technology and communications, but also a methodology for the formation of accounting and information resources. We should accept the view that the attention of specialists should be focused not only on the final information product, but on the whole process of its creation [6]. The objective of this paper is the advancement of theoretical issues of the functioning of the accounting system of an economic entity, including the development of a subject area of the accounting informatization.

2 Methodology

The study is based on a systems-based approach. To achieve the study objective, structural method and the case study method were also used. Results of scientific studies of Russian and foreign authors dedicated to the study of informatization of accounting processes in an enterprise served as the informational basis of the study. Information support of the study is represented by secondary qualitative information (results of scientific studies).

3 Results

As a result of the study, we propose the concept of a functionally oriented accounting system that makes it possible to determine the place of each of the areas of accounting activity in it in accordance with its subject area, objectives, principles and methods of implementation (Fig. 1). This approach ensures the focus on the substantive aspects of informatization of accounting processes over the technical aspects of its implementation.

As of today, we can talk about a significant differentiation of accounting activities of enterprises. The needs of users for accounting information depend, firstly, on the functional area in which they specialize and, secondly, on their position in the management hierarchy. It is these two aspects that determine the structure of the accounting system of an enterprise in a substantial way. As the functional areas of the company's accounting activities, we consider:

1. Accounting support for management activities within the scope of the core business process of the enterprise (CBP). The CBP includes activities on the development of new and improvement of existing products; promotion of products on the market and their sales; material and technical support; manufacturing process; reproduction of means of production and labor resources; financial activities. Accordingly, the company implements accounting for sales operations, supply and procurement, production records, personnel records, technical records, etc.
2. Accounting support for the implementation of management decisions within the scope of special branches of management. The latter include: tax management, environmental management, investment management, innovation management, operational management, risk management, quality management, social management, crisis management, marketing, etc. Accordingly, the following accounting areas can be implemented in an enterprise: tax accounting, marketing accounting, environmental accounting, social accounting, investment accounting, innovation accounting, quality management accounting, business process improvement accounting, risk management accounting, etc.
3. Accounting support for activities of top management in the coordination of individual management areas. To manage the activities of an economic entity, complex accounting information is required, ensuring a balanced development of

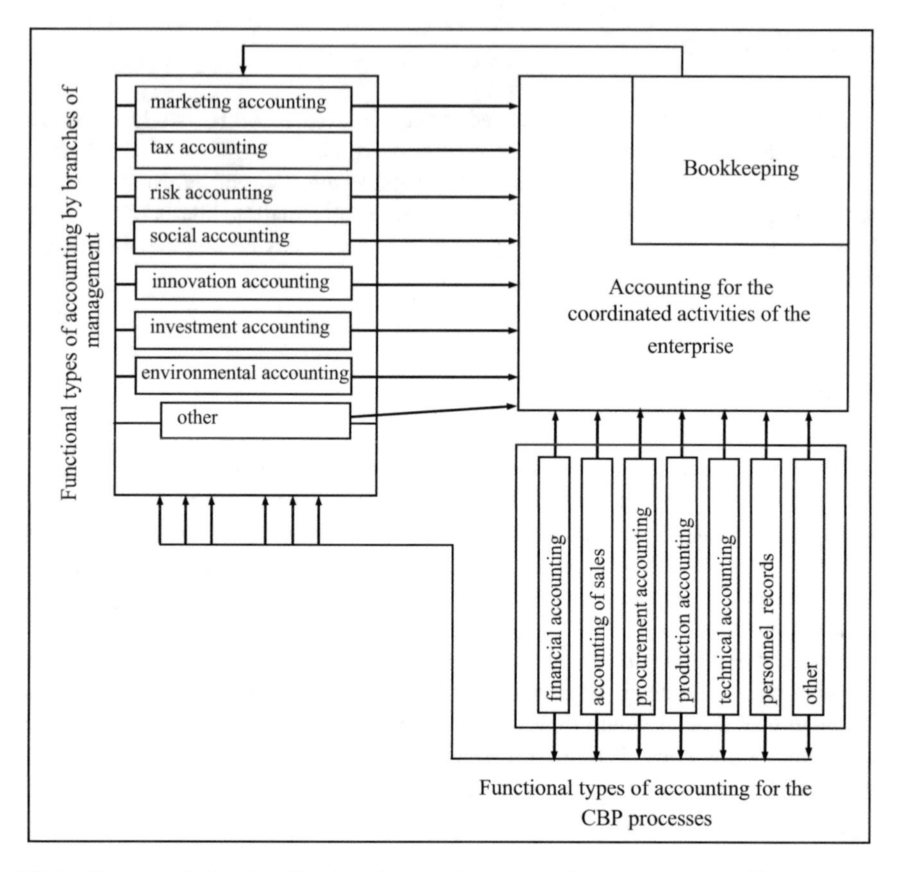

Fig. 1 Structure of a functionally oriented accounting system of an economic entity (*Source* author)

the enterprise. Bookkeeping serves as the core of such information: it is the bookkeeping information that provides a systematic quantitative and qualitative assessment of the integrated activities of an economic entity. This makes bookkeeping mandatory for an economic entity not only on the basis of the tasks of forming statutory information resources, but primarily on the basis of the need to generate internal consistent accounting information.

Accounting information fields of the levels of the accounting system shown in the figure differ from each other in the amount of data included in them. They are characterized by differences in the purpose and origin of this data, as well as the presence of different groups of users of outgoing accounting information. The totality of accounting information of functional types of accounting forms a potential field of accounting data for accounting of agreed activities and, consequently, for bookkeeping, which is a backbone component of an economic entity's accounting system.

Functioning of the accounting system, organized according to the functional feature, is directly interconnected with the management activity of the enterprise: management needs determine the structure and relationships of elements of the accounting system; in turn, the quantity and quality of accounting information determines the dynamics and quality of managerial decisions in certain areas of operations of the enterprise. This interdependent interaction of the accounting system and the enterprise's management system makes it possible for us to define the accounting system as the most important internal factor in ensuring sustainable and effective development of socio-economic systems.

The proposed functionally oriented structure of the accounting system provides for flexible restructuring of the accounting activities of the enterprise, in particular, making it possible to plan increasing or reducing the complexity of accounting activities by changing the number of functional types of accounting. Functionally-oriented approach to building the structure of the accounting system provides for ordering and identifying information flows of accounting data. The proposed structure organizes the algorithm of information interaction between the components of the accounting system in the course of its informatization.

4 Discussion

Modern accounting information systems provide the primary registration and processing of data on facts of business operations directly in the functional units of the enterprise that provide the main business process. Personnel of functional departments collects accounting information within the scope of their professional competence and based on the tasks solved by each specific functional department. Then the received accounting information is transferred to the bookkeeping system. Thus, certain accounting functions are performed outside the bookkeeping department. Some experts consider this a sign of the extinction of bookkeeping. We think otherwise, since the existence of bookkeeping is determined not only by its functions, but also by its purpose, methodology and informational result. The transfer of the functions of oversight, collection, registration, and partial processing of accounting data to functional department cannot fully ensure generation of standardized integrated accounting information structured in accordance with regulatory requirements.

Functional types of accounting create information opportunities for managing and supervising certain types of activities specifically. Comprehensive accounting information is required to manage the operations of the entire company, providing coherence and balance in the development of an economic entity. We define this type of accounting and information support as accounting in the field of coordination of the functioning of all types of activities (referred to as "accounting for coordinated activities"). Bookkeeping becomes the key component of this type of accounting, since it is precisely its information that provides a quantitative and qualitative assessment of integrated activities of an economic entity. Experts believe that bookkeeping

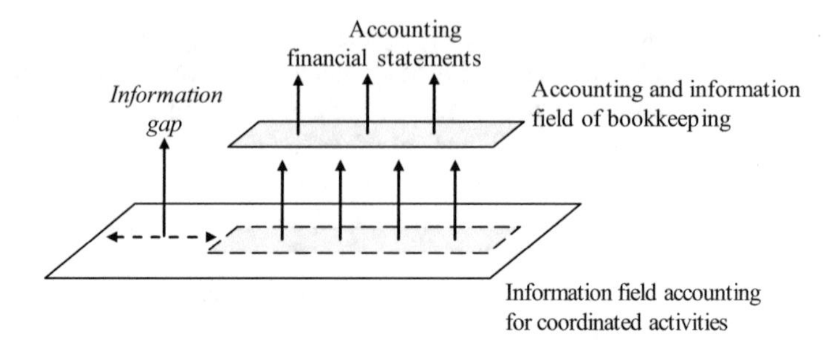

Fig. 2 Relations of accounting information fields of components enterprise accounting system (*Source* author)

is a well-designed system that brings various facts of economic life together into a single logical system [8].

Certain criteria for the selection of accounting data, their interpretation and processing lead to a narrowing of the accounting and information field of bookkeeping in comparison with the field of accounting of coordinated activities of the company (Fig. 2).

The main reasons for the gap in the volume of accounting information between the above components of the accounting system should be the availability of accounting information of a non-financial nature, inconsistent criteria for the recognition of accounting subjects and their characteristics for various functional types of accounting. The development of modern information technologies and accounting methodology provides for narrowing of this information gap. In actual accounting practice, minimizing this information gap is achieved through the formation of integrated reporting, supplementing the financial statements with information on the environmental and social activities of an economic entity, its intellectual potential, synergistic effects, etc.

Accounting activities carried out by an enterprise is a prerequisite for the formation of relevant management information. Changing factors of the external and internal business environment, the dynamic nature of modern conditions for the development of socio-economic systems are bound to result in the emergence of an increasing number of focal points for monitoring the facts of economic life, and, as a consequence, to a significant complication of accounting activities. We propose a structure of functionally oriented accounting system that explain the structure of interaction of a large number of accounting and information flows, making it possible to optimize it in the course of informatization.

5 Conclusion

The emergence of information technology in the accounting sector has a positive impact on the results of management activities of an enterprise. This process contributes to the improvement of the quality and efficiency of accounting information, ensures its transparency and relevance for concerned users. Active development of digital technologies expands the range of solvable problems in the field of accounting and reporting, and also makes it possible to establish new ones, associated with the complexity of management.

Informatization of the accounting system is a complex process, innovative in its nature and focused on optimizing objectives, methods, content and forms of accounting and information resources. The concept of a functionally oriented accounting system proposed in the paper can serve as a methodological basis for the informatization of the accounting of an enterprise. The use of a functionally oriented structure provides for modeling of the accounting system in accordance with the characteristics of the enterprise, providing for integrative interaction of various intra-company areas of accounting activities. Accordingly, the informatization of the accounting system based on the prepared methodology can bring accounting and information support of management processes to a new level of quality, accelerate the enterprise's adaptation to changes in the business environment, and accordingly ensure its long-term sustainability.

References

1. Akmarov PB, Knyazeva OP (2017) Prospects and problems of use of information technology in automation of accounting. Polythematic Online Sci J Kuban State Agrar Univ 130:139–154
2. Al-Htaybat K, von Alberti-Alhtaybat L (2017) Big Data and corporate reporting: impacts and paradoxes. Account Audit Account J 30(4):850–873
3. Betancur HD, Morales AMC (2019) Meta-theoretical approaches to the construction of accounting systems. Contabilidad Y Negocios 14(27):22–43
4. Griffin PF, Wright AM (2015) Commentaries on Big Data's importance for accounting and auditing. Account Horiz 29(2):377–379
5. Knudsen D (2019) Elusive boundaries, power relations, and knowledge production: a systematic review of the literature. Int J Account Inf Syst. https://doi.org/10.1016/j.accinf.2019.100441. Accessed 02 Feb 2020
6. Lyubenko AM, Kundrya-Vysotska OP, Rudnytska OV (2019) The development of praxeological action grammar in the modern accounting and analytical system in the context of supply of information requests. Financ Credit Act Probl Theory Pract 2(29):178–187
7. Meng F (2018) Characteristics of financial accounting transformation to management accounting in big data environment. In: Liu L, Ke G, Davis H (eds) Proceedings of the 2nd international conference on economics and management, education, humanities and social sciences. Advances in social science, education and humanities research, vol 151. Atlantis Press, Paris, pp 206–210

8. Mukhametzyanov RZ, Nugaev FS, Muhametzyanova LZ (2017) History of accounting development. J Hist Cult Art Res 6(4):1227–1236
9. Quattrone P (2016) Management accounting goes digital. Will the move make it wiser? Manag Account Res 31:118–122
10. Troshani I, Locke J, Rowbottom N (2019) Transformation of accounting through digital standardisation tracing the construction of the IFRS taxonomy. Account Audit Account J 32(1):133–162

Managing the Financial Stability of an Enterprise in a Digital Economy

F. T. Betaneli, N. V. Nikitina, and P. Zhelev

Abstract The purpose of this work is to develop an information and analytical model for analyzing and forecasting the financial stability of an enterprise. This model will allow you to quickly analyze the financial stability of the company. Discusses the theoretical justification for the need for industrial enterprises to conduct an analysis of financial stability in order to effectively manage it in a digital economy. Carrying out such an analysis in practice makes it possible to identify the strengths and weaknesses of the financial and economic activities of the enterprise under study, also to understand the possibilities for the further development of this enterprise, and to identify the threats and risks of various scenarios of further development of the company. The practical application of financial analysis methods made it possible to understand that the use of digital technologies in such an analysis not only reduces the time it takes to conduct the analysis itself, but also allows you to quickly and efficiently formulate recommendations to overcome threats and gain new development opportunities and, ultimately, make an effective management decision in a timely manner.

Keywords Financial stability · Analysis · Digitalization · Analytical model · Management

1 Introduction

Analysis and improving the financial stability of the enterprise is an important and urgent task facing the enterprise in the current conditions of digitalization of the

F. T. Betaneli · N. V. Nikitina (✉)
Samara State University of Economics, Samara, Russia
e-mail: Nikitina_nv@mail.ru

F. T. Betaneli
e-mail: Filipp-betaneli74@mail.ru

P. Zhelev
University of National and World Economy (UNWE), Sofia, Bulgaria
e-mail: pzhelev@unwe.bg

S. I. Ashmarina and V. V. Mantulenko (eds.), *Current Achievements, Challenges and Digital Chances of Knowledge Based Economy*, Lecture Notes in Networks and Systems 133, https://doi.org/10.1007/978-3-030-47458-4_31

economy [7, 12]. In the current competitive environment, the basis for a stable and sustainable position of the enterprise in the market is its financial stability. An enterprise with stable financial stability, being solvent, may have the opportunity to prevail over its competitors in the industry, have advantages in attracting bank loans, investments, in the selection of suppliers, as well as in the selection of personnel. The more stable the financial situation of the enterprise is, the more independent of changes it is in external market, political and legal conditions, and the lower the risk of bankruptcy is [1].

2 Methodology

The digital economy makes certain adjustments to the activities of modern enterprises and organizations. Ensuring competitive position on the market requires constant updating, the use of innovative technologies, which in turn is associated with attracting investment resources, and this is possible only in the case of stable, stable work. The more stable the financial and economic situation of an enterprise, the more it is independent of changes in external market, political and legal conditions. Also, high financial stability significantly reduces the risk of bankruptcy. Analysis of the financial and economic sustainability of the enterprise is very important for any industrial enterprise, as it allows you to identify your own advantages and strengths of competitors, the same applies to the shortcomings, weaknesses of the enterprise and its competitors [2].

In modern conditions of the digital economy, analysis of not only financial coefficients that are static in nature (however, there are coefficients confirming the dynamic nonrandom nature of the conclusions drawn from the values of these coefficients), but analysis of cash flows from operating, financial and investment activities is becoming more important. Such an analysis allows not only the cash flow budget (BDS) calculated directly, but also the BDS calculated indirectly using data obtained from the company's balance sheet [11]. At the same time, it does not matter according to which financial reporting system (international or Russian) this balance is drawn up. Also to assess the total cash flow allows an indicator such as net profit secured by cash flow calculated on the balance sheet. An assessment of the factors affecting this indicator allows us to fully determine the degree of influence on the company's cash flow of its financial result for the period, accounts receivable and payable, inventories, results and costs from the investment and financial activities of the company [9].

To quickly and qualitatively analyze the financial stability of the enterprise in the digital economy, we developed an analysis model in the EXCEL software product. The developed model allows not to calculate each coefficient of individuality, but simply "scoring" the company's balance sheet in the created form to get the calculated financial ratios for the analyzed period, deviations (from the plan, from the standards and in dynamics), also allows you to analyze the values obtained in the balance sheet models of analysis of financial stability and draw conclusions on automatically generated BDDS indirect method and net profit provided by cash flow and factors affecting it [4].

The developed model includes several analytical sheets, each of which represents a separate independent semantic part of financial analysis: analysis of financial stability; analysis of economic sustainability; analysis of liquidity and solvency; profitability analysis; analysis of turnover and business activity; analysis of tax burden; the balance sheet model of financial stability by calculating own working capital; the balance sheet model of financial stability by calculating and grouping assets and liabilities by liquidity and maturity, respectively; cash flow budget calculated using the indirect method; net profit secured by cash flow and factors influencing it.

3 Results

A study of the financial sustainability of the enterprise allows us to identify the strengths and weaknesses of the company and its competitors, to develop short-term and long-term strategy of enterprise development with consideration of the scenario activities. The developed information and analytical analysis model, which includes all stages of analyzing the financial stability of the enterprise, helps in this analysis. Conducting a consistent analysis of financial stability serves as an evidence-based and justifying basis for making management decisions in all areas of the company's business activities (operating, investment and financial). The value of current liquidity ratio in 2015–2019 more than 1.5: this suggests that the analyzed enterprise has more current assets than short-term liabilities: the company has a high ability to pay off its obligations in the short term [5].

The dynamics of the absolute liquidity ratio suggests that the company is not able to pay immediately obligations at the expense of cash. Nevertheless, the funds in the current account increased, but short-term liabilities also increased. The quick ratio for the analyzed period shows a fairly high value [5]. However, most of the liquid assets are accounts receivable, therefore, this value of the indicator cannot be considered sufficient; measures to reduce receivables are needed. The financial situation of the financial independence ratio deteriorated in 2018 compared with 2017, but remains higher compared to competing enterprises in the building materials industry.

The financial stability ratio during the analyzed period shows a positive trend. This trend suggests that the company is financially independent and improves its solvency over the long term [6]. The financial leverage ratio in 2019 increased compared to 2018, which is a negative trend. This value of the coefficient shows that more net profit remains at the disposal of the enterprise, as well as higher investment attractiveness compared to other enterprises in this industry [8]. Also evaluate the financial stability of LLC "Samara Stroyfarfor" BDDS, calculated by an indirect method, allowed us to assess the impact of the proposed activities on the amount of cash at the end of the planned 2019: a decrease in accounts receivable in the amount of 259 948 thousand rubles.

The increasing effect was exerted by a decrease in reserves in the amount of 250,868 thousand rubles [10]. The increasing effect was exerted by an increase in accounts payable in the amount of 14,819 thousand rubles. The cash balance at the end of the planned year 2019 amounted to 676,398 thousand rubles, which is 653,259 thousand rubles, which will certainly have a positive impact on the financial and economic stability of Samarsky Stroyfarfor LLC. Also, to assess the financial and economic stability of Samarsky Stroyfarfor LLC, the net profit secured by cash flow (PE ODP) was calculated.

The value of the state of emergency of ODP at the end of 2019 is planned at 699,240 thousand rubles, which is 562,556 thousand rubles. more Net income from the statement of financial performance. This suggests that the net profit of the company Samarsky Stroyfarfor LLC is fully secured by the cash flow and the analyzed enterprise has this money "really". It also makes it possible to assert that large wholesale consumers of Samarsky Stroyfarfor LLC pay on time for the products shipped by them and timely pay off their debts to Samarsky Stroyfarfor LLC (this is due to the planned decrease in accounts receivable). Thus, the financial and economic stability of Samara Stroyfarfor LLC has grown significantly in 2019 according to our assessment of this article.

4 Discussion

It seems necessary to consider the main opinions of scientists involved in the problems of financial stability of economic entities. The study of this aspect revealed two of the most popular approaches. Proponents of the first approach consider the concept of "financial stability" broader than the concept of "financial condition". Gannon noted that the financial condition characterizes the placement and use of the funds of the enterprise [6]. It is due to the degree of implementation of the financial plan and the measure of replenishment of own funds from profits and other sources, as well as the speed of turnover of production assets and especially working capital. According to these authors, the financial condition is manifested "in the solvency of enterprises, in the ability to timely meet the payment requirements of suppliers of equipment and materials in accordance with business agreements, to repay loans, pay wages to workers and employees, to make payments to the budget". With this approach, the concept of "financial stability" is embedded even more broadly than in the concept of "financial condition of the enterprise". Based on the content of the definition of financial condition, given G. Bannock, R.E. Baxter, R. Rees, we can make the following fundamental conclusion that the financial condition of an enterprise is expressed in:

- rationality of the structure of assets and liabilities, i.e. company funds and their sources,
- efficient use of property and profitability of products,
- the degree of its financial stability,
- level of liquidity and solvency of the enterprise [2].

According to Dessler, financial stability is one of the most important characteristics of the financial condition of the enterprise [3]. But on the other hand, the classification of types of financial condition uses the concept of financial stability. They distinguish four types of financial condition in which the enterprise can be: absolute financial stability; normal financial stability; unstable financial condition; crisis financial condition. From this we can conclude that the meaning of the concept of "financial stability" is wider than representatives who support the classical point of view, since the type of financial stability determines the type of financial condition; one concept is expressed through another.

It should also be noted that a number of learned economist, for example, Garbie share the approach of Garner [7, 9]. Of course, the coefficients calculated; the liabilities of the balance sheet are the main ones in this unit of assessing the financial stability of the enterprise, however, this characteristic with the help of such indicators is unlikely to be complete—it is important not only where the funds were raised from, but where they are invested in, what is the structure of investments. Therefore, it is necessary to draw a conclusion about the importance of the indicators calculated for the asset balance, which have a significant impact on the financial stability of the enterprise.

Proponents of the second approach consider the company as financially sound if it has the financial resources to pay off its financial obligations by the due date. Therefore, financial stability can also be understood in a broad sense. With this approach the definition of financial stability of an enterprise is close to the concept of its solvency. Solvency in international practice means the sufficiency of liquid assets to repay at any time all its short-term liabilities to creditors. The excess of liquid assets over liabilities of this type means financial stability.

5 Conclusion

Using the information model in the analysis can significantly reduce both the time of analysis itself and the time of making a management decision based on the conclusions of this analysis. Thus, the digital model developed by us can find a worthy practical application in industrial enterprises. It can also be further commercialized. In most real situations, the problem of product selection is still solved by simple expert methods. To improve the objectivity of decisions-making about the choice of suppliers and consumers, we can use models for organizing complex examinations based on the idea of decision matrices. This assessment is convenient for any market situation of buying and selling, and is especially important for complex technical complexes and software products.

References

1. Andreeva SV, Popova EE, Tarasova TM (2019) Integration of the internal control system for financial stability of the organization. In: Mantulenko V (ed) Global challenges and prospects of the modern economic development, the European proceedings of social and behavioural sciences, vol 57. Future Academy, London, pp 113–125
2. Bannock G, Baxter RE, Rees R (2015) The Penguin dictionary of economics. Penguin Books, Harmondsworth
3. Dessler G (2015) Personnel management. Modern concepts and techniques. Reston Publishing Company Inc., Virginia
4. Dowries J, Goodman JE (2017) Dictionary of finance and investment terms. Barron's, New York
5. Fitch TP (2018) Dictionary of banking terms. Barron's, New York
6. Gannon MJ (2015) Management: an integrated framework. Little, Brown and Company, Toronto
7. Garbie I (2016) Sustainability in manufacturing enterprises. Springer, Cham
8. Garmire BL (2017) Local government police management. International City Management Association, New York
9. Garner BA (2015) Dictionary of law. Peter Collin Publishing, Great Britain
10. Griffin RW (2015) Management. Houghton Mifflin Company, Boston
11. Oxford University (2016) Concise dictionary of business. Oxford University Press, Oxford
12. Vishnyakova AB, Golovanova IS, Maruashvili AA, Zhelev P, Aleshkova DV (2020) Current problems of enterprises' digitalization. In: Ashmarina S, Mesquita A, Vochozka M (eds) Digital transformation of the economy: challenges, trends and new opportunities. Advances in intelligent systems and computing, vol 908. Springer, Cham, pp 646–654. https://doi.org/10.1007/978-3-030-11367-4_62

Application of Information Technologies in Tax Administration

O. L. Mikhaleva and M. Vochozka

Abstract The relevance of the problem is determined by the results of the analysis on the use of information technologies in tax administration. Developed and implemented advanced digital technologies have allowed raising tax administration to a quite high level. The tax authorities of Russia, applying modern technologies in their work, are building a new strategy of interaction with taxpayers. The article investigates consequences of using advanced digital technologies in taxation. Modern technologies have a great potential for qualitative changes in the tax administration in the conditions of the economy digitalization. The use of technologies has conceptually changed the approach of tax authorities to control and analytical work. The practical significance of the article is to assess consequences of the use of information technologies for participants in tax relations.

Keywords Tax administration · Big Data · Blockchain · Artificial intelligence

1 Introduction

Over the period of the tax system existence in the Russian Federation, a certain mechanism of tax administration was gradually formed. In modern conditions, one of the main tasks of tax administration is to respond to changes in the economy and tax relations in a timely manner. The study of tax administration problems is directly related to improving the efficiency of the tax system [1]. High-quality administration of all taxes enables to make the process of collecting them more transparent, as well as reduce the volume of the shadow economy, which is characterized by non-payment of taxes. Effective operation of the tax system is possible only with the

O. L. Mikhaleva (✉)
Samara State University of Economics, Samara, Russia
e-mail: Mikhaleva2007@yandex.ru

M. Vochozka
Institute of Technology and Business in České Budějovice, České Budějovice, Czech Republic
e-mail: vochozka@mail.vstecb.cz

S. I. Ashmarina and V. V. Mantulenko (eds.), *Current Achievements, Challenges and Digital Chances of Knowledge Based Economy*, Lecture Notes in Networks and Systems 133, https://doi.org/10.1007/978-3-030-47458-4_32

application and use of modern information technologies: Big Data, user recognition technologies, Blockchain or distributed registries, and Internet of things tools.

2 Methodology

When studying tax issues, the greatest importance is shown in the development of empirical research and the application of theory to practical solutions. Applied research in the taxation field is determined by interest from both the state and taxpayers. The contribution investigates the role of advanced technologies in improving the efficiency of tax administration. The author tries to assess the transparency and fairness degree of taxation [8]. In particular, the comparative analysis allows structuring information, identifying areas for further research, and in practical terms—determining the most significant problems that need to be studied using theoretical tools.

The development of applied research, in turn, requires a variety of methods used. In the research process, the author used the following methods: theoretical (dialectical logic, rational cognition, etc.); diagnostic (diagnostic analysis of the research object condition and causes); empirical (description of facts, measurement and generalization of research results).

3 Results

The idea of creating a “household income register” is not new, as it already exists in many foreign countries. In the coming years, the sphere of tax administration is expected to see information innovations. Due to the large amount of information available to tax authorities, it will become more difficult for taxpayers to avoid paying taxes. According to the new head of the Federal Tax Service (FTS) D. Egorov, a key objective of the FTS is not to collect more taxes, and “clear the market”, but to make the tax payment process “inevitable and at the same time as comfortable as possible” [6].

The application of a whole set of digital systems into the FTS practice allows controlling taxes. The automated control system VAT-2 allows specialists to quickly detect gaps in the payment and VAT refund. With the help of VAT-2, it became possible to reduce gaps in the VAT payment chain from 8% of all transactions at the beginning of 2016 to less than 0.6% in November 2019.

Since the end of 2019, the automated information system “Tax-3”, which combines all electronic services and programs of the Federal Tax Service, as well as many state databases and registers, has been fully operational. The data set is huge and includes all the basic information about all taxpayers, their property, and taxes accrued and paid.

Online cash registers are related to the Internet of things technology (a computer network of tangible objects connected through the interaction with the external environment by providing special identifiers that can be recognized by technological devices). The Internet of things technology can also be attributed to the applied system of product labeling.

One of the latest FTS advances is a mobile app for self-employed people. The main activities for self-employed individuals are: transportation of passengers, renting of flats, construction services, tutoring, marketing, etc. About 70% of the participants of the special tax regime previously did not declare incomes from their business activities. With this app the accounting and tax calculation are made by tax authorities. Thus, completely remote interaction is provided through this application. By the end of 2019, more than 330 thousand Russians registered themselves as self-employed. The total amount of income tax on professional incomes exceeded 1 billion rubles.

Initially, in 2019, the experiment on taxation of the self-employed was conducted in Moscow, Tatarstan, and the Moscow and Kaluga regions. Since January 1 2020, 19 other regions have joined this tax regime, and since mid-2020, it is planned to expand the special tax regime for the entire Russian Federation. In the future, a similar principle will be gradually implemented in other tax systems.

Measures to improve the tax administration provided the main growth in tax revenues in 2019. Therefore, the total volume of all tax revenues is growing, despite the decline in oil prices and the introduction of a refundable excise tax to balance the prices of petroleum products in the domestic market. The share of tax administration in the total growth of non-oil and gas revenues was 17% in January–November 2019.

The use of Big Data technology has conceptually changed the approach to conducting control checks, minimizing the impact of the human factor. Digital services and pre-verification analysis have reduced the number of on-site tax audits. Employees of tax authorities began to go to the on-site inspection to collect evidence if they are sure of violations facts. With a reduction in the number of checks, the amount of additional charges has increased significantly. If in 2016, on average, inspectors added 13.6 million rubles per inspection, the amount reached 36.2 million rubles in the first 9 months of 2019.

The Federal Tax Service has transferred almost all services online—in the taxpayers' personal accounts, they can calculate and pay taxes themselves, and submit a tax return. In the foreseeable future, some of the functions of the tax administration will be transferred to artificial intelligence (AI). Currently, AI is used in the personal account of an individual taxpayer. This is done with the help of a robot that helps solve certain issues related to the tax administration. Examples of using AI in the field of taxation include: processing tax notifications; forecasting account balances to calculate the tax base for taxes; determining the amount of deductions; filing a tax return; and consulting on tax issues. The FTS digital platform of Russia has made it possible to create electronic services that help solve any tax issue online from anywhere in the world.

Currently, the issue of electronic data exchange via open communication channels with the use of cryptographic protection of tax information with the CIS countries is being discussed. Exchange of information on tax issues in the electronic form will become an additional tool for the tax administration.

4 Discussion

In order to work effectively, the entire tax administration system should necessarily be managed. The level of manageability is determined by how well processes of collecting, processing and analyzing information for organizing the work of tax authorities are organized. Every day, the Russian Federal Tax Service works with a huge array of data. In total, four petabytes of data are stored and processed in the information system of the Russian FTS.

Big Data refers to the use of heavy-duty computers and high-tech software to collect, process, and analyze huge amounts of data with rapid changeability. This technology, radically expanding information sources available to the organization online, as well as its collection, storage, processing and analysis, turns into a valuable and unique resource due to its characteristics [5]. Cloud storage will also be developed, which will provide easy access to a huge amount of information. It contributes to more effective communication between employees [3].

The advantage of Big Data lies in the almost unlimited potential for information accumulation, as well as in the operation speed [7]. According to Tapscott and Tapscott, blockchain is a perpetual digital distributed journal of economic transactions that can be programmed to record not only financial transactions, but almost anything that is valueable [9]. Blockchain technology in tax administration will allow creating distributed data storage systems with identification of each user and access security.

Using Big Data increases the transparency of tax processes. And together with the Blockchain technology, it becomes possible to track large transactions with high VAT, tax refunds and cross-border transactions.

Another trend of the Federal Tax Service towards digitalization will also affect individual taxpayers. One of today's priorities is to create a "household income register". After that, according to the Russian Prime Minister M. Mishustin, every income will be tracked from beginning to the end [4].

The idea of creating a common information base about Russian citizens first appeared more than 10 years ago. It was assumed that the state will not collect additional information about citizens, but will bring together in one system everything that it collected before. In 2016, its first practical implementation appeared—the data of all records were consolidated into a single register. So, there was a database that contains names, surnames, registration addresses, dates of birth and death of all Russian citizens. A new draft law is being discussed, according to which the information base should be updated with new data: information about education; compulsory insurance; registration as individual entrepreneurs and self-employed, etc.

5 Conclusion

In recent years, the policy of the Federal Tax Service of Russia has been aimed at digitizing the tax system. The main task was to transfer information electronically between various agencies in order to quickly obtain complete information about taxpayers. As a result, the country's most comprehensive database of taxpayers has been created and will continue to be updated. Thus, the tax administration process becomes faster and less time-consuming, since information about taxpayers can be obtained online. The conducted analysis confirms that there is a need for changes between the participants of tax relations arising in the tax administration.

The digital economy needs a review of many established approaches to taxation. Modern technologies have a great potential for qualitative changes in the tax administration in the conditions of the economy digitalization. World Bank experts consider the digital economy (in the broad sense of the word) as a system of economic, social and cultural relations based on the use of digital information and communication technologies [10]. Digitalization also raises new, fundamental questions, such as privacy protection, the future of work, cybersecurity, the market power of digital platforms, who has access to data, what information is reliable, how new technologies can be used ethically, and how to ensure that people and companies can keep up with this transformation [2]. Further use and development of digital technologies will allow tax authorities to get more benefit from existing data, to fight tax evasion and fraud.

References

1. Aleshkova DV, Greshnova MV, Smolina ES, Popok LE (2020) Research of efficiency of tax stimulation of innovative entrepreneurship. In: Ashmarina SI, Vochozka M, Mantulenko VV (eds) Digital age: chances, challenges and future. Springer, Cham, pp 80–84
2. Government of the Netherlands (2018) Dutch digitalisation strategy. http://www.government.nl/documents/reports/2018/06/01/dutch-digitalisation-strategy. Accessed 11 Feb 2020
3. Larkin J (2017) HR digital disruption: the biggest wave of transformation in decades. Strat HR Rev 16(2):55–59
4. Meduza (2020) Mishustin wants to create a register of all citizens and their income. Will they control us in China? https://meduza.io/feature/2020/01/18/mishustin-hochet-sozdat-reestr-vseh-grazhdan-i-ih-dohodov-nas-budut-kontrolirovat-kak-v-kitae. Accessed 11 Feb 2020
5. Nazarov MA, Mikhaleva OL, Chernousova KS (2020) Digital transformation of tax administration. In: Ashmarina SI, Vochozka M, Mantulenko VV (eds) Digital age: chances, challenges and future. Springer, Cham, pp 144–149
6. RBC (2020) The new head of the Federal Tax Service named the key task in working with business. https://www.rbc.ru/politics/20/01/2020/5e24da7e9a79473d1ed60549. Accessed 11 Feb 2020
7. Reese H (2017) Understanding the differences between AI, machine learning, and deep learning. Tech republic. https://www.techrepublic.com/article/understanding-the-differences-between-ai-machine-learning-and-deep-learning/. Accessed 11 Feb 2020

8. Revina SN, Paulov PA, Sidorova AV (2020) Regulation of tax havens in the age of globalization and digitalization. In: Ashmarina SI, Mesquita A, Vochozka M (eds) Digital transformation of the economy: challenges, trends and new opportunities. Springer, Cham, pp 88–95
9. Tapscott D, Tapscott A (2016) Blockchain revolution: how the technology behind bitcoin is changing money, business, and the world. Portfolio: Penguin, London
10. World Bank (2016) World development report 2016 digital dividends. https://openknowledge.worldbank.org/bitstream/handle/10986/23347/210. Accessed 11 Feb 2020

Information Technologies in Increasing Competitiveness of Small and Medium-Sized Construction Business

E. V. Volkodavova and S. M. Petrov

Abstract The unification of small and medium-sized construction enterprises into business networks creates a favorable environment for the development of healthy competition, and most importantly, stimulates the introduction of modern technologies, which can significantly reduce the innovation and technological gap with respect to world best practices, improve the administrative environment, and strive for accessibility and reliability of their financing, develop the infrastructure and service component. The fulfillment of this purpose is provided for small and medium-sized construction enterprises in business networks in both domestic and foreign markets using information technology. In the current business environment of small and medium-sized construction enterprises in business networks, the practice of using information technology is not sufficiently developed. This study considers information technology that can significantly increase the effectiveness and competitiveness of small and medium-sized construction enterprises in business networks and their value for potential customers.

Keywords Small and medium-sized construction enterprises · Business networks · Information technology · Scientific and technological progress

1 Introduction

Currently, within the framework of business networks (hereinafter referred to as BN) of small and medium-sized construction enterprises (hereinafter referred to as SMSCE), information technologies are one of decisive factors for increasing their

E. V. Volkodavova (✉)
Samara State University of Economics, Samara, Russia
e-mail: Vev.sseu@gmail.com

S. M. Petrov
Samara State Academy of Construction and Architecture of Samara State Technical University, Samara, Russia
e-mail: pesm@mail.ru

S. I. Ashmarina and V. V. Mantulenko (eds.), *Current Achievements, Challenges and Digital Chances of Knowledge Based Economy*, Lecture Notes in Networks and Systems 133, https://doi.org/10.1007/978-3-030-47458-4_33

competitiveness, since they allow entrepreneurs to quickly transfer relevant information over long distances, to cover significant number of suppliers and consumers, to build business management systems and to predict the economic situation in the construction market.

Computer-aided design systems greatly simplify the work of architects and designers. At present, AutoCAD, Allplan packages, which are constantly being developed and modernized based on the needs of the construction market and modern design as a whole, are widely used for designers in software.

Modern data processing capabilities allow you to maximize the ideas of planners and architects, and turn them into real models. In addition, modern software allows automating the creation of drawings, as well as using modules to convert future construction objects from drawings into specific volumes of labor and materials needed for the construction of buildings and structures.

There are many programs ("construction calculators") that allow you to perform calculations on the consumption and costs of materials, services and work in those specific areas of construction and architectural work, which is very promising for SMSCE in BN, as it allows extracting maximum profit without hiring additional staff.

At the moment, the use of 3D printers is gaining great popularity, allowing you to produce fairly complex, multi-level figures that cannot be produced in another way or which can be produced by other methods in a lot of time and will cost much more. This applies primarily to the production of prototypes, piece building products. At the same time, the cost of this high-tech tool and its size are constantly decreasing, which allows the use of printers in small offices, which is quite relevant for representatives of small and medium-sized construction enterprises.

2 Methodology

The methodological scheme for studying the competitiveness of construction production of SMSCE in BN using information technologies includes the following scientific and technical hypothesis: the system-technical coordination of all participants of small and medium-sized construction enterprises within the framework of construction production for a system analysis and assessment of the quality of construction and installation works should significantly increase the reliability of constructed objects based on modern information and computing technologies, as well as the competitiveness of business networks in segments of the construction market.

The methodological and theoretical basis of the study is:

1. A method of complex analysis, which made it possible to cover a wide range of issues, problems and solutions in the field of formation, organization and activity of small and medium-sized enterprises in business networks.

2. A system analysis method which considers the influence of information technology on interconnections and interdependencies of improving the quality and competitiveness of professional, managerial and entrepreneurial functions of Russian small and medium-sized enterprises.
3. The concept of rational behavior of market entities, the principles of which made it possible to analyze and evaluate the impact of information technology on improving professional results of small and medium-sized enterprises operating in business networks and their competitiveness in the construction market.

3 Results

In the formation and functioning of business networks, the use of effective information technologies is of great importance. As practice shows, only a small number of small and medium-sized construction enterprises use a wide range of IT in BN. Table 1 presents technologies that must be applied in business networks.

When the business network operates, the competitive activities of enterprises included in this network and the activities of the network itself largely depend on information technologies used, presented in Fig. 1.

Table 1 Digital technologies in business networks

Stage	Method	IT
1. Formation of the core of potential participants in business networks	Questioning, survey, analysis	Use modern videoconferencing systems, such as Polycom or CISCO, which even with a very narrow communication channel (for example, organized using ADSL technology) can organize business communication between SMSCEs in BN without their personal presence in the office They ensure legal, economic, technological, social databases so that the information necessary to solve managerial problems is available as soon as possible, at any point, to any MEOSAR in BN

(continued)

Table 1 (continued)

Stage	Method	IT
2. Development of the overall strategy	SWOT, PEST, SNW-analysis	Use IP-telephony in BN, which allows providing remote construction sites with telephone communications at the earliest possible time and at economical prices (you can combine all remote sites via VPN with PBX in central offices and provide employees with internal corporate numbers). Thanks to this, you can call colleagues on internal communication free of charge, and this allows construction companies to save on communication costs, and the more remote the construction is underway, the greater the savings They provide hardware and software, telecommunications systems in BN, information resources and access to them, including the storage, processing, transformation and transfer of information and knowledge between SMSCEs in BN
3. Development of pilot projects	SWOT, PEST, SNW-analysis	Use Microsoft Exchange to synchronize email, calendars, and contacts on mobile devices. In this case, all important business information is stored on a server that provides backup Due to this, they ensure the priority development of structures, and the production and reproduction of information and knowledge in BN

(continued)

Table 1 (continued)

Stage	Method	IT
4. Development of strategic projects	SWOT, PEST, SNW-analysis	Use IP for the construction business: on the 1C platform, called "Management of the construction organization" Develop and implement organizational and methodological foundations and programs for the consistent, focused and effective implementation of information technology in the management system of SMSCEs in BN
5. Self-regulation	Calculation and analysis of economic efficiency	SMSCEs in BN use electronic trading of Internet sites (Sberbank-AST, Roseltorg, MICEX, RTS-tender, Zakazrf, etc.) The advantages of using computer technologies between SMSCEs in BN at this stage are based on achievements of telecommunication technologies and distributed information processing. The ultimate goal of IT is not just to increase the efficiency of data processing and help SMSCEs in BN, but to create a highly efficient construction production. Applied IT should help SMSCEs in BN to withstand the competition and gain a competitive advantage

Source authors

The results of the implementation of information systems in SMSCE in BN are the following:

1. Improve the efficiency of business networks through electronic design documentation, increasing the efficiency of intra-network production and information interaction.
2. Implement distributed data processing when the workplace has enough resources to receive and analyze information.
3. Create developed communication systems when production facilities are combined to send messages as quickly as possible.

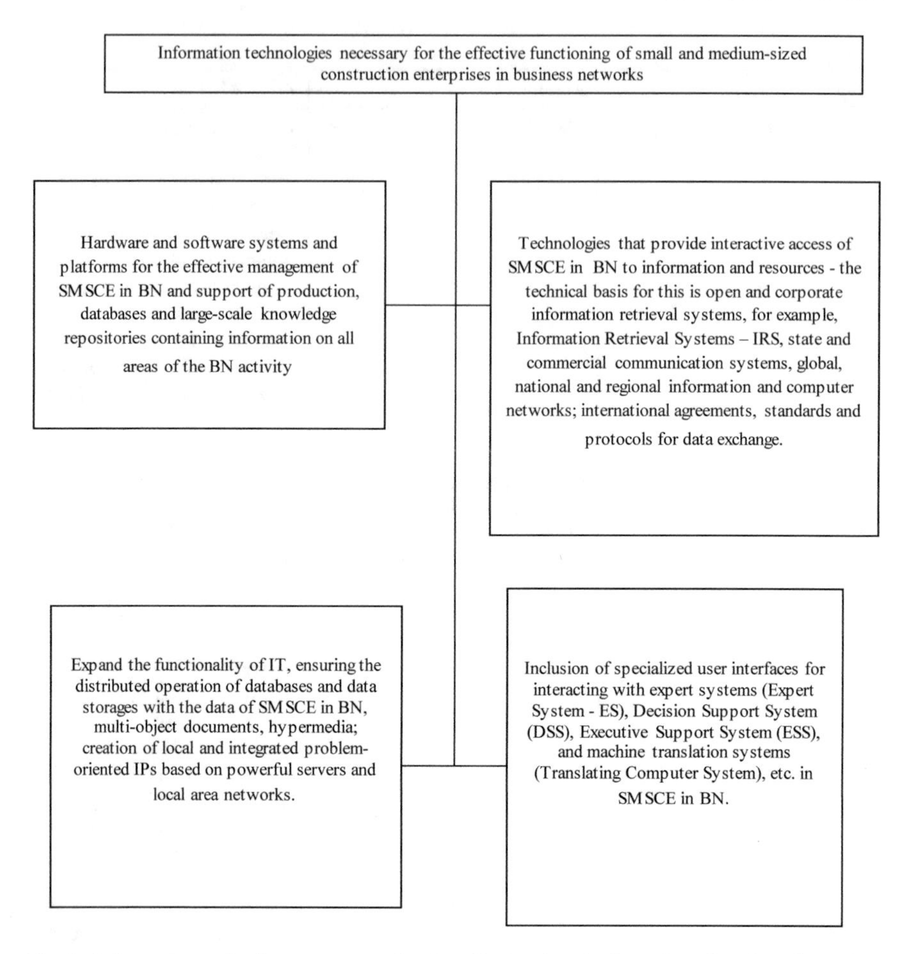

Fig. 1 Information technology necessary for the effective functioning of small and medium-sized construction enterprises in business networks (*Source* authors)

4. Eliminate interference in the system of integration of SMSCE in BN—external environment, direct access to world information flows.
5. Create and develop systems of electronic orders for construction products.

The main criteria for success in business are professional management, the ability to ensure the effective work of personnel, to correctly identify, design, implement and improve business processes, to effectively conduct organizational, administrative and economic activities. In these conditions, modern information technologies and integrated information systems created on their basis become an indispensable tool in ensuring the achievement of strategic goals and sustainable development of companies and organizations.

4 Discussion

Information technologies are used by SMSCE in various areas of activity: to search for office space; to develop automation systems; to improve the quality of buildings (using an information-intellectual environment), in pouring concrete (3D laser technology), in architecture and construction (laser scanning systems), in design and construction (electronic modeling), etc.

It should be noted that at present, one of the most important links in the management of construction production, especially for SMSCE is the quality control of construction production. One of the main tasks of control is to detect deviations from the planned course of organizational and technological processes, but the main control chain is preventive and corrective, i.e. as much as possible to prevent possible shortcomings, and in case of detection to correct them in a timely manner. In this case, the application of information and computing technologies to assess the quality of construction and installation works is carried out.

One of the most important elements of SMSCE is the introduction of the electronic document management system and design management systems to ensure the registration, storage and issuance of electronic design documents, which are original of the transmitted design documentation. Modern integrated workflow systems are complex software and technology complexes designed to organize the storage and transmission of various structured data over networks, including text documents in various formats.

Today, design and technological preparation of production is the same document management system, but it is more knowledge-intensive and requiring. In addition to computer science knowledge, it masters a huge number of subject disciplines. However, very often the concept of "automation of design and technological preparation of production" comes down to the well-known phrase "computer-aided design system" (CADS) in its simplified sense—we need systems for obtaining drawings, technological processes and control programs [5, 6].

The authors conducted research among the leaders of SMSCEs in the Samara region from September 2016 to November 2019 within the framework of the Scientific and Technical Center "Organization of Work in Construction" at Academy of Construction and Architecture of the Samara State Technical University, and concluded that due to insufficient comprehensive digitalization and automation of business processes, many SMSCEs leave the market, unable to withstand the competition of more experienced rivals. For the same reason, in the course of activities, SMSCEs face such problems as increased time and cost of design work (20% of respondents), increased costs that deprive them of existing and new customers (35% of respondents). We should agree with the opinion of the authors presented in [1–4, 7] that the implementation of information technologies will help to solve the tasks in a quality and efficient manner, without additional unforeseen expenses.

5 Conclusion

Today it is impossible to carry out successful entrepreneurial activities without information systems. Each year, the volume of information received, processed and stored increases. Therefore, SMSCEs will need to use the most modern and advanced technologies as part of BN. One of the main competitive advantages of SMSCE in BN is a quick response to any market changes and the ability to overcome them with minimal losses for the enterprise. And it is impossible to carry out these actions without information technology. The implementation of this development vector is determined by information technologies. Due to them, the competitiveness of SMSCEs will be ensured and increased in BN both in domestic and foreign markets.

References

1. Bogdanova AA, Demenkova AO, Kiselev AV (2016) Information technology for small and medium-sized businesses. In: Emelyanov SG, Chervyakov LM, Lapina TI, Sazonov SYu, Khalin YuA (eds) Information systems and technologies materials of reports of the II international scientific and technical correspondence conference. Southwestern State University, Kursk, pp 44–47
2. Kasyanova A (2009) Formation of the innovative economy as the most important task of modern Russia. INION RAS, Moscow
3. Marchenkov EI (2012) Server virtualization in information technologies for small and medium-sized businesses. In: Belousov LA (ed) Russia on the way to a post-industrial society: forecasts and reality. St. Petersburg State University of Economics and Finance, St. Petersburg, pp 42–43
4. Odintsov BE (2013) Information resources and technologies in the economy. Infra-M, Moscow
5. Volkodavova EV, Zhabin AP, Goryacheva T, Tomazova O (2018) Problems encountered in adapting the industrial policy to the new economic realities. Helix 8(2):3237–3247. https://doi.org/10.29042/2018-3237-3247
6. Volkodavova EV, Zhabin AP, Yakovlev GI, Khansevyarov RI (2020) Key priorities of business activities under economy digitalization. In: Ashmarina S, Vochozka M, Mantulenko V (eds) Digital age: chances, challenges and future. ISCDTE 2019. Lecture notes in networks and systems, vol 84. Springer, Cham, pp 71–79. https://doi.org/10.1007/978-3-030-27015-5_9
7. Zverev VV (2009) Information infrastructure of entrepreneurship: theory and practice of development: monograph. Volgograd Scientific Publishing House, Volgograd

Development Approaches of Gaming Applications Under Conditions of Innovative Economy

L. E. Popok, A. V. Mantulenko, and M. O. Suraeva

Abstract The purpose of this work is to analyze existing approaches and requirements to the development of gaming applications. The authors consider development approaches of gamified applications, the rules for their script designing and principles of the users' involvement into the game process under conditions of innovative economy. The article includes some practical examples for main phases of the software design with full or partial gamification, conclusions about the feasibility of using mixed approach in gamified software development and recommendations given for developers and IT-customers to increase the commercial effectiveness of gamified software.

Keywords Gaming applications · Gamification · Application design

1 Introduction

In the context of innovative and information economy, gamification (among other innovative technologies) has gained popularity in various areas: education, psychology, economics, manage, management etc. One of the most common practical applications is the design and development of fully gamified applications (including mobile ones) or applications with some gamified elements. Effective involvement into the game process, the developer (other application owner) will be able to get more commercial income than its competitors by attracting a larger audience, capture its attention for longer periods of time, etc. However, in order to develop effective

L. E. Popok
Kuban State Agrarian University (named after I. T. Trubilin), Krasnodar, Russia
e-mail: lpopok@gmail.com

A. V. Mantulenko (✉)
Samara University, Samara, Russia
e-mail: mantoulenko@mail.ru

M. O. Suraeva
Samara State University of Economics, Samara, Russia
e-mail: marusyasuraeva@mail.ru

S. I. Ashmarina and V. V. Mantulenko (eds.), *Current Achievements, Challenges and Digital Chances of Knowledge Based Economy*, Lecture Notes in Networks and Systems 133, https://doi.org/10.1007/978-3-030-47458-4_34

gamification software (from a marketing point of view), you need to choose an appropriate approach to development of such a system.

2 Methods

Theoretical and methodological basis of this article are works of Russian and foreign researchers who studied issues of gamified applications development and commercialization in conditions of innovations. General scientific and private methods of cognition for studying the research problem were used. These methods allowed exploring the conceptual framework of the gamified applications development for their better commercialization in the modern economic conditions. Private research methods were used at the stage of processing, analyzing and summarizing the obtained scientific information.

3 Literature Review

The concept of "gamification" was developed at the beginning of the 2000s, but real interest to the phenomenon has emerged only in the last decade in academic (scientific and business societies) [1, 5, 6, 10, 22]. To date, two approaches to the analysis of this phenomenon have been clearly identified.

The first concept of "gamification" is associated with the definition of "game", it allows researchers to treat gamification as an integral part of the overall cultural process, where games and gaming experience are considered as an essential component of society [9]. One of the main ideas in this concept is that the game process (according to its in nature) is not associated with rational activity, so it can be characterized in the logic of adaptation processes (as an instinctive reaction of the organism to environmental challenges) [8].

The second approach to gamification was formed in the sphere of economics and management design. Gamification is considered as an instrumental element of the environment design as a whole [21]. Gamification is given the ability to change and transform reality [12]. Gaming technology is becoming a key tool for solving different problems of the real world due to the opportunity to "reprogram" participants. Since the game is a model of reality, the gamification can be seen as a project for the transformation of our society, the acquisition of new knowledge, etc.

It is worth noting that a number of authors express doubts about the legitimacy of the "gamification" concept. They call the term itself as a problematic one [4], controversial [2, 3], paradoxical [14], emphasizing the lack of theoretical elaboration of this concept [6, 7, 11, 13, 15]. The consequences of its poor scientific development are difficulties with the implementation of gamification projects in practice [3, 7, 13, 20].

4 Results

According to Werbach, an expert in the field of gamification, in designing gamified applications, it is necessary to consider three basic rules: "journey" of a player, "balance" and giving "impressions" ("experience") [18].

The "journey"-rule assumes the need to perceive the target audience as *players*. While working with the gamified app, the player goes along the path comparable to a journey that allows getting unique gaming experience. For the developer, it is necessary to design such a journey, male it integral, structured and, last but not least, interesting, so you need to pay attention to all the elements: the beginning, the basic structure and the process of moving to a goal. Thus, at the beginning, it is necessary to help the user to get engaged in the game process, as quickly and simply as possible—for this purpose, we can use tooltips, limit some functional features that make the start of work complex and difficult, and the impossibility of losing at the early stages. For the basic structure design, we should pick up one or more training mechanics that will allow the player to overcome the above difficulties to eliminate the possibility of confusion or "stuck".

The next rule in gamification apps—the "balance"-rule—originates from the concept of balance in any game: at each stage, the game should be balanced in all its elements to remain interesting for each user. The last basic rule implies that the gamified application should enrich the user's experience, turn the current activity process into a richer and more extensive experience, without fundamentally changing the process itself.

Usage of the rules discussed above will help the developer to create a captivating and engaging gamified application that is able to turn on the player's emotions. This emotional component of the experience gained from winning, finding solutions, research and other game elements will allow the app developer to guide players to the mutual goals without losing the users' motivation.

Of course, before planning and implementation of a project, we should identify the target audience in detail: age, sex ratio, interests, and behavior, needs etc. Werbach classifies players as winners, researchers, socialists and aggressors. The behavior of the winners is determined by their desire to achieve a goal not taking into account means and obstacles [18]. Researchers primarily study the environment, trying to fully understand the presented possibilities. Socialists pay the main attention to interaction within the game, it is important for them to collectively perform tasks, create teams. Aggressors are those players for whom the enjoyment of the game is more dependent on rivalry and fierce competition [19]. This classification is used to determine needs of potential users and helps to select the necessary game elements and techniques.

When designing and directly creating gamified products, for the developer, it is important to understand the direction of the product development and maintain the focus of the game process. In addition, we should understand what experience and emotional response the player will get while using the application: the entire system should be built exactly around the user. It should be borne in mind that the game elements are only *means* of forming the experience.

In addition to the above mentioned game balance, it is necessary to create and maintain the balance between analytical and creative thinking of the player: the rationality must be harmoniously combined with the creativity to effectively gamify a particular process (especially if this process is routine one). This approach allows you to strengthen the involvement of the player in the gaming activity.

In the design of gamified applications, one of the most rational and universal solutions is the iterative development [17]: the creation of a software product in parallel with the permanent testing of a prototype by the control group and subsequent analysis of the usage results. After this analysis, the previous stages of direct development are adjusted according to the identified shortcomings, giving rise to a new conditional cycle "planning-implementation-verification-evaluation" until getting the final product. The main advantage of the iterative approach for creating gamified applications is the availability of real-time feedback from testers, which allows evaluating the project success at the early stages of application development.

As for gamified application scenarios, "activity loops" play an important role there. There are two types of such "loops" [16]. The "loop of involving" is a cyclical process consisting of three interrelated stages. This loop operates at the level of the users' activity. The stages of the activity loop are *motivation* (it is designed to encourage the user to perform a certain action), the *action* itself (the activity performed by the user, entailing a clear and effective feedback), *feedback* (the reaction of the application, which forms the basis of the next motivation, generating a new cycle).

The second type of the activity loop is the "progress loop". The progress loop is characterized by two basic principles. Firstly, the achievement of the goal in the game should be realized in small stages (steps), each of which should be clear to the player (clearly demonstrate the goal of doing each step and further development of the game). Secondly, the player development from basic (simplest) level "newbie" to the highest level "master" should include alternation of effort (work) and rest phases (such kind of balance allows preserving and maintaining an already established level of the player's involvement without overloading it).

In addition to the above mentioned approaches and principles of gamified applications development, there are two approaches to gamification, based on the difference in human perception. So, Werbach conditionally defines the first approach as "active" and the second as "sensitive" ones [18].

In the context of the active approach, the concept of "game" is considered only as something that corresponds to all the formal features of a game, and the concept of "gamification"—as a set of points, awards and leaderboards. In this approach, the incentive and reward system is used as a mechanism for users' involvement, and, for the users, the main value of the gamification itself is to meet their needs and desires. For the active approach, the growth of the player's status and self-esteem is chosen as an incentive, and the rewards are the players' motivation. In this approach, the development of game elements and mechanics is the main and primary aspect when creating the game apps. The main purpose of the gamification in this approach is to encourage users to commit a certain act (action).

On the contrary, the sensitive approach puts into the concept of "game" everything that is perceived by a person as a game, regardless of formalities, and the gamification for this approach is puzzles, tasks, "challenges" and training. The mechanisms of users' engagement, in contrast to the active approach, are positive impressions and experience, and the main value of gamification is the users' pleasure. In the sensitive approach, intrinsic motivation is used as a reward: new knowledge, the growing importance of the player to others. The players' motivation is the demonstration of their growth and progress as players. In this approach, the primary thing by creating an application is the development of game principles and rules, the idea itself, and game mechanics and elements are secondary. The main goal of gamification is to stimulate users to self-development.

5 Conclusion

As an optimal and appropriate solution for most developers of gamified applications in the innovative economic conditions, it is recommended to use a moderate approach, combining elements of both active and sensitive approaches, depending on the identified needs of players and the developer himself. From the point of view of information technology and development, it is strongly recommended to include in one of the initial stages of software design a description of the screen map (forms) of the application—a scheme reflecting all the screens (forms) of software, transitions and interactions between them, scenario loops. Such a scheme will help to identify the shortcomings of the scenario, design a complete database, and understand the logic of the application for further implementation. The screen map formed on the example of the mobile application "Navigation guide on the University campus with gamification elements" is shown in Fig. 1. The iterative development method, which is recommended for gamified software, implies revision and possible modification (addition) of the navigation map.

When designing a database, the developer also needs to consider the possibility of the future development of the software product, expanding and supplementing this database, since the involvement and gamification are long-term continuous processes that may evolve over time and require more storage options. Thus, when developing fully or partially gamified software, it is important to find an optimal approach that reflects the goals of using gamification elements, ensure a continuous process of creating engaging scenarios created by means of the above mentioned rules with the help of the iterative development and taking into account psychological and other characteristics of the target audience. It is also important to pay attention to the design phase of the navigation map and application database.

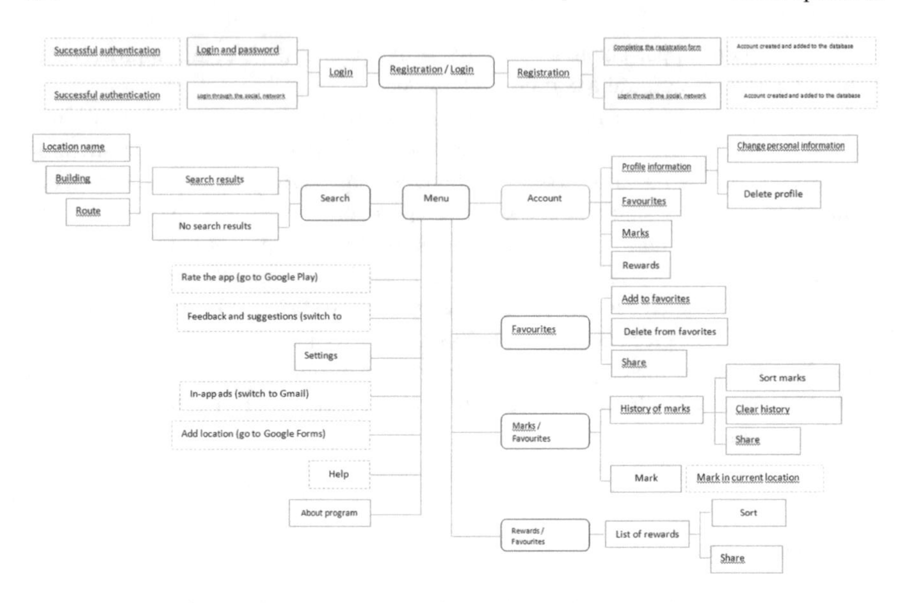

Fig. 1 Example of a gamified screen map mobile application (*Source* authors)

References

1. Babaeva AV, Klyuev AA (2017) Characteristics of gamification working with human resources: philosophical aspect. Vestn Minin Univ 4(21):13 (in Russian)
2. Bogost I (2011) Gamification is bullshit. My position statement at the Wharton gamification symposium. http://bogost.com/blog/gamification_is_bullshit/. Accessed 17 July 2019
3. Bogost I (2015) Why gamification is bullshit. In: Walz SP, Deterding S (eds) The gameful world: approaches, issues, applications. The MIT Press, Cambridge, pp 65–80
4. Fuchs M, Fizek S, Ruffino P, Schrap N (2014) Rethinking gamification. Meson Press, Lüneburg
5. Hamari J, Koivisto J, Sarsa H (2014) Does gamification work? – A literature review of empirical studies on gamification. In: Sprague RH (ed) Proceedings of the 47th Hawaii international conference on system sciences. IEEE Computer Society, Washington, pp 3025–3034
6. Hamari J, Koivisto J (2015) Why do people use gamification services? Int J Inf Manag 35(4):419–431
7. Huotari, K, Hamari, J (2017) A definition for gamification: anchoring gamification in the service marketing literature. Electron Mark 27:21–31. https://doi.org/10.1007/s12525-015-0212-z
8. Kayua R (2007) Games and people. Articles and essays on the sociology of culture. OGI Publisher, Moscow (in Russian)
9. Kheyzinga Y (2015) Homo ludens. Playing man. Izdatel'stvo Ivana Limbakha, St. Petersburg (in Russian)
10. Kurylev IN (2017) What is gamification? Game mechanics in business. http://gamification-now.ru/wtf/. Accessed 07 Sept 2019 (in Russian)
11. Landers RN (2015) Developing a theory of gamified learning: linking serious games and gamification of learning. Simul Gaming 45(6):752–768
12. McGonigal J (2011) Reality is broken: why games make us better and how they can change the world, 2nd edn. Penguin Group, New York
13. Seaborn K, Fels DI (2004) Gamification in theory and action: a survey. Int J Hum Comput Stud 74:14–31

14. Stenros J (2015) Behind games: playful mindsets and transformative practices. In: Walz SP, Deterding S (eds) The gameful world: approaches, issues, applications. The MIT Press, Cambridge, pp 201–222
15. Stenros J (2015) Playfulness, play, and games: a constructionist ludology approach. University of Tampere, Tampere
16. Valerina LP (2014) Gamification in education. Educ Pedagog Sci 6(2):314–317
17. Werbach K, Hunter D (2012) For the win: how game thinking can revolutionize your business. Perseus Distribution Services, Philadelphia
18. Werbach K (2014) Gamification: short course. https://www.coursera.org/learn/gamification. Accessed 07 Sept 2019 (in Russian)
19. Werbach K, Hunter D (2015). Engage and conquer: gaming at the service of business. Mann, Ivanov i Ferber, Moscow. Accessed 17 July 2019 (in Russian)
20. Whitton N (2009) Learning with digital games: a practical guide to engaging students in higher education. Taylor & Francis Group, New York
21. Zichermann G, Cunningham C (2011) Gamification by design: implementing game mechanics in web and mobile apps. O'Reilly Media, New York
22. Zikermann G, Linder J (2014) Gamification in business: how to overcome noise and attract the attention of employees and customers. Mann, Ivanov i Ferber, Moscow (in Russian)

Information and Analytical Support for Enterprise Business Management

O. A. Naumova and M. A. Tyugin

Abstract The lack of necessary data, using irrelevant or unreliable information can lead to ineffective management decisions. Analysis of business activity allows obtaining the necessary, timely and reliable quantitative and qualitative characteristics of it, which is currently the most important tool for successful business management that increases its efficiency. In the article, the business activity of the organization is considered in 8 areas: marketing, investment, innovation, production, finances, ecology, socio-economics and management. The main sources of information for assessing the identified areas of business activity are identified and the matrix of the information system of the business activity of the enterprise under the conditions of automated information collection and processing technologies is composed. Next, an assessment of the activities of five enterprises of the Samara region in the areas of activity. For evaluation, indicators developed by the authors were used. Based on the calculations and weighted average values of the indicators, the main trends in business activity in the digital economy are identified.

Keywords Economic analysis · Assessment of business activity · Information support for management decisions

1 Introduction

In the business world, the need to continuously build up the economic potential of each business entity becomes more and more obvious. Economic growth, especially in the context of global crisis phenomena, is impossible without the informational and analytical rationale for managerial decisions in certain areas of business activity that meets modern requirements, followed by an integral assessment of their

O. A. Naumova (✉) · M. A. Tyugin
Samara State University of Economics, Samara, Russia
e-mail: naumovaoa@gmail.com

M. A. Tyugin
e-mail: tyuginmaxim@gmail.com

S. I. Ashmarina and V. V. Mantulenko (eds.), *Current Achievements, Challenges and Digital Chances of Knowledge Based Economy*, Lecture Notes in Networks and Systems 133, https://doi.org/10.1007/978-3-030-47458-4_35

effectiveness. The methodology of business activity analysis, which allows obtaining the necessary, timely and reliable quantitative and qualitative characteristics of it, is currently the most important tool for successful business management, increasing its effectiveness. However, the lack of necessary data, the use of irrelevant or inaccurate information can lead to ineffective management decisions. The processes of adaptation of domestic accounting and reporting to international standards, the implementation of the provisions of the concept for the development of accounting and reporting in the medium term, as well as changes in the legislation have identified a range of actual problems for studying the information base for analyzing business activity indicators and methodological issues. At present, methodological and practical problems arising from the need to take into account the relationship of all areas of business activity and an integrated approach to the formation and analysis of its indicators are being updated and require further research. Existing methods for calculating these indicators do not fully reflect the overall picture of the work of the entire enterprise. Global factors of international economic instability reduce the reliability of the results of a strategic analysis of these indicators without additional adjustment calculations. Therefore, the obtained characteristics of the enterprise and industries' performance do not always affect even the short term.

The purpose of the study is to improve the methodology of analysis and the integrated assessment of business activity, which provides the necessary level of argumentation of managerial decisions taking into account the objective interests of the main stakeholders. To achieve this goal, the following tasks are defined: (1) to study the content of the information base of the analysis of the business activity of organizations to determine ways of forming the necessary and sufficient range of initial indicators, (2) form a system of indicators of business activity and determine its place in production management, (3) to assess the areas of business activity of enterprises according to the data of production enterprises and formulate the main trends.

2 Methodology

In our opinion, the modern definition of business activity should take into account the challenges of the time. Therefore, in business activity, it is appropriate to distinguish a broad (general) and narrow (depending on the purpose) interpretation. In a broad sense, this is, first of all, a set of actions aimed at achieving entrepreneurial goals, among which the most relevant are capital gains, increasing the intensity of its use in all types of activities, with a stable increase in economic profit and the competitiveness of the business entity. In a narrower sense, in order to analytically substantiate management decisions, business activity is characterized by an increase in the volume and (or) intensity of the company performance.

Integral assessment of business activity involves a comprehensive and systematic assessment of the organization. Thus, indicators of business activity are considered not only as one of the characteristics of the financial condition of the organization

Table 1 Information support of business activity analysis by external reporting forms in the directions

Business areas	Sources of information from financial statements
Marketing	Statement of Profit or Loss, Statement of Cash Flows, Notes to Financial Statements
Production	Statement of Profit or Loss, Notes to Financial Statements
Investment	Statement of Profit or Loss, Notes to Financial Statements
Innovative	Notes to Financial Statements
Financial	Statement of Profit or Loss, Statement of Cash Flows, Notes to Financial Statements, Statement of Financial Position
Labor, socio-economic, managerial	Part of the sustainability report

Source authors

but also expanded as an integral part of the system of indicators for its balanced development. Based on the functional composition of the types of business activity, information support for the analysis of business activity with financial statements is presented in Table 1:

Obviously, most of the data is non-financial in nature and cannot be reflected in the financial statements. In our opinion, as far as possible, the explanations should include information, for example, on funds aimed at environmental safety and the implementation of social programs. This will increase not only the analyticity of external reporting but also the investment attractiveness of the company. In the modern world, when the social responsibility of business is becoming increasingly important, information on the direction of funds for the implementation of social and environmental programs increases the company's business reputation and market position. This is confirmed by the work of researchers on social and environmental activity [1, 2, 8].

It is appropriate to present such information in tabular form in Notes to Financial Statements. Data on funds allocated for environmental purposes become more analytical if separate mandatory (required in accordance with current legislation) and voluntary directions of their movement are highlighted. It is also advisable to divide cash flows serving social programs into those that cover social guarantees for own employees (optional) and external sponsorship.

For the effective management of business activity, it is necessary to create a relevant information field within the organization. A well-built information support system makes the analysis process of the highest quality. To increase the interest of the main stakeholder groups and increase the analyticity of information, it is necessary to supplement the financial statements with the sources listed in Table 2.

Thus, it seems possible to build a scheme of the information support system for the analysis of the business activity of enterprises (Fig. 1).

Taking into account the principles of comprehensiveness and systematic economic analysis, the author has developed a system of indicators for the integrated analysis of the business activity of an enterprise in nine areas. The system contains two types

Table 2 Information support of the analysis of business activity by internal sources in the directions

Business areas	Formable sources of internal information
Marketing	Sales budgets and reports on their implementation, dynamics, and structure of advertising costs
Production	The production program and the report on its implementation, the plan for increasing production volumes, the implementation of the plan under the contracts, the plan for the modernization of production and the results of its implementation
Investment	Investment plan of the company, Report on the profitability of investment projects, the structure of the investment portfolio of the company, the amount of investment received
Innovative	Statement of R&D expenditures, report on R&D efficiency by type, movement, and availability of intangible assets, the production modernization plan
Financial	Financial budgets and reports on their implementation, a report on the financial and economic situation of the company, on the formation of financial results and their use, an expanded report on cash flows
Labor	Report on the movement and composition of personnel, data on labor productivity and its changes
Socio-economic	Plans and reports on environmental safety, a report on the implementation of social guarantees to employees, personal data of employees
Managerial	Data on the activities of management personnel, reports on rational proposals put forward and their application, reports on project implementation and the quality of their implementation

Source authors

of indicators: calculated, determined by mathematical actions, and expert, based on rough estimates. Using information from the enterprises of the electrical industry of the Samara region (Middle Volga Electrotechnical Company, Volgoenergokolshlekt, Samara Cable Company, Sevkabel-Volga Region, Electroshield). During the study, the weighted average values of the main indicators for the areas of business activity of these business entities for the period from 2014 to 2018 were calculated. The average level of the values of indicators characterizing the direction of business activity is determined as the weighted average. In this case, the share of sales in the group total proceeds is used as a weight:

$$\bar{\Pi} = \sum_{i=1}^{5} \mathrm{d}_i \times \Pi_{\mathrm{ij}} \quad (1)$$

where:

Π-the level of the value of the j-th indicator of the i-th enterprise;
d-the relative weight of sales of the i-th company in total sales;
j-name of indicator;
i-serial number of the investigated enterprise.

Accounting information		Accounting financial statements, financial statements of competitors, accounting policies	Accounting and management financial statements, operational accounting data
Extra-accounting information	Legislative	Legislation governing the methodology for conducting an external analysis	Internal regulatory documents and organization standards
	Other	Stock market data, advertising information, media reports, statistical information	Any information on the activities of the organization's services required for analysis
		External	**Internal**

Fig. 1 Matrix of information system for business analysis (*Source* authors)

3 Results

Based on the data obtained on the results of the work of five enterprises in the Samara region, the data were obtained in the areas of their business activity. They are shown in the Table 3.

The values of the obtained indicators by types of activity can be used as a comparison base for assessing the business activity of the subject of analysis. Based on the growth rate of its indicators in comparison with the weighted average values, we distinguish the following gradations of levels of business activity: zero, low, average, increased, high.

The results of a retrospective analysis of the areas of business activity according to Table 3 confirm that the economic and political crisis of 2014–2015 had a significant impact on the business activity of the studied enterprises. At this time, the level of key performance indicators was so significantly reduced that for some, pre-crisis values have not yet been reached. In this regard, it is proposed to carry out quarterly monitoring of the level of business activity on the basis of the developed methodology in assessing the financial condition of companies and developing regular and promising anti-crisis measures.

Currently, there is practically no environmental activity at enterprises, which may subsequently negatively affect their investment attractiveness and competitiveness of manufactured products. In this regard, expert estimates are summarized and it is determined that it is necessary to direct at least 1% of the proceeds to the implementation of environmental programs.

Table 3 Indicators of the integrated assessment of business activity of enterprises of the Samara region in 2014–2018

Business areas	Indicator	2014	2015	2016	2017	2018
Marketing	The share of advertising in the total cost of sales, %	0.83	0.58	0.62	0.44	0.32
	Turnover of finished products, number of turnovers	9.92	11.24	17.72	60.61	49.13
	The growth rate of sales, %	83	137	72	168	111
Production	Return on assets, RUB	12.8	10.5	6.90	10.76	11.2
	Turnover of materials, turnover	4.97	19.9	6.75	9.20	9.7
	The share of reject in the total amount of costs, %	5.11	3.01	4.32	5.12	5.41
Investment	Update rate, %	32	11	16	16	19
	NPV for active investment projects (>0/<0)	+	+	+	+	+
Innovational	The share of R&D put into production, %	0.00	79	0.00	65	0.00
	R&D expenses growth rate, %	159	212	100	3	100
Financial	Profitability on sales, %	9.44	5.82	4.52	5.76	3.41
	Profitability on total capital, %	6.54	2.28	−4.69	3.99	10.51
	The financial stability rate, %	0.36	0.38	0.29	0.74	0.44
	Net profit growth rate, %	138	41	267	190	304
Labor	Labor productivity, thousand rubles	6040	7237	5312	8356	9104
	Payroll, cop.	4	3	4	3	3
Socio-economic	The share of profits directed to social needs, %	0.15	0.05	0.04	0.04	0.05
Managerial	The level of costs for environmental safety, %	0.01	0.01	0	0	0.01

Source authors

Table 4 Approaches to the definition of the essence of the concept of business activity

	Traditional (Resource) approach	Structural approach	A complex approach
Approach features	Business activity is considered as a characteristic of the intensity and efficiency of use of the organization's resources	The business activity of an organization is an integral characteristic of its performance by various types of activity	Business activity characterizes economic potential, and its development is represented by the positive dynamics of general performance indicators
The authors	Efimova [4] and etc.	Revutsky [9] and etc.	Kovalev [5] and etc.

Source authors

4 Discussion

The issues of analysis of the business activity of an economic entity have always been in the field of scientific interests of leading scientists in the field of accounting, analysis and audit. The term business activity can be considered from three perspectives: a person, an enterprise (micro-level); countries (macro-level). The business activity of an individual is its psychological feature and is used in specific economic analysis. Business activity at the macro-level is characterized by the pace of economic development of the country's economy and can be represented as a set of business efforts of various institutional entities and individual business entities. The latter represents a micro-level of business activity. This level has a basic value, predetermines high activity in the economy of the whole country. Therefore, most of the work on financial analysis is devoted to this level. The results of the study of existing approaches to determining business activity in the domestic scientific and practical use are graphically presented in Table 4.

In the works of foreign researchers, unlike Russian, business activity is considered from the point of view of various directions. That is, there is a structural approach. The turnover of labor resources and productivity can be compared with labor activity [7]. Investment activity is considered in studies from both business angels and financial investments of enterprises [3, 10]. Innovative activity in the conditions of modern "factories of the future" and the development of technology is considered as the main driver of development not only of universities but of the business as a whole [6].

5 Conclusion

At present, the use of data from existing forms of financial statements for the analysis of business activity is possible only in its individual areas, assessment of social, environmental, and labor activity on this information is impossible. Increasing the analyticity of financial statements is proposed on the basis of additions, explanations to the balance sheet with information on expenses for environmental purposes and social programs.

In the course of the study, opportunities to improve the analysis of the results of activities of economic entities were found on the basis of the indicator of the total income of enterprises borrowed from IFRS and introduced into the Russian reporting since 2011. This indicator allows separating the financial result obtained in the course of business activity from the result associated with changes in accounting estimates.

It was also revealed that the information system of the analyzed enterprises contains a large flow of unsystematized and even redundant (repeated) documentation for various management and control purposes. For the purpose of analyzing business activity in directions, the author proposes to optimize the composition and content of existing forms of internal reporting at electrical enterprises, as well as more fully, taking into account the developed methodology for assessing the acceleration of business activity, use the available forms of statistical reporting on business activity. The work systematized sources of internal information on the activities of the enterprise. It is determined that the analysis of its business activity requires data on costs and results in the context reflected in the management reporting of individual structural units. An analysis of the marketing activity of the company is proposed to be performed according to the accounting management reports containing information on sales costs, classified by type, as well as sales budgets and reports on their implementation. In the dissertation, a form of a summary sheet of source data for the analysis of marketing activity is developed.

In the context of the digital economy, it seems possible, based on the application of the proposed structural approach, at the level of firms aggregate data into an information database used to calculate and analyze macroeconomic indicators. This was made possible thanks to the emergence of technologies for the automated collection and processing of information (primarily Weka, RapidManer, Knime, which allow unloading financial and non-financial indicators and using machine learning technologies to justify management decisions). Thus, the proposed methodology allows balancing the interests of key participants in economic relations in the digital economy.

A comprehensive, integrated assessment of the business entity is necessary for its effective management and justification of the choice of the investment object by investors. One of the main components of comprehensive and dynamic characteristics of entrepreneurial activity and resource efficiency is a business activity. The results of the study can be used by auditors and analysts in the process of evaluating the financial and economic activities of companies in order to increase their effectiveness.

- recommended the composition and content of the information support of the analysis and integrated assessment of business activity of enterprises based on the generated digital reporting,
- a methodology has been developed for the economic analysis of the business activity in general and in its individual areas, providing the required degree of detail in substantiating management decisions,
- identified the main trends in the transformation of areas of business activity in the digital economy in the studied enterprises.

References

1. Cheng B, Ioannou I, Serafeim G (2014) Corporate social responsibility and access to finance. Strat Manag J 35(1):1–23. https://doi.org/10.1002/smj.2131
2. Cheng J, Liu Y (2018) The effects of public attention on the environmental performance of high-polluting firms: based on big data from web search in China. J Clean Prod 186:335–341. https://doi.org/10.1016/j.jclepro.2018.03.146
3. Clercq D, Meuleman M, Wright M (2012) A cross-country investigation of micro-angel investment activity: the roles of new business opportunities and institutions. Int Bus Rev 21(2):117–129. https://doi.org/10.1016/j.ibusrev.2011.02.001
4. Efimova O (2018) Financial analysis: modern tools for management decision-making: textbook. Omega Li, Moscow
5. Kovalev V (2001) Analysis of economic activity of the enterprise. Prospect, Moscow
6. Li X, Tan Y (2019) University R&D activities and firm innovations. Financ Res Lett. https://doi.org/10.1016/j.frl.2019.101364. Accessed 11 Feb 2020
7. Pereira da Rocha L, Pero V, Corseuil C (2019) Turnover, learning by doing, and the dynamics of productivity in Brazil. EconomiA. https://doi.org/10.1016/j.econ.2019.11.001. Accessed 11 Feb 2020
8. Peng X, Tang P, Yang S, Fu S (2020) How should mining firms invest in the multidimensions of corporate social responsibility? Evidence from China. Resour Policy 65. https://doi.org/10.1016/j.resourpol.2019.101576. Accessed 11 Feb 2020
9. Revutsky L (2002) Production capacity, productivity and economic activity of the enterprise: assessment, management accounting and control. Perspektiva, Moscow
10. Yuan X, Nishant R (2019) Understanding the complex relationship between R&D investment and firm growth: a chaos perspective. J Bus Res. https://doi.org/10.1016/j.jbusres.2019.11.043. Accessed 11 Feb 2020

Prospects for Promoting a Tourist Product Using Virtual Information Space Technologies

O. V. Petryanina

Abstract The article introduces a comprehensive study of the introduction of digital technologies in the tourism business. These technologies cause significant changes in the activities of tourism enterprises. It is concluded that virtual and augmented reality actively contribute to the growth of demand for travel services, and a virtual presentation significantly increases the likelihood of a client buying a travel product. The concepts of virtual and augmented reality technologies are revealed. The main characteristics of the virtual tour are described and analyzed as an innovative marketing tool that allows to show an object to the potential consumer more expressively and to get objective information about a product or service. In this regard, there is a high potential for using virtual tours to promote tourism services. The author focuses on the identification and research on the economic effect and economic efficiency of the use of innovative means and activities in the tourism business sector.

Keywords Digital technologies · Tourism business · Virtual reality · Augmented reality · Virtual tour · Economic efficiency

1 Introduction

In the digital age, consumer behavior is changing rapidly, and their expectations are also growing quickly. In recent years the basis of the most consumer trends is information technology, which not only provided new ways to obtain information about a product and service, but also significantly modernized consumer behavior models, forms of customer interaction with the seller, and, of course, ways to promote goods or services.

Tourism as an industry characterized by high competition and, at the same time, high information richness implies a quick response to changes in consumer preferences and expectations. Therefore, in the context of the active development of electronic channels for promoting tourism products, an urgent solution on the part

O. V. Petryanina (✉)
Samara State University of Economics, Samara, Russia
e-mail: Petryaninaolga@rambler.ru

S. I. Ashmarina and V. V. Mantulenko (eds.), *Current Achievements, Challenges and Digital Chances of Knowledge Based Economy*, Lecture Notes in Networks and Systems 133, https://doi.org/10.1007/978-3-030-47458-4_36

of tourism industry entities is to strengthen the role of the technological component in the organization of their activities and the transformation of the formed information support system. The use of modern technological solutions in organizing the activities of subjects of the tourism industry contributes to the transformation of the traditional management of organizational structures, changes the competencies of personnel, and solves various problems in the field of management, implementation of innovations, and digitalization of activities [1, 6].

However, the main task of the strategic development of the modern tourism industry is not only to create a competitive and popular tourist product, but also to make it affordable. In this regard, the relevance and importance of programs aimed at implementing the process of transition of tourism to a new niche—affordable tourism, is greatly increasing. Tourism is currently becoming more popular among people with disabilities. Many states indicate the need for the integration of people with disabilities into society as one of the policy directions, and many of them recognize their unpreparedness for this process. Affordable tourism, ensuring the availability of tourism, destinations, products and services for all people, regardless of their physical limitations, disabilities or ages, should be realized accordingly only with the interaction of all interested parties: governments, international agencies, tour operators and end users.

2 Methodology

The main feature of promotion in the Internet space is the focus on the formation of consumer engagement. The development of information technologies and virtual means of communication leads to an increase in the intensity of communicative exchange and a change in the form of information messages and perception of the text. Information technology is changing the ways and forms of interaction with consumers of tourism services, without changing their essence. From the point of view of marketers, virtual reality technology (VR) is a suitable tool for improving the competitiveness of travel companies, which makes it easier for customers to take decisions and increase conversions [11].

A seemingly promising approach to the application of VR technologies to promote a tourism product has not yet become too widespread due to insufficient research on the effectiveness of such an application. There is still no clear answer to the question of how the impact of a virtual tourist environment can affect the intrapsychological processes leading to the purchase of a tourist product [8]. Most of the relevant developments are based on laboratory experiments, seeking to ensure the internal reliability of the results. This often occurs due to a distortion of validity, because laboratory conditions are not observed in a real environment. Consequently, there is an urgent need for empirical research on the development of virtual environments, which is supposed to lead to a better conceptualization of the role of VR, which forms the attitude towards the tourist product and the behavior of tourists.

In this work, the author sets the following tasks: to analyze the modern development of information technologies in the tourism industry, to study the impact of virtual reality on the emergence of innovative tourism products, to identify the most promising and competitive innovative forms of tourism services, to identify and consider the cost-effectiveness of introducing innovative techniques in the development of tourism business. As a research methodology, theoretical and comparative analysis, the method of analytical information processing, as well as the calculation method were used. The solution of these problems determines the main goal of the study—analysis of the use of the possibilities of a virtual tour as an innovative way to attract tourists and a new marketing tool.

The theoretical and methodological basis of the article was made by the works of such scientists as Bogomazova et al. [2], Cherevichko and Temyakova [4] researching digital technologies in tourist systems. Issues of innovative development are also reflected in the foreign research papers by Fereidouni and Kawa [5], Pikkemaat et al. [7], Xiang [14], Shafiee et al. [9]. The properties of an innovative tourist product and the features of its promotion are described by Kedrova and Kitsis [6], Reichstein and Härting [8].

Neither practice nor theory has yet developed a universally accepted definition of "virtual reality". This term is actually a comprehensive phrase to refer to completely different types of diving. Depending on the features of the VR tool and the goals of its operation, the possible consequences for consumer behavior can vary greatly.

In practice, VR often describe one of the new media technologies that creates a feeling of complete immersion in a purely virtual world and overcome the lack of physical tangibility of travel, which is important for a better demonstration of travel products, for making a purchase.

Studies show that virtual experience really increases the likelihood of a purchase, speeds up decision making and increases the number of customers. The sense of presence when using virtual reality technologies makes them predictors of behavioral activation. Providing interactive elements and performing multimedia functions, appealing to a whole complex of human feelings, an exciting virtual experience surpasses brochures or other "traditional" types of media used for tourism marketing. By integrating sensory experience into their communication strategies, tourism marketers could more effectively support the search for information and the customer decision-making process [4].

Focusing on the marketing component, we can define VR: virtual reality is a computer environment that provides users with "tips" for the sensations of reality in order to facilitate their customer choice. The main features of VR include:

- the ability to navigate in virtual space, interact with and manipulate content (for example, click on a button to see additional information or move an object);
- the provision of "rich content", which means the degree to which the VR application uses a combination of text, audio, video, graphics, formats to transmit information to the user.

3 Results

The development of forms and trends in the tourism industry stimulates the development of both niche and mass tourism. But, despite the active integration of digital technologies in the activities of many companies, there are many problems in the development of the tourism industry and, in particular, affordable tourism. The solution of problems associated with the industry's unwillingness to provide services to interested categories of people involves technological and innovative equipment and consists in identifying the level of demand for a possible product and service. According to international statistics, the growth rate of the number of persons with disabilities is 10% worldwide, which in quantitative terms is equal to 650 million people. The contribution of tourists with disabilities to the US market annually reaches nearly 117 billion US dollars. For example, in 2012, the direct gross value added of tourism for people with disabilities in the EU amounted to 150 billion euros. According to the European Commission in 2014, the total gross turnover of affordable tourism in the EU amounted to 786 billion euros [14].

Within the framework of rapidly developing innovations, prerequisites have been created for the development of three areas that are particular important for the tourism industry. They are highly relevant for people with disabilities and people with limited mobility. These are virtual reality, augmented reality and artificial intelligence.

Today, in the foreign and domestic tourism industry, the use of a virtual tour is considered an actual way to attract tourists in the development of both traditional and barrier-free tourism.

In practice, virtual tours are widely used in marketing strategies as a tool to promote products and services and attract customers, as they are more expressive, reliable and effective than photos and videos. The main advantages of 3D panoramas from ordinary photos or videos are that they give the most complete picture of the object being represented. Thanks to such interactive technologies, a person becomes better acquainted with the place being examined, as if he had been there. And this effect of a familiar place is invaluable when a person is faced with the choice of a particular route, excursion, hotel, etc. [12]. For example, the Barcelona Company Omnipresent invites people to explore the distant expanses of our planet through avatars equipped with a camera. In addition, the #Get Teleported system, developed by VFX Frame store and Marriott Hotels, offers the ultimate in virtual travel experience. Her system is equipped with an Oculus Rift virtual reality helmet, which creates a 4D effect. The British travel agency Thomas Cook offers its customers a virtual tour using a 3D helmet. The virtual travel application created by the Czech company Panoramas shows various places in the world using imaging technology—360° virtual tours created using GPS.

The creators of new technologies for virtual tourism do not set themselves the goal of finding an alternative to real travel. They expect to create an addition that will provoke people to travel even more.

A virtual tour provides an opportunity to increase the number of potential visitors due to the openness and accessibility of a tour product or service without time limits, reducing the time for choosing a route or excursion, and the tour product or service itself receives a real competitive advantage. In a comprehensive study, which was attended by more than 10,000 travelers, 31% of respondents said they would book a vacation after passing a virtual tour [2].

There are many examples of travel companies that have successfully integrated innovative programs in their activities. The French company Club Med offers its customers viewing videos in a 360-degree overview. The company's turnover amounted to 1527 million euros in 2017. The indicator grew by 3.8% compared to last year, and profit by 20%. In 2015, Marriot (the chain of the largest hotel brands) launched Vroom Service. Three exciting 3D travel stories about Chile, Rwanda and China have been uploaded to the virtual reality headsets provided to visitors. 51% of new guests who tested this service said that they would like to stay more often at Marriot hotels [5].

Another important trend of our time is the massive introduction of augmented reality technology, the unlimited scope of which (from advertising, marketing, tourism, museums, exhibitions, the magazine business, computer games to medicine and pedagogy) and the expansion of perception opportunities broadcast by them allows us to talk about the approach of a new a qualitative leap in the development of modern civilization.

Augmented reality is called so because of the addition of certain imaginary (virtual) objects, usually of an informative and auxiliary character, to the perceptions coming from the real world. With the improvement of technology capable of real-time overlaying digital data on the image in the cameras of various mobile devices, thanks to the special programs built into them, active adaptation of augmented reality to society has begun. Travel services are to some extent virtual and intangible: it is hard to imagine what a trip to an unfamiliar place would look like. Augmented reality in this sense can help make a trip decision. Even before departure to the destination country, you can see the expected landscapes and attractions.

Thanks to augmented reality, a unique opportunity to travel in time has appeared. Now you can see how buildings, streets and cities looked like many years ago. Plunging into the atmosphere of past eras, tourists become full participants in various events of the past (Augmented Reality mobile application). Such applications are not only informative, but also entertaining in nature. So, moviegoers can meet with their favorite movie fragments (Augmented Reality Cinema mobile application), and animal lovers—take a picture, for example, next to a tiger or panda (Arboretum guide mobile application) and so on.

Augmented reality will help you navigate in an unfamiliar city, understand the complex metro scheme, and get to the right stop. Now it replaces the thickness of paper maps and guides (mobile applications "Florence guide" from tips LTD, Metro AR Pro). The application for finding the starting point will be indispensable for travelers. A tourist can put a marker on a car, bicycle or public transport stop, and then boldly set off to explore the city. Using augmented reality, he can easily find his way back (My Augmented Reality mobile app). In order not to get lost in the vast space of the airport, you can use augmented reality—Flightradar24 application. There is no need to look for signs and signs, special indicators will make it clear where to drink coffee, check in your luggage, buy souvenirs, and you can also instantly buy a ticket and receive information about bonus points from airlines. So, technological solutions of virtual and augmented reality began to play a significant role in marketing and promoting tourism and are expected to become even more popular in the coming decades. In the light of fierce competition and rapidly changing markets, tourism campaigns should intelligently use new technologies to better understand consumer incentives and build a program to attract consumers to travel services [10].

The pace of implementation of such innovations depends on the reasonable cost of applying VR technology in the tourism business. Currently, there are approximately 72 million users of VR technologies in the world. Analysts predict in 2020 the growth of VR users to 200 million. One of the main conditions for user growth is the reduction in the cost of VR gadgets and their introduction to the mass market [3]. In addition, virtual technology is an obvious new step in borderless tourism. Virtual and augmented reality for a group of people with limited mobility who cannot afford a vacation, adapted to their capabilities, due to other personal difficulties, is a great chance to see the world.

4 Discussion

In modern conditions, one of the determining factors in the growth of the country's economy, improving the efficiency and competitiveness of the enterprise are innovative technologies. Tourism is considered one of the important areas of using new innovations, such as VR. The use of virtual reality applications and tools provides significant advantages for marketing purposes, in the design and production of a tourist product, its promotion and improving the quality of company services [13].

Virtual tours as interactive means of presenting information are becoming, at the same time, an increasingly promising tool for creating a positive tourist image and competitive tourist services.

To ensure the effective implementation of innovative tourism services, a tourism company should conduct a set of activities that are reflected in the formation of a marketing strategy. The development and implementation of a sales strategy involves solving the following fundamental issues: the choice of sales channels, the choice of intermediaries and the definition of an acceptable form of work with them. The identification of attractors is also considered the key one. It enables to form a competitive tourist product.

The process of introducing a virtual tour, as a rule, is based on the principle of functional integration, which consists in the need to form the functions of managing tourist destinations, in accordance with the goals of tourism development in the region and resources, taking into account tourist flows. It should be noted that the commercial goal of any tourism enterprise should be to profit in a competitive environment, and the goal of destination management should be to ensure sustainable functioning and development based on increasingly high-quality and diverse customer services (it is a necessary condition for ensuring mass tourism and obtaining the desired income as a source of development destinations) [9].

Consider an example of the cost-effectiveness of implementing a virtual tour in the Voronezh State Nature Biosphere Reserve named after V.M. Peskov. Despite the fact that the state budget continues to remain the main funding for the activities of many protected areas (specially protected natural territories) in Russia, the reserves can carry out economic activities and have profits in addition to budgetary allocations. In this context, the reserve becomes an ecotourism enterprise. Using the natural potential for economic benefits is usually the key one to the survival of the system of protected natural areas.

The innovative technology implementation program actually consists of several stages. At the first stage, a preliminary examination of the innovation proposed for implementation, its susceptibility, compliance with the goals and objectives of the organization, an analysis of the readiness of the enterprise to accept this new product and to implement relevant reorganization measures. The second stage is directly related to the creation of an innovative project.

Due to the fact that many tourists before visiting the Voronezh Reserve look for any information about the park on the Internet, it was decided to give them the opportunity to "walk" virtually through its territory. Key attractors were chosen. These are travel indicators: tourist attractions and crossroads. Then an interactive video presentation was developed, which is a visualization of a tourist panoramic video route (on foot or transport). In our case, the cost of creating such a tour was 33,000 rubles (direct costs) = 15,000 (creating 15 spherical panoramas) + 18,000 (editing and multimedia design of the tour).

The next step in creating a virtual tour is the promotion of goods on the market. Priority in choosing the channels for implementing this project was given to the Internet space. The Internet environment enables not only to track the latest changes in the market of tourist services, but also to respond to them in time: make new special offers, change directions according to the current situation. The main option for hosting a virtual tour is the website of the reserve—http://zapovednik-vrn.ru.

The results of the study showed that the virtual tour promotes the site. As you know, search engines use behavioral factors in the formation of SERPs. One of these factors is the average length of stay on the site. The presence of a bright and realistic virtual tour increases the time spent by users on the site.

In addition, official social media accounts VKontakte and Instagram interact with the target audience and attract new customers. In social networks, the reserve is represented by accounts on Facebook (1351 subscribers) and VKontakte (8 thousand subscribers), designed using the symbols of the organization. They contain information about the tourist offers of the organization, there are video materials, photographs and illustrations, a unit for interactive work with clients and feedback is designed. Content placed on the data pages of Internet sites helps to popularize the site not only among users of the global network, but also with robots that organize the issuance of links when entering a search query from a user. These tools allowed a successful integration of SMM-marketing into the overall marketing strategy of our company, thereby increasing the conversion from social networks.

It should be noted that the increase in the number of potential visitors was also realized thanks to partnerships with travel companies that provide faster and more efficient delivery of tourism products to the final consumer. The organization laid down an estimate of 18578 rubles for organizing an advertising campaign (illustrated brochures and booklets, leaflets, advertising and informational videos).

The final stage of the investment project is an assessment of the economic efficiency of the new tourist product, identification of its economically weaknesses, and development of measures to increase the economic efficiency of the tourist product or refusal to use the development.

To assess the economic efficiency of the tour and the appropriateness of its implementation, it is necessary to carry out calculations and find out the value of such financial indicators as profitability of sales, profitability of costs and perform the calculation. The simplest formula for calculating the return on investment is: ROI = (Income-Cost)/(Amount of investments) * 100%.

As a result, the choice of effective marketing tools led to an increase in demand for visiting the selected tourist destination. As a result, the average income of the organization received from tourism activities 2 months after the introduction of the interactive resource has increased and amounted to about 65,400 rubles per month. The amount of investments is 51578 rubles. Based on the obtained data, the ROI of ROI was calculated: ROI = 13822/51578 * 100% = 26.8%.

Thus, analyzing the value of the profitability indicator of 26.8%, the author of the article makes a conclusion about the economic efficiency of the introduced technology. From each invested ruble, the organization receives about 27 kopecks. extra profit. This simple example shows that a one-time investment in creating a full-fledged virtual tour paid off in the first months, and it will take a very long time to bring dividends. This technology has, first of all, an economic effect, as it serves as a source of equity formation when a new flow of tourists arrives. An increase in the level of recognition of even one destination in this or that territory can give a significant impetus to a rise in the number of foreign and domestic tourists, which will positively affect the economy of the definite region and the country as a whole.

5 Conclusion

Thus, the issues of effective promotion of tourist destinations and information support in tourism are relevant today and are the most important factor in increasing the competitiveness of Russian destinations in the domestic and inbound tourism market. The formation of the image of a tourist product has a direct impact on attracting investment, consumers, determines the further program of measures for the direction of development of the object. This theme is in the development trend of the digital economy and new "smart" technologies.

A virtual tour is a unique method of presenting an object. It makes possible to present your services interactively, creating a complete sense of presence for the viewer, representing the opportunity to stand out from competitors and become more visible to potential customers and partners [7].

A high degree of visualization and interactivity of virtual tours are the main criteria for their effective development and implementation in the activities of tourism enterprises. Presentations, demonstration videos and educational films created on the basis of modern interactive technologies allow us to ensure high rates of involvement of the Internet audience in the information support of the travel agency.

The feasibility of developing systems of tourist virtual information space is justified on the other hand by the economic efficiency of their implementation, which was proved in our work on a practical example in which a virtual tour has become a new source of increasing the profitability and profitability of a tourist site.

To sum up, we can conclude that innovative technologies in the tourism industry are a requirement of the time, which leads not only to improving the quality of services, but also to the rational use of all available resources for both tourists and tourism business owners. The cost of virtual reality technology is gradually becoming cheaper, accessibility is increasing. Great prospects in this direction are opened by the development and implementation of mobile applications with augmented reality. We believe innovative technologies in the tourism industry will also develop that will open up new opportunities for innovators and make tourism accessible to different categories of the population.

References

1. Aleksushin GV, Ivanova NV, Solomin IJ (2020) The use of technology of digital economy to create and promote innovative excursion products. In: Ashmarina S, Mesquita A, Vochozka M (eds) Digital transformation of the economy: challenges, trends and new opportunities. advances in intelligent systems and computing, vol. 908. Springer, Cham, pp 404–410. https://doi.org/10.1007/978-3-030-11367-4_39
2. Bogomazova IV, Anoprieva EV, Klimova TB (2019) Digital economy in the tourism and hospitality industry: trends and prospects. Serv Russ Abroad 13(3(85)):34–47. https://doi.org/10.24411/1995-042X-2019-10303
3. Bolodurina MP, Mishurova AI (2018) Information support of digitalization of the activities of tourism industry entities. Econ Anal Theory Pract 17(9):1710–1728
4. Cherevichko TV, Temyakova TV (2019) Digitalization of tourism: forms of manifestation. News Saratov Univ. New Ser. Ser. Econ. Control Right 19:59–64. https://doi.org/10.18500/1994-2540-2019-19-1-59-64
5. Fereidouni MA, Kawa A (2019) Dark side of digital transformation in tourism. In: Nguyen N, Gaol F, Hong TP, Trawiński B (eds) intelligent information and database systems, vol 11432. Springer, Cham, pp 510–518. https://doi.org/10.1007/978-3-030-14802-7_44
6. Kedrova EV, Kitsis VM (2019) Trends in the promotion of consumer behavior as the basis for promoting a tourism product. Mod Probl Serv Tour 13(2):21–33. https://doi.org/10.24411/1995-0411-2019-10202
7. Pikkemaat B, Peters M, Bichler BF (2019) Innovation research in tourism: research streams and actions for the future. J Hosp Tour Manag 41:184–196. https://doi.org/10.1016/j.jhtm.2019.10.007
8. Reichstein C, Härting R-C (2018) Potentials of changing customer needs in a digital world – a conceptual model and recommendations for action in tourism. Procedia Comput Sci 126:1484–1494. https://doi.org/10.1016/j.procs.2018.08.120

9. Shafiee S, Ghatari AR, Hasanzadeh A, Jahanyan S (2019) Developing a model for sustainable smart tourism destinations: a systematic review. Tour Manag Perspect 31:287–300. https://doi.org/10.1016/j.tmp.2019.06.002
10. Shamlikashvili VA (2015) Virtual tourism and the virtual space of the museum: the relationship of phenomena. Creat Econ 9(5):617–628
11. Turkaya B, Dıncer FI, Dincer MZ (2019) An evaluation of new values in economy and their impacts on future transformation in tourism. Procedia Comput Sci 158:1095–1102. https://doi.org/10.1016/j.procs.2019.09.151
12. Tussyadiah LP, Wang D, Jung TH, tom Dieck MC (2018) Virtual reality, presence, and attitude change: empirical evidence from tourism. Tour Manag 66:140–154
13. Vishnevskaya EV, Klimova TB (2017) Prospects for the development of virtual information space in the tourist industry. Sci Result Bus Serv Technol 3(1):22–33. https://doi.org/10.18413/2408-9346-2017-3-1-22-33
14. Xiang Z (2018) From digitization to the age of acceleration: on information technology and tourism. Tour Manag Perspect 25:147–150. https://doi.org/10.1016/j.tmp.2017.11.023

Features of Digitalization of the Business Sector in the Russian Economy

O. V. Trubetskaya

Abstract The relevance of the study is to identify the features and problems of the business sector development in Russia in the context of economic digitalization. The author considers the problem of adapting Russian companies to changing business conditions in the process of digital transformation. The purpose of the study is to identify the features of digitalization of the business sector in the Russian Federation. The following research methods were used: data comparison method, analysis and synthesis methods, statistical data method. Based on the index of readiness of Russian companies to work in the digital economy, the author made a conclusion on the degree of digitalization of the business sector, identified problems that impeded the further sustainable development of organizations.

Keywords Digital economy · Entrepreneurship · Business sector development

1 Introduction

The economic development of any state and the achievement of sustainable economic growth is impossible without the development of entrepreneurship. Small and medium-sized businesses affect the structure of the gross national income of the country, improve the quality of products, providing the necessary competitive environment, create new jobs, fill the state budget through taxes and duties. Economic digitalization leads to a change in the norms and rules of the business sector, also creates new development opportunities for business. Therefore, the interaction between the digital economy and the business sector is relevant. The ongoing digitalization in Russia involves the introduction of technological advances in all branches of the economy by 2025, in accordance with National Program "Digital Economy of the Russian Federation" approved by the government of the Russian Federation [6]. At

O. V. Trubetskaya (✉)
Samara State University of Economics, Samara, Russia
e-mail: olgatrub@gmail.com

S. I. Ashmarina and V. V. Mantulenko (eds.), *Current Achievements, Challenges and Digital Chances of Knowledge Based Economy*, Lecture Notes in Networks and Systems 133, https://doi.org/10.1007/978-3-030-47458-4_37

the same time, digitalization changes information and communication technologies in the interaction of economic entities, and also influences the behavior of buyers and firms in the market.

The development of new information technologies has led to the introduction of mobile devices, social networks, artificial intelligence technologies and working with data on the market. If prior to digitalization, purchases were made in stores and the client had to choose what enterprises offered him, now most of the trading has gone on-line. Digitalization has reduced the cost of services, and the services have become more affordable. All products and services can be modified to meet the needs of a specific client, which leads to increased competitiveness of companies, reduces advertising costs, makes goods and services more affordable, and increases customer satisfaction [2].

The information that entrepreneurs receive from consumers is of great value. For example, according to soft cameras, we can conclude which products and offers the buyer is interested in, which makes it possible to carry out a variety of marketing campaigns and offers more efficiently. Thanks to digital technologies, it is possible to quickly respond to customer questions and complaints through appropriate applications. Information has become a key resource now, and to process huge information fields, the business sector needs services implemented on a cloud platform. Thus, the entrepreneurial sector in the digital economy primarily considers the needs of the client, all services provided have high speed and mobility.

2 Methodology

Using the methods of analysis and synthesis, the author assesses the readiness of the Russian business sector for the digital economy. It was revealed that in a number of areas Russian enterprises are not ready for digital transformation. For organizations to become a driving force for economic development in the context of economic digitalization, we need to pay more attention to the skills of personnel in the field of digital technologies, information security issues and the promotion of goods and services created by enterprises via the Internet.

3 Results

In January–February 2019, Otkritie Bank and the NAFI Analytical Center conducted a study on the degree of implementation of digital technologies in Russian business. The result of this analysis was the development of the Business Digitalization Index (BDI), which considered the following five elements: channels of transmission and storage of information used by the business sector, internet tools used by

Fig. 1 Distribution of the business digitalization index by main elements (*Source* author based on [4])

business to promote goods and services, introduction of digital technologies in business processes, provision of information security, digital training of personnel in organizations.

The BDI index can range from 0 to 100% points. According to the study, for 2019, the Digitalization Index of Russian business is 34% of 100%, which indicates the lack of readiness of Russian companies for digital transformation.

Moreover, the distribution of values by index elements is as follows (Fig. 1): firms use digital technologies for transmitting and storing information—43%, integrate digital technologies into the business process 39%, use Internet tools to promote goods and services—38%, the information security index is 33%, the digital competencies development index is only 15%.

Based on the data presented, the following conclusions can be drawn: Russian companies are quite actively using modern digital technologies for data transfer and storage, as 78.3% of organizations use broadband Internet, 47.4% of enterprises actively use mobile broadband Internet, 22.6% use cloud services in their work. The most popular corporate technology is the intranet, companies are actively implementing Enterprise Resource Planning systems (ERP systems)—19.2%, Customer Relation Management systems (CRM systems)—13% and Supply Chain Management systems (SCM systems) [1]. Less than half of companies use the Internet as a tool to promote their products and services. 44% of organizations have a website with detailed information about the company and its activities. Messengers to communicate with customers are used by 35% of organizations. Many firms use pages on social networks to promote goods and services (Fig. 2).

The most popular networks are Vk and Instagram—70 and 44% of organizations respectively have information about their company there. The least popular social network is Twitter. Only 3% of the total number of enterprises represent information about the company and its goods and services.

The entrepreneurial network correspondent of the Russian Federation does not pay enough attention to information security issues: only 34.9% of companies have software for automating the process of analyzing and monitoring the security of computer systems, 26% of organizations use specialized programs to protect their

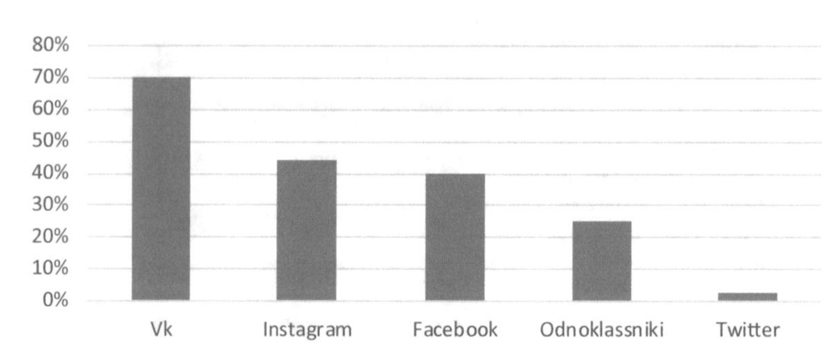

Fig. 2 The number of entrepreneurs (in%) having a page on social networks (*Source* author based on [4])

businesses, 30.7% of companies have the ability to back up data to carriers located outside the territory of organizations, and only 2% of enterprises have a network attack monitoring center. This attitude to security issues of Russian companies is associated with a lack of financial assets to purchase and maintain special software and equipment.

Russian companies devote very little attention to continuing education in the digital field. So, 94% of companies have never carried out professional development in this area for their employees, while many business leaders believe that educational programs in the field of ICT are needed only for those directly connected with digital technologies. The analysis shows that at present, the Russian business sector cannot ensure sustainable development of the Russian economy in transiting to digital.

4 Discussion

The problems of the functioning and development of business in the digital economy are reflected in the works of many modern economists. So, the OECD report says that the development of digital technologies has led to significant changes in the work of organizations in the field of entertainment, news, advertising and retail. The first large digital players appeared in these industries, which transformed traditional business models into digital technologies [8, 9].

Some scholars believe that economic digitalization will lead to further enlargement of companies, and in the end only gigantic digital and technically multinational companies will remain on the market [3]. Others pay attention to such a feature of doing business in the digital age as the ability of any small company or start-up to become a major exporter of goods or services through electronic acquisition along with such major giants as Alibaba, Amazon, eBay [5]. R. Strange and A. Zucchella in their work examine the impact of new digital technologies on international business practices. The main issue of their research is the process of creating new value within individual links of the international corporation [10]. Some economists say that the digital economy can create not only new business opportunities, but also lead to several problems: destroy established traditional companies, reduce jobs, transform markets [8]. Russian scientists in their works consider the issues of transformation of enterprises, their development in the digital environment [11, 12], the ideas of digitalization of business, as well as motives for digital transformation of the business sector [7].

5 Conclusion

For Russian business to be able to develop steadily in the digital economy, it is necessary to solve a few problems that cause difficulties in switching companies to digital:

- purchase and use high-tech equipment, but most small and medium-sized businesses are not able to do this due to high costs of equipment. At the same time, a possible solution to this problem may be the provision of discounts and special conditions for companies manufacturing such equipment to firms purchasing it,
- low level of professional development of employees in the field of digital technologies. This problem can be solved by increasing the number of specialists graduating from higher and secondary professional educational institutions, as well as by conducting special trainings for company management aimed at explaining the impact of digital transformations on Russian business, the competitiveness of organizations and advanced training of employees in the digital field,
- weak development of data storage and transmission services due to the reluctance of Russian business to spend money on cloud storage, data centers, and because businessmen do not understand the usefulness of such services. The solution to this problem is to explain the features and benefits of working with cloud services to company leaders, as well as providing test modes for working with them.

References

1. Abdraxmanova GI, Kuzminov YI, Sabelnikova M (eds) (2019) Indicators of the digital economy statistical digest. National Researched University of Higher School of Economics, Moscow
2. Betancourt RR (2017) Distribution services and the digital economy: implications for GDP measurement, productivity and household welfare. http://economics.fiu.edu/events/2017/seminar-roger-betancourt-2/betancourt2017.pdf. Accessed 01 Feb 2020
3. Bolwijn R, Casella B, Zhan J (2019) International production and digital economy. Int Bus Inf Digit Age Prog Int Bus Res 13:39–64. https://doi.org/10.1108/s1745-886220180000013003
4. Digitalization Index of SMEs (2019) NAFI. Research Center. https://nafi.ru/projects/predprinimatelstvo/indeks-peremen-gotovnost-rossiyskikh-kompaniy-k-tsifrovoy-ekonomike. Accessed 11 Feb 2020

5. McKinsey Global Institute (2016) Digital globalization: the new era of global flows. MGI, New York
6. National Program "Digital Economy of the Russian Federation" approved December 24, 2018. http://government.ru/rugovclassifier/614/events/. Accessed 11 Feb 2020
7. Nuretdinova Yu, Ometova D, Morozova M (2019) Transition of Russian enterprises to the digital economy: problems and solutions. Bull Altai Acad Econ Law 4:138–143
8. OECD (2014) Addressing the tax challenges of the digital economy. OECD Publishing, Paris. https://doi.org/10.1787/9789264218789-7-en
9. OECD (2018) Implications of the digital transformation for the business sector. Conference summary. http://www.oecd.org/sti/ind/digital-transformation-business-sector-summary.pdf. Accessed 01 Feb 2020
10. Strange R, Zucchella A (2017) Industry 4.0, global value chains and international business. Multinatl Bus Rev 25(3):174–184. https://doi.org/10.1108/MBR-05-2017-0028
11. Ustinova N (2019) Digital economy and entrepreneurship: Issues of interaction. Bull Saratov State Socio-Econ Univ 3(77):32–37
12. Volkodavova EV, Zhabin AP, Yakovlev GI, Khansevyarov, RI (2020) Key priorities of business activities under economy digitalization. In: Ashmarina S, Vochozka M, Mantulenko V (eds) Digital age: chances, challenges and future. ISCDTE 2019. Lecture notes in networks and systems, vol 84. Springer, Cham, pp 71–79. https://doi.org/10.1007/978-3-030-27015-5_9

Electronic Document Management in International Carriage: Russian Experience of Railway Business

R. V. Fedorenko, O. D. Pokrovskaya, and E. R. Khramtsova

Abstract The article addresses the issues of electronic document management support in organization of international goods carriage. Sophisticated technologies of international cargo transportation and a large number of the process participants objectively require digitalization of their relationships. The problem of a successful "one-click" business in the context of the 4th industrial revolution calls for applied, effective solutions. A state-wide information system is necessary for the implementation of end-to-end integrated technology of network import transportation. The article discusses and analyzes the Russian Railways Holding experience, which, in light of the digital transformation of business, has formed a global information space for participants in rail transportation. The methods of operations research, process approach, logistics, and customer focus are used. A target model of digital business is proposed. The electronic document management technology implemented in the Russian railway industry is considered.

Keywords Digitalization · Integration · International carriage by rail · Electronic document management

1 Introduction

The success of regional business integration into the system of world economic relations is largely due to the possibility of reducing transaction costs. The possibilities of international carriage intensification are largely limited by the logistics infrastructure state [5]. The importance of a competent assessment of the transport and

R. V. Fedorenko (✉) · E. R. Khramtsova
Samara State University of Economics, Samara, Russia
e-mail: fedorenko083@yandex.ru

E. R. Khramtsova
e-mail: romel06@mail.ru

O. D. Pokrovskaya
Emperor Alexander I St. Petersburg State Transport University, Saint Petersburg, Russia
e-mail: Insight1986@inbox.ru

S. I. Ashmarina and V. V. Mantulenko (eds.), *Current Achievements, Challenges and Digital Chances of Knowledge Based Economy*, Lecture Notes in Networks and Systems 133, https://doi.org/10.1007/978-3-030-47458-4_38

storage infrastructure current state was studied in earlier works of the authors [12, 13]. Modern research emphasizes the importance of digital technology application in various parts of the international supply chain. Digitalization capabilities are being explored in ports infrastructure [8], freight terminal management [10], road development [4], and air transportation [15]. The present article contemplates the Russian railway transport as a key element of the interregional and international trade logistics infrastructure.

Russian Railways, like any other business sector, actively implements digital technologies to facilitate business processes speed and information exchange at one-click-speed. Today, one of the key aspects of Russian railway industry digital transformation, as in most countries of the world, is the development of electronic document management, digital carriage and electronic signature. The positive impact of the transition to electronic document management in international trade is registered in a number of modern scientific publications [3].

The structure of Russian Railways business, involved in electronic document management, includes over 3700 stations, more than 14500 enterprises, as well as over 130,000 users. At the same time, an average monthly 2100000 documents are processed with electronic signature.

Consider the elements of a phased technology implementation:

1. In the "domestic" segment:
   - increase in the electronic documents share, including electronic accounting, primary documents and records,
   - replicating to the entire network the electronic signature application of those responsible for the goods placement and securing.
2. In the "import" segment:
   - start of transportation at a proof ground with VR Group (2019),
   - replicating of import and transit transportation supported with electronic documents (Q3 2019—Q2 2020).
3. In the "transit" segment:
   - integrating into the electronic transit zone the railway carriers of Mongolia (2019–2020), Azerbaijan and China (2021–2022),
   - integrating into the electronic transit zone the railway carriers of Latvia, Kazakhstan, Estonia (2019–2020).
4. In the "export" segment:
   - replicating export transportation (1st half of 2019).

Target indicators for the electronic document management implementation (in accordance with the implemented integrated scientific and technical project "Digital Railway") are presented in Fig. 1:

Stages of the 2016–2020 Electronic Train project, which the Russian Railways completed by 2020, are shown below:

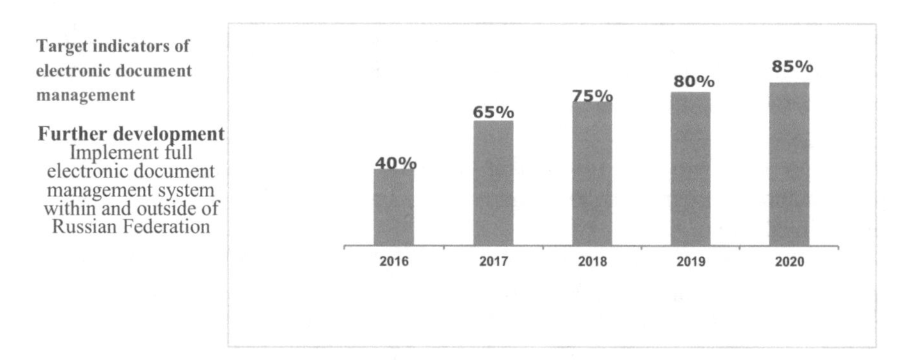

Fig. 1 Target indicators of international transit digitalization (*Source* authors)

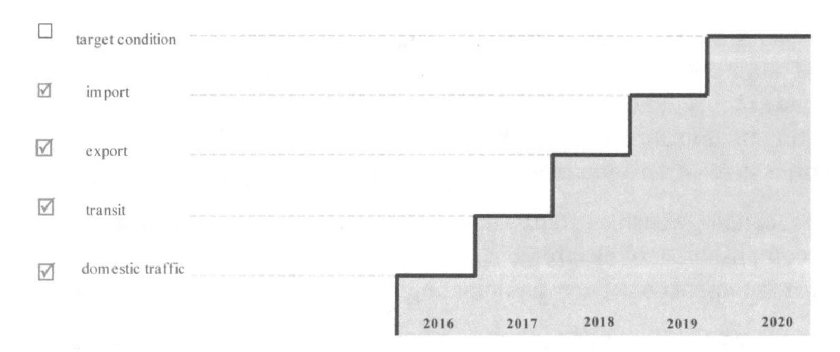

Fig. 2 Stages of electronic documents management implementation (*Source* authors)

1. Transportation of empty wagons in international traffic using paperless technology with electronic signature.
2. Transportation of export cargo using paperless technology.
3. Transportation of wagons in transit using electronic documents.
4. Transportation of imported goods using paperless technology in cooperation with federal executive bodies.

The promising and most important task for 2020 is to prepare a register of laws and regulations and make the necessary changes to them in order to effectively implement international trade in Russia using an electronic signature.

Figure 2 demonstrates stages of electronic documents management implementation by periods:

2 Methodology

In development of the principles and stages of international carriage organization applying a standard integrated electronic technology, the authors used a system and process approach, applied the theory of transport and storage systems, terminalistics, a synergetic and cluster approach, operations research, as well as tools for system analysis, dynamic programming and economic modeling.

Integrated electronic technology for processing export shipments implements a network-wide information system that integrates all participants in both the transportation and economic processes into a shared digital field with the possibility of online interaction.

In particular, on July 11, 2019, the Ministry of Justice of Russian Federation registered a joint order of the Federal Customs Service of the Russian Federation and the Federal Tax Service of Russia on the approval of electronic transport document formats provided by taxpayers to confirm the right to 0% VAT for export of goods by rail in electronic form.

The stages of implementation are as follows:

- improving the regulatory framework and software maintenance (2015–2016),
- pilot shipments in the Russian-Finnish direction (up to 37% of the total export volume in this direction)—2016–2018,
- making modification to the legislation of the Russian Federation—2018,
- current state—transition to replication of electronic export technology.

The technology enables preparation of the full set of electronic export documents and accomplishment of electronic documents exchange [2].

In particular, for consignors participating in the foreign economic activities (FEA):

- increase profitability of goods export not only through the tariff preferences but through the export cost decrease as well,
- reduce operational expenses of conducting FEA,
- optimize headcount involved in carriage administration, interaction with customs and tax authorities, engaged in accounting reporting due to the possibility of remote administration of export transportation from a single center.

For other foreign economic activities participants:

- leverage the Russian Railway and consignors operations through the digital servicing,
- reduce cost of operational personnel maintenance,
- provide online access to original electronic documents,
- provide digital interaction with tax authorities when submitting tax reporting.

Currently (at the beginning of 2020), more than 414 thousand waybills (1.429 million wagons) were issued in Russia with full electronic documentation for goods destined for Finland, Latvia, Estonia, Lithuania, Belarus, Kazakhstan and Russian ports for export cargo involving ocean shipping.

3 Results

Consider the basic model structure of organizing interaction with federal executive bodies through the system of interagency electronic interaction system (IEIS), Fig. 3.

The successful railways operation is largely determined by the information field coherence and mobility. The key tool, as demonstrated by the business practice of the Russian Railways Holding, can and should be electronic document management, which is called in Russia “paperless transport technology”.

As seen on the graph, the interaction is accomplished according to carriage process control levels in the interactive mode of electronic document exchange in the end-to-end information field of interaction. The electronic services structure that is currently being implemented by Russian Railways Holding to serve customers based on the digitization and customer focus principles can be represented as an information field “development” of Russian Railways field operations as the largest player in the transport and logistics market, Fig. 4.

Due to the implementation of the railway industry digital transformation project, in particular, the comprehensive scientific and technical project “Digital Railway”, a whole set of applied tasks was completed within the project framework of the electronic transit through the international transport corridors network, namely:

- integration of Kazakhstan, Estonia, Latvia railway carriers into the electronic transit zone (2019–2020),
- integration of Finland, Mongolia railway carriers into the electronic transit zone (2020),
- integration of Azerbaijan and China railway carriers into the electronic transit zone (2021–2022) [14].

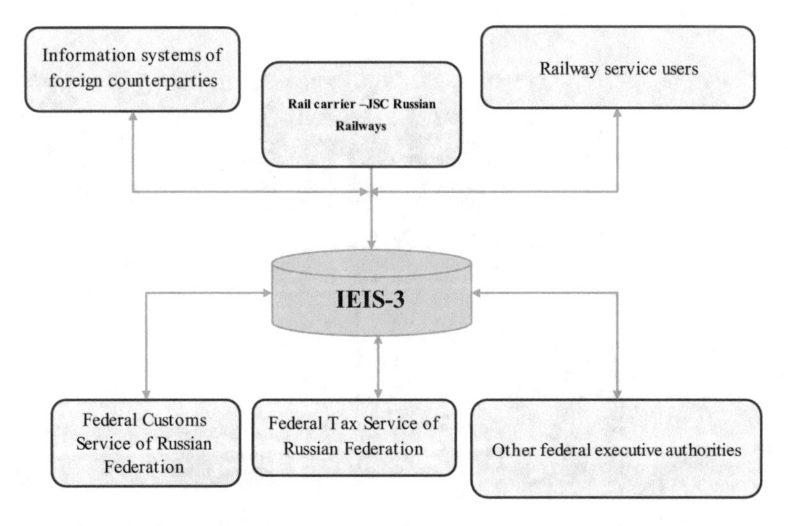

Fig. 3 Basic structure of interaction of JSC Russian Railways with international goods exchange process participants (*Source* authors)

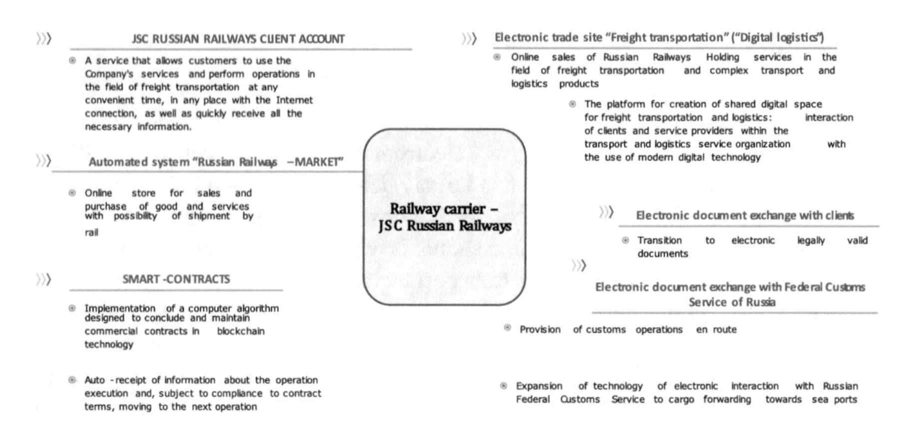

Fig. 4 Electronic services structure, implemented in Russian Railways Holding (*Source* authors)

Figure 5 shows a chart of the logistic links that are administered through electronic document management.

It can be assumed that the systematic, phased implementation of electronic document management in the Russian railways network will result in a multifunctional, poly-transport system for the cargo delivery in international traffic.

Figure 6 shows a general view of the conceptual flow chart for the goods "digital multimodal delivery" in international traffic. This figure shows a formal map of possible multimodal delivery routes. Introduction of electronic document management enables simplification of rail, sea and road transport interaction, as well as speeds up the international transit processing. Integrating transport companies, shippers and logistics service providers into a shared information field enables the regional transport and logistics centers development.

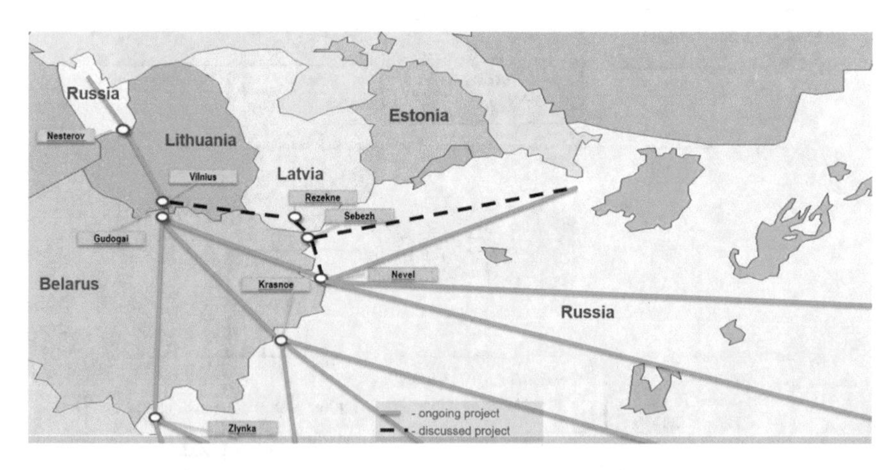

Fig. 5 Logistics links of JSC Russian Railways, administered through the electronic document management (*Source* authors)

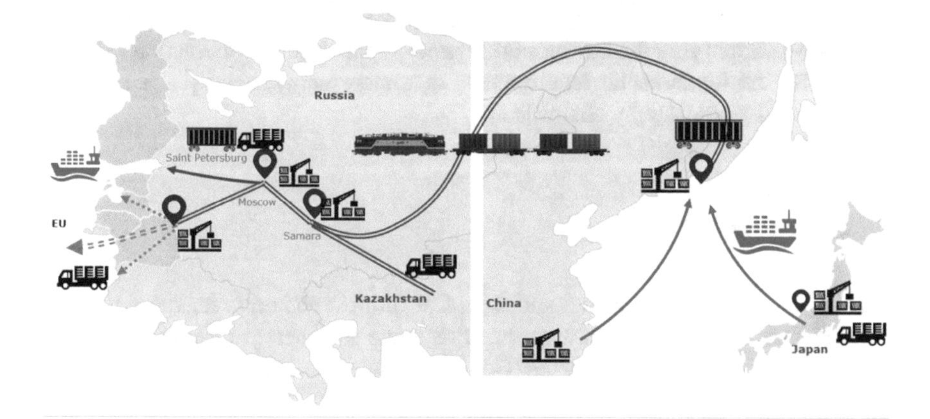

Fig. 6 Variant of goods multimodal transportation organization supported by electronic documents with the participation of Russian Railways (*Source* authors)

The presented figure shows the transport and logistics terminal in the Samara region as an example. Such terminals can operate in various regions through which significant transit cargo flows pass. The development of a unified electronic document management system will allow not only to attract additional volumes of transit deliveries, but also to accumulate export freight flows of neighboring regions. This will create additional opportunities for successfully integrating Russian regions into the world economic relations system.

4 Discussion

In today's world, more and more opportunities for digitalization of processes associated with the international goods carriage appear. Therefore, in the study of A. Jugović, J. Bukša, A. Dragoslavić, and D. Sopta, the possibilities of using blockchain technologies in goods delivery processing are considered [9]. In the article by R. Abduljabbar, H. Dia, S. Liyanage, S. Bagloee, the outlook of deploying artificial intelligence in the transport sector are analyzed [1]. Thus, it can be argued that the transition to electronic document management is a condition for the Russian transport industry survival in the context of toughening competitive environment.

In international trade, the transition to electronic document management is of great importance, due to the need for customs clearance among other reasons. Currently, electronic declaration is a ubiquitous customs clearance method [6]. The execution of electronic documents is becoming generally accepted in international container freight turnover. Engaging modern digital technologies allows foreign trade transactions participants not only to simplify the registration process, but also to increase the safety of delivery [7]. The introduction of electronic document management principles allows simplifying the transportation process participants' interaction in accordance with the fundamental requirements of transparency, security and customer orientation in global supply chains [11].

5 Conclusion

In conclusion, a number of first and foremost future-oriented tasks should be formulated that the Russian Railways need to solve in order to ensure full-fledged interaction with the Federal Customs Service of Russia and the Federal Tax Service of Russia in electronic format:

(1) In the field of customs regulations:

- exchange of electronic invoices, certificates, registers with customers (as of 2019, more than 2,300 clients work in the Russian Railways digital field, more than 110,000 invoices and 17,000 certificates of services acceptance and compendia have been issued),
- exchange of additional electronic accounting and source documents (expanding the electronic documents list used in freight transportation to account reconciliation statement for penalties payment, certificate of railway train services provision, etc.),
- transference of transportation documents electronic registers with electronic signature to the Federal Tax Service of Russia,
- verification of FEA participants and JSC Russian Railways the right for 0% export VAT,
- modification of Tax Code [16] and regulations of the Russian Ministry of Finance,
- full-fledged tax reporting exchange,

(2) In the field of tax regulation:

- electronic provision of preliminary information,
- electronic execution of customs operations: registration of arrival/departure, goods placement for temporary storage and delivery,
- an experiment on processing and completion in electronic form (Naushki—Brest railway line),
- obtaining electronic marks on the railway bill of lading,
- placement of goods under customs regime procedures and their completion,
- customs operations en route,
- formation of monitoring and information exchange analysis system,
- declaration of imported goods related to their release,
- expanding the technology of electronic interaction with the Federal Customs Service of Russia to cargo forwarding to seaports.

The solution of the above listed will allow transport companies to simplify the process of cargo clearance, which will create additional opportunities for Russian companies to successfully enter international markets.

Acknowledgements The reported study was funded by RFBR and FRLC according to the research project № 19-510-23001.

References

1. Abduljabbar R, Dia H, Liyanage S, Bagloee S (2019) Applications of artificial intelligence in transport: an overview. Sustainability 11(1):189. https://doi.org/10.3390/su11010189
2. Aptekman A, Calabin V, Klintsov V, Kuznetsova E, Kulagin V, Yasenovets I (2017) Digital russia: a new reality. McKinsey. https://www.mckinsey.com/~/media/mckinsey/locations/europe%20and%20middle%20east/russia/our%20insights/digital%20russia/digital-russia-report.ashx. Accessed 12 Jan 2020
3. Bakhtyar S, Mbiydzenyuy G, Henesey L (2016) A simulation study of the electronic waybill service. In: Al-Dabass D, Romero G, Orsoni A, Pantelous A (eds) Proceedings of the EMS 2015: UKSim-AMSS 9th IEEE european modelling symposium on computer modelling and simulation, vol 7579846. IEEE Computer Society, Los Alamitos, pp 307–312. https://doi.org/10.1109/ems.2015.53
4. Cao A, Fu B, He Z (2019) ETCS: an efficient traffic congestion scheduling scheme combined with edge computing. In: Xiao Z, Yang L, Balaji P, Li T, Li K, Zomaya A (eds) Proceedings of the 21st IEEE international conference on high performance computing and communications, 17th IEEE international conference on smart city and 5th IEEE international conference on data science and systems, HPCC/SmartCity/DSS 2019, vol 8855522. IEEE Computer Society, Los Alamitos, pp 2694–2699. https://doi.org/10.1109/hpcc/smartcity/dss.2019.00378
5. Fedorenko RV (2019) Modern issues of development of the customs and logistics infrastructure of the international north-south transport corridor. In: Ashmarina S, Vochozka M (eds) Sustainable growth and development of economic systems. Contributions to economics. Springer, Cham, pp 63–75. https://doi.org/10.1007/978-3-030-11754-2_5
6. Fedorenko RV, Persteneva NP, Konovalova ME, Tokarev YA (2016) Research of the regional service market in terms of international economic activity's customs registration. Int J Econ Financ Issues 6(5S):136–144
7. Grzelakowski AS (2019) Global container shipping market development and its impact on mega logistics system. TransNav 13(3):529–535. https://doi.org/10.12716/1001.13.03.06
8. Inkinen T, Helminen R, Saarikoski J (2019) Port digitalization with open data: challenges, opportunities, and integrations. J Open Innov Technol Mark Complex 5(2):30. https://doi.org/10.3390/joitmc5020030
9. Jugović A, Bukša J, Dragoslavić A, Sopta D (2019) The possibilities of applying blockchain technology in shipping. Pomorstvo 33(2):274–279. https://doi.org/10.31217/p.33.2.19
10. Pokrovskaya O (2018) Terminalistics as the methodology of integrated assessment of transportation and warehousing systems. In: Abramov AD, Manakov AL, Klimov AA, Khabarov VI, Medvedev VI (eds) Proceedings of X international scientific and technical conference "polytransport systems". MATEC web of conferences, vol 216. EDP Sciences, p 02014. https://doi.org/10.1051/matecconf/201821602014
11. Pokrovskaya O, Fedorenko R (2018) Evolutionary-functional approach to transport hubs classification. In: Murgul V, Pasetti M (eds) Proceedings of the 20th international scientific conference on energy management of municipal facilities and sustainable energy technologies. Advances in intelligent systems and computing, vol 982. Springer, Cham, pp 356–365. https://doi.org/10.1007/978-3-030-19756-8_33

12. Pokrovskaya O, Fedorenko R (2020) Assessment of transport and storage systems. In: Popovic Z, Manakov A, Breskich V (eds) Proceedings of the VIII international scientific Siberian transport forum. TransSiberia 2019. Advances in intelligent systems and computing, vol 1115. Springer, Cham, pp 570–577. https://doi.org/10.1007/978-3-030-37916-2_55
13. Pokrovskaya O, Fedorenko R, Khramtsova E (2019) Methods of rating assessment for terminal and logistics complexes. IOP Conf Ser Earth Environ Sci 403(1):012199. https://doi.org/10.1088/1755-1315/403/1/012199
14. Russian Railways (2017) The concept of the implementation of the comprehensive scientific and technical project "Digital Railway". https://www.irgups.ru/sites/default/files/irgups/science/document/koncepciya_realizacii_kompleksnogo_nauchno_tehnicheskogo_proekta_cifrovaya_zheleznaya_doroga.pdf. Accessed 11 Jan 2020
15. Sukhorukov A, Koryagin N, Sulyagina J, Ulitskaya N, Eroshkin S (2019) Digital transformation of airline management as the basis of innovative development. In: Popovic Z, Manakov A, Breskich V (eds) Proceedings of the viii international scientific Siberian transport forum. TransSiberia 2019. Advances in intelligent systems and computing, vol 1115. Springer, Cham, pp 845–854. https://doi.org/10.1007/978-3-030-37916-2_83
16. Tax Code of the Russian Federation July 31, 1998 N 146-FZ. http://www.consultant.ru/document/cons_doc_LAW_19671/. Accessed 01 Feb 2020

Automation Problems of ABC Costing in Russia

O. Potasheva

Abstract For a long time, it was the weak technical component of the automatic processing and systematization of information that prevented the use of ABC costing in the activities of Russian companies. In the context of the digitalization of the economy, the cost of processing accounting information is reduced, which leads to an increase in the possibilities for companies to use advanced methods of costing. The author conducted a survey of 30 enterprises using ABC costing for accounting support for management decisions. The survey results allowed us to establish that the most common partial ABC costing tool used by small and medium-sized assumptions is Microsoft Excel. 23% of the organizations examined use the Process Cost Analyzer product, OROS ABC, AllFusion Process Modeller and IDEF0.EM TOOL products are used by individual companies, which is associated with the relatively high cost of these software products. It was found that the greatest difficulties in setting up and using ABC costing by Russian organizations are caused by the modeling of ongoing business processes for the subsequent allocation of basic and auxiliary operations in order to distribute costs. In this regard, the author has developed a conceptual approach to the implementation of this stage.

Keywords ABC costing · Implementation of a costing system · Decision making · Computer program for cost accounting

1 Introduction

Modern business conditions lead to an increase in the share of overhead costs of companies, that requires their correct distribution between types of products. With the development of the economy, we also observe an increase in the range of products manufactured by manufacturers, which greatly complicates the assessment of the cost of each of them. A. Almeida, J. Cunha in their study of cost accounting methods by manufacturing companies indicate that accurate calculation of the cost

O. Potasheva (✉)
Samara State University of Economics, Samara, Russia
e-mail: olgakuzmina0212@gmail.com

S. I. Ashmarina and V. V. Mantulenko (eds.), *Current Achievements, Challenges and Digital Chances of Knowledge Based Economy*, Lecture Notes in Networks and Systems 133, https://doi.org/10.1007/978-3-030-47458-4_39

of manufactured products is a strategic task that determines the very possibility of doing business [1]. This task, according to these researchers, is complicated by the fact that the production process includes a number of actions common to various manufactured products, characterized by their own characteristics and various production requirements. A. Villalva-Cato, E. Ramos-Palomino, K. Provost, E. Casal emphasize the importance of the correct distribution of overhead costs not only at the level of one company, but also on the scale of efficient implementation of supply chains [7].

These circumstances determine the incorrect distribution of costs in cases where companies use traditional methods of cost accounting, which leads to a distortion of the cost of production, preventing the adoption of correct management decisions and business development.

ABC costing is an alternative to traditional cost accounting methods. His concept involves the classification of operations into main processes and the auxiliary operations required for their implementation. The types of activities for which the company collects information on costs are cost pools, and the factors that cause them to appear and change are drivers. As a result, based on the driver absorption rates per unit for all allocated cost pools, the company allocates costs based on the proportional consumption by the products of the operations performed. Thus, this approach allows a reasonable calculation of the cost of products.

Y. Anzai, M.E. Heilbrun, D. Haas, L. Boi, K. Moshre, S. Minoshima, R. Kaplan, V.S. Lee in their study emphasize the importance of using process maps to correctly account for company costs and the importance of automating settlement procedures for correct results [3]. For the practical application of ABC costing, it was important not only the development of the concept itself, which appeared in the 1980s, but also the formation of appropriate software for automating the process of cost accounting and calculating the cost of production by companies.

For a long time, it was the insufficient technical component that prevented large-scale automatic processing and systematization of information, which made it impossible to widely disseminate ABC costing in the activities of Russian companies. In the context of the digitalization of the economy, the cost of processing accounting information is reduced, which leads to an increase in the possibilities for companies to use advanced methods of costing.

2 Methodology

Automation of ABC costing is carried out by Russian companies based on the use of a large number of software products—from Excel to full-fledged ERP systems. When choosing products, companies take into account the complexity of production processes due to industry specifics and available resources (both financial for acquiring programs and labor).

To analyze the impact of cost accounting automation on the use of ABC costing by Russian companies, the author conducted a survey of 30 small, medium and large enterprises using ABC costing for accounting support for management decisions. Questioning of personnel of the surveyed organizations was carried out on the basis of a questionnaire presented in Table 1.

In order to systematize the results, fixed answers to questions were provided. To assess the level of automation of ABC costing, open-answer questions have also been added to enable a brief description of the software products and accounting processes implemented with their help.

The study was carried out in the 4th quarter of 2019, respondents' answers were analyzed and structured by the author, which allows using the obtained results to form a reasonable opinion on the state of ABC costing automation based on current practice as of the end of 2019.

Table 1 Questionnaire for assessing the impact of cost accounting automation on the process of using ABC costing by Russian companies

Question	Option № 1	Option № 2
1. Is ABC costing automated in your company?	Yes. What software product is used?	No. How is the application of the method organized?
2. Does the automation program cover all stages of calculations necessary for the application of ABC costing?	Yes. What software product is used?	No. Detail how calculations are carried out at stages that are not subject to automation
3. Does the software used support cost calculation?	Yes. What software product is used?	No. How is business process modeling carried out?
4. Is the mechanism for calculating data by cost drivers automated?	Yes. What software product is used?	Not. How are driver information evaluated and entered to perform the calculation?
5. Is the applied system for ABC costing integrated with the base of automation of accounting processes of the organization?	Yes. What software product is used and what are the benefits of integration?	No. What are the reasons for the lack of integration in this area?
6. What accounting operations cause the greatest difficulties in using ABC costing in your company?	Open response	–
7. What factors have a key influence on the choice of ABC costing automation system in your company?	Open response	–

Source author

3 Results

As a result of processing the respondents' answers, it was found that when using ABC costing, the surveyed companies use a number of software products, in particular, OROS ABC, Process Cost Analyzer, AllFusion Process Modeller, IDEF0.EM TOOL. The results of the evaluation of the method of automation of ABC costing application process by Russian companies presented in Table 2.

As we can see, the most common ABC costing tool is Microsoft Excel, which is used by 47% of the surveyed organizations. In questions with an open answer by respondents, it was indicated that the use of this product allows you to perform separate accounting actions in the process of distributing the organization's overhead costs and the final cost estimate. This massive use of Microsoft Excel is determined by its low cost and configuration flexibility.

But the respondents emphasized that the main problems of using ABC costing in this embodiment of the calculation are the high risks of errors in the numerical indicators and calculation algorithms, as well as the long processing time of large data sets. The specified tool is acceptable for the use of ABC costing in a relatively small number of business processes of small and medium enterprises with simple mechanisms for calculating the cost of products, but it is unsuitable for large production companies.

During the questionnaire, it was found that 23% of the organizations examined use the Process Cost Analyzer product, which can be described as a professional, narrowly focused software product. It allows for automated calculation of the cost of the main and auxiliary business processes, but requires the calculation of the final indicators of the cost of production manually, which is also associated with the human factor and, as a consequence, possible errors.

OROS ABC, AllFusion Process Modeller and IDEF0.EM TOOL products are used by individual companies, which is associated with the relatively high cost of these software products. Among the presented programs, it is OROS ABC that makes it possible to fully automate the use of ABC costing by a manufacturing enterprise. AllFusion Process Modeller and IDEF0.EM TOOL have similar functionality that allows to carry out operations to calculate the cost of business processes, but do not

Table 2 The results of the evaluation of the method of automation of ABC costing application process by Russian companies

Applicable product	The proportion of companies using this method %
Partial Automation with Excel	47
Process Cost Analyzer	23
AllFusion Process Modeller	13
OROS ABC	10
IDEF0.EM TOOL	7

Source author

imply automation of the calculation of numerical indicators for cost pools and drivers (manual entry is required).

Studying the respondents' practical experience in the field of ABC costing automation allows us to formulate the stages of this process. In particular, the process of automation of the cost accounting and calculation system is carried out by performing the following steps below.

1. Identification of the problem of the lack of automation of cost accounting and calculation. Based on the analysis of the information obtained at this step, the problem is formulated.
2. Making decisions on the need to solve the problem of automating cost accounting.
3. Analysis of the problem. A comprehensive study of the problem of automating cost accounting and calculation, evaluating the means at the company's disposal for its solution (financial and labor resources, organizational conditions).
4. Prediction of ways to solve the problem by identifying formal methods for automation of cost accounting.
5. Development of solutions to the problem by forming a list of software products available for purchase with an indication of their functionality.
6. Analysis of alternative solutions to the problem.
7. Choosing the best option for automating cost accounting in accordance with the evaluation criteria established by the company's management.
8. The acquisition and implementation of the selected solutions in the field of automation.

According to the results of the survey, it was found that the greatest difficulties in setting up and using ABC costing by Russian organizations are caused by the modeling of ongoing business processes for the subsequent allocation of basic and auxiliary operations in order to distribute costs. In this regard, the author has developed a conceptual approach to the implementation of this stage.

The general set of business processes that make up the activities of a manufacturing company and ensure the possibility of its normal functioning is based on the systematic implementation of a set of basic and auxiliary operations. From a methodological point of view, business processes implemented by the company can also be classified into existing (actually executed) and required (necessary for implementation). The degree to which the list of ongoing business processes matches the needs of the production system determines the quality of its functioning and is a fundamental factor in the sustainable development of the organization.

Modeling of the production system is carried out by transforming it into a conditional model consisting of a set of business processes. The basis of this transformation is the abstraction process, during which the business processes necessary for the production, storage and sale of each type of product produced by the organization are identified.

The working procedures carried out during the modeling of business processes are carried out in a certain logical sequence. The structure of the modeling of the production system includes:

- description of business processes and their grouping,
- determining the sequence of business processes,
- validation of the selection of business processes,
- graphic image of the logical sequence of the selected processes in the form of a test model.

Achieving reliability and completeness in the process of allocation of business processes is carried out through the implementation of a group approach, that is, the involvement of qualified specialists of various profiles. The correct formulation of the business processes of the facility is the basis for setting ABC costing in the company and is a prerequisite for its effective application. Validation of the grouping and distribution of selected business processes is carried out based on the use of deterministic logic techniques.

Establishing a functional sequence of business processes implementation involves identifying a set of targeted areas, which means a logical sequence of processes, the implementation of which leads to the production of specific types of products manufactured by the company.

In order to describe the relationships of business processes arising in the production of various products, they are compared and jointly analyzed by constructing special matrices. To build matrices, the technological features of the organization's products are analyzed. Subsequently, these matrices are used to develop options for the functional improvement of the business process system.

The functional organization of the system of ongoing business processes should be understood as a comprehensive description of the production system, reflecting the degree of its perfection in terms of concentration of business processes, their compatibility and flexibility.

To assess the functional organization of the system of ongoing business processes, it is advisable to test on the following aspects:

- the relevance of the business processes allocated to the production goals of products,
- a clear certainty of the specifics of business processes in order to concretize their content,
- strict consistency of selected business processes.

Respondents' answers about the key factors that determined the choice of a software product for ABC costing automation testify to the effect of many different conditions. The author concluded that a reasonable choice of the ABC costing automation option can be made by the company by solving multi-criteria optimization problems. The main areas of using multicriteria optimization methods are the following problem situations:

- the need to evaluate the analyzed process in terms of several indicators,
- the analyzed process is a component of several processes, each of which may have its own optimality criterion.

In accordance with the various strategic goals of the automation option companies, ABC costing can satisfy one of the following criteria:

- ensuring the greatest correlation of the beneficial effect of automation and its costs,
- achieving the greatest effect from automation ABC costing,
- ensuring the desired effect from the automation of ABC with minimal total cost of this process.

For the practical solution of this problem, prioritization should be made. The solution to this problem is carried out in the economic space, represented by a system of restrictions in the form of specific resources available to the company.

At the same time, in the process of choosing the optimal ABC costing automation option, one should also take into account the influence of the managerial decision made on economic indicators of management efficiency and the subsequent change in the company's market value compared to the relative growth of the industry market.

4 Discussion

In their study J.B.C.N. Araujo, A.N. Souza, M.S. Joaquim, L.M. Mattos, I.M. Lustosa Junior emphasize the serious impact of industry-specific features of the company on the specifics of ABC costing automation [4]. D.T. Cremonese, G.D.E. Tomi, M.R. Neves, confirming a similar position, concluded that the creation of an individual software product for the application of ABC costing is a profitable investment for the company, since in the future such an information system will show profitable and unprofitable products taking into account the specifics of the business that will contribute to its sustainable development [6].

Z.Y. Zhuang and S.C. Chang proposes the use of a mixed integer programming model based on a time cost accounting system. Using a time driver characterizing the relationship between a resource and an object of costs, taking into account the numerous resource limitations, the model allows you to search for a global optimal solution for a company. This avoids some of the possible limitations of the programming modeling approach when using the theory of constraints or activity-based costing [8].

Z.Y. Zhuang and S.C. Chang substantiate their theoretical approach using a numerical example. Based on the budget profit and loss statement, the results for the formulated models based on the time cost accounting system, the theory of constraints and the cost-based assessment of activity were used to compare profit indicators. As a result, it was established that the time-based cost-accounting model proposed by the researchers supports the formation of the most accurate information for making decisions on the profitability of the product range in comparison with other approaches that are implemented in practice [8].

Based on an analysis of prevailing cost accounting practices, C. Anghelache, S. Capușneanu, D.I. Topor, A. Marin-Pantelescu conclude that Microsoft Office Access is widely available for solving applied tasks of information support for managerial decision-making in terms of costs with detailed product range [2]. The results of their research indicate the possibilities of solving functional tasks using Microsoft Office Access for generating data on target cost indicators by product type and creating separate information blocks for information support for managing a small and medium-sized enterprise. The indicated functionality, in their opinion, can be the basis for automation of information support for managing targeted cost accounting.

V. Chouhan, G. Soral, B. Chandra argue that by improving the quality of information through automation of ABC costing, a company can improve the quality of its management decisions by using reliable data on the value of direct material costs and the value of available stocks [5].

5 Conclusion

Sustainable business development in modern economic conditions is based on high-quality information support for the management decision-making process. Ubiquitous digitalization leads to the spread of advanced methods of cost accounting and calculation, reducing due to automation the probability of accounting errors and the cost of obtaining information.

Based on the results of the survey, it was found that the following key requirements for ABC costing software are presented by Russian companies:

- purposefulness, characterizing the correspondence of the generated information to decision-making goals,
- completeness and reliability, determining the availability of information in the amount necessary and sufficient for management purposes,
- concreteness, that is, the exclusion of unnecessary and redundant information,
- efficiency, implying a reduction in time spent on the collection, processing, systematization and presentation of information,
- the balance of costs for the implementation and maintenance of software and the quality of information results of its use.

ABC costing in the context of digitalization is becoming an affordable and effective tool for solving the problems of the correct distribution of overhead costs in the context of the growth of their share and the complexity of the activities of business entities.

References

1. Almeida A, Cunha J (2017) The implementation of an activity-based costing (ABC) system in a manufacturing company. Procedia Manuf 13:932–939. https://doi.org/10.1016/j.promfg.2017.09.162
2. Anghelache C, Capușneanu S, Topor DI, Marin-Pantelescu A (2019) Target costing and its impact on business strategy computer program for cost accounting and administration. In: Oncioiu I (ed) Network security and its impact on business strategy. IGI Global, Hershey, pp 20–43. https://doi.org/10.4018/978-1-5225-8455-1.ch002
3. Anzai Y, Heilbrun ME, Haas D, Boi L, Moshre K, Minoshima S, Kaplan R, Lee VS (2017) Dessecting costs of CT study: application of TDABC (time driven activity-based costing) in a tertiary academic center. Acad Radiol 24(2):200–208. https://doi.org/10.1016/j.acra.2016.11.001
4. Araujo JBCN, Souza AN, Joaquim MS, Mattos LM, Lustosa Junior IM (2020) Use of the activity-based costing methodology (ABC) in the cost analysis of successional agroforestry systems. Agrofor Syst 9:71–80. https://doi.org/10.1007/s10457-019-00368-6
5. Chouhan V, Soral G, Chandra B (2017) Activity based costing model for inventory valuation. Manag Sci Lett 7:135–144. https://doi.org/10.5267/j.msl.2016.12.003
6. Cremonese DT, Tomi GDE, Neves MR (2016) Cost modelling of the product mix from mining operations using the activity-based costing approach. Revista Escola de Minas 69(1):97–103. https://doi.org/10.1590/0370-44672015690137
7. Villalva-Cataто A, Ramos-Palomino E, Provost K, Casal E (2019) A model in agri-food supply chain costing using ABC costing: an empirical research for Peruvian coffee supply chain. In: Bilof RS (ed) Proceedings of the 7th international engineering, sciences and technology conference (IESTEC). IEEE, Piscataway. https://ieeexplore.ieee.org/abstract/document/8943583. Accessed 11 Feb 2020
8. Zhuang ZY, Chang SC (2017) Deciding product mix based on time-driven activity-based costing by mixed integer programming. J Intell Manuf 28:959–974. https://doi.org/10.1007/s10845-014-1032-2

Role of Losses in the Financial Statements of Digital Companies

U. E. Monastyrsky

Abstract Nowadays, there is a problem that financial reporting does not fully meet current realities and features of digital companies, where intangible assets occupy a special position: developments, brands, organizational strategy, networks of partners and suppliers, client and social relations, computer data and software, as well as the human capital. With the growth of total investments, companies are increasingly faced with a paradox of "digitalization" and do not get the expected profit. The main purpose of this article is to determine how income and losses affect the value of the company, as well as to determine the main criteria for evaluating intangible assets of digital companies. The article presents results of analyzing the role of losses in the financial statements of digital companies. The author reveals the existence of an inverse relation between investments in the company's technological development and the growth of its losses in the financial statements.

Keywords Digitalization · Controlling persons · Bankruptcy · Liability of controlling persons

1 Introduction

The total capitalization of one hundred of the world's largest companies in 2019 amounted to more than 21 trillions US dollars [1], seven of the ten most expensive organizations are technology and online retail companies—Microsoft, Apple, Amazon, Alphabet, Facebook, Alibaba, and Tencent. This fact is a consequence of the digital revolution, the main result of which was the growth of the number of technology and digital companies around the world. The main feature of these companies is not providing physical goods and services, but intangible goods: information, digital goods and services, software, customers and brands [4].

In the study of Capgemini Consulting and MIT Sloan School of Management "Embracing Digital Technology: A New Strategic Imperative", based on the analysis

U. E. Monastyrsky (✉)
Moscow State Institute of International Relations (MGIMO University), Moscow, Russia
e-mail: monastyrsky@mzs.ru

S. I. Ashmarina and V. V. Mantulenko (eds.), *Current Achievements, Challenges and Digital Chances of Knowledge Based Economy*, Lecture Notes in Networks and Systems 133, https://doi.org/10.1007/978-3-030-47458-4_40

results of activities of 400 large companies from different industries, expectations from the use of new technologies and management techniques and the dependence of financial indicators were determined. For example, companies that actively use technologies and new management methods are on average 26% more profitable than their competitors; organizations that actively invest in digital technologies, but pay little attention to management, have financial indicators 11% lower; companies that improve only management receive plus 9% of their profits, but can potentially acquire three times more using digital technologies; companies that do not choose a digital development strategy have negative financial indicators compared to other market players—minus 24% [5].

Digitalization of business models has also brought its own adjustments to monetization strategies, ensuring production efficiency. Meanwhile, with the growth of total investments, companies are increasingly faced with a "digitalization" paradox and do not get the expected profit. As the company's investment in digitalization increases, the paradox becomes more likely, and only a few companies actually achieve high returns that meet their expectations [9].

However, with the rapid growth of the number of digital companies, the assessment of their profitability becomes particularly important [1]. For investors, information about profits and their driving forces is without a doubt the most valuable information disclosed by digital companies. Losses and profits have remained the main indicators of financial reporting over the past decades [16]. However, current trends indicate the inefficiency of traditional correlations between income and loss in the enterprise value assessment in relation to digital companies.

2 Methodology

The methodological basis of this research is a systematic approach to the study of modern practice of providing financial statements of digital companies. When processing the actual material, the author used traditional scientific methods such as dialectical, logical, scientific generalization, content analysis, comparative analysis, synthesis, source studies, etc. Their application allowed ensuring the validity of the analysis, theoretical and practical conclusions.

3 Results

Traditional financial indicators of the company profitability demonstrate that they are not applicable to digital companies. A prime example is Uber, whose net loss in 2019 reached 8.51 billion U.S. dollars, and in 2018, the company made a profit of $997 millions, and spent $172 millions on compensation to shareholders [17]. Despite significant losses, the company held a successful IPO in April 2019. In total, more than 90 investors have invested in this business. According to some experts, Uber has conducted 23 investment rounds since 2009 [6].

The situation was similar with Twitter and LinkedIn in 2013 and with WhatsApp in 2014. A reasonable question arises: why are technology companies that show a negative balance in demand in the market? Digital transformation exacerbates problems and difficulties for taking financial decision, making financial reporting, and accounting recommendations on which investors and market participants base their capital market decisions. The main reason lies in the goods/services produced by these companies. Valuation of assets of a company that produces physical goods is quite simple—fixed capital and its depreciation in a straight line over several years are taken as a basis. Thus, there is a direct link between the company's losses and the value of its shares.

In another situation are digital companies, a significant part of corporate balance sheets of which currently consists of intangible assets compared to physical assets. Why do investors react negatively to losses in financial statements for an industrial firm, but ignore such losses for a digital company?

4 Discussion

According to experts, financial reporting annually demonstrates less efficiency and applicability to the assessment of the current state of affairs of digital companies [3]. As it was noted earlier, for an industrial company, the profit and loss statement provides a complete description of the state of production assets. However, digital companies often have assets that are intangible in their nature, and many of them have ecosystems that extend beyond the company. Many digital companies do not have physical products and inventory to report. Therefore, the balance sheets of physical and digital companies present completely different pictures [11].

Digital companies' assets are based on intangible assets such as research and development, brands, organizational strategy, networks of partners and suppliers, customer and social relationships, computer data and software, and the human capital.

The Harvard Business Review published results of a survey of chief financial officers of leading technology companies and senior analysts at investment banks who monitor technology companies. Some of these ideas contradict traditional financial thinking, while others seem to be highly controversial.

It is assumed that financial capital is virtually unlimited, while some types of human capital are in short supply. For digital companies, the time of scientists, software developers, and product development teams is the company's most valuable resource. They believe they can always raise financial capital to cover a funding shortfall, or use the company's shares or options to pay for acquisitions and employee salaries. Thus, the main goal of the CEO is not to allocate financial capital wisely, but to direct valuable scientific and human resources to the most promising projects and to return and reallocate these resources in a timely manner when the prospects for specific projects dim. Traditional valuation models view risk as an undesirable feature. Digital companies, by contrast, are chasing risky projects that have lottery payouts. An idea with uncertain prospects, but at least some possible chance of reaching a billion dollars in revenue, is considered much more valuable than a project with a net present value of several hundred million dollars, but no chance of massive growth [10].

At the same time, for digital companies, in contrast to companies in the traditional manufacturing sector, there is a certain paradox: financing in its intangible assets is not capitalized as assets and is considered as an expense when calculating profits [12]. Thus, the more a digital company invests in building its future, the higher its reported losses are [15]. However, when investing in intangible assets, digital companies have the same goal as non-digital companies. Thus, investors have no choice but to ignore profit in their investment decisions [2, 7].

Intangible assets also have another distinctive feature. Their value increases as more people use a product or service, while physical assets depreciate as they are used [13]. Thus, the fundamental idea behind the success of digital companies (increasing returns from the scale) goes against the basic principle of financial accounting (assets are depreciated with use).

As digital companies become more visible in the economy and physical companies become more digital in their operations, there is a persistent trend that intangible investments have surpassed fixed assets, and understanding fixed assets and equipment as the main way to create capital is no longer applicable [9].

However, there is no place in financial accounting for the concept of network effects or increasing the value of a resource by using it. This actually means a negative depreciation expense in accounting language [8].

In such circumstances, the search for effective methods of evaluation of intangible assets becomes particularly important. It is interesting to use methodologies for generating income and cash flows that are specifically designed to assess the value of intangible assets.

According to S. Krishnan, when a company acquires another company, it is required to distribute the purchase price separately for all assets acquired for financial reporting purposes. As part of the purchase price allocation process, methods such as royalty exemption and excess profit method are used to evaluate intangible assets, which is very convenient when assessing the value of digital companies. Valuation of intangible assets based on these methods gives an accurate picture of where the real value is and what the cost factors are for these digital companies [14].

5 Conclusion

The current financial reporting structure, although necessary and authorized by the financial authorities, does not necessarily reflect incomes from intangible assets, and as a result, levels the value of digital companies, which is a huge disadvantage.

Digital companies generate a significant portion of their value, brand value, and value from intangible assets. The more a digital or technology company invests in growth and development in the form of intangible investments, the more it simultaneously makes itself less profitable in the form of higher operating costs.

The existing inverse correlation between investments in the company's technological development and the growth of its losses in the financial statements, as well as the lack of information about the growth in the value of intangible assets with their use, determines the search for additional information sources about factors underlying them. Such non-financial indicators are becoming more and more significant in the context of the growth of the digital companies number.

Companies' finance directors are also aware of the growing limitations of the current financial reporting model. As a way out of this situation, some modern digital companies disclose information about users, pricing, and downloads in addition to other performance indicators. Financial as well as non-financial disclosures show investors the real picture of affairs and the value of digital companies.

References

1. AICPA (2018) Audit and accounting guide: investment companies. Wiley, Hoboken. https://doi.org/10.1002/9781119563976.ch7
2. Alkhatib E, Ojala H, Collis J (2019) Determinants of the voluntary adoption of digital reporting by small private companies to companies house: evidence from the UK. Int J Account Inf Syst 34:100421. https://doi.org/10.1016/j.accinf.2019.06.004
3. Baruch L, Feng G (2016) The end of accounting and the path forward for investors and managers. Wiley, Hoboken
4. Büttner R, Müller E (2018) Changeability of manufacturing companies in the context of digitalization. Procedia Manuf 17:539–546. https://doi.org/10.1016/j.promfg.2018.10.094
5. Capgemini (2013) New research from Capgemini consulting and MIT Sloan management Review reveals why organizations are struggling to drive digital transformation. https://www.capgemini.com/news/new-research-from-capgemini-consulting-and-mit-sloan-management-review-reveals-why/. Accessed 11 Feb 2020
6. CBInsights (2019) Uber is going public. Here are the investors that stand to gain the most. https://www.cbinsights.com/research/uber-ipo-investor-analysis/. Accessed 11 Feb 2020
7. Duțescu A (2019) Financial accounting. An IRFS perspective in Romania. Palgrave Macmillan, London. https://doi.org/10.1007/978-3-030-29485-4
8. Charles R, Chaffin EdD (2019) Financial planning competency handbook, 2nd edn. Wiley, Hoboken. https://doi.org/10.1002/9781119642497.ch7
9. Gebauer H, Fleisch E, Lamprecht C, Wortmann F (2020) Growth paths for overcoming the digitalization paradox. Bus Horiz. https://doi.org/10.1016/j.bushor.2020.01.005. Accessed 11 Feb 2020
10. Govindarajan V, Rajgopal Sh, Srivastava A (2018) Why we need to update financial reporting for the digital era. Harv Bus Rev. https://hbr.org/2018/06/why-we-need-to-update-financial-reporting-for-the-digital-era?referral=03758&cm_vc=rr_item_page.top_right. Accessed 11 Feb 2020
11. Govindarajan V, Rajgopal Sh, Srivastava A (2018) Why financial statements don't work for digital companies. https://hbr.org/2018/02/why-financial-statements-dont-work-for-digital-companies. Accessed 11 Feb 2020
12. Kogdenko V (2018) Methodology of the financial analysis of digital companies. Account Anal Audit 5:94–109. https://doi.org/10.26794/2408-9303-2018-5-3-94-109

13. Kokh L, Kokh Yu, Prosalova V (2019) Disclosure of intellectual capital by digital companies. In: Ilin IV (ed) Proceedings of the international conference on digital technologies in logistics and infrastructure. Atlantis Press, Amsterdam, pp 193–198. https://doi.org/10.2991/icdtli-19.2019.36
14. Krishnan S (2015) Digital companies and the valuation of intangible assets. Toptal. https://www.toptal.com/finance/valuation/valuation-of-intangible-assets. Accessed 11 Feb 2020
15. Machado CG, Winroth M, Carlsson D, Almström P, Centerholt V, Hallin M (2019) Industry 4.0 readiness in manufacturing companies: challenges and enablers towards increased digitalization. Procedia CIRP 81:1113–1118. https://doi.org/10.1016/j.procir.2019.03.262
16. Rozhnova O, Atazhanova D (2016) Methodological aspects transparency financial statements of construction companies. Proc Voronezh State Univ Eng Technol 4:386–390. https://doi.org/10.20914/2310-1202-2016-4-386-390
17. Uber Investor (2019) Uber announces results for fourth quarter and full year. https://www.sec.gov/Archives/edgar/data/1543151/000154315120000005/uberq419earningspressrelea.htm. Accessed 11 Feb 2020

Automation of Design and Technological Preparation of Repair Works

A. Kornilova, E. Acri, and N. N. Pronina

Abstract The priority direction of the state policy in the sphere of technical operation of housing and communal facilities is the transition to the principle that implies the use of the most modern information technologies and the growth of the technical equipment level of engineering facilities. As a result, it increases the comfort of living conditions and energy efficiency of housing and communal facilities as a whole. Creating a single information space and a network management structure allows you to maximize the efficiency of services, coordinated work of all automated engineering systems of the building and objective information about the state of all equipment, a high level of resource management. The main advantages of implementing automation and dispatching systems for organizations that operate housing and utilities facilities are the detection and recognition of emergency situations at early stages and minimizing their consequences, reducing the cost of maintenance and operation of the building by up to 30%. These main advantages of IT application are considered in this work.

Keywords Housing and communal complex · Automation · Computer-aided design of works (CAD) · Sustainable development of organizations

1 Introduction

At the moment, in the field of technical operation of housing and communal facilities, there are many interrelated problems that lead to high accident rates, inefficient use of resources, high costs for the operation of various systems and a weak information

A. Kornilova (✉)
Samara State University of Economics, Samara, Russia
e-mail: adkornilova@yandex.ru

E. Acri · N. N. Pronina
Samara State Technical University, Samara, Russia
e-mail: kotay80@mail.ru

N. N. Pronina
e-mail: pronina_natalya@mail.ru

S. I. Ashmarina and V. V. Mantulenko (eds.), *Current Achievements, Challenges and Digital Chances of Knowledge Based Economy*, Lecture Notes in Networks and Systems 133, https://doi.org/10.1007/978-3-030-47458-4_41

base necessary for the development of programs for major repairs of the common property of apartment buildings.

The management system of apartment buildings, which implies providing favorable and safe living conditions, appropriate maintenance of the common property of apartment buildings and the provision of public services, requires changes that will be aimed at improving its quality.

One of the priority directions of the state policy in the sphere of technical operation of housing and communal facilities is the application of modern information technologies that are able to ensure the growth of the technical level of engineering facilities. As a result, information technologies improve the living conditions and the energy efficiency of housing and communal facilities [1–5]. In this regard, the authors consider a problem of automation of technical operation processes of housing and communal complex objects as relevant one.

2 Methodology

The authors determined that the management of housing and communal complex facilities should be based on such basic principles that will ensure the construction of an understandable and effective management system, regardless of the chosen form of real estate management:

1. *Object-based real estate management.* The control is carried out for each object based on the condition of the property, the extent of its accomplishment, as well as the status of the land assigned to it, and elements of landscaping and gardening located on it.
2. *Target management.* Goals should be specific and relevant for a specific moment, and be measurable by certain criteria and indicators that are specific to these goals and achievable during the periods for which they are set.
3. *Management efficiency.* This principle consists in achieving the set management goals (a certain qualitative result of management activities or the qualitative state of the object of management) with optimal costs.
4. *Professionalism in management.* The level of qualification and business qualities of the workforce is the most significant factor for improving the efficiency of the real estate management. It is extremely important that the financial condition of entities engaged in management activities depends on the results of their activities.
5. *Systematic management.* It implies that an effective system of management of housing and communal facilities is achieved through the joint efforts of owners of management facilities and management organizations.

3 Results

Based on the results of the analysis of modern automation systems, it is proposed to create a computer-aided design system for the production of repair work, which provides the design of the organization, technological preparation and repair work. The proposed model takes into account unscheduled events and helps the expert in making operational decisions aimed at saving material, time and human resources [6]. By presenting repair work plans (planned and unplanned), repair work, the composition of construction materials, products, machines and mechanisms (hereinafter—the nomenclature) that participate in the repair work, as well as structural elements that are being worked on in the form of separate entities, the design system provides the necessary "flexibility" [7].

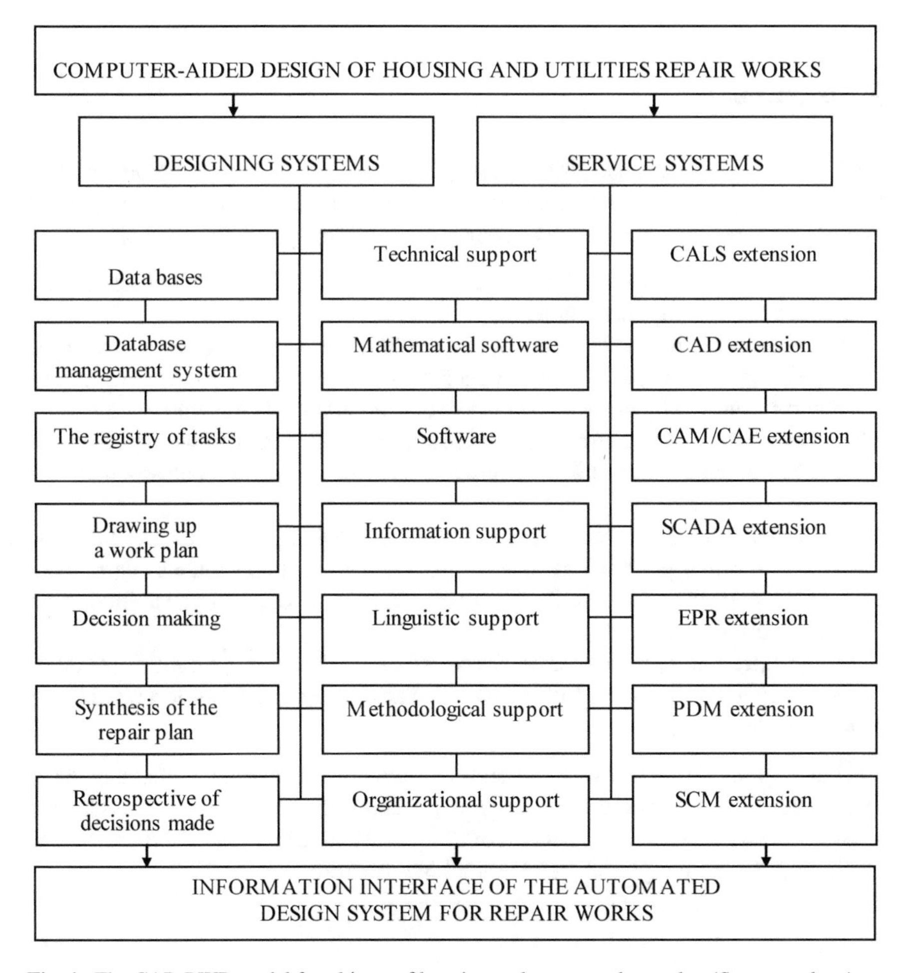

Fig. 1 The CAD RWP model for objects of housing and communal complex (*Source* authors)

The design subsystem of a computer-aided design system for the repair work production (hereinafter referred to as CAD RWP) is provided by: a database, a database management system, registers of the main tasks, preparation of a work plan (unplanned and planned), a decision-making subsystem, a subsystem for the synthesis of repair plans, a subsystem for analysis of solutions (retrospective of solutions) (Fig. 1).

The algorithm of decision-making in case of emergency situations in various elements of housing and communal complex objects is also proposed. An algorithm for synthesizing repair plans (emergency and planned) is developed and an organizational and technological scheme for conducting the retrospective process is proposed. The decision-making algorithm given in the article is necessary in case of emergency situations in the structural elements of considered objects and can help an expert in choosing a method of unscheduled repair work.

4 Discussion

Based on the analysis of modern technologies of the interactive and automated systems, the authors determined basic system requirements to enable the automated system of design and technological preparation of design work production:

- access to the system should be provided via a broadband Internet channel, with appropriate security restrictions,
- the system should be able to handle large volumes of structured and unstructured data storage (at the same time, access from outside, directly to the data storage should be physically closed, and the system should enable the work with data),
- the system should provide a set of services that interact in parallel with the complex of data and engineering infrastructure (a set of services should correspond to the specifics of each of the participants),
- the system should provide access to its services using public interfaces defined in the Extensible Markup Language (XML): the interaction should be carried out in accordance with the described formats, based on XML, and transmitted using Internet protocols,
- the system should provide the ability to use its services using a public user interface, which in turn should fully provide the ability to solve tasks described in the methodology.

These limitations allow us to determine a range of technologies that can solve these tasks. Cloud computing can become a technology platform that provides the ability to create an automated system for designing and technological preparation of repair work for buildings and engineering infrastructure.

5 Conclusion

The presented algorithm for making decisions in the emergency case is necessary when accidents occur in the structural elements. This model could help an expert by choosing a method of performing unscheduled repairs: partially or completely carry out emergency repairs as part of planned repairs or start unscheduled repairs. If there is an accident or breakage on a structural element, the expert should make a plan for emergency repairs, according to the CAD model of the for the repair work production.

References

1. Czegledy T, Fedorenko RV, Zaichikova NA (2019) Modelling of software producer and customer interaction: nash equilibrium. In: Ashmarina S, Vochozka M, Mantulenko V (eds) Digital age: chances, challenges and future. Lecture notes in networks and systems, vol 84. Springer, Cham, pp 298–307. https://doi.org/10.1007/978-3-030-27015-5_36
2. Komkov NI (2018) External and internal challenges and prospects for the modernization of the Russian economy. MIR (Mod Innov Dev) 9(1):12–24
3. Yakhneeva IV, Agafonova AN, Fedorenko RV, Shvetsova EV, Filatova DV (2019) On collaborations between software producer and customer: a kind of two-player strategic game. In: Ashmarina S, Mesquita A, Vochozka M (eds) Digital transformation of the economy: challenges, trends and new opportunities. Advances in intelligent systems and computing, vol 908. Springer, Cham, pp 570–580. https://doi.org/10.1007/978-3-030-11367-4_56
4. Chudaeva AA, Mantulenko VV, Zhelev P, Vanickova R (2019) Impact of digitalization on the industrial enterprises activities. In: Mantulenko V (ed) Proceedings of the 17th international scientific conference "problems of enterprise development: theory and practice" 2018. SHS web of conferences, vol 62. EDP Sciences, Les Ulis, p 03003. https://doi.org/10.1051/shsconf/20196203003
5. Nikitina NV, Chaadaeva VV, Chudaeva AA (2019) Effective development mechanism of companies in the communal services sector in modern conditions. In: Ashmarina S, Vochozka M (eds) Sustainable growth and development of economic systems. Contributions to economics. Springer, Cham, pp 335–348. https://doi.org/10.1007/978-3-030-11754-2_24
6. Semenyuk O, Beloussova E, Nechay N, Listkov V, Kurbatova V, Niyazbekova S, Abdrashitova T (2018) The influence of ecology and economic factors on eco-architecture and the design of energy efficient buildings. World Trans Eng Technol Educ 16(2):186–192
7. Belanova NN, Kornilova AD, Sultanova AV (2020) Target indicator and directions for the digital economy in Russia. In: Ashmarina S, Mesquita A, Vochozka M (eds) Digital transformation of the economy: challenges, trends and new opportunities. Advances in intelligent systems and computing, vol 908. Springer, Cham, pp 111–118. https://doi.org/10.1007/978-3-030-11367-4_11

Augmented Reality, Artificial Intelligence and Big Data in Education and Business

Artificial Intelligence in the Contemporary Digital Environment

J. D. Ermakova

Abstract The article is focused on stimulating the procedure of smooth functioning the latest advanced technologies using artificial intelligence in various areas of social, economic, educational and other spheres of the modern citizens. The author is studying the problems and possible implementation up to date scientific and technological researches not in the near future, but right now today. To make such embodiment more successful at the regional level, the system of contemporary economic and political relations certainly needs fundamental reconstruction. First of all, it is necessary to expand local export and import products by means of the modern software. And that is a real challenge. First of all, the society have to avoid misleading issues which usually take place when obsolescent software strategies take root in new technological generation and prevent to use it more efficiently. The most common areas where artificial intelligence could be used more successfully include economics, education, agriculture, medicine, aerospace industry, etc.

Keywords Artificial intelligence · Software · Educational environment · Economic development

1 Introduction

Nowadays generation, so called Generation Z, could hardly imagine educational environment without up-to-date advanced digital technologies. First ideas about computing intelligence appeared in the middle of the 20th century and they were represented by British mathematician Turing A., who raised the question of whether machines could think. He developed a simple heuristic to test his hypothesis: can a computer conduct a conversation and answer questions in such a way as to deceive a suspicious person into thinking that was the computer actually human? Then C. Shannon suggested creating a machine that could be taught to play chess. He made a proposal that the machine could be trained by brute force or by evaluation a

J. D. Ermakova (✉)
Samara State University of Economics, Samara, Russia
e-mail: ermjul@yandex.ru

S. I. Ashmarina and V. V. Mantulenko (eds.), *Current Achievements, Challenges and Digital Chances of Knowledge Based Economy*, Lecture Notes in Networks and Systems 133, https://doi.org/10.1007/978-3-030-47458-4_42

small set of strategic moves of the enemy [2]. Many scientists consider this research project the birth of artificial intelligence (AI). While AI research has steadily progressed over the past 60 years, the promise of early AI promoters turned out not to be really successful. This led to funding cuts and decrease interest in AI research in the 1970s. New growing trend to AI came with advances in computing power, which was produced in the 1990s. Data storage developed to the point where complex tasks became feasible. In 1995, AI took a big step forward with the development of an artificial linguistic Internet computer by R. Wallace, which was able to conduct small talks and simple conversations. Also in the 1990s, IBM developed a computer called Deep Blue, who played against world chess champion Garry Kasparov. Deep Blue could look forward six steps or more and calculate 330 million positions per second. In 1996, Deep Blue lost to Kasparov, but won a rematch a year later. In 2015 DeepMind released the alphabet software to play the ancient game of Go vs the best players in the world. He used an artificial neural network that was trained on thousands from human amateur and professional games to learn how to play. In 2016, Alpha Go beat the best player in the world at the time, Lee Sedol, four games to one. Then the developers of Alpha Go let the program play with itself by trial and error, starting with completely random ones play with a few simple guidelines. The result was a program (Alpha Go Zero) that taught itself is faster and was able to beat the original Alpha Go by 100 games to 0. Completely out of independent game—without human intervention and without the use of historical data—Alpha Go Zero surpassed all other alpha versions run for 40 days [2].

Over the past few years, the availability of Big Data, cloud computing, and related services computing and memory capacity and breakthroughs in artificial intelligence technology called "Machine learning" (ML), dramatically increased the power, availability, growth, and impact of AI. Ongoing technical progress is also leading to improved and cheaper sensors that capture moreover, it provides reliable data for use by artificial intelligence systems. The amount of data available for artificial intelligence systems continues to grow as these sensors become smaller and cheaper to deploy [6].

Result is a significant advance in many major areas of AI research, such as:

- natural language processing,
- autonomous vehicles and robotics,
- computer vision,
- language learning.

Some of the most interesting AI developments are outside of computer science in various fields for example, healthcare, medicine, biology, and Finance. In many ways, the AI transition resembles a way to distribute computers from several specialized enterprises to the broader economy and it also recalls how Internet access has expanded beyond multinational firms for the majority of the population in many countries in the 2000s, the economy will be more and more we need a bilingual sector. These are people who specialize in one area, such as Economics, biology or law, but also skilled in AI technique. They focused on applications that are used or projected in the short and medium term, rather than possible long-term developments, such as artificial general intelligence, do not have a generally accepted definition of AI. In November 2018, the AI group of companies OECD experts (AIGO) created a subgroup to develop a description of the artificial intelligence system [9]. The description must be clear, technically accurate, technologically neutral, and applicable to short-term and long-term time horizons. It is wide enough to cover many of them. AI definitions are widely used by scientific, business, and political circles. In addition, he informed about the development of the OECD Council's recommendation on Artificial intelligence [7].

This description of the artificial intelligence system is based on the conceptual representation. This view is consistent with the widely used definition of AI as "the study of computations that allow perceive, reason, and act" and with similar general definitions. The conceptual view of AI is first presented as a high-level structure of the general AI system (also denoted as "intelligent agent") [5].

2 Methodology

The artificial intelligence system consists of three main elements: sensors, operational logic, and actuators. Sensors collect raw data from the environment, while the drives act to change the state of the environment. Key strength of the artificial intelligence system it resides in its operational logic. For a given set of goals and based on input data from sensors, operating logic provides an output for actuators. Thus recommendations, forecasts, or decisions that may affect the environment.

To cover different types of artificial intelligence systems and different scenarios, the program separates the process of building a model (for example, ML) from the model itself. Model building is also separate from the model interpretation process, which uses the model to make predictions, recommendations, and decisions; Executive mechanisms use these results to influence environment. The environment in relation to the AI system is the space observed through perception (through sensors) and act through actions (via actuators). Sensors and actuators are either cars or people. The environment is either real (for example, physical, social, mental) and usually only partially observed, or virtual (for example, board games) and, as a rule, completely observable. An artificial intelligence system is a machine system that can, for a given set of human-defined goals, make predictions, recommendations, or decisions that affect real or virtual environments. It does this using machine and/or human input data to:

- perceive the real and/or virtual,
- abstract such representations in the model through analysis in an automated method (for example, using ML or manually),
- use model inference to formulate options information or action. Artificial intelligence systems are designed to work with different levels of autonomy, AI model, model construction, and model interpretation.

3 Results

The core of an artificial intelligence system is an artificial intelligence model that represents all or part of the system, an external environment that describes the structure and/or dynamics of the environment. Model can be based on expert knowledge and/or data obtained by humans and/or automated tools (for example, ML algorithms). Goals (such as output variables) and performance indicators (such as accuracy, resources for learning, the representativeness of the dataset) direct the build process. Model inference is the process by which people and/or automated tools get results from model. They take the form of recommendations, forecasts, or decisions. Goal-setting and performance indicators guide the implementation. In some cases, (for example, deterministic rules), the model can offer a single recommendation. In other cases, (for example, probabilistic models), the model can offer a variety of recommendations. These recommendations are related to various levels, for example, performance indicators, such as confidence, reliability, or risk. In some cases, in the process of interpretation, it is possible to explain why some specific make recommendations. In other cases, the explanation is almost impossible.

Credit scoring system, for instance, illustrates a machine system that affects the environment (whether loans are issued to people). It provides recommendations (credit rating) for this product. It does this using both machine inputs (historical data on people's profiles and whether they have repaid loans) and human resources (set rules). Using these two sets of input data, the system perceives the real environment (whether it is people repay loans on an ongoing basis). It abstracts such perceptions in the model automatically. The credit scoring algorithm can, for example, use a statistical model. Finally, it uses an output model (credit scoring algorithm) to formulate a recommendation (credit assessment) of possible outcomes (granting or refusing a loan).

Assistant for the visually impaired an assistant for visually impaired people illustrates how the machine system is affected his entourage. It gives recommendations (for example, how a visually impaired person can avoid obstacle or street crossing) for a given set of goals (moving from one place to another). It does this using machine and/or human input data (large image databases with labels, objects, written words, and even human faces) for three ends.

- first, it perceives images environment (the camera captures an image of what is in front of the person and sends it to application),
- second, it automatically abstracts such perceptions in the model (object recognition algorithms that can recognize a traffic light, car, or obstacle on the sidewalk),
- third, it uses the output model to recommend options for results (providing audio description of objects found in the environment) so that a person can decide how to act and thus affect the environment.

Artificial intelligence provides more easily accessible prediction. From an economic point of view, recent advances in artificial intelligence (AI) either reducing the cost of forecasting or improving the quality of available forecast [1]. Many aspects of decision-making are separated from forecasting. However, it has improved, low-cost and widely available AI prediction can be transformative because prediction it is a contribution to a large part of human activity. As the cost of AI prediction has decreased, there are more opportunities to use prediction. It appeared, as with computers in the past. The first AI applications were long recognized as forecasting problems. For example, machine learning (ML) predicts credit defaults and insurance risk. As their cost declines, some human activities are reinterpreted as forecasting problems.

In medical diagnostics, for example, the doctor uses data about the patient's symptoms and fills in missing information about the cause of these symptoms. The process of using data for the complete missing information is a prediction. Object classification is also a prediction problem: a person's eyes perceive information in the form of light signals, and the brain fills in there is no information about the label. AI, thanks to less expensive forecasting, has a large number of applications, because forecasting this is a key contribution to the decision-making process. In other words, prediction helps you make decisions, and decision-making happens everywhere. Managers make important decisions related to hiring, investments both strategy and less important decisions about which meetings to attend and what to say at that meeting [12]. Judges make important decisions about guilt or innocence, procedures and sentencing, and smaller decisions on a specific item or petition [4]. Similarly, individuals constantly make decisions—from whether to get married, to what to eat or what to sing.

Overcoming uncertainty is a key issue in decision-making. Because the prediction reduces uncertainty, it is a contribution to all these solutions and can lead to new opportunities. The machine replaces the human prediction. Another important economic concept is substitution. When the price of the product (such as coffee) falls, people not only buy more of it, they also buy fewer substitute products (for example, tea). Thus, as machine prediction becomes less expensive, machines will replace for people in forecasting tasks. This means that the reduction of labor associated with forecasting the key will be the impact of AI on human performance. Just as computers meant that few people now perform arithmetic as part of their job, AI this will mean that fewer people will have forecasting tasks [10]. For example, transcription-conversion from spoken words to text is a prediction in the sense that it fills in the missing information on the set characters that correspond to spoken words. AI is already faster and more accurate than many people whose work is related to transcription. Data, actions, and judgments complement the machine's prediction. When the price of a product (for example, coffee) falls, people buy more of its components (for example, cream and sugar). Thus, identifying additions to the forecast is a key task in relation to the latest achievements in the field of artificial intelligence. While forecasting is a key contribution to decision-making, forecasting this is not a solution in itself [8].

Other aspects of the solution complement AI: data, action and judgment. Data is information that is included in the forecast. Many recent developments in the field of artificial intelligence depend on large amounts of digital data for artificial intelligence systems to predict based on past examples. In general, the more examples from the past, the more accurate the forecasts are. Thus, access to a large the amount of data is a more valuable asset for organizations because of AI. Strategic plan the value of the data is thin, as it depends on whether the data is useful for predicting something important for the organization. The value also depends on whether only data is available historically, or whether an organization can collect continuous feedback over time. That the ability to continue learning with new data can create a sustainable competitive advantage [1].

4 Discussion

Newer problems arise from other elements of the solution: actions and judgments. Some actions are inherently more valuable when they are performed by a person rather than a machine. (for example, the actions of professional athletes, child caregivers, or salesmen). Perhaps the most important this is the concept of judgment: the process of determining the reward for a certain action in the special environment. When AI is used for predictions, the person must decide what to do to predict and what to do with the predictions. Implementing AI in an organization requires additional investment and the process changes like computing, electrification, and the steam engine, AI can be considered a universal device technology [4]. This means that it has the potential to significantly improve performance in a wider range of sectors. At the same time, the AI effect requires investments in a number of additional funds input. This can lead an organization to change its overall strategy.

In the context of AI, organizations need to make a number of additional investments. Previously, AI had a significant impact on productivity. These investments are related to infrastructure for continuous data collection, specialized workers who know how to use the data, and changes in processes that take advantage of new opportunities resulting from reduced uncertainty. Many processes in each organization exist to make the best use of the situation in the face of uncertainty instead of serving customers in the best possible way. Airport waiting rooms, for example, make it convenient for customers while they wait for their plane. If passengers had accurate forecasts of how long it will take to get to the airport and through security, rest rooms may not be necessary [11]. The scope of opportunities offered by the best forecasts is expected to vary across companies and industries. Google, Baidu and other major digital platform companies are well positioned benefit from large investments in AI. On the supply side, they already have systems inside a place to collect data. On the demand side, having enough customers to justify a high fixed the costs of investing in technology are at the early stages of its development. Many other businesses do not have they have completely digitized their workflows and are not yet able to apply artificial intelligence tools directly to existing processes. However, as costs decrease over time, these businesses will recognize the available opportunities. This is possible by reducing uncertainty. Guided by their needs, they will follow the industry leaders and invest in AI. Private investment in AI startups Investment in AI in general is growing rapidly, and AI is already having a significant impact on business. According to MGI estimates, between $ 26 billion and $ 39 billion were invested in AI around the world in 2016. Of this amount, internal corporate investments amounted to about 70%, investments in AI startups-about 20%, and AI acquisitions—about 10% [5]. Major technology companiės have made three-quarters of these investments. Outside the technology sector, AI implementation is at an early stage; only a few firms have deployed.

AI startups attracted about 12% of all global direct investment in the first quarter of 2018, a sharp increase from just 3% in 2011. All countries analyzed we have increased our share of investments in AI-oriented start-ups. About 13% of the investment U.S. and Chinese startups had AI startups in the first half of 2018. Most dramatically, in Israel, the share of investments in AI startups jumped from 5% up to 25% between 2011 and the first half of 2018; autonomous vehicles (AVS) are represented 50% of investments in 2017 [3].

5 Conclusion

Efforts are being made to develop empirical AI indicators, but they are disputed other issues include definition issues. Clear definitions are crucial for making accurate and comparable measures. The Institute for Innovation and competition (MPI) offers a three-way approach aimed at measuring:

- AI achievements in science reflected in scientific publications;
- technological developments in the field of AI, confirmed by patents;
- development of AI software, in particular, open source software.

This approach involves using expert advice to identify documents (publications, patents, and software) that are uniquely related to AI. These the documents are then used as a reference for evaluating the degree of affinity of the AI with other documents. Scientific publications have long been used to evaluate the results of scientific research and achievements in science.

References

1. Agrawal A, Gans J, Goldfarb A (2018) Prediction machines: the simple economics of artificial intelligence. Harvard Business School Press, Brighton
2. Anyoha R (2017) The history of artificial intelligence. Harvard University Graduate School of Arts and Sciences Blog. http://sitn.hms.harvard.edu/flash/2017/historyartificial-intelligence/. Accessed 07 Feb 2020
3. Breschi S, Lassébie J, Menon C (2018) A portrait of innovative start-ups across countries. OECD science, technology and industry working papers, no. 2018/2. OECD Publishing, Paris
4. Brynjolfsson E, Rock D, Syverson C (2017) Artificial intelligence and the modern productivity paradox: a clash of expectations and statistics. NBER working paper, no. 24001. http://dx.doi.org/10.3386/w24001. Accessed 07 Feb 2020

5. Dilda V (2017) AI: perspectives and opportunities. http://www.oecd.org/going-digital/ai-intelligent-machines-smart-policies/conference-agenda/ai-intelligent-machines-smart-policies-dilda.pdf. Accessed 12 Jan 2020
6. Gringsjord S, Govindarajulu N (2018) Artificial intelligence. The Stanford Encyclopedia of Philosophy Archives, Stanford
7. Ilyas A, Engstrom L, Athalye A, Lin J (2018) Blackbox adversarial attacks with limited queries and information. In: Bierbaum M (ed) Proceedings of the 35th international conference on machine learning, Stockholmsmässan. OECD Publishing, Stockholm, pp 2142–2151
8. Knight W (2017) 5 big predictions for artificial intelligence in 2017. MIT Technol Rev. https://www.technologyreview.com/s/603216/5-big-predictions-for-artificial-intelligence-in-2017/. Accessed 12 Jan 2020
9. OECD (2017) OECD digital economy outlook 2017. OECD Publishing, Paris
10. Skorodumov P, Badanin D (2015) Construction of a virtual environment of economic modeling. Quest Territ Dev 4(24):1–12
11. Somu N, Kirthivasan K, Sriram V (2017) A computational model for ranking cloud service providers using hypergraph based techniques. Futur Gener Comput Syst 68:14–30
12. Syntetos AA, Kholidasari I, Naim MM (2016) The effects of integrating management judgement into OUT levels: in or out of context? Eur J Oper Res 249(3):853–863

Application of Big Data in the Educational Process

Yu. I. Efremova

Abstract Scientific progress and automation of many processes contribute to the fact that in the modern world it is necessary to operate with big data. The education system has always generated a significant amount of data. We see the problem only in how to implement Big Data in Universities, because this process associated with gravest cost and time, and how to start working with this data at the system level: analyze it, make decisions based on it. The purpose of the study is to consider issues related to the possibility of implementing big data in the learning process. To achieve this goal, we have studied the theoretical and practical aspects of obtaining and using big data in education; analyzed the main types of Big Data; identified problems and formulated proposals for their elimination; and considered the prospects for implementing big data into the education system.

Keywords Modern technologies · Methods · Forecasting · Education · Digital society

1 Introduction

Big Data refers to a specific technology that allows you to analyze and extract new knowledge from arrays of unstructured data. This phenomenon is now known in the broadest professional circles. If 10 years ago, huge volume of information often could not find their application due to the complexity of searching a universal data structure, now it is possible to extract useful information from them. And this is increasingly used in a variety of fields of activity. An example of implementing Big Data analysis is the Skillsoft platform, which provides educational materials online to more than 20 million users. By using the results of Big Data analysis, it was possible to personalize the content provided to the level of training of a particular client. As a result, the quality of online education has improved.

Yu. I. Efremova (✉)
Samara State University of Economics, Samara, Russia
e-mail: yul-efrem@yandex.ru

S. I. Ashmarina and V. V. Mantulenko (eds.), *Current Achievements, Challenges and Digital Chances of Knowledge Based Economy*, Lecture Notes in Networks and Systems 133, https://doi.org/10.1007/978-3-030-47458-4_43

Along with the emerging source of new information in the form of Big Data, the Internet makes its own adjustments to the development of these processes due to the high speed of information exchange. The number of people who regularly use the Internet is constantly increasing. Therefore, the number of devices and the amount of data transmitted by these devices is growing rapidly. In this regard, approaches to creating and storing information also apply to education.

The use of Big Data in the education system is becoming a hot topic every day. With their help, you can transform the educational process and bring it to perfection. It is possible to speed up the solution of various pedagogical tasks, namely: scientific, research and educational, etc. Today, there are special "references" where you can put data and analyze it [10].

Effective implementation of Big Data in a higher education institution will depend on successful collaboration between different departments in a particular institution. The significance of big data is based on the skill to jointly create management structures and deliver more progressive strategies. To date, the implementation of Big Data is difficult due to errors in determining the scope of work, technical difficulties (data processing and analysis). This project has not yet passed many years of verification.

2 Methodology

The following methods were used to implement the tasks set in the work: theoretical—analysis of literary sources and pedagogical experience in the aspect of the studied problem. The methodological basis of our research is the formalization of Big Data technology aimed at the development of educational systems.

Research in Russian and foreign literature on the use of Big Data in the education system is quite scattered. Therefore, it is worth starting the study of this topic in modern education with the specification of the concept itself. The term Big Data appeared in 2008 and is associated with the name of Clifford Lynch, editor of the journal Nature. Currently, there are a number of definitions that can be combined as follows: big data is a set of approaches, tools, and methods designed to accumulate and analyze unstructured data in order to extract knowledge. This concept reflects the increase in the volume of processed data and technological perspectives concerning the likely transition "from quantity to quality". The characteristics of Big Data include a number of factors, namely: high-speed processing and obtaining results, reliability, visualization, value, etc. [8]. Thus, Big Data is understood as a specific amount of data, and as methods of processing them. With their help it is possible to process not only huge data arrays, but also small volumes. Due to the constant growth of data and the need to process it, it is necessary to clarify the principles of working with Big Data. These include:

- horizontal scalability (any system must be extensible),
- fault-tolerance (it is necessary to take into account possible failures in operation),

- data locality (if possible, you must process the data on the computer where it is stored).

To structure Big Data management processes in education, five interrelated groups of processes can be identified:

1. Goal setting: defining the goal (for example: expert and analytical assessment of the personnel potential of educational organizations in the Samara region) and research objectives (to identify age characteristics of teachers, to assess the qualifications of existing teachers).
2. Planning: selection of information sources, procedures for obtaining data, algorithms for processing information (annual form of federal statistical observation, questionnaire form for obtaining missing information from educational organizations).
3. Data collection: organization of data collection in a single database (comparison of the allocation of teaching staff in the region by age in comparison with the normal distribution in the context of the taught subject).
4. Analysis of indicators: analysis of the received data, determination of ways to present results.
5. Adjustment: development of practical regulatory measures.
6. Completion: fixation of patterns (fixing the identified problems in the final report) [6].

V.V. Utyomov and P.M. Gorev formulated a number of properties of the collected database, allowing to increase the efficiency of using Big Data in education [9].

Partial independence. There is local data management on each database segment. At the same time, each segment is a component of the entire database, but can be considered as a separate small database with its own set of procedures and rules.

Uninterrupted operation. The ability to retrieve data from any database segment, even if the data on that segment is already used for other processes.

Access transparency. If you have access rights to data, the analyst should not take into account the parameters of the physical location of the information. Data delivery is performed automatically by built-in tools.

Multiplication. The multiplication capability should allow data from different systems to integrate with each other.

Distributed query. The collected data must be able to be retrieved via distributed queries, i.e. parallel queries to multiple database segments.

Free tool. Any software and hardware solutions can be used as data processing tools.

Thus, Big Data in education as a technology is characterized by distinctive features, structuring of management processes, a system of data collection directions, as well as properties of the collected database that allow increasing the efficiency of using Big Data in education.

3 Results

Economic, political, and social development of society are factors in the development of the Big Data idea. Big Data can predict not only the image of the consumer, but also model the future regarding business, education, etc. The application of Big Data in the educational process is necessary in order to improve the quality of education. To do this, on the one hand, you need to use a huge amount of accumulated information that needs to be analyzed and systematized. On the other hand, Big Data makes it possible for each student to build their own individual educational trajectory in a new way, evaluate the quality of education in an educational organization, and choose an acceptable way of learning for themselves. In this regard, consideration of the possibilities of using big data to evaluate and improve the quality of education is relevant.

No one denies the role of a teacher, but information systems can be very effective and serve as an additional means of learning to perform educational tasks. At the very least, they will help you save time searching for information. People with disabilities will also be able to use Big Data. This will help eliminate the concept of "educational inequality". They also allow you to provide information about the best training methods and control of knowledge, skills and competencies acquired in various educational organizations or independently. Methods of objective data analysis that form the basis of our algorithms allow us to calculate the patterns that arise in the learning process. This, in turn, will help optimize the learning process and make it more fun.

4 Discussion

In the field of education, there are five main types of big data analysis: personal data; data on students' interaction with electronic learning systems and with each other (electronic textbooks, online courses, bounce rates, page viewing speeds, page returns, number of links, distance of links, number of page views per user, etc.); data on the effectiveness of educational materials (what type of student interacts with what part of the content, interaction results, educational results, etc.); administrative (system-wide) data (attendance, absences due to illness, number of lessons, etc.); forecast (expected) data (what is the probability of a student's participation in a particular activity, what is the probability of completing a task, etc.) [5].

It is obvious that many educational institutions in Russia do not have a special electronic environment that would have a large amount of online content and, as a result, a large number of users of this content and interactions with each other regarding it. Therefore they have to work with small data.

Big Data, like any technology in education, does not save the teacher from interacting with the student. The ability of a person to empathize and motivate is always important, and this function is not available to computers. Their advantage is that they help make the teacher a super teacher [9]. How does this happen? For example, the system can analyze hundreds of thousands of texts on the Internet and choose the one that contains the right number of new words. This is something that a person is not capable of, but a machine can do. Using big data, you can do, relatively speaking, three important things: create methods that are adapted to a large number of students; personalize content; and select a learning mode. It should be noted that Big Data will soon change the technology of higher education, making it possible to make students' training more individual: not only to choose their own course program, but also to give a separate homework assignment, as well as to ensure that the content is assimilated [7].

Let's consider the possibilities of effective use of big data in education.

1. Data mining and data analytics can be implemented on the basis of open and Big Data in the future, which will provide students and teachers with quick feedback. Collective and large-scale data can predict which students need more help from the education system to avoid failure in learning courses. This, in turn, will lead to the search for new pedagogical approaches in which students with special needs are particularly interested.
2. In recent years, online education has played an important role in the world of education. Students will be able to receive educational material without leaving home. Large-scale use of open educational resources will allow the education sector to accumulate significant amounts of personal information, which will help to improve the quality of education.
3. In addition, big data can make education more personal and improve the efficiency of learning materials. The new information will help research teachers identify hidden relationships between students and their success.

In Russia, you can list a number of barriers that prevent the introduction of Big Data in the educational process, namely: data distribution and inconsistent access to it (data can be password protected, stored on personal computers and, therefore, are not available for aggregation); fear of data loss; privacy (data security problems). The personnel issue is no less important when implementing this project. Big Data in higher education is increasing, but most of it is scattered across desktops, departments, and presented in various formats, making it difficult to extract it [2].

To resolve these issues, you must:

1. Distribute information flows according to the degree of importance, speed of its recovery, and availability.
2. The content of the information should be divided according to the following points, for example, operational data, working time data, archive data, etc.
3. It is necessary to compare the means of ensuring the security of information and its cost.
4. Project requirements must be met in accordance with the standard and the law.

One of the ways to use big data tools in higher education is to evaluate the skills of some students and make a personalized curriculum that corresponds to their specific areas. When used effectively, big data can help educational institutions improve the learning process and improve student performance, reduce dropout rates, and increase the number of graduates. Successful implementation of Big Data is associated with the use of three data models (descriptive, predictive, and prescriptive), as well as the usefulness of each of them for more effective decision-making [1].

Descriptive Analytics

Descriptive Analytics is aimed at describing and analyzing historical data about applicants such as: enrollment in higher education and graduation, as well as obtaining a degree. Using descriptive analytics, educational institutions can also examine data within learning management systems, examining the frequency of logins, page views, monitoring completed courses of students, who took which courses, what content they attend, and so on [4].

Predictive Analytics

Predictive Analytics aims to assess the likelihood of future events by examining trends. It allows you to pay more attention to lagging students and predict the degree of completion of the course for a particular direction.

Prescriptive Analytics

Enables universities to combine analytical results from both descriptive and predictive models to look at the assessment and identify new ways of working to achieve the desired results. Prescriptive Analytics allows decision-makers to look into the future and see opportunities (problems) for solving different tasks, and also presents the best plan of action.

It is also important to note the benefits of using big data for the administrative staff of higher education institutions. Academic performance, attendance, scholarships, and other personal information about students are subject to continuous collection, processing, and analysis. Working with this amount of data requires considerable effort. Automation will lead to savings in financial and human resources both in individual educational organizations and in many other organizations [3]. In our opinion, it is necessary to address issues related to the prospect of implementing Big Data in the educational process:

1. Early professional orientation. Big Data can help students choose certain educational products that best match their personality characteristics (behavioral characteristics) and social needs. This will potentially allow them to choose educational organizations more meaningfully, and provide training and development of abilities in accordance with their innate inclinations.
2. New adaptive educational trajectories.
   The analysis of Big Data will allow determining at an early stage the abilities and inclinations of the student, on the basis of which educational trajectories can be created that best contribute to the development of the necessary competencies of specific students, taking into account their abilities, motivation and needs of both the society and the students themselves.
3. Control of career path. Big Data analytics can allow an educational organization to track the professional success of its graduates.

4. Openness and transparency of education. The free and unrestricted circulation of unstructured data and their availability for analysis and processing will open up opportunities for deeper involvement of students in educational processes.

Summarizing the above, the analysis of big data and its free circulation (subject to privacy) can serve as a basis for qualitative changes, the formation of a new modern and dynamically developing education system. At the same time, pedagogical science and practice itself will receive a tool whose scale of change and effectiveness can only be compared with the introduction of vaccination in health care. We cannot predict all the consequences of the mass introduction of free-flowing information analysis technologies in all areas of society. They can have an impact on society and its institutions that is difficult to predict. When considering issues related to the use of Big Data, it is impossible not to mention their negative consequences of widespread implementation. For example, Big Data will inevitably reduce the number of jobs. Radical changes can affect not only operators of various services, but also doctors and laboratory technicians, and perhaps even some teachers who are engaged in simple translation of knowledge.

It is important to understand that Big Data can actually cause a change in the approach to information at the level of all mankind, and not just in the areas of marketing services and education. There is a high probability that all fields of activity will adopt new methods and tools.

5 Conclusion

Big Data analysis technologies will soon change higher education and help solve a number of problems in education in Russia. They are able to identify useful data and turn it into useful information. They will make the training more individual: you can not only choose your own course program for everyone, but also give a separate homework assignment. Students will receive more detailed recommendations. Predict how well the course will be completed before the start of training. The system will also help teenagers choose a University: it is assumed that robots will choose the best places for future students themselves, they will not even have to apply. By the end of the university study, each student will have a digital portfolio, and it will be easier for young professionals to navigate when choosing a career. There will be fewer laggards, as technology will allow you to identify students who may be at risk in advance. Teachers will be able to better help lagging students, as the program will indicate which areas have problems. In the future, the analysis of Big Data collected by organizations, as well as open data, will help to implement a mechanism that provides effective interaction between teachers and students in real time, which will allow for a deep and comprehensive study of learning models implemented by educational organizations, and to optimize them taking into account the new knowledge available through Big Data analysis.

References

1. Elia G, Solazzo G, Lorenzo G, Passiante G (2019) Assessing learners' satisfaction in collaborative online courses through a Big Data approach. Comput Hum Behav 92:589–599. https://doi.org/10.1016/j.chb.2018.04.033
2. Galin LR (2019) Strategic use and application of big data technologies in higher education: the potential of Big Data for the development of higher education in Russia. Stud Electron Sci J 30(74):37–40
3. Isaac O, Aldholay A, Abdullah Z, Ramayah T (2019) Online learning usage within Yemeni higher education: the role of compatibility and task-technology fit as mediating variables in the is success model. Comput Educ 136:113–129. https://doi.org/10.1016/j.compedu.2019.02.012
4. Kim YH, Ahn J (2016) A study on the application of Big Data to the Korean college education system. Procedia Comput Sci 91:855–861. https://doi.org/10.1016/j.procs.2016.07.096
5. Kondrateko AB, Kondrateko BA (2017) Opportunities for using Big Data in education in the digital age. Bull Kaliningrad branch St. Petersburg Univ Ministry Intern Aff Russia 4(50):112–115
6. Mamedova GA, Zeynalova LA, Melikova RT (2017) Big Data technologies in e-learning. Open Educ 6:41–48
7. Shakhovska N, Boyko N, Zasoba Y, Benova E (2019) Big Data processing technologies in distributed information systems. Procedia Comput Sci 160:561–566. https://doi.org/10.1016/j.procs.2019.11.047
8. Shadroo S, Rahmani AM (2015) Systematic survey of Big Data and data mining in internet of things. Comput Netw 139:1947. https://doi.org/10.1016/j.comnet.2018.04.001
9. Utyomov VV, Gorev PM (2018) Development of educational systems based on Big Data technology. Concept 6:449–461. https://doi.org/10.24422/MCITO.2018.6.14501
10. Wassan JT (2015) Discovering Big Data modeling for educational world. Procedia Soc Behav Sci 176:642–649. https://doi.org/10.1016/j.sbspro.2015.01.522

Artificial Intelligence as an Object of Legal Protection

T. Kazankova

Abstract In the modern world, questions related to artificial intelligence are gaining more and more popularity. Investments are increasing, and the market for artificial intelligence technologies is growing rapidly. In the current reality, robotics is used to create various types of intellectual property items. All this creates an urgent need to rethink the basic values that are so familiar to us, which constitute the object of intellectual rights, and need innovations in the field of their protection at the legal level. In this regard, there is a well-founded question that addresses the issue of whether it is appropriate to protect the results that are created in the course of artificial intelligence activities. As well as the problem of the procedural nature of their protection. Who will be the legal owner of the rights to them.

Keywords Artificial intelligence · Intellectual property · Protection of property rights · Product of creative activity

1 Introduction

Artificial intelligence is the subject of scientific activity of many research centers and companies specializing in the development of research areas. Rapidly developing information and communication technologies create an urgent need for research and legal improvement of the legal status of artificial intelligence. The modern world, where industry is actively using computer technologies, applying their work in various areas of production and scientific and technical activities, is gradually replacing "live" work with "smart" machines. The result of such activities can be works of art, musical compositions, as well as new results in the scientific and technical field. From the point of view of intellectual property law, this means using artificial intelligence systems to create new results of intellectual activity. The problem of legal regulation of legal protection is most acute when artificial intelligence systems create

T. Kazankova (✉)
Samara State University of Economics, Samara, Russia
e-mail: tatianaok78@yandex.ru

S. I. Ashmarina and V. V. Mantulenko (eds.), *Current Achievements, Challenges and Digital Chances of Knowledge Based Economy*, Lecture Notes in Networks and Systems 133, https://doi.org/10.1007/978-3-030-47458-4_44

potentially protected intellectual property products autonomously, that is, without the direct participation of human content.

Thus, there is a need to rethink the values and concepts underlying the institution of protection of intellectual property rights, especially the concept of defining the concept of "creative activity". This raises a number of legitimate questions, such as the legal protection of results produced by artificial intelligence, whether there is a need for it at all, and how it should be implemented. Whether the results of artificial intelligence activities are subject to intellectual property rights protection and who will own the rights to them. In the scientific literature, both domestic and foreign, there is no single concept on this issue yet. Many authors, such as Denisov [3], Tamm [7], Kafiev et al. [4] reflect only the technical component of this problem in the issues of artificial intelligence.

Researchers of the legal regulation of artificial intelligence activities discuss existing or possible difficulties in protecting the use of results created by artificial intelligence systems [8]. However, possible alternatives to change the current legislation are not offered. The lack of proper legal regulation can be a serious obstacle to the development of relations in the field under consideration and, as a result, can be a deterrent to scientific and technological progress. In the absence of legislative regulation of problems related to the legal protection of results created by artificial intelligence systems, the general provisions of intellectual property law alone are not enough to satisfy the practical interest and need of all interested parties in these relations.

In connection with the above, there is an urgent need to adjust the domestic legal framework regarding intellectual property rights. It is necessary to bring it closer to the realities of modern trends in the field of innovative and technological development. If the current situation does not change in the near future, it will create dangerous preconditions in which the relationship between the parties concerned about the creation and use of the results of artificial intelligence activities will become stable, and the legislator will have to accept the existing "rules of the game" by legalizing the existing practice. Timely response and adaptation of the current legislation to modern realities will not only avoid negative consequences, but also increase the country's prestige as one of the flagships in the development of legislation in the field of intellectual property, and create a favorable ground for the investment climate.

All of the above allows us to highlight the main goals of this study: to form a scientific understanding of the features of the legal regime of products created by artificial intelligence, to contribute to the development and improvement of the concept of their civil protection. To do this, it is necessary to solve several main tasks: to analyze current problems of legal protection and address the issue of ownership of rights to products created by artificial intelligence in both Russian and foreign legislation, to analyze and systematize existing theoretical developments on this issue; to develop proposals for improving the current legislation in the field of legal regulation of public relations arising.

2 Methodology

When choosing the methodology of this study, the analysis of existing publications on the problems of legal regulation and improvement of artificial intelligence was applied. A detailed analysis of the legal norms regulating public relations affecting the legal protection of intellectual property results is carried out. General scientific methods of scientific knowledge, such as dialectical, historical, logical, and system methods, as well as special legal research methods, such as comparative legal, formal legal, and legal modeling methods, were used in the research process.

3 Results

In the course of the research, the author comprehensively considers the problematic issues of the legal regime of results created by artificial intelligence systems in the context of the current state and prospects of scientific and technical development from the position of intellectual property law. The foreign experience of this issue and current trends in this area have been studied. Insufficient elaboration of this problem at the theoretical level, the lack of comprehensive regulatory and law enforcement practice creates a number of problems and leaves a number of unresolved issues related to the legal protection of intellectual property objects created in the course of artificial intelligence systems.

It is established that the results created by artificial intelligence systems cannot be included in the area of legal regulation and protection. This is naturally due to the absence in this chain of the author-an individual-of the mandatory component necessary for providing legal protection to the results of intellectual activity, as a result of whose creative activity the results of intellectual activity were created.

It is proposed to amend the provisions of article 128 of the civil code of the Russian Federation, including as objects of civil rights the results created by artificial intelligence systems and status equate them to protected results of intellectual activity, in respect of which exclusive rights will be recognized without granting the right of authorship and other personal non-property rights. As well as in this regard, expand the provisions of paragraph 1 of article 1225 of the Civil Code of the Russian Federation in terms of increasing the list of protected results of intellectual activity [2].

4 Discussion

In the conditions of modern realities, with the development of scientific and technological progress, products created by artificial intelligence systems are increasingly included in our daily life. From the point of view of intellectual property rights, this means using artificial intelligence systems in the process of creating potentially protectable intellectual property products.

Any product of intellectual activity is impossible without creative efforts of human potential. Human labor, as a priority link in the protection of intellectual property rights, acts as an obstacle to the implementation of the concept of legal protection of the results of artificial intelligence. The creative component is a mandatory feature of the results of intellectual activity and a necessary condition for granting them legal protection as objects of intellectual rights. The definition of "creative activity" and the corresponding measures of protection capacity are directly dependent on a positive decision to recognize a particular person as the author of the result of intellectual activity and provide him with legal protection. According to paragraph 1 of article 1228 of the Civil Code of the Russian Federation, the author of the result of intellectual activity can only be recognized as an individual, in the course of whose creative work it was created [2]. The above-mentioned approach is the basis of traditional views of the Russian right-wing doctrine. Neither the world scientific concept nor the legislation defines creative activity from the point of view of intellectual property rights, as well as a single methodology for evaluating the criteria for its protection. The resolution of this issue is increasingly transferred to the analysis of the activity of creating a particular result for its independence, originality, novelty and inventive level, depending on the type of object of protection and the dominant theory. The interpretation of such a concept is carried out all the time in court.

Despite the different approaches to understanding the concept of "creative activity", as well as the measure of assessing the protection capacity of products created as a result of such activity, there is still some common opinion on the issue of ownership of authorship rights. They are always recognized as an individual. In order to be able to protect intellectual property rights, the product of intellectual activity must contain the peculiarity of the author's personality, emphasize its exclusivity and individuality. Thus, according to the legislation of the Russian Federation, as well as the legislation of many foreign countries, only those products of intellectual activity that were created as a result of human creative detail are protected as a general rule. In the content of the designated question, it seems logical to designate the problem of "machine creativity". That is, those products, created by machines, are capable of such activities without the participation of human potential. Despite certain common features inherent in human creativity, "machine creativity", in view of its lack of intelligence, cannot be called an intellectual activity. On this occasion, we come to the conclusion that with regard to the creative activity of machines, this study will not focus on the results of intellectual detail, since such a term can be applied exclusively to human creativity, but on the results that are potentially protected as objects of intellectual property rights. In addition, it makes sense to talk about the legal validity of artificial intelligence only if they are recognized as legal entities at the legislative level. However, there is still an exception to this rule, meaning the recognition of the rights of the humanoid robot with artificial intelligence Sofia, which in October 2017 received citizenship of Saudi Arabia [6].

On the issue of recognizing the copyright of artificial intelligence systems for the results they create, most researchers do not find this practical need [5]. And what machines create is essentially public domain. At the same time, taking into account the rapidly growing information technologies and the emergence of an urgent need for legislative modernization of existing provisions of domestic legislation, the question of revising certain concepts of legal regulation in the field of intellectual property for their subsequent effective adaptation in the field of rapidly developing artificial intelligence technologies is quite justified. Our state's obligations to the World trade organization in terms of ensuring and guaranteeing the implementation of intellectual property rights norms involve promoting technological progress and the dissemination of technology to achieve mutual benefit of producers and users of technical knowledge, contributing in all possible ways to social and economic well-being and achieving a balance in terms of rights and obligations [1].

In the process of starting to modernize domestic legislation, it is advisable to change a number of legal norms. Thus, the norms of article 128 of the Civil Code of the Russian Federation are proposed to be expanded to include as objects of civil rights the results created by artificial intelligence systems and to equate them with protected results of intellectual activity, in respect of which exclusive rights will be recognized without granting the right of authorship and other personal non-property rights. The expansion of the list of objects of civil rights in this regard seems to us natural and consistent with current trends in the technological development of high-tech technologies. The amendments to article 128 of the Civil Code will entail the extension of the provisions of paragraph 1 of article 1225 of the Civil Code in terms of increasing the list of protected results of intellectual activity [2].

Taking into account the above, the domestic legal doctrine should follow the global trends in the modernization of approaches and technologies within the traditional legal model of protection of intellectual property results and protection in addressing the issue of ownership of exclusive rights to products created by artificial intelligence systems.

5 Conclusion

There is no doubt that in the foreseeable future, artificial intelligence will have an impact on traditional concepts in the field of intellectual property. Timely and efficient use of legal tools in solving this issue, as well as a comprehensive approach to the legislative regulation of the results of artificial intelligence activities are not only of theoretical importance, but also of great practical value. In its entirety, the solution of these problems will have a significant impact on innovation and economic development, will contribute to the solution of these issues at the legislative level, which will increase the image of Russia as one of the potential flagships in the field of development of legislation regulating intellectual property, and will also create a favorable ground for potential investments will help attract additional funds for the development of the modern information technology sector.

References

1. Agreement on trade-related aspects of intellectual property rights (trips Agreement). http://www.wipo.int/wipolex/ru/details.jsp?id=12746. Accessed 20 Feb 2020
2. Civil Code of the Russian Federation. Part four: Federal law No. 230-FZ of December 18, 2006: adopted by the State Duma on 24 November 2006. http://www.consultant.ru/document/cons_doc_LAW_64629. Accessed 21 Feb 2020
3. Denisov E (2019) Robots, artificial intelligence, augmented and virtual reality. Ethical Leg Hyg Issues 98(1):5–10
4. Kafiev IR, Romanov PS, Romanova IP (2019) The selecting of artificial intelligence technology for control of mobile robots. In: Mansurov VZ (ed) Proceedings of the international multi-conference on industrial engineering and modern technologies, FarEastCon. IEEE, New Jersey, pp 8602796
5. Morhat PM (2017) Artificial intelligence: a legal view. Buki Vedi, Moscow
6. NTV (2017). Humanoid robot Sofia received citizenship of Saudi Arabia. http://www.ntv.ru/novosti/1945500/. Accessed 22 Feb 2020
7. Tamm M (2015) A semiotic theory of cultural memory: in the company of Juri Lotman. In: Kattago S (ed) The ashgate research companion to memory studies. Tallin University, Tallin, pp 127–142
8. Yarichin EM, Gruznov VM, Yarichina GF (2018) Intellectual paradigm of artificial vision from video-intelligence to strong artificial intelligence. Int J Adv Comput Sci Appl 9(11):16–32

Application of Information Technologies in Physical Education in Samara Higher Educational Institutions

O. A. Kazakova, L. A. Ivanova, and V. S. Martynenko

Abstract The authors analyzed the State educational standard of higher professional education in the "Physical culture and sport" discipline, the working program on "Physical culture and sport" discipline in two universities in Samara and studied the scientific literature of local and foreign authors based on the research topic. The purpose of this study was to analyze the social significance and social functions of information and computer technologies in the educational process in the "Physical culture and sports" discipline. The study used the following empirical methods: observation, survey, questionnaire, practice tests. The following conclusions: the questionnaire offered to the students of these two universities confirmed the presence of the students' interest in the introduction of information and computer technologies in educational process on physical training, which will increase the motivation to learning and doing sports activities in universities; authors marked significant social function of ICT in the process of "Physical culture and sport" discipline in a particular university.

Keywords Physical culture and sports · Information and computer technologies · Forms of physical culture · Interactive forms · Electronic applications

O. A. Kazakova
Samara National Research University named after Academician S. P. Korolev, Samara, Russia
e-mail: kazakova.kpn@gmail.com

L. A. Ivanova (✉)
Samara State University of Economics, Samara, Russia
e-mail: kfv2012@mail.ru

V. S. Martynenko
Volgograd Academy of the Ministry of Internal Affairs of the Russian Federation, Volgograd, Russia
e-mail: lidiya66@mail.ru

S. I. Ashmarina and V. V. Mantulenko (eds.), *Current Achievements, Challenges and Digital Chances of Knowledge Based Economy*, Lecture Notes in Networks and Systems 133, https://doi.org/10.1007/978-3-030-47458-4_45

1 Introduction

Global informatization of the world community and the rapid development of scientific and technical progress lead to new standards of management and business development [1], market competition becomes more rigid, therefore, the education has followed the trends and the development of informatization of the education system [2].

The Internet has become an integral part of modern society. Informational technologies are being constantly updated in all spheres of activity, and are actively involved in education: they develop e-learning resources and new interactive forms of education such as distance learning, etc.

In modern society, they observe the significant interest in computer technology of higher education system in recent years, defined by the objective reasons primarily connected with the fact that gamification of the educational process is able to give high growth of learners' motivation towards acquiring knowledge and self-development. The creation and implementation of electronic applications into educational programs is comparable to modern video games, which offer sets of different quests and missions, and the completion requires the acquisition of new knowledge, that is probably one of the best way to create motivation for students to study [5].

Nowadays a specialist in any sphere should easily be able to obtain, process and use the flow of information using computers and other means of information technology. The discipline of "Physical culture and sport" is not an exception.

The aim of our study was to analyze the social significance and social functions of information and computer technologies (ICT) in educational process of "Physical culture and sport" discipline.

The main objective of the study was to prove the possibility of application of information and computer technologies in the development of a working program of "Physical culture and sport" discipline and the definition of their practical application for bachelor and specialist students.

2 Methodology

Main forms of ICT in the physical education and sports, which are used in the educational process of the University, were analyzed. 300 students of the second and third years of Samara National Research University (Samara University), Samara State Economic University (SSEU) took part in the research.

Research methods include observation, survey, questionnaire. In addition, they analyzed basic regulatory tests and physiological state of students with the use of information computer technologies.

Informatization in the pedagogical process at the department of physical education in universities helps to use modern technology in various directions [3, 4, 6]:

- creating databases of various documents for educational and training process,

- technical and organizational services of a competition: computer timing, software, e-mark, etc.,
- the use of electronic textbooks, educational courses, information retrieval and help systems,
- the use of computer aided training programs,
- computer-based training courses,
- use of computerized control of knowledge (e-zine), etc.

In addition, specialists from the physical education departments of both universities developed and applied multimedia complex on the basis of the "Physical culture and sport" discipline, which include theoretical, practical and methodical assignments which greatly facilitate the learning process according to a mandatory educational program.

Especially there is a positive effect on students with health problems, they have an opportunity of distance-learning examination of theoretical and practice-teaching material for this discipline and receive a credit. All university students of the 1st to 3rd years use this multimedia system for passing the theoretical and methodological classes, which leads to greater efficiency in the absorption of the material and uses less time, because the learning process takes place in a convenient place at convenient time and in convenient and more efficient mode.

Also, at the moment, computer technologies are widely used in Samara University, for example, the data bank of indicators of health and physical qualities, where in a few minutes of entering the required data of the student, they can monitor physical and functional state of these students. Unfortunately, these resources are not always used at other universities in a sufficient degree.

3 Results

During the study we conducted surveys of 300 students in different faculties from the above mentioned universities (150 each), which include the "Physical culture and sport" discipline. The questionnaire consisted of questions about ICT implementation in the educational program of this discipline in the University, the attitude of students, how often they had to use ICT tools during the classes, and whether ICT contributes to a more rapid assimilation of the program material at the lessons of physical culture and sports, etc. Figures 1 and 2 reflect the results of questionnaire of students of the above mentioned universities.

The blue field shows the percentage of students who think that ICT in the "Physical culture and sport" discipline are needed, and they are happy to use them. The red field reflects the students' attitude towards ICT as positive, but they have never used them in the classroom in this discipline. The green field shows a negative attitude towards ICT used in physical education classes, they note that the best time spent on the use of ICT would be used for the practical training in physical culture and sports.

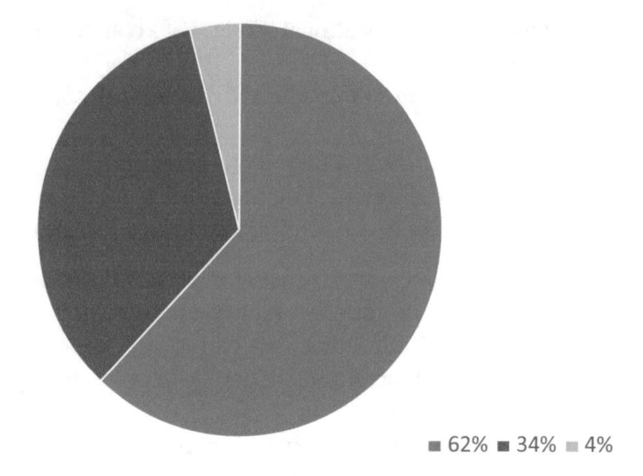

Fig. 1 Attitude of students of the Samara University to the use of ICT in educational process of "Physical culture and sport" discipline (*Source* authors)

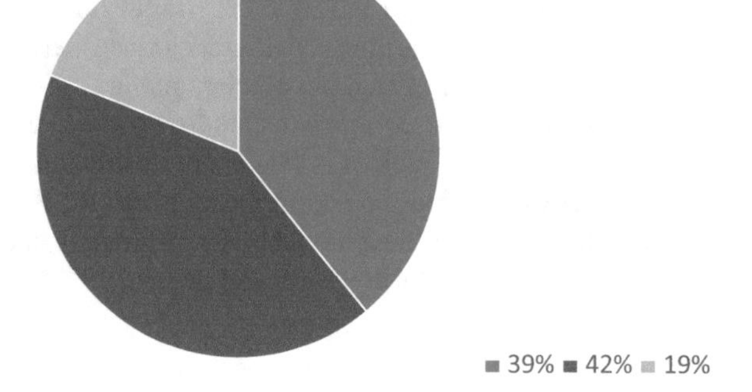

Fig. 2 The attitude of SSEU students towards the use of ICT in the "Physical Culture and Sports" classes (*Source* authors)

The analysis of the questionnaire showed that the majority of respondents in both universities have a positive attitude towards the use of ICT in physical education and sports (62% and 39%), and they understand that ICT improves the assimilation of the studied practical and theoretical material and they undoubtedly increase motivation and interest in classes. Although this conscious attitude to ICT is more expressed at Samara University than at the SSEU, in our opinion, this can be explained by specifics of the university, Samara University (with a technical grade) constantly uses ICT tools in all disciplines, unlike the SSEU.

A neutral position on the use of ICT in physical education and sports is supported by 34% and 42%, which is an approximately equivalent result. And the negative attitude towards ICT by 4% and 19% of respondents is due to the fact that athletes who do not want to be distracted from practical training sessions and use ICT tools to gain knowledge, they are used to trust the knowledge of a trainer-teacher.

4 Discussion

We agree with the opinion of some researchers, that information and computer technologies mean not only work with a computer, but also a person's work with any information in any forms [1, 2, 4, 7]. The specificity of classes in the "Physical Culture and Sports" discipline is motor activity, that is why in such classes it becomes difficult to use information technology directly during the study of practical material. But, nevertheless, the need of use of information and computer technologies to master full knowledge in the educational process of students' physical education is caused by the need of the quality improvement of higher education. Information and computer technologies contribute to creation of the educational process at a new higher quality level and provide more complete and, most importantly, rapid assimilation of educational material by the students.

5 Conclusion

To sum up, the social significance and social functions of ICT in the course of classes in the "Physical Culture and Sport" discipline are increased, as in other disciplines of universities. Information and computer technologies will become one of the main aspects of students' motivation in the field of physical education and sports, and they are already being analyzed from the perspective of improving the quality of modern education as a social institution, and very soon, these technologies will be used by all the sports coaches and university teachers.

References

1. Aleksina AO, Chernova DV, Ivanova LA, Aleksin AY, Piskaykina MN (2019) The main directions in informatization of the sphere of physical culture and sports service. In: Ostrovskaya VN, Popkova EG (eds) Proceedings of the 5th national scientific and practical conference on perspectives on the use of New Information and Communication Technology (ICT) in the modern economy. Advances in intelligent systems and computing, vol 726. Springer, Cham pp 473–479
2. Volkova LM, Volkov V Yu (2014) Modern information and diagnostic technologies in the practice of physical education. Phys Cult Sport Health 23:17–20
3. Ivanova LA, Kazakova OA, Aleksin AY (2018) Predictive response analysis in context of boxers' mental progress. Theory Pract Phys Cult 11:4
4. Nikolayev PP, Nikolaeva IV (2017) Integral analysis of model morphofunctional characteristics of girls – representatives of mass categories in power triathlon. Theory Pract Phys Cult 9:104
5. Robertso A, Munkler M (1994) New health promotion movement. Health Educ 21(3):295–312
6. Saifullina AF, Aleksina AO (2019) Application of innovative technologies in aerobics classes in modern condition. Trends Dev Sci Educ 1(50):70–73
7. Shikhovtsov YuV, Kolemanova IV, Gryaznov SA (2002) Using modern information technologies to control the level of tactical readiness of volleyball players. Vestnik Samara State Econ Acad 1(7):272–275

Applied Pattern of Artificial Intelligence and Big Data in Business

A. V. Loshkarev

Abstract This article explores the applied pattern of artificial intelligence and Big Data in business. Now, during the rise of digital technology, companies are actively attracting IT technology, including artificial intelligence and Big Data. Companies using these technologies have certain advantages over those that do not use these technologies. The relevance of the topic is reflected in the growing interest of both large companies and states in the topic of artificial intelligence. Big Data is one of the tools of artificial intelligence. The author considers the most popular areas of application of artificial intelligence and Big Data in business as well as the problems associated with their application.

Keywords Artificial intelligence · Big Data · Business · Digital technology · IT technology · Digital economy

1 Introduction

Nowadays digital technologies such as artificial intelligence (AI) and Big Data have a huge impact, transforming every area of public life. With the advent and development of these technologies, people quickly appreciated the possibilities of their application in business and analytics. In this regard, in recent years, the need for machine learning and big data specialists has increased significantly. Researchers point out that Big Data attracts the attention of many enterprises which consequently face many problems when adapting to new technologies [3]. One of the most pressing tasks in the field of IT-technologies today is the creation of a powerful computer that can think like a human. The term "artificial intelligence" was first formulated by A. Turing prize winner John McCarthy in 1956 [7]. He defined artificial intelligence as the ability of intelligent systems to perform creative functions that are traditionally considered the prerogative of a person [7]. Big Data is one of the tools of artificial intelligence.

A. V. Loshkarev (✉)
Samara State University of Economics, Samara, Russia
e-mail: 2482337@mail.ru

S. I. Ashmarina and V. V. Mantulenko (eds.), *Current Achievements, Challenges and Digital Chances of Knowledge Based Economy*, Lecture Notes in Networks and Systems 133, https://doi.org/10.1007/978-3-030-47458-4_46

Russian Federation actively supports development in the field of artificial intelligence and other cross-cutting technologies. In the framework of the national project "Digital Economy of the Russian Federation", approved by the Presidium of the Presidential Council for Strategic Development and National Projects on December 24, 2018, plans were developed on artificial intelligence technologies, robotics, Big Data, distributed registry systems, quantum technologies, new production technologies, industrial Internet, wireless communications, virtual and augmented realities [11].

Despite the government's policy aimed at introducing artificial intelligence tools into the economy of our country the technology is not widely used among Russian companies. According to the results of a joint survey of the All-Russian Center for the Study of Public Opinion of the department on implementation of the national program "Digital Economy" of the Analytical Center under the Government of the Russian Federation 91% of respondents are aware of AI technology, 43% do not use AI in their company and do not plan to implement it in the near future [14]. As part of our study we examine why the direction of machine learning and Big Data is so popular in business, for what purposes these technologies are used, what problems exist in this area.

2 Methodology

When forming the methodological basis of this study we used general scientific methods: analysis, synthesis, generalization of information and a comparative method. The study is based on the conclusions and proposals of Russian and foreign scientists, studies of civil society institutions, and the rule of law.

3 Results

Artificial Intelligence and Big Data technologies are widely used in business industry. A study conducted by PwC shows that by 2030, artificial intelligence will provide an increase in world GDP of 14%, which is about 15.7 trillion dollars [8]. Now, during the rise of digital technology, companies are actively attracting IT technology, including artificial intelligence and Big Data. Companies using these technologies have certain advantages over those that do not use these technologies.

Benefits of using artificial intelligence and Big Data are a new level of interaction with customers (the use of chat bots (based on AI) allows you to provide round-the-clock customer support, as well as respond to their requests as soon as possible); improving business efficiency (artificial intelligence quickly processes a large amount of data, which accelerates the production process, increases the number of sales and has a positive impact on productivity).

Artificial Intelligence and Big Data can take the business industry to a whole new level. Let's consider some areas of application of these technologies. One of the areas of application of artificial intelligence, which is in great demand and mentioned above, is virtual assistants or interlocutors (chat bots). Companies use chatbots to help customers and perform routine operations. For example, in 2017, The Royal Bank of Scotland launched the Luvo virtual assistant, who communicates with bank customers through speech recognition using digital devices. Luvo answers customer questions and performs standard operations like money transfers.

In retail organizations, the main direction of the development of artificial intelligence technologies is customer service, logistics optimization, inventory, cost reduction and demand forecasting. For example, the company "M. Video" introduced the artificial intelligence technology, which analyzes the behavior of the client on the website, transitions between pages, section views, after which the system prepares personal offers of goods and sends it to the client by e-mail. The company also introduced a mechanism that determines the optimal time for interaction with the client. As a result of these and other measures to optimize the operation of the store's website, online sales increased.

Artificial intelligence is actively used in the banking sector. Mostly banks use various scoring tools that allow them to optimize decision-making processes. For example, Sberbank plans to shift decision-making on issuing loans to artificial intelligence. Artificial intelligence will compare the biometric data of the client, the credit history of income, expenses and decide independently. Even now, in some companies, a preliminary interview with candidates is conducted by a robot that asks questions depending on the situation and if the candidate is suitable, then switches him to a person—an employee of the HR service.

4 Discussion

The introduction of new technologies requires legal regulation. This is a difficult task since the legislator must ensure balance between the interests of society on the one hand, which wants to use new technologies and extract useful properties from them, and, on the other hand, the need to minimize the negative consequences of using these technologies. In the case of AI and Big Data, the question is most acute. These technologies are developing rapidly, and law is far behind. Further on we move to consider the problems associated with the use of artificial intelligence, which we believe should pay attention to the states sponsoring the development of AI.

1. The dilemma of responsibility for the damage caused by artificial intelligence: a designer, a developer who created the algorithm or an employee, an accountant, an operator who did not check the result?
2. Protection of personal data. How will the safety of personal data be ensured if artificial intelligence analyzes data, including gender, age, health, recognition of faces, without obtaining consent from individuals?

3. Problems with the source data. Obstacles to the use of artificial intelligence can be low quality and rapid obsolescence of raw data.
4. Errors associated with the human factor. Artificial intelligence requires human control. And a person is imperfect and can make mistakes which can be very expensive.
5. Job cuts. The introduction of artificial intelligence automates production, and accordingly, many people will be left without work, because the machine copes with certain tasks better than a person.

In our country, the main regulatory acts on relations related to the use of artificial intelligence and Big Data are the Federal Law "On the Protection of Personal Data" dated July 27, 2006 No. 152-FZ (hereinafter—the "Federal Law on Personal Data") and the Federal Law of July 27, 2006 N 149-FZ "On Information, Information Technologies and Information Protection" [12, 13]. There is no specific regulation for Big Data in Russian law. The legislator approaches the regulation of Big Data from the perspective of personal data [9]. In addition, the legislator has not established legal definitions of artificial intelligence and big data, has not settled the issues of their legal status as a commercial product, and has not established the basis for compensation for possible damage caused by artificial intelligence. We agree with researchers who believe that governments need to take structural measures to create, use and develop the advantages of a digital economy to ensure increased competitiveness both domestically and abroad [6].

The problem of the lack of legal definition of AI is relevant not only for the Russian Federation. At the moment this situation persists in all countries. This is due to the lack of a unified approach to establishing the essential characteristics of this technology. The creators of the European Civil Law Rules on Robotics attribute this to the presence of different types of robots. They believe that it is necessary to approach the study of the latter casuistically, considering each robotic system separately [10]. Their views are shared by American researchers who distinguish four types of artificial intelligence, each of which requires an individual approach: Super machines (for example, the Deep Blue chess computer from IBM that defeated G. Kasparov in the early 1990s), the main feature of this type is the inability to accumulate and analyze data on already implemented solutions; Systems with limited memory that can use past experience for future decisions (for example, some functions of unmanned vehicles); Intelligent machines—"reasonableness" in this context implies the understanding that others have their own beliefs, desires and intentions that influence the decisions they make; Systems with artificial self-awareness, which can form a picture of oneself.

In addition, researchers from St. Petersburg pointed out the relevance of developing a legislative framework governing the circulation of big data and noted the possibility of worldwide uncontrolled use of personal data in social networks in 2018 [2]. Currently, the possibilities of regulating the technical aspects and managerial abilities in big data analytics are seriously studied [5]. Meanwhile, the potential for using big data is hard to overestimate. For example, Russian scientists offer solutions for personnel management in manufacturing enterprises based on the use of big data. It is proposed to collect information on personnel productivity in the form of events and process it in real time using parallel computing technologies [4]. It should be noted that the issues of the application and legal status of artificial intelligence are also conducted by foreign scientists [1]. We believe it is necessary to discuss the problem of the need for a unified approach to the concept of AI, the share of responsibility between many parties involved in the development or use of artificial intelligence technology. In this regard, the urgent tasks facing legal scholars in the field of AI, in our opinion, are: to formulate the concept of legal regulation of artificial intelligence; to determine the framework of the legal personality of AI and the possible responsibility for violations in this area; to study legally significant problems that have arisen in connection with new developments of artificial intelligence, as well as those associated with the use of existing types of autonomous intelligent systems.

5 Conclusions

We examined the areas of application of artificial intelligence and Big Data in business, analyzed the main problems associated with the use of these technologies. Currently, no large company can do without artificial intelligence. The use of artificial intelligence optimizes the production process, minimizes costs, maximizes revenues. That is why it is very popular now. Despite the increased interest in artificial intelligence, there are not enough specialists in this area. A lot of attention is paid to this problem: machine learning and Big Data faculties have appeared at universities, a lot of literature on this topic is published, and there are a lot of training courses on the Internet. It seems that soon the problem will be solved. In addition, issues with legal regulation of artificial intelligence, big data, robotics, cryptocurrencies still remain. Although dome of modern business is no longer possible without artificial intelligence, AI will not be able to do without human help for a long time.

References

1. Alpkan L, Belgemen M, Şenel AA (2019) Discussion about the effects of artificial intelligence on the social life. In: Zehir C, Erzengin E (eds) 7th international conference on leadership, technology, innovation and business management. European proceedings of social and behavioural sciences, vol 75. Future Academy, London, pp 287–296
2. Boldyreva E, Grishina N, Duisembina Y (2018) Cambridge analytic: ethics and online manipulation with decision-making process. In: Chernyavskaya V, Kuße H (eds) Professional culture of the specialist of the future. European proceedings of social and behavioural sciences, vol 50. Future Academy, London, pp 91–102
3. Hazirbaba N, Yalcintas M (2019) Designing strategy dimension of the organization based on big data analytics. In: Özşahin M (ed) Proceedings of the 15th international strategic management conference. European proceedings of social and behavioural sciences, vol 71. Future Academy, London, pp 299–309

4. Ivaschenko A, Simonova M, Sitnikov P, Shornikova O (2019) Big Data analysis for HR management at production enterprises. In: Mantulenko V (ed) Global challenges and prospects of the modern economic development. European proceedings of social and behavioural sciences, vol 57. Future Academy, London, pp 463–471
5. Karaboga T, Zehir C, Karaboga H (2018) Big Data analytics and firm innovativeness: the moderating effect of data-driven culture. In: Özşahin M, Hıdırlar T (eds) Joint conference: 14th ISMC and 8th ICLTIBM-2018. European proceedings of social and behavioural sciences, vol 54. Future Academy, London, pp 526–535
6. Korobeynikova E, Ermoshkina N, Kosilova F, Sheptuhina I, Gromova T (2019) Digital transformation of Russian economy: challenges, threats, prospects. In: Mantulenko V (ed) Global challenges and prospects of the modern economic development. European proceedings of social and behavioural sciences, vol 57. Future Academy, London, pp 1418–1428
7. McCarthy J (2007) What is artificial intelligence?. Stanford University, Stanford
8. PwC (2020) PwC's global artificial intelligence study: exploiting the AI revolution. What's the real value of AI for your business and how can you capitalise? https://www.pwc.com/gx/en/issues/data-and-analytics/publications/artificial-intelligence-study.html. Accessed 02 Feb 2020
9. Sosnin K (2019) Legal regulation of Big Data: foreign and domestic experience. Intell Prop Court J 25:30–42
10. Shestak V, Volevodz A (2019) Modern needs of the legal support of artificial intelligence: a view from Russia. Russian J Criminol 13(2):197–206
11. The national program "Digital Economy of the Russian Federation" approved by the Presidium of the Presidential council for strategic development and national projects on 24 December 2018. http://government.ru/info/35568/. Accessed 02 Feb 2020
12. The Federal Law "On the Protection of Personal Data", 27.07.2006, No. 152-FZ. http://www.consultant.ru/document/cons_doc_LAW_61801/. Accessed 10 Feb 2020
13. The Federal Law of July 27, 2006 N 149-FZ "On Information, Information Technologies and Information Protection". http://www.consultant.ru/document/cons_doc_LAW_61798/. Accessed 10 Feb 2020
14. VTsIOM (2019) Artificial intelligence and business: is there contact? https://wciom.ru/index.php?id=236&uid=10068. Accessed 02 Feb 2020

Artificial Intelligence Technologies in the Field of Legal Services: Relevant Aspects

N. V. Deltsova

Abstract The article is concerned with the issues surrounding the application of artificial intelligence technologies in the provision of legal services. The relevance of the study is conditioned upon the introduction of new service delivery forms in the digital environment and their impact on the implementation of the professional skills of a lawyer. The purpose of this research is to study the artificial intelligence technologies used by lawyers to provide legal services, as well as some problems associated with their implementation. The following methods were used in the research process: general scientific—analysis, synthesis, comparison, generalization; partially-scientific technical legal method, comparative law method. It is concluded that even absent a special legal regulation, artificial intelligence technologies are actively used in the provision of legal services, both in the Russian Federation and abroad. AI-technologies and robotic systems are property, therefore, they are an object, not a subject of law. Consequently, liability issues are resolved with entitled persons (owners, title holders, copyright holders, etc.). AI-technologies can greatly simplify the work of lawyers, save them from performing technical functions, solving routine tasks, but are not able to entirely replace a lawyer as a specialist.

Keywords Artificial intelligence · Field of legal services · Robotic technologies

1 Introduction

The current socio-economic development stage is characterized by digital transformations, which has affected many sectors of the economy and business areas. The field of legal services is no exception. The activities of a modern lawyer are associated with the active use of information technologies that allow you to search for the necessary regulatory and legal material, law enforcement practice, quickly process the information received, and therefore make effective and informed decisions. Moreover, the classical approaches to the legal business organization have changed

N. V. Deltsova (✉)
Samara State University of Economics, Samara, Russia
e-mail: natdel@mail.ru

S. I. Ashmarina and V. V. Mantulenko (eds.), *Current Achievements, Challenges and Digital Chances of Knowledge Based Economy*, Lecture Notes in Networks and Systems 133, https://doi.org/10.1007/978-3-030-47458-4_47

significantly under the influence of digitalization. The provision of legal services increasingly goes beyond the office of a lawyer and is carried out through Legal-Tech services. Information technologies used in the legal sphere make it possible to offer legal services online, providing advice and preparation of legal documents on the website pages. The above processes are based on the use of automated information data systems, which are among the information technologies, constituting a complex including computer and communication equipment, software, linguistic tools, information resources designed to collect, prepare, store, process and provide information, as well as system personnel supporting the information model to meet the information needs of users and for decision-making purposes [8]. Among them, special attention should be given to automated information data systems that operate on the basis of artificial intelligence technology. The latter are actively used in such areas of legal activity as the expert and consulting, and are also used to prepare various types of documents (agreements, complaints, statements of claim, etc.). The scientific literature emphasizes that the use of these technologies will inevitably affect the further development of the lawyers' professional skills, their functions and methods of providing professional services [18]. Having regard to the above, the relevance of research on the application of artificial intelligence technologies in the provision of legal services is beyond dispute. The purpose of this research is to study the artificial intelligence technologies used by lawyers to provide legal services, as well as some problems associated with their implementation.

2 Methodology

The following methods were used in the research process: general scientific—analysis, synthesis, comparison, generalization; partially-scientific technical legal method. The research base is scientific efforts, publications of Russian and foreign scientists and lawyers studying the problems of using artificial intelligence technologies in legal activities. The study also used analysis findings on issues related to the implementation of the above technologies in the field of practical jurisprudence in the provision of legal services.

The study of the problem was carried out in two stages: First stage: the existing scientific and analytical literature on the subject of research, as well as legislation on the use of artificial intelligence technologies in the provision of legal services, was analyzed; the problem, purpose, and research methods were highlighted. Second stage: the conclusions were drawn from the analysis of scientific and analytical literature and legislation; the publication was prepared.

3 Results

The term "artificial intelligence" was proposed in 1956 at a conference held in Dartmouth, devoted to the study of the manifestation of various computing abilities of the mind and the creation of intelligent algorithms and systems that can independently solve their tasks, make decisions. This term was the basis of a whole line of scientific research aimed at the development and implementation of computer and other information technologies related to human intelligence, capable of reasoning, learning and solving problems. The introduction of robotic systems (robots) that have the indicated capabilities and are controlled by artificial intelligence technologies is considered as the relevant and most promising direction for the development of most socio-economic activities [20].

It should be noted that to date, no legislative act in the Russian Federation contains a definition of artificial intelligence. Separate regulatory elements are present only in strategic and industry documents. In the industry document—GOST R 43.0.5-2009. National standard of the Russian Federation. "Informational ensuring of equipment and operational activity. Information exchange processes in technical activities. General Provisions", approved by Order of the Federal Technical Regulation and Metrology Agency on December 15, 2009, No. 959-st [12], artificial intelligence is defined as a simulated (artificially reproduced) intellectual activity of human thinking.

The development of artificial intelligence as the main development trend of information and communication technologies is mentioned in Decree of the President of the Russian Federation dated May 9, 2017 No. 203 "On the Strategy for the Development of the Information Society in the Russian Federation for 2017–2030" [5], Decree of the President of the Russian Federation dated December 1, 2016 No. 642 "On Strategies of scientific and technological development of the Russian Federation" [6], etc. The above brings us to the conclusion that the current legislation defines artificial intelligence as an information technology that contains elements (algorithms) that simulate the cogitative (intellectual) activity of a human being.

The scientific literature identifies several main areas of artificial intelligence technologies application in the field of practical jurisprudence: provision of legal professional activity, legal proceedings, legal advice, rule-making, criminal intelligence, preliminary investigation, criminalistics [19].

The article considers the scope of legal services only; the remaining areas of legal activity are not included in the research subject. Legal services in this article shall mean any services of a legal nature, which include verbal and written consulting on legal issues, drafting of contracts, letters before action, claims, complaints, statements, appeals, transaction support, legal representation, representation of the client's interests in government agencies and municipalities, corporate regulation, etc. [17].

The legal basis for the provision of legal services is formed by the Constitution of the Russian Federation [4], which guarantees everyone the right to receive competent legal assistance (Article 48), and by the Federal Law "On the advocacy and the bar in Russia" dated May 31, 2002 No. 63-FZ, which governs the procedure for the provision of qualified legal assistance by special persons—barristers [10]. In addition, the current legislation allows the provision of legal services by other persons and entities—private lawyers and commercial organizations on the grounds of a fee-based service agreement.

The legal services market competition has greatly accelerated the process of introducing artificial intelligence technologies (AI-technologies) in the field of legal services both in Russia and abroad. In the Russian Federation, this sphere is developing in the following directions: standard legal services automation, growth of legal online services for clients, creation of solutions based on artificial intelligence [25]. In the context of the designated use during the provision of legal services, external and internal AI technologies are distinguished. External AI technologies are systems and services that are provided by lawyers to clients for a fee or for free. Such technologies include design tools for legal documents, web platforms intended to search for lawyers, chatbots, and online consultations. Internal AI technologies are technologies that help lawyers themselves to automate various legal tasks. Among these are platforms for recording their legal cases, CRM systems (for example, Law CRM, AmoCRM), electronic document management, reference and legal systems [16].

Artificial intelligence technologies in the field of legal services, along with other information technologies, are subject to the Federal Law No. 149-FZ of July 27, 2006, "On information, information technology, and information protection" [9]. The norms of this legislative act determine the legal basis for the search, receipt, transfer, production and dissemination of information, the use of information technology and the protection of information.

Besides, according to some authors, the field of artificial intelligence is also regulated by Part IV of the Civil Code of the Russian Federation in terms of creating computer programs, protecting algorithms as know-how, impossibility of recognizing authorship rights for robots, etc. [20, 23]. However, the lack of a clear conceptual framework and special regulation makes it difficult to resolve some practical issues related to the use of these technologies in real practice.

In the field of providing legal services, these are, first of all, questions about responsibility for the actions and decisions of artificial intelligence technology. This aspect is especially pronounced in the legal service business, based on the use of robotic systems—chatbots, video bots, as well as online consultations on specialized sites. The scientific literature actively discusses the matter of responsibility in the case of the artificial intelligence technologies application, as well as the question of the robot's legal personality [2].

This is largely due to the foreign practice of adopting conceptual legal acts that create the basis for the robots and robotic technologies' legal standing. For example, in 2008, the "Intelligent Robots Development and Distribution Promotion Act" [15] was adopted in South Korea; in February 2017—the European Commission's resolution "Civil Law Rules on Robotics" [24]; in 2017, Notice of the State Council issuing the "New Generation of Artificial Intelligence Development Plan" [21] was adopted in China.

It is worth paying attention to Russian initiatives in the field of legislative development. To wit, in December 2016, Grishin Robotics Foundation proposed the Concept of the Law on Robotics [3], where the robot could be a subject of civil matters and be vested with elements of legal personality.

Based on the legislatively established design of a legal entity as a subject of civil-law relations, the developers of the Concept of the Law on Robotics propose amendments to the Civil Code of the Russian Federation [23] to include norms on robotic agents. According to the developers' theory, robots created to participate in civil circulation are subject to inclusion in a special register and only from this moment they can have elements of legal capacity [11]. At the same time, the robot is also considered as an object of law, since it belongs to the owner or possessor.

Legal personality, as a characteristic of the subject of legal relations, implies the ability to have rights, obligations and bear the responsibility for the actions committed. It seems that today it is premature to talk about vesting robots with legal personality. The subject of legal relations must have consciousness and will. At this stage in the development of robotics, the formation of these robot's qualities is not possible without the constant assistance of humans. A robot or a robotic system is an object of law—a type of property, therefore, the risks associated with their use are borne by an authorized person (owner, title owner, copyright holder, user, etc.).

Another important issue that provokes the discussion concerns the role and functions of lawyers in connection with the mainstreaming of AI technology in their daily activities since some types of functional duties and services can be performed by these technologies without human input.

First of all, it concerns the solution of typical and standardized tasks. In particular, chatbots and video bots that use legal information materials in the program's memory are able to provide answers to standard or most frequently asked questions, and to draft standard legal documents. Such legal platforms as Rocket Lawyer in the UK and LegalZoom in the USA, pravoved.ru in Russia are used to provide legal services.

In Russia, the services of creating legal documents (personnel documents, contracts, powers of attorney, etc.) are offered by the popular online design tool Freshdoc.ru. Sberbank, the largest bank in the Russian Federation, uses the chatbot to draft typical claims. It should be recognized that these technologies are largely able to simplify the work of a lawyer, save him from performing technical functions, solving routine tasks, but are not able to fully replace a lawyer as a specialist with legal thinking, knowledge of not only norms, but principles of law, and with a feeling for law and order, allowing to interpret legal norms containing evaluation categories. Moreover, a lawyer with the skills, knowledge, and abilities in the field of legal psychology, conflict resolution and rhetoric certainly has an advantage over robotic systems in the field of legal representation and mediation. It seems that with the introduction of AI-technology in the provision of legal services, the role of a qualified lawyer will only increase.

4 Discussion

The problems of using AI-technologies and robotic technologies are actively discussed in scientific and analytical publications in the Russian Federation and abroad. We should take note of the interest in the challenge of using these technologies in the legal sphere in general and in the provision of legal services in particular.

General issues of the "artificial intelligence" concept are investigated in the works of Morhat [19], Nagrodskaya [20]. The problems of organizing business processes in the legal sphere are considered in an article by Tung [26].

Rietveld et al. [22] investigate the use of technology-based artificial intelligence assistant to analyze and evaluate judicial decisions. Alarie et al. [1] write about the impact of artificial intelligence tools on the legal profession. About artificial intelligence technologies and their application in law argues Gowder [13]. The creation of an AI lawyer is discussed in an article by Dervanović [7]. Greenleaf, Mowbray, Chung write about technologies for providing free legal advice using artificial intelligence [14]. These publications and the discussions that arise on their basis show the relevance of research on the use of artificial intelligence technologies in the provision of legal services.

5 Conclusion

Artificial intelligence technologies are actively used in the provision of legal services, both in the Russian Federation and abroad. However, to date, there is no clear legislative regulation in the field of application of these technologies and robotic systems. There are only conceptual legal acts recognizing the possibility of using these technologies. The problem of responsibility for the actions and decisions of artificial intelligence systems also remains unresolved, since the issue of the legal personality of these technologies has not been settled. In Russia, taking into account the existing legal regulation, it is fair to say that AI-technologies and robotic systems are property, therefore, they are an object, and not a subject of law. Consequently, liability issues are resolved with entitled persons (owners, title holders, copyright holders, etc.).

AI technologies are largely able to simplify the work of a lawyer, save him from performing technical functions, solving routine tasks, but are not able to fully replace a lawyer as a specialist with legal thinking, knowledge of not only norms, but principles of law, and with a feeling for law and order, allowing to interpret legal norms containing evaluation categories. It seems that the introduction of AI-technology in the provision of legal services will help to increase the role of a qualified lawyer.

References

1. Alarie B, Niblett A, Yoon AH (2018) How artificial intelligence will affect the practice of law. Univ Toronto Law J 68:106–124
2. Arkhipov VV, Naumov VB (2019) On some issues of the theoretical grounds for the development of legislation on robotics: aspects of will and legal personality. Law 5:157–170
3. Arkhipov VV, Naumov VB (2017) The concept of the law on robotics. http://robopravo.ru/proiekty_aktov. Accessed 22 Feb 2020 (in Russian)
4. Constitution of the Russian of Federation of 15 December 1993. http://www.consultant.ru/cons/cgi/online.cgi?req=doc&ts=150167684009108327130112795&cacheid=6D95DFD64FE024A3E816930244333AA3&mode=splus&base=LAW&n=2875&rnd=8D46E21450CF868D587DECE31B521602#1ta9gw689rf. Accessed 18 Feb 2020 (in Russian)
5. Decree of the President of the Russian Federation No. 203 of May 9, 2017 "On the Strategy for the Development of the Information Society in the Russian Federation for 2017–2030". http://www.consultant.ru/cons/cgi/online.cgi?req=doc&ts=60663846408650616445053736&cacheid=B985E9138DDAF3E3BEC8BA60FC17E562&mode=splus&base=LAW&n=216363&rnd=0.8049608825822591#2owyl08fkr4. Accessed 10 Feb 2020 (in Russian)
6. Decree of the President of the Russian Federation No. 642 of December 1, 2016 "On Strategies of scientific and technological development of the Russian Federation". http://www.consultant.ru/cons/cgi/online.cgi?req=doc&ts=60663846408650616445053736&cacheid=7DB7C3C878809698698883B21D0FD7F9&mode=splus&base=LAW&n=207967&rnd=0.8049608825822591#2expz0567wz. Accessed 10 Feb 2020 (in Russian)
7. Dervanović D (2018) I, inhuman lawyer: developing artificial intelligence in the legal profession. Perspect Law Bus Innov 2018:209–234
8. Dzhatdoev A (2018) Information technologies in jurisprudence. Young Sci 6:20–24
9. Federal Law No. 149-FZ of July 27, 2006 "On information, information technology and information protection". http://www.consultant.ru/cons/cgi/online.cgi?req=doc&base=LAW&n=339396&dst=100144&rnd=42DEEAF95EEFE6057CB515371FE28609#05098454103320023. Accessed 18 Feb 2020 (in Russian)
10. Federal Law No. 63-FZ of May 31, 2002 "On the advocacy and the bar in Russia". http://www.consultant.ru/cons/cgi/online.cgi?req=doc&ts=60663846408650616445053736&cacheid=6DDBDE74EBFB0F33512E965271DA6A7A&mode=splus&base=LAW&n=339270&rnd=0.8049608825822591#14426a6mi41. Accessed 18 Feb 2020 (in Russian)
11. Filipova IA (2018) Legal regulation of artificial intelligence: regulation in Russia, foreign research and practice. State Law 9:79–88. https://doi.org/10.31857/S013207690001517-0
12. GOST R 43.0.5-2009. National standard of the Russian Federation. Informational ensuring of equipment and operational activity. Information exchange processes in technical activities. General Provisions, approved by Order of the Federal Technical Regulation and Metrology Agency No. 959-st of 15 December 2009. http://www.consultant.ru/cons/cgi/online.cgi?req=doc&ts=60663846408650616445053736&cacheid=CDDE3A3556825C5FFF8206AD91FD1112&mode=splus&base=STR&n=13472&rnd=0.8049608825822591#114i15alhx6 Accessed 18 Feb 2020. (in Russia)
13. Gowder P (2018) Transformative legal technology and the rule of law. Univ. Toronto Law J. 68:82–105
14. Greenleaf G, Mowbray A, Chung P (2019) Legal information institutes and AI: free access legal expertise. Front. Artif. Intell. Appl. 317:199–211
15. Intelligent Robots Development and Distribution Promotion Act No. 9014 of 28 March 2008. http://robotunion.ru/files/Robotics_SOUTH-KOREA-LAW.pdf. Accessed 23 Feb 2020
16. Khoroshilov A (2018) Automation of law: can artificial intelligence cope with legal tasks? https://zakon.ru/blog/2018/3/20/avtomatizaciya_prava_smozhet_li_iskusstvennyj_intellekt_spravitsya_s_yuridicheskimi_zadachami. Accessed 22 Feb 2020. (in Russia)
17. Lenkovskaya RR (2015) The concept and the legal nature of the legal and law services. Gaps Russ Law 4:36–38

18. McGinnis JO, Pearce RG (2019) The great disruption: How machine intelligence will transform the role of lawyers in the delivery of legal services. Actual Probl Econ Law 13(2):1230–1250
19. Morhat PM (2018) Opportunities, features and conditions for the use of artificial intelligence in legal practice. Administrator Court 2:8–12
20. Nagrodskay VB (2019) New technologies (blockchain/ artificial intelligence) in the service of law: Instructional guide. Prospect, Moscow
21. Notice of the State Council issuing the "New Generation of Artificial Intelligence Development Plan" №35 of 27 July 2017. http://www.gov.cn/zhengce/content/2017-07/20/content_5211996.htm. Accessed 18 Feb 2020
22. Rietveld R, Rossi J, Kanoulas E (2019) Distilling jurisprudence through argument mining for case assessment. Eng Comput Sci 1:19–221
23. The Civil Code of the Russian Federation (part 4) No 230-FZ of 18 December 2006. http://www.consultant.ru/document/cons_doc_LAW_64629/#dst0. Accessed 22 Feb 2020. (in Russia)
24. The European Commission's resolution "Civil Law Rules on Robotics" 2015/2013(INL) of 16 February 2017. https://www.europarl.europa.eu/doceo/document/TA-8-2017-0051_EN.html. Accessed 20 Feb 2020
25. Tsvetkova I (2017) Artificial intelligence in a court, bot lawyers and crowd-funding of legal disputes—how the LegalTech revolution begins. https://rb.ru/opinion/legaltech/. Accessed 20 Feb 2020. (in Russia)
26. Tung K (2019) AI, the internet of legal things, and lawyers. J Manag Anal 6(4):390–403

Augmented Reality Technology in Russian Business

A. V. Levchenko, Yu. V. Shikhovtsov, and A. V. Timoshenko

Abstract The purpose of this work is to analyze trends in the use of augmented reality technologies in Russian organizations. The authors consider the problem of implementing AR technologies in the economic processes of enterprises in the conditions of Russian economic, legislative, social and technical barriers. The relevance of this problem is confirmed by the high demand of Russian enterprises for AR projects, while the Russian consumer market for augmented reality technologies is lagging behind the world market. This article analyzes the materials of analytical and consulting companies, as well as current data and indicators that characterize consumers of augmented reality projects and the market of AR technologies. As a result, opportunities to overcome barriers to the introduction of augmented reality technologies and prospects for the development of the Russian consumer market for these technologies were identified.

Keywords Augmented reality · Digitalization of the economy · Innovations · Information technologies · Russian industry

1 Introduction

The fundamental element of scientific and technological progress that affects the development of the world economy as a whole is information technology, through which the processes of production, promotion and sale of products and services are transformed. Currently, digital technologies are developing at a high rate, allowing businesses to constantly improve. However, to achieve business goals and increase the

A. V. Levchenko
Samara State Social and Pedagogical University, Samara, Russia
e-mail: lavsport_67@mail.ru

Yu. V. Shikhovtsov (✉) · A. V. Timoshenko
Samara State University of Economics, Samara, Russia
e-mail: shikhovtsovy@bk.ru

A. V. Timoshenko
e-mail: timoshenkoall@mail.ru

S. I. Ashmarina and V. V. Mantulenko (eds.), *Current Achievements, Challenges and Digital Chances of Knowledge Based Economy*, Lecture Notes in Networks and Systems 133, https://doi.org/10.1007/978-3-030-47458-4_48

competitiveness of organizations, visible reality is no longer enough. Corporations around the world are actively using augmented reality technology to attract customers and increase profits. Augmented reality is becoming a technology that is changing modern society, especially in the field of business and Economics [8].

Augmented reality (AR-technology) is a virtual environment that integrates into the real physical world to improve the perception of information about the surrounding reality by adding sensory data to the perceptual field and further modeling using computer processing of elements, which allows you to visualize many difficult to perceive objects or processes [3]. AR AR devices are equipped with an increasing number of features, SUCH as MICROPHONES, speakers, or systems with multiple cameras [7]. Thus, AR technology it is a new interface for human-computer interaction that takes it to a qualitatively new level.

The USE of AR technologies is not limited to just one production area, but extends to almost all business processes of the organization (Table 1).

Currently, some representatives of Russian business have begun to actively research and apply the business opportunities of AR technologies. The relevance of the introduction of augmented reality systems for domestic companies is confirmed by the creation Of the Association of augmented and virtual reality (ADVR) in 2015. The main goal of ADVR is to bring the domestic AR industry to the leading positions of the world market, which requires the joint efforts of representatives of

Table 1 The use of AR-technologies in the processes of economic

Business process	Possibilities of using AR-technologies
Production	Production workers using augmented reality glasses receive the necessary information about the object, in particular, properties, tips on assembling parts and potential problem areas [7]
Training	Using tablets and personal computers with augmented reality software will help employees more easily learn the training material. The use of AR-technologies in staff training can be noted in almost any field, from production to trade [4]
Logistics	The application of augmented reality technology is carried out when choosing the parameters of cargo transportation, taking into account factors affecting the implementation of transportation and loading and unloading operations; the preparation of the route and other operations to automate the processes of internal warehouse logistics [5]
Retail	Using augmented reality, buyers have the opportunity to obtain the necessary information about the product (product composition, shelf life, method of use, etc.) using the camera of their mobile device with Internet access [9]
Security	AR-technology for the recognition, detection and tracking of objects and people will facilitate security services to prevent emergencies [5]
Marketing	Marketing services organizations can use AR technology to provide contextual links between their consumer offerings, online resources, and points of sale [9]

Source authors based on [4, 5, 7, 9]

this industry and support in the development and monetization of AR technologies [2]. The need to achieve this goal arose in the context of existing barriers to the introduction of AR projects and the lagging Russian consumer technology market augmented reality. That is why, in this work, the authors set themselves the task of determining how to overcome the constraints.

2 Methodology

To achieve the research objectives, a set of interrelated research methods was used: methods of induction and deduction, synthesis and analysis, and a systematic approach. In particular, the analysis of materials of analytical and consulting companies working in this field, as well as current data and indicators that characterize consumers of augmented reality projects and the market of AR technologies was carried out.

3 Results

The Russian market of AR technologies is 1–2 years behind the world market, in particular, a number of areas that are actively developing abroad are currently pilot projects in Russia. However, in General, there is a strong growth of the Russian art industry. Thus, in 2016–2018, the volume of sales of AR technologies increased almost 5 times, and the number of organizations involved in this industry increased from 60 to 200 [5]. The areas of application of augmented reality technologies in Russia are shown in Fig. 1.

Thus, most of the Russian enterprises that already use AR technologies represent the real sector of the economy: construction, mechanical engineering, mining and processing, energy, etc. thus, using augmented reality in construction and architecture are carried out visualization of projects, building three-dimensional models of future structures; in the automotive industry-control of production, logistics and Assembly quality; in mining companies-modeling of rock conditions and geological exploration; in energy and industry-maintenance and repair of machines and equipment, monitoring the operation of installations; in education-visualization of the educational process, reading books on a virtual screen; in trade-simplification of product selection with the provision of all necessary characteristics [5].

In 2019, the Association of augmented and virtual reality of Russia and the CIS conducted a study on the Use of augmented and virtual reality technologies [AR/VR] in industrial enterprises, which was attended by 50 Russian enterprises (PJSC Gazprom Neft, JSC Russian Railways, PJSC KAMAZ, etc.) [2]. The results of the study showed exactly what tasks AR technologies solve in industrial enterprises in Russia (Fig. 2).

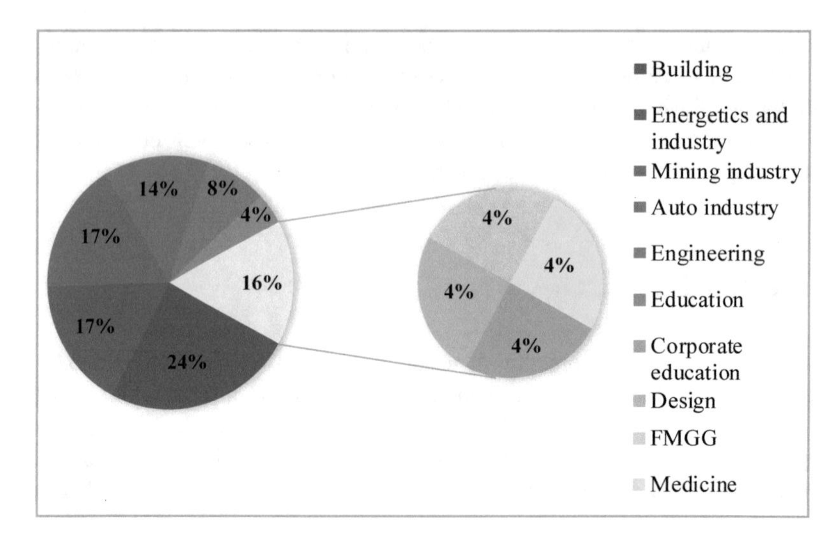

Fig. 1 Structure of AR technology consumption in Russia in 2018, % (*Source* Authors based on [5])

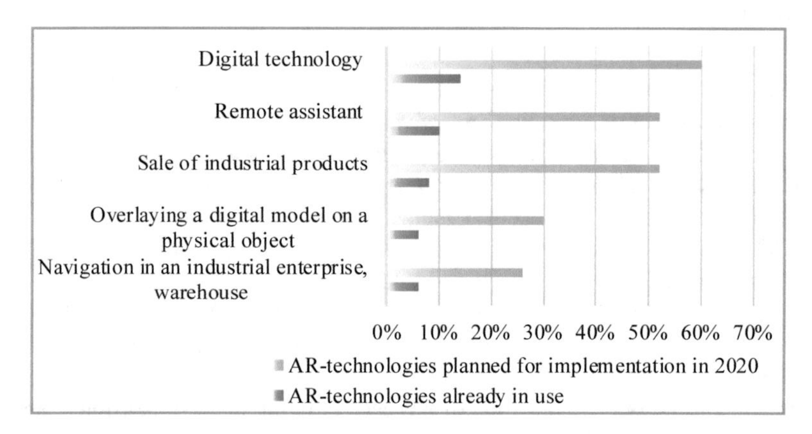

Fig. 2 AR AR-technologies (*Source* authors based on [2])

Thus, five most relevant cases were identified:

1. Digital instructions are consistent recommendations that help the worker perform Assembly, maintenance, and repair operations.
2. A remote assistant is a remote specialist who sees the same thing on his computer monitor as a worker using a camera on glasses, while voice communication is set up between them and the ability to place the necessary information in space using digital interfaces.
3. Superimposing a digital model on a physical object, which makes it possible to identify errors in production or construction by comparison.

4. Navigation through an industrial enterprise and warehouse allows you to improve logistics within the organization, as well as identify dangerous areas of production.
5. Sale of products of an industrial enterprise. In the framework of this case study is a demonstration of future products to the customer via the interactive animated 3D models. This technology is used in cases of technical complexity and difficult transportation of the product [2].

For practical implementation of these tasks, industrial enterprises use Microsoft HoloLens mixed reality glasses, GMT-1Z1 realwear augmented reality glasses and Epson Moverio Pro BT-2200, iOS and Android operating systems, smartphones, tablets and headsets that can render virtual models in a physical environment using visual bindings for augmented reality, as well as their own developments [1, 2].

4 Discussion

The use of these technologies can increase the efficiency of economic processes in the enterprise, namely:

- to increase productivity,
- reduce production cycle time,
- to ensure the growth of operating efficiency in logistics,
- reduce costs and the number of staff errors,
- speed up the process and improve the quality of staff training [3].

However, in the process of implementing AR projects, Russian industrial enterprises face a number of barriers (Table 2).

Thus, the mass proliferation of AR-technologies is hindered by a number of factors, among which technological disadvantages occupy a separate place [7]. Companies themselves are not able to influence the removal of technical barriers, as the

Table 2 Obstacles to using augmented reality technologies

Type of barrier			
Economic barriers	Regulatory barriers	Social barriers	Technical barriers
• High costs for the formation of infrastructure for the implementation of AR-technologies • High cost of AR-technologies • Lack of a clear correlation with economic efficiency [2]	• Inability to certify educational AR courses • The need to develop clear industry standards [2]	• Resistance to technology adoption by management or employees • A long process of approval and approval at industrial enterprises • Conservatism of the industrial sector [3]	• High weight of devices • A small viewing angle in glasses • Low autonomy • Limited computing power and limited flexibility of the interface platform of AR devices [7]

Source authors based on [2, 3, 7]

developers themselves are responsible for this. These current trends in the creation and development of new information technologies, it can be expected that significant shortcomings will be eliminated in the near future. It should also be noted that, in conjunction with the parallel development of other digital technologies (virtual reality, Blockchain, artificial intelligence, etc.), there is a likelihood of the formation by 2030 of a platform for the active development and improvement of augmented reality technologies. If the developers of AR-technologies do not overestimate consumer expectations, and potential consumers are not disappointed in AR-technologies, then by 2025 it will be possible to argue about their transition to the stage of stable commercial implementation [3]. To overcome social barriers, it is important that Russian companies share their experience in using AR technologies in various business processes, and AR projects are covered in the media and on the Internet. Also, not only technology developers and heads of companies using AR projects, but also ordinary users and employees of organizations should discuss potential possibilities of implementing augmented reality technologies [3].

In the Russian Federation is actively working to overcome regulatory and legislative barriers, in particular, has developed a national program "Digital Russia", which is part of the Federal project "Information security". In this project, the task is by 2022 to develop, adopt, harmonize and implement standards for information security systems, virtual and augmented reality and artificial intelligence technologies [6]. Another important factor hindering the spread of AR technologies in Russian companies, are economic barriers such as high cost and complexity of implementing these technologies in conjunction with non-obvious benefit from their use [3]. The cost of implementing AR technology depends on a number of parameters of task complexity, used devices, etc. However, the study, "Use of augmented and virtual reality [AR] on industry" highlights some average benchmarks. Thus, the timing of piloting account for about 3 months, the price of the pilot project—1 to 3 million rubles, the implementation time ranges from 6 months to a year, the price of implementation is in the range of 10–30 million rubles [2]. Under these conditions an effective funding tool AR projects can be public (development institutions, for example, the SKOLKOVO Foundation) and private (business angels) funding.

Thus, at present, is actively working to eliminate these barriers. There is a growing demand for augmented reality technology, with the result that there is the prospect of the development and spread of AR technologies in the Russian business [5]. For example, in "Gazprom oil" developed a road map for the development of AR technology by 2025, according to which by 2025 the training on labor protection and industrial safety with the use of augmented reality will take place every employee of the company, which will provide an opportunity to reduce risks by 30%. And by the end of 2021 in "Gazprom oil" with the aid of VR will be accepted until 90% of the engineering models that would reduce engineering time by 10% [2].

5 Conclusion

Thus, augmented reality technology is becoming an effective tool for the Russian industry, as it is capable of launching a wave of technological changes behind itself, which is part of the overall process of digitalization of the company. With the proper use of AR-technologies, enterprises have the opportunity to increase profits due to the growth of personnel productivity, optimization of production and other business processes, attracting new consumers and customers [3]. The use of AR-technologies is a significant potential for expanding the business and increasing the competitiveness of industrial goods and services [9].

References

1. Bahremand A, Nguyen L, Harrison T, Likamwa R (2019) Hololucination: a framework for live augmented reality presentations across mobile devices. In: Bruder G, Anderson M, Hiyama A, Kitazaki M (eds) Proceedings – 2019 IEEE international conference on artificial intelligence and virtual reality, vol 8942234. IEEE, Piscataway, pp 243–244
2. Degay E (2019) Research «Application of Augmented and Virtual Reality Technologies [AR|VR] in industrial enterprises». http://4industry.ru/ar-vr-research. Accessed 11 Feb 2020
3. Ivanova AV (2018) Technologies of virtual and augmented reality: opportunities and obstacles of application. Strateg Decis Risk Manag 3(106):88–107
4. Lovcheva MV, Konovalova VG, Simonova MV (2020) Development of corporate digital training. In: Ashmarina S, Vochozka M, Mantulenko V (eds) Digital age: chances, challenges and future. Lecture notes in networks and systems, vol 84. Springer, Cham, pp 473–479
5. MegaResearch (2019) Augmented reality market analysis in Russia: the industry is just emerging. https://www.megaresearch.ru/news_in/analiz-rynka-dopolnennoy-realnosti-v-rossii-otrasl-tolko-formiruetsya-1437. Accessed 11 Feb 2020
6. National program «Digital Economy of the Russian Federation» (approved by the Council Presidium of the Presidential of the Russian Federation for Strategic Development and National Projects. Protocol dated 04.06.2019 N 7). http://www.consultant.ru/document/cons_doc_LAW_328854/. Accessed 11 Feb 2020
7. Popper UmJ, Ruskowski J (2018) Modular augmented reality platform for smart operator in production environment. In: Correa FR (ed) IEEE industrial cyber-physical systems, ICPS 2018. IEEE, Piscataway, pp 720–725
8. Roxo MT, Brito PQ (2018) Augmented reality trends to the field of business and economics: a review of 20 years of research. Asian J Bus Res 8(2):94–117
9. Vilkina MV, Klimovets OV (2020) Augmented reality as marketing strategy in the global competition. In: Popkova E, Sergi B (eds) The 21st century from the positions of modern science: intellectual, digital and innovative aspects. Lecture notes in networks and systems, vol 91. Springer, Cham, pp 54–60

Big Data Analytics in the Model "Cargo Flow—Transport and Logistics Infrastructure"

E. V. Bolgova, V. A. Haitbaev, and S. A. Nikishchenkov

Abstract The absence of a model combining the flow of goods and infrastructure facilities into a single system is limiting the effectiveness of companies and the spatial and socio-economic development of Russian subjects. Big Data Analytics (BDA) technologies adapted for Supply Chain Management (SCM) allow overcoming this limitation. For the development of the model «cargo flow—transport and logistics infrastructure», the authors: described the structure of BDA in the elements of advanced analytics, in the composition of Descriptive analytics, Predictive analytics, Prescriptive analytics; validated BDA taxonomy; classified different types of data characterizing the cargo flow; proposed models and developed step-by-step algorithms for modelling transport and logistics infrastructure according to the parameters of cargo flows linked to the application of different types of data in BDA.

Keywords Big Data Analytics · Supply chain management · Model · Cargo flow · Logistics · Infrastructure

1 Introduction

Large business players who accept Big Data as a new management paradigm are oriented towards digital transformation of their business and increased operational efficiency. In Supply Chain Management (SCM), successful logistics models based on Big Data attract the attention of both practitioners and researchers and frequently appear in news headlines. Amazon uses Big Data to track the delivery of 1.5 billion commodities distributed over 200 warehouses around the world. Amazon's logistics

E. V. Bolgova (✉)
Samara State University of Economics, Samara, Russia
e-mail: elena_bolgova@rambler.ru

V. A. Haitbaev · S. A. Nikishchenkov
Samara State Transport University, Samara, Russia
e-mail: haitbaev@samgups.ru

S. A. Nikishchenkov
e-mail: nikishchenkovs@samgups.ru

S. I. Ashmarina and V. V. Mantulenko (eds.), *Current Achievements, Challenges and Digital Chances of Knowledge Based Economy*, Lecture Notes in Networks and Systems 133, https://doi.org/10.1007/978-3-030-47458-4_49

is based on a predictive "pre-emptive delivery" analytics that is able to predict when a customer will purchase a product and send it to a warehouse close to its final destination in advance. Wal-Mart handles over a million customer transactions every hour, transmits information of over 2.5 petabytes using the radio frequency identification (RFID) product markings, which generate 100 to 1,000 times more data than traditional barcode systems. The deployment of telematics by the UPS courier service in the segment of cargo flow has enabled to transform itself into a multi-billion-dollar corporation. The ability to offer its clients from 200 countries a variety of ways to synchronize freight, information, and financial flows has enabled UPS to take the lead in supply chain management. However, modern business is so swamped with data that it collects it more than it knows what to do with it [5]. This is also true for logistics companies, which are considered to be a benchmark in the management of freight flows. Reality shows that most organizations have high expectations of Big Data Analytics (BDA), but the actual use of BDA is limited by the lack of a model combining the flow of cargo and the transport and logistics infrastructure into a single system to increase the operational efficiency of business, the concentration of transport and logistics infrastructure objects in the economic space of a region. The overall objective of this research is to fill the knowledge gap between data science and supply chain management (SCM) by linking Big Data and modelling applied to the characteristics of the flow of goods defining its quantity, geography, capacity and the technical equipment of transport and logistics infrastructure. In particular, this article will include: (1) the results of a study of the BDA concept for SCM; (2) the BDA taxonomy, which classifies the different types of data that characterize the traffic; (3) the development of models and algorithms for modelling transport and logistics infrastructure in accordance with the load flow parameters associated with BDA, which proved to be very effective in solving complex SCM tasks.

2 Methodology

BDA in SCM is a subject of debate, interdisciplinary and covers different areas of management. It is understood by the authors that management decisions are seldom taken in a high degree of certainty and realized as an ideal problem with complete data. But even where completeness is not in doubt, there are practical difficulties in providing data to support solutions for a particular problem. The data blocks themselves increase the quality of decisions and the value of business. Gunasekaran et al. described this field as "new and emergent" [3]. Barratt and Oke recognized the relevance of finding new BDA practices in SCM [1]. From these preconditions follows the recognition of a gap in the BDA concept in relation to SCM, the need to fill this gap and to define BDA (1) in its own terms, (2) in relation to SCM research and practice. The method of review of BDA-specific terms exhaustively describes the composition of BDA, the method of contextual review of BDA in SCM creates a solid conceptual basis for the subsequent stages of study; the modelling method results in a set of working models «cargo flow—transport and logistics infrastructure» and an algorithm for modelling transport and logistics infrastructure linked to the cargo flow and BDA determining its parameters.

2.1 The Method of Review of BDA-Specific Terms

The review of BDA-specific terms systematized by Varela Rozados and Tjahjono has revealed the notion of «information flows in SCM» —an extended supply chain functioning both as a relationship system and as a flow of materials, information, resources [11]. Unlike centuries-old commodity information, which was stored and transmitted together with the goods themselves as physical documents, the extended supply chain involves technologies synchronized with SCM, but managed as a separate stream at the stages of data collection, processing, analysis, storage, data exchange. The Expanded Supply Chain is a multi-level system which links organizations and enables cooperation and integration, not competition between individual firms.

The concept review method leads to the understanding that among the stages of SCM information flow, BDA focuses on analysis, and the analytical tool is *Advanced analytics (AA)*—the scientific process of transforming data into the basis for making the best decisions. As a formal discipline, the advanced analytics has grown into a mathematical field of operational research, and Chae, Sheu, Yang, Olson have proposed a classification of the advanced analytics into three main sub-types [2]:

- *Descriptive analytics (DA)* contains an analysis of data collected to describe past business situations in such a way that their trends and patterns become apparent. This level of analytics answers the question «what happened» and is a way to get an idea of the best approach to the future. The DA method is described in detail by Zeng and includes (1) standard reports and standard package information; (2) special reports: queries configured by the end-user in the package interface; (3) query details: first level of data mining, allowing for complex multi-dimensional information from databases; (4) alerts: formed in any one of the descriptive analytics groups by selecting the user-priority data; when a parameter crosses a reference value; (5) visualization: data in any graphic form, illustrating the acts performed and events [15],
- *Predictive analytics (PA)* analyzes real-time data and scenario data for the development of forecasts and estimation of probability of future events. In Siegel's view, these include technologies based on algorithmic methods, which include: (1) time-series methods and extended forecasting, widely used in SCM for marketing, sales forecasting, insurance reserve estimation; (2) controlled training includes regression and algorithms of statistical analysis; (3) method of «decision tree» which uses a hierarchical sequence of structures; (4) clustering, advanced density-based analysis technique [10],
- *Prescriptive analytics (PrA)* uses predictions based on data that are capable of offering sets of actions that make it possible to achieve (or avoid) a specific result. PrA includes studying the problem of variability of expected results with the help of scenario analysis «what/if» or mathematical theory of games. PrA is associated with optimization and modelling and is particularly relevant in uncertain environments where it is not possible to use deterministic algorithms.

2.2 The Method of Contextual Review of BDA Terms in SCM

The definition of BDA in SCM is the combination of two inextricably linked disciplines: Big Data (BD) and advanced analytics (AA). Formally, there is no single definition of the term «Big Data» , but Ward and Barker's review suggests a characterization based on fact that Big Data is the next milestone for innovation, competition, and performance ensured by data properties: (1) *volume*—data set, data quantity, a factor allowing, in particular, to group companies. Intel believes that organizations that create about 300 terabytes of data per week are part of a group of large data generators; (2) *velocity*—data transmission from batch to real-time mode. According to IBM, 2.5 quintillion bytes of data are created every day in the world, and 90% of the data in the world have been created in the last two years; (3) *variety*—multiple formats, structure, data storage standards in organizations [14].

The use of Advanced Analytics (AA) in SCM is based on supply chain analytics, multi-dimensional database analysis technology that allows consolidation, measurement, tracking, prediction and management of SCM business data. Some authors identify BDA in SCM as a tool of business process re-engineering, others believe that BDA aims to improve SCM «transparency» in the whole chain. The speed of processing converted BDA's use into the complex pattern extraction method in SCM, increased the availability and speed of updating of information, which led to a change to the term «chain analytics» (Advanced Supply), describing a new paradigm, where models should be proactive to data rather than reactive. The general definition of BDA, specialized and adapted for SCM, is presented in Table 1.

The contextual study enables to conclude that BDA in SCM may be viewed as a process, but at the same time as a method, an information review, an operational shift, in a complex transforming SCM and allowing to state that «The better BDA, the more accurate the management algorithms».

2.3 The Methods for Modelling Transport and Logistics Infrastructure in Accordance with the Cargo Flow and BDA Characterizing Its Parameters

The fundamental principle of modelling taken into account by the authors is the consideration of cargo flow parameters for which the transport and logistics infrastructure is an external factor. As an external environment for the cargo flow, transport and logistics infrastructure—a complex economic facility, unique in its properties, unmanageable at the level of individual companies, diverse and variable. The influence of transport and logistics infrastructure on cargo flow is carried out in the direction of «time—transport costs»; «cycle—delivery schedule». In addition to the impact of transport and logistics infrastructure on the cargo flow, there are impacts of the flow on the development of transport and logistics infrastructure. For countries with large territories, such as the Russian Federation, the second type of impact is

Table 1 The contextual definition of BDA in SCM

Keyword	Definition/source
Process	of extracting and presenting information on SC for measuring, monitoring, forecasting and managing SC [4]
Process	by which individuals, companies and their organizational units use supply chain information through the ability to measure, control, forecast and manage the business process associated with SC [12]
Methods	of mathematics and statistics applied to large amounts of data [9]
Methods (quantitative and qualitative)	of different disciplines combined with SCM theory to solve current problems and predict results, taking into account data quality and availability issues [13]
Methods (quantitative)	for obtaining forward-looking conclusions from the data, for a better understanding of what is happening «upstream» and «downstream», being able to evaluate the operational effects on SC [7]
Review	of the whole supply chain to show the complete product, provide a single view on SC, provide a high KPI analyst [8]
Operational shift	in management models that can help SC professionals analyse all big data sets using proven analytical and mathematical methods [6]

Source authors

extremely important. The approach to infrastructure facilities based on their spatial location across the country in such a way as to overcome the high level of regional differentiation requires cargo flow to be considered as a factor and infrastructure as a result in SCM tasks. Thus, the problem of cargo delivery is solved by creation of transport-logistics infrastructure in the model «cargo flow—transport-logistics infrastructure» .

3 Results

BDA process understanding in SCM is associated with four stages of the supply chain: procurement–marketing–transportation–warehouse and warehouse operations. At the transportation stage, BDA content is presented by the data units describing cargo traffic (Table 2).

The dependence of the transport and logistics infrastructure on the parameters of cargo flow presented in BDA Advanced analytics enables to form the models of

Table 2 BDA taxonomy according to the data types characterizing cargo flow

Advanced analytics (AA)	BDA content
Descriptive analytics (DA)	Operational cargo position data
Predictive analytics (PA)	Diagram of the transport network, the route of the vehicles with the indication of the starting and destination points; transport costs and other static data of the route to create alternative routes in real time
Prescriptive analytics (PrA)	Operational data on direct tracking of shipments, volume of transit stocks, estimated time for ordering in specific traffic conditions and weather variables; real-time marginal costs for different delivery channels and types of freight flows; accessible indicators of smart transport systems, mass delivery networks

Source authors

transport and logistics infrastructure, linking parameters "time—cost of transportation", "cyclicity—schedule of delivery" with quantity, geography, capacity, technical equipment, technology of operation of infrastructure objects (Table 3).

Accepted model abbreviations: **IM**—inventory management; **MS**—mass service; **NPM**—network planning and management; **DP**—dynamic programming; **EM**—econometric models; **GT**—game theories; **LP**—linear programming; **BS**—balance sheet; **FM**—functional; **MA**—morphological analysis; **HA**—heuristic algorithms; **MHA**—meta-heuristic algorithms; **CA**—combinatorial algorithms.

Table 3 Transport and logistics infrastructure models in BDA Advanced analytics

Advanced analytics (AA)	Model selection criteria						
	Suitability	Universality	Cost reflection	Time	Digital technologies	Fullness	Adaptability
IM	++	+++	++	++	+	–	+
MS	+	++	+	++	+	++	–
NPM	+	+	+	++	+++	+	+++
DP	++	++	++	–	+++	+	+++
EM	++	++	+++	++	+++	–	+
GT	+	++	–	–	+	+	+
LP	+++	+++	+	+	+++	++	+++
BS	+++	++	++	–	++	+	+++
FM	++	+	++	+	++	++	+++
MA	+++	+	–	+	+	+	+
HA	++	++	+	+	++	+	++
MHA	++	+++	+	+	++	++	+++
CA	+++	+++	++	++	+++	++	+++

Source authors

As criteria for selecting a model based on BDA Advanced analytics, Table 3 shows: Failure to implement the model (−); Low degree (+); Leverage (++); High Realization (+++).

The model of transport and logistics infrastructure formation chosen according to the given criteria determines geography, quantity, type, distances, capacity, technical equipment, transport and engineering communications of an infrastructure object. It is obvious that the implementation of the model is effective when there is an opportunity to use already existing objects, an established infrastructure of cargo flow. However, the achieved versatility of the model must be supplemented by flexibility, the ability to change according to the characteristics of the cargo flow, the requirements of the consumer to the place, time, volume, intensity and cost of cargo, the movement of goods to new markets, new suppliers and customers. We have developed a step-by-step modelling algorithm to achieve the criteria for the effectiveness of models (Table 4).

Table 4 BDA transport and logistics infrastructure modelling algorithm in SCM

Iteration	The action of the step
Step 1. Identifying the purpose of modelling	to define the target characteristics of cargo consumption by place, time, volume, quality, price
Step 2. Calculation of cargo flow parameters	to define the trajectory, the starting, intermediate and final point of the cargo flow; speed, time, intensity, mass, volume of the delivery unit; costs, form of movement (warehouse, transit)
Step 3. BDA organization	to form Advanced analytics blocks as part of Descriptive analytics, Predictive analytics, Prescriptive analytics
Step 4. Planning the transport and logistics infrastructure	to calculate the number, determine the type, capacity, technical equipment, geography of the location of the logistics object
Step 5. Cargo allocation regulations	to develop a scheme of consolidation and distribution of cargo to flow in the supply chain links
Step 6. Transport plan	to develop a route, select the type of a vehicle
Step 7. Inventory management system	to calculate parameters of inventory management in subjects of transport and logistics infrastructure, to organize internal cargo processing

Source authors

4 Discussion

In the modern sense, the infrastructure should follow the flow of cargo, not the flow of cargo follows the infrastructure, which, however, should not be understood literally. It is important to define points of formation of perspective flows of cargoes, their vector, speed, volume, and in accordance with these parameters to form the strategy for developing transport and logistics infrastructure. Otherwise, the need to "put" cargo flow into the existing infrastructure facilities scheme leads to an increase in the operating costs of companies, industry imbalances of the economy and high differentiation of socio-economic development of the territories. This situation is described in a number of works on cybernetics and system analysis, characterized by centralized management, high rigidity of the structure, lack of plasticity and low adaptability to changes. These properties may turn out to be fatal for the system, since the contradictions between the constant structure and the changes caused by evolution can grow to global dimensions and require such radical and sharp restructuring as is no longer possible within the framework of this structure and leads to its destruction.

5 Conclusion

In a situation where Big Data determines management efficiency, BDA in SCM allows to develop qualitative models and precise algorithms for the development of transport and logistics infrastructure. The proposed methods of research, models "cargo flow—transport and logistics infrastructure" and the modelling algorithm implement the author's approach and the principle of priority of cargo flow to the development of infrastructure. The research resulted in the systematization and refinement of the BDA concept in relation to SCM. The study of BDA-specific terms exhaustively describes the structure of BDA in the Advanced analytics elements, as part of Descriptive analytics, Predictive analytics, Prescriptive analytics. The method of contextual analysis allowed to define BDA in SCM as a combination of two inextricably linked disciplines: Big Data (BD) and Advanced analytics, to justify its understanding as a process, but, at the same time, a method, an information review, an operational shift, as a whole transforming SCM practice. The author's taxonomy of BDA classifies various types of data that characterize the flow. The method of modelling transport and logistics infrastructure—which is an external factor to the load capacity—made it possible to develop models and an algorithm for modelling the transport and logistics infrastructure in accordance with the parameters of cargo flows linked to the application of BDA, to highlight the criteria of the model efficiency in various BD units. In solving complex tasks of SCM, the authors' proposals are of practical significance in analyzing and assessing the parameters of cargo flows, in modelling transport and logistics infrastructure, in developing sectoral and territorial documents of strategic planning, strategies for spatial development of the subjects of the Russian Federation and corporate strategies of transport and logistics companies.

References

1. Barratt M, Oke A (2007) Antecedents of supply chain visibility in retail supply chains: a resource-based theory perspective. J Oper Manage 25(6):1217–1233
2. Chae B, Sheu C, Yang C, Olson D (2014) The impact of advanced analytics and data accuracy on operational performance: a contingent resource based theory (RBT) perspective. Decis Support Syst 59(1):119–126. https://doi.org/10.1016/j.dss.2013.10.012
3. Gunasekaran A, Papadopoulos T, Dubey R, Fosso S, Stephen W, Childee J, Hazen B, Akter S (2017) Big Data and predictive analytics for supply chain and organizational performance. J Bus Res 70:308–317. https://doi.org/10.1016/j.jbusres.2016.08.004
4. Marabotti D (2003) Build supplier metrics, build better product. Quality 42(2):40–43
5. McAfee A, Brynjolfsson E (2012) Big Data: the management revolution. Harvard Bus Rev 90(10):61–67
6. O'Dwyer J, Renner R (2011) The promise of advanced supply chain analytics. Supply Chain Manag Rev 15(1):32–37
7. Pearson M (2011) Predictive analytics: looking forward to better supply chain decisions. Logist Manag 50(9):22–26
8. Rozados IV, Tjahjono B (2014) Big Data analytics in supply chain management: trends and related research. In: Pujawan IN, Vanany I, Baihaqi I (eds) Proceedings of the 6th international conference on operations and supply chain management. Sepuluh Nopember Institute of Technology, Surabaya, pp 1096–1107. https://doi.org/10.13140/rg.2.1.4935.2563
9. Sahay BS, Ranjan J (2008) Real time business intelligence in supply chain analytics. Inf Manag Comput Secur 16(1):28–48
10. Sanders NR (2014) Big Data driven supply chain management: a framework for implementing analytics and turning information into intelligence. Pearson Education, Inc., Pearson
11. Siegel E (2013) Predictive analytics: The power to predict who will click, buy, lie, or die. Wiley, Hoboken
12. Smith M (2000) The visible supply chain. Intell Enterp 16:44–50
13. Waller MA, Fawcett SE (2013) Data science, predictive analytics, and Big Data: a revolution that will transform supply chain design and management. J Bus Logist 34(2):77–84. https://doi.org/10.1111/jbl.12010
14. Ward JS, Barker A (2013) Undefined by data: a survey of Big Data definitions. http://citeseerx.ist.psu.edu/viewdoc/download?doi=10.1.1.705.9909&rep=rep1&type=pdf. Accessed 01 Feb 2020
15. Zeng X, Lin D, Xu Q (2011) Query performance tuning in supply chain analytics. In: Yu L, Wang S, Lai KK (eds) Proceedings of the fourth international joint conference on computational sciences and optimization. IEEE, Piscataway, pp 327–331. https://doi.org/10.1109/cso.2011.212

Digital Education Influence on Students' Intellectual Development and Behavior

G. N. Alexandrova, M. V. Cherkunova, and J. S. Starostina

Abstract The article analyses the development of cognitive, creative and communicative abilities of the youth in the modern digital era. The authors consider the problems of acquiring new knowledge, developing critical thinking, and improving socializing skills by young people. The article investigates the correlation between Internet addiction and creativity, school education and successful study at University. The authors use the methodology of diagnosing testing and questioning. The challenge of modern education is to adapt students to changing educational environment, to make them more flexible and confident in their future. University curricula should envisage what soft and hard skills would be necessary in the nearest and furthest future. More attention should be paid to activities, inspiring face-to-face communication among students. Another problem is lacking of well-equipped schools and qualified teaching staff.

Keywords Cognitive abilities · Creativity · Digital education · Critical thinking · Knowledge acquisition

1 Introduction

We can't no longer imagine our society without modern gadgets, devices, various software, that help us think, or do not think how to do a lot of chores around the house, a lot of work in the garden, in an office, with a button press we send letters, invitations, warnings and so on [3]. The development of digital technologies has embraced and captured the whole planet in its smart networks. Some people are

G. N. Alexandrova (✉)
Samara State University of Economics, Samara, Russia
e-mail: Alexandrova.gn@rambler.ru

M. V. Cherkunova · J. S. Starostina
Samara National Research University, Samara, Russia
e-mail: m.cherkunova@mail.ru

J. S. Starostina
e-mail: juliatim@mail.ru

S. I. Ashmarina and V. V. Mantulenko (eds.), *Current Achievements, Challenges and Digital Chances of Knowledge Based Economy*, Lecture Notes in Networks and Systems 133, https://doi.org/10.1007/978-3-030-47458-4_50

completely lost in a familiar city, in which they have lived for more than a dozen years, if in a hurry, and we are always in a hurry, they ran out of the house without a phone. Everything around us is rapidly developing, changing, accelerating and improving. Listening to lectures delivered by a famous professor from a European university, using educational online platforms, traveling through time and space,—everything is possible in digital epoch. There are unique opportunities for cultural self-development,—you can visit your favorite art galleries, listen to music, travel, draw, compose music or write novels, design buildings and even entire cities [2].

The relevance of this study is initiated by the rapid spread of information technology and the widespread of society digitalization. Digitalization affects the social, cultural, and political spheres of life [1]. People gain more knowledge, their social life is improving, they show greater awareness of other people lifestyles, traditions, and customs. All these lead to a more tolerant perception of other values and world outlook. Modern people spend more and more time on the Internet, solving production issues and domestic problems [7]. Their total involvement in the virtual environment of communication changes the usual way of life and poses many questions for scientists and researchers regarding the possible consequences of this phenomenon for the full development of social, physical and mental skills of younger citizens.

The development of digital services in the fields of education, healthcare, and economy has allowed the population to receive huge amount of information they need for professional growth, running successful business, and solving urgent problems [4]. In addition, along with the positive aspects of such easy and affordable way of obtaining information, studies appear about the negative impact of modern technology on the mental and cognitive abilities of children and teenagers. Transferring interpersonal communication to the virtual plane leads to problems of socialization and loss of empathy. People are growing apart from each other, family and friendship ties are weakening, institution of family and marriage disappears. Similar troubling arguments appearing in the media require careful analysis and research [9].

The purpose of the study is to analyze the development of cognitive, creative and communicative abilities of the youth in the modern digital era. The object of the study is cognitive, analytical and communicative skills of students of economic specialties in modern conditions of digitalization, who learn with the help of digital technologies [12]. The subject of the study is transformation of secondary and higher education, as a result of the use of digital educational platforms and the impact of digital resources on the level of education and personal growth of students of economic specialties.

2 Methodology

The methodological basis of this scientific research was the achievements of cognitive linguistics, psycholinguistics, pedagogy, the theory of management of modeling of socio-economic systems and neuro-linguistic programming studies. The methodological base of this is formed by the general scientific principles of a systematic and logical approach [10]. The authors used the following methods: diagnostic testing, directed situational experiment, analysis of the intellectual activity of the students during their studies at the university.

This research was conducted in five schools and two Universities. The research on cognitive, creative, and communicative skills was conducted among students during their study at the university. Taking into account the ongoing research in schools, we have developed a methodology for diagnostic testing of test groups. We developed the following tests: test to examine long-term and short-term verbal memory; test to examine long-term and short-term digital and temporary memory; test to examine skills and a tendency to research activities; test to examine desire and ability to work in a team; test to examine availability of academic writing skills and presentation of research results; test to examine the ability for self-criticism, verification of their research results and probabilistic forecasting.

In addition to these tests, we compiled questionnaires to find out the degree of ability of students to search for and find necessary information on the Internet, the degree of enthusiasm for computer games, the number of interlocutors in social networks living near the house, outside the city, outside the country. For this purpose, we selected two sample groups of students in the first year: the first group with high USE (United State Examination) scores (100 – 90 points) and, besides, this group of students was getting a second additional education in the field of professional communication. At the end of their study, they got a second diploma and a second profession—a translator of economic and legal texts. The group consisted of 40 people—20 students from Samara University of Economics and 20 students from Samara National University. The second group also consisted of 40 people—20 students from Samara University of Economics and 20 students from Samara National University with lower scores (90 – 65 points).

3 Results

The study of students' desires and requests began 10 years ago in connection with the introduction and development of new curricula for continuing education at Samara State University of Economics. It was necessary to determine more accurately the range of interests and needs of students in the acquisition of skills beyond those, offered by the curriculum. As a result, linguistic departments, at which translators in the field of professional communication are being trained, were opened in two higher educational institutions of Samara,—the University of Economics and the Samara National University. The research started exclusively for practical purposes has grown into an interesting scientific topic, in which two large Samara universities are involved. Since the profession of a translator is purely humanitarian and requires a wide outlook and a large amount of extra-linguistic knowledge, it has become necessary to check what knowledge students have about the world around them, namely literature, art, geography, and politics.

It turned out that students who graduated from high school, pursued the same educational programs, had the same ambitions and goals, demonstrated the huge difference in knowledge. More or less the same knowledge was in those subjects that were necessary for admission to the university. In Russia, there are no entrance examinations to higher educational institutions. Universities accept applicants according to the results of a United State Examination, which they take at the end of school education. At the next stage of the study, we decided to investigate the methodology of teaching compulsory subjects, additional training programs, extracurricular activities, technical equipment, staff and qualifications of teachers in secondary schools in Samara region. For this purpose, we selected five schools from the lowest to the highest level of student exam results, geographically located in various parts of Samara region: two schools—in the center of Samara, one school—in the outskirts of the city, one school—in a regional center town and the last one—in a small village.

In addition, we decided to analyze the degree of involvement of school students in the educational process, their ability to work on a computer, to find necessary information on the Internet, to use digital technologies for study and for personal communication with real and virtual friends. Additionally, we focused on the composition of the family, the social status and financial situation of parents, full and single parent families, and the participation of parents in raising children. In order to obtain complete and truthful results, we compiled questionnaires. These questionnaires were designed in such a way that the answers received could be double-checked, using other questions.

Two out of five sample schools are schools with in-depth study of a number of school subjects. In the first one, the students benefited from in-depth study of exact sciences, in the second one—from the humanities. Digital technologies are widely used in these schools. All classrooms are equipped with electronic whiteboards and computers. We would conditionally identify these schools—school number 1 and school number 2. The third school is also located in Samara, the level of digitalization is average, the curriculum includes a minimum of what is recommended by the Ministry of Education. Two other schools are located outside Samara, one of them (school number 4)—in the regional center, the other (school number 5)—in a small village. In the regional center, the equipment of the school with electronic devices is average, electronic whiteboards are not used, computers often break down, and there is a lack of technical personnel. In a remote village, the Internet connection does not always work, but the equipment of the classes with technical facilities is not bad.

We also focused on how high school students use the Internet and personal digital devices (home computers, laptops, mobile phones). We considered the strategic planning of the methodology of the entire educational process, the development and use of lecture materials, and artificial intelligence to be very important. From 2009 to 2013, the time spent by schoolchildren communicating with friends in the virtual space was no more than 2–3 h a day, by 2019 this time increased to 5–6 h a day. Moreover, 70% of high school students admitted that they did it during classes at school. However, the time of face-to-face communication reduced: for example, many students, strolling with friends in the park and, at the same time, socializing with someone on the network. Along with this, there is an alarming trend of increasing suicidal cases among adolescents and the creation of such groups on the Internet, as evidenced by official statistics. 95% of high school students almost do not watch

TV, sometimes they watch entertainment programs, and their watching time is less than one hour per week. Only 5% watch TV 3–4 h a week. 7% follow political events, happened at home or abroad, but they read this information on the Internet. 23% of high school students are not searching for additional information and materials for preparing homework, 35.8% are looking for information only if they are given individual tasks at school, for example, an essay or a report. Almost all respondents answered that they prefer ready-made essays or reports from the Internet that they could simply rewrite. At the same time, many students admitted, that it was difficult for them to read and understand information on history, politics, science, literature, and culture due to the huge amount of material and different points of view of the authors.

It should be pointed out, that the situation with the use of personal digital devices in the first two schools is completely different. Despite the fact that the students prefer virtual communication with peers, school activities compensate for the lack of personal communication. The students also perform many creative tasks during their lessons at school and at home, their self-study is monitored and verified by a teacher. They rarely search for relevant information with the help of keywords, and use reliable sites and certain educational platforms.

Over the course of ten years of research in these schools, changes have occurred only in terms of updating technical equipment and increasing the paper work of teachers. Since we have been introducing digital educational platforms at our universities, planning and monitoring students' independent work, we need to have an idea of how to draw up curricula in order to be sure that students will master the material and acquire necessary soft and hard skills.

Students who have high scores on the results of a United State Examination do not miss classes, actively participate in class and seminar discussions, enroll additional training programs, engage in scientific and public work, and do not ignore creative and cultural events. Over the past 10 years, according to the results of the study, the number of such students has been stable and amounted to 10% (plus or minus 1.7%) of the total number of applicants accepted for the first year. We can conclude that they have better motivation to study.

In this article, we present the results of studies of sample groups that studied at our universities from 2014 to 2018. Since all students studied in non-humanitarian faculties, we felt that it would be interesting to compare the results of students receiving additional humanitarian education with those, who did not receive additional education. Diagnostic test results showed that 12% and 72% of students in the first and second sample groups, respectively, experienced difficulty in self-searching for the necessary scientific information, they were distracted by the ads appearing on the monitor, had difficulty reading long and complicated texts, and had problems with long-term memory. 9% and 36% students, respectively, could not work in a team, showed signs of intolerance to an opposite opinion, did not know how to bring the discussion within the framework of the chosen theme and sound convincing when defending their point of view. 3% and 18% students respectively admitted that they spent more than 3 h a day playing computer games and often could not force themselves to stop playing. 9% and 19% students, respectively, prefered to communicate with peers virtually.

According to the results of academic performance and various achievements recorded by students in the portfolio, it can be seen that the greatest number of scientific achievements were among students who initially entered the university with better results. That is, from the first group—this is 98% of students, from the second group—23%. According to the results of the final test, very few students experienced problems in self-searching for the necessary scientific information—0% and 2%, respectively, but it should be clarified that information search was supervised by the supervisor. Tutors recommended their students to look through a list of reliable sites and educational platforms. The percentage of students who could not work in a team, showed signs of intolerance towards an opposite opinion, did not know how to participate in discussions and be convincing when defending their point of view decreased—2% and 9%, respectively.

4 Discussion

In order to develop cognitive abilities, critical thinking, and increase the share of students' research work using digital technologies, the university is developing electronic courses with the elements of artificial intelligence, the number of tasks encouraging the development of research and scientific work skills is increasing, a special academic course has been introduced in the university curriculum. It is planned to pay more attention to reading and analyzing scientific texts while supervising students' research work, since diagnostic testing showed that 78% of students in the second sample group were not able to analyze and critically interpret difficult texts. In the first group, there are 0% of such students. Unfortunately, teachers cannot control the time students spend on the Internet being distracted by unnecessary advertising or useless information. By the end of the fourth year of study, most students mastered the curriculum, acquired the skills necessary for their future job, they greatly benefited from a new practice of close cooperation between business and our universities, and gained essential experience. From the point of view of developing communication skills, of particular importance is the involvement of students in extracurricular activities, especially in volunteer projects, aimed at helping people in difficult situations, assisting in restoration of monuments, and training the elderly to increase computer literacy.

Digital technologies have a significant impact on the organization and development of the entire education system at home and abroad. Changes occur both in the field of preschool, school, and, especially, higher education [8]. With the undoubted benefits and enormous advantages of the modern education system, serious challenges arise that concern both teachers and students. In addition to a traditional duty of a teacher to give knowledge and teach, another function, that is becoming more and more necessary and important, is to teach students how to learn. In other words, it is necessary to teach students how to choose the right information in this rapid information flow, to highlight those facts or arguments that are necessary for the students at this stage of training, which would be understandable and useful to them. Students should develop a desire to work on their own in the network information space. In addition, students should develop critical thinking skills and analysis of acquired knowledge [6].

The primary task of teaching students at the University is to teach them how to work with digital products, to develop independent thinking, stimulate creativity, manage time and self-control, and raise efficiency of their work on the Internet [11]. However, we would like to draw attention to the considerable difficulties arising from the use of the Internet and various software. Unfortunately, in both groups the percentage of students who spend time playing computer games and chatting with friends in social networks has increased—10% and 26%, respectively. What is more, most of employers consistently mention that being able to use various software and artificial intelligence are the most important and valued skills in today's work world [5].

5 Conclusion

Thus, according to the results of the study, we can conclude, that the cognitive abilities of students have improved, communication skills have decreased. We have not identified any changes in student' creative abilities. The next stage of the study, we believe, will be to find out the reasons for the decrease in communicative abilities, the lack of development of creative abilities and ways to solve these problems. At the beginning of the study, we thought that cognitive abilities should improve significantly, but this has not yet happened. In our opinion, it has not happened for several reasons: students need to read too much information; students are not always able to understand which information source is reliable, especially if they have to search for information on their own; it is difficult for students to evaluate this amount of information on their own. In addition, the amount of information is steadily growing; some students show a disturbing trend for Internet addiction. We should reorganize and adapt the whole system of higher education to new digital epoch challenges. Forecasting and introducing new professional competences into universities curriculum in advance is a must for new educational environment.

References

1. Beech D (2019) Art and postcapitalism: aesthetic labour, automation and value production. Pluto Press, London. https://doi.org/10.2307/j.ctvr0qv2p.10
2. Bellucci S, Otenyo E (2019) Digitisation and the disappearing job theory: a role for the ILO in Africa? In: Gironde C, Carbonnier G (eds) The ILO @ 100: tackling today's dilemmas and tomorrow's challenges. Brill, Leiden and Boston, pp 203–222. https://doi.org/10.1163/j.ctvrxk4c6.17
3. Cayley J (2020) Reading language art in digital media: reconfigurations of experimental practices. In: Colby G (ed) Reading experimental writing. Edinburgh University Press, Edinburgh, pp 185–204. https://doi.org/10.3366/j.ctvss3z34.14
4. Hassan R (2020) The condition of digitality: a new perspective on time and space. In Hassan R (ed) The condition of digitality: a post-modern marxism for the practice of digital life. University of Westminster Press, London, pp 73–96. https://doi.org/10.2307/j.ctvw1d5k0.6
5. Karam B (2019) The digital divide and film. In: Mutsvairo B, Ragnedda M (eds) Mapping digital divide in Africa: a mediated analysis. Amsterdam University Press, Amsterdam, pp 153–172. https://doi.org/10.2307/j.ctvh4zj72.12
6. Lethbridge J (2019) Identifying diverse sources of expertise. In: Lethbridge J (ed) Democratic professionalism in public services. Bristol University Press, Bristol, pp 55–84. https://doi.org/10.2307/j.ctvhktk4b.8
7. Pollitzer E (2018) Creating a better future: four scenarios for how digital technologies could change the world. J Int Aff 72(1):75–90. https://doi.org/10.2307/26588344
8. Skorodumov P, Badanin D (2015) Construction of a virtual environment of economic modeling. Questions Territorial Dev 4(24):1–12
9. Strutt D (2019) Dynamic digital spaces, bodies, and forces. In Strutt D (ed) The digital image and reality: affect, metaphysics and post-cinema. Amsterdam University Press, Amsterdam, pp 113–158. https://doi.org/10.2307/j.ctvx8b78q.7
10. Yakhneeva IV, Agafonova AN, Fedorenko RV, Shvetsova EV, Filatova DV (2020) On collaborations between software producer and customer: a kind of two-player strategic game. In: Ashmarina S, Mesquita A, Vochozka M (eds) Digital transformation of the economy: challenges, trends and new opportunities. Advances in intelligent systems and computing, vol 908. Springer, Cham, pp 570–580. https://doi.org/10.1007/978-3-030-11367-4_56
11. Zechner M, Rinne P (2018) Social work education in Finland. In: De los Santos M, Agudo M, Millan L, Cobos M, Perez S, Nieto-Morales C, De Martino Bermúdez M (eds) Social work in XXI century st.: challenges for academic and professional training. Dykinson, S.L., Madrid, pp 147–168 https://doi.org/10.2307/j.ctv6hp39z.12
12. Ziegele F, Mordhorst L (2019) Competition, collaboration and complementarity: higher education policies in Europe. In: Pritchard R, O'hara M, Milsom C, Williams J, Matei L (eds) The three Cs of higher education: competition, collaboration and complementarity. Central European University Press, Budapest and New York, pp 11–26. https://doi.org/10.7829/j.ctvs1g987.6

The Main Ethical Risks of Using Artificial Intelligence in Business

E. L. Sidorenko, Z. I. Khisamova, and U. E. Monastyrsky

Abstract The article reveals ethical foundations underlying the creation and application of systems with artificial intelligence (AI) in the interests of business. The main problem today is the lack of internationally accepted principles regulating a responsible attitude to processing and use of personal data used for training systems with AI, as well as the use of these systems in the interests of business. The main research goal is to find the best ways to solve existing problems in the field of ensuring the confidentiality of personal data and ensuring interests of business and the state in the development of advanced machines. The authors note the need for an early adoption of global initiatives at the international level to regulate a responsible attitude to personal data processing in order to train AI and create data ethics. In addition, the authors justify the need to distinguish AI ethics as a separate direction in the field of data ethics.

Keywords Artificial intelligence · Data ethics · Business interests · Ensuring balance · Ethical principles

1 Introduction

About 5 years ago, various researchers made predictions about the possibility of automation and digitization of production processes by about 50% in the next 20 years [6]. The MIT Technology Review notes that AI experts often make mistakes in

E. L. Sidorenko (✉)
Moscow State Institute of International Relations (University) of the Ministry of Foreign Affairs, Moscow, Russia
e-mail: 12011979@list.ru

Z. I. Khisamova
Krasnodar University of the Ministry of Internal Affairs of Russia, Krasnodar, Russia
e-mail: zarahisamova@gmail.com

U. E. Monastyrsky
Monastic, Zyuba, Stepanov & Partners, Moscow, Russia
e-mail: monastyrsky@mzs.ru

S. I. Ashmarina and V. V. Mantulenko (eds.), *Current Achievements, Challenges and Digital Chances of Knowledge Based Economy*, Lecture Notes in Networks and Systems 133, https://doi.org/10.1007/978-3-030-47458-4_51

such predictions [5]. We see a confirmation of this everywhere. The technological development pace significantly exceeds all existing forecasts. In its 2018 research, the analytics company Gartner identified artificial intelligence as a key technology and technological trend [3]. This is confirmed by data on the volume of global investments in the field of AI development, exceeding $5 billions [14, 17]. The key areas of active AI implementation in the next 10 years will be:

- medicine where the effectiveness of AI has already been clinically proven. A team of researchers from the University of Nottingham has developed four machine learning algorithms to assess the risk of cardiovascular disease in patients, the accuracy of the algorithm exceeded 76.4%, while the standard diagnostic procedure implemented by the American College of cardiology provides an accuracy of only 72.8% [11],
- finance and insurance, where the neural network, unlike a person, is able to assess and anticipate all possible risks and fraud attempts, for example, the Japanese insurance company Fukoku Mutual Life Insurance has replaced its employees with a neural network to check medical certificates, as well as accounting for the number of hospitalizations and operations carried out by the insured person to determine the terms of insurance for customers. The neural network is expected to increase the insurance productivity by 30%. For similar purposes, AI is also used by the PayPal payment system [23],
- online retail, commerce and media. The main purpose of using AI for e-commerce is to ensure interests of potential customers and offer them the most likely and desirable product for purchases. For these purposes, machine learning algorithms analyze the behavior on the site and compare it with millions of other users, which provides trading platforms with up to 35% of sales, and streaming services with up to 70% of views of certain resources [4, 13]. AI is also actively used in other business areas, such as marketing, transportation, law, consulting, agriculture, and even security [22].

Artificial intelligence technologies in marketing and advertising are becoming ubiquitous and invisible, and imperceptibly for a person constantly reduce the space of supposed freedom of choice for each individual consumer. According to analysts, the global GDP growth due to the application of AI could reach 14% by 2030, and provide an additional $15.7 trillions, that will allow the AI to become a leader among technological drivers of the economic growth. It is expected that the greatest benefits from the AI implementation will be felt in China (+26% of GDP by 2030) and North America (potential growth of 14%) [19]. However, with the growth of investments in AI, a number of problems that are inextricably linked to it are also of increasing concern. First of all, these are problems of an ethical nature.

2 Methodology

The methodological basis of this research is a systematic approach to the study of ethical foundations for the use of AI systems. During the research, the main risks associated with the use of AI systems in the interests of business were classified and analyzed. Traditional scientific methods such as dialectical, logical, method of scientific generalization, content analysis, comparative analysis, synthesis, source studies, etc. were used in the processing of factual material. Their application made it possible to ensure the validity of the analysis, theoretical and practical conclusions and developed proposals.

3 Results

The research work identified the main ethical problems associated with the use of artificial intelligence in business. These problems are divided into two blocks: the first block is related to the collection and processing of personal data for the training purpose of systems with AI, the second block of problems is caused by the ethics of decisions made by AI and their compliance with generally accepted morality. The authors summarize that in the conditions of rapid development and implementation of AI systems, it is necessary to adopt global initiatives aimed at the responsible attitude to the handling of data in the development of AI systems, as well as the direct application of such systems.

4 Discussion

For training systems with AI, a huge array of data is used, which can only be collected using data collection and processing technologies, so the relation between artificial intelligence and data accounts for almost 100%. The main question today is: how to use maximum data with minimal risks? Thanks to modern computing power, AI technologies can analyze huge amounts of data and find complex and deeply hidden relations. However, there are still many questions about the ethics of using data, including personal data, for training AI technologies applied in business [1, 7, 10]. Data ethics is developing and becoming more relevant, as it is evidenced by the relevant documents of both public associations and international organizations. Defining boundaries of ethical access to data is a complex issue that affects various stakeholders: citizens, the state, business corporations, government agencies, etc. and requires a comprehensive solution [12].

Data ethics as a kind of applied ethics has appeared relatively recently and does not have a generally accepted definition yet. Data ethics is a new branch of applied ethics that describes value judgments and approaches we use when generating, analyzing, and distributing data. This includes deep knowledge of data protection legislation and another relevant legislation, as well as the appropriate use of new technologies, which requires a holistic approach to applying best practices in computing, ethics, and information security [9]. Ethical issues related to the use of AI can be divided into 2 groups: problems related to the collection, analysis and processing of digital data; problems related to AI decision-making based on generalized data. Ethical difficulties are caused primarily by the collection, analysis and processing of citizens' digital data, Big Data, social and personal data. As it was already mentioned, businesses need them for AI training, online advertising, and online commerce, while the state needs them for making administrative decision, interacting with citizens, and ensuring the national security. Thanks to the collection and analysis of Big Data using AI, technology giants are able to build correlations that people themselves cannot determine themselves [20]. This also raises a question of maintaining a balance between ensuring the right to protect personal data and interests of science, business and our society in general, interested in the widespread use of AI technologies. In a number of countries today, there is a fierce debate between the use of citizens' personal data by government (and not only) organizations to ensure the public safety, training systems with AI and ensuring the right to privacy.

According to some experts, total surveillance is more dangerous than anything it should protect us against. Questions are raised about the ethics of using AI systems to control a socially approved behavior. For example, the Chinese government uses the artificial intelligence to create a social rating, using all information about the network users: their behavior, purchases, credit history, movements, social circle, interests. The basis of the European legislation since 2018 is the General Data Protection Regulation (GDPR) [8]. The main purpose of the regulation is to create a legal regime in the society, in which any citizen can be sure of the confidentiality of their personal data, including the ability to respond to cases of using his personal data without consent—to delete posted information.

At the same time, even despite the measures taken, a large amount of information remains in the network, which allows you to identify a person (using some methods of processing). So, using PCs, smartphones with mobile applications, most people without any hesitation give their consent to the smartphone use of geo-location data, access to the phone book and calls, to search queries and search history, to photo ads, etc. And the thesis that our data has become a product has become more relevant than ever. The lion's share of the world's most technology companies use a business model based on the collection and processing of users' personal data. And here the issue of ethics not only of collecting such information, but also of using it is on the agenda.

The second problem is related to the ethics and humanity of AI decisions. Systems with AI that are capable of learning go through a training phase where they "learn" to detect correct patterns and act according to the data entered. Once the system is fully trained, it can go into the test phase, where it will be given various examples and we can see how it copes with them. However, the experience of recent years shows that training is not enough, and despite the lack of feelings and emotions in AI systems, they cannot be considered as impartial and ethical ones for a number of reasons.

First, the very concept of ethics is very conditional, and can differ significantly within a single society and generation, not to mention over longer time periods. Second, it is the "level of bias and purity" and relevance of the data used for AI training. It seems extremely unlikely, even improbable, to find a large enough group of people in the society with zero bias to be able to use their data for AI training. As a result, we have cases of AI discriminating against people based on race or gender. Undoubtedly, programmers and large companies are trying to immediately eliminate such manifestations of "racism, chauvinism or xenophobia of AI".

In their research, PwC talked about how to prepare data for AI training, extract maximum profit from the technology and not lose the trust of customers. According to analysts, the best basis for implementing AI is the competence center for artificial intelligence. Business and IT representatives should develop uniform regulations for working with data for the entire corporation and monitor their compliance; define technical standards for relationships with suppliers; manage the intellectual property; and evaluate the implementation level of the artificial intelligence. All tasks of artificial intelligence training should be solved by the AI competence center [17, 18].

However, the rapid development and use of AI technologies makes such control almost impossible. In 2018, New York city passed a law aimed at preventing discrimination through algorithms used by public services. It became the basis for the creation of a public group of experts who analyze legal and ethical aspects of the city automated decision-making systems, as well as the creation of a position in the city mayor office that is dedicated to combating biases in algorithms and increasing the responsibility for their decisions [16].

However, it should be noted that excessive restrictions on access to data can slow down the development of AI technologies. Well-thought-out legislation will allow maintaining a balance between regulating the volume and degree of anonymity of personal data without imposing numerous prohibitions. But such legislation, unfortunately, has not been created in any country in the world yet. It should be noted that this issue will become the main topic of discussions in the world arenas in the coming years [25]. A number of countries have already adopted separate acts aimed at creating a legal regime for the application of AI systems [2]. However, from the point of view of the ethical component, the most interesting is the self-regulation of the industry. Thus, in 2017, developers and researchers in the AI field adopted the Asilomar AI Principles [10], which formed the basis of ethical aspects spelled out in the Montréal Declaration for Responsible Development of Artificial Intelligence, adopted in December 2018 [24]. The responsibility principles in the development of advanced AI systems, personal data privacy, freedom and privacy, as well as security are at the heart of these ethical memoranda and a number of other private initiatives [21].

5 Conclusion

In general, it can be stated that as the number of systems with AI that actively process citizens' personal data increases, the question of the ethics of their use for various purposes, including commercial ones, is increasingly raised. AI ethics can be identified as a new independent area of data ethics. Without forming an ethical framework for the use of AI systems and the principles of personal data use in their processing, further progress is impossible. Public legal institutions are already actively involved in the ethical regulation of artificial intelligence.

In our view, the integration of ethical principles should be regulated by national or international organizations [15]. It should be noted that this sphere has become a key issue on the agenda for discussion on the world arenas. It is important to maintain a balance between interests of individuals and businesses. Well-thought-out legislation and its enforcement will allow maintaining the balance of regulating the volume and degree of anonymity of personal data without imposing numerous prohibitions. But such legislation, unfortunately, has not been created yet.

References

1. Bostrom N, Yudkowsky E (2011) The ethics of artificial intelligence. Cambridge University Press, Cambridge
2. Carrillo MR (2020) Artificial intelligence: from ethics to law. Telecommun Policy. https://doi.org/10.1016/j.telpol.2020.101937. Accessed 09 Jan 2020 (in press)
3. Cearley D, Burke B, Searle S, Walker M (2018) Top 10 strategic technology trends for 2018. Gartner. https://www.gartner.com/doc/3811368. Accessed 09 Jan 2020
4. DTI Algorithmic (2017). Neural networks: how artificial intelligence helps in business and life. https://blog.dti.team/nejroseti/. Accessed 09 Jan 2020 (in Russian)
5. Emerging Technology from the arXiv (2017). Experts predict when artificial intelligence will exceed human performance. https://www.technologyreview.com/s/607970/experts-predict-when-artificial-intelligence-will-exceed-human-performance/. Accessed 09 Jan 2020
6. Frey CB, Osborne MA (2013) The future of employment: how susceptible are jobs to computerisation? www.oxfordmartin.ox.ac.uk/downloads/academic/The_Future_of_Employment.pdf. Accessed 09 Jan 2020
7. Geis JR, Brady A, Wu C, Spencer J, Ranshaert E, Jaremko J, Langer SG, Kitts AB, Birch J, Shields WF, van den Hoven van Genderen R, Kotter E, Gichoya JW, Cook TS, Morgan MB, Tang A, Safdar NM, Kohli M (2019) Ethics of artificial intelligence in radiology: summary of the joint European and North American multisociety statement. J Am Coll Radiol 16(11):1516–1521. https://doi.org/10.1016/j.jacr.2019.07.028
8. General Data Protection Regulation (GDPR). https://gdpr-info.eu/. Accessed 09 Jan 2020
9. Government Digital Service (2018) Guidance data ethics framework. https://www.gov.uk/government/publications/data-ethics-framework/data-ethics-framework. Accessed 09 Jan 2020
10. Koering D (2019) Ethics, AI, and human beings. Unpublished. https://doi.org/10.13140/rg.2.2.25131.80169. Accessed 09 Jan 2020
11. Krittanawong Ch, Zhang HJ, Wang Zh, Aydar MT, Kitai T (2017) Artificial intelligence in precision cardiovascular medicine. J Am Coll Cardiol 69(21):2657–2664

12. Lee M, Cochrane L (2020) Issues in contemporary ethics: AI warfare. https://www.researchgate.net/publication/339434471_Issues_in_Contemporary_Ethics_AI_Warfare. Accessed 09 Jan 2020
13. Market Research Future (2019) Video streaming market research report-global forecast 2023. https://www.marketresearchfuture.com/reports/video-streaming-market-3150. Accessed 09 Jan 2020
14. MarketsandMarkets Research Private Ltd. (2019) AI in fintech market by component (solution, service), application area (virtual assistant, business analytics & reporting, customer behavioral analytics), deployment mode (cloud, on-premises), and region—global forecast to 2022. https://www.marketsandmarkets.com/Market-Reports/ai-in-fintech-market-34074774.html. Accessed 09 Jan 2020
15. Mittelstadt B (2019) Principles alone cannot guarantee ethical AI. Nat Mach Intell 1:501–507
16. NYC (2019) The New York City automated decision systems (ADS) task force. https://www1.nyc.gov/site/adstaskforce/index.page. Accessed 09 Jan 2020
17. PwC (2018) Risk management for sustainable growth in the age of innovation. https://www.pwc.ru/ru/riskassurance/publications/assets/pwc-2018-risk-in-review-russian.pdf. Accessed 09 Jan 2020
18. PwC (2019) How AI will transform the CFO's role. https://www.pwc.com/gx/en/issues/data-and-analytics/artificial-intelligence/cfo-artificial-intelligence.html. Accessed 09 Jan 2020
19. PWC (2019) Sizing the prize. What's the real value of AI for your business and how can you capitalise? https://www.pwc.com/gx/en/issues/analytics/assets/pwc-ai-analysis-sizing-the-prize-report.pdf. Accessed 09 Jan 2020
20. Saariluoma PI, Leikas J (2020) Designing ethical ai in the shadow of hume's guillotine. In: Ahram T, Karwowski W, Vergnano A, Leali F, Taiar R (eds) Intelligent human systems integration 2020. Advances in intelligent systems and computing, vol 1131. Springer, Cham, pp 594–599. https://doi.org/10.1007/978-3-030-39512-4_92
21. Shahriari K, Shahriari M (2017) IEEE standard review. Ethically aligned design: a vision for prioritizing human wellbeing with artificial intelligence and autonomous systems. In: Umiversity R (ed) IEEE international humanitarian technology conference. IEEE, Piscataway, pp 197–201. https://doi.org/10.1109/ihtc.2017.8058187
22. Tadviser (2019) Artificial intelligence (Russian market). http://www.tadviser.ru/index.php/%D0%A1%D1%82%D0%B0%D1%82%D1%8C%D1%8F:%D0%98%D1%81%D0%BA%D1%83%D1%81%D1%81%D1%82%D0%B2%D0%B5%D0%BD%D0%BD%D1%8B%D0%B9_%D0%B8%D0%BD%D1%82%D0%B5%D0%BB%D0%BB%D0%B5%D0%BA%D1%82_(%D1%80%D1%8B%D0%BD%D0%BE%D0%BA_%D0%A0%D0%BE%D1%81%D1%81%D0%B8%D0%B8). Accessed 09 Jan 2020 (in Russian)
23. Terry H (2018) Fukoku mutual—insurance firm to replace human workers with AI system. The Digital Insurer. https://www.the-digital-insurer.com/dia/fukoku-mutual-insurance-firm-to-replace-human-workers-with-ai-system-2/. Accessed 09 Jan 2020
24. The Montréal declaration for responsible development of artificial intelligence. https://www.montrealdeclaration-responsibleai.com/. Accessed 09 Jan 2020
25. UN & COMEST (2017) Report of COMEST on robotics ethics. https://unesdoc.unesco.org/ark:/48223/pf0000253952. Accessed 09 Jan 2020

Actual Problems of Industry Digitalization in Russia

A. B. Vishnyakova, D. A. Bondarenko, and M. A. Agisheva

Abstract One of the main components of the Russian economy today is industrial production. After all, such indicators as GDP, GNP, national income, the level of provision of the population with necessary goods and services depend on it. The deplorable state of the material and technical base and the level of innovation implementation require the development of state measures to improve the efficiency of machine-building enterprises. But poorly developed investment activity slows down industrial modernization. The relevance of digitalization development is presented not only at enterprises, but also at the state level. Every year, new projects and programs are created, and funds are allocated from the state budget for the innovative development of industrial production and services. The article investigates some problems of industrial development in Russia and ways of overcoming them. The authors consider informatization as a way to overcome the production backwardness and the only way to ensure sustainable development of the country.

Keywords Informatization · Industrial production · Mechanical engineering · Digital economy

1 Introduction

In general, digitalization is understood as a socio-economic transformation initiated by the mass application and development of digital technologies, that is, technologies for creating, processing, exchanging and sending information. The need for modernization in the domestic machine building industry is changing every year, because

A. B. Vishnyakova (✉) · D. A. Bondarenko
Samara State University of Economics, Samara, Russia
e-mail: Angelina8105@yandex.ru

D. A. Bondarenko
e-mail: dasha172223@gmail.com

M. A. Agisheva
Kazan National Research Technological University, Kazan, Russia
e-mail: kalen7979@mail.ru

S. I. Ashmarina and V. V. Mantulenko (eds.), *Current Achievements, Challenges and Digital Chances of Knowledge Based Economy*, Lecture Notes in Networks and Systems 133, https://doi.org/10.1007/978-3-030-47458-4_52

Table 1 Investments to the program "Digital economy of the Russian Federation", billion rubles

Directions of the program «Digital economy Russian Federation»	Sources of investment from 2018 to 2020	
	From the budget	Off-budgettools
Informational infrastructure	100	336
Information security	22.3	11.7
Formation of research competencies and technological reserves	48	2
Normative regulation	0.9	0.3
Total	**171.2**	**350**

Source authors

of the technological gap from the Western countries for more than 20 years [7, 8]. Digitalization is one of the main directions of the economy. If this goal is achieved, the effective use of digitalization will be the key to the country's competitiveness in the near future.

Educational programs that are responsible for "digital literacy" will become particularly important. The national program "Digital economy of the Russian Federation" was created to integrate digitalization in such areas as economy, entrepreneurship, healthcare and public administration (Table 1). The main result is the creation of communication networks and infrastructure. Digitalization provides many opportunities for organizations: saving time and financial resources, the company will be able to respond faster to changes in the markets and prevent the loss of a competitive position [14].

2 Methods

In the framework of this analytical study the researchers used general scientific methods. The authors analyzed theoretical sources on the considered issues (articles, monographs, conference proceedings and dissertation thesis), as well as documentation (national programs, strategies, reports etc.). Data obtained was systemized and presented in this work compared with similar aspects of other countries. Current trends of global economic development were taken into account.

3 Results

It seems that Russian enterprises are not ready for informatization yet, since costs of more than half of them for the modernization and development of information technology does not exceed 2% of their budget [12]. Only 7% of enterprises spend about 6% of the budget, while Western countries spend at least 8%. In 2017, a study was conducted to assess enterprises' infrastructure availability, which showed that their IIoT equipment is estimated at only 95 billion rubles.

The main condition for digitalization is that the company is equipped with numerically controlled equipment. Unfortunately, in Russia only 14% of factories have half of this equipment [15]. The largest number of CNC machines is in: aircraft industry (30%), instrumentation (10%) and more than 10% in the machine tool industry (for comparison, the percentage of equipment with CNC machines in other countries: Japan (78%), Germany and the United States (73%), China (31%)). Unfortunately, the automotive and heavy machinery industries cannot reach 10% in this aspect [12]. The companies plan to purchase about 80% of the machines in three years. Is it possible to implement this plan in such a period? If only 10% of Russian enterprises have an innovation director.

We should admit that the current state of machine building in Russia is deplorable, as the dynamics of output since 2014 is negative. This is because of a decline in the domestic consumption. The machines that are located at enterprises have been working for more than 20 years, almost half of them (about 46%) are physically worn out and must be written off. The process of reducing machines is in full swing, but the staff remains unchanged. On average, there are about 4.5 workers per machine, while in European countries there are 0.8 workers per machine. This means that the labor productivity in the EU is much higher. Therefore, Russian products will not be able to withstand the onslaught from European products.

In order to achieve sustainable and stable development of the economy, it is necessary to implement a number of measures to modernize it. The basis for the modernization is the use of the latest technologies and developments of scientific and technological progress, this process plays a very important role. Unfortunately, this function is poorly developed in Russia. Our country is on 46th place in the Global innovation index 2019 (comparing with 2018, the position has not changed). In the UNESCO ranking, the Russian Federation is in the top 10 in terms of investment in R&D (this is an important share of all investments in innovations), experts estimated its investment to this sphere at $40.5 billion in purchasing power parity. However, Russia is not among the top 15 of UNESCO countries in terms of investments to GDP, estimating the level at 1.2% of GDP (the global average is 1.7%, and for North America and Western Europe, it is 2.6%) [3].

Legislation cannot provide a set of benefits for enterprises that carry out innovative activities, this fact negatively affects the pace and scale of scientific and technological progress, and the corruption of society affects not only the economy, but also its modernization. Excessive paperwork, which the entrepreneur does not want to get involved with, slows down the application of innovations. One of the solutions to this issue is SKOLKOVO, a scientific and technological complex for creating new technologies. The national project "Science" with a budget of 637 billion rubles should solve the problem by 2024. Hopefully, this complex and the national project will contribute to the modernization.

If Russia takes an active path of digitalization, it will get:

- flexibility of the production process, which leads to the rapid implementation of the modernization plan,
- reducing the cost of goods and improving their quality, which contributes to high competitiveness in both the domestic and foreign markets,
- effective and high-quality solution of tasks not only in the sphere of production, but also of environmental safety, creation of new business opportunities,
- digitalization will help you analyze sales, inventory, and the state of the production capacity; this will increase the interaction level with suppliers and customers.

So, let's imagine that the Russian production has abruptly become on the path of modernization, then the economy will be hit by a number of irreversible consequences:

- across the country, a wave of massive staff cuts that have accumulated over the past 20 years will pass, millions of people will be without work and not able to find a new job, as the reduction in one industry will not be compensated by other industries,
- in addition, digitalization in Russia will be uneven: some regions will get more technologies and intellectual resources, such as Moscow, Saint Petersburg, Kazan, and Nizhny Novgorod,
- increase in digital fraud and piracy.

The main documents for the transformation of the Russian society are the "Strategy for the development of the information society in the Russian Federation for 2017–2030" and the program "Digital economy of the Russian Federation". These projects pay special attention to goals, objectives and measures for the information technologies use, as well as the national economy formation. According to analysts, the program budget may exceed 3.5 trillion rubles [1]. Russian Experts estimated the costs of the Russian Federation for the development of the digital economy (Table 2).

The process of digitalization in Russia has its own trends:

unstable domestic expenditures on research and development, as a percentage of the gross domestic product (Table 3), but in turn, by 2030, the GDP is expected to grow, due to digitalization (Table 4);

- the decline in the innovation activity not only in certain activity types, but also in the country as a whole,
- the growing need for innovation (evidenced by costs of organizations).

Digitalization is a new stage of the revolution that has affected all spheres of the human activity, but most of all, high-quality updates are expected in the industry. The importance of a state policy in this issue is great. Digitalization will help to solve a number of topical issues for modern Russia: speeding up the production process, creating import-substituting goods and increasing the labor productivity.

Table 2 Structure of internal costs for the development of the digital economy by type of costs, %

Sector of the economy	Total
Internal expenditures on research and development in the priority area of science, technology and technology development "Information and communication systems"	2.4
Purchase of computer and office equipment	13.6
Acquisition of telecommunications equipment	9.8
Purchase of software	9.5
Employee training related to ICT development	0.2
Payment for services of third-party organizations and specialists related to ICT (except for communication and training services)	14.5
Payment for telecommunication services	35.0
Other ICT costs	3.7
The acquisition of digital content	11.3

Source authors

Table 3 Internal research and development costs, as a percentage of GDP for the Russian Federation (RF)

Year	2010	2011	2012	2013	2014	2015	2016	2017	2018
Total for RF	1.13	1.01	1.03	1.03	1.07	1.10	1.10	1.11	1.0

Source authors based on [6]

Table 4 Assessment of the digitalization contribution to the GDP growth by accumulated total, %

Year	GDP growth due to other factors	The contribution of the information industry	Contribution of digitalization of economic sectors
2017	0,4	0.1	1.0
2018	0.9	0.2	2.2
2019	1.3	0.3	3.0
2020	1.9	0.4	4.3
2021 (forecast)	3.5	0.6	5.6
2025 (forecast)	10.2	1.3	11.1
2030 (forecast)	17.8	2.3	18.4

Source authors based on [5]

4 Discussion

Expanding markets, increasing intra-industry competition, and increasing the competitiveness of individual countries' industries on world markets are consequences of digitalization. The McKinsey Global Institute has made the following forecasts: the Chinese GDP will increase to 22% by 2025 due to Internet technologies; and in the United States, the increase in value due to digital technologies by 2025 will be 1.6–2.2 trillion U.S. dollars. In Russia, the GDP increase will be 4–8 trillion rubles due to digitalization by 2025 [2].

These results can be achieved by applying new technologies and business models, such as a digital platform, a digital ecosystem, Industry 4.0 technology, and the Internet of things. The fourth industrial revolution ("Industry 4.0.") includes robotization of production, modeling and forecasting based on advanced analytics and Big Data technologies. Many global enterprises are moving to work with "Industry 4.0", but the implementation of these principles at Russian enterprises is questionable, since the information base is outdated, corporate information systems cannot be implemented because of the lack of technology. According to some experts [3], the Russian economy is not ready to implement the principles of "Industry 4.0", since at present the Russian industry is a combination of new, high-tech machines and the equipment that was created by Soviet specialists. Unfortunately, all the latest production machines are located in the sectors that bring the most profit, for example, in the gas and oil industries. This means that one of the most important problems of the Russian industry is the lack of domestic innovations and technologies that can be used to improve the efficiency of other enterprises. The entire production process depends on foreign suppliers of know-how, on technological progress that is rapidly developing on the territories of European developed countries.

To overcome this problem, the state should create conditions for training young professionals who can withstand tough competition, as well as have creative thinking that can find a way out of a crisis situation and help solve complex problems. But not only investments in higher education institutions are needed to train young professionals, important is also their motivation. The main mass of emigrants from Russia in 2018 are people at the working age, up to 33 years old. At the same time, the number of people leaving the country with higher education has increased rapidly: in comparison with 2012, it is more than 4 times [10].

Qualified professionals leave for other countries for the sake of profit. It is not surprising, because in developed countries they pay several times more than in Russia. The solution to this problem is to create a program of financial support, as well as to attract foreign masters to adopt their knowledge. The income of domestic specialists should not differ from the income of foreign colleagues. These actions should motivate new workers and possibly attract new ones from developed countries.

But there are also opposite views, where experts believe that Russia is ready to fight for a leading position in the world market and overcome the existing problems (for example, the industrial production index in 2019 (2.4%) in Russia slowed significantly compared to 2018 (2.9%). No matter how sad the Rosstat (the Federal State Statistics Service) news sounds, there is no way to hide these figures. Oil and gas production is slowing, although in 2019 a post-Soviet record for oil and gas production was set, and LNG (liquefied natural gas) production increased by almost 47% [11].

By 2035 Russia will take a leading position in several markets, each of them will contain 100 milliards U.S. dollars [4]. "Industry 4.0." opens up new opportunities for all sectors of the economy, including industrial production. Over the past

Table 5 Rating of technologies based on their investment attractiveness

Type of technology	Currently, %	In three years, %
«Internet of things»	73	63
Artificial intelligence	54	63
Robotics	15	31
3D printers	12	17
Additional reality	10	24
Virtual reality	7	15
Drones	5	14
Blockchain	3	11

Source authors based on [9]

10 to 15 years, countries with a digital economy have undergone mass modifications of production in the direction of artificial intelligence, and at the moment they have a fairly digitalized, high-tech and robotic production, connected to production networks [16].

Analyzing the development of Industry 4.0, we note the following forecasts:

1. Automation of professions by 60% by 2030.
2. The creation of 30 million jobs in the IT sector in 2030.
3. Retraining of people, mastering new skills (about 370 million people).
4. With the development of artificial intelligence, the economy will receive about 14 trillion dollars.
5. About 40 million jobs will be taken by machines, that is, every second employee will be affected.
6. By 2022, 43 million devices in Russia will be connected to the Internet of things.

Currently, the Internet of things is actively developing in the fields of economy, industry, agriculture, and healthcare. IoT is in the first place in the ranking of technology investment (Table 5).

Internet of things in the industry has a positive effect through the following aspects: automation of monitoring and remote control of equipment, troubleshooting (by 10%); reduced downtime by 10% [16]; creating effective relations between the organization and its participants; improving the product quality and compliance with the requirements; creation of new service business models.

The application of IIoT will increase the labor productivity and ensure the GDP growth. There are some financial estimates of the IIot development:

- stage from 2017 to 2023, this market will grow at an average annual rate of 14.36% and by 2023 its size will be $700.38 billion,
- by 2025, the world market for IIot will reach 484 billion Euros,
- by 2030, the contribution of IIot to the world economy in monetary terms will be more than $14 trillion, including up to $6 trillion in the USA and more than $70 billion in Germany [13].

The data obtained show that economic, managerial and social spheres are developing due to the digitalization. The effective use of remote control by the company "General Electric" has gained a benefit of $100 billion. In Russia, devices that are engaged in the automated data collection have taken root. For example, "Rosenergoatom" Concern has upgraded the equipment operation function at units 1 and 2 of the Smolensk Nuclear power plant (NPP) over the past 2 years. The eSOMS system (manufacturer – ABB) has been implemented, which implies issuing terminals to employees that suggest suitable routes of circumvention and allow real-time transmission of data to the central information system. There, all the information is combined, analyzed and transmitted to the station operators together with advice and instructions, in case of deviations. The implementation of this system allowed to reduce the time spent on working rounds by almost 20 times, reduce the paper workflow constructively, and increase the research quality. The financial effect of reducing labor costs is estimated at 45 million rubles per year [13]. By 2025, the effect of implementing the Internet of things in Russia will be 2.9 trillion rubles.

5 Conclusion

In general, Russia has a number of prerequisites for further development of the digital economy, which contributes to improving not only enterprises, but also the social sphere and public administration. One of the problems is a lack of active and effective measures on the part of the state and the lack of a regulatory mechanism.

The application of digital technologies is slow, this is not enough for further development, the country needs to acquire the status of "rapid development" instead of "traditional" one. Undoubtedly, there are some prospects in the economy, but first of all, company managers need to take a risk, get out of their comfort zone and move towards the innovative development. In Russia, the concept of "smart manufacturing" is used in the transport, aircraft, rocket and space industries. At the largest enterprises, the use of standardized solutions, automated quality control systems, and energy consumption management has become a trend.

References

1. Abdrakhmanova GI, Gokhberg LM, Kovaleva GG, Suslov AB (2019) Digital economy. Internal costs for the development of the digital economy. https://issek.hse.ru/data/2019/06/05/1499451712/NTI_N_131_05062019.pdf. Accessed 27 Feb 2020
2. Aptekman A, Kalabin V, Klitsov V, Kuznetsova E, Kulagin V, Yasenvets I (2017) Digital Russia: a new reality. http://www.tadviser.ru/images/c/c2/Digital-Russia-report.pdf. Accessed 21 Feb 2020
3. Belkin A (2019) How innovations develop in Russia. https://www.vedomosti.ru/partner/articles/2019/10/09/813027-razvivayutsya-innovatsii. Accessed 25 Feb 2020

4. Borovkov AI, Klyavin OI, Maruseva VM, Ryabov YuA (2016) Digital factory of the Institute of advanced production technologies of SpbPU. http://fea.ru/news/6387. Accessed 25 Feb 2020
5. Dryanev YuYa, Kuchin II, Fadeev MA (2018) The contribution of digitalization to the growth of the Russian economy. https://issek.hse.ru/data/2018/07/04/1152915836/NTI_N_91_04072018.pdf. Accessed 25 Feb 2020
6. Federal State Statistics Service (2019) Science and innovations. https://www.gks.ru/folder/14477. Accessed 25 Feb 2020
7. Karpanova LD (2018) The digital economy in Russia: its state and prospects of development. Econ Taxes Right 2:58–69. https://doi.org/10.26794/1999-849X-2018-11-2-58-69
8. Korovin GV (2018) Problems of industrial digitalization in Russia. Proc Ural State Univ Econ 19(3):100–110
9. PWC (2017) «Internet of things» (IoT) in Russia. The technology of the future, available now. https://www.pwc.ru/ru/publications/iot/iot-in-russia-research-rus.pdf. Accessed 25 Feb 2020
10. Radio Freedom (2019) "Project": Rosstat six times underestimates the number of people who left Russia. https://www.svoboda.org/a/29713531.html. Accessed 27 Feb 2020
11. RBC (2018) The Ministry of industry and trade assessed the readiness of Russian enterprises to digitalize. https://www.rbc.ru/technology_and_media/03/07/2018/5b3a26a89a794785abc9f304. Accessed: 27 Feb 2020
12. RBC (2020) Rosstat has found a growing dependence of the Russian economy on oil and gas. https://www.rbc.ru/economics/17/02/2020/5e4a79d49a79471aa1e28c38. Accessed 27 Feb 2020
13. Tadviser (2018) Industrial Internet in Russia. http://www.tadviser.ru/a/410570. Accessed 27 Feb 2020
14. The national program "Digital economy of the Russian Federation" dated 24 December 2018. http://government.ru/rugovclassifier/614/events/. Accessed 27 Feb 2020
15. Tolkachev SA (2017) Industry 4.0 and its impact on the technological foundations of Russia's economic security. Bull Finan Univ 1(25):86–91
16. Vishnyakova AB, Golovanova IS, Aleshkova DV, Maruashvili AA, Zhelev P (2020) Current problems of enterprises' digitalization. In: Ashmarina S, Mesquita A, Vochozka M (eds) Digital transformation of the economy: challenges, trends and new opportunities. Advances in intelligent systems and computing, vol 908. Springer, Cham, pp 646–654

Factor Analysis as a Modeling Tool in the Digital Economy

O. A. Dinukova and N. N. Gunko

Abstract Digitalization implies availability of tools for strategic planning and sustainable development of socio-economic systems in regions and at enterprises. The application of digital tools allows enterprises to make better decisions in strategic planning, improve the efficiency of production processes using Big Data analytics. To identify hidden reserves, enterprises need an objective assessment of their resource potential. The research problem is based on the fact that in practice, enterprises do not fully use available tools of financial analysis: modeling and analysis of factor models to justify conclusions about results of their activities. By the analysis of production factors, production capacity and material resources are considered, and the impact of labor resources is underestimated. To identify production reserves and justify management decisions, the authors built a three-factor model of production and sales, conducted a deterministic factor analysis based on regional enterprise data, and identified areas of organizational development.

Keywords Labor factors · Deterministic factor analysis · Labor productivity · Number of employees · Product sales

1 Introduction

The functioning of any enterprise as a socio-economic system occurs when a complex of factors, both external and internal, interact with each other. The ratio of supply and demand, purchasing power, level of competition and market saturation will be the basis for determining optimal production volumes in terms of external factors. Internal sources of analysis for an economic entity are production capacities, material and labor resources, the effective management which can ensure an increase in the competitive advantages of the productions.

O. A. Dinukova (✉) · N. N. Gunko
Samara State University of Economics, Samara, Russia
e-mail: odinukova@yandex.ru

N. N. Gunko
e-mail: gunko_nn@mail.ru

S. I. Ashmarina and V. V. Mantulenko (eds.), *Current Achievements, Challenges and Digital Chances of Knowledge Based Economy*, Lecture Notes in Networks and Systems 133, https://doi.org/10.1007/978-3-030-47458-4_53

All phenomena and processes of the economic activity of enterprises are interdependent. Among many forms of natural relations, an important role belongs to cause-and-effect (deterministic) ones, in which one phenomenon generates another. With unlimited demand and limited production capacity, the production volume will be decisive. In the case of increased competition or as the market becomes saturated, the volume of sales will be the basis for planning production at the enterprise.

Based on goals and needs of an economic entity in a specific time period, it is necessary to analyze internal factors that the company can influence and focus on when making management decisions. The application of digital tools in operational activities will allow enterprises to quickly collect information about physical indicators and transit it into digitized data for further processing, exchange information in the electronic form, process information using machine learning and artificial intelligence to obtain qualitatively new conclusions.

Among the priority tools related to the use of information technologies are: mobile applications and services; automated system for storing documents, locations for stores, raw materials management, Big Data; virtual cloud infrastructure; digital HR transformation; machine learning, etc. [1].

By combining various technologies, enterprises get tools that allow them to increase the output of finished products, reduce the level of defects and the material consumption, and increase the availability of equipment. One of these tools is factor analysis, which aims to reduce the number of variables and determine the structure of relations between variables.

Deterministic factor analysis is used for a comprehensive systematic study and measurement of the impact of labor factors on the volume of production and sales of products. The use of deterministic factor analysis allows to select from the whole set of factors the main ones, better understand causes of changes in the studied phenomena, evaluate the place and role of each factor in shaping the values of effective indicators.

Deterministic factor analysis is widely used in the analysis of such economic indicators as financial stability, profit [3], in particular, this method allows assessing the cost efficiency and profitability of credit institutions [2]. It is reasonably applicable to factors affecting the economic security [8]. A number of articles is also devoted to problems of the labor productivity under conditions of changes in the production structure and the volume of final demand [5]. The authors of the article investigated a possibility of using the above method to assess the labor potential of enterprises and its impact on the volume of sales of products.

2 Methodology

The main idea of the factor analysis is that the existing relations between a large number of initial variables are determined by the existence of a much smaller number of hidden or latent variables called factors. The relevance of this research is based on the natural need of any enterprise to increase its financial performance with an optimal ratio of available internal resources.

The methodology of the factor method is based on determining the role of each of the factors in changing the performance indicator. Mapping relations describes specific dependencies of the input indicators in the form of mathematical formulas, the quantitative parameters of which are determined based on factual data [2].

Construction of a three-factor model of production and sale of products, deterministic factor analysis, implementation of accounting procedures for analyzing the model was carried out according to the reporting data of a regional enterprise. Conducting a deterministic factor analysis of this study involves the following procedures:

- construction of an economically justified deterministic factor model that links the cost of products sold, the number of employees and the labor productivity:

$$y = f(x_1, x_2, \ldots, x_n),$$

Where y is the effective attribute, x-factor features;

- determining the amount of influence of individual factors on the growth of the effective indicator by the method of absolute differences.

When using it the size of the impact factors is calculated by multiplying the absolute growth values of the investigated factor to base (planned) values of the factors that are to the right of it, and the actual values of the factors to the left of it in the model;

- formulation of conclusions based on the results of the analysis and definition of recommendations.

3 Results

To build the model, linking sales with the availability of labor, organization inputs parameters of the average number of employees, the average number of working hours, the average hourly output per worker, according to the report of the regional enterprise for the 2017–2018 (Fig. 1).

As a result of the analysis, it was found that the value of the population index was stable during the study period, the indicators of the average number of working hours and the average hourly output changed abruptly.

The next input construction factor is the annual change in the volume of produced and sold products under the influence of labor factors, which involves an analysis of the dynamics of produced and sold products, absolute indicators of the number of labor resources and their production, indicators that characterize the efficiency level of the organization's production capacity (Table 1).

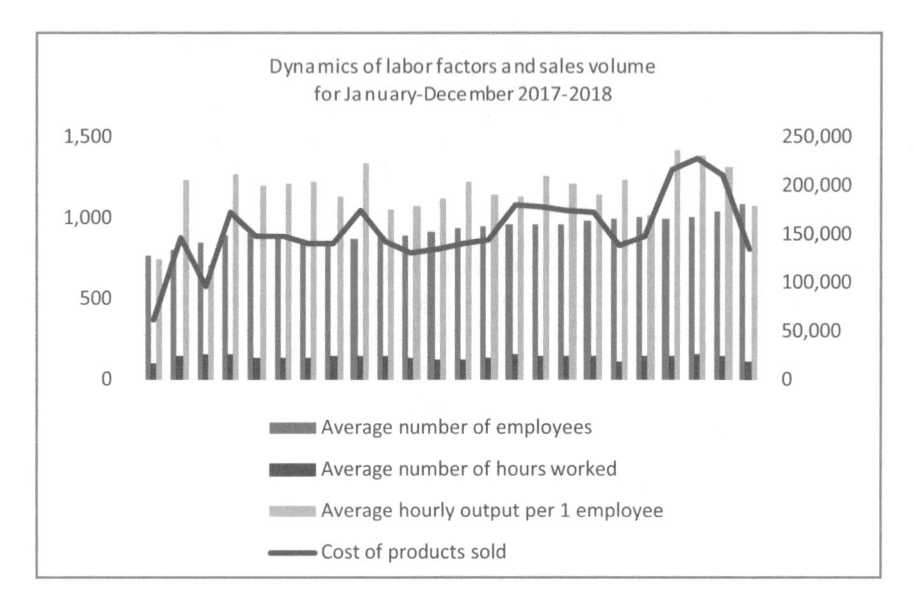

Fig. 1 Dynamics of labor factors and sales volume for January–December 2017–2018

Table 1 Data for factor analysis of the volume of produced and sold products

Indicators	2017	2018	Absolute deviation (+, −)	Growth rate, %
Cost of manufactured products, thousand rubles	1634247	2084780	450533	127,57
Cost of products sold, thousand rubles	1632846	2063694	430848	126,39
Average number of employees, people	861	991	130	115,10
The number of working person-hours	1460047	1690593	230546	115,79
Average number of hours worked	1695,76	1705,95	10,19	100,60
Residual value of fixed assets, thousand rubles	661690	599899	−61791	90,66
Return on funds, rubles	2,47	3,48	1,01	140,89
Capital productivity, rubles	768,51	605,35	−163,16	78,77
Average hourly output per 1 employee, rubles	1118,35	1220,69	102,34	109,15

Source authors

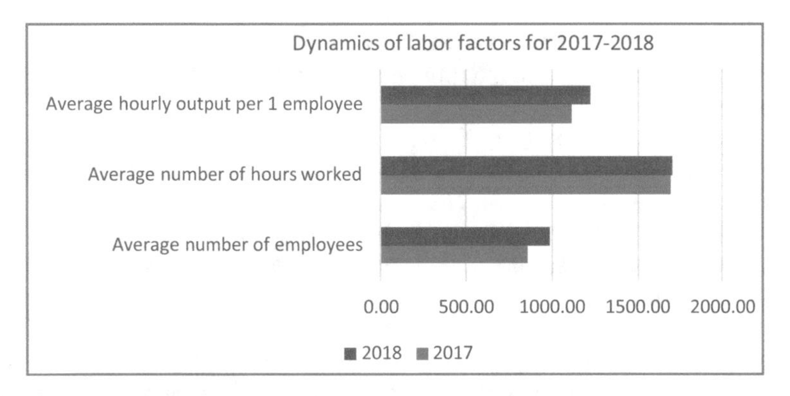

Fig. 2 Dynamics of labor factors for 2017–2018 (*Source* authors)

The impact of labor factors on the increment of sales in the amount of 430848 thousand rubles is calculated. Its change was influenced by the following factors:

- increase in the number of employees,
- increase in the average number of hours worked,
- increase in average hourly output per 1 employee (Fig. 2).

Using the method of absolute differences, we calculate the influence of factors on the growth of the effective indicator. To do this, we multiply the absolute increase in the value of the studied factor by the basic values of the factors to the right of it, and by the actual values of the factors to the left of it in the model.

The impact of changes in the number of employees was 246538 thousand or 57%, the impact of changes in the average number of hours worked was 11292 thousand rubles or 3% and the impact of changes in the average hourly output per employee was 173023 thousand rubles or 40%, respectively. As a result of the analysis, it was revealed that the increase in the sales volume was carried out by 40% due to the production intensification, i.e., an increase in the labor productivity, and 57% due to an extensive factor—an increase in the number of employees. In addition, we also noted the growth of the labor productivity with an increase in the capital productivity and a decrease in the index of capital equipment.

4 Discussion

The success of enterprises depends a lot on their ability to effectively process large amounts of information and analyze the obtained indicators. One of the methods of data processing is deterministic factor analysis, which allows us to understand causes of changes in the studied phenomena.

The scope of epy factor analysis is wide enough to enables a choice between applying basic psychometric approaches, which can be complex and time-consuming, and applying simple and fast, but less accurate approaches, such as sum or average score. Moderated nonlinear factor analysis provides more accurate estimates than traditional methods [4].

The use of the factor analysis allows selecting main factors from the whole set, and processing data from questionnaires [6]. The use of this method is not limited to the physical features of processes. The factor analysis is both a powerful method for monitoring processes and is applicable to the design of systems of various purposes. In medicine, the factor analysis has been used for more than 10 years in various fields of healthcare [9].

In psychology, this method is used to study personality properties and behavior by answering questions from psychological tests. In sociology, the factor analysis is actively used to process survey results, which is one of the main tools for studying the public opinion [3].

In the work of B. Xu, J. Sandra-Garcia, Y. Gao, and X. Chen, it is shown that labor factors have a significant impact on all aspects of the overall factor productivity. Improving the effectiveness also varies in the same direction as the increase in the labour productivity. Human resource management may be the main reason for poor performance [11].

There are many studies in the scientific literature that examine these factors and their relation to the labor productivity. In the study of A. Kazaz, S. Ulubeyli, T. Acikara, and B. Er, factors were classified into various groups and ranked according to their levels of importance [7].

In their work M.-A. Tarancon, M.-J. Gutierrez-Pedrero, E. Callejas, and I. Martines-Rodriges, investigate a relation between the labor productivity and the production efficiency, which also helps to identify key sectors for improving the efficiency, taking into account the influence of labor factors [10]. This article examines the relation between the labor productivity and the degree of production efficiency for 24 countries within the European Union (EU) over a period of 17 years.

To identify production reserves and justify management decisions, it is possible to use a three-factor model of production and sales. Assessing the impact of labor factors allows you to identify a direction of organizational development.

5 Conclusion

The functioning of the enterprise as a socio-economic system is based on the interrelation of all components, which determines the necessary activity result. Reducing the uncertainty of forecasting management processes is achieved by using economic and mathematical models, which allows us to justify the concept of improving the labor efficiency and identifying reserves in modern production facilities.

The very definition of the term “digital economy” means an increase in the efficiency of enterprises based on analysis results of processing large amounts of information. The use of IT technologies for processing large amounts of data and identifying the most relevant areas leads to the transformation of the business model, the definition of a new format for conducting business. Digital transformation of enterprises is an expansion of opportunities for managers and employees through new technologies, an opportunity to increase productivity and business development.

As a result of the analysis, the dependence of changes in factor and performance indicators in the initial model was revealed. The extensive direction of organizational development is determined, as a result of which there is an increase in resources. Thus, labor factors influence the volume of production and sales of products. Assessing the degree of their influence and influencing them, it is possible to increase the efficiency of the enterprise as a whole.

References

1. Bakanach O, Proskurina N, Kornev V (2019) Deterministic factor analysis of financial results of credit organizations. Bull Samara Municipal Inst Manag 2:96–104
2. Fomina EE (2017) About the possibility of applying of factor analysis to social data processing. Modern Scientific Research and Innovation. http://web.snauka.ru/issues/2017/01/77488. Accessed 22 Feb 2020
3. Gottfredson N, Cole V, Giordano M, Bauer D, Hussong A, Ennett S (2019) Simplifying the implementation of modern scale scoring methods with an automated R package: automated moderated nonlinear factor analysis (aMNLFA). Addict Behav 94:65–73
4. Hudcovsky M, Labaj M, Morvay K (2017) Employment growth and labour elasticity in V4 countries: structural decomposition analysis. Prague Econ Pap 26(4):422–437
5. Jalehimanesh F, Soltani M, Dastoorpoor M, Moradi N (2019) Factor analysis of the Persian version of the voice disability coping questionnaire. J Voice. https://doi.org/10.1016/j.jvoice.2019.06.001. Accessed 20 Feb 2020 (In Press)
6. Kabitova EV (2018) Development and application of an individual model of discriminant analysis and a deterministic factor model for assessing the financial stability of an economic entity. Econ Sustain Dev 4(36):346–350
7. Kazaz A, Ulubeyli S, Acikara T, Er B (2016) Factors affecting labor productivity: perspectives of craft workers. Procedia Eng 164:28–34
8. Korshakevich I, Pashkovskaya O (2017) Using factor analysis in medical information systems. Curr Probl Aviat Cosmonautics 2(13):363–364
9. Krasnova OV, Shkabienko AS (2017) Application of factor analysis for the study of economic security indicators. In: Lavrikova YuG (ed) Development of territorial socio-economic systems: issues of theory and practice. Institute of Economics of the Ural branch of the Russian Academy of Sciences, Ekaterinburg, pp 99–102
10. Tarancon M-A, Gutierrez-Pedrero M-J, Callejas E, Martines-Rodriges I (2018) Verifying the relation between labor productivity and productive efficiency by means of the properties of the input-output matrices. The European case. Int J Prod Econ 195:54–65
11. Xu B, Sandra-Garcia J, Gao Y, Chen X (2020) Driving total factor productivity: capital and labor with tax allocation. Technol Forecast Soc Change 150:119782

Digital Platforms and Sharing Economy

Waste Management and Circular Economy in the Public Discourse in Russia

B. Nikitina

Abstract The waste management system, badly organized in Russia, finally came to the critical point with wide public protests. While EU develops its New Waste Framework Directive, working on extended producer responsibility and waste prevention activity, Russia making faked system of new waste taxation without changing of core waste treatment institutions. While EU countries create waste prevention programs using power of crowdsourcing, the Russian population was set aside from the discussion about planned transformations. Development of circular and sharing economy as a way of waste prevention and reduction is hardly discussed in Russia. Russians have a lot of spontaneous practices of resources savings, when they use Internet platforms to organize different activity to reduce, reuse and recycle municipal waste, but there is no legislative support or State investments in such kind of projects. Civil society in Russia is again ahead of its state in that field. The article makes a contribution to analyze the public discourse about significance of Circular and Sharing economy methods in the field of Municipal Waste management sphere in Russia.

Keywords Circular and sharing economy · Municipal Waste management · Public discourse about waste management

1 Introduction

Until recently, the problem of municipal waste was on the periphery of the attention of the Russian population. However, over time, the amount of waste increased so much that it was necessary to expand the territories for landfills. As a result, there was a movement of "garbage protests", especially active in those regions that are close to the maximum producing municipal waste produsing capital cities- Moscow and St. Petersburg. Garbage that can no longer be placed in the vicinity of Moscow is being taken further and further from the Moscow region, then to the Yaroslavl and

---

B. Nikitina (✉)
Samara State University of Economics, Samara, Russia
e-mail: belanik@yandex.ru

S. I. Ashmarina and V. V. Mantulenko (eds.), *Current Achievements, Challenges and Digital Chances of Knowledge Based Economy*, Lecture Notes in Networks and Systems 133, https://doi.org/10.1007/978-3-030-47458-4_54

Kaluga regions, which have caused powerful social and environmental protests there [12]. Similar victim of poor management of the capital's garbage is the Arkhangelsk region, which became famous for the powerful protest of the population against the construction of the so-called Ecotechnopark Shies [16].

What is happening in the field of waste management has been brewing for a long time, but the Russian government postponed the decision until the last minute, but starting in 2019, the "garbage reform" was announced. Finally, an accelerated modernization of the waste management sphere was announced, but its essence in practice was reduced to "renaming" the existing household service without real, understandable and noticeable changes for Russian citizens. In some cases, the result of the "garbage reform" was a change in the location of garbage sites. Sometimes boxes or nets for collecting plastic or paper that were installed by someone in the previous period completely disappeared, sometimes they changed the design and name of the owners, and in some places the schedule for collecting garbage changed. The most noticeable change for the citizens was a big change in the appearance of garbage trucks, which received a new coloring and unusual logos and images. However, in large cities, there were no major changes for residents of Central areas.

As for the outskirts or rural areas that are closer to landfills, the problems for residents have only increased, since garbage from natural dumps has been brought to landfills. As a result of clearing natural areas, landfills began to overflow, the smell increased, especially when fires began to occur. The expansion of landfills has contributed to the deterioration of the landscape, as well as the reputation of the area, which has affected the prices of apartments or houses on the outskirts near waste management facilities.

The most serious consequences of the proliferation of landfills were brought to rural residents when new landfills were built near their place of residence, or when overcrowded old landfills that were left unattended and properly reclaimed began to burn. It is obvious that desperate residents of these territories began to protest against environmental discrimination in defense of their right to life and health [16].

To some extent, this social protest can be called a primitive and selfish activity, driven by the ideology of "Not in my backyard", defending only its own interests. But what is noticeable here is that protesters often switch to eco-friendly practices in their own homes. Some representatives of protest movements began to turn to the implementation of spontaneous practices of circular economy, such as separate collection of waste paper, plastic, etc.

However, such activities often face a lack of proper infrastructure, i.e. a business that can organize the collection, transportation and processing of separately collected waste. The move to incineration, proposed by the authorities as an alternative means of solving the problem—and in fact only for a small reduction in the volume of waste-was a trigger for the inclusion of new social groups in the protest, who quickly began to understand that their personal and family health could be seriously affected by air pollution. As a result, the discontent of the population is growing, the situation is becoming more tense and alarming for the local population, the continuation of the previous waste management is categorically rejected by them, but there is no understanding of how to change the situation, where to move and develop correctly.

The reasons for this impasse are that the problem of waste generation is very complex, related to how society as a whole functions—from managing the household of the most disadvantaged population to the work of the most high-tech institutions related to the field of waste management.

It should be recognized that this area is so complex, multi-layered and heterogeneous that its full understanding is beyond the power of individual narrow specialists without the participation of various stakeholders, and only collective discussion and awareness by entire significant social groups that have influence on managerial decision-making can bring Russian society closer to solving this problem.

At the same time, we should pay attention to the specific situation when the implementation of the principles of circular and sharing economy, which have long been recognized as effective and necessary in European countries, is not considered in the official Russian discourse and even critical discourse as a mainstream direction.

As for the state policy in the field of waste management, it can be described as insufficiently environmentally consistent and focused on resource-intensive and corrupt waste management models. To a certain extent, it can be contrasted with public discourse that focuses on a more socially-economically and environmentally acceptable way of developing the garbage industry. However, the question is the degree of maturity of such a discourse, the availability of a clear and accessible to the public theoretically justified system of waste management, which could act as a clear real alternative to what the state offers. To solve this problem, it was decided to conduct a study of critical discourse on the sphere of waste management in one of the most radical Russian media—radio stations "Echo of Moscow" and "Echo of St. Petersburg".

2 Methodology

Based on these key points, an analysis of the content of radio broadcasts was carried out to find out which ways to solve the problem are lobbied by special stakeholders, and which of them are discredited, which stages of waste management are considered as critical, and which are unfairly ignored within the critical public discourse of modern Russian society.

3 Results

During the analysis and coding of texts, comments related to the above-mentioned stages of waste management were highlighted. Then all the encoded text fragments were numbered (they could contain several codes) and the weight of a particular code was calculated as a percentage. Encoding and distribution of the text volume by codes in accordance with the waste management stages in accordance with the Ad Lansink "conception of Ladder" was performed as the initial stage of the analysis.

This distribution of codes by weight is shown below. Social and political aspects and interpretations of these processes will be given in the results discussion section.

Level 1—The Collection (12%)

Most often, guests of all radio programs discuss the topic of insufficient infrastructure development for separate waste collection, as well as insufficient incentives for the population to separate waste. At the same time, some experts point to the lack of administrative regulations, while others mention the lack of a control system. The most frequently mentioned is the economic aspect, i.e. the lack of economic incentives to reduce the amount of waste generated by each individual resident. When discussing the issue of waste generation, the issue of consumer culture and environmentally friendly product selection is also discussed. The issue of plastic as a source of pollution and Vice versa—the potential of plastic as the most suitable material for recycling is actively discussed. As for waste collection companies, it is not profitable to reduce the volume of waste.

Level 2—The Accumulation of Waste (3%)

The idea that waste accumulation is a fairly important stage of competent logistics organization in the field of waste management is voiced in isolated cases. Guests of radio programs rarely mention the need to create infrastructure facilities of regional and municipal scale, where recyclable waste fractions can accumulate. Only a few brief references to the experience of European countries were found about the possibilities of collecting, accumulating and redistributing unused and unclaimed items that are potentially waste. There is no discussion about the importance of creating such sites in Russia. Accordingly, problems related to the possibility of storing separately collected waste fractions in residential areas are more often mentioned. The possibility of forming a business that accumulates recycled materials, as well as resources that can be used as a kind of "second-hand" for "cross-consumption" is critically evaluated.

Level 3—Transportation (8%)

According to the guests who participate in the radio programs we are studying, the formation of the Institute of regional operators in the case of monopolization of the competitive garbage truck market is a big problem. After the announcement of the "garbage reform", the number of small businesses that have gradually developed in cities has significantly decreased due to the identification of profitable commercial niches associated with the circular economy. Many experts note that the growth of circular practices "from below" was actually suspended by the introduction of state transport monopolists (regional operators) "from above". These monopolists are interested in maximizing transportation, rather than reducing the volume of waste at the place of its formation by removing secondary raw materials from their composition. Thus, the principal depravity of the waste management system proposed from above is revealed. At the same time, paradoxically, it is the regional operators who should be responsible for the removal of waste, the rate of formation of which is determined without taking into account the possibility of allocating recyclable components. The corruption nature of this legislation is one of the most common issues heard in radio programs about "garbage reform" on the Echo of Moscow radio station. In particular, the refusal of a regional operator to accept waste based on the actual volume of its formation may lead to a collapse: since the conclusion of a contract with a regional operator in accordance with Russian law is mandatory, those who have not signed a contract with a regional operator will not be served. Another problem that is clearly recognized by the guests-experts of radio broadcasts is the lack of waste processing enterprises in the regions, as a result of which the organization of separate waste collection on the ground can lead to an increase in tariffs.

Level 4—Processing (23%)

In the framework of the analyzed array of speech transcripts, much attention is paid to the waste treatment process itself, technological solutions, and their cost. It is often mentioned that the absence of a system for separate collection of household waste at the place of its formation leads to an increase in the cost of processing, which leads to an increase in tariffs. The cost of high-tech waste processing complexes is also discussed. There are also concerns that the cost of building them will not always cover the revenue from the sale of the received materials and goods due to the relatively low depth of processing. In turn in the absence of waste processing plants in the region the cost of processing and transportation of sorted waste makes this type of business unprofitable.

Level 5—Disposal (34%)

The most frequently discussed idea in the speeches of radio guests is a critique of existing statistics on waste generation and the capacity to remove and recycle them. It is clear that the existing waste management infrastructure does not cover the entire volume of waste generated. In critical discourse, we are talking about the need to find the most successful locations for waste landfills. The necessity of forming the waste processing industry as a tool for reducing the volume of waste generation is discussed. From the same point of view, the possibility of reducing the volume of waste due to incineration is considered. In this issue, most of the discourse expressed fear of the consequences of the existence of waste incineration plants, a closely related topic is the low profitability of incineration. The lack of financial resources for the development of the processing industry in the regions and the unfavorable investment climate significantly hamper the implementation of projects for the construction of plants that meet modern environmental safety requirements.

A number of experts focus on the economic aspects of waste management. Waste disposal for private businesses at the stage of formation in the absence of associated infrastructure (communications, stable volumes of raw materials supplies, sales market) is unprofitable without adequate support from the state. The system of collection of recycling and environmental payments intended for the development of waste disposal infrastructure has a number of drawbacks: since 2014, waste producers and importers have no alternative to organizing their own disposal; the amount of collection does not always adequately reflect the real costs of waste recycling; not always taken into account the interests of domestic producers of technical equipment and its industries; there is no Fund utilization and lack of transparency in the allocation of funds; there are gaps in the law when establishing relationships between importers, manufacturers, consumer goods and associations, representing their interests in the sphere of payment of environmental charges and self-organization of waste management; there is no description of the mechanism for independent organization by the manufacturer or importer of the collection and processing of goods and packaging.

Level 6—Decontamination (11%)

A separate topic that occurs periodically in discussions is the determination of the amount of payment for waste disposal in terms of fully determining the cost of reclamation of waste disposal facilities. Often underestimating this cost leads to subsequent environmental problems for the local population.

Level 7—Placement (9%)

The inclusion of fees for waste disposal in a single tariff deprives regional operators of incentives to reduce the negative impact of waste disposal facilities on the environment, contributing to the redistribution of responsibility between waste owners. Most landfills and landfills are not equipped in accordance with environmental safety requirements, and soil protection measures are not systematic. The above-mentioned thematic dominants are most often found in discussions about the organization of the waste management system—we deliberately excluded political interventions regarding the system's features from the analysis.

According to recent studies [2, 5, 8, 9, 13], the listed thematic dominants identified in the coding of radio broadcasts partially coincide with the opinion of specialists in the development of waste management systems, but many topics are not mentioned in radio broadcasts.

4 Discussion

The coding results presented above clearly demonstrate that the socio-political discourse of modern Russian society is focused on the most pressing issues of waste disposal (which take the second biggest part of encoded texts), with an emphasis on the definition of new waste disposal sites and the quality of remediation of old landfills. In close conjunction with the discussion of the locations of new landfills, the issue of creating waste sorting complexes is being discussed. The possibility of separate waste collection is also actively discussed, but in this topic, garbage sorting is considered only on the basis of the "either—or" principle, and this is despite the fact that it has already been proven that the most effective sorting is carried out after the population produces the primary sorting.

Incineration wastes attract a lot of attention as a separate topic, but the issue of the quality of incinerated garbage is also very poorly addressed. It becomes clear that there is no clear understanding in public discourse that only sorted garbage is burned in developed countries. In the Russian discourse, much greater hopes are associated with miracle technologies that will free the population from the need to participate in the waste management process.

In this context, the very small importance attached to the issue of the allocation of organic waste as a valuable fraction becomes noticeable. In fact, a very small part of the participants in radio broadcasts spoke about composting, which indicates a low level of biological culture of the modern urban population.

In general, the current public discourse on the implementation of the reform in the field of waste management is dominated by a critical attitude towards the effectiveness of the government's activities. First of all, the element of corruption is emphasized, when large projects are assigned in one way or another to structures affiliated with the country's top—ranking leaders-from the President to the Prosecutor General. Corruption is seen in shifting goals from terminal goals to instrumental goals. At the same time, concerns are expressed on both occasions—both the failure to achieve the goals set, and the consequences if they are achieved. If the goals are met, experts anticipate both negative environmental and economic consequences.

Informants are also critical of the process of forming legislation in the field of waste management. However, none of the invited experts show deep and comprehensive knowledge of the legislation.

Problems in the field of waste management are identified by experts and guests of radio broadcasts starting from the stage of development of territorial waste management schemes. In practice, most of these documents were not discussed with the public in the time frame established by the state—from 2017 to 2020. Conflict situations are associated with both the location of new waste management facilities and the nature of their impact on the environment. Most discussions are chaotic due to the fact that the terminology itself is undeveloped or developed incorrectly. In particular, the type of activity of enterprises that are intended for waste processing is now unclear, since 2019, the legislation considers waste incineration as a form of garbage processing.

The very quality of the development of territorial waste management schemes for each region of the Russian Federation causes complaints. There is no complete understanding of the quality and quantity of waste produced.

Experts, participating in radio-programs, see positive prospects in the introduction of more advanced technologies. However, there are two directions: some experts focus on pyrolysis and thermal gas processing, while in their opinion the problem in the field of waste management system will be radically different. Most experts suggest that it is necessary to develop recycling, which will reduce the volume of waste generation. For this purpose, the emphasis is on the introduction of separate waste collection. Much attention is paid to the fact that each person is able to change the amount and quality of waste generated in the household, provided that the waste is collected separately, as well as changing the style of consumption.

Thus, circularity as a category is now considered in most statements as something positive, as a prospect of lifting the crisis. However, the depth of consideration of this issue is insignificant. Some informants point to the lack of waste processing capacity, while others point to the lack of raw materials for such enterprises. At the same time, the issue of the quality of recyclable materials and the market for such enterprises is considered much less frequently. There are also no questions about the need for the state to encourage products with the inclusion of recyclable materials, which is a political tool for interfering in market relations. There are no claims against the state here. Also, the question of the main, fundamental question of who should be responsible for sorting waste remains out of focus. If this responsibility is transferred from the consumer to any other actor—whether it is the owner of a waste sorting or waste processing enterprise, the manufacturer of goods or the manufacturer of packaging—in all these cases, the style of consumption itself will not be sufficiently controlled by the society, will not be subjected to its pressure-due to insufficient awareness and interest of the population. In this regard, it should be pointed out that the level of criticism on this issue clearly does not correspond to the degree of its seriousness.

The question about recycling as a core moment of waste management system doesn't still have a simple interpretation [1], Despite the fact that on average in the European Union countries in 2018, almost half of the waste (47%) is recycled [3], the economy of this region is still not circular in the full sense of the word. Indeed, the European policy in the field of waste management has long been based on the principles of consistency with the hierarchy of waste A. Lansink, identifying the possibility of using not only the most effective waste management, but also to prevent their occurrence [14]. At the same time, the complexity of waste prevention calculations is obvious, which does not eliminate the need to take this important factor into account [4].

Even before the transition to the circular economy model, the question was raised in European countries that we should not just study the environmental and economic consequences of an irrational approach to resources and their ruthless transfer to the category of waste or even recyclable materials. Environmentally concerned public formed the very discourse of the need to move to a specific policy in the field of consumption, which would stimulate the most rational and careful consumption, would stimulate it both morally and economically [7].

Note that environmentally friendly consumption is seen in European countries as more socially oriented, when the goal is not to reduce opportunities to meet existing needs, but to transform them, i.e., to give up excessive desires imposed by the society of "consumption" [10]. In some cases, such refusal entails not only moral satisfaction, but also a healthy lifestyle [11].

Understanding that the circular economy cannot be based solely on technological solutions leads to the need to organize a special infrastructure for implementing socially friendly actions that allow achieving environmental goals at the same time [6]. Experts are also critical of the population, constantly pointing out the inability of citizens to maintain cleanliness and responsible attitude to the separation of waste. At the same time, there were also suggestions of positive role models, but they were mainly reduced to individual handling of waste and refusal to buy disposable items.

It should be noted here that the ideas of the sharing economy are very poorly represented in critical public discourse. Although there are calls to donate your own unclaimed items to charity, no more detailed discussion of this issue took place during the year within the framework of the "Ekho of Moskow" radio station. It also notes a very eloquent fact-there are no ideas about changes in lifestyle, about the need for an independent transition to secondary consumption, to the use of things that were in someone's use or are your own inheritance.

To a lesser extent, informants address the issue of changing consumption patterns. An interesting fact is that the examples given regarding the Japanese leaving home food in favor of eating in restaurants in order to avoid problems with waste disposal are perceived with laughter and considered as a joke.

It is interesting that informants do not actually reach the level of system understanding of the situation. If informants start to talk about the situation from an economic point of view, then most often the economy of construction and operation of polygons is considered. At the same time, the economy of housing communities is completely out of focus. The most important thing is to completely ignore the role of the municipality in the process of organizing waste management, despite the fact that a few years ago it was the municipal authorities that played the main role in this issue. The only function that is left at this level of government is the purchase of garbage containers. At the same time, participants of radio broadcasts do not pay special attention to what is purchased for waste management. Thus, the issues of convenient and understandable infrastructure and everything else is left to the discretion of the population, which is then also accused of a low level of environmental culture. These conclusions, by the way, are completely inconsistent with the latest polls [15].

5 Conclusion

Thus, today the socio-political discourse in the field of waste management, presented in such an oppositional media as Echo of Moscow (including its branch Echo of St. Petersburg), is quite broad, representing almost the entire range of problems in the waste management system. At the same time, rational ideas, representatively articulated here, scattered chaotically. A systematic view of the nature of the crisis and how to get out of it is rarely presented, as a rule, in some radio broadcasts programs the emphasis is on some part of the waste management system, not at the system as whole. There are bulk of different critical comments, but there are only few substantive proposals that claim to fundamentally solve the problem of waste management. Such proposals are rare and go beyond the ideology of the development of circularity. The rest of the proposals are mainly limited to the development of separate waste collection and subsequent recycable waste processing, but they do not claim to be a final and complete solution to the problem.

As for the proposals related to the development of the sharing economy as one of the ways to resolve the problem of waste management, this topic is discussed very poorly. While a lot of attention is paid to the discussion of the plastic problem, in particular, the issues of avoiding the use of disposable tableware, tubes and plastic bags are discussed, there is little discussion of the need to change the packaging material as such, and there is almost no discussion of changing the mode of functioning of households that generate certain types of waste.

Quite often it is said that you should not throw things away and should give them to charity, but the topic of the need to become acceptors of things that were in use, almost does not rise.

Thus, there is no complex discussion about the waste management system at several levels in a public critical discourse at the most popular Russian broadcasts station.

1. There is no understanding of such a basic level as the primary waste collection system.
2. The culture of waste management is considered as something that naturally appears at the time of formation of a particular waste management system. At the same time, in a situation where the technology of waste collection, processing and disposal is not obvious to its creators, which is clearly demonstrated by the state of territorial waste management schemes, the population cannot correctly "guess" the correct behavior. And even more so, it is impossible to guess it in the absence of economic incentives, which should support the correct behavior. In a situation when ecologists trying to implement more environmentally friendly behavior in the field of waste management and even they found it is difficult and inconvenient, it is unlikely to expect changes in behavior population at whole.
3. There is no wide discussion about ideas that the waste management system is part of the economic system as a whole, that waste is generated in the life support system, which is formed depending on the nature of the economy.

Acknowledgements The research conducted for this article is part of the project "Readiness of local communities to develop joint consumption and management of solid waste through the development of IT as a strategic factor influencing the socio-economic development of Samara", funded by Competition of fundamental research conducted in 2018 by RFBR together with the Subjects of the Russian Federation, grant number 18-411-630003.

References

1. Bartl A (2014) Moving from recycling to waste prevention: a review of barriers and enables. Waste Manag Res 32(9):3–18
2. Ermolaeva Yu (2019) Modernizing Russia's waste management industry: the scope of expert analysis. Vestnik Sociol Inst 10(3):131–150. https://doi.org/10.19181/vis.2019.30.3.596 (in Russia)
3. Eurostat (2019) Recycling rate of municipal waste. http://ec.europa.eu/eurostat/web/products-datasets/product?code=sdg_11_60. Accessed 17 Mar 2019
4. Haupt M, Vadenbo C, Hellweg S (2016) Do we have the right performance indicators for the circular economy? Insight into the Swiss waste management system. J Ind Ecol 21(3). https://onlinelibrary.wiley.com/doi/full/10.1111/jiec.12506. Accessed 17 Mar 2019
5. Hobson K (2019) Small stories of closing loops: social circularity and the everyday circular economy. Climatic Change. https://doi.org/10.1007/s10584-019-02480-z. https://link.springer.com/article/10.1007/s10584-019-02480-z. Accessed 14 June 2019
6. Hult A, Bradley K (2017) Planning for sharing – providing infrastructure for citizens to be makers and sharers. Plan Theory Pract 18(4):597–615

7. Hultman J, Corvellec H (2012) The European waste hierarchy: from the sociomateriality of waste to a politics of consumption. J Ind Ecol 44:2413–2427
8. Kudryavtseva O, Solodova M, Korenevskaya D, Kutubaeva R, Tishkova A, Shchevyeva L (2018) Perspectives of the solid waste management in Moscow. Scientific research of faculty of economics. Electron J 10(2):64–87 (in Russia)
9. Latypova M (2018) Analyzing the development of the municipal solid waste system in Russia: challenges, prospects and evidence from Europe. Natl Interests Priorities Secur 14(4):741–758 (in Russia)
10. Michael K, Ho P (2020) Unravelling potentials and limitations of sharing economy in reducing unnecessary consumption: a social science perspective. Resour Conserv Recycl 153:104546
11. Morone P, Falcone P, Morone E (2018) Does food sharing lead to food waste reduction? An experimental analysis to assess challenges and opportunities of a new consumption model. J Cleaner Prod 185(1):749–760
12. Nikitina B (2019) "Zimnyaya Vishnya" and the "Yardovo" landfill as two sides of the same coin: difficulties in comprehending the obvious. Vestnik Instituta Sotziologii 10(1):28–57 (in Russia)
13. Nikitina B, Korsun M, Sarbaeva I, Zvonovsky V (2019) Development of the practice of sharing economy in the communicative information environment of modern urban communities. In: Ashmarina SI, Mesquita A, Vochozka M (eds) Digital transformation of the economy: challenges, trends and new opportunities. Springer, Cham, pp 376–394
14. Pires A, Martinho G (2019) Waste hierarchy index for circular economy in waste management. Waste Manag 95:298–305
15. Shabanova M (2019) Citizens' socio-economic practices as a resource to alleviate the waste issue in Russia. Sociol Stud 6:50–63
16. Steshenko A (2018) Analysis of protest activity of the local population against landfills in environmental conflicts in the Moscow region, Scythian. Stud Sci Issues 6(22):15–20

Digital Platforms and Banks

G. Panova

Abstract The article presents some issues of innovative modernization of banks and banking. It investigates debatable issues of theoretical understanding and practical use of digital platforms and banking ecosystems. The article shows how technologies change the financial market and the perception of bank customers, stimulate entrepreneurial activity and change the appearance of traditional financial markets in the process of implementing innovative business models (so-called ecosystems of modern banks). The author assesses the role of banks in the process of technological transformation of financial markets in Russia. The main research objectives are: to identify and quantify customer needs, as well as obstacles to the technological transformation of banks in Russia; to analyze market conditions and regulatory support for banking activities and determine its further development direction; to evaluate policy options for banks in terms of ecosystem development. The main purpose of the article is to identify the current problem of the bank transformation and show directions of its further development. In general, the research was aimed at bridging the knowledge gap between technology developers, bank employees and other financial market participants, as well as at developing the optimal policy of the Russian financial regulator in modern conditions.

Keywords Digital platform · Ecosystem · Financial market · Bank · Financial intermediation

1 Introduction

All economy sectors are subject to shocks. According to various estimates, currently about 5%–10% of the traditional economy is affected by technological changes or technological explosion. And for 90% of the economy, change is still ahead. This is a huge field of opportunities for businesses and entrepreneurs in all sectors of the economy without exception. The transformation of financial and credit institutions

G. Panova (✉)
Moscow State Institute of International Relations (MGIMO University), Moscow, Russia
e-mail: gpanova@mail.ru

S. I. Ashmarina and V. V. Mantulenko (eds.), *Current Achievements, Challenges and Digital Chances of Knowledge Based Economy*, Lecture Notes in Networks and Systems 133, https://doi.org/10.1007/978-3-030-47458-4_55

in the context of the fourth industrial revolution feeds discussions of scientists and practitioners who argue about the banks' prospects in the near and distant future. Traditional banks are concerned that 23% of their business may be at risk due to the development of financial technologies. FINTECH companies expect that they will be able to capture 33% of the traditional banking business [5]. According to Citibank's forecasts, the growing influence of financial technologies may lead to job losses for 30% of bank employees by 2025 [2]. In these conditions, banks need to develop cooperation with FINTECH companies. On the other hand, large banks can open up huge opportunities for FINTECH companies and give them access to global financial markets.

A futuristic view of the future of banks is being formed today. Banks are becoming more customer-oriented, and digitalization of their operations allows them to provide the best, emotionally pleasant customer service in terms of speed, convenience and security of operations in digital channels [10]. According to many experts, Russia is currently at the forefront of digitalization of financial services, being one of the world leaders. Moscow, for example, is one of the top three cities in the world for the implementation level of digital technologies in urban infrastructure. The Russian retail banking business has also long established itself as one of the most advanced in the world.

Under the influence of technological innovations, banks and their businesses are changing dramatically [4]. Those banks that have great technological potential use their capabilities to search for fundamentally new business models. They analyze activities of aggregators, marketplaces, ecosystems; create technological services for customers, etc. Banks operating in this new paradigm are seriously transforming their activities and transforming their business models.

2 Methodology

In the research work, the author uses methods of comparison, benchmarking of market practices and recommendations of international consultants. The study was conducted to determine the current situation and prospects for the development of digital platforms for banks and the potential market response, taking into account the impact of systemic and individual risks (organizational; technological; market conditions; regulatory; operational risks, etc.). Examples of technological and innovative state support for banks were studied in order to identify and determine various forms of bank behavior in the market. We also analyzed comparable approaches in other countries (for example, in Germany, Japan, and the United States) in order to understand their orientation and find out how to apply their best practices and experience in Russia. The analysis was based on reports and other materials from the World Bank, the Bank for International Settlements, PWC, the Association of Russian Banks, Sberbank, VTB Bank, Tinkoff Bank (Russia), Citibank, and others.

3 Results

3.1 The New Digital Reality: Russian Experience

An alarming global trend of recent years is a decline in customers' confidence in banks, as it is evidenced by public opinion polls. Sociologists have even created a special index—the Millennial Disruption Index—to assess potential changes in the world economy. In the context of a fundamental transformation of the banking activity, the survey results showed a record low level of confidence in banks: 73% would prefer to use the new financial services of large technology companies, rather than deal with traditional banks. At the same time, according to the World Bank, in many regions of the world, 2.5 billion adults (it is about half of the working-age population) do not have access to the global financial system [9]. Russia is developing a bank-oriented model of financial markets. The banking sector is one of the leaders in applying the achievements of the digital economy. In these conditions, large banks occupy a special place in the country's economy.

The Bank of Russia, as a mega-regulator of financial markets, develops financial infrastructure and coordinates actions of financial market participants and government agencies. The Bank of Russia's strategy for the development of financial technologies identifies legal regulation, information infrastructure, and information security as key issues. "The main directions of financial technology development for the period 2018–2020" [1] are synchronized with the national program "Digital economy of the Russian Federation" [3] and other projects in the field of financial technology development [5, 8, 11]. Implementation of projects under this strategy in the field of financial technology contributes to the digitalization of the financial sector and improves the availability of services of financial-credit institutions to the population in all regions of the country.

The main goals of the Bank of Russia in implementing the program documents on the development of financial technologies include promoting the competition in the financial market, increasing the availability, quality and range of financial services, reducing risks and costs in the financial sector, as well as increasing the level of competitiveness of Russian technologies in general. The Bank of Russia considers the financial industry in the future as a driver of the economic growth and development of the financial services sector, which will create a broad infrastructure that will integrate other (including public) services. The development of financial technologies for the megaregulator of financial markets (the Bank of Russia) and for market participants is becoming a fundamental direction for changing their business models and developing the financial market as a whole.

3.2 Digital Technology. New Challenges for Banks

Digital technologies affect the economy with an emphasis on organizational changes in financial markets and the use of digital platforms in Russia. Technologies are radically changing the demand for financial services and the availability of services for consumers [6]. The most promising in terms of interests of financial and credit institutions and their clients are Big Data technologies, artificial intelligence, robotics, biometrics, cloud technologies, open interfaces (APIs) and distributed registry technologies. The Bank of Russia uses them in its own interests, including to organize the interaction between banks, clients and government agencies (operators of state information system).

The key elements of the digital financial infrastructure are: platforms for remote customer identification, fast payments, platforms for the marketplace of financial products and services, as well as new platforms based on distributed ledger technologies and cloud technologies. The key element in the financial market infrastructure will be open interfaces (Open APIs).

The platform for remote identification of clients of financial organizations based on the database of the unified identification and authentication system and biometrics allows banks (and other organizations) to provide remote services to the population.

The fast payment platform allows the population to make money transfers to individuals and legal entities, including state and municipal entities, using the maximum number of identification methods (for example, mobile phone number, QR code, and other identifiers such as messengers).

The marketplace allows customers to get the necessary financial services in a "single window" mode with clear and transparent pricing. The Moscow exchange develops a platform that analyzes customer needs, selects the most suitable offers from financial institutions and offers them via chatbots.

The platform for registering financial transactions allows you to maintain a unified register of transactions in the financial market; the launch of this platform will support the implementation of the marketplace.

Payment platforms. Based on the unified payment infrastructure, a promising payment system is being created that will allow banks to manage their liquidity more effectively. The Bank of Russia is expanding the list of services provided by the national payment card system. For example, it is planned that the MIR card will be integrated with non-financial applications (campus, social, student, etc.). The implementation of innovative services such as MIR-Accept (with independent authentication of the issuer's client) will continue. The national payment card system continues to be promoted among the member states of the Eurasian Economic Union (EEU).

The payment platform for transmitting financial messages is created on the basis of distributed registry technology and is intended for transmitting financial messages in the SWIFT format. At the initial stage of its existence, this platform will only work on the territory of Russia. However, in the future, the Bank of Russia expects to organize an inter-system interaction of the EEU countries on its basis.

The Central Bank of the Russian Federation actively encourages the application of new technologies, including the creation of a so-called end-to-end client ID and a platform for cloud services.

Distributed Ledger technology is mainly represented in the strategy by the Masterchain platform, which the Bank of Russia is developing together with members of the FINTECH Association. Currently, pilot projects are being implemented in the field of accounting for electronic invoices, digital letters of credit and digital bank guarantees.

The creation of a new financial infrastructure allows creating equal conditions for all market participants and contributes to the development of the competition in the financial market. At the same time, the development of digital technologies is accompanied by an increase in cyber threats, which require financial market participants and mega-regulators to be able to quickly prevent or minimize them. For this purpose, the Bank of Russia pays special attention to the legal regulation of the use of breakthrough technologies and their direct implementation and development in the financial market of Russia and the EEU. At the same time, the Russian regulator is testing new digital tools and services on the RegTech platform (regulatory sandbox—test platform), which reduces the time required to ensure the legal regulation for the use of new technologies and allows selecting projects that are significant for the population and the national economy.

Other banks are using new technologies to develop new products and services, as well as to improve their compliance with the Bank of Russia's regulatory requirements and risk management. Customer identification, suspicious transaction detection, reporting automation, and compliance control are areas where new technologies (biometrics, Big Data, and Blockchain) have found their application.

The application of SupTech (Supervision Technology) allows the Bank of Russia to more actively analyze the affiliation of borrowers, determine the current demand for cash and its prospects, analyze and maintain the financial stability of credit institutions, and detect fraudulent transactions.

At the same time, there is a transition to electronic interaction between government agencies, financial market participants, and their clients. The Bank of Russia aims to improve the security of using new products and services, creating conditions for improving the financial literacy of the population.

3.3 *Organizational Changes in the Business Models of Russian Banks*

Analysis of the available data showed that even in large banks, the digital infrastructure needs to be improved. Customer needs are constantly being improved, which requires an adequate offer from banks. Bank clients are interested in increasing the availability, simplicity, and efficiency of operations that should be performed in one click, which determines the need to use digital channels. Modern banks are obliged not only to ensure the safety of the client's money, but also to provide him with uninterrupted remote round-the-clock access to it.

The constant transformation of activities of classical banks and financial intermediation in general (especially under the influence of a fierce competition from FINTECH companies) dictates the need to update not only the range of banking operations and services, but also the strategy for the development of banking activities and business models of banks. For example, VTB Bank in 2019 adopted a new strategy, which includes a goal of making from retail services for customers a convenient omnichannel service with a high share of digital technologies. By 2022, 70% of the bank's operations and 50% of VTB's sales are planned to be performed exclusively in digital channels. First of all, banks start performing daily operations in a digital mode (paying for mobile communications and the Internet, renting premises, utilities, etc.), and then they start processing new products (opening accounts, accepting deposits and issuing loans, issuing plastic cards, etc.).

In general, Russians are ready to use different online services. They have mastered payment tools and realized that it is safe, reliable and profitable. Therefore, banks continue to develop their offline processes to provide full-fledged digital customer service. At the same time, with the development of remote banking services, banks not only do not abandon the network of physical offices, but also open new ones. For example, VTB bank currently has more than 1,500 offices, and their number has been increasing in recent years, since there are operations that cannot be performed without the participation of a bank employee (depositing cash by a client to a bank account, receiving advice on complex structured products, renting safe deposit boxes, etc.). At the same time, the bank's offices are functionally adapting to the global digitalization of the banking industry; the transition to the format of paperless branches is underway, where the client confirms the transaction and signs everything in the bank's mobile app in the presence of an employee who forms the document itself and its digital image.

In recent years, banks have been actively using a system of financial incentives for their clients, including loyalty programs, where cashback, bonuses, miles and other preferences are combined in one mobile app and on the bank's website. The Bank provides its clients with these incentives for active use of its programs and those of its partners. The bank's client saves time and money by getting on-line information about products, services, and services that they need at any given time, and also has the opportunity to purchase them remotely. Thanks to modern technologies, some of the information is provided to the client remotely in a visual format. For example, using virtual reality glasses, bank clients can see their future apartment or other property, and then remotely apply for a loan, open the own personal mortgage office, get loan approval, and then the bank will transfer the funds to the borrower's account in a non-cash manner.

3.4 The Marketplace and the Banking Ecosystems

The trend of recent years is the development of banking marketplaces. The largest and most successful banks create so-called ecosystems. A number of large Russian banks, whose share in the banking sector assets is 66%, declare their development strategy as development "in the ecosystem paradigm". These banks are leaders in the development of new financial technologies.

In the modern economic literature, there are two main types of banking ecosystems: 1) ecosystems that meet a wide range of customer needs in various goods and services (Sberbank), and 2) ecosystems that meet needs of customers in one particular segment (Rosselkhozbank, VTB) [7, 9, 12].

For example, the country's largest bank, Sberbank, began to create its own ecosystem in 2017, when (at the shareholders meeting) its head Hermann Gref called the main competitors of the bank the IT giants (such as Google, Yandex, Alibaba). And the shareholders approved a strategy for gradually transforming the bank into an ecosystem while the technology platform and the bank's ability to aggregate an increasing number of services are created. Since then, the bank has gradually become an open ecosystem. In 2017–2019, Sberbank spent 1 billion U.S. $ (or 3% of its profit during this period) on purchasing various assets (financial and non-financial) to create an ecosystem. Currently, the Sberbank ecosystem includes companies that provide various services to individuals and legal entities, including: cellular communications, credit bureaus, recruitment services, real estate portal, logistics, online movie theater, and others. In 2019, Sberbank added a number of digital assets to its business, including a joint venture with Mail.ru and a block of Rambler shares, as well as media assets "Gazeta.ru" and "Lenta.ru". Sberbank's ecosystem includes food delivery companies (Delivery Club) and taxi companies (Citymobil), restaurant businesses (the SberFood platform and R-keeper—ERP for restaurants). As an important educational project, the bank considers the establishment of a joint venture with the company "Soyuzmultfilm". It is expected that the capitalization of non-banking business will eventually be equal to the capitalization of Sberbank itself, which exceeds 5 trillion rubles. In the future, Sberbank aims to become the largest technology company.

VTB Bank announced its intention to create marketplaces and a digital banking ecosystem in January 2019. The creation of a marketplace allows the bank to integrate many services. For example, VTB is currently developing a so-called "salary" marketplace, creating housing and automotive ecosystems, and developing VTB's multi-service for online commerce. The "Housing ecosystem" project is currently under active implementation. Its goal is to meet all the housing needs of clients in a single window mode, including: choosing a suitable apartment or another real estate object, conducting a transaction, evaluating and registering, insuring, creating a design project for the object to be purchased, getting help with moving and servicing the residential area, ordering small services for the house, repairs, etc.; getting mortgage and consumer loans on the marketplace, etc. An open partner ecosystem of VTB is being created that offers new services to all market participants, primarily construction companies that sell their objects, professional realtors, and other banks.

In August 2019, VTB launched a new website for the sale of collateral non-core assets—“CommissiON”. In 2020, it is planned to open a marketplace on its basis. Cars and real estate received from debtors will be put up for sale, and options for purchasing and registering the purchased property will also be presented here.

“VTB mobile”, the mobile operator of VTB, is represented on the market as a project of the bank and Tele2. SIM cards of this system will work throughout Russia and abroad. Moreover, the joint use of banking and telecommunications products will provide the maximum effect in terms of benefits and convenience. Communication can be absolutely free for customers when they actively use banking products. Special attention is paid to security issues, as there are cases when customers of traditional mobile operators have faced the fact that their SIM cards were replaced by fraudsters who gained access to Internet banking and withdrew funds from personal accounts of citizens. In VTB, all operations to replace SIM cards are performed only in the bank offices with the identity verification, which minimizes the risk of fraud. The client’s gadget is also identified. In this case, if the card is replaced in some way, then when you install it on someone else’s device, banking operations using this device will be blocked.

Among the Russian systemically important banks that are building their ecosystems focused on customer service in a particular segment of the financial market are: Rosselkhozbank, Rosbank and Alfa-Bank. In 2020, the Russian agricultural bank (Rosselkhozbank) plans to create a digital ecosystem for small businesses in rural areas, which will include services for hiring seasonal workers, receiving veterinary care, digital farm management, expanding sales markets, and agrotourism. Rosbank plans to develop an ecosystem for supporting mortgage lending: providing a mortgage loan, access to the services of realtors, notaries, insurance companies, as well as services for repairing and purchasing furniture. Alfa-Bank has been developing its “ecosystem” since 2013 on the principle of buying financial companies and startups (Alfa-stream, Alfa-Leasing, Alfa-Finance).

In these conditions, banks are constantly working to improve technologies, trying to bring them to a new higher level in order to provide optimal customer service in terms of speed, convenience and range of services in digital channels. Obviously, customers will only stay with the banks that can solve these tasks.

4 Discussion

The main idea of the work was to determine the effectiveness, efficiency and directions for further development of the relevant strategies of banks. The study of the digitalization experience of banking activities allowed us to reflect the current realities of the development of Russian financial market entities, identify existing barriers and determine the main development directions for banks’ policies in the era of the fourth industrial revolution. In particular, new approaches of the Bank of Russia as a megaregulator of financial markets often cause discussion among economists. The analysis of the theory and practice of applying the latest technologies allowed us to find an explanation for organizational changes in the banking industry. These changes were confirmed by the rapid development of so-called "digital platforms" and "ecosystems" of banks, since participants in these systems perform intermediary financial transactions for households and companies significantly faster and cheaper than traditional banks. At the same time, together with advantages of using innovations that increase the efficiency of financial and credit institutions, it is also necessary to take into account a significant increase in systemic and other risks, which are based on cyber risks.

5 Conclusion

Increasing the competitiveness of Russian banks on national and global markets is impossible without applying technological innovations, using digital platforms and banking ecosystems. The main trend in the banking industry today is the emergence of new opportunities to solve the transformation problem, updating the practice of functioning of traditional banks. In these conditions, it is necessary to overcome a knowledge gap between technology developers, bank employees and other financial market participants, as well as to develop optimal policy options for different groups of banks: the regulator, large systemically important banks, as well as small and medium-sized banks, whose needs for digitalization are determined by the reality.

References

1. Bank of Russia (2017) The main directions of financial technology development for the period 2018–2020. https://www.cbr.ru/StaticHtml/File/36231/ON_FinTex_2017.pdf. Accessed: 11 Feb 2020
2. Citibank (2016) Digital disruption: how fintech is forcing banking to a tipping point. https://www.ivey.uwo.ca/cmsmedia/3341211/citi-2016-fintech-report-march.pdf. Accessed 11 Feb 2020
3. Government of the Russian Federation (2018) The national program "Digital economy of the Russian Federation". http://government.ru/rugovclassifier/614/events/. Accessed 16 Mar 2020
4. King B (2020) Bank 4.0, banking everywhere, never at a bank. Wiley, New Jersey
5. PWC (2019) Blurring boundaries: how financial technology companies affect the financial services sector. http://www.pwc.by/ru/publications/other-publications/edges-blurring.html. Accessed 11 Feb 2020
6. Sidorenko EL (2020) Stablecoin as a new financial instrument. In: Ashmarina SI, Vochozka M, Mantulenko VV (eds) Digital age: chances, challenges and future. Lecture notes in network and system, vol 84. Springer, Cham, pp 630–638. https://doi.org/10.1007/978-3-030-27015-5_75
7. Simonia NA, Torkunov AV (2016) The impact of geopolitical factors on international energy markets (the US case). Polis 2:38–48

8. Sologubova GS (2019) Components of digital transformation: Monograph. Yurayt Publishing House, Moscow
9. The World Bank (2013) Financial inclusion: helping countries meet the needs of the under-banked and under-served. http://www.worldbank.org/en/results/2013/04/02/financial-inclusion-helping-countries-meet-the-needs-of-the-underbanked-and-underserved. Accessed 11 Feb 2020
10. Tofanyuk E, Uskov N (2019) G. Gref: the transformation of the savings bank is an eternal process. https://www.forbes.ru/biznes/387895-german-gref-transformaciya-sberbanka-eto-vechnyy-process. Accessed 11 Feb 2020
11. Trushina KV, Smagin AV (2019) The main trends in the development of the banking system of the Russian Federation at the present stage. Bank Serv 12:7–11
12. World Bank Group (2019) World development report 2019: the changing nature of work. World Bank, Washington, D.C

Sharing Economy and Legal Barriers to Its Development

E. L. Sidorenko

Abstract The paper deals with the sharing economy development. Special attention is paid to the definition of subject boundaries of the sharing economy, its main economic and legal features, functioning principles, indicators and development trends. The data of major expert centers proving the viability and prospects of the development of the sharing economy are presented, its individual branches are considered with the indication of key performance parameters. Using the example of individual projects of the economy of shared consumption, the author reveals legal aspects of this new phenomenon and describes the main legal barriers that prevent its development. The paper defines the main directions of the legislation development in terms of implementing sharing in the traditional economy, namely, changing the model of tax administration, distributing responsibility for goods and services among participants of digital turnover, determining the legal status of sharing platforms, etc.

Keywords Digital economy · Sharing economy · Economy of shared consumption · Digital platforms · Legal responsibility · Taxation

1 Introduction

In the conditions of active development of digital technologies and digital platforms, the question of adaptability of the traditional economy to the drastic change in consumer behavior of citizens comes to the fore. The application of models of direct communication between consumers of goods and services has led to the development of a fundamental new economic phenomenon—the sharing economy, or the economy of shared consumption.

In 2015, the volume of the sharing economy in the European Union reached 28 billion euros, and in the same year it was estimated that this amount would increase to 335 billion euros by 2025 [18]. According to experts, in China, as a leader in the

E. L. Sidorenko (✉)
Moscow State Institute of International Relations (University) of the Ministry of Foreign Affairs, Moscow, Russia
e-mail: 12011979@list.ru

S. I. Ashmarina and V. V. Mantulenko (eds.), *Current Achievements, Challenges and Digital Chances of Knowledge Based Economy*, Lecture Notes in Networks and Systems 133, https://doi.org/10.1007/978-3-030-47458-4_56

development of the economy of shared consumption, the development of sharing will contribute to the GDP growth in the range of 1.5–2% and by the end of 2020 may generate about 10 giant firms [27]. In the report "Economy of joint consumption in Russia" (2018) by the Russian association of electronic communication (RAEC), it was noted that by 2024, the development of joint consumption projects in Russia can lead to the following results:

1. Increase the urban environment quality index by 20% through carsharing, bike sharing and scooter sharing projects.
2. Reducing the level of air pollution due to bikesharing and scooter sharing by 40 tonnes of emissions.
3. Providing favorable conditions for business and self-employed activities for 2.5 million people through exchanges of service providers and coworking.
4. Accelerate the implementation of digital technologies and create a comprehensive system for financing projects for the development of platform solutions due to the appearance of about 7,000 non-gaming applications.
5. The decline in the share of roads operating in the overload mode by 15% through the development of carpooling and others [25].

Among the main advantages of the sharing economy are:

- increase the speed of service delivery due to the speed and flexibility of request processing. Excluding intermediaries from the producer-consumer chain allows reducing the cost of services by 10–20% on average and increasing the speed of their delivery by more than 40%,
- increasing the availability of goods and services, flexible consumer logistics system. Unlike traditional production models, the sharing economy is based not on statistical patterns, but on the individual consumer needs,
- a possibility of additional earnings when selling or renting out unused property,
- the efficient use of resources and significant reduction of waste. For example, the application of a foodsharing model in Russia allows saving 1 million tons of food annually for a total of more than 85 billion rubles [26],
- reducing transaction costs by eliminating intermediaries, creating a new model of trust between consumers.

As individual projects of shared consumption, key principles of the sharing economy began developing:

- the principle of equality of exchange participants (the principle of peering),
- the principle of self-regulation, or voluntary division of responsibility,
- the principle of rationality and openness of information,
- the principle of environmental friendliness,
- the principle of evangelization, evaluating your economic behavior not only from the position of your own benefit, but also the benefit to the society as a whole [22].

There are several sharing economy models depending on the principles of activity and the product that is in common consumption. According to the first criterion, the following models are distinguished: #We provide (renting out property, providing services on request); #We share (centralized rental of things or space from companies); #We give (giving, exchanging or selling unnecessary things); #We unite (co-financing and cost sharing) [25].

Based on the second criterion, we have sharing of digital content, sharing of physical goods, and sharing in the context of social projects.

Despite some common features associated with the trust business model and the exclusion of intermediaries, each of these models has legal specifics and excludes a stencil approach to the analysis of the sharing economy as a homogeneous phenomenon. With the growth of economic indicators of the sharing economy, the conclusion about its heterogeneous nature becomes more and more obvious.

In this regard, significant methodological and legal issues related to the definition of subject boundaries of the sharing economy and the search for its "weak links" that need point-to-point legal regulation come to the fore.

2 Methodology

The methodological basis of this work is general scientific and private scientific methods of knowledge, methods of empirical and theoretical research. The dialectical method allows us to consider the sharing economy through the prism of the laws of transition quantitative changes in qualitative and negation of negations etc.

The deterministic approach, concretizing such paired categories of dialectics as "necessity—chance", "possibility—reality", "cause—effect", allowed not only to study the cause-and-effect relations in the development of the sharing economy, but also to develop an approach to its classification.

The following general scientific methods of cognition were used as the main ones: ascent from the abstract to the concrete, modeling, historical method, comparison, analysis and synthesis, induction and deduction. The latter were supplemented by empirical methods, including statistical (statistical observation, summary and grouping, statistical analysis), sociological (study of documents, surveys in the form of questionnaires and interviews, observation) and socio-psychological (testing). The use of these approaches, along with the formal legal method, allowed us to conduct a comprehensive study of the legal aspects of the sharing economy.

3 Results

The study investigates legally significant features of the sharing economy and identifies the main problems associated with the legal support of digital projects. Legal issues of the economy of shared consumption are considered in the context of comparison with the main activity stages of participants in the sharing economy, taking into account current legislation and judicial practice. Based on the analysis of problems of determining the legal status of digital platforms, sharing participants and guarantees of providing goods (services) of appropriate quality, the author forms a number of recommendations aimed at developing the current legislation, and justifies the importance of their compliance to minimize legal risks of joint consumption.

4 Discussion

Despite the active development of digital projects of shared consumption, neither at the international nor at the national level, a single definition of the sharing economy has been developed. This significantly complicates the search for its legal basis, since it prevents the development of common legal features of the sharing economy and a strategy to minimize its legal risks.

Along with the definition of "sharing economy" in English-language literature, the terms "gig economy", "freelance economy", "collaborative consumption", "digital economy", "peer economy", "access economy", "crowd economy", "platform economy", "on-demand economy", and others are used [2, 5, 7, 10, 13, 21–24]. Despite their obvious terminological similarity, these concepts focus only on certain legally relevant features of the sharing economy. Thus, the term "peer economy" emphasizes the absence of intermediaries in economic exchange, and "crowd economy" focuses on the use of Internet platforms to meet the interests of large masses of consumers (Uber and Airbnb platforms) [23]. The term "collaborative consumption" emphasizes the fact of joint consumption of goods and sees this as an independent legal system [6].

Speaking of the sharing economy, experts see it as a situation where individuals and legal entities provide each other on a paid basis access to their assets, which they themselves do not fully use [6], a set of online platforms that provide a combination of supply and demand, through the use of three methods: peer-to-peer sales, peer-to-peer leasing and crowdfunding [11]; the use of digital platforms to provide access, rather than ownership of tangible and intangible assets [16].

Quite interesting is the approach to defining this term by the Australian tax service. It refers taxi services, car sharing, renting out real estate and movable things, as well as providing freelance services to the sharing economy. At the same time, it is specifically stipulated that sharing is not identical to online trading, cryptocurrency infrastructure, and crowdfunding platforms [1].

Analysis of existing approaches from the perspective of the completeness of generalization of the legal characteristics of the sharing economy suggests that the most appropriate legal content is the following legal definition: sharing economy is a business model in which the production, exchange and consumption of material and non-material benefits are realized through the use of consumer digital platforms.

Thus, the key legal features of the sharing economy are:

- a specific set of participants in legal relations: digital platform providers—persons who provide resources (goods, their working hours, services) on a regular or professional basis. As a rule, material resources are provided for temporary use and do not change owners; users of digital resources (individuals and legal entities); intermediaries—online platforms that provide interaction and conclusion of transactions between providers and users;
- rights of participants: mainly the product passes to the user for temporary use. However, in the #We provide and #We give away models, the range of user permissions can change significantly, up to and including the rights to dispose;
- a special participant in legal relations is a digital platform. Unlike a traditional intermediary or agent, it performs the function of searching for a consumer provider and ensuring the transaction security. However, a unified approach has not yet been developed as to whether the platform should have a duty to provide guarantees of use, and if so, to what extent they should be implemented;
- a paid form of providing goods (services). With the development of the sharing economy, the universality of this term is increasingly questioned because of the emergence of projects related to charity, free transfer of things or products at the request of users, free provision of housing, transport services, etc.

Knowledge of the main legal features of the sharing economy allows us to see legal barriers that hinder its development.

Participant status. Individuals who rent their property to the sharing system (for example, homeowners within the framework of Airbnb) receive income, but cannot be considered as employees of the corresponding service or independent entrepreneurs [5, 14, 19].

A separate issue is raised regarding the status of individuals who actually work for the sharing platforms. So, in California, there was a case brought against the company Lyft (functionally similar to Uber), in which employees demanded that they be recognized as employees under all guarantees of the labor law, and not just as temporary contractors. As a result, the company paid employees $27 million, but retained the right to treat them as simple contractors [10].

Even more unclear is the legal status of sharing platforms themselves and people who organize their work. Their mediation functions are anonymous, automated, and hardly fully regulated by classical civil law models of contracts. Modern digital platforms do not fully fall under the status of agents. The basis of their legal activity is mainly user agreements, use programs and (or) loyalty, taking into account the principle of the contract freedom. This means that rights and obligations of platforms can only be restricted by law if they directly contradict it. The operation of digital platforms based on the freedom of contracts, on the one hand, gives the sharing economy the necessary flexibility, but on the other hand, significantly restricts the rights of users, since when concluding an offer agreement, their rights to change the terms of the agreement, changes in time and price are effectively nullified.

Responsibility of participants in sharing projects. Because of the unsettled status of participants, the issue of liability between these persons and in their relations with third parties remains unresolved. A partial solution to this problem in the United States is section 230 of the Communications Decency Act (CDA) of 1996, which states that the platform owners are not responsible for services, products, or content distributed by users through this platform [4].

However, even on this issue, disputes persist. For example, in San Francisco in 2014, a local law was passed requiring all landlords to register and pay for the right to rent out their homes. In 2016, local authorities filed a lawsuit against Airbnb and HomeAway for working with unregistered homeowners. The court found the companies guilty, but in the end the case was resolved by a settlement agreement. Another case took place in 2015, when lawsuits were filed against Uber in connection with two cases of harassment of taxi customers by Uber taxi drivers. However, the case was resolved before the trial began [10].

As an example may be considered the case of Elena Graschenkova against Yandex.Taxi. A client had an accident while in a taxi. Initially, the court did not recognize the victim's legal relationship with Yandex.Taxi as a contract of affreightment and refused to collect monetary compensation from the company. But later, the court still recognized that the woman was a customer of this digital service. And since the service organized the trip, it should be responsible for it [17].

Right of disposal. In the sharing economy, there are and will continue to be products that citizens may have quite controversial rights to. Thus, the BetrSpot system [3] provides an opportunity to organize the trade of places in queues—a phenomenon whose turnover, obviously, is not regulated by law in any state.

In addition, platforms dealing with the circulation of prohibited goods and services began to appear. In this case, the status of a seller and a buyer is not in doubt and should be qualified under the criminal law, and the question about liability of the platform organizers does not have a clear answer, largely because of the inability to prove an intent to participate in a criminal scheme.

Consumer protection is directly related to the status of platforms and their participants. The possible difficulty is that this protection is only possible within the framework of the consumer-entrepreneur relationship.

This point, in particular, is highlighted on the website of the government of Western Australia [9]. Economic and theoretical understanding of the problem of consumer protection in the framework of sharing platforms was undertaken by Koopman and co-authors in the work "The sharing economy and consumer protection regulation: The case for policy change" [12].

Risk minimization and quality standards. The sharing economy involves replacing several large centralized business owners (for example, hotels) with many small decentralized owners. As a result, many of the rules and standards that apply to former "official" market participants (for example, on residential safety, sanitation, and customer identification) do not actually apply to new "unofficial" participants. This generally undermines the possibility of power control over the entire sector of the economy and can potentially lead to new threats [11]. Thus, carsharing can give access to cars for people who do not have or are deprived of a driver's license, and in homes, there is often no fire alarm [15]. Issues of mandatory insurance, including liability, also remain unresolved.

Competition law. Some platforms, such as Uber, independently set prices for services provided by individuals working within their framework, thereby excluding price competition between these individuals. It is unclear whether this system can be considered a violation of antitrust law. So, in the United States, there was a court case in which the court found Uber not guilty of monopolistic activities [20].

Taxation. The issue of taxation in the sharing economy is ambiguous only for the reason that each platform can provide different products, different commercial models and bring different income to different individuals, so creating uniform tax standards for all these platforms can be difficult.

However, the issue of taxation is the only internationally developed aspect of the sharing economy. For example, the OECD prepared a report in March 2019 [16], which made recommendations to member states on this issue:

- the need for cooperation between tax authorities and platforms in order to better inform platform participants about their tax rights and obligations;
- minimizing the bureaucratic burden for the platforms participants;
- international cooperation of tax authorities in order to form common understanding of the procedure for establishing revenues within the framework of platforms and their taxation;
- standardizing requirements for various platforms, including reporting requirements.

It is important to note that the issue of tax administration of digital platforms is already being considered in detail at the level of individual states. For example, there is already a detailed set of rules in Australia [1], the United Kingdom [9], as well as Greece, Denmark, Hungary, France, Ireland, Spain, Singapore and many other countries studied by OECD analysts [8].

At the same time, general recommendations on taxation are useless when addressing private issues of sharing. In particular, when an organization transfers goods to charity, these goods are considered sold, and therefore affect the determination of the market value in income and the value of the tax base for the income tax. In addition, the company that transfers something for sharing is the last in the value chain and therefore has to pay VAT. Thus, taking into account the specifics of the country, the total tax burden for gratuitous projects (for example, for foodsharing) can be up to 50%.

Appeal. The issue of appealing the actions of participants and challenging transactions concluded in the context of sharing platforms can be quite acute in those countries where there is a possible waiver of the right to appeal to the court (especially in the USA) as a number of sharing platforms directly provide a duty of everyone to resolve related disputes exclusively through special arbitration mechanisms available on the platform [20]. These types of rules restrict users' rights to a full and comprehensive review of the case by authorized courts.

5 Conclusion

In conclusion, it is important to note that the active development of sharing models in the economy should be accompanied by a deep and high-quality study of issues of their legal support. In particular, at the level of international organizations, unified principles of the sharing economy operation should be developed. We also need to define its legally significant signs and consider the issue of the responsibility scope of the digital platform, its providers and users. In addition, the issue of tax administration of digital platforms and their providers also needs a deep legal analysis. In part, a differentiated approach to taxation should be developed depending on the remuneration of turnover, characteristics of goods and services, as well as the legal status of providers and users.

References

1. Australian Government (2019) The sharing economy and tax. https://www.ato.gov.au/General/The-sharing-economy-and-tax/. Accessed 20 Feb 2020
2. Belk RW (2014) Sharing versus pseudo-sharing in web 2.0. Anthropologist 18:7–23
3. BetrSport (2019) World is crowded, trade up a space. http://www.betrspot.com/index.html. Accessed 20 Feb 2020
4. Communications Decency Act (CDA) of 1996. https://www.eff.org/issues/cda230. Accessed 20 Feb 2020
5. Erickson K, Sørensen I (2016) Regulating the sharing economy. Internet Policy Rev, 5(2). https://policyreview.info/articles/analysis/regulating-sharing-economy. Accessed 20 Feb 2020
6. Frenken K, Meelen T, Arets M, van de Glind P (2015) Smarter regulation for the sharing economy. The Guardian. https://www.theguardian.com/science/political-science/2015/may/20/smarter-regulation-for-the-sharing-economy. Accessed 20 Feb 2020
7. Görög G (2018) The definitions of sharing economy: a systematic literature review. Management 13(2):175–189
8. Government of UK (2017) Sharing economy: user characteristics and tax reporting behavior. https://www.gov.uk/government/publications/sharing-economy-user-characteristics-and-tax-reporting-behaviour. Accessed 20 Feb 2020
9. Government of Western Australia (2018) The sharing economy for consumers. https://www.commerce.wa.gov.au/consumer-protection/sharing-economy-consumers. Accessed 20 Feb 2020
10. Hofmann S, Sæbø Ø, Braccini AM, Za S (2019) The public sector's roles in the sharing economy and the implications for public values. Govern Inf Q 36(4):101399
11. Ivantsov SV, Sidorenko EL, Spasennikov BA, Berezkin YuM, Sukhodolov YaA (2019) Cryptocurrency-related crimes: key criminological trends. Russian J Criminol 13(1):85–93 https://doi.org/10.17150/2500-4255.2019. (in Russian)
12. Koopman C, Mitchell MD, Theirer AD (2014) The sharing economy and consumer protection regulation: the case for policy change. J Bus Entrepreneurship Law, 8(2). https://papers.ssrn.com/sol3/papers.cfm?abstract_id=2535345. Accessed 20 Feb 2020
13. Migai C, de Jong J, Owens J (2018) The sharing economy: turning challenges into compliance opportunities for tax administrations. eJournal Tax Res. 20:285–317
14. OECD (2015) Digital economy outlook 2015. https://www.oecd.org/internet/oecd-digital-economy-outlook-2015-9789264232440-en.htm. Accessed 20 Feb 2020
15. OECD (2017) Model tax convention on income and on capital: Condensed version. Retrieved from: https://dx.doi.org/10.1787/mtc_cond-2017-en. Accessed 20 Feb 2020

16. OECD (2019) The sharing and gig economy: effective taxation of platform sellers: forum on tax administration. https://www.oecd-ilibrary.org/taxation/the-sharing-and-gig-economy-effective-taxation-of-platform-sellers_574b61f8-en. Accessed 20 Feb 2020
17. Oparin M (2017) Muscovite against Yandex.Taxi: what will the court decide? https://www.vesti.ru/doc.html?id=2954245. Accessed 20 Feb 2020 (in Russian)
18. PWC (2015) The sharing economy: sizing the revenue opportunity. http://www.pwc.co.uk/issues/megatrends/collisions/sharingeconomy/the-sharing-economy-sizing-the-revenue-opportunity.html. Accessed 20 Feb 2020
19. Rogers B (2015) Employment rights in the platform economy: getting back to basics. Harv Law Policy Rev 10(2):480–519
20. Schmitt R (2017) The sharing economy: can the law keep pace with innovation? Stanford Lawyer, 96. https://law.stanford.edu/stanford-lawyer/articles/the-sharing-economy-can-the-law-keep-pace-with-innovation/. Accessed 20 Feb 2020
21. Schor JB, Walker ET, Lee CW, Parigi P, Cook K (2015) On the sharing economy. Contexts 14(1):12–19
22. Shved VV (2017) Theoretical aspects of the sharing economy. Econ Forum 1:26–31
23. Sundararajan A (2016) The sharing economy, the end of employment and the rise of crowd based capitalism. The MIT Press, Cambridge
24. Sutherland W, Jarrahi MH (2018) The sharing economy and digital platforms: a review and research agenda. Int J Inf Manag 43:328–341
25. RAEC (2018) Sharing economy in Russia 2018. https://raec.ru/upload/files/sharing-economy2018.pdf. Accessed 16 Mar 2020 (in Russian)
26. TiarCenter (2019) Food sharing in Russia: report. https://foodsharing.ru/wp-content/uploads/2019/10/foodsharing_tiar.pdf. Accessed 20 Feb 2020 (in Russian)
27. Zhong N (2017) Report says China's sharing economy to grow 40% annually. China Daily. https://www.chinadaily.com.cn/business/2017-03/23/content_28647692.htm. Accessed 20 Feb 2020

Development of Sharing as a Factor in the Tertiarization of the World Economy

V. A. Perepelkin

Abstract The article studies the distribution of sharing relations in consumption in the process of forming a post-industrial economy. A quantitative analysis of tertiarization in the 2000s was carried out at the level of the entire world economy and its individual parts and also the specificity of tertiarization of the Russian economy was revealed. It is proved that the digital organization of the access-based consumption which expressed in the replacement of material goods by the provision of services has become a new factor of tertiarization. Besides, the article reveals the content of the economic and social aspects of sharing and the stages of its development and shows the dependence or the intensity of post-industrial cooperation in consumption on the level of maturity of social reletions and the richness of the country's human potential.

Keywords Tertiarization · Sharing · Access-based consumption · Services · Collaborative consumption · Online platforms

1 Introduction

Under the influence of digitalization, the content and forms of manifestation of a wide range of social relations are changing, including the organization of the consumption process. The structural effect of digitalization is expressed in promoting the emergence of new industries and industries, modernization of traditional ones. Embracing the spread of electronic information and communication technologies in the economy, digitalization is not equally intensive in its different parts. Due to the immaterial nature of the products created, the third sector (service sector) is more susceptible to digital technologies than the commodity-producing sectors. Perhaps for this reason, the terms «digital economy» and «post-industrial economy» are often used as synonyms. A number of scientific publications substantiate that the

V. A. Perepelkin (✉)
Samara State University of Economics, Samara, Russia
e-mail: slavaap@rambler.ru

S. I. Ashmarina and V. V. Mantulenko (eds.), *Current Achievements, Challenges and Digital Chances of Knowledge Based Economy*, Lecture Notes in Networks and Systems 133, https://doi.org/10.1007/978-3-030-47458-4_57

post-industrial economy is a generic concept in relation to the resulting terminology array of definitions of the modern stage of economic development.

The general direction of forming the structure of the post-industrial economy is tertiarization, which is quantitatively described by increasing the share of services in the total values of production, consumption and employment [5]. There is an opinion about the need to take more account of the new qualitative characteristics of tertiarization, demonstrated mainly by developed national economies [14]. Having accumulated significant reserves of human capital and advanced productive capacity, they ensure further expansion of services by restructuring the tertiary sector by increasing the share of knowledge-intensive services with progressive productivity in their creation and high efficiency in consumption [11]. In this context, it is relevant to study sharing, which has a service-oriented nature of development with a high susceptibility to the use of digital technologies. Based on Internet communications and processing of large data sets (big data), sharing relationships have penetrated into the sphere of consumption of economic agents, forming the structure of their demand already on post-industrial principles. The transition from possession of goods to access to them on the right of use in modern conditions is accompanied by the replacement of consumption of material goods with services that provide an opportunity to obtain a similar consumer effect with lower costs. Such a post-industrial shift in consumption indicates the emergence of a new factor of tertiarization, which allows us to predict the continuation of this trend of structural development.

2 Methodology

The theoretical and methodological basis of tertiarization factor analysis has become a three-sector Fisher-Clark's model that structures the economy into primary, secondary and tertiary sectors which represent the economic base of theory of development of post-industrial society. Expressing itself through a long-term shift in the structure of social reproduction from material goods to services the tertiarization is associated with significant changes in the system of economic relations. One of this changes is a change in consumer preference in favor of services which is stimulated by the emergence of digital-based organization of the consumption process [8]. The approach that was described predetermined the setting for the disclosure of the content of sharing as a consumption method through the access to goods without owning them in the context of the formation of post-industrial social relations [10]. In the framework of the presented topic of the article, the social side of the research object was studied as well as the economic one. The methodological apparatus that was applied included a set of General scientific methods and some special methods of scientific knowledge such as system, analytical, functional and graphic methods.

The information base of the research was the results of scientific research, reports of intergovernmental economic organizations, international analytical agencies, sociological surveys and state statistical services. Based on the data collected using the methods of grouping and comparisons as well as constructing rows of speakers shows the development of tertiarization of the world economy as a whole and its separate parts and justifies the factor influence of sharing propagation on the investigated trend of the post-industrial structural transformation.

3 Results

The existence of a long-term trend in the growth of the services or tertiary sector, resulting in an increase in its share of gross product and aggregate employment generated by national and global economies, has been confirmed for many decades by empirical studies [12]. Tertiarization continued to serve as a universal structural shift as part of the process of transforming mature industrial economies into post-industrial and into the 21st century. Having exceeded half of the world's gross value added (GVA) in the late 1960s, the tertiary sector's share in the global economy was already 62.185% in 2000 and 69.092% in 2018.

The graphs presented in Fig. 1 show the superiority in the level and sustainability of the progress of the tertiarization of the economies of the developed countries of the Organization for Economic Cooperation and Development (OECD) over the similar world averages and indicators of the group of least developed countries. The picture clearly shows the dependence of the growing share of the tertiary sector on the level of development achieved by the economy. The tertiarization of the economies of

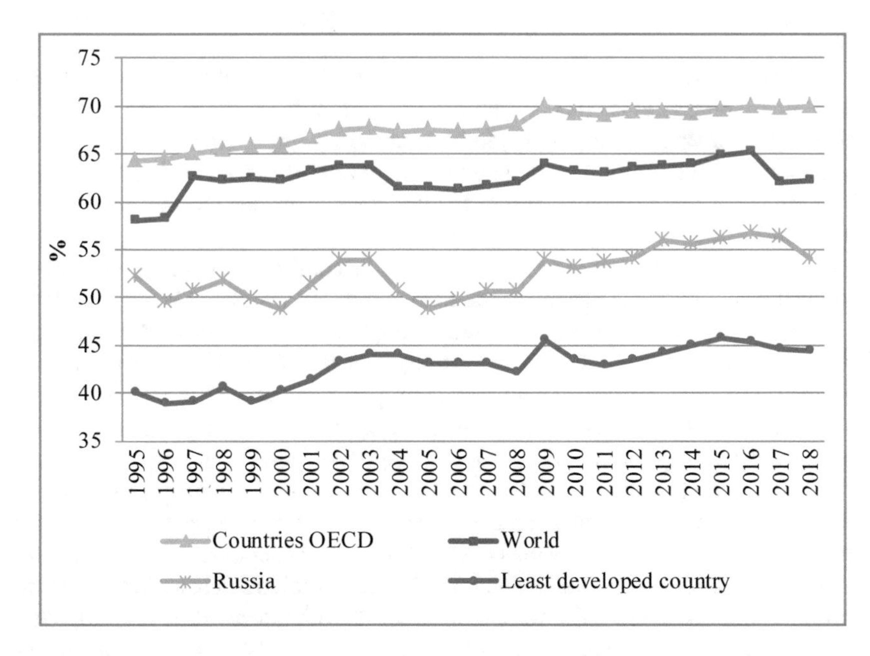

Fig. 1 The share of tertiary sector in Gross domestic product, in % (*Source* authors based on [15])

the least developed countries of the world has not yet reached the world average values of the middle of the last century and during the period 1995–2018 the lag from most national economies in the share of the tertiary sector has not decreased, but has increased (from 17.98% in 1995 to 20.66% in 2018). At the same time, global averages were slightly closer to those of OECD countries (instead of 6.24%, the lag was 4.79%). The dynamics of the Russian economy in this regard were inferior to all the groups of countries represented in the figure: the lag from the world average and OECD level increased (from 5.83% to 10.97% and from 12.07% to 15.76%, respectively), and the advance compared to the least developed countries decreased (from 12.16% to 9.70%).

The situation with the level and dynamics of tertiarization of the Russian economy, which in 2018 occupied the 59 largest gross national income (GNI) per resident in the world (26470 US dollars in purchasing power parity with 17842 US dollars on average worldwide), is represented unusual. In theory, it is considered to be the main factor in the expansion of the service sector. The amount of income of the population affected the Russian economy less than usual. This has also been the case in almost all the largest oil exporting countries in the world, indicating the direction of the search for the reasons for the existence of the detected feature of the sector structure of the economy.

Statistically gradually losing the decisive influence on the rate of tertiarization of the world economy, developed countries retain their role as initiators of qualitative changes in production and consumption patterns, leading to further absolute and relative expansion of the tertiary sector. One of the most significant of these in recent times appears to be the emergence of the so-called sharing economy in all the diversity of its inherent consumption practices.

Initially, in a market economy, resources were made available for loan through leases, in which relationships between little or no familiar persons arise within the framework of private property rights, and consumption is preceded by the provision of market-based services. Rent is charged in cash, in most cases, the economic interest of the recipient is not to compensate for the costs of ownership, but to generate additional income.

Aimed at realizing private interest, renting in the context of the formation of post-industrial social relations has given life to new ways of accessing benefits without the right to own them. Clearly, this transformation was evident in the transition from car rental to car wool. It is noteworthy that the original idea of carsharing was borrowed from the pro-social practice of group use of cars owned by several households (for which it was called "cooperative or social carsharing"). By expanding the scale of coverage from strictly local to regional, transferring communication links from offline to online, as well as commercializing the scheme of separation of use of vehicles, entrepreneurs launched a fundamentally new service for passenger transport. As a result, car borrowing between friends and neighbours was replaced by short-term rental with market prices and an entrepreneurial structure as a coordinating centre. The first commercial sharing car rental service originated in 2000 in the United States (Zipcar), in 2001 a similar one appeared in Germany (RentMyCar), and today car sharing has spread around the world. The main obstacle to the continuation of the rapid expansion of personal car sharing was the imperfect mechanism to ensure the required level of trust between the owner of the car and the user, achieved through verification of formal (driver's license number, driving experience) and informal information about the latter (feedback on the electronic platform on the results of participation in previous car sharing contacts). As a result, personal car sharing with access to a driverless vehicle has given way to Uber-impersonated transport services on demand by freelance and aggregator specialists. Success in the market was due to a number of obvious advantages of auto-sharing over auto-lease from the point of view of the consumer: access to the vehicle is as close as possible to a convenient place for the user; The contract is concluded remotely and once, at the first circulation, after which the car is reserved without other formalities via the Internet; Per-minute payment based on actual travel time, taking into account mileage; Practical virtual communications with the owner of the car.

As a post-industrial regime of access-based consumption, sharing has a number of characteristics: temporary access, monetization of underutilized resources owned by suppliers, market impersonal connections, self-service of consumers through the transfer of part of operational activities to them, dominance of private interests among participants over group and public interests, reliance on the possibilities of digitalization. The desire to save on the costs of ownership, utilitarianism, supported by profit-making, demonstrates the primacy of the economic component in the conduct of activities. It is no accident that sharing exchange systems according to the B2C model (business-consumer) are often identified with short-term rent. The shift to online communications within the consumer community has made it possible to share benefits more effectively, unite and transfer them. In addition to the cost-effectiveness, speed, flexibility of the circulation of goods, publicity about their properties, the benefits of consumption without possession consist in increasing the total efficiency of the use of economic resources by society through intensification of their reuse, in reducing the physical values of waste [7].

The next stage in the development of the sharing economy was the development of collaborative consumption. Unlike the term "sharing" (derived from the English verb "to share"), which indicates how resources are allocated, the meaning content of the phrase "collaborative consumption" is closer to targeting. In addition, unlike other options for the implementation of sharing relations, the study of joint consumption deserves special attention to the identification of priority goals and tasks among the groups of participants. In order to maximize the integral benefits of consumer cooperation, they act on the grounds of greater convenience and benefit from temporary access to goods rather than their continued possession. The notion of the importance of owning things becomes obsolete as the availability of these same goods increases through the spread of the latest forms of co-operation in consumption. There is a real prospect of improving the quality of life at a faster rate of growth in consumption than in the production of benefits. A survey of participants in the market of sharing services in the USA showed that 43% of respondents covered by its estimate property as a burden. Temporarily unutilized goods in increasing quantities are supplied to the market, thereby increasing the competition from which end-users benefit. Post-industrial changes in consumer attitudes are reflected in a relative decrease in demand for durable items (e.g. automobiles), with increasing demands on their functionality and durability [2].

The intensity of post-industrial cooperation in consumption depends on the level of maturity of social relations achieved. In most countries of the world, shared consumption practices are mainly found in communities of relatives or people linked to each other by their local material living conditions (apartment building or country partnership). A system of informal values has been developed in developed countries, facilitating the cooperation of even familiar people who take responsibility not only for their immediate surroundings but also for the situation in the locality and place of residence. And even if the desire to save or earn additional income by joining the new consumption system prevails, its implementation depends on the readiness to interact with other societies on equal, mutually beneficial and trusting terms.

Sharing extra to become freer and richer is an informed choice dictated not by altruism but by the benefits acquired. In order to negotiate in a virtual environment the temporary transfer of irregularly used things to people with little knowledge, it takes a certain trust in them. It arises from the developed reputation of a person in the online community, reviews in which replace the experience of personal communication. Of course, the authors of reviews do not always have the information and skills to give a correct assessment to their former partners, but the accumulation of a reputational database helps to overcome this problem. Thanks to the Internet, trust is formed at the global level and young people who have grown up in a virtual world without borders are more confident in online contacts. Good psychological compatibility with what is happening in the Internet environment, as well as the inherent desire for new and limited funds of young people, explains the most active participation in the "sharing economy" of people of the age of beginning of work and family life, noted by sociologists. Along with youth, sociological research indicates a high level of education as a personal characteristic, increasing the desire to "share" or share benefits in consumption, as well as high incomes - as a reducing circumstance.

The reasons for joining communities of cooperation in consumption, on the one hand, are determined by considerations of economy and convenience, and on the other, altruism, values of communication and new dating [1]. At the stage of the formation of such communities, they were dominated by a social component, they were in the form of cooperatives created for mutual assistance. In Western Europe, young citizens began to join clubs in their territorial location at the end of the last century, sharing cars (carsharing) or housing (couchsurfing) with each other on a non-profit basis. The sharp strengthening of the economic component occurred after the transformation of the social initiative for cooperative use into business models by entrepreneurs. Prior to this, people agreed on collective alternate access to available resources directly, the aim being to increase the efficiency of the consumption process by increasing the usefulness of things while reducing the unit costs of participants. Exchanges were made offline, accompanied by rather close personal communication. The introduction of online platforms created and regulated by business structures into the exchange mechanism has made it much faster and more convenient. Electronic platform calculations have made it possible to monetize the economic benefits by bringing commercial content. Acting in a consumer to consumer exchange model

(C2C) by intermediaries, or as part of a model with B2C providers, business structures bring market relations to the end-user community, changing the dominant main target of interaction from social attractiveness to financial advantage. The result of the exacerbation of contradictions between the social and economic motives of consumer participation in sharing relations is the so-called "heart and purse paradox" [4].

Since online platforms have taken a central place in the mechanism under consideration to ensure cooperation in consumption, in the seat of their ignorance, better-digitalized services displace goods from these electronic trading and exchange sites. A clear illustration of the reality of such changes is the so-called "uberization," when, similar to Uber, even industrial manufacturers increasingly offer customers to buy from them not real goods in possession, but services of their use. Exchanges of information and knowledge are the most promising. The expansion of benefits only at the time of their use gives added impetus to tertiarization, which is reflected in the growth of the services sector ahead of the rest of the economy, making the reproduction process, including consumption, more post-industrial.

According to estimates of international consulting companies, the growth of consumption on the basis of access looks impressive: in 2014 the volume of the world market of sharing services amounted to 43 billion dollars, in 2017—18.2 billion dollars, in 2021 Juniper Research predicts 40 billion dollars, and by 2025 PricewaterhouseCoopers expects 335 billion dollars. With US leadership in this area, thanks to London's role as a global financial technology centre, the UK's sharing market has grown faster than continental Europe's, competing with the explosive growth of a similar market in China. The volume of transactions of the Russian "access economy" increased by 30.4% in 2018, reaching 510.78 billion rubles, which is quite comparable to Chinese values in dynamics, but significantly less when measured in absolute values [6]. According to the assessment of researchers of the Russian Association for Electronic Communications (RAEC), in Russia the largest share (72.4%) was accounted for by C2C-sales of things and, with a significant lag, followed by services of freelancers (19.2%), ridesharing (2.69%) and car sharing (2.54) [13]. However, the results of the international comparisons given on the degree of development of sharing are rather conditional due to country differences in the interpretation of the nature and forms of access-based consumption and resulting differences in the quantification of the latter.

The link between the increase in the volume of sharing in the economy and the decrease in the share of total consumption in real goods is obvious. The object of the sharing transaction is no longer goods, but the possibility of temporary use of the name. Replacement of the ownership of benefits with access to them on the right of use leads to a relative reduction of the need for real benefits, while the need for services increases due to the service nature of the sharing relations. Demand is shifting from goods to services, to which the sector structure of production and employment responds, demonstrating continued tertiarization.

4 Discussion

Factor analysis of tertiarization shows the variety of forces that affect it, given the country specifics of their action. Thus, while remaining the strongest on the demand side, the factor of household income does not universally determine the degree of tertiarization of national economies. In particular, in most of the largest oil-exporting countries, the share of the service sector in the generated gross public product and total employment is significantly less than in countries with similar average incomes of residents, but a different sectoral structure of the economy [12].

Publications on the development of the tertiary sector often focus on progressive changes in its structure, associated with an increase in the share of production of business intellectual services. Their interpretation as a driver of growth of the entire tertiary sector and even the economy as a whole looks quite reasonable. However, in a market economy, supply can only determine economic dynamics by interacting with demand. Among the latest factors influencing tertiarization on the demand side, the most powerful today is the sharing economy, which has created a mechanism for implementing post-industrial changes in people's consumer preferences in practice. Consistently occurring for many decades, the replacement of the possession of material goods by the provision of services similar to the results of consumption has received economic and institutional support in the form of the practice of sharing goods.

When considering sharing as a factor of tertiarization, it is important to take into account the state of the country's human potential as a characteristic of its economic development. The greatest distribution sharing relationship has received in the developed countries (without a comma) among the urban educated youth. Sociologists describe a typical participant in the "sharing economy" as a young person (18 to 30 years old) with an income close to the national average, having a family with children and living in a large city. The functional prerequisites for participation in access-based consumption are considered to be mobile Internet connectivity and online payments, along with having a good education, supported by the desire to master electronic information technologies and progressive social practices to solve emerging problems. Within this "core" of sharing participants, new practices of consumer behavior were born, implementing the ideas of joint consumption in the conditions of formation of a post-industrial society.

The spread of sharing relationships can significantly reduce in the future the amount of material goods necessary for humanity to live comfortably. In particular, this will help to mitigate the severity of the problems that accompany the process of motorization. For the consumer, the quantitative assessment of the latter should not consist in determining the number of cars, but the distances covered by them. According to a study by Morgan Stanley, if in 2012 car sharing accounted for 4% of the total mileage of cars in the world, by 2030 this share may reach 26% [9]. Taking into account expert estimates that one car sharing car replaces 5–6 privately owned cars it is easy to predict the possible consequences of the rapid growth of car sharing in the global economy for the automotive industry. This example suggests that the cumulative effect of reducing the specific consumption of material goods from all forms of sharing can significantly enhance the process of tertiarization.

Now the circle of participants in access to goods without ownership of them covers, as a rule, personally unfamiliar people. Contacts between specific participants are usually isolated, and trust in the counterparty arises from the presence of his profile, verification and rating. This was made possible by the spread of the Internet, which created a functional infrastructure for the expansion of access-based consumption on a global scale. Along with the channels of communication, the composition and structure of the goods provided have changed: three decades ago, exchanges of cars, clothing, books, music and video recordings prevailed, and now almost any movable or immovable property is the object of consumption on the basis of access. Receiving benefits only for the time necessary to use them is attractive both economically, due to cost savings, and socially, because it allows you to join the comfortable and prestigious consumption of those who are not able to claim it in traditional sales.

Acting as a socio-economic phenomenon of post-industrial society, access-based consumption has evolved from personal offline exchanges (such as children's clothing or music recordings of a favorite artist) driven by intra-group mutual assistance of end-users to online commercially oriented C2C models (such as the eBay online auction) and B2C (such as Amazon and AliExpress online stores) [3]. At the next stage, a more organic combination of social and economic is expected, since the P2P model, which is recognized by many experts as the most promising, built in a peer-to-peer system of relations, along with the equality of the parties, assumes the presence of clearly expressed economic interests among the participants. Like other innovations in access-based consumption, the first application of this model was prosocial. Initially, it was practiced within groups of people United in social networks on thematic areas of consumption (for example, borrowing money or driving cars). Then the model was positively perceived by entrepreneurs and transferred it to specialized online platforms working on the P2P model, as happened during the transformation of spontaneous hitchhiking into organized ridesharing. On the basis of such platforms, there are now services for searching for car companions or ridesharing (in particular, French BlaBlaCar), mutual lending or crowdlanding (for example, American Lending Club, British Zopa), exchange of places of residence among travelers or couch surfing (for example, American CouchSurfing), social car sharing (such as American Turo, French Drivy, German Tarnuca). Time will show the effectiveness of electronic platforms built as decentralized overlay computer networks, in which the functions of the client and server can be performed by any node. Improving the digital organization of consumption based on access will expand the possibility of replacing material goods with services for the use of them, which will be an additional impetus for the passage of tertiarization.

5 Conclusion

Departure from the traditional model of consumption of property-acquired goods leads to tertiarization because of a replacement of material goods by the provision of services. Thus, being released from investments in the purchase of technical devices, the user additionally transfers the increasingly complex and expensive process of servicing them to its owner and pays only for services received. As a result, the total costs of the consumer are reduced while the efficiency of using the consumer's resources is increased. The action of the considered factor of tertiarization has the economic side as well as the social side. The formation of network communities of cooperation in consumption is facilitated by the values which are cultivated by the Internet environment for expanding the circle of communication and more intensive information exchange with certain opportunities for collecting data about specific members of the community.

The observed expansion of sharing relations in consumption as a new reason for the continuation of tertiarization fits seamlessly into the context or the formation of post-industrial economy and society. The speed and success of this socio-economic transformation of the economy largely depends on the country's human potential which allows sufficiently realize the opportunities for further development.

References

1. Böcker L, Meelen T (2017) Sharing for people, planet or profit? Analysing motivations for intended sharing economy participation. Environ Innov Soc Tr 23:28–39. https://doi.org/10.1016/j.eist.2016.09.004
2. Godelnik R (2017) Millennials and the sharing economy: lessons from a 'buy nothing new, share everything month' project. Environ Innov Soc Tr 23:40–52. https://doi.org/10.1016/j.eist.2017.02.002
3. Gollnhofer J (2017) The legitimation of a sustainable practice through dialectical adaptation in the marketplace. J Public Policy Market 36(1):156–168. https://doi.org/10.1509/jppm.15.090
4. Guyader H (2018) No one rides for free! Three styles of collaborative consumption. J Serv Mark 32(6):692–714. https://doi.org/10.1108/JSM-11-2016-0402
5. Esposito P, Patriarca F, Salvati L (2018) Tertiarization and land use change: the case of Italy. Econ Model 71:80–86. https://doi.org/10.1016/j.econmod.2017.12.002
6. Lovata (2019) Sharing economy and collaborative consumption: definition, examples, market size and statistics 2018. https://lovata.com/blog/sharing-economy-collaborative-consumption-definition-statistics-examples.html. Accessed 11 Feb 2020
7. Mitchell A, Strader T (2018) Introduction to the special issue on "Sharing economy and on-demand service business models". IseB 16(2):243–245. https://doi.org/10.1007/s10257-018-0373-3
8. Montalban M, Frigant V, Jullien B (2019) Platform economy as a new form of capitalism: A régulationist research programme. Camb J Econ 43(4):805–824. https://doi.org/10.1093/cje/bez017
9. Stanley M (2019) Shared mobility on the road of the future. https://www.morganstanley.com/ideas/car-of-future-is-autonomous-electric-shared-mobility. Accessed 11 Feb 2020
10. Pais I, Provasi G (2015) Sharing economy: a step towards the re-embeddedness of the economy? Stato e Mercato 105(3):347–378. https://doi.org/10.1425/81604:y:2015:i:3:p:347-378

11. Perepelkin V, Perepelkina E (2017) Reduction of an economy's raw material dependence and the human capital of a country. Comp Econ Res 20(1):53–73. https://doi.org/10.1515/cer-2017-0004
12. Perepelkina E, Perepelkin V, Abramov D, Vlezkova V (2018) Comparative intercountry analysis of tertiarization as a structural component of post-industrial development. Helix 8(6):4550–4557. https://doi.org/10.29042/2018-4550-4557
13. Tiar Center (2019) Economics of collaborative consumption in Russia 2018. Russian association of electronic communications. https://raec.ru/upload/files/raec-sharing-economy-nov2018.pdf. Accessed 09 Feb 2020
14. Souza KB, Bastos SQA, Perobelli FS (2016) Multiple trends of tertiarization: a comparative input-output analysis of the service sector expansion between Brazil and United States. EconomiA 17(2):141–158. https://doi.org/10.1016/j.econ.2015.10.002
15. World Bank & OECD (2020) Services, value added (% of GDP). https://data.worldbank.org/indicator/NV.SRV.TOTL.ZS. Accessed 11 Feb 2020

Digital Platforms in Modern Enterprise Management

S. A. Chevereva and E. S. Popova

Abstract In this article, digital platforms are considered as the basis for the digitalization of modern society. Thanks to their implementation, companies increase the efficiency of their activities, automate the production process, establish a relationship between the client and the company itself and reduce their costs. The degree of influence of digital platforms on management activities is considered, the criteria for choosing a particular platform or a combination of the many options are considered. This is an innovation that will help companies reach a new level and integrate into the digital environment.

Keywords Digitalization · Management · Digital platforms · Enterprise · Optimization

1 Introduction

In the digital age, companies are trying to actively incorporate advanced technology into their workflow. Otherwise, they may be defeated in the race for the client. There is a transformation of the organizational process, as companies are introducing digital platforms that can significantly improve the quality of work and significantly optimize their activities. So what is a digital platform? The digital platform is a basis for the transformation of all sectors of the economy, a system of mutually beneficial relations between participants in digitalization in order to reduce costs through the use of a package of digital technologies. This is an absolute innovation that undermines the usual foundations of doing business [1].

S. A. Chevereva (✉) · E. S. Popova
Samara State University of Economics, Samara, Russia
e-mail: chevereva@yandex.ru

E. S. Popova
e-mail: popovaks07@yandex.ru

S. I. Ashmarina and V. V. Mantulenko (eds.), *Current Achievements, Challenges and Digital Chances of Knowledge Based Economy*, Lecture Notes in Networks and Systems 133, https://doi.org/10.1007/978-3-030-47458-4_58

2 Methodology

The authors have considered digital platforms as a key element in the development of the company and optimize management activities, consider the criteria for choosing a digital platform and assess the degree of their impact on the business, identify the practical application of digital platforms in global companies. To address these issues, existing companies were studied, the main digital platforms used in them were examined. The practical results of the companies' activities were examined by considering their final indicators. World ratings were reviewed to assess the impact of digital platforms on business optimization.

3 Results

Today, the digital transformation is taking place along the way with "Uberization". This is a phenomenon of integration into a mobile payment system and a digital marketing platform for business. The Plazius platform, thanks to which the restaurant client can simply get up and leave the restaurant, having pre-paid his lunch in the application. This is a marketplace where business and consumers find and help each other. This application is used by about 1,500 restaurants and shops and more than 4.5 million customers. That is, the platform itself attracts a customer base and retains it by analyzing the audience and the degree of its involvement in the shopping process. This leads to an increase in the frequency of arrival of guests by 33–65%, and the average check is increased by 25% [5].

Digitalization does not bypass industrial enterprises that are actively introducing digital platforms. So, PJSC Gazprom Neft set itself the task of reaching a fundamentally new level of flexibility, security and efficiency by creating a digital platform. It will lead to total integration of processes and functions within the company, full integration into the digital economy, high speed of changes in the business model, the rapid introduction of new technologies and lower costs for the company's digitalization. This digital platform will combine application and infrastructure platforms, which include capital construction, production, logistics, sales of b2b and b2c, as well as the digital work environment in general, digital models, cloud resources, cyber security and others. These are all the services needed to develop digital business solutions. The success of the platform is the higher, the more companies use it. Therefore, Gazprom Neft aims to develop the EvOil platform. It is an open continuous production management platform. How it will look in practice.

Exact information in real time gets into the digital system of remote online monitoring (GPN-BM Central Office). Metrological data are sent there, which contain all the production parameters available to the operator in the system. Here, the full automation of production management takes place with the complete absence of people in the danger zone. Updating the production plan itself is available online. All this is possible under control using the EvOil platform, which will solve the problem of the fragmentation of control models and makes information available constantly, and not once a month upon request. In addition, the company's logistics will be optimized with the help of smart BitumMAP, where the necessary parameters are agreed in the personal account on the part of the consumer and GPN-BM (representative of Gazprom Neft—Bituminous Materials). Complex automation of production will be available at the BitumPLANT site, where there is the possibility of remote monitoring. Robotic packaging and filling are also added to help the logistics process, which greatly improves efficiency. As well as the analytical system of quality management BitumLAB, where the quality of produced BMs (bitumen materials) is significantly optimized. As a result, there will be prompt processing, shipping and delivery of materials with personalization of the proposal in the personal account of the counterparty. The changes will affect the sales area, where there will be annual and monthly forecasting of demand and prices based on neural networks based on market analysis and competitive activity. All this is accompanied by the use of smart contacts in interaction with customers.

Not so long ago, Huawei introduced the digital platform, the digital platform, which will lay the foundation for the digital world. The company believes that it is necessary to integrate the digital and physical world, so their platform is able to horizontally integrate new ICTs, artificial intelligence, cloud, converged communications, as well as vertically connect devices, peripherals, network and cloud. This platform will help businesses reach a new level, as well as experience the benefits of adaptive implementation of new technologies. About 2 billion rubles will be directed to the development of industrial digital platforms, the Ministry of Industry and Trade spoke about this as part of the implementation of the Digital Economy 2019 program. Project authors will receive subsidies for up to 50% of their projects. A total of 178 applications were reviewed, of which 61 were approved.

4 Discussion

There are different types of digital platforms, among which are operating (Uber, Yandex, Get), innovative (Android, IOS), integration (Apple), investment (Kickstarter). In addition to the above platforms, there are aggregated (Alibaba), social (Facebook), mobilization (CRM-systems) and training (YouTube) platforms. All of them form the digital infrastructure of the market, provide communication delivery of content, and manage users by processing the data array [6]. They also distinguish instrumental, infrastructure, and application platforms, which we will consider in more detail.

The tool platform consists of a software product that is designed to create software and hardware solutions for applications. Participants are platform and solution developers. The main result will be a product (software or software and hardware) for processing information. Examples are Java, Amazon, Bitrix, Intel, IOS. The infrastructure platform brings together the digitalization market to accelerate market entry and provide consumers with solutions for automating the activities of companies using end-to-end digital technologies. The product of such a platform will be

IT—service and information for decision making. The participants are information providers, platform operators, developers (platforms and IT services) and consumers. Examples include Predix, ArcGIS, CoBrain, public services. Finally, the application platform, which is a business model for providing an algorithmized exchange of values between independent market participants by conducting transactions in a single information environment. The result is a transaction, that is, a transaction fixing the exchange of goods or services. Participants in this activity: suppliers of resources and goods (services), consumers, platform operator and its regulator. Examples of application platform: Uber, Avito, Airbnb, Booking.com, Yandex.

There is a relationship between digital platforms, which is as follows: the WebGL tool platform provides a software product for the infrastructure platform, where Google maps and Yandex maps are created, which, in turn, form the application platform Uber and Yandex taxi. Accordingly, the platforms themselves are both participants and consumers of their products. So, Uber users get a quick taxi service, and drivers receive a stream of orders that can increase taxi utilization up to 90%.

It is important to note that the diversity of digital platforms is due to their evolution. Initially, questions were solved on the implementation of applied software and hardware solutions, then a need arose for a communication infrastructure for the delivery of content, which subsequently led to the formation of a digital market infrastructure that implements innovative models. And finally, a digital market infrastructure has been formed that manages users based on the processing of large data arrays [2].

In practice, there are five main users of the platforms: operator (responsible for maintaining the platform's operability and manages the development process of the functional itself); supplier (provides services or products sold through this platform); consumer (the main buyer of a product or service); service provider (creates functional modules); regulator (the body that monitors compliance with the legal framework). Each company is guided by certain criteria when choosing a digital platform. An important aspect is the presence of a single information environment in which the interaction between participants and the information and technological infrastructure will take place [3].

Platform participants work on the principle of "win-win", that is, the basis is the mutually beneficial relationship. The number of participants also determines the scale of the activity and is a significant criterion. The platform should provide the effect of reducing transaction costs in the interaction of various participants. Moreover, all processes must be fully algorithmic [4].

5 Conclusion

In modern conditions, the digital platform is becoming a key element in the development of the company, as it allows you to integrate all processes together, reduce the cost of digitalization and increase the speed of changes in business processes [5]. It is important to follow all the criteria when choosing a digital platform that will allow you to structure tasks, create technological unity and a single information space, automate processes as much as possible, increase the predictability of malfunctions and possible accidents, and identify data that is of particular interest to the company. Digital platforms are the future, so it's especially important to keep up with the times and become part of the digital wave.

References

1. Allen DK, Irnazarow A, McLauglin F (2019) Practice, information and the development of a digital platform. Proc Assoc Inf Sci Technol 56(1). https://doi.org/10.1002/pra2.101. Accessed 11 Dec 2019
2. Gribova VV, Velichko AS, Kolmogorov AV (2019) Digital platform concept for business activities. Inf Technol 8(25):502–511. https://doi.org/10.17587/it.25.501-511
3. Khelfaoui M, Sedkaoui S (2020) Digital platforms and the sharing mechanism. In: Sharing economy and big data analytics, pp 51–60. https://doi.org/10.1002/9781119695035.ch4
4. Radjenovic T, Janjić I (2019) The importance of managing innovation in modern enterprises. Economics 65(3):45–54. https://doi.org/10.5937/ekonomika1903045J
5. Sedkaoui S, Khelfaoui M (2020) Sharing economy and big data analytics. Wiley, Hoboken. https://doi.org/10.1002/9781119695035.ch4
6. Zutshi A, Nodehi T, Grilo A, Rizvanović B (2019) The evolution of digital platforms. In: Shrivastava AK, Rana S, Mohapatra AK, Ram M (eds) Advances in management research. CRC Press, Boca Raton. https://doi.org/10.1201/9780429280818-3. Accessed 02 Feb 2020

On the Development of an Integrated Information System of Municipal Finance Management

A. A. Petrogradskaya

Abstract The article covers problems and prospects for the development of an integrated information system for managing municipal finances. Using this system is the most important way to increase the efficiency of municipal budget management. Noting the successes in the development of the Electronic Budget system at the federal and regional levels, the author indicates that at the municipal level, the development of integrated financial management information systems is complicated by a number of problems, among which there is not only a lack of the necessary technical equipment and software, but also proper interaction within municipal finance management systems.

Keywords Electronic Budget · Integrated information system · Municipal finance · Local budget · Intergovernmental regulation

1 Introduction

World experience and the practice of the functioning of the domestic economic system indicate that information systems occupy an important place in the socio-economic space of most civilized countries. For this reason, one of the functions of a modern state is the accessibility and transparency of information and the provision of access to it to the population. One of the signs of democracy and the exercise of constitutional rights of citizens is the provision of information on the functioning of the state budget, the opportunity to take part in its formation.

Of particular note is the problem of access to information on the state of local budgets and their management. The relevance and importance of this problem is due to the fact that it is local budgets that are the instrument of economic growth of municipalities, provide the most efficient spending of local finances, which ultimately contributes to the growth of investment attractiveness of municipalities and improve

A. A. Petrogradskaya (✉)
Samara State University of Economics, Samara, Russia
e-mail: petrogradckaya@yandex.ru

S. I. Ashmarina and V. V. Mantulenko (eds.), *Current Achievements, Challenges and Digital Chances of Knowledge Based Economy*, Lecture Notes in Networks and Systems 133, https://doi.org/10.1007/978-3-030-47458-4_59

the quality of life of its residents. The purpose of this study is to analyze the problems and prospects for the development of an integrated information system for managing municipal finances.

2 Methodology

The main method used in this study is general scientific dialectic method of cognition. In addition, the following research methods were used in the study: the method of formal logic, the comparative legal method and other. Using this research method revealed the most common patterns of development of the municipal finance management system. Also, to write the article, an analysis of the specialized literature and legal acts on the topic of the study was carried out.

3 Results

The scientific novelty of the study is an objective analysis of the problems of developing an integrated information system for managing municipal finances, comparing it with the regional level, identifying the conditions for the effectiveness of the functioning of the studied system, and proposing solutions to the problems identified. The lack of proper interaction within the municipal finance management system happens not only with electronic budget system, but also with other information systems of federal operators. Such interaction problems naturally lead to an increase in additional costs on the part of municipalities. At the same time, the federal center does not take the necessary actions to form interagency cooperation within the public finance management system. Moreover, study results show there is a rather strict accountability of local government to the federal center, both with the use of levers of financial control and intergovernmental regulation, and simultaneously with the use of centralized (federal) information systems.

4 Discussion

One way to improve the efficiency of municipal finance management is to create an integrated information system [10]. Local finance management issues are reflected in the “Concept of Regional Informatization” approved by the Government of the Russian Federation [5]. In this legal act it is noted that it is necessary to ensure the integration of municipal finance management information systems of state authorities of the constituent entities of the Russian Federation and the local government with the “Electronic Budget” system at the regional level.

The “Electronic Budget” system was developed as part of the “Public Finance Management and Regulation of Financial Markets” program, approved by the Government on April 15, 2014 [4]. The developers of the Electronic Budget system studied the experience of more than a dozen countries, met with the creators and users of similar systems in France and Brazil, in the countries of the Asia-Pacific region, consulted with leading World Bank specialists, and studied the regional experience and experience of corporate financial management systems.

The analysis of the “Concept for the creation and development of the state integrated public finance management information system “Electronic Budget””, approved by the Government of the Russian Federation on July 20, 2011, by decree № 1275-p [3], showed that its general scheme has much in common with the scheme of integration of the French project processes Chorus (Chorus—the state financial information system for managing expenditures, non-tax revenues and state accounting in accordance with the provisions of the budget reform in legislation of France. Developed and tested in 2008–2011, put into operation in 2012) [8].

The electronic budget in the Russian Federation is an integrated information system for managing public finances aimed at the transparency, openness and accountability of the efficiency of the chief administrators of budgetary funds in exercise of budgetary powers, as well as the creation of a single information space as a tool for improving financial management. The creation of the system is aimed not only at the collection and processing of information for subsequent analysis and management decision-making, but also in order to form and maintain a document flow at all stages of the budget process. “Electronic Budget” state information system is one of the elements of cloud technology, which allows to translate almost all of the accounts of participants in the budget process into the “cloud”. New control systems allow receiving online data based on the analysis of large volumes of data, while increasing the speed of decision-making, instantly responding to changes in the environment, and they also focus on specific users of the system.

Since the system incorporates uniform registers, classifiers, reference books, all business operations carried out by institutions in conducting financial and economic activities, for the budget system as a whole, the application of a unified budget execution methodology should lead to a reduction in duplication of information, since all reference data and documents are entered into the system once. Consequently, the federal electronic system “Electronic Budget” is aimed at eliminating paper workflow, standardizing procedures in order to increase the efficiency of budget expenditures and saving budget funds.

The system allows you to create a single software product in the public domain on the Internet [7]. To date, the concept of the “Electronic Budget” system has been developed and approved at the federal level, and the system itself has been in stage of development for several years.

In the Decree of the Government of the Russian Federation dated June 30, 2015 № 658 “On the State Integrated Public Finance Management Information System “Electronic Budget” there is a paragraph five that follows in recommendation to state authorities of the constituent entities of the Russian Federation and local government bodies to use the service subsystems of the Electronic Budget system for the implementation of budget relations in accordance with the Regulation approved by the decree [6].

In a number of regions (Tambov, Moscow, Tula, Kaliningrad regions, the Republic of Udmurtia, etc.) regulatory and administrative documents on the creation, development and implementation of state information systems called "Regional electronic budget" are adopted. In addition, leading software developers such as NPO Krista and BFT Company in their marketing publications also point to the creation of systems with the same name.

In some cases integrated solutions used in the process of managing regional finances provide up to 90% of the functional capabilities declared under the Electronic Budget. Therefore, even now, such decisions can be called regional segments of the "Electronic Budget", which provide the basic principles proclaimed by the Ministry of Finance of the Russian Federation:

- openness, transparency and accountability of the activities of state authorities and local authorities,
- improving the quality of financial management of organizations in the state and municipal government sector,
- creating conditions for the most efficient use of budget funds and assets of public law entities.

However, despite the results achieved, much remains to be done in developing the regional segments of the Electronic Budget and preparing mechanisms for integration those with the federal level of the Electronic Budget in terms of implementing business process standards, classifiers and forms, information data flows.

As for the municipal level despite the large number of financial management facilities, the development of an integrated information system for managing local budgets is very difficult, and from January 1, 2020, all municipalities should already work fully in the Electronic Budget system and post a large amount of data on a certain unified portal for budget system. Information must be generated both as a whole for the district, and specifically in the context of each urban and rural settlement. The gradual implementation of the "Electronic Budget" will help solve the problem of the lack of awareness of citizens about municipal spending in various areas of life.

An analysis of the problems of developing an integrated information system for managing municipal finances in the Russian Federation revealed a number of problems in this area. So, the serious problem of the implementation of the Electronic Budget system in municipalities is the lack of the necessary technical equipment, as well as software. The transition to new software is always difficult for specialists in institutions. It is necessary to competently organize staff training at the implementation stage and create a high-quality support system [11].

The next problem in the development of an integrated information system for managing local budgets is the fact that each municipality has its own specifics, develops and functions in its own way [2]. In this regard, it is obvious that a unified system of municipal finance management for all will not be effective. Currently, it is undeniable that the "Electronic Budget" in municipalities should be really in demand on the ground, and for this the system should solve the administrative tasks of municipalities, as well as take into account the specifics of each of them. To do this, it should be adapted to the needs of end users, in this case, to the needs of municipalities. In our opinion, this kind of adaptation will be most effective if specialists from municipalities deal with it.

One of the most important problems in the development of an integrated information system for managing municipal finances is the problem of the interaction of the Electronic Budget with municipal financial management systems.

The analysis of the municipal budget management system made it possible to formulate requirements for the functional characteristics of the implemented budgeting information system:

- support and control of budget data exchange regulations,
- scenario planning and scenario analysis,
- consistency control at the data entry stage,
- correct work with time series,
- use of mathematical and financial functions,
- obtaining an array of planned data for each operation,
- the ability to determine exchange rates, budget assumptions and targets,
- use of hierarchically organized budget measurements,
- formation of a set of budget reports,
- means of budget execution control.

In addition to taking into account the functional characteristics of software used in the field of budgeting, an important aspect is the observance of the requirements for computer budgeting programs. Budgeting information technologies should:

- adapt to the specifics of the budget process of the municipality,
- adapt to the financial structure and its possible changes,
- have moving regulations that allow for sliding [9].

The system should be aimed at ensuring the integration of preparation and execution of the budgets processes, accounting, as well as integrate preparation of financial statements and other analytical information of public law entities, state and municipal institutions [1]. Here it is necessary to agree with Yerzhenin, who claims that it was necessary for the local government to organize information exchange, and not to connect the local government specialists directly to a system that they cannot affect the performance of. This actually happened in 2018 when a number of municipalities were not able to connect to the system due to their lack of technical capabilities [12].

5 Conclusion

Thus, in the current situation, it is clearly early to say that the integrated information system for managing municipal finances is developing adequately. However, a solution of mentioned problems will allow the both adequate development and effective functioning of the municipal electronic budget as an independent public law and institutional unit in the digital economy of Russia. So, at present, an integrated information system is a very promising way to improve the efficiency of municipal finance management. The results of its implementation at the federal and regional levels already now indicate an increase in openness, accountability of the activities of local authorities, as well as an improvement in the quality of financial management of organizations in the municipal government sector. At the same time, it is hardly possible to say at the municipal level that the integrated financial management information system is developing properly.

References

1. Aguzanova KA (2016) Increasing the efficiency of budget expenditures in the field of state (municipal) procurement. Bull Tomsk State Univ Econ 4(36):110–121
2. Belousov YV, Timofeeva OI (2017) Rating of regions by the level of budget openness as a tool to improve the efficiency of public finance management. Issues State Munic Manag 4:139–157
3. Decree of the Government of the Russian Federation of July 20, 2011 № 1275-r (as amended on December 14, 2018) On the Concept for the Creation and Development of the State Integrated Public Finance Management Information System "Electronic Budget". http://base.garant.ru/55171780/. Accessed 11 Feb 2020
4. Decree of the Government of the Russian Federation of April 15, 2014 № 320 (as amended on March 29, 2019) On Approving the State Program of the Russian Federation "Public Finance Management and Regulation of Financial Markets". http://base.garant.ru/70644234/. Accessed 11 Feb 2020
5. Decree of the Government of the Russian Federation dated December 29, 2014 №. 2769-r (as amended on October 18, 2018) "On Approving the Concept of Regional Informatization". Retrieved from: http://pravo.gov.ru/proxy/ips/?docbody=&nd=102366731. Accessed 11 Feb 2020
6. Decree of the Government of the Russian Federation of June 30, 2015 №. 658 (as amended on December 14, 2018) On the State Integrated Public Finance Management Information System "Electronic Budget" (together with the Regulation on the State Integrated Public Finance Management Information System "Electronic Budget"). https://rulaws.ru/goverment/Postanovlenie-Pravitelstva-RF-ot-30.06.2015-N-658/. Accessed 11 Feb 2020
7. Grover R, Walacik M, Buzu O, Gunes T, Yildiz U, Raskovic M (2019) Barriers to the use of property taxation in municipal finance. J Financ Manag Prop Constr 2:166–183
8. Mitchell D, Thurmaier K (2016) (Re)defining the disarticulated municipality: budget accountability for networked governance. Public Budg Financ 1:47–67
9. Portnyagina AV, Tyushnyakov VN (2016) Analysis of the requirements for the municipal finance management information system. Int Stud Sci Herald 4(4):605–606
10. Sidorenko EL, Bartsits IN, Khisamova ZI (2019) The effectiveness of digital public administration: theoretical and applied aspects. Issues State Munic Adm 2:93–114
11. Simpson NP, Shearing CD, Simpson KJ, Cirolia LR (2019) Municipal finance and resilience lessons for urban infrastructure management: a case study from the Cape Town drought. Int J Urban Sustain Dev 3:257–276
12. Yerzhenin RV (2019) Municipal electronic budget: it is impossible to execute pardon. State Power Local Gov 3:59–63

Industrial Digital Transformation and Ecosystem Formation Based on Advanced Digital Platforms

O. Pudovkina and E. Ivanova

Abstract Entering a new phase of industrial development under the influence of the 4th industrial revolution, there is a problem of digital technological transformation (it is important to take urgent measures to implement breakthrough development in order to overcome the lag behind world leaders). One of the main roles in solving this problem belongs to the industrial production, whose overall contribution to Russia's GDP is quite high. In this regard, it is necessary to develop an innovative approach to managing the development of the industrial sector under reindustrialization conditions. Using this approach would ensure sustainable development and competitiveness of Russia for the long term, which indicates the relevance of this research. The article is aimed at developing recommendations for the formation of a digital ecosystem of industrial cooperation based on the use of digital platforms. The main research method was a content analysis. This method made enabled to identify the most functional information resources needed for this study. This article presents an approach to the transformation of the industrial complex in the conditions of a deep penetration of digital technologies in the economy. The research result is a digital ecosystem based on the theory of a new industrial society.

Keywords Digital ecosystem · Digital platform · Reindustrialization · Industry · Digitalization · Advanced technologies

1 Introduction

Digital economy is shaping a new industrial development paradigm. There are two areas that emerge from judgments in different professional communities. The first direction claims that the digital economy is a digital platform system that provides

O. Pudovkina (✉)
Syzran Branch of Samara State University of Economics, Syzran, Russia
e-mail: olechkasgeu@mail.ru

E. Ivanova
Michurinsky State Agrarian University, Michurinsk, Russia
e-mail: ivanova_ev@list.ru

S. I. Ashmarina and V. V. Mantulenko (eds.), *Current Achievements, Challenges and Digital Chances of Knowledge Based Economy*, Lecture Notes in Networks and Systems 133, https://doi.org/10.1007/978-3-030-47458-4_60

communication between producers and consumers of goods and services. The second direction is broader. The concept of the digital economy also includes processes and results of activities, as well as communications that take place when performing activities. Digitalization of the economic system entails fundamental changes in the processes of creating the added value [8].

In order to build a digital economy, it is necessary to move to a new paradigm of managing socio-technical systems based on the use of digital platforms. The level of development of the digital economy directly correlates with the development level of the material sphere [1].

Digital platforms in the industrial production are a key digitalization tendency, a global trend and a prospect of digital transformation for the Russian economy. The IT industry is an industry that anticipates the economic growth. The digital transformation of the industrial production is a process that reflects the transition of the industrial sector from one technological mode to another through digital technologies in order to increase the efficiency and competitiveness of enterprises and, in general, the Russian economy.

Digital technologies are transforming business models and influencing how people and companies interact with each other. A new world is being formed around us, and data is becoming the most important raw material of the XXI century. New ways to work with data create new opportunities and change business processes. The major economic trend is the "upérisation"—the process of destruction of traditional industries and companies. Digital platforms that collect large amounts of data are coming to the fore in the new economy. They form their own ecosystems. Full integration into the world economy is possible for the country only through innovations, reaching the world level in its development, and even exceeding it for innovative enterprises.

Today, humanity has entered the digital stage of its development. Digital technology affects all life spheres. Digitalization leads to the disappearance of entire sectors of the economy, businesses and jobs, it changes people's social behavior, influences labor and property relations. Business models of digital platforms are built today on the basis of collecting global user data, which allows us to improve content, services, attract more users and quickly increase capitalization with the help of breakthrough technologies, including artificial intelligence. At the same time, large digital companies become owners of data, and the product is often not specifically the service provided by such platforms, but the data itself.

Russia has breakthrough technologies for implementing the most ambitious tasks in the field of digitalization. We have our own digital business leaders providing global ecosystems, developing a direction that will change our world in the very near future. At the same time, we not only possess them, but also implement them in real life. The country's digital development will follow the path of integrating contactless services into the digital ecosystem of people and companies. Digital technology has become a key element of the global competition, and achieving leadership positions requires defining more global business objectives.

The analysis of directions of the economic and innovative growth, production and social infrastructure [2, 4, 10, 12] and directions of scientific research revealed the importance of consolidating resources of organizational systems for doing business in the digital economy for industrial enterprises. An important step in solving the existing problems of consolidating resources of organizational systems is to ensure interoperability, as the ability of two or more information systems or components to exchange information and use this information [11].

The essence of a proposed concept is to create a digital ecosystem based on the use of a single model of a digital platform for industrial enterprises. The fundamental condition that determines the effective prospects for implementing a digital platform project within the framework of cooperation is that all industrial enterprises are "built and operate" according to the same standards, and use the same standards for presenting information about the life cycle of products (Standards for systems of product development and commissioning, Unified system of design documentation, Unified system of technological documentation, Unified system of technological preparation of production, etc.). From this point of view, the key conclusion for defining the digitalization strategy is that there is a fundamental possibility to develop a model of the digital ecosystem of industrial cooperation that can be configured for specific features of the economic activity [9]. Thus, the article offers a promising model of the digital ecosystem of industrial cooperation based on advanced digital platforms in the reindustrialization conditions.

2 Methodology

To solve research problems, the authors used a set of scientific methods. The theoretical and methodological basis was the basic provisions of the study of processes and phenomena: system analysis, abstraction, analysis and synthesis. The main research method was a content analysis. This method made it possible to identify the most functional information resources needed for this study.

3 Results

Digital platforms are a foundation for digital transformation towards Industry 4.0. Digital platforms are technologies that allow firms to unify, edit, and distribute data on an unprecedented scale [5]. At the moment, a number of digital platforms are used in the Russian industry: at the macro level, they are federal state information systems (SIS), in particular, GIS for industry; at the meso level, they are basic and analytical digital platforms such as Cerebra, Watson IoT, Xively, Relayr, Davra, Everyware, Motiware Melody One, and others. There are several cloud computing platforms available for use in the industrial Internet of things and Big Data analytics [7]. At the corporate level, these are the industrial Internet of Things platform Smart Factory, industrial Internet of Things, industrial IoT, and others.

Let's consider the digital transformation and digital platforms in the industrial production in more detail. The evolution of the production management system in the industrial sector is a set of PDCA cycles, i.e. a decrease in the discreteness of planning and feedback time (Table 1). Every five years a person is squeezed out of the PDCA cycle (from any segment), that is, all world studies show that 70% of the inefficiency in production is because of the mismatch of people and equipment, i.e. people and the material world.

Gradually, people squeezed out of the control system, these functions began to move to computers, when the fact of production was somehow taken into account. For the next five years, people began to be squeezed out of intra-monthly planning, that is, when the computer began to plan not monthly, but weekly. In the world, this is called MRPII. The classic MRPII approach is weekly planning (the industrial production in the USA the 70s, Russia—in the 90s). The next step: gradually the person began to be squeezed out of the dispatching function, when the plan started to be recalculated and if there were problems in production, they had to be re-planned.

Starting in 2017, using the concept of the APS 4 On-line digital platform, when the computer itself optimizes production, a person was squeezed out of virtually all PDSA cycles. So, we can state the fact that the algorithm of the digital platform APS 4 On-line is an unmanned production. Of course, there are old-style machines at Russian enterprises, but the main industry has undergone technical re-equipment.

The Smart Factory digital platform is an integration of the production management system and the execution system. The relation of the production management system is an automatic, human-free recalculation of the production plan for a short interval. In other words, if certain deviations of the production delay occur, no dispatcher or production manager is needed, these functions are performed automatically.

The next digital platform for industry is the Industrial Internet of Things. This is a PDCA-based technology. What happens in the system using this digital module? All Russian enterprises had two production management cultures that saw little of each other. The first culture was called ASU, the second—ASUTP. They both talked about managing production, but they hardly understood each other. Why? Because some people were focused on managing information and material flow, others spoke about managing the production process in a given time and almost did not pay any attention to the information and material flow. With the help of the industrial Internet of Things digital technology, these two production management cultures merge. Now the system from the APS4 On-line digital module issues a control effect on the equipment, reads data in the on-line mode, performs production dispatching if necessary, and then the entire PDSA cycle turns without any human intervention. This is how the control effect occurs in an enterprise with a universal type of production.

The next digital technology, Industrial IoT is a new digital module of the system that receives a task, turns data flow diagrams into an information impact on the equipment, reads information from sensors, and turns data flow diagrams into business processes. The production staff in the resulting scheme is needed not to provide control actions, but to look after business processes.

With the help of the PPMM 2/0 digital module, the business process element is deployed in the data flow diagram of the industrial Internet of Things. In fact, if a

Table 1 PDCA cycles of production management system evolution—reduction in the discreteness of planning and feedback time

Stage	Before 1975	1987 RCCP	1991 MRP II	2001 > APS 1	2004 > APS 2	2014 APS 4 & Mobile	2017 APS 4,5 On-Line
Plan	Person monthly	Computer monthly	Computer weekly	Computer weekly	Computer by the day	Computer On-Line at the request of a person	Computer On-Line independently
Do	Person	Person	Person	Person	Person	Computer On-Line/Person	Computer On-Line/IIoT
Check	Person	Person	Person/Computer weekly	Person/Computer by the day	Person/Computer by the day	Computer On-Line	Computer On-Line
Act	Person	Person	Person	Person by the day/Computer weekly	Person by the day/Computer by the day	Person by the day/Computer several times a day	Computer On-Line
Registrar	Person & Paper	Person & Paper	Person & Paper & Computer	Person & Paper & Computer	Person & Paper & Computer	Person & Tablet/Smartphone/Data collection terminal	IIoT
The frequency of the control	Month	Month	Week	Day	Day	Within a day, on request of a person	On-Line

Source authors

command is written at the stage of the business process to complete the end of the task, this is all passed to the data flow diagram, which generates a certain information impact and starts or stops the equipment.

The new digital product Smart Eam is a system for maintenance and repair of equipment. This is a completely new product that answers a simple question: do you want to manage repairs or do you want to increase the technical readiness of equipment? These are two different aspects. This digital module is designed to increase the coefficient of technical readiness of the equipment. The ideology came from the aviation industry.

The next digital technology is Smart Manager. In fact, it is an extension of business process management. It allows you to build not only business processes, but also to establish a system for executing business processes, in other words, to establish control and distribution within a non-production group.

Thus, having carried out a theoretical review of modern digital platforms for the industrial production, we can talk about using a set of business applications of various complexity in order to create a digital ecosystem of the industrial sector, on which the entire line of products and digital modules discussed above is built. This is a complete ecosystem, a trend in the Industry 4.0, with which it is possible to get ready-made business solutions. Industry 4.0, proposed in 2011, is a kind of product equivalent to the consumer-oriented Internet of Things [3]. Under these conditions, the digital ecosystem cooperates with technological, functional, and infrastructure platforms in the form of industrial, logistics, financial, regulatory, and other digital platform pools (Fig. 1).

In our view, digital platforms for industrial cooperation should be created in the form of digital twins, combined on a sectoral basis with the nature of a specific industry interaction, which can serve as a source of local progress. The digital twin of the industrial sector is a single model that reliably describes all parameters, processes, and interdependencies, both for an individual industrial enterprise and for the entire industrial sector as a whole. The emergence of "digital twins" was a logical result of the development of the concept of digital production and the industrial Internet of Things [6]. Many integrated and interconnected digital platforms that interact within the ecosystem will always benefit from a single digital platform in terms of efficiency and will provide the maximum value from switching to working with it.

The authors developed a process for implementing digital platforms for industrial cooperation within the digital ecosystem (Fig. 2). The fundamental stage is the formation of a competence center for each industrial sector, whose functions should include the cooperation of platforms that assemble industry consortia (digital platform operators) from suppliers of technological solutions, expert organizations, and industrial associations.

The implementation process of the Industrial cooperation competence center within the digital ecosystem should allow the implementation of about 300 digital platforms by 2024. The total number will be higher, but because of integration problems, not all platforms will be able to enter the digital ecosystem. The introduction of digital platforms into the industrial cooperation guarantees an annual acceleration of GDP growth by at least 1%. The application of digital platforms in

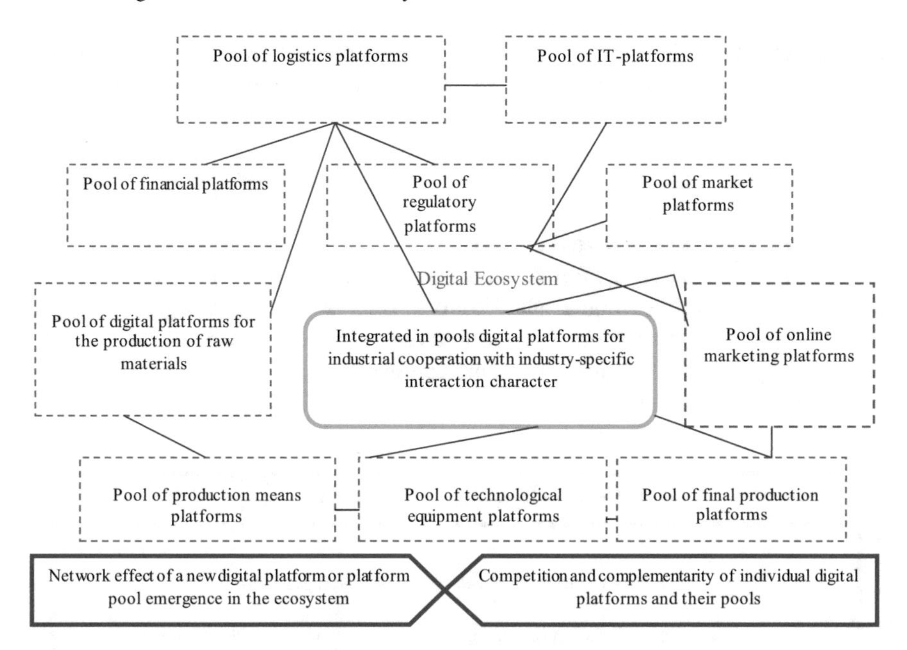

Fig. 1 Digital ecosystem of integrated in pools digital platforms in the industrial production (*Source* authors)

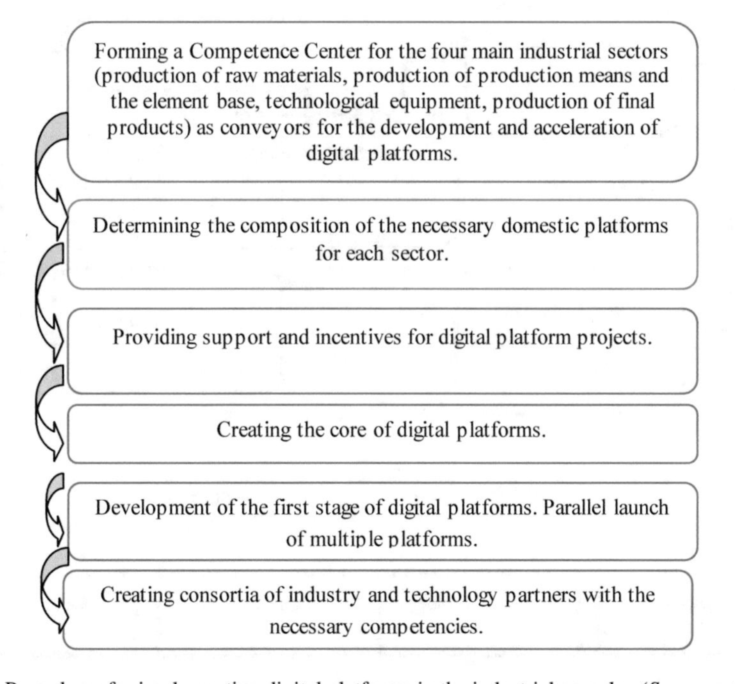

Fig. 2 Procedures for implementing digital platforms in the industrial complex (*Source* authors)

the industry will entail a digital transformation—the transition to a new paradigm of business and production, changes in business processes, competencies, and the entire system of economic relationships. The relations and interdependencies in the process of managing data, knowledge, competencies, projects, changes, and challenges that arise during the implementation of the digital transformation program for industrial enterprises are shown in Fig. 3.

It should be noted that digital transformation should be implemented at the stage of maturity in five areas: the formation of the regulatory framework, advanced technology creating a single digital platform or a pool of integrated digital platforms on the corporate infrastructure, training and retraining of personnel for work in the digital ecosystem, the search and implementation of new technological solutions, the transition to a new business paradigm.

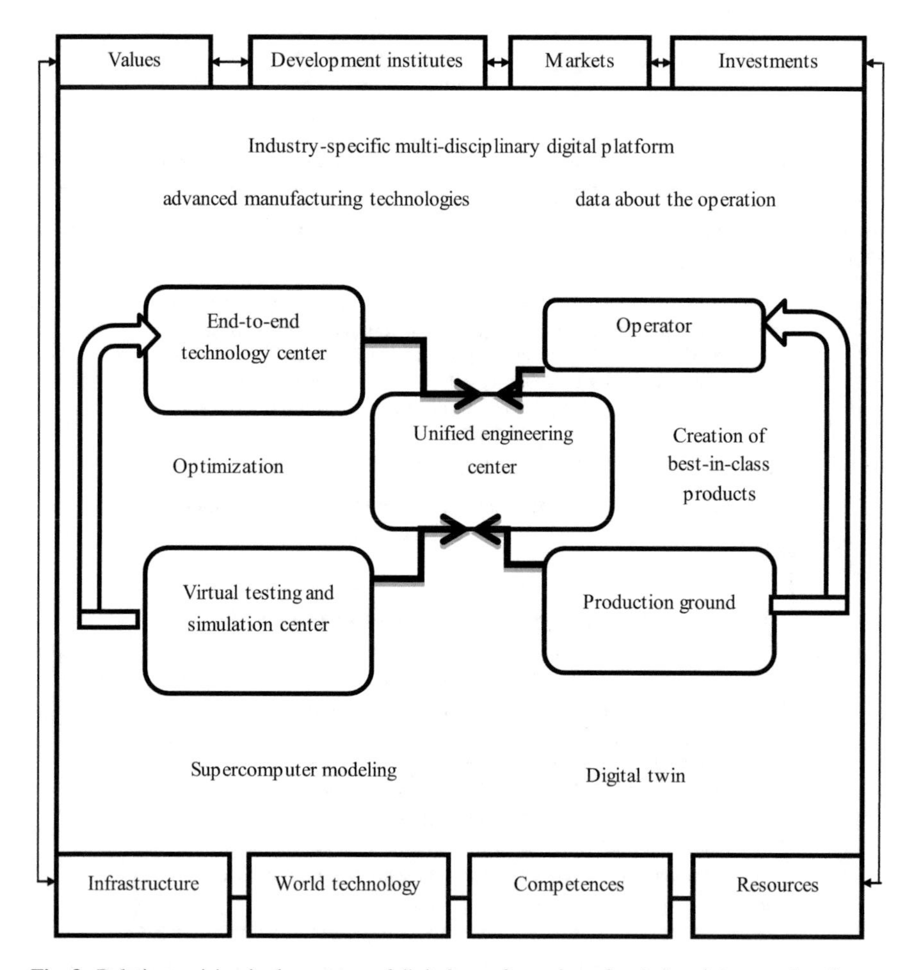

Fig. 3 Relations arising in the process of digital transformation of an industrial enterprise (*Source* authors)

The key technology for digital transformation of industrial enterprises is the creation of digital twins. The digital transformation program should be carried out by the top management of industrial enterprises and requires significant support, qualified personnel, corporate culture and proper reporting. The projected economic effect of the digital transformation of industrial enterprises is as follows: a 10–50% reduction in costs, reduction in production time by up to 4 times, profit growth up to 2 times, an increase in the number of new products by 50–70%, and a 7–15% reduction in the number of equipment units.

4 Discussion

In the prevailing trends of digitalization, most enterprises that want to work effectively should go through a process of digital transformation. In practice, this means creating a system of interconnected business processes that can be described as a digital business ecosystem. The process of transition to a new industrial evolution has passed a number of stages from the 0-th industrial revolution to the structure of changes in industry 4.0. Figure 4 shows the scale of changes that characterize the industrial revolution.

Research has shown that a key factor in the industrial development in the new digital economy is a "digital platform". The authors define a digital platform as a set of integrated tools based on modern digital technologies, the use of which simplifies the

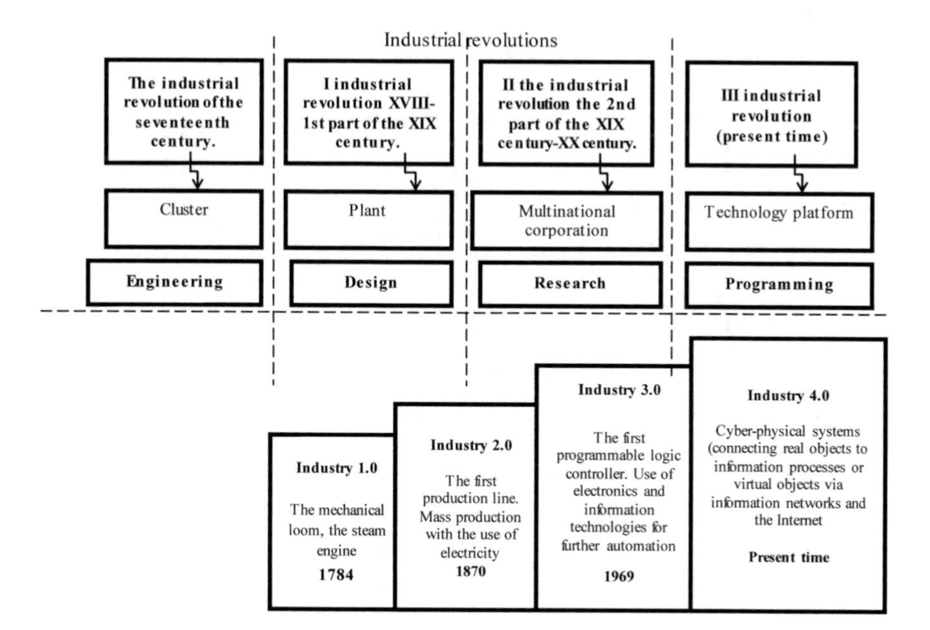

Fig. 4 Stages of the industrial revolution (*Source* authors)

management of functioning and interaction both inside and outside a socio-economic system. Digital platforms set new standards for managing organizations, develop competition, and generate dynamic ratings for industry participants. Therefore, the transition to the fourth industrial revolution is an inevitable innovation process that will result in a fully automated digital production with the prospect of merging into a global industrial ecosystem.

The basis of "Industry 4.0" is technology convergence—a process of convergence and mutual penetration of industrial technologies, in which the boundaries between individual industrial technologies are blurred, and many interesting results arise in the context of interdisciplinary work at the intersection of areas. This leads to structural and spatial transformations of industrial systems. Technologies begin to use the same generalized resource, and the selection leads to a quantitative improvement in the efficiency of already established types of technologies or management forms within the new technical and economic paradigm.

The study showed that the transition to digital business models should be carried out iteratively, with maximum use of the synergy effects, scale and network development. The transition from digital platforms to the ecosystem is shown in Fig. 5.

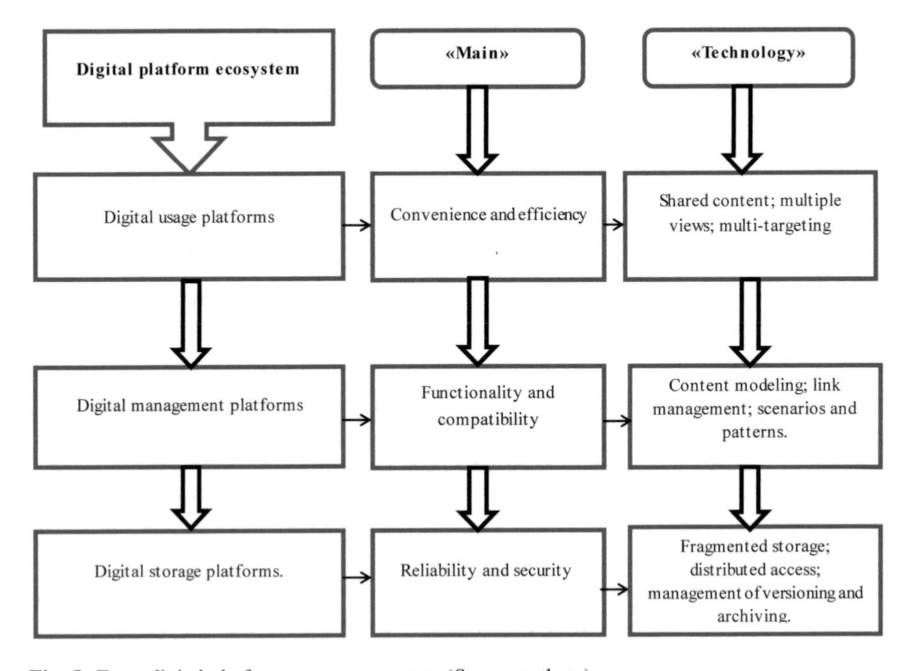

Fig. 5 From digital platforms to an ecosystem (*Source* authors)

5 Conclusion

The results of the study are of great economic importance, which is to justify the formation of conceptual provisions for the development of the industrial complex in the conditions of reindustrialization, and which can serve as a factor for increasing its competitiveness.

The article investigates stages of the industrial revolution and digital transformation of the industrial production, characterized by innovative digital technologies. The authors came to the conclusion that in the conditions of re-industrialization and the implementation of breakthrough development of the Russian economy it is advisable to use a digital platform of industrial cooperation through the formation of digital ecosystems, creating a common ground for the implementation of the digital agenda, which aims to revitalize and strengthen business in obtaining its high performance.

References

1. Akberdina VV (2018) Transformation of the industrial complex of Russia in the digital economy. Izvestiya Ural State Econ Univ 19(3):82–99
2. Aleksandrov AA, Larionov VI, Sushchev SP (2015) Uniform methodology of the risk analysis of emergency situations of technogenic and natural character. Herald Bauman Moscow State Tech Univ Ser Nat Sci 1(58):113–132
3. Bobrova VV, Berezhnaya LYu (2019) Digitization of the transport industry in Russia: problems and prospects. In: Nazarov A (ed) Proceedings of the 1st international scientific conference "Modern management trends and the digital economy: from regional development to global economic growth". Advances in economics, business and management research, vol 81. Atlantis Press, Paris, pp 174–177
4. Burkaltseva DD (2017) Points of economic and innovative growth: a model for organizing the effective functioning of the region. Mod Innov Res 1(8):8–30
5. Cenamor J, Parida V, Wincent J (2019) How entrepreneurial SMEs compete through digital platforms: the roles of digital platform capability, network capability and ambidexterity. J Bus Res 100:196–206
6. Gromova EA (2019) Digital economy development with an emphasis on automotive industry in Russia. Espacios 40(6):27
7. Kabugo JC, Jämsä-Jounela SL, Schiemann R, Binder C (2020) Industry 4.0 based process data analytics platform: a waste-to-energy plant case study. Electr Power Energy Syst 115:105508
8. Molotkova NV, Khazanova DL, Ivanova EV (2019) Small business in digital economy. In: Mantulenko V (ed) Proceedings of the 17th international scientific conference "problems of enterprise development: theory and practice". SHS web of conferences, vol 62. EDP Sciences, Les Ulis, p 04003
9. Pudovkina O, Sharokhina S (2019) Digital platform of industrial cooperation – innovative direction of regional industry development. In: Mantulenko V (ed) Eurasia: sustainable development, security, cooperation. SHS web of conferences, vol 71. EDP Sciences, Les Ulis, p 04016
10. Radzievskaya TV, Mishina AV (2016) Model of the quality evaluation of management of the "Hsuman capital – innovative technologies" system in the presence of counteraction to economic development. Bull Voronezh State Univ Econ Manag Commun 2:5–12

11. Zatsarinnyy AA, Shabanov AP (2019) Model of a prospective digital platform to consolidate the resources of economic activity in the digital economy. Procedia Comput Sci 150:552–557
12. Zatsarinnyy AA, Gorshenin AK, Volovich KI, Kolin KK, Kondrashev VA, Stepanov PV (2017) Management of scientific services as the basis of the national digital platform "science and education". Strateg Prior 2(13):103–114

Global Challenges and Infrastructure Background of the Economic Security of Regional Development

V. A. Noskov

Abstract The urgency of the analyzed issue is determined by the critical role of infrastructure sectors, particularly a transport complex and a system of science and education, digital economy of the economic security in the sustainable economic development of territories in the modern global economy. The purpose of the article is to make a conclusion about the role of the formation of modern city agglomerations in the economic security (on the example of the polycentric Samara-Togliatti city Agglomeration—STA), on the basis of the formation of their scientific, educational, information, innovative and transport frameworks. The effectiveness of economic functions in education and science is implemented and determined taking into account the institutional and organizational characteristics of the territory, it forms the structure and quality of the scientific and educational frame of territorial branch clusters. It seems appropriate to consider economic clusters of the Samara region developed and mature with established research, educational and information and transport frame. The important prerequisite for the implementation of competitive advantages and the economic security of the Samara region, which has a system of high-tech economic clusters and their system of scientific and educational and information frameworks.

Keywords Infrastructure sectors of the economy · Economic security · Scientific and educational complex (SEC) · Scientific and educational framework of the sectoral cluster · Samara-Togliatti city agglomeration

1 Introduction

The need of the Russian economy for the development of agglomerations is very up-to-date. In many countries the city agglomeration is the main form of living rather than in cities. The optimization of labor resources, attraction of the population within the borders of Samara-Togliatti agglomeration will contribute to the transformation of

V. A. Noskov (✉)
Samara State University of Economics, Samara, Russian Federation
e-mail: noskov1962@inbox.ru

S. I. Ashmarina and V. V. Mantulenko (eds.), *Current Achievements, Challenges and Digital Chances of Knowledge Based Economy*, Lecture Notes in Networks and Systems 133, https://doi.org/10.1007/978-3-030-47458-4_61

the city districts in agglomerations with intensively populated suburbs which provide the chain formation of large urbanized systems—industrial, scientific and cultural centers in the structure of settlements of the Samara region. The agglomeration development as the territory with attractive environment for living and working is becoming very actual, because of the perspective decrease of workforce [13].

The Samara region has powerful Samara-Togliatti agglomeration (the third place for the number of the population in the country—more than 2.2 million people), more than 85.3% of the population of the region. In the Samara region there are the most powerful and competitive sectors—automotive sector, aerospace, petrochemicals, energy and agriculture complexes. They form clusters, the core ones of which are large industrial enterprises, scientific and educational organizations [12]. The development of infrastructure sectors of the regional economy in the context of globalization is an important tool for achieving the social, economic, foreign policy, the economic security and other objectives, enhances the quality of life [6]. A special role among infrastructure sectors in the sustainable development of the modern economy plays the transport sector and the education system and science, digital economy. On the territory of the Samara region there was formed the largest transport system in the Volga Federal District that provides transport links of the federal and regional levels in all directions.

A favorable geographical location of the region also contributes to this: the proximity of sales markets with neighboring countries, the presence of junctions connecting traffic flows in the direction of international transport corridors (ITC). If there is the sustainable development, the growth rate of turnover, energy consumption and employment should be lower than the rate of the GRP growth. In the formation of market relations, the approaches to the further development of the transport planning must be based on the determination of final results of this development. One of indicators, acting as the final result for the industry, is a cargo intensity of the economy.

A characteristic feature of both Russia, as a whole, and the Samara region is a relatively high value of the cargo intensity compared with the global rate. Thus, the level of the traffic load on the GDP, expressed in ton-kilometers of transportation per unit of the parity GDP, at the beginning of the XXI century Russia was inferior more than 3 times to the United States, 4 times—to China, 29 times—to Germany and 143 times—to Japan. It is obvious that the industrial growth is impossible without the transportation growth. However, their growth should be less than the GDP growth. This is confirmed by a retrospective analysis of changes in the cargo intensity of the GDP: the transition of developing countries to the developed category, and the overall social progress is accompanied by a decrease in the transport burden on the economy. In the literature the criterion of the region's transition from a developing to a developed one by the cargo intensity of the GRP: 1–2 ton-km./US dollars [15]. This new quality of the regional economy will provide a highly developed scientific and educational complex (SEC) and it's forming high technologies in economic clusters. SEC of the region as well as regional economic clusters unite enterprises, universities, scientific research institutions, laboratories, information centers, technological parks located on the territory of STA.

The tendencies of the economic development in the regions of Russia are more and more defined by the quality and quantity of labor resources. The gradual transition of the economy to the postindustrial phase increases the impact of information technologies on all spheres of life, intellectualization of production, the dynamics of which depends on the quality of workforce and makes an increasing impact on the labor potential. Due to the innovative quality of the labor force it is possible to increase the competitiveness in all areas of the economy and to create breakthrough technologies. The decisive role in this process is played by the educational system which is the basis of the formation and development of qualitative characteristics of the labor force. Qualitative and quantitative characteristics of the labor force are closely related, quantitative indicators are the basis for planning the structure of labor force quality, providing human resources for all areas and levels of economic activities. The most visible indicator of the evolution in labor quality is the level of education in the region. Vocational education in the modern world economy acts as a catalyst of many social and economic processes, gradually moving from the service function to the function of forming environment and shaping the future and the economic security of society [5].

The development of STA, on the territory of which it is expected to form high-tech industries, will promote productive jobs creation and it will develop the educational system of the Samara region. SEC of the Samara region is represented by more than 30 higher educational vocational institutions. These are universities including the research ones, academies, and institutes functioning on the territory of STA: 17 state and municipal higher institutions and 13 non-state higher institutions and their research and scientific commercial sub-divisions.

Evaluation of the SEC potential of the Samara region on the structure of its subdivisions allows making a conclusion that the generating (scientific and educational) function of SEC is performed more actively in technical universities. The leader is Samara State University (SSU) since March 2016, and it has in its structure 7 scientific research institutes, 8 branch scientific research laboratories, 17 scientific research laboratories, 2 scientific research groups, 6 scientific research centers and the university got the status of the research university in 2009. In classical and humanitarian universities the generating function activity is substantially lower which adversely affects the quality of the scientific, commercial and educational activities. The effectiveness of economic functions implementation aimed at achieving a maximum economic growth and development of the economic area of the Samara region by university ensures its sustainable competitive position and, therefore, the survival in the long run.

The household functions of some universities and scientific institutions are individualized and can be performed independently from their realization by other territorial scientific and educational organizations, but economic functions are inherent at universities and institutes as part of the interrelated complex with equal participation in solving the problems of the regional development. This uniting characteristics of economic functions of education and science is an objective basis for the formation of the scientific and educational and information framework of territorial-industrial clusters and innovation framework in the region.

2 Methods

It is necessary to use an institutional approach in the research of the fundamental role of education in the economic growth. Economic functions of science and education are concretized by economic functions of the scientific and educational frame of territorial and branch clusters of the Samara region [4]. Taking into account the economic security, the economic growth and qualities of the economic space improvement of the Samara region as the target priorities and institutional peculiarities of the territory and organizational characteristics of significant conditions we can make the following assumption:

The effectiveness of economic functions in education and science is implemented and determined taking into account the institutional and organizational characteristics of the territory, it forms the structure and quality of the scientific and educational frame of territorial branch clusters—units of the scientific and educational system of the region and links realizing its ability to be the driving force of innovations. The efficiency increase of economic functions in education and science makes them more effective for the economic security, the economic development and evolution of the territory, the modern human capital development of the region by the regional branch economic clusters development. These processes are accompanied by:

First, expansion of the functional field within the scientific and educational and information framework—the increase in the number of functions performed by the units of the scientific and educational frame in innovative processes of the cluster. Expansion of the functional field in education and science provides the innovative development of the regional economy and its economic growth.

Secondly, the complexity of the structure of the internal (between the links of the scientific and educational frame) and external (between the links of the innovation and transport frame: its scientific-educational and educational-production parts) connections. Complicating the relations between the links of the scientific and educational frame implements the task of integrating scientific, educational, information and industrial space, and increases the competitiveness of the territory. All these processes are based and take place in the Samara region, first of all, on the territory of STA defining its formation and development.

3 Results

3.1 Ways of Reducing the Cargo Intensity of the Regional Economy

In 2015 the cargo intensity of the Samara region, excluding the pipeline transport was slightly higher than the value of this indicator in Russia on average, but it was lower than in such regions of the Volga Federal District as Tatarstan and the Kirov region. Despite the fact that the cargo intensity of the Samara regional economy significantly exceeds the national average, it is worth noting that from 2011 to 2017, in a relatively economic growth, there was a negative trend to its increase (Table 1). The changes of this trend will contribute to a new quality of the economic growth.

There are a number of countries with extremely high decrease in the cargo intensity. They include India (the first years after independence) and Japan (this country does not have its own natural resources and it is highly dependent on their transportation), and in recent decades—the United Kingdom. In the UK the system regulating the cargo intensity index is very important. In this country, as in any other, the GRP growth is outstripping the cargo growth [16]. At the same time the main source of reducing the cargo intensity becomes a decrease in the average length of traffic by a transmitting short line (up to 250 km, but not up to 150 km as it was considered in Russia before) from the railway to road transport, which in turn implies the priority development of the road network. The main source of reducing the cargo intensity of the GRP is to reduce the transportation distance, despite the fact that its value will rise (after growing economy). It is estimated that 77% of the cargo intensity could be shorter or it is not needed [1]. According to forecasts and assessment of the Ministry of Economic Development, Investments and Trade of the Samara Region (MEDT SR) by 2040 the value of the cargo intensity of the economy could fall to 1.05 ton-km./US dollars, that displays the Samara region by this indicator on a par with the most developed countries. Today, the transport costs for many sectors of the economy are a burden. The commercial speed of commodities' transportation from a producer to a consumer in Russia is 2–3 times lower than in Europe and the United States. The share of transportation costs in the cost of goods for production purposes ranges from 5 to 35%, with average values in the country: in agriculture—7%, in the food industry—8.9%, in capital construction—13%, in logging—22% [9]. In the USA, for example, the transport component varies on average from 4 to 8%, and in only five sectors it exceeds more than 10%: in the pulp and paper industry (11%), in furniture (12%), in food (13%), in timber (18%), and petrochemicals industries (24%) [15].

Table 1 The cargo intensity of the Samara regional economy

Indicators	2011	2012	2013	2014	2015	2016	2017
The GRP (million rub.) (in comparable prices of 2017)	865122	935197	970734	780470	864810	896772	941611
The cargo turnover by modes of public transport, billion ton-km							
In total	164,8	162,4	163,7	146,3	142,7	148,5	146,3
The cargo intensity (ton-km./rub., the GRP)	0,19	0,17	0,17	0,19	0,17	0,17	0,16

Source author based on [14]

3.2 Problems of the Passenger Growth Capacity of the GRP

The dynamics of the passenger turnover capacity in the Samara region does not correspond to global trends: in 2017 there was a significant decrease in this indicator compared to 2000, despite the economic growth in 2000—years, both in Russia as a whole, and in the Samara region (Table 2). However, the analysis of the population mobility shows that such negative trend in the passenger capacity can be explained, first of all, by a faster pace of the GRP growth compared to the passenger turnover. In addition, it requires analyzing the issues of increasing the people mobility in connection with the increase in motorization of the population and the possible unreported trips by public transport. The absolute value of the passenger capacity in the Samara region in 2015 was lower than in Russia as a whole and than in many countries [10, 11]. It is known that the higher the level of the industrial and social and economic development of the country is, the higher its rate of mobility and the people mobility are. In the Western countries, only 35–40% of the passenger turnover accounts for passenger trips (in Russian—60% or more).

Certain studies have shown that the transport accessibility significantly affects the increase in per capita of the GDP (GRP), while the GDP growth in itself does not guarantee the transport accessibility growth [7]. The population mobility of the population in 2003 was 4300 km for 1 person per year. For the Soviet Union in 1990 it was 8700 km, whereas in developed countries, primarily in Western Europe, this figure now exceeds 10 thousand km for a person per year (in Germany—10.7 thousand km). In the Samara region in 2017 it was equal to 1388 km for 1 person per year. Relatively to other regions of Russia, the Samara region has the average population mobility. The Russian transport strategy provides for the growth in mobility and the population mobility by 2020. It is primarily due to the growth in the vehicle fleet and to the increase in the share of passenger cars in the passenger transportation from 33% to 53% [14]. Every adult resident of the Samara region does less than 2000 km per year with social and cultural purposes. The development strategy of the unified

Table 2 The passenger capacity of the GRP of the Samara region

Indicators	2000	2002	2005	2015	2016	2017
The GRP, billion rub.	140,4	206,3	402,3	694,8	896,8	941,6
The passenger transportation burden (by public transport, thousand people)						
In total (including urban electric transport)	1110,7	1032,7	442,4	302	264	296
The passenger turnover (by public transport, million passenger-km.)						
In total (including urban electric transport)	10247	9804	6670	4914	4397	4462
The passenger capacity (passenger-km./1USD, the GRP)	2,3	1,4	0,4	0,25	0,19	0,16
Mobility (passenger-km./1citizen)	3202	3073	2101	1526	1368	1388

Source author based on [14]

tariff net of the Samara region, within which the passenger transport will develop, guarantees the mobility growth of 1 resident with social and cultural purposes at least 3 times by 2040.

The growth of quality of life is accompanied by the growth of the population mobility, primarily due to trips with social and cultural purposes. In the last 15 years the population mobility of the Samara region has decreased in all branches of public transport and modes of public transport. This negative trend has not corresponded to the development strategy of the region and should be overcome in the long term. We believe that by 2020–2025 the passenger capacity of the GRP of the Samara region can be increased 1.5–2 times under any scenario of the GRP growth. It's enough to reach the rate of the population mobility, which was in the country in 1990.

3.3 The Scientific and Educational Complex as the Most Dynamic Sector of Modern Economy

Education and science, digital economy as the factors for the stable development and the economic security of the regional economy are the most important modern branches [2, 8]. The links of the scientific and educational and information framework of the territorial branch economic cluster in the region are the combination of scientific and educational and information resources and integration ties which are prerequisites for achieving the targets of region's economic and spatial development. The practical differentiation orientation of household and economic functions in education is fundamentally different by the orientation on various development indicators and by the institutional apparatus of the regional economy and the characteristics of institutions linked to the objective of reforming the education system in Russia. The problem solving of productive use of education and science in Russia is connected with the recognition of their economic functions as priority ones. The contradiction between the developed system of education in Russia and low productiveness of its functioning reveals itself in the disparity between the high coverage of higher education and a low growth rate of its GDP (GRP).

The task of expanding economic functions within the scientific and educational and information framework of territorial and sectoral economic clusters requires a focused, efficient and full use of already formed resources, the effective use of which should be realized in the economic growth, the economic security and competitiveness of the territory. In the Samara region the reserve of such growth is the development of STA. With regard to the regional economic growth the scientific and educational and information framework acts as a producer of new knowledge—a tool for the growth and the source of human development and capital. In this sense, being the tool of the fundamental economic resource of the modern economy, education and science is the driving force for the sustainable economic growth [3].

The concentration of innovation technologies in the economic space of the region forms territorial innovation and information saturation. Organizational and institutional abilities of the scientific and educational frame in regard to innovation and intellectual saturation of the territory leads to the additional increase of the GRP due to the increase of capitalization of the regional economy [16]. This development of the scientific and educational and information frame improves organizational characteristics of the territorial and branch clusters and the institutional construction of the whole economic space of the Samara region. At the same time the formation of branch clusters in the region is the necessary prerequisite for these processes and the efficiency of the cluster forming in its turn depends on the company's effectiveness of internal and external ties determining the qualities of scientific and educational, innovation and information and transport frame structures of the territorial and branch economic clusters.

3.4 Compliance with the Scientific and Educational and Information Complex and the System of the Regional Economic Clusters

Table 3 compares the structure of the scientific and educational complex of the Samara region and the structure of its territorial and branch clusters. It allows revealing the ability of the regional system of science and education to fulfill the functions of economic clusters in the region [11]. Table 3 shows that the structure of the scientific and educational complex of the Samara region is actually formed in accordance with the structure of industrial clusters of the territory and follows the structure of STA formation. The previous periods of the system development of science and education in the Samara region contributed to the fact that the links of the scientific and educational potential of the region correspond to the sectoral focus of its developing economic clusters and make a good potential to realize the functions of their scientific and educational frame. In a number of economic clusters, their scientific and educational, innovative and information and transport framework has already been developed to a high degree, in others—they are only in the process of formation and they require a significant effort for their development both from the governing bodies of the Samara region, and in the self-development of universities and scientific organizations.

This, in our opinion, the most important problem to develop a more innovative and competitive economy of the region must be taken into account in the formation of the concept of STA development. The analysis of Table 3 shows that practically all economic clusters and their scientific and educational frames spread on all the territory of STA making it economically, scientifically and educationally uniform. It seems appropriate to consider economic clusters of the Samara region developed and mature with established research, educational and information and transport frame. In STA they include: petrochemical, aerospace, automotive, energy and agro industrial

Table 3 The interrelation in structures of territorial and branch clusters and the scientific and educational complex of the Samara region

Branch clusters of the Samara region	Structure elements of the scientific and educational frame of branch clusters
1. Oil-producing (petrochemical) cluster	
Core–oil-producing companies—the main company is JSC "SamaraNeftegaz" (Rosneft); JSC "Saneko", JSC "Samarainvestneft"; -petrochemical companies: JSC "Kuibyshevazot", "Togliattikauchuk" Ltd., JSC "Novokuibyshevsk Petrochemical Company"	The core—Samara State Technical University (SamSTU); Togliatti State University (TSU), Samara State University of Economics (SSEU), JSC "Samaraneftegeophizika", "SamaraNIPIneft" Ltd., JSC "Samaraneftehimproject"
2. Aviation and space cluster	
The core—"Central Design Bureau Progress", JSC "SNTK named after Kuznetsov N.D."; JSC "Motorostroitel", JSC "Aviakor-Aviation Plant", JSC "Aviaagregat", JSC "Hydroavtomatika", JSC "Metallist-Samara"	The core—Samara State University (SSU); The structure: SamSTU, Volga State University of Telecommunications and Information Technologies (PSTITU), SSEU
3. Automobile cluster	
The core—JSC "AvtoVAZ"; JSC "GM-AvtoVAZ", JSC "AvtoVAZagregat", JSC "VAZinterservice", JSC "Samara-Lada", JSC "Motor-Super"	The core—TSU; SamSTU, SSU, SSEU, Volga Academy of Water Transport Engineers (Samara branch) (VA WTE)
4. Energy cluster	
The core is being formed "Srednevozhskaya Gas Company" Ltd., JSC "Samara City Electric Chains", JSC "Volzhskaya TGK", JSC "RusHydro"	The core—SamSTU; The structure: SSU, TSU, SSEU, Samara Branch of Physical Institute of the Academy of Science
5. Agroindustrial cluster	
The core is being formed. JSC "Agroprogress", Brewery "Baltika", "Danon", "Nestle", "Wim-Bill-Dan", closed JSC "SV-Povolzhskoye", closed JSC "Severnyi Kluch", closed JSC "Alikor-Trade", AIC "Krasnyi Kluch" Ltd., closed JSC "Lunacharsk"	The core—Samara State Agricultural Academy (SSAA); The structure: TSU, SamSTU, SSEU, Research institutes: JSC Scientific Research Institute named after Tulaikin S.M., The Institute of Ecology of the Volga Basin
6. Transport and logistics cluster	
The core-Kuibyshev Railroad—the branch of JSC "RR"; JSC International Airport "Kurumoch", airlines making flies via airport "Kurumoch"; JSC "Samara River Port"	The core is being formed. Samara State University of Railways (SamSRU), SSU, SSEU, VA WTE, TSU, SamSTU
7. Tourism and recreation cluster (potential)	
The structure is being formed	The core is being formed. SSCAA, SSU, SSEU
8. IT cluster (potential)	
The structure is being formed	The core is being formed. PSTITU, SSU, SSEU, SSTU, SSU, TSU

Source author

clusters. In their structure there are universities and research organizations which educational and scientific activities are fully focused on the needs of the cluster.

In addition, the structure of their scientific and educational framework has the core—Profile University [11]. Potential territorial and branch clusters of the region have not got a clear scientific and educational and information frame or a core yet—the leading profile university that is oriented towards the needs of the cluster.

4 Discussion

On example of the development of the Samara-Togliatti agglomeration we examined the role of the transport sector, education and science, the digital economy, leading universities and research centers in the region, in the formation of modern urban agglomerations. Having determined the importance of the services of the most important infrastructure sectors in the economic development of the territories in the global economy, we conclude about the new crucial role of the transport complex, education and science, digital economy, leading universities and research centers in the development of the regions in the transition to a modern industrial society. This role is manifested through the development of regional sectoral economic clusters, on the basis of the formation of their scientific, educational, information, innovative and transport frameworks, the growth of their modern human capital. Thus, it seems that the presence of the leading specialized university, surrounded by effective scientific infrastructure, forming the core of the scientific and educational and information frame and focused on the needs of the cluster, is the most important criterion for the successful development, prospects and competitiveness of the cluster.

The important prerequisite for the implementation of competitive advantages and the economic security of the Samara region, which has a system of high-tech economic clusters and their system of scientific and educational and information frameworks, is the further development of the transport and logistics framework of the STA. The problems of practical implementation of the factors that strengthen the competitiveness of the economy in the Samara region, on the one hand, especially in the face of globalization, and on the other hand, in the face of economic instability and economic sanctions, should be the subject of the further serious research.

5 Conclusions

An important background for the successful formation of STA, the realization of competitive advantages of the Samara region with the system of a high technological economic cluster and a developed system of scientific and educational and information frames is the development of the transport and logistics system of STA. It must become the subject of a further research of the STA development. The improvement of transport services leads to significant improvements in health outcomes (preservation of human capital) of regions, reduction of unemployment, crime and deaths from traffic accidents and it will increase the migration attractiveness in the global economy.

In the transition to a predominantly stable type of the economic growth, the cargo turnover growth in the Samara region will lag behind the growth in the GRP and to an even greater degree it will lag behind the passenger turnover growth. This will determine the way in which the economy of the Samara region will gradually move into the post-industrial development stage, accompanied by a decrease in the contribution of traditional economic sectors in the GRP and increasing the population mobility.

We can assume further that in the process of moving towards the new economy based on knowledge, the role of leading universities in respective economic clusters will increase, and they will increasingly play the role of the main centers in the formation and development of territorial and sectoral clusters. In this sense, STA development as a single economic and scientific and educational and information complex will play an integral ever-growing positive role, and will soon become indispensable for rapid capitalization of the region, promoting the economy of the Samara region in the direction of more the economic security and sustainable development.

References

1. Bolgova EV, Noskov VA, Noskov IV (2011) The infrastructure framework of economic space in the region. Monograph. Publishing House of the Samara State University of Communications, Samara (in Russian)
2. Chekmarev VV (2003) Economic relations in the production of educational services. Publishing House of the SSC RAS, Novosibirsk (in Russian)
3. Dahlman CJ, Routti J, Ylä-Anttila P (2006) Finland as a knowledge economy: elements of success and lessons learned. World Bank, Washington, D.C.
4. Fredman M (1953) The methodology of positive economics. In: Fredman M (ed) Essays in positive economics. University of Chicago Press, Chicago, pp 3–43
5. Karlsson C, Johansson B, Stough RR (eds) (2012) Entrepreneurship, social capital and governance: directions for the sustainable development and competitiveness of regions. Edward Elgar Publishing, Cheltenham. https://doi.org/10.4337/9781781002841
6. Kinnear S, Ogden I (2014) Planning the innovation agenda for the sustainable development in resource regions: a central Queensland case study. Resour Policy 39(1):42–53. https://doi.org/10.1016/j.resourpol.2013.10.009
7. McKinnan AC (2007) Decoupling of road freight transport and economic growth trends in the UK: an exploratory analysis. Transp Rev 27(1):39–46
8. McRae H (1996) The world in 2020. Power, culture and prosperity: a vision of the future
9. Metyolkin PV (2008) The comparative analysis of the transport systems development in Russia, the USA, China and the European Union. Bull Transp 10:34–42
10. Noskov VA (2003) Fundamentals of the formation of the market for higher education services. Econ Sci 5:91–98 (in Russian)
11. Noskov VA, Noskov IV (2015) A role of the transport system in expansion of economic space and a sustainable development of the Samara region in conditions of world economy globalization. Monograph. Samara State University of Economics, Samara (in Russian)

12. Noskov VA, Vlezkova VI (2017) Problems of competition in the market of banking services in the context of globalization (on the example of the world economy, the economy of the Russian Federation and the Samara region). Monograph. LAP Lambert Academic Publishing, Beau-Bassin (in Russian)
13. Sedlacek S, Gaube V (2010) Regions on their way to sustainability: the role of institutions in fostering the sustainable development at the regional level. Environ Dev Sustain 12(1):117–134. https://doi.org/10.1007/s10668-008-9184-x
14. The Territorial Body of the Federal State Statistics Service of the Samara Region (2018) The transport of the Samara region. Statistical yearbook. Samarastat, Samara (in Russian)
15. The White House (2014) An economic analysis of the USA transportation industry. ARTBA, Washington, D.C.
16. Tonn BE (2004) Integrated 1000-year planning. Futures 36(1):376–382. https://doi.org/10.1016/S0016-3287(03)00068-5

Smart City and Living Standards: Significant Correlation

E. Repina

Abstract In contemporary Russia the significant inequality in municipalities' living standards is caused by differentiation in their socio-economic development. The breakthrough in economy and social sector could improve both the living standards and the demographic environment. This paper attempts to determine the correlation of "smart city" digitization process and the living standards in municipal districts of Samara *oblast*. A combination of descriptive, multidimensional and distribution-free methods was used in the data analysis. The aggregated index for living standards of the municipal districts was calculated. The framework for the "smart city" digital platform was described. Statistically significant correlation between "smart city" digitization process and living standards determinants was ascertained.

Keywords Smart city · Living standards · Correlation

1 Introduction

Significant differentiation features the socio-economic development of municipalities in the RF regions. This affects the living standards of the territories. A notable difference in the level of economic development of municipalities is likely to contribute to the implementation of a comprehensive policy aimed at combating poverty, increasing the social well-being of citizens and sustainable development of the country at whole. Currently the transformation of sustainable development domain, which is based on the theory of environmental modernization, is being carried on globally.

Previous research findings into economic development differentiation of municipalities have been addressed by Glinsky et al. [6]. T. Egorova and A. Delakhova investigated differences in the development of the territories of the North-East Russian macro-region [2]. A slowdown in the business activity of institutions was revealed, which significantly affects the socio-economic development of the territory. J.V. Ragulina, A.V. Bogoviz, S.V. Lobova, A.N. Alekseev, V.I. Pyatanova studied the

E. Repina (✉)
Samara State University of Economics, Samara, Russia
e-mail: violet261181@mail.ru

S. I. Ashmarina and V. V. Mantulenko (eds.), *Current Achievements, Challenges and Digital Chances of Knowledge Based Economy*, Lecture Notes in Networks and Systems 133, https://doi.org/10.1007/978-3-030-47458-4_62

post-crisis development of the Russian economy in the period 2009–2018. They concluded that innovative development based on digital technologies is a source of global competitiveness of the country, as well as a determinant for the surge of living standards [10]. The issue of differentiating the development of member countries is also relevant for the European Union. According to Eurostat data, the value of GDP per capita varies significantly across the European Union: Luxemburg's index (maximum in the European Union, 32,220 euros) exceeds that of Serbia (minimum in the European Union, 9690 euros) by 3,325 times [3]. Spatial and socio-economic differences are also common in other regions of the world, such as China. Fang and Yu [4] investigated different perceptions of globalization changes and the effects of digitalization in 20 urban agglomerations in China.

It should also be noted, that small administrative entities-municipal districts, urban districts of one territorial entity of the country are subject to considerable differences in socio-economic development.

The country's national development goals until 2024 are set out by the Decree of the President of the Russian Federation No. 204 of May 07, 2018. The decree states the direction aimed at ensuring sustainable growth of citizens' incomes, increasing life expectancy, improving housing conditions, and accelerating the introduction of digital technologies in the economy and social sector [13]. At the highest state level, a policy has been declared to improve the living standards of the population with accelerated digitalization. The use of digital technologies can neutralize significant differences in the socio-economic development of territories, which will improve the living standards of the population. This study was designed to investigate the potential impact of "smart city" digital technologies on the living standard growth. The study is based on the case of municipal urban districts of Samara *oblast* and takes into account the differentiation in their inequality. The standard of living in municipality is seen in this paper as the three-component synergy, namely the levels of socio-demographic and economic development together with the level of living environment. This approach is in agreement with the component system of living standard aggregated index, which was proposed in Mkhitaryan and Bakumenko [8]. Conceptually this approach is compatible with the methodology for calculating the living standard by Yakovenko et al. [14].

2 Methodology

The statistical data array to calculate the living standard aggregated index (quality of life, *QL*) for Samara *oblast* municipalities was based on the open statistics from the Federal State Statistic Service of Russia [5]. The period under study is from 2017 to 2018 ($n = 10$), that represents an objective evaluation and is in line with the complex methodology for determination of the living standard level [7].

The system of statistical indicators is comprised of three sets:

(1) Human capital (HR)—11 indicators: birth rate; death rate; migration growth; number of registered marriages/divorces per 1,000 people; average monthly wages in organizations; official unemployment rate; number of doctors of all specialties/secondary medical personnel, per 1,000 people; the capacity of out-patient clinics per 10,000 people; the number of places in organizations that carry out pre-school educational activities, per 1,000 people under the employable age.
(2) Economy (E)—5 indicators (per capita): shipped goods of own production, self-performed works and services; investments in fixed capital; local budget revenues; total volume of all food products sold within the city district, monetary by value for the financial year; number of individual entrepreneurs per 1000 people of working age.
(3) Comfortable environment (CE)—6 indicators: commissioning of residential buildings per capita; the percentage of roads in compliance with regulatory requirements, in the total length of roads; number of sports facilities per 10,000 people; the number of service providing companies per 10,000 people; the number of places in collective accommodation facilities per 1000 persons; total area of residential premises, on average per inhabitant (at the end of the year).

The units of measurement among those indicators are different. The pattern method with the use of transformation was applied to put the statistical data in comparable form: $X^* = \frac{X}{X_{best}}$, где X_{best}—the best value based on common sense and theoretical proposition of economics. The mathematical description of the living standard aggregated index calculation QL:

1. The calculation of the aggregated index constituent for the sets 1–3 for each year under study:

$$HR_i(E_i, CE_i) = \frac{\sum_{j=1}^{m} X_j^*}{m}, \quad (1)$$

Where, i—urban district, j—statistical indicator of the corresponding set;
2. The calculation of the integrated index for each urban district for every year of the period under study:

$$QL_i = \sqrt[3]{(HR_i \times \alpha_{HR}) \times (E_i \times \alpha_E) \times (CE_i \times \alpha_{CE})}, \quad (2)$$

Where, $\alpha_{HR} = 0.5$; $\alpha_E = 0.4$; $\alpha_{CE} = 0.1$—the share of every set of statistical indicators in the integrated index QL formation based on the expert estimations, with the sum of 1 [11].

The method of comparative dynamic analysis was applied to study the dynamics of the living standards in the urban districts in Samara *oblast* during the period under study. The method of systematization and descriptive analysis were used to define the concept of the “smart city” model. Nonparametric statistical methods were used to estimate the correlation between the constituents of the aggregated index QL and the index of urban environment standards, which is seen as a tool to aggregate the potential efficiency of "smart city" technologies integration.

3 Results

The calculated values of the aggregated index QL were derived for 10 urban districts of Samara *oblast* and are demonstrated in Table 1.

The highest index of living standards in 2018 was recorded in Samara (city with more than 1 million inhabitants) and Otradnoye (a small city with a population of up to 50,000 people). The worst QL index belongs to Oktyabrsk (2018). Comparative dynamic analysis of the QL index allowed stating (Fig. 1) that the standard of living increased in Togliatti (+3.02%), Syzran (+6.11%), Otradnoye (+5.44%), and Oktyabrsk (+24.74%). The leader of the anti-rating was the urban district of Zhigulevsk (−30.87%).

The decline in the living standards in 6 out of 10 urban districts of the Samara region requires strategic decisions to address this extremely important problem. In 2019, national projects aimed at providing financial support for the growth of the territory's economy and the living standards of its residents, including the "Housing and Urban Environment" project, began to be implemented in Samara *oblast*. The introduction of smart city technologies represents the certain stage of the implementation of this national project.

There are 4 directions in which the "smart city" concept can be implemented:

1. Smart communal service. The national project "Housing and Urban Environment" defines it as a digital platform "Digital Twin City", integrated with the "Active Citizen" platform [9]. There are the following functional features: automation of metered values collection, control of equipment operation and

Table 1 The values of QL index for 2017–2018

City districts of Samara *oblast*	2017	2018
Samara	0.1791	0.1766
Togliatti	0.1272	0.1310
Syzran	0.1185	0.1257
Novokujbyshevsk	0.1446	0.1443
Chapaevsk	0.1065	0.0937
Otradny	0.1409	0.1485
Zhigulevsk	0.0906	0.0626
Oktyabrsk	0.0467	0.0583
Kinel	0.1234	0.1122
Pokhvistnevo	0.1445	0.1334

Source author

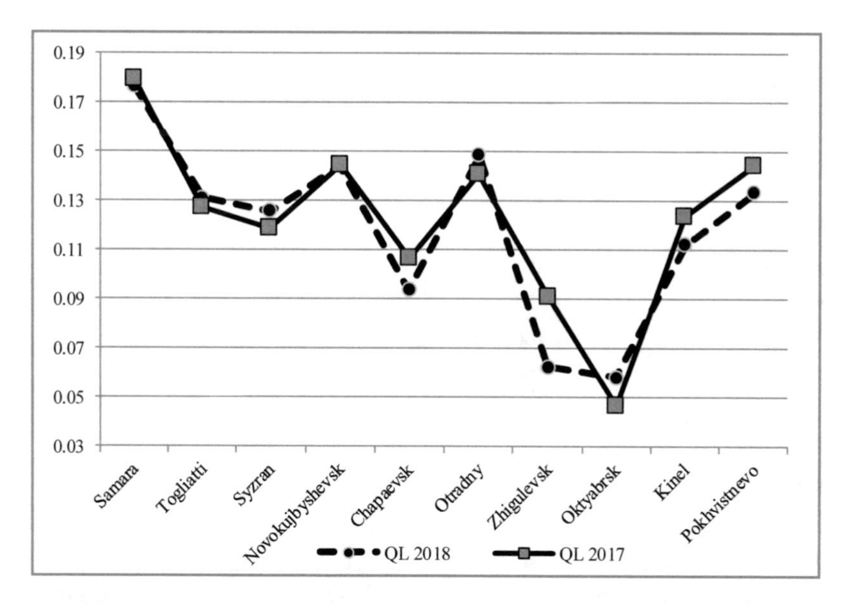

Fig. 1 Dynamics of the aggregated index QL for 2017–2018 (*Source* author)

rapid response to emergency situations, identification losses, energy passports for real estate objects, financial monitoring of public utilities.

2. Smart residential sector. It implies information and communication platform integrated with mobile applications, software, and sensors inside residential buildings. The building information model is provided for the whole life cycle of the building;
3. Smart urban transport. It aggregates the digital platform for car parking administration with photo and video recording of road accidents, a system for monitoring the state of the road surface, artificial intelligence for managing traffic flows, a network of smart bus stops.
4. Smart urban environment. It synthesizes the capabilities of interactive urban lighting management (including architectural lighting), monitoring of road and municipal services, car sharing system, public Wi-Fi, "guest card" service, intelligent fire safety systems, online environmental monitoring (air, water) on a single digital platform.

Urban environment monitoring has been implemented in Russia since 2018. The rating is calculated based on 36 indicators; each of them is evaluated in the interval (0, 1). The maximum value of the urban environment quality index is 360 points. The entire set of index indicators is divided into 6 sets that are evaluated according to 6 criteria. There are the following sets: housing and surrounding spaces, street and road network, green spaces, infrastructure, public space with the following criteria: safety, comfort, health ecology, identity and diversity, urgency and relevance, management efficiency.

Table 2 UES and the rank of QL constituents (2018)

City districts of Samara *oblast*	*UES*, point	*UES* rank	HR rank	*E* rank	*CE* rank
Samara	163	5	1	1	2
Togliatti	181	1	6	3	9
Syzran	152	7	4	6	5
Novokujbyshevsk	173	2	3	2	6
Chapaevsk	136	10	7	9	10
Otradny	155	6	2	4	1
Zhigulevsk	166	3	10	7	7
Oktyabrsk	166	3	9	10	8
Kinel	145	9	8	8	3
Pokhvistnevo	147	8	5	5	4

Source author

The urban environment standards (UES) of city districts in Samara *oblast* (http://индекс-городов.рф) and the rank of the constituents of the aggregated index QL are shown in Table 2.

The order variables Rank *UES*, Rank *HR*, Rank *E*, Rank *CE* were used to calculate the multiple coefficient of rank correlation (concordance coefficient) with the absence of "linked" ranks by the formula:

$$W = \frac{12S}{m^2\left(n^3 - n\right)}, \tag{3}$$

Where, $m = 4$ (the number of potentially correlated variables), $n = 10$ (the number of ranks combinations), S—the deviation of the rank squares sum from the average of rank squares.

The empirical value of the concordance coefficient was $W = 0{,}501$, that marks the noticeable correlation between indices. The Pearson criterion was used to test the statistical significance of the coefficient $\left(\chi^2_{obs} = 22{,}03;\ \chi^2_{obs}(\alpha = 0{,}05;\ k = 9) = 16{,}92\right)$. The coefficient is statistically significant at 5% level.

4 Discussion

One of the possible ways to increase the living standards of the population is the implementation of "smart management" of urban districts the development. It is based on innovative technological solutions-digitalization of urban management elements and creation of digital information platforms. The advantage of such platforms is the speed of making management decisions based on big data analysis.

The global trend of implementing management based on "smart cities" technologies is considered in a host of academic research. A. Sharida, A. Hamdan, and M. AL-Hashimi analyzed the impact of "smart city" technologies on the standards of living in Bahrain [12]. Transparency of management decision-making, increased communication between residents and authorities are recognized by the authors as the driving forces of sustainable development and improvement of living standards and environmental situation. The increasing need to assimilate rural migrants is caused by the acceleration of the urbanization process. S.A. Almagashi, S. Lomte, S. Almansob, A. AL-Rumaim, and A. Jalil have investigated the issue of information and communication technologies application in this context [1]. D. Zanchett, J. Junior, A. Monteiro, D. Haddad, and L. Assis studied the phenomenon of ShareBus (which is the information system of urban mobility and represents a highly specialized domain of the "smart city") [15].

The term "smart city" implies such urban planning and management solutions in city management that would integrate the capabilities of digital intelligent systems, Internet technologies and other areas of development of information and communication tools. It is possible to distinguish the functioning principles of the "smart city". Among them there are the following: high-tech, based on information and platform principle; flexibility; synergetic approach; convenience for all participants; efficiency.

In Russia, the project of digitalization of urban entrepreneurship "smart city" is being implemented within the framework of the national project "Housing and Urban Environment" and the national program "Digital Economy". The Ministry of Construction, Housing and Utility Services approved the basic and additional requirements for the cities participating in this project [9] and the goals of its implementation. Pursuing those goals will lead to an increase in the living standards of the population. This will definitely have a positive impact on improving the demographic situation in the country. In Samara *oblast*, certain urban districts can apply for participation in the urban digitization project and receive federal subsidies for its implementation.

5 Conclusion

The research has confirmed the hypothesis about correlations of components of the living standard aggregated index and the success of the "smart city" project. One of the methods for evaluating the success of the implementation of digital urban management technologies is the urban environment quality index, calculated for all urban districts of Russia. The growth in the urban spaces quality, as well as the introduction of digital services in all spheres of citizens' life is likely to contribute to the living standards improvement and positively impact the demographic processes in society.

References

1. Almagashi SA, Lomte S, Almansob S, AL-Rumaim A, Jalil A (2019) The impact of its in the development of smart city: opportunities and challenges. Int J Recent Technol Eng 8(3):1285–1289. https://doi.org/10.35940/ijrte.b3154.098319
2. Egorova T, Delakhova A (2020) Regional peculiarities and differentiation of socio-economic development of the north-east of Russia. Smart Innov Syst Technol 138:272–282. https://doi.org/10.1007/978-3-030-15577-3_27
3. Eurostat (2020) Main GDP aggregates per capita. Database. https://appsso.eurostat.ec.europa.eu/nui/. Accessed 10 Feb 2020
4. Fang C, Yu D (2020) China's urban agglomerations. Springer, Cham. https://doi.org/10.1007/978-981-15-1551-4_4
5. Federal State Statistic Service (2020) Database of indicators of municipalities. https://gks.ru/dbscripts/munst/munst36/DBInet.cgi. Accessed 12 Jan 2020
6. Glinsky VV, Serga LK, Bulkin AM (2016) Differentiation of municipalities as a factor of economic development of territories. Stat Issues 8:46–51
7. Kozlova OA, Gladkova TV, Makarova MN, Tukhtarova EH (2015) Methodological approach to measuring the quality of life of the region's population. Reg Econ 2:182–193. https://doi.org/10.17059/2015-2-15
8. Mkhitaryan VS, Bakumenko LP (2011) Integrated assessment of the living standards of the population of the Republic of Mari El. Stat Issues 6:60–70
9. Order of the Ministry of construction and housing and utility services of the Russian Federation, October 31, 2018 No. 695/pr «On approval of the passport of the departmental project of digitalization of urban economy "Smart city"». https://www.minstroyrf.ru/docs/17594/. Accessed 10 Feb 2020
10. Ragulina JV, Bogoviz AV, Lobova SV, Alekseev AN, Pyatanova VI (2020) Strategy of increasing the global competitiveness of Russia's economy and Russian's becoming a new growth vector of the global economy. In: Popkova E (ed) Growth poles of the global economy: emergence, changes and future perspectives. Lecture notes in networks and systems, vol 73. Springer, Cham, pp 203–210. https://doi.org/10.1007/978-3-030-15160-7_20
11. Repina EG, Guseva MS, Polyanskova NV (2019) Integrated statistical assessment of the level of socio-economic development of municipal districts of Samara oblast. In: Science of the XXI century: current directions of development, vol 1–2, pp 148–154
12. Sharida A, Hamdan A, AL-Hashimi M (2020) Smart cities: the next urban evolution in delivering a better quality of life. In: Hassanien A, Bhatnagar R, Khalifa N, Taha M (eds) Toward Social Internet of Things (SIoT): enabling technologies, architectures and applications. Studies in computational intelligence, vol 846. Springer, Cham, pp 287–298. https://doi.org/10.1007/978-3-030-24513-9_16
13. The Decree of the President of the Russian Federation No. 204 of May 07, 2018. http://www.kremlin.ru/acts/bank/43027. Accessed 11 Oct 2019
14. Yakovenko NV, Didenko OV, Safonova IV (2019) Socio-ecological well-being of the population (the regions of the Central Federal District are example). IOP Conf Ser Earth Environ Sci 272(3):253–259. https://doi.org/10.1088/1755-1315/272/3/032035
15. Zanchett D, Junior J, Monteiro A, Haddad D, Assis L (2019) Collaborative information system to find efficient routes using public transport. In: dos Santos J, Muchaluat-Saade D (eds) Proceedings of the 25th Brazillian symposium on multimedia and the web, WebMedia 2019. Association for Computing Machinery, Inc., Cham, pp 473–476. https://doi.org/10.1145/3323503.3361717

Smart City: Ensuring Economic Security of Administrative Center of the Russian Entity

I. A. Svetkina

Abstract The article discusses certain issues of economic security (ES) at the level of the administrative center of a constituent entity of the Russian Federation (ACERF) in the context of the development of the "smart city" system (ESSC—the economic security of the "smart city"). The general structure of ESSC includes qualitative and quantitative indicators defined by the main characteristics of a large city, which may vary depending on location (climate, geographical location, population, industry, infrastructure, etc.). Smart city services are usually recommended for implementation in cities with a population of over 100 thousand people. The analysis of extensive statistical and practical material made it possible to classify the main (specific) "smart risks", "smart threats" and "smart challenges" for the ESSC of a large city. Therefore, the mechanism for ensuring economic security at the level under consideration should include diagnostics, monitoring of risks, challenges and threats, assessment of the state of the ESSC of the administrative center of the subject of the Russian Federation.

Keywords Risks · Threats · Challenges · Economic safety · Smart city

1 Introduction

The concept of Smart Cities can be understood as a holistic approach to improve the level of development and management of the city in a broad range of services by using information and communication technologies. It is common to recognize six axes of work in them: (1) Smart Economy, (2) Smart People, (3) Smart Governance, (4) Smart Mobility, (5) Smart Environment, and (6) Smart Living [2, 13].

Particular attention in the implementation of the concept of "smart city" in different countries is paid to sustainable energy issues [3, 5, 6], the analysis of large amounts of data for the formation of a smart urban environment [7, 11, 12, 15], as well as environmental issues [8]. Note that the "smart city" (SC) is a combination

I. A. Svetkina (✉)
Samara State University of Economics, Samara, Russia
e-mail: svetkinairina@yandex.ru

S. I. Ashmarina and V. V. Mantulenko (eds.), *Current Achievements, Challenges and Digital Chances of Knowledge Based Economy*, Lecture Notes in Networks and Systems 133, https://doi.org/10.1007/978-3-030-47458-4_63

of multivariate information services that ensures the quality of life of the population and effective city management.

There are 85 territorially diversified entities of the federation (regions) in Russia, the administrative centers of which are most often their large cities. The Smart City projects have been launched in many ACERFs in Russia (Moscow, St. Petersburg, Kazan, Yekaterinburg, Krasnoyarsk, Novosibirsk, Ufa, Sochi, Perm, Rostov-on-Don, Voronezh, Chelyabinsk, Nizhny Novgorod, Omsk, Volgograd, Samara). And the list of cities is constantly expanding, as the SC system has a tendency to develop, and after a few years, innovation will affect every city.

It is impossible to build an ESSC system in each ACERF according to a single template, as large cities differ significantly in their characteristics (economic, social, geographical, environmental, etc.). The complex internal structure of the electronic system of each ACERF should take into account the requirements of SC (stability, the ability to self-development and progress). But with the introduction of SC in ACERF, "smart risks", "smart challenges" and "smart threats" of electronic banking arise, since this process captures all aspects of human life and a large city as a living organism.

2 Methodology

The main areas of ESSC include: focusing on people, ensuring the manufacturability of urban infrastructure, improving the quality of urban resources management, a comfortable and safe environment, economic efficiency of the digital and engineering technologies used [11].

SC development drivers are state and administrative resources. For example, the city administration provides the conditions and resources of such areas as: the provision of public services in electronic form, the management of urban transport, the rational use of electricity and water, the development of the urban health system, the use of innovations, and efficient waste management.

The information base for adapting SC tools to the individual characteristics of a large city is:

- experience of leading cities considered in international studies [10, 13, 14];
- key international and domestic ratings and standards [10];
- various indicators and indicators, bibliometric and patent analysis in the field of smart cities [10];
- information portals and statistical databases, international standards, etc.

In the process of providing ESSC, certain data sources are examined to identify potential risks and threats that can be divided into two categories:

1. Main: statistical databases of federal, regional and municipal levels, GIS, a survey of representatives of authorities at various levels, news sources, information resources, web portals, open data on the Internet.

2. Additional: social networks (digital footprint), data from telecom operators, data from the industrial Internet, data from digital service owners, providers and sellers.

Given the openness of the economic system of a large city, we note the unattainability of complete economic security by a separate economic entity, since its isolation from challenges, risks and threats is impossible.

Approaches to the analysis of the level of economic security (high, medium, low) imply the implementation of the following steps: determining the source of the negative impact; identification of the nature of challenges; challenge classification for forecasting their development; identification of objects prone to challenges; determination of hazard levels for objects, etc. Within the framework of ESSC, it is necessary to classify "smart challenges" into groups in order to identify "smart risks" and "smart threats" more accurately in the future (Table 1).

Since the "smart city" system is an integral part of the urban environment, risk-generating factors act on it (Table 2).

It should be borne in mind that when SC elements change or communication between them is broken, the state of the SC system is destabilized and a "smart threat" to economic security arises. "Smart threats" (internal, external) can destabilize the

Table 1 Classification of "smart challenges" ESSC

Characteristics	Groups
Social and economic interests of population	Direct Indirect
The level of negative impact possibility	Treats Risks
Sources of origin	Institutional Organizational-structural Contract
The level of manageability	Manageable Less manageable Uncontrollable
The level of criminal situation	Uncriminal Half-criminal Criminal
The length of negative impact	Short-term Middle-term Long-term
The area of origin	Infrastructural Social and demographic Credit Legislative Ecological Fiscal

Source author

Table 2 Risk-generating factors influencing ESSC

Feature	Groups
Focusness	Internal (broken connection of several elements) External
Impact direction	Professional Purposeful Accidental
Source of impact	Anthropogenic Technogenic Climatic
Strength of impact	Severe damage (from 50% to 100% the whole system or the main part of the system) Medium damage (less than 50% to 20% of the entire system or individual subsystems) Low damage (less than 20% to 3% of the entire system or individual subsystems
Frequency of impact	One time Periodic
Type of impact	Resource Technological Networking

Source author

ACERF economic system, and "smart risks" can lead to damage to interests in the economic sphere when the corresponding threat is realized.

In the "smart city", digital technologies meet residents at every step, forming a single ecosystem and responsible for all aspects of human life, creating, transmitting and analyzing information coming from various urban sources. Of course, there are dangers and risks of introducing intelligent technologies into public life, among which are cybercrime, electronic inequality, dehumanization of public space, a decrease in the level of creativity of a society with the development of a technocratic lifestyle.

Residents of the city understand that a lot depends on themselves, and part of the population is ready to make their own efforts for the benefit of the development of the city, to establish a dialogue between the city authorities and residents, since joint efforts can achieve good results [4]. To involve citizens in city management, in the quality control of various SC services, appropriate tools are needed. For the real work of these tools, investment in the infrastructure of the ACERF is required.

The volume of received, stored and processed data on subjects and objects, processes, activities of the ACERF is constantly increasing. Note that at times today, dependence on energy sources is increasing. Uninterrupted water, gas, heat and energy supply should be provided. For example, there is a threat of a power outage at a distance or a change in the mode of consumption of electricity, which can lead to data loss, stopping the operation of various devices and mechanisms, etc.

In addition, data privacy and security must be ensured. An improperly configured SC system or the intervention of intruders can lead to millions of losses. Thus, remote access to infrastructure facilities can lead to serious risks and consequences, for example, if attackers gain access via the Internet to regulate the operation of road transport systems or emergency services. Therefore, a comprehensive ESSC provides a state of security of the economy of a large city from threats and risks that impede sustainable reproduction and effective development. Also, ESSC mechanisms should be aimed at reducing the risk of emergencies, at preserving the health of people, reducing the size of environmental damage and material losses in case of their occurrence.

3 Results

The "Economic Security Strategy of the Russian Federation until 2030" [9] lists 40 indicators of the state of economic security. We believe that it is necessary to insert a generalized indicator "smart city" in the ACERF indicator system, which is disclosed for each large city depending on individual characteristics, but the conceptual apparatus must be adopted at the state level and adjusted with the international names of SC indicators in order to conduct a comparative, factor analysis and various analytical studies. For example, in the context of the introduction of the "smart city" system in the ACERF, additional indicators that are developed by various scientific institutions of the country are included in the electronic circuit. For example, NIITS JSC developed 26 SC indicators in 7 key areas of a smart city (smart economy, smart management, smart residents, smart technologies, smart environment, smart infrastructure, smart finance) [1].

In the draft preliminary standard "Information Technology. Smart city. Indicators." indicators are distributed in a number of categories (economy, education, energy, environmental change, health, safety, transport), which in total include 38 indicators [10]. Expanding the composition of indicators does not mean improving the quality of assessing the level of EB in SC conditions, since the more indicators, the more difficult it is to track them. Moreover, individual safety factors are not constant and may vary depending on the specific situation and socio-economic, political and legal, ethnic, environmental and other problems. Conducting a study of the level of ESSC and socio-economic indicators, we can conclude that in the system of "smart city" there are both positive and negative aspects of development.

In providing the ACERF, the technological approach to ESSC is applied (the application of specific modern technical solutions) and an integrated approach (study and monitoring of indicators of the level of development and integration of various spheres of life of a large city and its population). Thus, the management bodies control the efficiency of using natural, labor, material, financial resources, achieving economic growth, and improving the quality of life of the population.

In addition to monitoring, it is necessary to monitor the operation of SC services. In our opinion, the SC safe development monitoring system should meet the following fundamental requirements: the monitoring base should be statistics; monitoring should be systematic. Therefore, the main scenarios implemented by ESSC are: (1) reactive, providing a passive response of the electronic system to the functioning of SC in normal mode; (2) leading, stipulating the initiation, creation and implementation of projects to increase the effectiveness of SC; (3) coordinating SC compliance with modern requirements and technologies.

4 Discussion

We ask ourselves the question: "smart city"—a cost point or a new income item for a large city? For each specific city, a separate calculation of the expenditure part and the revenue part by category SC is required. It is incorrect to compare the capitals of leading countries [15] with the ACSRF. Since currently the city of Moscow is in first place in the implementation of SC technologies in Russia (93%), and the remaining large cities account for only 7%. The gap is huge, but this indicator is not negative, because the population of the city of Moscow also far exceeds the population of other large cities of Russia.

The economic component of ESSC looks no less important than the technological. One of the serious barriers to the development of the sphere of smart cities, experts consider the lack of funding. In other words, the cost-effectiveness of smart city systems is difficult to calculate, especially in the cost/benefit ratio. But here it is hardly possible to offer a general solution, because various tools are used.

The implementation of the concept of "smart city" runs into a number of significant difficulties. First: in almost all cases, the main investor is the state. To achieve the expected economic effect of the implementation of the SC system, it is necessary to create a number of conditions, for example, to ensure a certain level of penetration of these technologies into the city.

Moreover, in a number of areas, the implementation of SC solutions in principle cannot be considered an investment, for example, in the field of public safety or in the field of social protection. For example, in many cities they find money as soon as they realize that warning an emergency, for example, is cheaper than eliminating it. Relatively speaking, it is more profitable to spend 1 billion rubles to create a monitoring system, to prevent fire or flood, than to spend 1.5 billion rubles to eliminate their consequences.

A huge number of SC services require both corresponding capacities for data storage, and computing power for processing flows of "smart information"; continuous data transmission (communication services), continuous maintenance, support, modernization of all devices and systems are necessary. All in all is the colossal operational cost of maintaining and servicing. The return on investment is ensured by the savings received by the city government from the introduction of a digital solution, or additional income. This, for example, photo-video recording systems on the roads, which allow to fill the city budget due to fines of motorists.

The second difficulty: the gap between those who bear the costs of implementation and those who benefit. The lack of interest of private investors in participating in Smart City projects is explained by the fact that often such investments do not become highly profitable. Even in the case when the costs of deploying systems are borne by a private company, and at the stage of implementation the state does not spend budgetary funds on it, all these start-up costs will certainly be taken into account in the future cost of servicing (rendered services), since a private company is created for profit and will not work at a loss.

In order to interest a private company in SC investments, it is possible to propose at the regional level a reduction in the tax base for the payment of income tax, personal income tax deduction for certain categories of developers, and the allocation of targeted subsidies and grants.

Third difficulty: the implementation of various SC services often ends in failure, because they are implemented suddenly, without preliminary preparation of the population for them, without any special desire on the part of the authorities to get a really positive effect from the innovation, that is, they are introduced because an order was received from above, without clarifying the need for them and explanations of their usefulness. For the full implementation of the SC mechanism, federal, regional, municipal funds, as well as investments by private investors, are required. For more efficient work of the city, it is required to systematically plan expenses for SC technologies, ensure their transparency and control, evaluate the effect obtained from each implemented project.

5 Conclusion

Based on the foregoing, it seems possible to make some generalizations. The study revealed the "smart challenges", "smart risks" and "smart threats" of the economic security of large cities associated with the implementation of the SC project. Possible threats, negative factors, negative values of the dynamics of some important indicators—all this provides the basis for the formation of methods and recommendations to reduce the share of these problems.

Undoubtedly, the process of expanding SC in conjunction with the digitalization of all aspects of human life, family, society, city, region, country cannot be stopped. Since only in the conditions of digitalization the properties of SC are maximally realized, which contribute to: improving the quality of life, mobility, urbanization, socialization of the population, personification, the introduction of smart technologies, virtualization of space, competent and optimal modernization of infrastructure.

Also social inclusion approaches are essential for enhancing the social sustainability of smart cities. It does so by highlighting how such a perspective is often ignored by discourses and visions that favor generalized and socially skewed ways of framing the "city" as well as the citizens who are expected to become "smart" and benefit from high technologies [1, 10, 15].

Information on the functioning of SC in different countries and regions of Russia is constantly updated and expanded. Systematic, coordinated processes of diagnosis, monitoring and evaluation of the implementation of SC functions in the entire available variety of possible situations and environmental conditions contribute to the detection of internal defects that threaten the catastrophic consequences for the economic security of the ACERF.

Timely identification and neutralization of threats, as well as crisis situations at the level of the ACERF, will significantly reduce the risk of regional and national security threats. A reliable, effective ESSC system can serve as a guarantee of uniform, stable and sustainable socio-economic development of the state.

Acknowledgements The reported study was performed in the frame of the international research and educational project "SmartCitiZens" funded by the Movetia fund (Switzerland).

References

1. NIITC (2017) Indicators of smart cities. http://niitc.ru/publications/SmartCities.pdf. Accessed 27 Feb 2020
2. Alba E (2016) Intelligent systems for smart cities. In: Friedrich T (ed) Proceedings of the 2016 genetic and evolutionary computation conference. Association for Computing Machinery, New York, pp 823–839. https://doi.org/10.1145/2908961.2927000
3. Almeida M, Mateus R, Ferreira M, Rodrigues A (2016) Life-cycle costs and impacts on energy-related building renovation assessments. Int J Sustain Build Technol Urban Dev 7(3–4):206–213
4. Aurigi A, Odendaal N (2020) From "Smart in the box" to "Smart in the city": rethinking the socially sustainable smart city in context. J Urban Technol 27. https://doi.org/10.1080/10630732.2019.1704203. Accessed 27 Feb 2020
5. Bohne RA, Huang L, Lohne J (2016) A global overview of residential building energy consumption in eight climate zones. Int J Sustain Build Technol Urban Dev 7(2):38–51
6. Cupelli L, Cupelli M, Ponci F, Monti A (2019) Data-driven adaptive control for distributed energy resources. IEEE Trans Sustain Energy 10(3):1575–1584. https://doi.org/10.1109/TSTE.2019.2897661
7. Gessa A, Sancha P (2020) Environmental open data in urban platforms: an approach to the big data life cycle. J Urban Technol 27. https://doi.org/10.1080/10630732.2019.1656934. Accessed 27 Feb 2020
8. Gagliano A, Detommaso M, Nocera F, Berardi U (2016) The adoption of green roofs for the retrofitting of existing buildings in the Mediterranean climate. Int J Sustain Build Technol Urban Dev 7(2):116–129
9. Presidential decree No. 208 of May 13, 2017 "On the economic security Strategy of the Russian Federation for the period up to 2030". https://www.garant.ru/products/ipo/prime/doc/71572608/. Accessed 27 Feb 2020
10. Tadviser (2020) Smart cities. http://www.tadviser.ru/index.php/Статья: Smart_city. Accessed 27 Feb 2020
11. Tomor Z, Meijer A, Michels A, Geertman S (2019) Smart governance for sustainable cities: findings from a systematic literature review. J Urban Technol 26(4):3–27

12. Udupi PK, Malali P, Noronha H (2016) Big data integration for transition from e-learning to smart learning framework. In: Kiran GR (ed) Proceedings of the 3rd MEC international conference on big data and smart city. IEEE, New Jersey, pp 1–4
13. Velinov E, Ashmarina SI, Zotova AS (2020) Participatory budgeting in city of Prague: boosting citizens' participation in local governance through digital tools (Case study). In: Ashmarina S, Vochozka M, Mantulenko V (eds) Digital age: chances, challenges and future. Lecture notes in networks and systems, vol 84. Springer, Cham, pp 189–197
14. Wiig A (2019) Incentivized urbanization in Philadelphia: The local politics of globalized zones. J Urban Technol 26(3):111–129
15. Yun HY, Zegras C, Heriberto D, Palencia A (2019) Digitalizing walkability: comparing smartphone-based and web-based approaches to measuring neighborhood walkability in Singapore. J Urban Technol 26(3):3–43

App-Based Multimedia in Foreign Language Teaching: Options for Language Learner

O. V. Belyakova and N. A. Pyrkina

Abstract The article aims to study the use of app-based multimedia in foreign language teaching. The authors consider the problem of mobile apps usage distinguishing its advantages and disadvantages. A comparative analysis of several popular mobile resources made it possible to identify success factors for further application of mobile devices in teaching foreign languages and enabled the authors to produce some recommendations. It was determined that the use of mobile devices for autonomous study of foreign languages enhances the participants' performance and helps to master students' knowledge of foreign languages.

Keywords Multimedia · Mobile applications · Foreign language teaching · Language learner · E-learning

1 Introduction

In today's world, digitalization penetrates into all spheres of activity, multimedia resources cover public administration, business, manufacturing, science, and education as well. Due to the specific nature of the subject (in teaching a foreign language), there is the need to create an artificial language environment for students, which implies the most flexible and widespread use of multimedia tools that open up new opportunities in education. This makes the study of multimedia resources (hereinafter referred to as MR) in teaching a foreign language very timely and relevant.

Multimedia have been studied in the works of both Russian and foreign researchers [1–3]. They view modern multimedia educational resources as hypermedia systems, where static components (images and texts) are interconnected with dynamic information blocks (audio, video and animation). In other words, multimedia is

O. V. Belyakova (✉)
Samara State University of Economics, Samara, Russia
e-mail: olgabel5893@yandex.ru

N. A. Pyrkina
Samara State University of Social Sciences and Education, Samara, Russia
e-mail: pyrkina@pgsga.ru

S. I. Ashmarina and V. V. Mantulenko (eds.), *Current Achievements, Challenges and Digital Chances of Knowledge Based Economy*, Lecture Notes in Networks and Systems 133, https://doi.org/10.1007/978-3-030-47458-4_64

the combination of various media (audio, video, images, text, animation, graphics) accessed by a computer. However, there are some discrepancies in the interpretation of the concept of "multimedia". On the one hand, it is a single digital space that represents different ways and types of information presentation. On the other hand, the term "multimedia" is also referred to the end product made on the basis of multimedia technologies and multimedia instrumental programs and shells, and modern computer equipment (the availability of DVD-ROM drive; a large memory and so on).

In general, a MR is a system of iconic and symbolic means of displaying educational information; a tool of storing and transmitting a large amount of information in a compact form; and the form of presentation of diverse educational content. MRs appeared in the middle of the 20th century, but the rapid growth in this area began only in the 1980s, when the first attempts to combine video and audio materials with computer programs were made, which eventually led to the development of interactive training programs. As a result, there is a wide range of multimedia that contain all the information essential for learning foreign languages. The study is aimed at providing tips for undergraduate students to use MRs. Today, students spend most of their time on the Internet, social networks, and use electronic media resources, which give them unlimited access to various educational resources and the opportunity to be in contact with the teacher all the time [4].

Despite the fact that the area of study connected with the use of multimedia tools for teaching foreign languages is sufficiently covered, there are still a few works focusing on identifying the features of MRs manifested in teaching the above-mentioned subject, and contributing to more effective mastering of the subject being studied, as well as developing recommendations for their use.

There are different formats for using multimedia tools. For example, they are used in formal online courses (those providing formal qualifications) offered by universities. Such courses usually take place in LMS and focus on developing all four types of language skills (reading, writing, listening and speaking). The latter are practiced in an asynchronous mode, while speaking is performed through videoconferencing (a synchronous mode). LMS MOODLE is a good example of an educational platform used by a number of Russian and foreign universities. Besides formal Language Massive Open Online Courses (LMOOCS) have proved to be very efficient these days. They do not offer formal qualifications [2] as university courses but demonstrate options for using multimedia (presentations, video lessons, audio and video lectures). A modern form of MRs are online language learning communities providing online space and offering live language classes to global audience. Teachers can be accessed when students are in need of them. One more alternative method of studying a foreign language is to make use of various applications that can be installed on your PC or smartphone.

A mobile app is software designed to work with smartphones, tablets, and other mobile devices. It consists of using adaptive learning to help learners to get through pre-packaged individually focused content associated with a particular language area [5]. The issue of using mobile apps in teaching languages is rather controversial so it needs further study. The most popular mobile apps are Duolingo with 10 million users, LinguaLeo and Memrise. Many websites have mobile versions in the form of mobile apps. Some mobile applications are complex (LinguaLeo, Duolingo), while others are highly specialized: apps for mastering spoken language and communication practice, apps for translation, and apps-dictionaries. Mobile apps are the central tool of mobile learning theories. In foreign research papers, the emphasis is on the personal role of the teacher and the mobility of students [5]. Thus, due to the variety of formats of MRs in foreign language learning, there is a need for a study that would highlight the features of multimedia in teaching foreign languages, consider its advantages and limitations, and make recommendations for using multimedia in the educational process. Since language apps for university-aged learners are not very well studied, we will focus on analyzing the apps' advantages and disadvantages.

2 Methodology

First and second-year undergraduate students at higher institutions in Russia are required to take English courses once or twice a week during three or four semesters. Depending on the curriculum of the university, the English course may last up to 250 academic hours approximately. Teacher-led (face-to-face) classes are targeted at teaching students the four basic skills: reading, writing, listening and speaking. The students are also supposed to master their grammar and vocabulary skills.

At the start of their study at university students are required to take a placement test to see what level of English they might fit in. As a rule students of mixed levels come together in one class (from elementary to intermediate). But nearly all the surveyed usually rank themselves as poor in English and express the desire to improve their language skills for different reasons: removing the language barrier when communicating with their English counterparts, studying abroad, travelling to Europe where English is widely used, searching a rewarding job, building a career, etc.

But learning a language is not easy and takes much time and effort on the part of the learner. Studies suggest that most of language learners opt for traditional teacher-student classes [6]. They are reluctant to take full responsibility for their education. With the traditional face-to-face interaction, it is the teacher who manages the class, provides learning materials for the students, selects digital resources in accord with the students' abilities and preferences and assesses their achievement.

But with limited hours of teacher-student English classes at university students are expected to self-study the language. With various multimedia available online today students may make use of numerous e-learning materials free of charge. The availability of mobile devices in recent years has contributed to the rise in the so-called 'mobile learning' among university-aged people all over the world. Findings suggest that modern students demonstrate high levels of e-readiness for learning which is connected with both demographic (age, nationality, technological accessibility, etc.) and non-demographic (motivation, learner autonomy, attitude towards online learning, language proficiency, computer literacy, etc.) factors [6].

Self-directed learners are free to select their online multimedia resources for studying English, but sometimes it is not an easy option for them, which source to choose. At the international IT language market there are lots of free online multimedia resources that are targeted at university students learning a language for general academic purposes. Since university-aged learners opt for mobile devices in their everyday life (though generally used for socializing), it stands to reason to encourage them using their tablets and smartphones for educational purposes. Young people studying English at universities can improve their language skills while utilizing one of the numerous mobile apps for language learning. Language apps use adaptive learning schemes and are primarily targeted at memorization, translation, multiple choice and listening comprehension exercises. An app identifies individual learner 'needs' (in grammar or vocabulary as a rule) and provides the learner with individually focused content. Special attention is focused on assessing students' progress.

Several of the most popular language learning apps are Duolingo, Memrise, Lingualeo. Though they share some common traits concerning their computer-generated algorithms, their content and methods of introducing the language material are different. To help university-aged learners navigate through these app-based multimedia we analyzed some of the most popular language apps in accordance with students' needs and compiled recommendations for them. We used an integrated approach which combined comparative analysis and synthesis of mobile apps.

First we gathered the data from open sources such as the corresponding websites of the above-mentioned mobile apps. We analyzed such points as easiness of navigation, the structure, typical activities, the target audience, skills to be developed. At the next stage of the study we compared the data obtained from three mobile apps to trace similarities and differences which enabled us to reveal the advantages and disadvantages of the studied mobile devices. At the final stage the findings were used to produce recommendations for language learners.

3 Results

The current study investigated that *Duolingo* by Google company is ranked the first among top popular apps. With Duolingo you can study five languages for Russian speakers and more than thirty languages for English natives. Studying English with Duolingo can be quite challenging if you are persistent in learning a language and fulfill all the activities provided by the course. The app is aimed at developing learners' four basic skills (reading, writing, listening and speaking).

The lessons are structured in the form of a game so it can be quite motivating to self-study a language with this app resource. Duolingo has its own currency (lingots), so if you fulfill all the lessons of a chosen course you will be awarded with a number of lingots. With them you will get some bonuses such as changing the image of a program, getting access to special term data bases (for example, business or negotiations terminology), etc. If you fulfill your lessons daily, you can get some extra lingots. To start working with the program you should get registered first. Press the 'Get started' key. To choose a course press the 'Course taught in English'. Then set your goal (the pace of studying a language). It can be 'Easy' that is 5 min of language practice a day, 'Average'—10 min of exercise or 'Advanced' (15 min of activities a day). To construct your own study route with Duolingo you should pass a special Qualification test.

The course is divided into units. You can choose either a vocabulary or a grammar unit. Each lesson comprises four types of exercises: (1) a translation exercise, (2) a listening comprehension exercise, (3) a vocabulary exercise (matching activity), (4) a speaking exercise (phonetics drill). Typical activities are 'Translate from foreign into native', 'Choose the right word or word combination', 'Pronounce the word' (The new word is viewed on the screen and voiced, the student is supposed then to repeat the word or phrase through a voicer (a microphone)). Sometimes the new words are highlighted for you to click on them and find the right translation. If you fulfill the exercise successfully, you will be able to go on with the following task. The learner has three the so-called 'lives' which means he/she can make not more than three mistakes. Otherwise he/she will have to restart the lesson. It will take you from 5 to 20 min approximately to complete one lesson. The program assesses the student's progress and puts a mark.

The advantages of this application are the following:

- it is free for users,
- it can be easily installed on your phone or tablet,
- its interface is easy to use,
- the new vocabulary is voiced and you can replay the words any time you wish,
- it teaches you vocabulary in stages,
- its key targets are vocabulary and grammar,
- it tracks your progress which is motivating.

The drawbacks of this application are as follows:

- it does not provide any grammar rules in the Grammar part of the course, the learner has to consult grammar reference books,
- the choice of words in the Vocabulary part of the course is rather subjective,
- it does not provide a natural voice system which is sometimes demotivating,
- it is difficult to develop speech habits with this app,
- it is efficient only for elementary and pre-intermediate learners, for more advanced students there are not many activities.

Though there are a number of limitations, all in all the application is a useful tool to improve elementary or pre-intermediate students' grammar and vocabulary skills. The developers modify and improve the app, so we are about to see new courses, functions and content.

Another popular language platform and mobile application is *Memrise* (available on iOS and GooglePlay). It is based on gamification methods. The study process looks like a space travel with learners being cosmonauts growing flowers, so the more words you learn, the more flowers you have.

The app is targeted at developing learners' vocabulary skills. The new word is introduced in English with its Russian translation. The meaning is also illustrated with a picture. The app provides four types of exercises to memorize the vocabulary:

- matching the Russian word with its English translation,
- matching the English word (or phrase) from a video with its Russian translation,
- matching the word (or phrase) from an audiotext with its translation,
- translating the word (or phrase) from Russian into English using the app keyboard or your phone keyboard.

The advantages of this application are the following:

- it is easy to navigate,
- it is gamified,
- its target is vocabulary acquisition,
- it uses various mnemotechniques,
- audiotexts and videoclips are voiced by native speakers.

The drawbacks of this application are as follows:

- it is not free for users,
- no grammar is drilled,
- there are no reading or speaking exercises,
- it is based on user-created content which is rather subjective and not every learner likes it.

Though there are a number of limitations, the application has proved to be a useful tool to improve students' vocabulary skills and develop their mnemotechniques.

One more popular online educational platform is *Lingualeo,* which is available as an application on iOS, Android and Windows Phone. The app was launched in 2010 in Russia and like the apps mentioned above is based on gamification. It is named after its mascot, the Lion (Leo). When starting work with this app the user takes a special placement test. Based on the results of the test the course providers compose an individual study route for the learner. He/she is exposed then to native audiotexts and videoconferences from TEDtalks, Coursera MOOC and other open sources platforms.

The advantages of this application are the following:

- it has a freemium model (the basic exercises are free but the user should pay for extra activities and courses),
- it is easy to navigate,
- it is gamified,
- it is targeted at training vocabulary, grammar, pronunciation and listening skills.

The drawbacks of this application are as follows:

- it uses learner-downloaded texts which is rather subjective.

To sum it up, the application has proved to be a useful tool to improve students' vocabulary and listening skills. All this enhances higher education learners' motivation and develops their language skills [7]. When using mobile apps learners focus their attention on the presented content, which encourages understanding, training and memorizing the language material.

4 Discussion

The results show that multimedia in general and mobile applications in particular are an effective means of teaching a foreign language, which is especially true for young people who are used to working with mobile devices. The study identified the main opportunities of multimedia in foreign language teaching, such as providing information in digital form and in different forms in the same resource; the visual, interactive nature that offers nonlinear language presentation; rapid feedback between the user and the mobile app; simplified control and verification of knowledge system; possibility to store and analyze any amount of data; suitability for independent study and for group work; saving time of both teachers and students, and finally, entertaining character that helps to relieve the psychological tension of language learners.

However, the study also identified organizational and technical problems, related to the probability of finding insufficient information on the studied issues, the new role of the teacher (from teacher-instructed lessons to self-directed study) and students' need to build up an autonomous educational paradigm. It is obvious that mobile apps cannot replace personal communication with a teacher, they only expand his/her capabilities [8]. Accordingly, it is necessary to make recommendations for using multimedia apps in language learning.

This study shows that the best results are obtained when mobile tools are used for independent study, both under the supervision and guidance of a teacher and extracurricular self-directed work. In the classroom it is advisable to use multimedia for practicing listening, reading and writing skills, since they can be checked after the completion in the form of statistics in the application [9]. You can also offer project work on topics set by the training program. In general, learner's autonomous work with multimedia devices and with mobile apps in particular is rather effective as the student manages his/her speed of learning and chooses the educational route which implements individual modeling of the study process [8].

The other success factor in the use of mobile devices in teaching a foreign language is the active role of a teacher. He should be responsible for the overall organization, for preparing tasks, consulting students, monitoring the pace of completion of tasks, motivating students to work with mobile devices outside the classroom, providing psychological security of the student's information interaction with other users. In general, all participants in the educational process need to be ready for mobile learning, but the teacher needs to do much more effort, since he bears the burden of organizing mobile learning. Also, the effect of using mobile applications in the educational sphere is determined by the psychological readiness, competence and technical expertise of both students and teachers.

The data obtained also indicate the urgent need for regular use of mobile applications both in the classroom and in the independent language practice of students, since only in this case a positive result can be obtained.

All these facts make it possible to compile some recommendations for language learners. First of all, educational establishments should provide all the necessary conditions and facilities for using mobile devices in foreign language classes. Secondly, teachers should be technically-minded and aware of the possibilities of mobile applications. Thirdly, the teacher must be ready for a huge organizational work, he must be able to explain how to work with a particular device. When preparing students to work with a mobile app, you should first start with a general introduction to the device, and then gradually move on to the language aspects of using it. Fourthly, it is necessary to clearly define the areas of use of mobile applications for teaching a foreign language. It is advisable to use them for independent work of students. It is recommended to use mobile apps first of all to enlarge vocabulary and train your grammar skills, though mastering reading, writing and listening skills is also possible. Finally, the information contained in the application must meet the interests of students.

In general, it is recommended to develop an instruction on how to use mobile devices in teaching foreign languages. These studies confirm the effect of app-based teaching and indicate the need for a consistent and thorough use of multimedia content within the above-mentioned limitations.

5 Conclusion

Nowadays modern MRs have recently become widely available among language learners. We all live in the world of the so-called "e-culture," so the use of MRs in general and mobile applications in particular becomes a relevant direction in foreign language learning. Mobile applications make it possible to personalize and differentiate learning, have important features such as visibility, interactivity and flexibility of submitting language material, and providing many opportunities for foreign language learners [10].

This is especially important for the youth with a high level of information competence, which makes multimedia an influential educational driver through simultaneous effect of audio and visual information. The research of the study indicates that mobile apps are particularly useful in organizing autonomous study, both within and outside the classroom. It is recommended to use mobile applications in teaching project activities, reading and writing skills, in improving vocabulary and grammar skills. A review of a number of mobile applications used to teach foreign languages has shown that their use gives students the opportunity to receive information in a format that is convenient and interesting to them and to have access to educational resources, both during their studies and in their spare time, which contributes to improving the quality of education. Multimedia has a great expressive potential which facilitates the learners' developing their creative skills through active ways of learning.

Multimedia linguistic environment is a factor of successful foreign language acquisition as teaching foreign languages becomes easier and more effective with the help of MRs. The current study has implications for developing and implementing new Multimedia devices in foreign language teaching preferably of mobile character.

References

1. Guan N, Song J, Li D (2018) On the advantages of computer multimedia-aided English teaching. Procedia Comput Sci 131:727–732. https://doi.org/10.1016/j.procs.2018.04.317
2. Bovtenko MA, Parshukova GB (2018) Subject MOOCS as component of language learning environment. In: Filchenko A, Anikina Zh (eds) Linguistic and cultural studies: LKTI 2017. Advances in intelligent systems and computing, vol 677. Springer, Cham, pp 122–127. https://doi.org/10.1007/978-3-319-67843-6_15
3. Ma B (2019) The practice of the multimedia courseware for college foreign language teaching based on the network resources. In: Atiquzzaman M, Yen N, Xu Z (eds) Big data analytics for cyber-physical system in smart city, BDCPS 2019. Advances in intelligent systems and computing, vol 1117. Springer, Cham, pp 1720–1726. https://doi.org/10.1007/978-981-15-2568-1_242
4. Zheng D, Liu Y, Lanbert A, Lu A, Tomei S, Holden D (2017) An ecological community becoming: language learning as first-order experiencing with place and mobile technologies. Linguist Educ 44:45–57. https://doi.org/10.1016/j.linged.2017.10.004
5. Perez-Paredes P, Guillamón CO, Van de Vyver J, Meurice A, Jimenez PA, Conole G, Hernández PS (2019) Mobile data-driven language learning: affordance and learners' perception. System 84:145–159. https://doi.org/10.1016/j.system.2019.06.009
6. Mehran P, Alizadeh M, Koguchi I, Takemura H (2017) Are Japanese digital natives ready for learning English online? Int J Educ Technol High Educ 14:8. https://doi.org/10.1186/s41239-017-0047-0
7. Cho M-H, Castañeda DA (2019) Motivational and affective engagement in learning Spanish with a mobile application. System 81:90–99. https://doi.org/10.1016/j.system.2019.01.008
8. Shadiev R, Hwang W-Y, Liu T (2018) Investigating the effectiveness of a learning activity supported by a mobile multimedia learning system to enhance autonomous EFL learning in authentic context. Educ Technol Res Dev 66:893–912. https://doi.org/10.1007/s11423-018-9590-1
9. Tarighat S, Khodabakhsh S (2016) Mobile-assisted language assessment: assessing speaking. Comput Hum Behav 64:409–413. https://doi.org/10.1016/j.chb.2016.07.014
10. Jurkovich V (2019) Online informal learning of English through smartphones in Slovenia. System 80:29–37. https://doi.org/10.1016/j.system.2018.10.007

Potential of Digital Footprints for Economies and Education

Digital Potential of Economic Education: Information Technologies in a Management University

F. F. Sharipov, T. Yu. Krotenko, and M. A. Dyakonova

Abstract The changes happening in the world these days imply the inevitable digital transformation of business, science and education. To undergo such a serious transition Russian companies of any business scope need to provide the full-scale readiness of the whole system. Digital transformation without the elaborated high-quality preparation of all system elements and connections may turn out to be an untimely expensive measure hindering company's development. There are analyzed the obstacles defined as part of the study and related to digital transformation of economy and education. Education is viewed as a leading factor of socio-economic development. Thus, the request for implementation of digital transformation is addressed primarily to education. The study raises a problem of lack of a well-founded psychological and pedagogical concept of digital learning which could be used by the subjects of education as the basic one. The results of the survey conducted among university students majoring in public management have shown the following: it is necessary to expand professional competences of modern managers needed for carrying out the activity in the epoch of digital revolution; there emerges the question about rational inclusion into public management university curricula the disciplines corresponding to real demands of the labor market.

Keywords Science · Public management education · Knowledge · Information technologies · Digital literacy · Challenges of digital transformation

F. F. Sharipov (✉) · T. Yu. Krotenko · M. A. Dyakonova
State University of Management, Moscow, Russia
e-mail: fanissh@rambler.ru

T. Yu. Krotenko
e-mail: krotenkotatiana@rambler.ru

M. A. Dyakonova
e-mail: marie.d@mail.ru

S. I. Ashmarina and V. V. Mantulenko (eds.), *Current Achievements, Challenges and Digital Chances of Knowledge Based Economy*, Lecture Notes in Networks and Systems 133, https://doi.org/10.1007/978-3-030-47458-4_65

1 Introduction

Nowadays one can talk about digital transformation more than just a cutting-edge trend. Digital transformation becomes an obligatory factor for success of any socio-economic system in the market, and education institutions are not an exception in this case. The adoption of the National program "The Digital Economy of the Russian Federation" and appearance of a number of projects in the framework of the markets of Russia's National Technology Initiative are testimony to understanding that digital transformation is something one cannot do without these days [14]. The implementation of such a large-scale project on a national level is possible only with its support on the part of science, business and education.

During the past five years the State University of Management has been holding international forums, conferences, workshops and roundtable discussions where hot issues related to digital transformation of education have been discussed. In 2019, the year of the 100th anniversary of the State University of Management, as part of the launch of the platform University's Tochka Kipeniya (Boiling Point) (aimed at designing educational trajectories for students and teaching staff and expanding the university's capacity to develop necessary students' digital competencies) the authors have conducted the study on the topic: «Resources of Business, Science and Education for Implementation of Digital Transformation Program».

The expert group of the study included representatives of the sphere of education, science and business, namely, scientists, university students, entrepreneurs in the field of technologies, members of professional communities, creators and users of online learning content. During the development of the study program there was firstly identified the scope of challenges which businessmen and entrepreneurs were suggested to evaluate with regard to their relevance in the modern business environment, including the following ones:

- companies' readiness for digital transformation,
- companies' application of the latest digital technologies,
- barriers which companies face on the way to digital transformation.

Secondly, creators and users of learning content, i.e. teaching staff and students engaged in online learning, were suggested to discuss the following issues:

- vectors of development of modern learning environment,
- today's understanding of what digital literacy is,
- abilities of modern education to provide human resources for economy's transition to a digital one.

Studying the processes of digital transformation in education, it is quite expected to discuss the building of a basis of digital literacy. Complication of media landscape and exponential growth of digital technologies requires from educational institutions not only establishment of fundamentals of basic literacy skills (reading, writing and counting), but development of much deeper knowledge and skills of application of information and communication technologies. The continuous progress inevitably leads to extension of the meaning of the word literacy.

Digital literacy nowadays means the knowledge of basics of computer programming, ability to create a content using digital technologies, having the skills of data processing, free communication of users with professionals [1]. At the same time one should all the time take into account the fact that this set of basic digital literacy skills will every hour and gradually be supplemented by new skills required for the subsequent new level of development. The environment of ongoing knowledge building and permanent processing of huge amount of diverse information should be useful for subjects of education, providing for their development rather than being the source of unceasing stress [3].

2 Methodology

Digital transformation which is developing as a national project, needs support on the part of science, education and business. From this perspective, we have made an attempt to identify the challenges of digital technologies application which face representatives of the above mentioned institutions. We have formulated the three hypotheses of the study.

Hypothesis 1. The major problem which hinders the implementation of information technologies in business, is a low level of digital literacy of business leaders, as well as insufficient development of digital competencies of managers and company experts.

Hypothesis 2. Professional competencies of managers of the future should be extended and supplemented with knowledge and skills in the sphere of digital technologies.

Hypothesis 3. The main obstacle for efficient use of information technologies at universities is the lack of a solid psychological and pedagogical concept of digital learning, which could serve a basis for the subjects of an educational process.

We have applied the methods of the study in the following sequence: (1) survey of representatives of companies and universities with further data processing and analysis; (2) content analysis of the answers to the questions where experts state their own position and personal point of view.

The questionnaire for the representatives of the real sector of economy contained a number of open questions concerning digital transformation in their companies:

1. Which digital technologies are already used in your company? Does the company have a plan of digital transformation?
2. Name the obstacles you had to face during the company's transformation.

The students of the State University of Management majoring in management were asked to answer one question: Name the digital technologies which you will need in your professional activity in the near future.

The questionnaire for the respondents who are creators of online content as well as for the State University of Management teaching staff included the following 4 open questions:

1. What factors do you think influenced the keen interest of people all over the world (including Russians) in digital learning?
2. In which directions are information technologies used in an educational process?
3. How do you evaluate the possible contribution of online learning into development of managers of the future?
4. Are digital technologies efficient just as they are implemented in education these days? What are the challenges? How to improve efficiency?

3 Results

Results of the Survey of Business Representatives. The applied content analysis of 55 questionnaires of businessmen helped to make the following conclusions. Large Russian companies have embarked on a course of digital transformation, understanding that implementation of digital technologies will allow to increase labor productivity, cover the expenses, prepare the inflow of investments, etc. [9]. However, the companies which have started a planned realization of projects in a digital sphere, are not numerous [10]. Leaders of the companies which have in view and implement pilot transformation projects, have enumerated the obstacles on their way (Fig. 1).

The respondents name such challenges as, naturally, the deficit of financial resources and a slow return of investments in digital projects (29% of those who operated with a questionnaire). According to entrepreneurs, there is no doubt that digital technologies can improve many aspects of business activity, such as client communication and advertising, product quality enhancing and improving the means of product delivery, automation of internal business processes, simplification of cooperation with state authorities [8]. At the same time some respondents claim that digital solutions do not correspond to already designed business processes (36% of respondents) and digital transformation plans contradict to strategic plans adopted by the company. In such cases there are needed more serious measures related to introducing changes into companies, which also requires investments.

The answers of the survey reflect some concerns regarding the risks of an experiment (9% of experts), low level of enthusiasm required to conduct efficient transformation (12% of respondents), ineffective horizontal collaboration (13%), week infrastructure which leads to assimilation of digital solutions (15%), low level of companies' economic and informational security (18%), as well as the lack of a complex program of company's digital transformation (19%). As it was expected at the beginning of the study, the respondents in their answers put emphasis on an

Fig. 1 Barriers to using IT in business (*Source* authors)

insufficient level of leaders' literacy in the sphere of information technologies (43%) and weak digital competences of some managers and experts (39%) [12].

Since the information was obtained during the study conducted at the university where one of the majors is management, it can be considered as an invitation to review educational programs of the university with the aim to supplement them with such courses which will allow students to be well-informed about the professional digital environment.

Student Survey Results. 193 students of the State University of Management (2-year, 3-year and 4-year students) majoring in Management and Anti-Crisis Management took part in this stage of the study.

When answering the question, "Name those digital technologies which you will need in your professional activity in the near future", the students mentioned a number of software programming languages, databases, organizational IT-systems, data analysis facilities, SMM tools, nonlinear tools for search, analysis and forecasting, Machine Learning facilities.

There was identified the following indicator of a conscious need of possessing competences in information technologies for conducting professional activity: the students underlined in their answers the fact that they make an attempt to study some digital tools on their own. Thus, there arises a reasonable question concerning the rational inclusion into the specified university curricula the disciplines corresponding to real demands of the labor market [4].

Results of University Teaching Staff Survey. Almost all of the respondents (69 teachers in higher education were interviewed) claim that digital technologies which are already present in many spheres of our life, are bursting into the learning environment. Internet access, common learning content, systematization of evaluation, work with smart devices, cloud computing, AR/MR, as well as just the dynamic character of university classes resonate both with students and many teaching staff members. But all the above mentioned amenities have their reverse side.

Since modern digital solutions displace from the industry those ones, whose labor is replaced loss-free by smart technologies, education should not turn into a wide-scale assembly line churning out exactly the same specialists. Such workers are no longer needed in the present era of figures. The economy ruthlessly gets rid of them already nowadays [15]. The topical issue now is to educate unique experts who will possess the following competences: high social and individual mobility, unique development trajectory, self-motivation skills, self-planning and self-analysis [11]. Can information and communications technologies (just as they are in an educational process these days) be the leading ones in a complex development of people with such competences?

Experts in the sphere of online development pay attention to the fact that the world-wide spreading of digital learning (and Russia is not an exception in this case) was influenced by the following four simultaneous conditions:

- development of computer science, engineering and cognitive psychology; representatives of these areas of science began to talk in concert about the application of a computer metaphor when describing machines' actions as well as human brain's activity,
- development of digital learning goes along the same way as computer-assisted learning, which means that nowadays we deal with reproduction of a technological approach to managing an educational process,
- development of mass production of personal computers, software and various digital devices,
- searching for the markets to sell digital amenities which were mentioned above, where the sphere of education for such a purpose is an unlimited space long wished for.

University teaching staff members acknowledge the fact that a computer is used in an educational process as: (1) a training device, when there is required a minimal systematization and just refining of the acquired skills or the covered material; (2) a tutor: the task is clear, the conditions are well-defined, there is only one correct variant; (3) a modelling tool, which sets a task and students are given an opportunity to solve the task independently. "Training devices" and "tutors" have quickly gained popularity at high schools and universities at the beginning of the era of computerization [7]. Such forms of using electronic devices for educational purposes significantly increase the pace of information processing. Although, one needs to admit that the "fast speed fever" quickly stopped when it was discovered that student's thinking started to lack quality improvements and students could no longer independently work and find non-linear solutions of difficult tasks [2]. Modelling (the third type of using computers) at present moment holds a gigantic variability, since a student acts on his own in a simulation model. This has a lot of positive features and has the edge over teaching without computer modelling. The mode of such a game definitely has a high potential, as it gives space for creative thinking in a designed learning environment.

The following opinions of experts are of great interest. In education there are often confused such two conceptually different notions as "information" and "knowledge". Any computer, as it exists these days, is not yet ready to transfer information into knowledge and meaning into sense. Such transformations are purely thinking processes explained by psychological patterns. They are typical only of human beings. Transfer of the information present in class (especially using electronic devices) into sense, and then into certain practical actions poses a huge problem. Successful transformation of such type are already a victory. This is why we should keep in mind the reasonable balance between using unique opportunities of computers and real-life communication, as well as the limits of application of a "computer metaphor".

The supporters of digital learning consider an ideal the fact of minimizing face-to-face communication of the subjects of an educational process. But should one strive to this so eagerly? The role of perception in sense-making no longer needs scientific evidence. However, the point of view of e-learning supporters regarding the fact that proper thinking process is possible without active reflection by a person of humans, events and situations, still needs to be scientifically proven. Degradation of living speech (in the conventional sense), fragmentary and discrete nature of a thought, absence of concentration on one object, depthlessness, unsubstantiated character of judgements, preference for visual messages over text ones, text messages over a live talk—all these are features a new digital generation psychology.

Moreover, experts assure that there have not been conducted any longtime observations related to the aftermath of a wide-scale introduction of electronic devices into educational institutions and there have not been developed any psycho-physiological, clinical, health protection or psychological standards of using gadgets in an educational process. There are no solid and sufficient justifications of the fact that electronic books, interactive devices and multimedia gadgets are not injurious to children's health. Eyesight and hearing deteriorate, as well as metabolic processes and the state of internal organs. Children develop early onset scoliosis, atony and psychosomatic illnesses. The consequences of thoughtless introduction of digital means into high schools and universities are not studied well enough. In fact, there is no well-founded psychological and pedagogical concept of digital learning which could be used by the subjects of an educational process as the basic one (48% of respondents talked about this).

Experts have been discussing the circumstances which led to the groundbreaking changes of learning environment which are observed nowadays. Educational institutions are united into complexes, the number of levels at primary and secondary schools, as well as at universities is increasing. There is the growth of competition between educational institutions for service consumers, orders for scientific and research developments and qualified staff. In addition to the above, learning platforms get updated and improved and already now they can stand competition with many universities regarding the quality of information resources and services. Artificial intelligence abilities improve and thus it has the full potential to get implemented into an educational process. The range of paid educational services is getting wider, there appear distant ways of learning [5]. Education is turning into a business: services are sold, high potential skills are bought with the purpose of further profitable capitalization. However, these tendencies should not be considered separately and emotionally, as they present a whole set which influences the situation in education and shapes a common vector of development.

4 Discussion

In the conditions of cyber socialization of all public institutions lifelong learning becomes a norm, and even the condition for human living. In a situation when educational content is actively digitized and there appear online learning platforms, experts are seriously concerned to what extent students and teaching staff are ready for transformations and whether there are needed inner resources and pedagogical tools for their development.

In general, the problems of digital learning can be conveniently classified into the problems of temporary nature and immanent problems typical only of this very type of learning. The development of electronics inevitably leads to educational market transformation. The main subjects will be those structures which generate new knowledge and are successfully involved in the development of fundamentally new educational amenities and staff training [6]. As a result of the "natural selection" the following institutions will come to the forefront: big universities, and later companies producing digital educational products, as well as leaders transmitting educational amenities to customers, i.e., global learning platforms which have proved to be powerful. And for the moment we deal with weak and/or dishonest operators. It is sometimes hard to distinguish quality products and services from their imitations on a vast and rapidly developing market of online learning (31% of respondents). This is connected with the lack of users' experience, skills of specialists taking part in creating the learning content, deficit of expert reviews and systematic monitoring, obtrusive advertising which ensures success and credibility (29% of respondents) (Fig. 2). Educational production is a typical kind of privileged goods. This means that a consumer has to hope to receive good quality, without possibility to evaluate the product as an expert. Reviewing can serve as a filter which can be set before the placement of a product on an online platform. One can also test artificial intelligence abilities.

Experts are talking about the problems typical only of online education. First of all, this is the craving for simulation of full-time education (15% of experts). As a result, a high-technology copy is often inferior to the original created in an offline mode with respect to the force of its impact on users. Similar to the way digitized

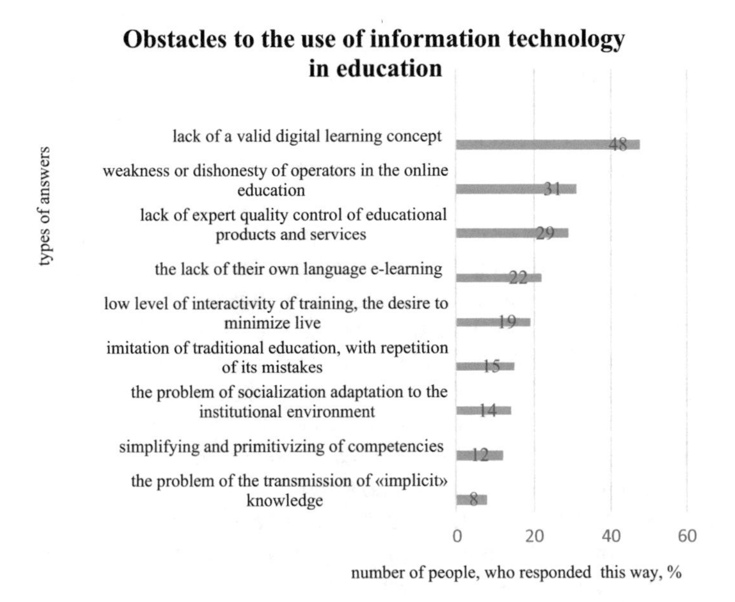

Fig. 2 Obstacles to the use of information technology in education (*Source* authors)

books, documents and movies lose touch with the time when they were created, so called impersonal digital twins in education turn out to be less interesting than their classical prototypes.

Weak interaction is one more sore spot of online education (19% of respondents). Here electronic learning repeats the mistakes of traditional education and rests hopes on artificial intelligence. At present moment the problems of activity of the subjects of learning are solved, for example, by means of specialized forums. And, in general, distant learning in an electronic environment is successful already nowadays, including in a real-time mode. But there arise the issues of social involvement and cultural inheritance.

Even if an online course has some interactivity, there remains the problem of reproduction of norms, rules and values accepted in a community. The problem of social involvement is debated a lot by experts (14% of respondents). Formal rules and regulations can definitely be transferred remotely, in an electronic form. However, this will not make it possible to adapt a neophyte to life and activity in conformity with the laws of an institutional environment. The transfer of conventional skills is possible only during real-life interaction—face to face, hand in hand. A student needs to have a teacher's powerful personal example, the potential of emotional contact and psychological support. It is necessary to play different roles, to live situations according to samples, to get acquainted with traditions which allow participants to understand each other and act together. The function of upbringing is likely to be placed on an online-tutor of the future. The problem of how to transmit information was solved long time ago, while the problem of sociocultural inheritance will get more and more acute [13].

The issue of transfer of tacit knowledge is similar to the previously mentioned problem of social involvement (8% of experts). Tacit or implicit knowledge (transferred only from a teacher to a student) is not just non-formalized and superfluous information, but rather a mandatory base for logical forms of knowledge. Although "peripheral knowledge", in the framework of Michael Polanyi classification, is not included in a discourse (as it is the product of sensational and intellectual intuition), it establishes the basis for scientific theories. Authors and holders of this useful knowledge are certain people. Such type of knowledge is a strategic nonmaterial resource of an institution, its intellectual capital, keeper, transmitter and the center of reproduction of the type of thinking through cooperative speculation. People also keep and transfer their vision of the world, techniques of information processing, their vision of the culture of holding a discussion, the art of creation and ideas generation. All this is inseparable from a man and is transferred only through a joint action with other people. Nowadays, as experts claim, there can be seen a meaningless contraction of the area of traditional reproduction of knowledge.

The problem of transfer of tacit knowledge has something in common with the problem of skills primitivization, or automatic simplification of competences (12% of respondents). Mass usage of electronic devices, such as spelling applications, calculators, navigators and others, which are supposedly meant to make people's work easier, in fact drastically decrease user's motivation for an independent search. Moreover, not having such helpers at hand, a person is unable to complete different tasks and make on-the-spot decisions.

The enumerated symptoms in a sum present a sad picture. Already now reformers in the sphere of education hold short-sighted discussions of separate education of such specialists as generators producing new knowledge and operators who can work in already created programs. And, in this sense, to design a society of ordinary consumers and manufacturers of an elementary range of goods and services, i.e. economy without competitive advantages, is the best option is a test type of education, regulating underperformance on a timely basis.

5 Conclusion

Nowadays there seem to be present the following several possible forms of digital transformation of a learning environment:

1. Transformation of already created learning materials into a digital environment. Many lecture courses, course books and study guides, books of problems, presentations, self-instructional materials, tools for testing, etc. have already been digitized.
2. Development of conceptually new teaching tools which do not copy the already known forms. Undoubtedly, in order to be perceived as a whole (in visual, audio, kinesthetic and digital forms) the learning content should be structured in a different way.

3. Creating an interactive digital space for beneficial interaction and communication of all the subjects of an educational process. For instance, these can be teachers' online accounts, online discussions and forums, webinars, survey forms, etc.
4. Creating brand new training resources which are aimed at transfer of techniques of thinking in a "here and now" mode, rather than at an ineffective way of transmitting facts about the techniques of thinking (the very thinking process is much richer than knowledge about it). This can be done, among other things, by using the achievements of gaming and simulation modelling.
5. Involvement of useful artificial intelligence abilities into an educational process.

The first three forms are successfully implemented to a different extent in a learning environment of high schools and universities. This makes many learning materials more accessible, reduces insignificant and routine work of teaching staff, expands the range of online educational services. The process of e-learning adaptation will have positive outcomes only in those educational institutions which will manage to offer distant learning services of continuously improving quality, with an original language of electronic education (22% of experts told that there was no such services). In a reverse situation, Russian system of education, taking a challenge of digital transformation, can remain on the fringes of global learning environment.

Business, it has to be noted, is quickly embracing digital technologies. Although, there are some obstacles, such as difficulties at the step-by-step planning of transformation and weak infrastructure for assimilation of new technologies. Representatives of a real economy sector consider the crucial one the problem of insufficient digital acumen and lack of literacy in the sphere of information technologies on the part of companies' leaders and specialists. Overall 82% of the respondents answering the questions of the survey talk about such kinds of barriers on the way to digital transformation. Therefore, we can consider the first hypothesis of our study to be confirmed.

The conducted survey of the students who are going to become future managers and administrators, allows concluding that acquiring digital competences is, evidently, of crucial importance for their present and future professional activity. It proves the second hypothesis. The today's world creates a totally new space for people's life and work. Digital technologies break into learning platforms. This presupposes a dynamic correction of educational standards regarding their appropriateness to modern time, application in practice, competitive ability, and most importantly (the thirds hypothesis), the development of a well-founded psychological and pedagogical theory which proves the efficiency of digital means application in an educational process at universities.

The presence of such a strategic benchmark is obligatory when designing university educational programs. Otherwise, there is a real danger which is in the following: education which claims its own priority and innovative nature will transform from the main factor of society development into a service structure meeting chaotically arising needs of economy in educational services.

References

1. Baskakova ME, Soboleva IV (2019) New dimensions of functional illiteracy in the conditions of digital economy. Issues Educ 1:244–263. https://doi.org/10.17323/1814-9545-2019-1-244-263
2. Bostic JD, Krupa EE, Carney MB, Shih JC (2019) Reflecting on the past and looking ahead at opportunities in quantitative measurement of k-12 students' content knowledge. In: Bostic J, Krupa E, Shin J (eds) Quantitative measures of mathematical knowledge: researching instruments and perspectives. Routledge, New York, pp 205–229. https://doi.org/10.4324/9780429486197
3. Davison RM, Ou CXJ, Martinsons MG (2018) Interpersonal knowledge exchange in China: the impact of guanxi and social media. Inf Manag 55(2):224–234. https://doi.org/10.1016/j.im.2017.05.008
4. Ellis RA, Bliuc AM (2019) Exploring new elements of the student approaches to learning framework: the role of online learning technologies in student learning. Act Learn High Educ 20(1):11–24. https://doi.org/10.1177/1469787417721384
5. Gibadullin AA, Karagodin AV (2019) Challenges of the digital economy in the field of personnel training. Curr Probl Econ Manag 22(2):33–42
6. Grant SB, Preston TA (2019) Using social power and influence to mobilise the supply chain into knowledge sharing: a case in insurance. Inf Manag 56(5):625–639. https://doi.org/10.1016/j.im.2018.10.004
7. Janelli M (2018) E-learning in theory, practice and research. Issues Educ 4:81–98. https://doi.org/10.17323/1814-9545-2018-4-81-98
8. Konstantinov V, Sakulyeva T, Makeeva V (2019) Development of economic tools for managing regional innovation clusters. J Entrep Educ 22(1S). https://www.abacademies.org/articles/development-of-economic-tools-for-managing-regional-innovation-clusters-7988.html. Accessed 03 Mar 2020
9. Lokuge S, Sedera D, Grover V, Dongming X (2019) Organizational readiness for digital innovation: development and empirical calibration of a construct. Inf Manag 56(3):445–461. https://doi.org/10.1016/j.im.2018.09.001
10. Ryazanova GN, Sazanova AA, Sazanova SL (2018) The impact of the digitalization of the economy on the activities of non-financial organizations. Management 2:52–56. https://doi.org/10.26425/2309-3633-2018-2-52-56
11. Salehan M, Kim DJ, Lee JN (2018) Are there any relationships between technology and cultural values? A country-level trend study of the association between information communication technology and cultural values. Inf Manag 55(6):725–745. https://doi.org/10.1016/j.im.2018.03.003
12. Sazonova SL (2019) Structural modeling of the institute of entrepreneurship in Russia. Econ Sci Modern Russ 84(1):34–48. https://doi.org/10.33293/1609-1442-2019-1(84)-34-48
13. Sergeeva M, Shilova V, Evdokimova A, Arseneva N, Degtyareva V, Zuykov A (2019) Future specialists socialization in the context of competence approach. Revista Praxis Educ 15(34):571–583. https://doi.org/10.22481/praxisedu.v15i34.5796
14. The national program «Digital Economy of the Russian Federation», approved by the order of the Government of the Russian Federation of July 28, 2017 No 1632-r. http://static.government.ru/media/files/9gFM4FHj4PsB79I5v7yLVuPgu4bvR7M0.pdf. Accessed 03 Mar 2020
15. Tinyakova VI, Morozova NI, Gunin VK (2019) Transformation of the system of professional training of personnel who are competitive in a knowledge-based economy. Econ Sustain Dev 37(1):242–245

Digital Footprint Interpretation in Vocabulary Training When Working with Electronic Dictionaries

A. L. Kuregyan and E. A. Pertsevaya

Abstract This article deals with the digital footprint interpretation in modern English vocabulary training when working with electronic dictionaries in the process of forming the translation competence. The authors provide a detailed analysis of the digital footprint as a part of learning processes digitalization in foreign language teaching. Some dictionaries existing in electronic form were compared for optimal use in the perspective of digital footprint analysis. The detailed analysis conducted by the authors of the paper revealed that the most preferable dictionary is the Lingvo dictionary.

Keywords Digital footprint · Vocabulary training · Electronic dictionaries · Multitran · Lingvo · Reverso

1 Introduction

The digital footprint as a set of search history according to the material studied is quite a new topic. The present research assessing digital footprint left by users when they request information focuses on the problem of identifying social media users. Scientists consider the possibility of applying a digital footprint to create a profile based on the personal characteristics of users of social networks [2], which will help in determining the preferred and negatively evaluated information. Any material obtained from the analysis of the user visits history can be collected, processed and interpreted in accordance with the specific task of the researcher [2], so it is necessary to raise the awareness of the digital footprint that users leave when searching for data [1]. It is logical to use the digital footprint in foreign language teaching, as dictionaries and full-text translation services are widely used in the process of learning new

A. L. Kuregyan
Samara State Technical University, Samara, Russia
e-mail: amleku@mail.ru

E. A. Pertsevaya (✉)
Samara State University of Economics, Samara, Russia
e-mail: kmilyutina@mail.ru

S. I. Ashmarina and V. V. Mantulenko (eds.), *Current Achievements, Challenges and Digital Chances of Knowledge Based Economy*, Lecture Notes in Networks and Systems 133, https://doi.org/10.1007/978-3-030-47458-4_66

vocabulary. When learning a foreign language, English in particular, the use of digital footprint has several advantages over the traditional system of learning, which has several disadvantages. For example, it is impossible to get feedback from the student, as well as to track and analyze the means used by the student to do the exercises. Analysis of the digital footprint as part of learning processes digitalization in foreign language teaching is possible. Consequently, it is necessary to consider the possible means of analyzing the digital footprint of foreign language learners, especially electronic dictionaries. In our work, we compared dictionaries existing in electronic form for optimal use in the perspective of digital footprint analysis.

2 Methodology

There are 3 main dictionaries used in Russia to find the meanings of the necessary word on the Internet. These are Multitran, Lingvo, and Reverso. The choice of the most suitable source for the research was carried out taking into account 12 main criteria. Currently, most of the sources are copyrighted and are not free of charge. Consequently, the first evaluation criterion was the fee for using the resource. Moreover, a number of educational resources have not only the site but also associated accounts on social networks, as well as applications that allow using it on a mobile device. So, these dictionaries have been reviewed for the availability of a mobile application. A third criterion was the industry's specificity of the dictionary, as English vocabulary had a significant multiplicity of meanings due to the conversion existing in the language. Then the dictionaries were evaluated by the number of represented industries reflected in the comments to the lexical unit. In addition, a number of sources allow the users themselves to add word meanings based on a possible one reflected either in the literature dedicated to the phenomenon or translated by a professional translator for the first time into Russian. So, the fifth criterion of evaluation is the possibility of adding a terminology list by the users of the database. The next aspect is the opportunity to listen to the correct pronunciation of lexical units in the English language. In a number of paper and electronic dictionaries, there are examples of usage; therefore the seventh criterion was connected with the presence of word usage examples. So, it was also considered in the paper. Moreover, it is necessary to take into account the possibility of the word translation in the text in Word when working with large passages. This is the criterion number eight. The ninth point was the aspect of having different programs and applications to memorize words. The next criterion is the existence of a forum where participants can discuss the correctness of the meaning chosen. The eleventh criterion is the ability to integrate with Microsoft Office to check the meaning and spelling of the word. The final aspect is the presence of a user-friendly interface that allows you to work with an electronic dictionary easily. The data comparison of existing dictionaries by the 12 abovementioned criteria is shown in Table 1.

Table 1 On-line dictionaries comparison analysis

Criteria	Online dictionaries		
	Multitran	Lingvo	Reverso
Price, RUB (per month)	Free	Free	Free
Application availability	+	+	+
Industry specification	+	+	+
A number of branches	+	–	–
Ability to add the terminology base	+	–	–
Availability to listen to the correct pronunciation	–	+	+
Examples of dictionary units usage	+	+	+
Availability of word translation in the text in Word	–	+	–
Programs and applications to memorize words	–	+	–
Forum	+	+	–
Integration with Microsoft Office	+	+	+
User-friendly interface	+	+	+

Source authors

3 Results

The processes of integrating digital technology into all areas of activity are closely connected with the foreign language learning, as they require constant clarification of the meaning of words and sentences in the language different from the mother tongue. An electronic dictionary is the most convenient way of information search about a specific lexical unit. The electronic version of the printed dictionary seems much more effective as it allows to save time while selecting the meaning, and also to search a word on its basis, offering options of spelling at basic letter combinations search. In studying a foreign language, students need not only to understand basic vocabulary units with translation and interpretation but also to constantly replenish their vocabulary with new lexemes related to modern issues.

Teaching a foreign language in the modern world should not only be a process of a message delivering, but also a process of getting students' feedback. Taking into account the information that a student receives in the course of English language learning, it is possible to understand the topics he/she is interested in and to highlight the words that are most frequently requested in different search systems. Students of non-linguistic majors of higher education institutions are in search of foreign words different from students of specialized areas because, mainly, the search for words of a foreign language does not require interpretation in the original language, and consists in the selection of the Russian equivalent.

The resources for learning a foreign language, most frequently used, include different sources for self-study and sources used by the teacher in the process of foreign language teaching. Both resources include online dictionaries (there is a number of such dictionaries, some of them are translation ones or explanatory dictionaries). It is possible to compile a list of the most frequently requested words of modern English to make the training more customizable and aimed at learning the most frequently used words taking into account students' involvement in the electronic environment of the higher education institution (this indicator is 100%). While searching the meaning of English words that are necessary to perform tasks related to the foreign language they are integrating into a dictionary system that interprets those words. Making a list of the words may be done by analyzing the search history (digital footprint) in open online search resources—namely, electronic dictionaries.

As we can see from Table 1, the most appropriate dictionary for all the criteria studied is Lingvo, which has the minimum number of restrictions—it does not specify the number of industries in which the word can be used and there is no possibility of adding content by users themselves. These aspects are not so important, but this possibility is present in the dictionary that is in second according to the number of missing criteria, i.e. the Multitran dictionary. Almost every user can offer his own variant of lexical unit translation, different from the versions listed in the dictionary. If such a variant is recognized, it is placed in the list of requested word analogues. Such an approach is interesting, but the translated variant is not always absolutely correct, and it gives, however, the possibility to search for the word interpretation among the meanings of lexical units. The Lingvo dictionary is updated by professional translators, making it more academic and accurate. However, in the Multitran dictionary, it is possible to find analogues of words and word combinations, which are used for the first time in the printed version and do not have any analogues in Russian. However, they are present in the texts and require an equivalent translation into Russian. This aspect is mainly related to the use of word combinations translation.

In the Multitran dictionary, it is not possible to listen to the correct pronunciation of the requested word, which is a significant disadvantage, because nowadays quite a few students specify the transcription of the requested words, and even fewer can read the transcription and thus use the correct variant of sound in speech. Most words are often read using a Latin-like pronunciation, which is relatively clear in medical and technical texts, but absolutely unacceptable in other areas because it distorts the meaning of the word. Accordingly, the availability to listen to the word pronunciation is a fundamentally significant feature that is not found in the dictionary involved.

In addition, the Multitran dictionary does not have the possibility to translate the text in the Word application, which is the most common version of text documents, as well as the application to remember and repeat new words, which is very effective when learning a foreign language.

The Reverso dictionary has another significant disadvantage, that is the lack of a forum where participants can exchange opinions on questionable vocabulary units and share successful translation variants. At present, the forum as a platform for information exchange is very important. Not having found the necessary meaning of the word many users prefer not to use other sources but to apply directly to the forum and ask for the right answer or ideas there.

Despite a number of drawbacks, all the dictionaries reviewed are free of charge, which means that they are easy to access for information retrieval and easy to use. All dictionaries take into account the specifics of the industry, as the same word can be used in different texts, acquiring different meanings or shades of meaning, and it can be used in the highly specialized and completely neutral texts. Also, all the dictionaries provide a more or less extended version of using a lexical unit in context, which allows minimizing grammatical and semantic disturbances in the unit use. All the dictionaries can be integrated with Microsoft Office and have a user-friendly interface that allows using them easily. Thus, all these dictionaries can be used in work when teaching English. Lingvo is the most preferred dictionary, but the others also have prospects for use in the educational process, as they allow analyzing the statistics of user's requests through the digital footprint left by the requested lexical units.

In practice, we used the following method of digital footprint analysis: students received a specialized text, all the unfamiliar words, and phrases meaning had to be checked. The search for unknown meanings was performed using an electronic dictionary, unfamiliar words were highlighted in the text and then sent to the teacher's e-mail. On the basis of the requested units analysis, it became possible to select necessary means of language studying and mastering the unknown vocabulary. Such work was carried out both taking into account the individual needs of the student and summarizing the data received from the student group.

It should be noted that the possible variants of studying the data of the digital footprint in the English language teaching are statistical accounting of the number of users, as well as the search history, that can be categorized by several parameters:

- by the word-formation principle (for example, the most frequently requested parts of speech),
- by semantics (what values are the most frequently requested),
- by vocabulary categories (for example, economic terms or common words).

On the basis of the data studied it is possible to predict the students' needs in the field of language learning and to fill the existing knowledge gaps. Moreover, these data can reveal to the teacher possible problems and areas in vocabulary learning. Teaching English to each group of students becomes more adapted and customizable, taking into account the specific tasks that are important in mastering a foreign language. In addition, based on the data obtained it is possible to create a dictionary in which either the students themselves or the teacher, introduces lexical units variants that are difficult to learn.

4 Discussion

The digital footprint left by foreign language learners while using electronic dictionaries has considerable potential for use. The use of traditional media, namely printed dictionaries, is currently being questioned as users prefer to use electronic media, applications, and devices [5]. The use of traditional media is very time-consuming, since searching for a word in a dictionary causes a long spelling analysis, which makes reading the text much more difficult. More and more applications, platforms and other language learning options are now available. In modern literature, there is also the term MALL (Mobile Assisted Language Learning), which covers all kinds of digital opportunities used in language learning [3]. Electronic dictionaries are a kind of these applications, so they can be studied in this aspect.

Search for words in a dictionary, as presented by Liu et al. [6] plays a significant role in the formation of general background knowledge of any foreign language and can contribute to the skills of English words spelling and general cognitive load. In the process of foreign language learning in modern realities two types of electronic dictionaries are involved: dictionaries in which you need to click on an unfamiliar word (Click-on) and dictionaries in which you need to print an unfamiliar word on the keyboard (Key-in). Both types of dictionaries have significant value in learning foreign language vocabulary, but the latter has more advantages in long-term learning and vocabulary memorization, although this dictionary is more time and effort consuming. The dictionaries that we have considered in the process of our research are of the second type. Moreover, dictionaries, in particular electronic dictionaries, are considered to be a source of continuous learning [4] because they allow us to acquire knowledge about new words we have never seen before. The analysis of dictionaries to determine the optimal source of digital footprint has identified promising areas of research.

5 Conclusion

In general, analyzing the digital footprint of English vocabulary is quite promising and makes learning more appropriate for each specific group of students. The most common means for learning English are electronic dictionaries of various types, which allow you to select the right word meaning. Our analysis of dictionaries revealed that the most preferable dictionary is the Lingvo dictionary. Analyzing the units requested while working with text, the prospects of processing data digital traces of students in lexical units search of modern English are:

- simulating a student profile for each specific group, taking into account the individual needs and vocabulary gaps of each category;
- categorization of the requested lexis into specific groups according to the semantics, part of speech, etc.;
- analysis of students educational needs in the English language vocabulary;
- identification of certain problem aspects that exist within the vocabulary learning;
- individualization of the foreign language learning process and adaption of the material to each students group;
- creating lexical units dictionary requested by a specific student group.

It should be noted that the analysis of the digital footprint can be done not only for English dictionaries but also for all the other languages, creating educational courses which take into account the needs of a particular student or a particular group of students in mastering the vocabulary.

References

1. Camacho M, Minelli J, Grosseck G (2012) Self and identity: raising undergraduate students' awareness on their digital footprints. Procedia Soc Behav Sci 46:3176–3181. https://doi.org/10.1016/j.sbspro.2012.06.032
2. Deeva I (2018) Computational personality prediction based on digital footprint of a social media user. Get Rights Content Personal Individ Diff 124:150–159. https://doi.org/10.1016/j.procs.2019.08.194
3. Hoi VN (2020) Understanding higher education learners' acceptance and use of mobile devices for language learning: a Rasch-based path modeling approach. Comput Educ 146:9–20. https://doi.org/10.1016/j.compedu.2019.103761
4. Jalaluddin NH, Zainudin IS, Ahmad Z, Mohamad Sultan FM, Radzi HM (2014) The dictionary as a source of a lifelong learning. Procedia Soc Behav Sci 116:1362–1366. https://doi.org/10.1016/j.sbspro.2014.01.398
5. Kukulska-Hulme A (2013) Mobile-assisted language learning. In: Chapelle CA (ed) The encyclopedia of applied linguistics. Blackwell Publishing Ltd., Hoboken, pp 3701–3709
6. Liu TC, Fan MHM, Paas F (2014) Effects of digital dictionary format on incidental acquisition of spelling knowledge and cognitive load during second language learning: click-on vs. key-in dictionaries. Comput Educ 70:9–20. https://doi.org/10.1016/j.compedu.2013.08.001

Prospects of Digital Footprints Use in the Higher Education

V. V. Mantulenko

Abstract The objective increase in the role of digital technology comes into a real contradiction with the real practice of its usage in the Russian educational system. In the pedagogical science and practice, educational opportunities of digital tools and resources, their didactic and educational potential are still underestimated. This article investigates possible directions of digital footprints use in the system of higher education in the Russian Federation. The main research methods are analysis, systematization and modeling. Based on them, the author identifies three promising application areas for digital footprint technology in the educational context, which are interdependent: ensuring continuity and integration of educational levels; organization and management of the educational process; management of the educational system as a whole (educational management).

Keywords Digital footprint · Education · Digitalization of the educational system · Higher education

1 Introduction

Overcoming the consequences of the global economic crisis and focusing on the innovative development required the modernization of the education system, which is a key factor in improving various spheres of the human activity. The Russian national project "Education" emphasizes that the global competitiveness of higher education in the Russian Federation can be ensured by cooperation with partners from the real economy sector to develop adaptive, practice-oriented and flexible higher education programs that provide students with necessary professional competencies [7]. It is important that these competences meet current requirements of the labor market, including in the field of digital economy, entrepreneurship, and project management in relation to the future areas of professional activity.

V. V. Mantulenko (✉)
Samara State University of Economics, Samara, Russia
e-mail: mantoulenko@mail.ru

S. I. Ashmarina and V. V. Mantulenko (eds.), *Current Achievements, Challenges and Digital Chances of Knowledge Based Economy*, Lecture Notes in Networks and Systems 133, https://doi.org/10.1007/978-3-030-47458-4_67

Today, digital technologies are a comprehensive platform for the development of all sectors of the economy, including education. The use of digital technologies in higher education has a didactic potential in the aspect of organizing the educational and cognitive process, providing new quality opportunities through the implementation of the principles of virtualization, mobility, adaptability, and instant feedback. Digital education sets new requirements for subjects of the educational process, to the content of the information and educational space, to the regulation of all participants' interaction in the educational process, and to methods and parameters for evaluating the educational and cognitive activity of students.

This study is consistent with the key directions and objectives of the development of the Russian education system identified in the national project "Education": updating the content, creating the necessary modern infrastructure, and the most effective mechanisms for managing the industry [7], the federal project "Digital educational environment" [6] and the Decree of the President of the Russian Federation from May 7, 2018, No. 204 in the aspect of creating a modern and secure digital educational environment by 2024 that ensures high quality and accessibility of education of all types and levels; modernization of professional education, including through the application of adaptive, practice-oriented and flexible educational programs [4].

In order to ensure the creation of a modern digital educational environment, the federal project "Digital educational environment" is aimed at the application of modern technologies (including virtual and augmented reality, "digital twins", etc.) by the implementation of basic educational programs. Effective use of these technologies requires scientific justification and development of specific mechanisms and models for their practical use for educational purposes.

A digital footprint is a huge and unstructured array of data that we leave in the global information network from any of our actions, and that can carry extremely useful information. In the field of education, a digital footprint is a student's written works, notes, tests, online courses, photos, etc. Modern technologies allow us to recognize faces, voice, translate speech into text and vice versa—and all this in seconds. Based on the analysis and special processing of this trace, we can give some advice to students, guide them, and make professional training more individually oriented. The digital footprint can allow educational institutions to better understand student's behavior and better prepare the students' assistance and mentoring in the direction of discovering and developing their abilities.

In the context of a new economic paradigm, an economy built on knowledge and information, lifelong learning becomes not just a necessity, an important prerequisite for the professional self-realization, but also a basic component of the spectrum of personal and professionally significant value orientations. Educational systems in many countries are undergoing changes not only in terms of the active usage of information and communication technologies in the educational management and organization of the educational process. In our opinion, changes related to ensuring the continuity of different educational levels (in particular, between schools and high educational institutions/HEIs), strengthening cooperation between universities and the business community in terms of creating joint training programs, courses, practices, etc. are more in demand and relevant at the moment. These transformations are extremely significant for the post-Soviet countries and countries with developing economies, as they lead to the creation of a new quality of human capital, which is in demand in the context of global digitalization to provide a new format for organizing life and professional growth.

In a world of growing mass customization, where some experts see a unique chance to realize the economic growth and create a new quality of human capital, digital technologies and data have an inexhaustible potential and play a crucial role in training unique specialists of the future.

All modern information and communication technologies and teaching methods based on them should ultimately be directed to ensuring a possibility of forming each student as a separate unique specialist who has passed his own trajectory of formation and development, and has accumulated his own combination of competencies. Soon, we may find ourselves in a future where there will be no departments or training areas, but everyone will have their own set of competencies (like a unique fingerprint) that can be developed, modified and supplemented throughout life.

In modern science, attempts have been made and are being made to find effective ways to use digital technologies in the educational environment (including artificial intelligence and augmented reality). In real pedagogical practice, there are already attempts to use some of the features of artificial intelligence and augmented reality in education, and there are numerous tools for analyzing and systematizing Big Data. However, there is still no scientific justification for the effective and legitimate use of such data as the "digital footprint", its didactic potential and opportunities for building individual educational trajectories have not been studied. Many recommendations become obsolete as quickly as technologies themselves change rapidly. Therefore, one of the objectives of this research is to analyze the experience of the educational space digitalization in different countries over the past decade, to identify problems and prospects that are specific for the Russian Federation in the context of using digital footprint in education.

2 Methods

In this research, the author used a complex of methods: analysis of scientific literature and pedagogical documentation; study and systematization of research sources; comparative-historical and historical-typological methods; quantitative and qualitative data analysis, modeling. The usage of these methods allowed achieving the research goal (determining some possible directions for application of digital footprint in higher education) and solving research objectives (reviewing documentation regulating the development of the Russian educational system until 2024, and analyzing modern research works on the issues of digital footprint use in the educational environment).

3 Literature Review

At the moment, there are very few scientific publications devoted to the use of Big Data and digital footprint in education, despite the high potential and importance of this area. Here are some attempts of Russian and foreign authors to evaluate the possibilities of these resources and use them for didactic purposes.

Structural and informational characteristics of the digital educational footprint as a way to evaluate results of educational and cognitive activity of students in the process of teaching mathematics were studied by Galimova et al. [5].

Based on the analysis of data related to the digital footprint, D. Azcona, I. Hsiao, and A.F. Smeaton developed their own methodology that allows them to automatically identify "risk groups" among students in the aspect of their failure to complete tasks on computer programming modules (courses) and simultaneously maintain adaptive feedback. This technique was implemented as part of a joint project of Irish and American scientists and described in the article "Detecting students-at-risk in computer programming classes with learning analytics from students' digital footprints" [1].

S. Suhonen tried to analyze the digital footprint that students leave on the Moodle learning management system and WhatsApp messenger in terms of understanding their own learning achievements [12]. The possibilities of a digital footprint for detecting social rejection and victimization of bullying in social networks were studied by Ophir et al. [10].

Noting that modern students are quite competent in the field of digital technologies, since they have already mastered their skills of creating and posting content on the Internet, a number of experts emphasize the importance of that the younger generation could realize the responsibility that it bears by voluntarily contributing to public or semi-public corners of the global information network. In this regard, issues of digital identity, personal development, social relations, and lifelong learning in the context of digitalization are becoming relevant. By focusing on "digital identity", some researchers try to better understand the nature of students' social and cultural experiences. The study in this area showed that the use of digital technologies is a key factor for the productivity of students' learning, however, this is not always associated with activities that are perceived by students themselves as "pleasant" ones [2].

In order to get a more holistic view of how different aspects of the context and interaction of students outside formal learning spaces affect their behavior and learning outcomes, E.Y.L. Ng, N. Law and A.H.K. Yuen consider the digital footprint left by students in social networks The researchers take into account the widespread use of digital technologies by students, especially outside of the structured learning time, and note the growing interest in understanding students through biographical and ethnographic methods [8].

In recent years, biographic and narrative research has been increasingly used in studies of various fields of knowledge related to the educational world. The inclusion of ICTs in the educational process and their huge impact on the lives of students and teachers encourages us to think about the technological imprints that mark their academic and personal experience. Analyzing the technological experience of university students in formal and informal contexts, a number of researchers come to the conclusion that it is necessary to further promote not only digital literacy, but also the critical and reflective content of media use in the educational (formal) and informal (personal) context [11].

Thus, in our opinion, in the field of considered issues, there is no comprehensive view of the problem. The scientific evidence of the potential of unstructured Big Data for the education system is needed. Important is also to identify some efficient ways of using digital footprint in the context of higher education.

4 Results

In our opinion, the potential of the digital footprint for the educational system lies in 3 different areas:

- ensuring continuity and integration of educational levels (for example, school—HEI),
- organization of the educational process (for example, creating individual educational trajectories),
- management of the educational system (educational management): for example, in the aspects of ensuring the quality of education, the competitiveness of HEIs (image, branding, etc.).

In order to solve the problem of ensuring continuity and integration of different educational levels, the author believes that the digital footprint potential is primarily in the aspect of monitoring the activity of students on the official web pages of HEIs, including in their social networks groups. In this function, the digital footprint helps to better understand needs, interests, expectations, and moods of the younger generation, on the one hand. On the other hand, the analysis of the digital footprint of potential applicants will ensure understanding of shortcomings of the HEI's own information policy in the field of branding and advertising, working with institutions of general secondary education. The results of analytical work in this area will be valuable for universities in terms of adjusting work programs, organizing scientific events aimed at the target group "applicants and their parents", developing joint programs with schools to prepare school graduates for entering the higher education, and in general, to develop/adjust the development strategy of the educational organization. Here we can see the connection with the third highlighted direction of digital footprint use.

One of the directions of using the digital footprint in the organization of the educational process is educational analytics. Learning analytics or "education based on Big Data" is a fixation, processing and analysis of the digital footprint in order to understand and optimize the educational process and the educational environment. It is used to create individual educational trajectories, improve teaching and learning at the HEI, and improve educational results. Learning analytics can be descriptive, predictive, and prescriptive.

Descriptive analytics is based on describing the current situation using available data to give an objective and most accurate assessment of what is happening. This description is based on visualization through charts, graphs, infographics, and so on. Visual tools play an important role in the aspect of transforming large digital arrays into clear, accessible, and easily perceived information. The value of descriptive analytics is to form a holistic view of what is happening at the moment. This information is essential for making certain decisions. An example of using descriptive analytics in the higher education is monitoring student engagement in the educational process. Based on such engagement indicators as frequency of work in the library, applications for optional training courses, attendance at lectures, use of electronic courses of the university and others, the HEI management can understand which aspects need special attention to improve the involvement of students in the learning process. This analytics is also useful for the student community, as it informs students about their own academic activity and allows them to compare their activity with fellows.

Predictive analytics aims to forecast a situation based on a comparison of data for the previous and current periods. In practice, this type of educational analytics is most often used to identify students at risk from the point of view of poor academic performance, non-attendance, etc. and provide them with advance support, mentoring, preventing the "drop-out" of the educational process of students potentially prone to this. It analyzes not only the academic history of students (the amount of time that students spend on completing certain educational tasks, types of tasks themselves, academic performance, and attendance of certain courses), but also demographic data.

Prescriptive analytics is aimed at finding recommendations for changing an existing or potentially possible situation. To do this, we use generalized information about the experience of previous users with similar characteristics. At the output, some algorithms and behavior patterns are created, based on which we can predict the actions of new students and direct them to certain changes in their educational trajectories. For example, based on data on student performance, academic history of past students with similar profiles, it becomes possible to choose courses that best match interests, abilities and curriculum of actual students. Another way to use prescriptive learning analytics is to improve students' achievements through adaptive learning environments (ALEs). These are feedback systems that allow students to monitor their own progress and develop individual learning paths [9].

The formation of individual educational trajectories of students based on their individual characteristics and previous experience, reflected in the "digital footprint", already has successful examples of implementation in the form of various practical solutions in Russian and foreign practice, but most often such forms are narrowly focused (in one area, for example, learning a foreign language, or for solving private, specific tasks (for example, to increase student attendance at training sessions). Building individual educational trajectories within the framework of bachelor's, master's, or post-graduate programs is still far from being implemented. According to experts, this is prevented by various barriers. In our opinion, the most relevant of these are: *a low quality of "digital footprint" data»* and *slowness of the "industrial" educational model.*

In order to update or create educational content, programs, and algorithms that are focused on specific educational needs, uniform and detailed data, it is required to take into account various individual characteristics and experiences. At present, there is little evenly and accurately recorded data on students' academic experience in Russian high educational institutions. As a rule, such information is scattered across different information systems, and some data has not yet been digitized. The available information is not always sufficient for correct data comparison. For example, data on the academic performance do not reflect specific criteria for obtaining a particular assessment, assessment criteria themselves are not always clear to both students and teachers. Universities rarely have detailed information about other metrics of students success in addition to the academic performance (for example, their employment, career trajectory, international mobility, additional education, etc.). It is impossible to build an effective educational trajectory without taking these aspects into account and processing them in a high-quality way.

According to I. Chirikov and I. Smirnov, Russian HEIs have developed an "industrial" educational model, in which universities act as "factories" for the production of qualified specialists from applicants who have entered them. "United in groups in the training areas, students move along educational trajectories pre-prepared for them. Based on this model, resources, teaching timetable, and classroom fund are planned. In practice, the ability to build individual educational trajectories is very limited [3]. Experts believe that this model has little space for developing students' independence and initiative, and digitalization of this model is not only not effective, but in some cases even harmful. It is difficult to agree with I. Chirikov and I. Smirnov in the aspect that the main difficulty lies in the need to change approaches to teaching that have developed in Russian HEIs, that teaching is rarely built based on the logic of educational results, and the content of previous courses is poorly synchronized with subsequent ones. In our opinion, the key problem is that the motivational component of teachers' readiness to work in the new (digital) educational model is poorly developed, including because teachers (as well as methodologists and university administration) do not see the expediency in developing individual educational trajectories, investing their working time, knowledge, efforts [13]. They hardly understand that this will not be implemented in practice, since new curricula, programs are needed for individual educational trajectories. The development of individual learning trajectories actually requires a qualitative restructuring of the entire educational process and changes in the existing educational standards. Therefore, this is a complex problem that requires consolidated efforts of not only HEIs, but also regulatory bodies (ministries, agencies, etc.).

5 Conclusion

International comparative studies on the higher education quality show that the problem of improving teaching methods is really growing in the Russian higher education. The solutions to this problem, however, do not necessarily have to lie in the digitalization sphere. There are many successful examples in the world practice of increasing student engagement through active learning and developing self-reliance and initiative. Many of them were implemented long before the emergence of artificial intelligence. Taking into account current technological conditions, betting on artificial intelligence can enrich contractors who implement it, but it is unlikely to significantly improve the quality of teaching and educational outcomes of students. In addition, in the context of declining investments in higher education, this can divert resources from really important initiatives. A significant part of them can rely on the "natural" intelligence of teachers and students. Therefore, the aspect of expediency of using digital technologies in education is the most important and should be a priority in the development of any initiatives of federal, regional or local level.

However, this does not mean that the didactic, organizational, and technological potential of digital tools should be ignored. In the age of information, this is hardly possible. Digitalization of education (at different stages and levels) is a necessary condition for the survival and development of schools and higher educational institutions today, in order to speak the same language with the younger generation, to better prepare them for life in the information society.

In this context, the author considers some possibilities of the digital footprint that seem to be quite interesting and promising for solving a number of urgent problems faced by the Russian system of higher professional education at the present stage of its development.

References

1. Azcona D, Hsiao I, Smeaton AF (2019) Detecting students-at-risk in computer programming classes with learning analytics from students' digital footprints. User Model User-Adap Inter 29:759–788. https://doi.org/10.1007/s11257-019-09234-7
2. Camacho M, Minelli J, Grosseck G (2012) Self and identity: raising undergraduate students' awareness on their digital footprints. Procedia Soc Behav Sci 46:3176–3181. https://doi.org/10.1016/j.sbspro.2012.06.032
3. Chirikov I, Smirnov I (2019) False digital footprint: 5 challenges for artificial intelligence in higher education. http://www.edutainme.ru/post/5-vyzovov-dlya-iskusstvennogo-intellekta/. Accessed 12 Mar 2020 (in Russian)
4. Decree of the President of the Russian Federation from May 7, 2018, No. 204 "On national goals and strategic objectives for the development of the Russian Federation for the period up to 2024". http://kremlin.ru/acts/bank/43027. Accessed 12 Mar 2020 (in Russian)
5. Galimova EG, Konysheva AV, Kalugina OA, Sizova ZM (2019) Digital educational footprint as a way to evaluate the results of students' learning and cognitive activity in the process of teaching mathematics. EURASIA J Math Sci Technol Educ 15(8):em1732. https://doi.org/10.29333/ejmste/108435

6. Ministry of Education of the Russian Federation (2019) Federal project "Digital educational environment". https://edu-frn.spb.ru/files/iiMBxQ4cNH1BCsaWn2WqDgFinWeU3rVYpmO6sd33.pdf. Accessed 12 Mar 2020 (in Russian)
7. Ministry of Education of the Russian Federation (2019) The Russian national project "Education". https://edu.gov.ru/national-project/. Accessed 12 Mar 2020 (in Russian)
8. Ng EYL, Law N, Yuen AHK (2018) Understanding learner lives through digital footprints. In: Story AL (ed) Proceedings of the technology, mind, and society conference (TECHMINDSOCIETY 2018), article no 26. ACM, New York. https://doi.org/10.1145/3183654.3183708
9. O'Farrell L (2017) Using learning analytics to support the enhancement of teaching and learning in higher education. Report. National Forum for the Enhancement of Teaching and Learning in Higher Education, Dublin
10. Ophir Ya, Asterhan ChSC, Schwarz BB (2019) The digital footprints of adolescent depression, social rejection and victimization of bullying on Facebook. Comput Hum Behav 91:62–71. https://doi.org/10.1016/j.chb.2018.09.025
11. Sonlleva Velasco M, Torrego González A, Martínez S (2017) It's crazy to live withouth Facebook or WhatsApp": the technological footprint in the teacher training. EDMETIC Revista de Educación Mediática y TIC 6(2):255–275. https://doi.org/10.21071/edmetic.v6i2.6935
12. Suhonen S (2019) Learning analytics: combining Moodle, Whatsapp and self-evaluation data for better understanding. In: Popma W, Francis S (eds) Proceedings of the 6th European conference on social media, ECSM 2019, vol 149414. Academic Publishing, Brighton, pp 410–413
13. Zotova AS, Mantulenko VV (2014) Assessment of readiness of the Russian higher education to the introduction of modern information and communication technology. Econ Entrep 6(47):248–252 (in Russian)

Digital Footprint Analysis to Develop a Personal Digital Competency-Based Profile

L. V. Kapustina

Abstract Social networks have become an integral and important part of younger generation's life today. The information that users place on their web pages on the social network is the product of their activities. It reflects the users' needs and values. The analysis of the digital footprint left on social media has undeniable potential to improve all processes in society in general, and the educational process in particular. The purpose of this study is to build a model for developing a personal digital competency-based profile through the analysis of web users' digital footprints left on the social network. According to the analysis report on digital footprints, the author of the study developed a model of the user's digital competency-based profile. Intellectual efforts of universities, scientists and trainers in order to create digital competency-based profiles and improve the personal educational trajectory can lead to educational system improvement and increase the chances for personal career growth under the conditions of the digital economy.

Keywords Digital footprint · Digital competency-based profile · Internet · Social media · Web users

1 Introduction

Currently, Internet tools such as blogs and social media provide millions of web users with the opportunity to express themselves. At the same time, they are young people who most actively use social media [11]. The processes taking place in the Internet space can have a profound impact on people's lives and the entire society outside the worldwide web. Social networks have become an integral and important part of younger generation's life today. The information that users place on their web pages on the social network is the product of their activities. It reflects the users' needs and values. The analysis of the user's digital footprint (DF) left on social

L. V. Kapustina (✉)
Samara State University of Economics, Samara, Russia
e-mail: lkap@inbox.ru

S. I. Ashmarina and V. V. Mantulenko (eds.), *Current Achievements, Challenges and Digital Chances of Knowledge Based Economy*, Lecture Notes in Networks and Systems 133, https://doi.org/10.1007/978-3-030-47458-4_68

media has undeniable potential to improve all processes in society in general, and the educational process in particular.

DF gives an idea of the psychological characteristics of web users, their value systems, professional and personal development, searching for new ways of socialization, and specific educational requests. To use the data analysis of this DF is now the request of the society. The world is rapidly changing, and many of the currently relevant information technologies will become irrelevant by the time of nowadays students' graduation. A personal digital competency-based profile is an extremely flexible tool that allows adjusting the educational trajectory to the society needs and labor market. The purpose of this study is to build a model for developing a personal digital competency-based profile through the analysis of users' DF left on the social network.

2 Methodology

A personal digital competency-based profile is an opportunity for a mentor, trainer, and employer to assess the potential of a particular person, his or her intelligence, personal characteristics, motivation, and knowledge of the subject. The personal digital competency-based profile provides a chance to notice, analyze, and correct trainees' weaknesses. Consequently, it is a chance to modernize the educational system.

In 2018–2019, the study was conducted on the basis of Samara State University of Economics to collect and analyze first-year students' DF, left on the social network "VKontakte", to develop their personal digital competency-based profiles. The study took into account only the information that social network users have made public even before the start of the study. Also, the study participants were asked for verbal consent to collect and process this information. To collect and analyze users' DF, the distinctive characteristic such as subscription to online communities in "VKontakte" was considered on their web pages. On the basis of the characteristic listed above, the following users' criteria were evaluated:

- high educational interest to the particular science,
- high level of intelligence, motivation, and creativity.

These criteria formed the basis of the model to develop the web user's personal digital competence-based profile. To determine high level of educational interest, the author of the study used heuristic analysis of users' subscriptions to online communities. A sample of marker communities that are thematically related to educational topics was created. The sample was classified by subject areas: business, sociology, management, law, etc. Frequency estimation of the subjects of users' subscriptions allowed the author to draw conclusion about high/low users' educational interest in a particular field of science and right/wrong choice of the faculty to study.

Users' creativity, intelligence, and motivation were determined through the following algorithm. The author of the study had a control group of students who were tested on the features listed above. Having analyzed the subscriptions of the users who did not pass the tests and compared them to ones of the control group, the author found some correlation and obtained the list of web users with a likely high level of intelligence, motivation, and creativity.

3 Results

The penetration of information technology in our lives has led to the complete transfer of certain aspects of human activity to the virtual space. Users' profiles on social media have become a projection of their real personal, creative, and business lives.

First of all, the author of the paper implemented the collection and analysis of DF. The development of methods for collecting and structuring heterogeneous digital data has made it possible to quickly search for information about a user on a social media and combine it for the analysis.

The study of users' DF allowed to carry out modeling of their physiological, psychological, and cognitive features and use this information to forecast, develop, and manage the educational process. According to the analysis report on users' DF, the author of the study developed a model of the personal digital competency-based profile (Table 1).

The average share of users' subscriptions to educational online communities was 5–10%. According to the author, only users with 60–80% share of subscriptions to the educational content have high level of interest to a particular subject. This is the first level—potential—of the personal digital competency-based profile.

Intelligence became the basis of the second level of the proposed model of the personal digital competency-based profile. It is called an instrumental level. It is the level of having tools in the topical area, from knowledge of existence of this tool to the use of the tool without assistance. The third, conceptual level, reflects the understanding of the topical area from the terminological awareness to the ability to transfer knowledge to others and generate new knowledge.

The forth level of the personal digital competency-based profile assesses the ability to perform effective and productive activities in a particular area, that is to say, create the final product. Motivation for effective and productive activities increases with

Table 1 Personal digital competency-based profile

Characteristic of the user's web page	Evaluation criteria	Level of the personal digital competency-based profile
Subscriptions to online communities in "VKontakte"	Interest	Potential
	Intelligence	Instrumental
	Creativity	Conceptual
	Motivation	Effective

Source author

increase in the amount of online educational communities on the social media where the user leads an active digital life.

Collecting and analyzing users' DF on the social network "VKontakte" resulted in building their personal digital competency-based profiles. Their professional qualification and value depends on the amount of levels of the personal digital competency-based model they have. Respectively, the educational system should be harmonized to guarantee the development of all four levels of the model proposed.

The author of the study also considers it possible to monitor the further professional career of users on the basis of their DF analysis in "VKontakte". The graduates also leave their DF on the Internet: where they studied, where they now work, what they do, and how successful they are today. That can help develop personal digital competency-based profiles of successful graduates. This will give universities the opportunity to search for applicants with the personal digital competency-based profile Similar to the one of a successful graduate.

4 Discussion

The technological revolution has provided many new opportunities in the field of education. In the educational environment, the advantages of the technological era are availability of information sources, information speed, and ability to use new knowledge from various fields [8]. In professional education, institutional electronic systems and personal electronic environments can serve as data sources to build a personal digital competency-based profile.

Institutional electronic systems include electronic information and educational environments (EIEE). EIEE include all the students at the university. These environments allow to store, collect, and analyze the following DF:

- ratio of the types of users' educational activities (viewing content, communication, and tasks) and comparison of these values with the average ones in the students group, which allows to determine an individual learning style and use this information to correct teaching methods,
- regularity of training activities,
- users' ability to self-organize,
- student electronic portfolio [2].

The implementation of the Priority Project "Current digital educational environment in the Russian Federation" will allow universities to work on the basis of a unified digital portfolio, containing information about the results of students' training at all online courses, which were included in the unified portal [9].

Personal electronic environments that accumulate DF are primarily popular social networks. According to the research, the university can identify up to 93% of its students on social media and clarify information about them [10]. Linguistic analysis of texts on the users' web pages on social media makes it possible to learn about the users' interests [5].

In order to analyze users' DF, Internet blogs have also proved their effectiveness. They can be used to increase learning motivation taking into account users' age and their cultural and gender aspects [7]. Students can also learn how to apply their new knowledge by analyzing real and synthesized networks with the help of individual summary of the network profile and research project. This supports users to learn the network analysis blocks [6]. Users also need to be trained to understand and analyze large dynamic systems and media. This can be done through online research courses, which are usually interdisciplinary [1].

However, the Internet data should be treated with a certain degree of skepticism. Internet technologies are gaining popularity in education, but the information presented on Internet platforms must be carefully analyzed [4]. Sometimes it is necessary to clarify the social media content, because it could be influenced by momentary feelings of the web user who posts some comments [3].

5 Conclusion

By analyzing web users' DF and building their personal digital competency-based profiles, universities will be able to support learners more effectively and train them for their entry into the labor market. High schools should create new services based on the analysis of such digital data. This will provide web users with the best individual trajectories in education, as well as allow trainers to find more effective solutions in the field of teaching under the conditions of digitalization, introduce innovations, and share best practices.

Previously, the quality of graduates' education was provided by natural selection through the expulsion of underachieving students, but now higher education is aimed at ensuring the success and competitiveness of each person in the labor market.

Intellectual efforts of universities, scientists and trainers in order to create digital competency-based profiles and improve the personal educational trajectory can lead to educational system improvement and increase the chances for personal career growth under the conditions of digital economy.

Acknowledgements The author expresses her gratitude to all those participated in this study for their kind cooperation, including Samara State University of Economics for the opportunity to conduct the research and publish the results in the form of this paper.

References

1. Arney C (2018) Network science undergraduate minor: building a foundation. In: Cramer C, Porter M, Sayama H, Sheetz L, Uzzo S (eds) Network science in education. Springer, Cham, pp 59–70

2. Babanskaya OM, Mozhaeva GV, Feshhenko AV (2014) Individualization in e-learning based on the “E-tutor model”. In: Gómez Chova L, López Martínez A, Candel Torres I (eds) Proceedings of the II international scientific and practical conference modern information and communication technologies in higher education: new educational programs, pedagogy with e-learning and improving the education quality. University La Sapienza, Rome, pp 91–96
3. Bharti SK, Pradhan R, Babu KS, Jena SK (2017) Sarcasm analysis on twitter data using machine learning approaches. In: Missaoui R, Abdessalem T, Latapy M (eds) Trends in social network analysis. Lecture notes in social networks. Springer, Cham, pp 51–76
4. De Medio C, Limongelli C, Marani A, Taibi D (2019) Retrieval of educational resources from the web: a comparison between google and online educational repositories. In: Herzog M, Kubincová Z, Han P, Temperini M (eds) Advances in web-based learning. Lecture notes in computer science, vol 11841. Springer, Cham, pp 28–38
5. Feshchenko A, Goiko V, Mozhaeva G, Shilyaev K, Stepanenko A (2017) Analysis of user profiles in social networks to search for promising entrants. In: Gómez Chova L, López Martínez A, Candel Torres I (eds) Proceedings of the 11th international technology, education and development conference. IATED Academy, Valencia, pp 5188–5194. https://doi.org/10.21125/inted.2017.1203
6. Gera R (2018) Leading edge learning in network science. In: Cramer C, Porter M, Sayama H, Sheetz L, Uzzo S (eds) Network science in education. Springer, Cham, pp 23–44
7. Korovina S, Pushkina A, Gurova N (2016) Online blogs in the process of development of students’ reading skills. In: Chis V, Albulescu I (eds) Education, reflection, development. The European proceedings of social & behavioural sciences, 4th edn., vol 18. Future Academy, London, pp 298–305
8. Manea AD, Stan C (2016) On-line communication. In: Chis V, Albulescu I (eds) Education, reflection, development. The European proceedings of social & behavioural sciences, 4th edn., vol 18. Future Academy, London, pp 317–323
9. Passport of the Priority Project “Current digital educational environment in the Russian Federation”. http://static.government.ru/media/files/8SiLmMBgjAN89vZbUUtmuF5lZYfTvOAG.pdf. Accessed 10 Jan 2020
10. Smirnov IB, Sivak EV, YaYa K (2016) In search of lost profiles: the reliability of “VKontakte” data and its significance for education researches. Educ Issues 4:106–119
11. Titov VV (2013) Internet as a place for transformation of system of communicative identities of the Russian society (based on the data of the World Internet project (WIP) international research). Public Opin Monit Econ Soc Changes 3(115):42–50

Academic Development in the Era of Globalization of Scientific Communication

P. M. Taranov and M. A. Taranov

Abstract Global competition in the scientific and educational services market make universities to stimulate academic development of scientific and pedagogical personnel on a systematic basis using modern digital solutions. The leadership in research (in important part of universities rankings) is impossible without development of new methods, tools of scientific communication, which requires constant modernization of the training system of researchers. Information technologies and the advanced use of information and analytical systems has become an integral part of research activities, which requires the development of scientific and methodological background for researchers' training. This article is aimed at analyzing experience of Russian universities, research and educational organizations on the academic culture development, substantiating methodological prerequisites and practical recommendations for the formation of effective scientific communication skills. The research methodology is based on the study and generalization of the accumulated international and Russian experience in the area using statistical and scientometric tools. The authors proposed the content of descriptors of research competencies on the example of a master course. It is concluded that the introduction of systematic academic development should take place at each stage of higher education, which will be a factor in increasing the global competitiveness of universities.

Keywords Academic development · Scientific communication · Scientometrics · Academic master degree · Research activities

P. M. Taranov
Don State Technical University, Rostov-on-Don, Russia
e-mail: taranov@inbox.ru

M. A. Taranov (✉)
Don State Agrarian University, Zernograd, Russia
e-mail: ma.taranov@gmail.com

S. I. Ashmarina and V. V. Mantulenko (eds.), *Current Achievements, Challenges and Digital Chances of Knowledge Based Economy*, Lecture Notes in Networks and Systems 133, https://doi.org/10.1007/978-3-030-47458-4_69

1 Introduction

International scientific cooperation is one of the most dynamic forms of international relations in the modern world. International integration in the field of scientific communication, based on Internet-connected information and analytical systems, has made universities competitive in the global market of research and educational services. Universities in developed and developing countries have followed corporations in the process of transnationalization, which requires form them to constantly modernize and adapt to current world economic conditions.

A large number of scientific papers are devoted to the study of the mission of an university in the post-industrial society. One of the main interpretations is that the current mission of the university is to perform an integration function, i.e. to realize the role of a mediator in the modern global society [15].

Highly qualified research and teaching staff remain a key factor in ensuring the global competitiveness of universities. Gaining leadership in the global high-tech markets requires an increase in the number and qualification of scientific personnel trained for research work as part of research teams and project teams.

The reproduction of scientific personnel in the leading countries of technological development is proceeding at a faster pace, which is reflected in the absolute and relative growth of the number of researchers (Table 1).

Russia is almost the only country among the twenty world leaders in publishing activity where the number of researchers per million of inhabitants is decreasing. This trend will not contribute to the improvement of the country's position in high-tech markets, or to the growth of the population's income (Fig. 1).

However, along with the declining number of researchers, the quality of training of scientific personnel remains an urgent problem. It should be noted that according

Table 1 Number of researchers in R&D, people (per million people)

Countries and groups of countries	1996	2006	2016
OECD countries, including	2472	3185	4047
Germany	2827	3432	4748
France	2647	3419	4307[a]
Great Britain	2489	4178	4350
The USA	3130	3795	4313
Japan	4874	5333	5173
BRIC countries, including			
India	153	139	231
Brazil	–	541	881[a]
China	438	921	1159
Russia	3796	3240	3122

[a]Data of 2014
Source authors based on [14]

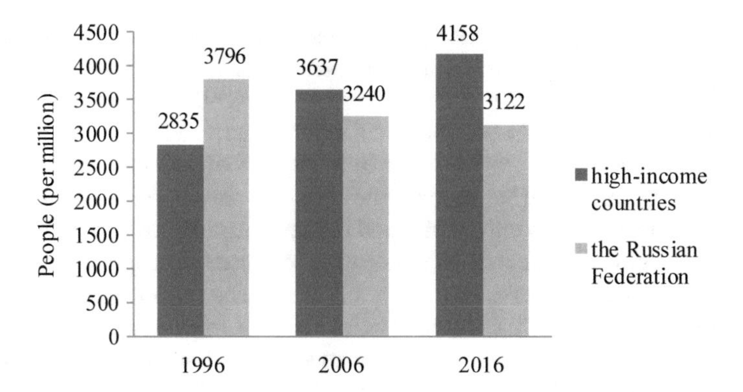

Fig. 1 Dynamics of the number of researchers in R&D, people (per million people) (*Source* authors based on [14])

to the Decree "On national goals and strategic objectives for the development of the Russian Federation for the period up to 2024", one of the key tasks in the field of science is called "formation of an integrated system of training and professional growth of scientific and scientific-pedagogical personnel" [3].

This task emphasizes the relevance of developing a modern scientific and methodological basis for ensuring the academic development of researchers. Modern scientific personnel should be able to use the current instrumental and methodological research base, dynamically integrate into international research projects, effectively use various tools of scientific communication, etc. The high level of academic development of the scientific personnel determines not only the international competitiveness of universities, but also the ability of the national innovation ecosystem to achieve leadership in new global high-tech markets.

2 Methodology

The authors of scientific research devoted to the academic culture issues note that the content of the concept of "academic development" can vary significantly. One of the reasons for differences in interpretations of this term is that academic development includes many activities aimed at adapting the academic community to the changing reality. The researchers note that academic development as an activity for the professional development of scientific and pedagogical personnel has quite a variety of tasks. Thus, some researchers see the main task in developing skills for using scientific communication methods, bibliometry, and academic writing [5]. Other scientists emphasize the need to adapt pedagogical training in the context of the transition to the model of continuous education and digital transformation of the education system [10]. Proponents of the entrepreneurial university model define academic development primarily through the acquisition of entrepreneurial skills by scientific and pedagogical personnel, taking into account the need to commercialize scientific developments [11]. Finally, some researchers draw attention to the need to develop the academic mobility in the context of the scientific globalization [7]. The most important trends are the transnationalization of activities of scientific and educational institutions, as well as the internationalization of markets for scientific and educational services, which has become one of the key reasons for the transformation of approaches to managing universities and research institutes as corporations in the spirit of the ideology of "new managerialism" and "academic capitalism".

The world university rankings (first of all, the Times Higher Education (THE), the Shanghai rating (ARWU), and the QS World University Rankings) play a fundamental role in the management of higher education, which today have a decisive impact on scientific, technical, and educational policies of countries [2]. This trend has determined the relevance of such tasks of the academic development as the development of modern methods of scientific communication and academic writing in terms of evaluating the effectiveness of research activities based on scientometric indicators [1].

3 Results

One of the key factors for strengthening the competitive position of key Russian universities in the global market of scientific and educational services is the level of the publication activity, which is directly related to the academic development level of researchers. The assessment of Russia's place among the leaders of the publication activity based on Web Of Science Core Collection (WoS CC) data shows that the share of our country in the global number of publications increased from 1.7% in 2008 to 2.7% in 2018 (Table 2).

Indeed, according to Russian researchers, after the recession in 2011–2014, our country's position in the global array of publications has improved, while Russia is in the Top 10 countries in 39 from the 252 subject categories of WoS CC [9]. In general, Russia's position in the ranking of leading countries over a ten-year period has risen from 15th place in 2008 to 13th place in 2018, i.e., despite significant expenditures of the state budget (including under the 5–100 Program), it is still not in the top ten.

There is no denying that academic readiness implies that researchers have the ability, on the one hand, to generalize and critically evaluate results of domestic and foreign research, and, on the other hand, to present results of the own research to the scientific community and—in an adapted form—to the general public, which requires possession of differentiated means of scientific communication.

The practice of developing a culture of scientific communication exists in developed countries, where the term "scientific communication skills" is firmly entrenched, and their formation takes place within the framework of the so-called

Table 2 Level of the publication activity in countries

Country	2008			Country	2018		
	Number of publications, thousand pieces	Place in the ranking	Country's share in the global number of publications, %		Number of publications, thousand pieces	Place in the ranking	Country's share in the global number of publications, %
The USA	561,6	1	27,9	The USA	760,1	1	24,9
China	181,8	2	9,0	China	512,0	2	16,8
Germany	131,0	3	6,5	Great Britain	199,9	3	6,6
Great Britain	129,5	4	6,4	Germany	177,7	4	5,8
Japan	112,3	5	5,6	India	132,9	5	4,4
France	90,3	6	4,5	Japan	130,2	6	4,3
Canada	82,2	7	4,1	Italy	119,3	7	3,9
Italy	75,7	8	3,8	Canada	118,6	8	3,9
Spain	59,1	9	2,9	France	117,3	9	3,8
Australia	54,7	10	2,7	Australia	110,3	10	3,6
Russia	34,5	15	1,7	Russia	83,4	13	2,7

Source authors based on Web of Science Core Collection

"scientific education" [12]. Thus, the academic development of scientific and pedagogical personnel and students is an urgent task for universities and scientific organizations, and is the focus of international and domestic scientific and educational policy.

An important area of academic development is the development of training methods and forms for young researchers. The formation of their research competencies is a complex multi-stage task that should be solved at every education stage. There is an approach according to which elements of the academic culture and scientific communication should be instilled not only in the framework of special disciplines [13]. The effectiveness of the multi-level development of academic readiness is confirmed by the practice of universities in developed countries [4], but a special role in shaping the scientific culture of a young scientist belongs to the master degree level [6].

The practice of scientific supervision and analysis of publication activity allow us to identify typical problems in the formation of research competencies among young scientists and formulate current educational tasks.

Academic development should, first of all, form understanding of ethical standards and reputational responsibility in science, as well as develop the following skills: the use of scientific information and analytical systems and full-text databases; the use of tools for systematizing scientific information; the use of modern solutions for organizing scientific research as part of a research team; the use of scientific communication techniques and academic writing.

Curricula of academic master programs usually contain one or more methodological disciplines (for example "Methodology of scientific research", "History and methodology of science"), but these courses are mainly theoretical in their nature and need to be supplemented with applied disciplines that could focus on practical aspects of the scientific creativity, including issues of scientific communication.

It seems that the development of scientific communication skills, academic writing and applied bibliometry should take place within the framework of each academic master program. However, this task requires the development of adequate solutions in the form of new disciplines and their methodological support. It should be noted that it can be considered as an established practice when issues of academic writing are studied within the framework of a separate discipline "Academic writing", using e-learning technologies [8]. The relevance of mastering tools of modern scientific communication is determined by the rapid growth of the number of scientific publications and the development of information technologies, which required solutions for comprehensive coverage and organization of the growing publication flow.

It should be noted that over the past ten years, the flow of scientific publications only within the Web of Science Core Collection database has increased by 1.5 times from 2014 thousand pieces in 2008 to 3050 thousand pieces in 2018. The number of scientific publications, even on fairly narrow issues, often exceeds the perception capabilities of experienced scientists, not to mention young researchers. Scientometry served as a methodological basis for the formation of instrumental solutions for the systematization of the publication flow—international and national information and analytical systems, which allowed to a certain extent to solve the problem of evaluating, analyzing and selecting current scientific publications.

The development of modern international research databases and software is a necessary condition for young scientists not only to achieve the best results in their scientific research, but also to be able to effectively present results of the own research work to the international scientific community.

Bibliometric databases have become a necessary tool for the research activity, which forms the need of young scientists for developing skills of using global information and analytical systems (e.g., Web of Science, Scopus), and their national counterparts (e.g., SCIENCE INDEX based on RSCI in Russia).

Russian universities and professional associations (Moscow State University, High School of Economics (HSE), Ranepa, NEICON, etc.) have relevant educational practices in terms of the academic development of researchers, which are implemented in the following forms:

- professional development programs (e.g., National Training Fund programs for research and teaching staff; professional development programs, NEICON conferences and webinars for journal editors and publishers),
- scientific seminars provided for in the master and postgraduate study plans (e.g., Moscow State University, Saint Petersburg State University),
- programs for the formation of personnel reserves (for example, Young Faculty Development Program in High School of Economics, in other universities of the 5–100 program, supporting universities),

- preparation of specialized periodicals (Information-analytical Bulletin of HSE "Windows of Growth", etc.),
- academic writing centers (e.g., Academic Writing Center in HSE, Academic Writing University Center in National Research University of Technology MISIS).

Not only universities, but other organizations are also involved in certain aspects of the academic development. For example, the international company Clarivate Analytics regularly conducts online seminars in Russian on all major products and services (Web of Science, InCites, etc.). NEICON and Training and consulting center (TCC) "Academy of ANRI" are also regularly engaged in professional development in the field of scientific communication and applied bibliometry, but the main target audience of these structures is librarians and editors of scientific journals.

In 2014–2019, the National Training Fund (NTF) implemented the project "Development of scientific communication in the academic sphere", in which NTF specialists conducted advanced training courses for scientific and pedagogic personnel, including in the regions. The seminars are aimed at the academic development of research and teaching staff, who should pass their knowledge and form practical skills by their students.

4 Discussion

Based on the conducted review, we can conclude that among existing formats of the academic development, there is clearly a lack of a regular master course that could be included in the educational programs of different training areas. Modernizing the training of modern researchers requires the development and implementation of applied disciplines aimed at developing practical skills in different research activities.

Creating a special master course is one of the most promising solutions for the formation of the academic culture among young scientists. It seems methodically expedient to separate the body of theoretical and applied knowledge, methodological guidelines, practical tasks and cases aimed at forming such academic culture elements as scientific communication skills, scientometric analysis, systematization of scientific information and organization of research activities into a separate discipline.

The practical nature of the discipline determines the need for extensive information and analytical support in the form of scientific citation databases, scientometric packages, full-text scientific databases, electronic library systems, project management software and text borrowing detection systems (Fig. 2).

It should also be noted that tasks in the scientific communication field are often solved in a cross-platform format, when a web service, mobile application, desktop application, and PC application can Simultaneously exist to perform approximately the same functionality.

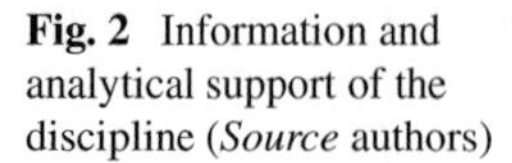

Fig. 2 Information and analytical support of the discipline (*Source* authors)

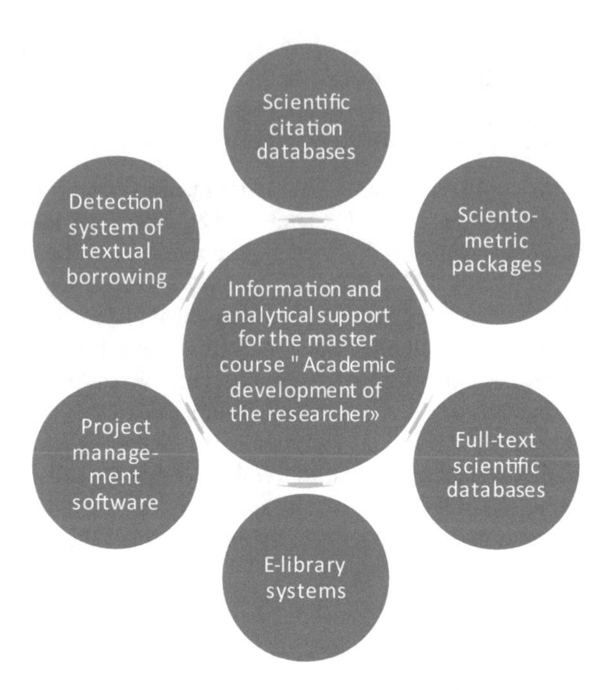

In the information-analytical and software support used for the educational process, we can select the main and additional tools. An example of the authors' distribution of information-analytical and software tools in the context of research competencies in the direction 38.04.01 "Economics" for the discipline "Academic development of a researcher" is presented in Table 3.

As a rule, the main tools are the most important for the development of the discipline, to which the university has either open access or subscription. For example, the Web of Science and Scopus scientific citation databases are used in the course as part of a national subscription, and the scientific electronic library eLIBRARY.RU—based on open access rights. Valuable scientometric packages InCites and SciVal, unfortunately, can only be studied in a review, because they are not included in the national subscription.

It should be noted that the master course "Academic development of a researcher" in terms of information-analytical resources and software focuses primarily on the study of universal solutions, i.e. the course is cross-program in its nature and can be taught in any academic direction of master programs. For example, systems of professional market analysis (e.g., Ruslana, Bloomberg, EMIS, Kommersant "KARTOTEKA", SPARK) will be in demand by undergraduates of the direction group "Economics and management", but they are unlikely to be of interest to students of other areas. At the same time, scientific citation databases, project management systems, and bibliographic managers will be in demand regardless of the training area.

Table 3 Information-analytical support of the master course in the context of research competencies in the direction of 38.04.01 "Economics"

Research competences (RC)	Information-analytical resources and software	
	Main	Additional
RC-1: the ability to generalize and critically evaluate results obtained by domestic and foreign researchers, identify promising areas, and develop a research program	*Scientific citation databases*: Web of Science Scopus eLIBRARY.RU *Open access resources*: Google Scholar *Full-text scientific databases*: ScienceDirect SPRINGER	*Scientometric packages*: InCites SciVal *Open access resources*: Cyberleninka Microsoft Academic Kopernio *Full-text scientific databases*: Wiley Online Library, JSTOR, DOAJ, NEICON
RC-2: the ability to justify the relevance, theoretical and practical significance of the chosen topic of scientific research	*Scientific citation databases*: Web of Science Scopus eLIBRARY.RU *Full-text scientific databases*: ScienceDirect SPRINGER	*E-library systems (ELS)*: ELS «University library on-line» ELS «IPRbooks» ELS «Znanium.com» *Full-text scientific databases*: Wiley Online Library, DOAJ, NEICON
RC-3: the ability to conduct independent research in accordance with the developed program	*Project management software*: MindMeister	*Project management software*: Mindjet Mindmanager, Microsoft Project, Trello
RC-4: the ability to present the research results to a scientific community in the form of an article or report	*Bibliographic Manager*: Zotero *Systems for detecting text borrowings*: Antiplagiat.ru	*Bibliographic Manager*: EndNoteWeb Mendeley *Systems for detecting text borrowings*: Grammarly Turnitin Plagscan

Source authors

The course "Academic development of a researcher" can form the research competence in the framework of the federal state education standard of the higher education in the training area 38.04.01 Economics (master level) based on the practices of Don State Technical University (DSTU). The content of level descriptors for the research competences RC-1 and RC-4 is presented below as an example.

For example, the implementation of the RC-1 (Table 4) requires the use of Web of Science, Scopus, and eLIBRARY.RU citation databases as the main tools, as well as familiarization with capabilities of scientometric packages InCites, SciVal. Google Scholar is used as the main open access resource, and additionally, the students study such web resources as Microsoft Academic and Cyberleninka.

Table 4 Descriptors of the competence levels for the RC-1 (according to the federal state education standard of the higher education in the training area 38.04.01 Economics)

RC-1: the ability to generalize and critically evaluate results obtained by domestic and foreign researchers, identify promising areas, and develop a research program	
Know	
Level 1	Domestic and foreign bibliographic, full-text and abstract databases
Level 2	Bibliometric indexes and their application areas; types and features of scientific publications
Level 3	Main domestic and foreign research networks; opportunities for scientific communication in research networks
Can	
Level 1	Carry out the examination and selection of domestic and foreign scientific publications
Level 2	Carry out the examination and selection of domestic and foreign scientific sources, authors, organizations
Level 3	Correctly draw up borrowings of various materials in scientific publications
Possess	
Level 1	Skills to determine the relevance and problems of scientific research for the preparation of the research program, including on the basis of bibliographic and full-text databases
Level 2	Skills to determine the development degree and prospects of the study for the preparation of a research program, including on the basis of bibliography and full-text databases
Level 3	Skills of determining goals, object, subject, and objectives of a scientific research for the preparation of a research program, including on the basis of bibliographic and full-text databases

Source authors

Skills for preparation of a research program are ultimately formed based on the study of current scientific publications through access to full-text scientific databases ScienceDirect, SPRINGER, JSTOR, DOAJ, Wiley Online Library, and NEICON. It should be noted that expanding the access of Russian universities to full-text scientific databases remains an urgent task, which in some part it would be advisable to solve in the format of a national subscription.

For example, the implementation of the RC-4 competence also involves the development of a number of skills by undergraduates (Table 5). Thus, the task of preparing and presenting the results of the research to the scientific community in full compliance with the norms and principles of scientific ethics cannot be performed without familiarization with modern systems for detecting text borrowings: Antiplagiat.ru, Grammarly, Turnitin, Plagscan, etc.

To develop skills for systematization of scientific publications and correct registration of borrowings, students are getting acquainted with the functionality of bibliographic managers—Zotero, EndNoteWeb, Mendeley, and others.

Table 5 Descriptors of the competence levels for the RC-4 (according to the federal state education standard of the higher education in the training area 38.04.01 Economics)

RC-4: the ability to present the research results to a scientific community in the form of an article or report	
Know	
Level 1	Main institutions and procedures for scientific publication activities
Level 2	Norms and principles of modern scientific ethics and reputation responsibility; bibliographic managers and standards
Level 3	Standard rules for registration of scientific publications, the procedure for reviewing publications in scientific journals and conference proceedings
Can	
Level 1	Correctly draw up borrowings of various materials in scientific publications, use bibliographic managers to systematize scientific information
Level 2	Prepare and present the research results to the scientific community in the form of a report at regional, national and international conferences
Level 3	Prepare and present the research results to the scientific community in the form of a research or review article in peer reviewed scientific journals
Possess	
Level 1	Skills of registration and promotion of scientific publications
Level 2	Methodology for building a publication strategy
Level 3	Complex skills of presenting the research results to the scientific community in the form of scientific publications and reports

Source authors

The presented examples of the content of competence descriptors indicate the interdisciplinary nature of the course "Academic development of a researcher", the main task of which is the formation of applied research skills of master students.

The initiative to develop the course "Academic development of a researcher" was recognized as promising at the Don State Technical University, and in 2019 the project received professional and grant support from experts of the Charitable Foundation named after V. Potanin, which confirms the high degree of relevance of this discipline. Currently, the course is implemented for master degree programs in an enlarged group of specialties 38.00.00 "Economics and management".

5 Conclusion

Systematic academic development of researchers should be implemented at every stage of higher education (bachelor, specialty, master, postgraduate levels), which can and should be a factor in increasing the research potential and competitiveness of the Russian universities.

The stated task can be performed by developing and including in the curriculum such courses that are focused on the formation of scientific communication skills—"Academic writing" and "Academic development of the researcher", but the introduction of these disciplines will require modern logistics, and information-analytical support.

The integration of the systematic academic development should take place not only in the form of separate specialized disciplines. The academic culture elements and scientific communication skills should be instilled in the study of each discipline and practice where research competencies are implemented. It should be a priority for universities to ensure the availability of modern tools for scientific communication for research and teaching staff and students. An important factor in modernizing the scientific communication and increasing the competitiveness of domestic universities is the expansion of the state support for ensuring access to the best international information, analytical and software resources.

Acknowledgements This research work was supported by the Vladimir Potanin Foundation, project ID GK GK190000020.

References

1. Akoev M, Moskaleva O, Pislyakov V (2018) Confidence and RISC: how Russian papers indexed in the national citation database Russian Index of Science Citation (RISC) characterize universities and research institutes. In: Costas R, Franssen T, Yegros-Yegros A (eds) Proceedings of the 23rd international conference on science and technology indicators, STI 2018 conference proceedings. Universität Leiden, Leiden, pp 1328–1338
2. Cakur MP, Acarturk C, Alasehir O, Cilingir C (2015) A comparative analysis of global and national university ranking systems. Scientometrics 103(3):813–848
3. Decree. On national goals and strategic objectives for the development of the Russian Federation for the period up to 2024. http://www.garant.ru/hotlaw/federal/1195467/. Accessed 12 Feb 2020
4. De Oliveira JRS, Queiroz SL (2015) Scientific communication in undergraduate chemistry courses: a review. Quim Nova 38(4):553–562
5. Korotkina IB (2017) Academic literacy and global scientific communication methods. Sci Editor Publ 2(1):8–13
6. Kuehne LM, Twardochleb LA, Fritschie KJ, Mims MC, Lawrence DJ, Gibson PP, Stewart-Koster B, Olden JD (2014) Practical science communication strategies for graduate students. Conserv Biol 28(5):1225–1235
7. Kuznetsov AYu, Vershinina EV (2017) Factors of development and transformation of academic mobility. High Educ Russ 10:144–148
8. Lin CC, Liu GZ, Wang TI (2017) Development and usability test of an e-learning tool for engineering graduates to develop academic writing in English: a case study. Educ Technol Soc 20(4):148–161
9. Meskhi BCh, Ponomareva S, Ugnich EA (2019) E-learning in higher inclusive education: needs, opportunities and limitations. Int J Educ Manag 33(3):424–437
10. Mokhnacheva YuV, Tsvetkova VA (2019) Russia in the global array of scientific publications. Vestnik Russ Acad Sci 89(8):820–830
11. Sigahi T, Saltorato P (2019) Academic capitalism: distinguishing without disjoining through classification schemes. High Educ. https://doi.org/10.1007/s10734-019-00467-4. Accessed 12 Feb 2020

12. Spektor-Levy O, Eylon BS, Scherz Z (2009) Teaching scientific communication skills in science studies: does it make a difference? Int J Sci Math Educ 7(5):875–903
13. Stanley JT, Lewandowski HJ (2016) Lab notebooks as scientific communication: investigating development from undergraduate courses to graduate research. Phys Rev Phys Educ Res 12(2):020129
14. The World Bank Group (2019) World Bank Group – International development, poverty, and sustainability. https://data.worldbank.org/indicator/SP.POP.SCIE.RD.P6?contextual=default&end=2016&locations=RU-CN-US-GB-DE-IN&start=1996&view=chart. Accessed 18 Dec 2019
15. Unger M, Polt W (2017) The knowledge triangle between research, education and innovation – a conceptual discussion. Foresight STI Gov 11(2):10–26

Learningmetry: Effectiveness E-Learning Measuring and Reflection of Educational Experience

N. A. Zaychikova

Abstract The task of accumulation and analysis reflective experience when students acquire knowledge at e-learning was considered. Effectively teaching the student and collecting digital reflective experience in e-learning is of particular relevance. The aim of the work is to consider the application of method of results measurement in e-learning as a means of analyzing educational reflective experience. The basis for the research is new method of results measurement in e-learning, which can be called learning results metric or learningmetry. The learningmetry is method for research the effectiveness of the results of cognitive learning on the scale of academic achievement-motivation, a necessary element of which is to obtain feedback in the learning process, a sufficient element is the data on the results of training and reflection on it. With the help of learningmetry, the analysis of the results of learning students is carried out, conclusions and forecasts are made on the progress of the educational discipline. For a teacher working with students to master the courses in e-learning, it is possible to draw up an effective learning program for both the group as a whole and an individual student.

Keywords Learningmetry · Effectiveness learning measuring · Academic achievement scale · Educational experience reflection · E-learning

1 Introduction

Recent studies have focused on the issue of learning effectiveness and the realization of educational goals in e-learning. This is because e-learning is taking up more and more space in the modern education. In the process of e-learning, the distant nature of learning leads to more independent student activity, to the possibility of choosing an individual educational path, and the teacher is transformed into a tutor. Various test systems are ubiquitous and are widely used in higher education institutions.

N. A. Zaychikova (✉)
Samara State University of Economics, Samara, Russia
e-mail: zajna@yandex.ru

S. I. Ashmarina and V. V. Mantulenko (eds.), *Current Achievements, Challenges and Digital Chances of Knowledge Based Economy*, Lecture Notes in Networks and Systems 133, https://doi.org/10.1007/978-3-030-47458-4_70

Therefore, it is important to create a methodology for analyzing the effectiveness of e-learning.

Having collected a real digital footprint in e-learning, having analyzed reflection, we can see the real state of affairs in the educational process: how students learn from the tests; which tests are best perceived; in which areas of learning activity is effective learning. Without collecting a digital trace, a vivid picture, confirmed by data, and not by the subjective opinion of experts, cannot be compiled. The cognitive experience of creating a digital footprint of educational experience is here [2].

To create a digital footprint of educational experience, feedback from participants in the educational process, the collection of information and the method for its processing and analysis are necessary. As such a technique, we offer learningmetry. The learningmetry is method for research the effectiveness of the results of cognitive learning on the scale of academic achievement-motivation.

The reason for creating a new method was the task of effectively teaching a student using tests in e-learning. The idea of the method arose on the basis of the sociometric method, developed by Moreno [6].

The material for research on the learningmetry method for measuring learning outcomes on the achievement-motivation scale was data on the work of students of the economic direction during the development of the course "Probability Theory and Mathematical Statistics" and "Econometrics" with test systems in e-learning environment. The description of the learningmetry method was considered earlier in the articles of the author [12, 13].

2 Methodology

Let us describe the research e-learning results procedure Similar to the sociometric procedure. Students who have been tested on a number of tests of a specific educational discipline by e-learning, for each test are invited to answer the following questions:

1. What is your test rating on a five-point scale?
2. Indicate how much you liked the test: what is your assessment of the test (on a five-point scale: 5 is a great test, 4 is a good test, 3 is a satisfactory test, 2 is a unsatisfactory test)?

It should be noted that in an electronic educational environment, test results by subjects are usually always recorded, and are available to the teacher, therefore, to collect data on learningmetry, it is necessary to organize a feedback form.

After the information is collected, its processing begins. Usually it has three stages: tabular, graphic and indexological.

Examples of completed matrices of questionnaires of personal data are presented in Tables 1, 2, which are convenient to work with in Microsoft Excel. Below, $i = 1..n$, $j = 1..m$.

Table 1 Personal data. Student rating

Student serial number, (i)	Test (j)					
	1	2	3	…	m	The average rating of the i-th student
1	s_{11}	s_{12}	s_{13}	…	s_{1m}	s_1
2	s_{21}	s_{22}	s_{23}	…	s_{2m}	…
3	s_{31}	s_{31}	s_{33}	…	s_{3m}	…
…	…	…	…	…	…	…
n	s_{n1}	s_{n2}	s_{n3}	…	s_{nm}	s_n

Source author

Table 2 Personal data. Test rating

Student serial number, (i)	Test (j)					
	1	2	3	…	m	The average rating for the tests of the i-th student
1	t_{11}	t_{12}	t_{13}	…	t_{1m}	ts_1
2	t_{21}	t_{22}	t_{23}	…	t_{2m}	ts_2
3	t_{31}	t_{31}	t_{33}	…	t_{3m}	ts_3
…	…	…	…	…	…	…
n	t_{n1}	t_{n2}	t_{n3}	…	t_{nm}	ts_n
Average rating of j-test	t_1	t_2	t_3	…	t_1	

Source author

To visualize student statuses and tests, we will use the graph of the ratio of the statuses of objects of research (students or tests), selected on various grounds. According to the traditional arrangement, the inner circle corresponds to the rating "excellent", the second—"good", the third—"satisfactory", the outer—"unsatisfactory". As in the sociometric procedure, young men are denoted by triangles, girls by circles, for tests we introduce the notation with a square.

The main graph of the learningmetry is compiled on the basis of the average student's assessment of the tests. For the graph for each individual test, arrows or segments can connect students with an identical opinion about the test passed. Thus, micro-groups are visible by motivational activity in learning on tests of a specific subject. A graph of student results for a specific test is compiled on the basis of the test rating. Further calculated by learningmetry indices that serve to identify quantitative characteristics on a scale of academic performance-motivation. Distinguish between personal and group indexes. The main personal indices are: Student Perfomance Index (SPI), Index of Student Motivational Activity (ISMA), Index of Test (IT), Learning Student's Index of Test ($LSIT_{ij}$), Learning Student's Status (LSS_i). The main group indices are: Group Performance Index (GPI), Group Motivational Activity Index (GMAI), Learning Group Index (LGI). Separately defined concepts related to tests, Learning Test Index (LTI_j). Below we describe the definitions of the main characteristics.

Student Perfomance Index (SPI) is a characteristic of the position of a member of the group on academic performance, calculated as the average score obtained after completion of the tests (Table 1).

Index of Student Motivational Activity (ISMA) is a characteristic of the position of a group member, showing generally the attitude to the tests and the desire to master the discipline material, calculated as the average rating of tests that were evaluated by students (Table 2).

Index of Test (IT) is a characteristic of the test position, obtained as the average rating set by a particular test by various members of the study group (Table 2).

Learning Student's Index of Test ($LSIT_{ij}$) is the ratio of the test score to the score received by the student upon completion of this test.

$$LSIT_{i,j} = \frac{t_{i,j}}{s_{i,j}},$$

where $t_{i,j}$ is the value of the rating set for the j-th test by the i-th student, and $s_{i,j}$ is the value of the rating received by the i-th student at the completion of the j-th test. By analogy with the coefficient of satisfaction with communication, this index is a kind of measure of satisfaction-dissatisfaction with the result of training in tests, the description of which is given in the Table 3.

It is known from the theory of general pedagogy and psychology that if motivation is too strong, the level of activity and stress increases, therefore, some discrepancies in activity and behavior, that is, work efficiency, especially with long-term projects, worsens. Then a high level of motivation is the reason undesirable emotional reactions (tension, excitement, stress, etc.), which leads to a deterioration in activity [11]. Too high motivation with insufficient success quickly leads to exhaustion. At the same time, a slight decrease in motivation in the long term gives the best result with stable work.

In the case of effective work on mastering the material of the discipline, the area that is optimal in terms of academic performance and positive motivation is shown in the Fig. 1. The optimum here is the optimum student satisfaction with the learning

Table 3 Characteristic values of $LSIT_{ij}$

Value group	Values of $LSIT_{ij}$	Description
I group	$LSIT_{ij} < 1$	Motivation below learning success, low need for achievement
II group	$LSIT_{ij} = 1$	Motivation is adequate to success in training, average need to achieve
III group	$LSIT_{ij} > 1$	Motivation above learning success, high need for achievement

Source author

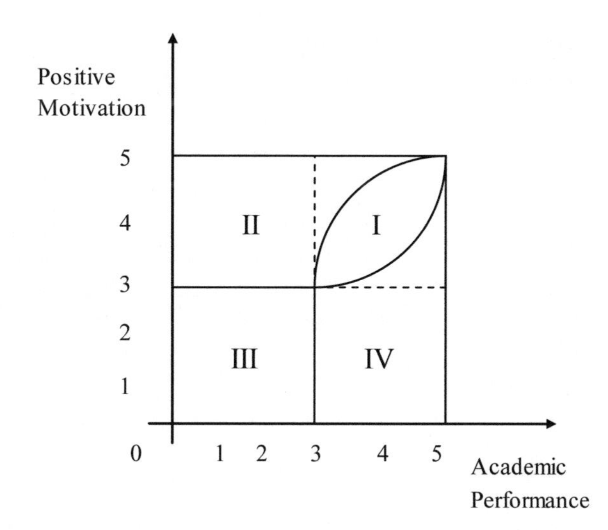

Fig. 1 Areas of mastering the material of the educational discipline in terms of academic performance-motivation (*Source* author)

outcome, when the desire and efforts made to master the material are proportional to the learning outcome. Development zones is quadrants I and II, quadrant III is the degradation zone, quadrant IV is the zone of stable (routine) work, are also noted. Since quadrant II is correlated with poor performance with high effort, quadrant II for long-term activities without changing conditions can be called a zone of increased risk of exhaustion, overload and emotional burnout.

Below are a few more indicators:

Learning Student's Index (LSI_i) is the average of the $LSIT_{ij}$, metrics for each test *j* = *1* .. *m*: $LSI_i = \sum_{j=1}^{m} LSIT_{ij}$ (for this index, you can describe a table similar to the Table 3);
Learning Student's Status (LSS_i) is a characteristic of the position of a group member, calculated as a rank depending on the LSI_i;
Group Performance Index, (GPI) is the average rating received by students of the group on completion the tests;
Group Motivational Activity Index (GMAI) is the average rating for discipline tests that were evaluated in this study group;
Learning Test Index (LTI_j) is the average of the $LSIT_{ij}$ for students of the study group *i* = *1* .. *n*: $LTI_j = \sum_{i=1}^{n} LSIT_{ij}$;
Learning Group Index (LGI) is average of *Learning Student's Index,* LSI_i, of students in a study group. For this index, you can describe a table similar to the Table 3.

The results of the calculation of learning metric indices can be presented in the table of the form of the Table 4.

Table 4 Index table view

Student serial number, (i)	LSIT(i, j)					
	1	2	3	…	m	LSI (i)
1	LSIT(1, 1)	LSIT(1, 2)	LSIT(1, 3)	…	LSIT(1, m)	LSI (1)
2	LSIT(2, 1)	LSIT(2, 2)	LSIT(2, 3)	…	LSIT(2, m)	LSI (2)
3	LSIT(3, 1)	LSIT(3, 2)	LSIT(3, 3)	…	LSIT(3, m)	LSI (3)
…	…	…	…	…	…	…
n	LSIT(n, 1)	LSIT(n, 2)	LSIT(n, 3)	…	LSIT(n, m)	LSI (n)
LTI (j)	LTI (1)	LTI (2)	LTI (3)	…	LTI (m)	*LGI*

Source author

3 Results

Educational experience in applying the teaching of the learningmetry was carried out in groups of the economic direction of undergraduate studies as part of the courses "Probability Theory and Mathematical Statistics" and "Econometrics" based on Samara State University of Economics.

Consider, for example, the results of an analysis of the work of the tests of several typical students.

Let us consider examples of the correlation between learningmetry indices obtained from experimental data in student groups. Suppose performance and positive motivation of student number 1 fluctuate around the diagonal of the 1st quadrant. Then the indicators of this student are in zone I of the optimal learning, and we can expect effective mastery of the academic discipline. Suppose student number 2 academic performance is below average group values, and positive motivation is high. Then the indicators of this student are in zone II of intense learning, and we can expect effective mastery of the courses with considerable effort.

Suppose student number 3 academic achievement is less than average group values, and positive motivation is low. Then the indicators of this student are in zone III, degradation. In this case, an unfavorable forecast is given in the progress of the education discipline: the student does not succeed in studying successfully, and there is no motivation to correct the situation.

Suppose student number 4 academic progress is above average group values, and positive motivation is low. Then the indicators of this student are in zone IV of routine work. Then, progress in the development of discipline, due to the lack of positive motivation and interest in learning, cannot be expected.

In general, for the group, the gradation shown in Table 3, if by teaching the group learningmetric index, LGI, is included in the I group of values with reduced motivation, then we need to work towards increasing positive motivation. If the group index corresponds to the optimal value, then a favorable forecast can be made in the development of discipline by the group as a whole. If the group index is in the third group, corresponding to high motivation and reduced success, then the forecast in mastering the discipline is favorable, but training and practicing in mastering the knowledge, skills in this group should be given special attention.

So, we can draw up the following research program of a teacher working with students on the development of educational discipline in e-learning. As the first stage of the study, we can single out the collection of statistical data, for the receipt of which a questionnaire is used for students who have passed a certain type of tests in the discipline under study in the e-learning. The second stage of the study is data processing, according to the described method of learningmetry. Then a matrix by learningmetry and status graphs, determining micro-groups by motivational activity in the study of academic discipline material, is constructed. At the third stage, the analysis of the results, the zone of proximal development of both an individual student and the group as a whole is determined, and a forecast is made on the effectiveness of learning the academic discipline material on tests and satisfaction with the results of educational activities.

4 Discussion

Various test systems are ubiquitous and are widely used in the e-learning environment in higher educational institutions. Much attention is paid to the new digital format of education, and in particular the problem of reducing motivation in e-learning. In this regard, it is relevant to measure the effectiveness of training, adjusted for motivation, in the process of mastering courses in the e-learning environment.

Papers deal with the issues related to the assessment of the efficiency and the quality of mathematical training of university students in conditions of electronic educational environment [10]. E-learning approach in teaching of mathematical disciplines for future economists are discussed in articles [3, 4, 7].

To assess learning activities in a digital educational environment, special tools are needed to collect and process information and analyze it. We need scales, parameters, indices and feedback from students. As data collection tools are considered an academic achievement test, motivation scale and attitude scale [1]. A feature of the teaching of the learningmetry method is the collection of information on the feedback of students as in other works [5, 9]. Constructing an assessment instrument with behavioral scales to evaluate university teachers and describing clear and unambiguous feedback was done in some works earlier [5].

It is noted that in recent years particularly important was the increase in attention to the current control of knowledge and the intensification of student contacts with the teacher [8], therefore, the task of determining the effectiveness of activities in the e-learning environment and its adjustment by the teacher is relevant.

5 Conclusion

In the educational process, there is a fundamental difference between teaching and learning. What the teacher wants to convey to students (and what, in his opinion, students can and should learn) does not always correspond to what they really learn and understand. And this is not about "knowledge" (or assigned information for active use or passive storage), but about "understanding" (reflexively embedding into the general semantic structure, building links with other thematic areas) and the practical mastery of the tools necessary for moving into activity. This becomes especially noticeable when using e-learning. Under these conditions, it becomes important to find a way to collect specific information (about the course of the learning process, about feedback), its processing and subsequent description. As such a convenient tool, a procedure called learningmetry is proposed. And if earlier in e-learning information was collected on the success of the educational trajectory in the form of ratings and grades obtained during testing, then when applying learningmetry, a necessary condition is the collection of reflection and feedback from the participants in the trainings. Using the learningmetry proposed in the article, one can not only process information, but also make a forecast for effective learning in e-learning, and create their own educational path for the student group and for students.

References

1. Ahmed AM, Osman ME (2020) The effectiveness of using wiziq interaction platform on students' achievement, motivation and attitudes. Turk Online J Distance Educ 21(1):19–30
2. Komissarov A (2019) Digital trace of educational experience. Inform Bureau 20.35. https://ntinews.ru/blog/publications/tsifrovoy-sled-obrazovatelnogo-opyta.html. Accessed 21 Feb 2020 (in Russian)
3. Lyko J (2018) New technologies vs. curricula of quantitative courses for students of economics. In: Chova LG, Martinez AL, Torres IC (eds) Proceedings of the 12th international technology, education and development conference. INTED proceedings. Iated-Int Assoc Technology Education & Development, Burjassot, pp 3087–3090
4. Massing T, Schwinning N, Striewe M (2018) E-assessment using variable-content exercises in mathematical statistics. J Stat Educ 26(3):174–189. https://doi.org/10.1080/10691898.2018.1518121
5. Matosas-Lopez L, Carlos A-FJ, Gomez-Galan J (2019) Constructing an instrument with behavioral scales to assess teaching quality in blended learning modalities. J New Approach Educ Res 8(2):142–165. https://doi.org/10.7821/naer.2019.7.410
6. Moreno JL (1951) Sociometry, experimental method and the science of society; an approach to a new political orientation. Beacon House, Beacon
7. Orszaghova D (2018) E-learning approach in mathematical training of future economists. E-learning 10:427–442
8. Ovsyannikova TL (2014) Foreign experience of distance learning mathematics. Sci Notes Oryol State Univ 1(57):389–392 (in Russian)
9. Tawafak RM, Romli AM, Alsinani MJ (2019) Student assessment feedback effectiveness model for enhancing teaching method and developing academic performance. Int J Inf Commun Technol Educ 15(3):75–88. https://doi.org/10.4018/IJICTE.2019070106

10. Toktarova VI (2019) Assessing the efficiency of teaching mathematics in the e-learning environment. In: Uslu F (ed) Proceedings of INTCESS—6th international conference on education and social sciences. Int Organization Center Acad Research, Istanbul, pp 428–431
11. Yerkes RM, Dodson JD (1908) The relation of strength of stimulus to rapidity of habit-formation. J Comp Neurol Psychol 18:459–482
12. Zaychikova NA (2018) Construction of the method of measurement by scale success-motivation of the results of work with test systems in e-learning. Mod Probl Sci Educ 4. https://doi.org/10.17513/spno.27902. http://www.science-education.ru/article/view?id=27902. Accessed 23 Feb 2020 (in Russian)
13. Zaichikova NA (2019) Application of learning science methodology in student groups in e-learning. In: 21st Century Science: Current Trends, vol 2, issue 1, pp 35–40 (in Russian)

Peculiarities of the Digital Generation in the Context of Education and Management

S. Grishaeva and E. Mitrofanova

Abstract The article takes a closer look at the most stable and widespread ideas about the socio-cultural peculiarities of the digital generation (the so-called "Y" and "Z" generations), which currently represents a significant population category with high standards and unique features. In terms of effective education and management, significant attention is drawn to issues related to the lifestyle and value orientations of the digital generation, formed in the era of the information revolution and the rapid development of computer and communication technologies. The purpose of the study is to analyze the ideas about the socio-cultural features of the digital generation and to show how these features should be considered in modern education and personnel management. The research methods used are the following: secondary analysis of data obtained from Russian and foreign studies, as well as the primary information collection methods: a survey of youth representatives (a questionnaire) and an expert interview. As a result of the research, the views and values typical for the digital generation are outlined, and the tools for using the socio-cultural peculiarities of the digital generation are methodically reasoned both for creating new digital educational process and for choosing the methods for personnel management in the company.

Keywords Socio-cultural peculiarities · Digital generation · Education · Management · Generation "Y" · Generation "Z"

1 Introduction

Culture is the main mechanism of social and individual transformation, it forms values, standards, and means of transmitting cultural patterns. It is values (objects that are given a special meaning in the individual or group conscience) that are the most important regulators of individual behavior (or of social one). The information

S. Grishaeva (✉) · E. Mitrofanova
State University of Management, Moscow, Russia
e-mail: grishaeva@bk.ru

E. Mitrofanova
e-mail: elmitr@mail.ru

S. I. Ashmarina and V. V. Mantulenko (eds.), *Current Achievements, Challenges and Digital Chances of Knowledge Based Economy*, Lecture Notes in Networks and Systems 133, https://doi.org/10.1007/978-3-030-47458-4_71

society produces its own values, which are formed under the influence of digital technologies development and differ quite significantly from the values of preceding industrial society. The values of personal freedom, individualism, "... forms of human existence are being formulated...Virtual space forms a "parallel world", which contains functioning ... equivalents of the real world, ... system of universal values is transforming into the triune unity—unity of physical, spiritual and virtual values" [16, p. 44]. According to scientific opinion, information technologies, on the one hand, expand the boundaries of "freedom of cultural choice for the subject", providing an opportunity to access various information sources. But at the same time, the opposite process is inevitable, as "many subcultures lead to non-freedom, because now the individual is deprived of a guide that determines value orientations" [1, p. 52].

The following features of the information value system are determined [15]:

- perception of the Internet as a source of absolute knowledge,
- formation of new public information spaces with a variety of content (often attractive for a consumer) and a complete lack of meaning and content control,
- accessibility and ease of mastering various cultural fields of almost all nations, eras and times relate to all aspects of modern life, which facilitates work (especially intellectual that), saves resources and time, and brings diversity to recreational activities, etc.,
- recently becoming traditional, a retrieval system searching "by request" and collecting data from many simultaneously operating sources of information gives rise to a special type of perception characterized by fragmentation, clip-thinking and shard-thinking, lack of a unified picture of the studied phenomenon, of the causal relations reflection and of the genesis of development.

Digitalization makes society more dynamic and flexible, changing the form and the content of many social categories: power, status, mobility, etc. The concept of "profession" is also changing its role in the information society—it becomes less stable, permanent, and unchangeable. Today, we are hearing more and more often about the dominance of interdisciplinary professional competencies over narrowly specialized ones (so-called soft skills over hard skills) [18]. Probably, no one can be sure that the profession he received after graduation will remain permanent throughout his working life. The understanding of classes in modern society is also changing: "the problem of a new aspect of the late capitalism social structure transformations, which is mostly associated not with the changes in production forces but with the changes in production relations" [9, p. 23]. The values of the information society are especially noticeable in the behavior of the youth—the digital generation representatives, which are closely investigated in this article.

2 Methodology

The authors base on the results of a secondary data analysis of Russian and foreign studies, as well as on the materials of a study conducted with the assistance of one of the authors [10]. In addition to the secondary analysis, primary information collection methods have been used: a survey of youth representatives (in the form of a questionnaire) and an expert interview. A nationwide selection was carried out across the federal subjects of Russia as a part of the survey, and 1,600 respondents were interviewed. As a result of the interviews, the data obtained during the survey was presented to the experts and supplemented by them. Scientific community representatives, as well as organizers of the youth work, were selected as experts.

3 Results

As a result of the study, the following values typical for the modern youth have been distinguished [10]: material values, love/relationships, friendship/social contacts/communication, patriotism/civil stance, cultural values/cultural identity, career, individualism, healthy lifestyle, self-realization, education/knowledge, social success/social status, family, pleasure, urge to change. Moreover, according to experts, top-priority values of modern Russian youth are the following: material prosperity, career, communication, self-expression, and individualism.

How are these values being brought to life? International survey of the third millennium generation, conducted among 13,416 representatives of generation Y from 42 countries and among 3,008 generation Z representatives from 10 countries [12] has demonstrated the following results:

- the level of economic, social and political optimism has fallen to a record low,
- a lack of trust in traditional social institutions, including media, has been detected,
- a pessimistic view on the possibility of social mobility,
- low life, financial situation, work, management, government, media, and personal data usage satisfaction,
- choosing a new experience and new sensations,
- traveling and socially useful activities are more important than starting a family or starting your own business,
- skeptical about the motives that drive the business,
- the respondents are quite critical about business in terms of companies' integrity, their public contribution, and their willingness to make the world a better place,
- in general, the representatives of generations Y and Z use the services of those companies whose stance coincides with their values. For example, many respondents indicated that they are ready to immediately reconsider or even break off their relations with a company if its activities, values and political sympathies contradict their beliefs.

The respondents have demonstrated a wide range of life goals in their answers [10]. For the majority of youth (77.7%), the main goal in life is to build a good family. 67.9% of respondents would like to live in prosperity. 62.7% indicated getting a good job as their main life goal. 59.3% of respondents chose getting their own apartment as a life goal. Just over a half of the respondents (53.5%) indicated getting a good education as a crucial life goal. Every second respondent (49.3%) indicated that his/her goal in life was "to do good to other people". 46% stated "living in harmony with oneself" as their life goal. 45.4% of respondents would like to make use of their talents. 42.8% of young people would like to earn a lot of money and 38.7% would like to establish their own business. 27.5% of participants are looking forward to becoming bright individuals. 27.4% of respondents would like to occupy a high social position. 15.5% indicated "power, the ability to lead other people" as their life goal and, finally, 15.2% would like to get universal recognition and fame.

As we can see, successful employment is quite important for young people. At the same time, it is necessary to consider specific features of the digital environment as of an environment of professional implementation, implying a specific cultural form—information culture—existing in that particular environment. Researchers propose to treat the information culture of an individual "…as a set of mechanisms that contribute to adaptation to changing social conditions. The need … to have knowledge about … working methods with information … gives…grounds to include them in the minimum … functional literacy of an individual" [13, p. 35]. Therefore, the digital environment imposes quite high requirements for the level of so-called "digital competencies" among employees. However, while it is believed that young people proficiently possess digital competencies, not all the studies confirm that [7]. One of the reasons for the lack of "digital competencies" development and, in particular, digital literacy, is digital inequality, referring to "… an unequal access of subjects to information technologies, to information, to the accumulated knowledge of society" [2, p. 17].

Another digitalization trend driven by value transformation is the possibility of "remoted" employment. Indeed, the values of individualism imply, among other things, the possibility of choosing the time, place and pace of work. The digital society provides such opportunities, making professional self-realization possible for those who previously found it difficult, for instance, for people with disabilities. As a result, there is a tendency to move from exclusion (not taking into account or not employing individuals who differ by some indicators) to inclusion and, consequently inequality associated with exclusion decreases significantly [17].

New technologies are also changing behavioral patterns for example, the method of communication via e-mail, which is quite common in modern organizations, is likely to lose its popularity among new generation employees. So, according to a study conducted by American consulting firms 8 × 8 and M. Nikolsky Research [20], "only 68% of young employees believe that in 5 years, e-mail will still be a relevant way of communication within the working processes, and only 19% actively use e-mail for such purposes. For older generations, the corresponding indicators are equal to 83% and 40%, respectively".

Digital environment can influence the formation of professional adaptation strategies for the Russian youth, as well as the strategies for sociocultural adjustment. These strategies may not always be constructive, for instance, there are:

- the effect of cognitive reduction: knowledge consumed from the Internet is uncritically evaluated by the recipients, as there are difficulties in distinguishing reliable information from the one that is not trustworthy; there are no competencies necessary for working with big data; the problem of overcoming information overload,
- the phenomenon of digital alarmism and gambling: much more flexible consciousness due to new technologies appearance; leaving offline environment for online communication; fake information presence; emphasis on user reviews in response to such a phenomenon as "life for a show" ("just for show life"); the prevalence of non-constructive, often socially dangerous outrage; illegal placement of personal data, Internet phishing (Internet fraud), online gaming addiction, inability to distinguish reality from virtual space,
- communicative destruction effects: loss of interpersonal business communication skills; blurring borders, barriers, and respect for communication; decrease in the level of communication depth and underdevelopment of emotional intelligence; prevalence of social adaptation problems in conditions of living space digitalization [8].

The presented data indicates the need to take specific values and views of the digital generation into account both when building a new digital educational process, and when choosing methods for managing this category of personnel in the company.

4 Discussion

4.1 Considering Sociocultural Peculiarities of the Digital Generation in Educational Process

First of all, the educational problems caused by the values and views of the digital generation presented above require correction of the negative sociocultural characteristics of this generation's representatives. It should be taken into account that these peculiarities not really represent a lack of personal qualities but mostly demonstrate a lack of social competencies, which can be compensated with the help of targeted educational technologies (Similar to "knowledge gaps").

Moreover, it is crucial to note that digital generation also disposes of a range of positive characteristics important for education. That are, primarily, fluency in the latest digital technologies, and the following aspects:

- in terms of cognitive development—a constant desire for novelty and self-improvement, creativity, an ability to synthesize different types of thinking,

non-linearity, an ability to process different information flows at the same time (multitasking), an aptitude to use different informational sources, high speed of information processing and decision-making;
- in terms of social development—desire for self-expression, preference of the "horizontal" (partner) type of relationship to the "vertical" (hierarchical) one, openness to cross-cultural and international communication, optimism and self-confidence [5, 6].

In general, the strategy of working with digital generation representatives should be based on the fact that it is almost impossible to integrate them into traditional educational process. It requires a significant transformation, the result of which is the construction of a new, digital educational process [19].

Digital technologies significant for education include: artificial intelligence; telecommunication technologies; electronic identification and authentication technologies; Big Data processing technologies; distributed registry technologies (including Blockchain); gamification, virtual and augmented reality, additive technologies, digital double technology, and others [4]. These digital technologies create new opportunities for building the educational process and solving a wide range of educational tasks, considering specific values of the digital generation.

4.2 Considering Sociocultural Peculiarities of the Digital Generation in Personnel Management

Choosing the personnel management methods suitable for the digital generation requires consideration of the following sociocultural peculiarities [14]:

- striving not for the long-term results, but for quick ones,
- significance of the meaningful aspect of professional activity,
- relevance of game aspects of professional activity,
- pursuing an informal style of communication, a comfortable psychological climate and free schedule,
- ideas of one's own about career prospects and the role in society,
- strong motivation for meaningfulness, interest at work, and moderate one—for money.

In accordance with these features, effective management methods for this particular generation representatives are the following:

- informal approach to any kind of activity: from the job opening description to interior design,
- free working schedule,
- detailing the goals and objectives of professional activity,
- considering candidate's interest in the proposed work when selecting staff,
- a corporate culture that allows employees to feel "included in the community",
- active use of verbal encouragement.

Thus, digital generation representatives will work actively if the company has a clear and interesting corporate culture, mission, values, and traditions. Tasks must be set specifically and clearly, indicating crucial features of the required goals.

Many european and some of Russian IT companies, such as Google, Yandex, Facebook, Euroset, Mail.Ru and others, have already successfully implemented this theory in practice in the sphere of personnel management. In particular, operations in Facebook's department of application development (the employees are mainly young people, aged from 17 to 27) are organized as follows [14]:

- there is no strict working schedule,
- an opportunity to deal with the business of one's own more than 90% of the working time,
- opportunities to play video games while working,
- collective informal discussion of the employees' ideas,
- conditions have been created for employees to live in the office if necessary and desired, etc.

The features mentioned above characterize the digital generation representatives, who are called generation "Y" or Millennials in the modern theory.

However, the "Y" generation is being replaced by new representatives of the digital generation "Z", who were born in the digital world. They cannot imagine themselves without a mobile phone or modern gadgets, they already know a lot and are able to perform many tasks. This is a generation growing at a rapid pace. They are brought up on the basis of information technology, they process information quickly and are guided by innovative development [14].

People born in 2003 and later are still too young to demonstrate their talents and professionalism. And it is yet difficult to say what will be their priority and what type of employees will they turn out to be. Moreover, according to the research carried out by the Millennial Branding company, the values of generation "Z" will be limited to free communication and enthusiastic world perception. The most dominant qualities of this generation will be a desire for continuous learning and self-education and creativity. It is assumed that generation "Z" will be interested in innovative technologies and science, biomedicine, art, and robotics [23].

The main activity area of the companies that will employ generation "Z" representatives will be innovative tasks, while creativity, freedom, and non-standard approach will stand in the foreground. The main labor motivation of generation "Z" is the opportunity to work on important and interesting projects, for example, creating unique technologies for solving global world problems [21].

In the sphere of management it is crucial to consider that generation "Z" representatives base on the "I want" concept, but there is no concept of "need" for them. For this generation it is fundamentally important to spend time with interest, travel and build a career for which "there will be no shame" [3].

In accordance with these features, the most effective management methods for representatives of this generation are the following [11]:

1. To provide a good career line that should correspond with a horizontal professional path, and not only in terms of finances. It should be interesting to work—it may include engaging projects and a large area of responsibility.
2. Potential travel opportunities, for example, business trips or training programs in foreign countries.
3. Another important motive is that generation "Z" representative wants to be an individual, to be different in some way (not for themselves, but for others). It is important to provide an opportunity to be unique so that others can see it.

5 Conclusion

Summing up the results of the study, we can draw the following conclusions. It is desirable to avoid excessive generalizations while predicting the behavior of young people in a transforming society. As it was fairly noted [7], "myths" about generation "Y", regardless of the actual professional qualities of its representatives, have already affected the attitude of employers and colleagues towards them, creating the effect of "reverse ageism" or discrimination against young professionals, as, for example, occurs in the United States [22], and may happen again with the representatives of "Z" generation. However, both education and business should be methodically and organizationally ready to cooperate with the digital generation representatives, focusing on the idea of sociocultural peculiarities of both generation "Y" and "Z", which has already been formed in theory and in practice and involves the following features:

- opportunities to play video games at the workplace,
- collective informal discussion of the employees' ideas,
- conditions have been created for employees to live in the office if necessary and desired,
- pursuing an informal style of communication, a comfortable psychological climate and free schedule,
- ideas of one's own about career prospects and the role in society,
- strong motivation for meaningfulness, interest at work, and moderate one—for money.

Considering these peculiarities will allow modern companies to successfully manage personnel of different generations and develop strategies for mutual interaction.

References

1. Afanasiev SV (2017) Conflicts of moral values in the culture of information society. Conflictology 2:48–60. https://doi.org/10.7256/2454-0617.2017.2.23383

2. Afanasyeva AS (2019) Digital inequality as a problem of introducing e-government in Russia. Soc Polit Econ Law 6:16–17
3. Anderson HJ, Baur JE, Griffith JA, Buckley MR (2017) What works for you may not work for (Gen) me: limitations of present leadership theories for the new generation. Leadersh Q 28(1):245–260. https://doi.org/10.1016/j.leaqua.2016.08.001
4. Baker Rosa NM, Hastings SO (2016) Managers making sense of millennials: perceptions of a generational cohort. Qual Res Rep Commun 17:52–59. https://doi.org/10.1080/17459435.2015.1088895
5. Baker Rosa NM, Hastings SO (2018) Managing millennials: looking beyond generational stereotypes. J Organ Change Manag 31(4):920–930. https://doi.org/10.1108/JOCM-10-2015-0193
6. Blinov VI, Dulinov MV, Esenina EYu, Sergeev IS (2019) Project of the didactic concept of digital professional education and training. Pero Publishing House, Moscow
7. Bogacheva NV, Sivak EV (2019) Myths about "Generation Z". HSE, Moscow
8. Brodovskaya EV, Dombrovskaya AYu, Pyrma RV, Sinyakov AV, Azarov AA (2019) Influence of digital communications on the formation of professional culture of the Russian youth: results of a comprehensive practical research. Public Opin Monit Econ Soc Changes 1:228–251. https://doi.org/10.14515/monitoring.2019.1.11
9. Buzgalin AV, Kolganov AI (2019) Transformations of the social structure of late capitalism: from the proletariat and the bourgeoisie to the precariat and the creative class? Sociol Res 1:18–28
10. Chuev SV (ed) (2017) Value orientations of Russian youth and the implementation of state youth policy: research results. Publishing House SUM, Moscow
11. Costanza DP, Finkelstein LM (2015) Generationally based differences in the workplace: is there a there? Ind Organ Psychol Perspect Sci Pract 8(4):308–323. https://doi.org/10.1017/iop.2015.15
12. Deloitte (2020) The international survey of the third millennium generation – 2019. https://www2.deloitte.com/ru/ru/pages/about-deloitte/articles/millennialsurvey.html#. Accessed 23 Jan 2020
13. Gavrilyuk VV, Sorokin GG, Farakhutdinov SF (2009) Functional illiteracy in transition to the information society. IUT, Tyumen
14. Gladkova T (2019) Generation X, Y, and Z in the labor market – how will personnel management change? HR-Portal. https://hr-portal.ru/blog/pokolenie-h-u-i-z-na-rynke-truda-kak-izmenitsya-upravlenie-personalom. Accessed 23 Jan 2020
15. Gnatyshina EV, Salamatov AA (2017) Digitalization and digital culture formation: social and educational aspects. South Ural State Humanit Pedagog Univ Bull 8:19–24
16. Kotova SA (2015) New values of post-industrial society. Bull Prikamsky Soc Inst 4(72):42–46
17. Lyons S, Urick M, Kuron L, Schweitzer L (2015) Generational differences in the workplace: there is complexity beyond the stereotypes. Ind Organ Psychol 8(3):346–356. https://doi.org/10.1017/iop.2015.48
18. McDonald NC (2015) Are millennials really the "go-nowhere" generation? J Am Plan Assoc 81(2):90–103. https://doi.org/10.1080/01944363.2015.1057196
19. Murphy W (2012) Reverse mentoring at work: fostering cross-generational learning and developing millennial leaders. Hum Resour Manag 51(4):549–574. https://doi.org/10.1002/hrm.21489
20. Nikolsky M (2018) How we lose generation Z: 6 errors in communicating with new people. Knife. https://knife.media/generation-z/. Accessed 23 Jan 2020
21. Parry E, Urwin P (2011) Generational differences in work values: a review of theory and evidence. Int J Manag Rev 13(1):79–96. https://doi.org/10.1111/ijmr.2011.13.issue-1
22. Raymer M, Reed M, Spiegel M, Purvanova R (2017) An examination of generational stereotypes as a path towards reverse ageism. Psychol-Manag J 20:148–175
23. Schawbel D (2015) Elite daily and millennial branding release landmark study on the millennial consumer. Millenial Branding. http://millennialbranding.com/category/blog/. Accessed 23 Jan 2020

The Role of Social Networks in the Adaptation of People with Disabilities

N. A. Ustina

Abstract The relevance of the problem is due to the low activity of patient non-profit organizations in using the Internet space to involve people with disabilities in social and professional activities. The purpose of the study is to identify the main trends in interaction of people's with disabilities in social networks and to develop methods for evaluating the effectiveness of managing their groups in social networks. Research methods include a quantitative analysis of the activity of groups in social networks engaged in promoting content for people with disabilities. Based on the results of the analysis of social networks content, a link was found between the nature of content posted in social groups of people with disabilities and their willingness to be engaged in communication, use social networks for socialization and professional self-determination. The result of the study was the development of a methodology for evaluating the communication formation process in social networks. This method allows determining the most effective ways to work in social networks that contribute to the formation of an active communication environment for people with disabilities.

Keywords Communication in social networks · Socialization of people with disabilities

1 Introduction

One of the most important trends in the field of socialization and social adaptation of people with disabilities is their growing need for access to an interactive communication environment that expands their understanding of the opportunities to lead a full life. At the same time, we can observe a low activity of non-profit organizations that protect the interests of people with disabilities in using media and the Internet space to inform about opportunities of people with disabilities, conducting interactive events to engage these people in interaction on issues of their socialization and self-realization.

N. A. Ustina (✉)
Samara State University of Economics, Samara, Russia
e-mail: nina_ustina@mail.ru

S. I. Ashmarina and V. V. Mantulenko (eds.), *Current Achievements, Challenges and Digital Chances of Knowledge Based Economy*, Lecture Notes in Networks and Systems 133, https://doi.org/10.1007/978-3-030-47458-4_72

The Internet space is currently the most accessible and effective platform: (1) for spreading information to the target audience of non-profit organizations including people with disabilities and their environment; (2) for organizing communications that increase the interest of people with disabilities in social and professional self-realization. By 01.10.209, the number of people with disabilities in the Russian Federation is more than 11 million people. At the same time, medical indications of a significant number of disabled people allow them to work and participate in society. But according to statistics, only a third of such citizens works. Only 11.2% of people with disabilities of working age assess themselves as capable of active professional activity and socialization. This situation is connected, on the one hand, with the stereotypical thinking of the society about the inability of people with limited health opportunities and people with severe chronic diseases to an active life, on the other hand, with the lack of motivation of people with disabilities themselves. The reasons for this attitude are the dependent passive lifestyle of people with disabilities, their lack of an adequate positive perception of life.

The problem can be solved by informing about the possibilities of an active life, creating confidence in this information, spreading the successful experience of social and professional self-realization of people with disabilities and involving them in the implementation of project initiatives aimed at their socialization.

The most important function of patient non-profit organizations and non-profit organizations that unite people with disabilities is to facilitate the involvement of people with disabilities in active social life and professional activities, including by providing access to information about their opportunities for self-realization. The Internet space is currently the most accessible and effective platform not only for spreading information to the target audience of non-profit organizations that unite people with disabilities, but also for organizing communications that increase the interest of people with disabilities in social and professional self-realization.

Active growth of information impact to destroy existing stereotypes is possible only if non-profit organizations working with this target group will produce high-quality interesting material in an accessible and interactive communication space. The development of communication platforms aimed at promoting the idea of an active life for people with disabilities, increasing their social and professional involvement in the social and economic life of our society is a task that can be solved within the framework of SMM-management and content management (CM) technologies.

2 Methodology

The developed methodology for evaluating the activity of non-profit organizations in social networks includes such indicators as:

- representation of a non-profit organization in social networks,
- popularity of a non-profit organization in social networks,
- publication activity of the organization in social networks,

- nature of publications in social networks,
- the nature of responses from members of the social group to publications.

The representation of a non-profit organization in social networks was evaluated in points (0 to 3). The maximum score was assigned to organizations represented in 4–5 of the most popular social networks in the Russian Federation (Vkontakte, Odnoklassniki, Facebook, Twitter, YouTube). Presence in 2–3 social networks was estimated at 2 points. Accordingly, the presence of a group in one of the five social networks was estimated at 1 point.

To determine the popularity of a non-profit organization in social networks, the number of participants in social groups was calculated. Organizations were awarded points from 0 to 3, depending on their place in the ranking by the number of participants. ABC-analysis was used to allocate the organization's place in the rating. Group A includes organizations that had the highest rates of participation (up to 80% of the total number of participants in social networks of all studied organizations). Group B includes organizations that cover 15% of all participants, and group C includes 5%.

The publication activity was determined by the number of posts published in all accounts of the non-profit organization during the studied period. A 3-point scale (from 0 to 3) was used to evaluate this indicator. Organizations received points based on the ABC-analysis. Group A includes 20% of organizations, group B—30%, and group C—50% of all organizations that posted publications. Organizations that did not publish at all during the studied period received 0 points.

The nature of publications was determined by their interactivity degree, i.e. their ability to elicit an active response from group members and their readiness for active actions. To determine the interactivity degree, all publications were divided into three groups. The first group includes publications that are only informative in their nature, the so-called informative messages [9]. The second group includes publications that contain problematic (transformative) information. This information changes the attitude of a member of a social group to different situations, updates individual problems, makes them experience the situation, and encourages them to form an attitude to it. The third group includes publications that encourage action (decision-making publications). This is the most active type of communication, which involves not only changing the attitude to the situation, but also changing the behavior of the communication participants. These publications include an invitation to participate in the event, a story about positive experience in solving the problem. If there are only the first type of publications in a social group of a non-profit organization, the group is assigned 1 point. If there are publications of the second group—2 points. Accordingly, the placement of publications that can be attributed to the third group gave 3 points.

The object of our study was the social network accounts of regional offices of the All-Russian public organization of disabled people with multiple sclerosis. Currently, the organization is represented by its offices in 54 regions of the Russian Federation. It is the most active member of the patient community in Russia, is a collective member of the All-Russian Union of patients, and effectively builds communications with authorities, insurance companies, medical institutions, and pharmaceutical companies. The presence of communication projects of this public organization allowed us to put forward a hypothesis that the leaders of this organization and its regional offices have a high level of competence in building communications with patients using social networks to form a unified communication environment for patients.

3 Results

The study of activities of non-profit organizations in social networks showed that at the moment only large organizations that deal with the problems of people with disabilities actively conduct an information policy and post professionally prepared materials in the Internet space.

From the 54 public organizations that represent the interests of patients with multiple sclerosis, only 13 have accounts in social networks, which is 24% of all regional offices of the All-Russian public organization of disabled people with multiple sclerosis.

From 13 organizations that have their own accounts in social networks, only three are represented in 2 or more social networks (Moscow, Omsk and organization of the Republic of Khakassia). The total number of participants in all social groups of regional offices of ARPO DPMS is 3644 people, the total number of publications for three months in these groups—397.

The results of the popularity rating and publication activity of organizations in social networks are shown in Fig. 1.

Analysis of the popularity of social groups in regional branches of the ARPO DPMS showed that the largest number of participants have accounts in the Leningrad branch, St. Petersburg, Moscow and the Moscow region. The least popular groups are those in the Samara, Voronezh regions and the Republic of Khakassia. At the same time, according to the number of publications, the rating of regional offices has a different type. The social groups with the largest number of participants have the lowest rates of the publication activity. Accounts of four regional offices had no publications for 3 months of observation. Conversely, groups that have a small number of participants show high publication activity. This effect can be interpreted from different positions. In our opinion, the lack of direct correlation between the publication activity of organizations in social networks and the number of participants is primarily determined by the quality of these publications, and the lack of skills of activists to form content that is interesting for members of these organizations.

Figure 2 shows the results of analyzing the quality of content posted in social groups of ARPO DPMS representative offices.

In general, most non-profit organizations that post information in their social media groups are limited to passive publications. They are not ready to use the modern features of the Internet space fully and do not try to create an interactive communication space that provides access to interesting information useful for the socialization of people with disabilities.

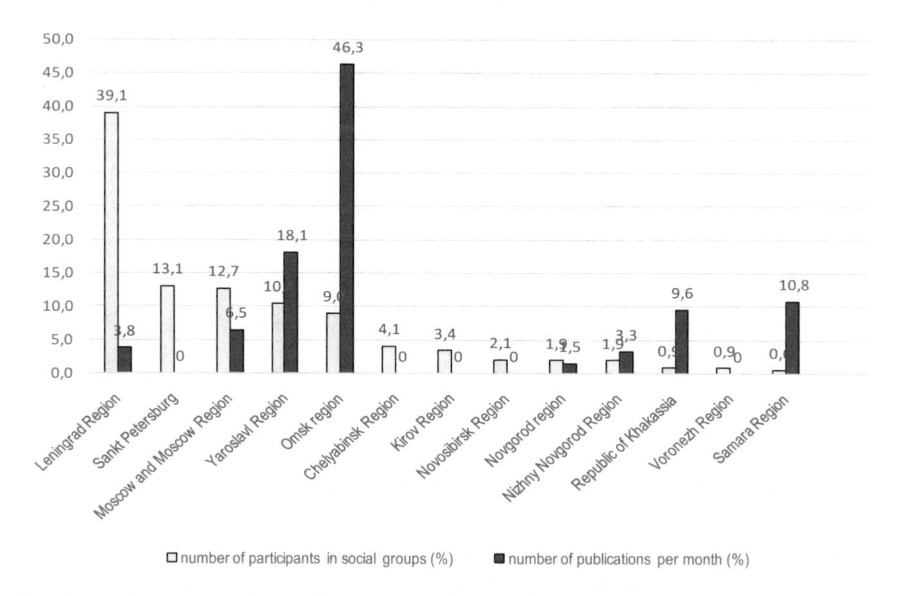

Fig. 1 Activity of non-profit organizations in social networks (% of the total number of participants and publications of regional organizations of ARPO DPMS) (*Source* author)

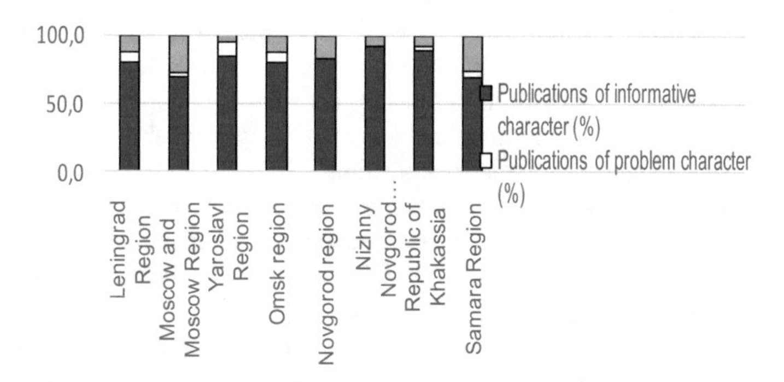

Fig. 2 Quality of content in social groups of regional offices of ARPO DPMS (% of the total number of publications) (*Source* author)

The passive nature of content posted on social networks in most cases does not cause an active response from participants. The total number of responses to all publications in social networks of regional offices of the ARPO DPMS is 1306 units. In the social groups of two organizations—Nizhny Novgorod and Novgorod, there were no responses to publications at all for three months. Figure 3 shows the ratio of responses of various types for social groups in each regional office.

Based on the results of the analysis of the activity of patient non-profit organizations in social networks, their overall rating was compiled, which was based on

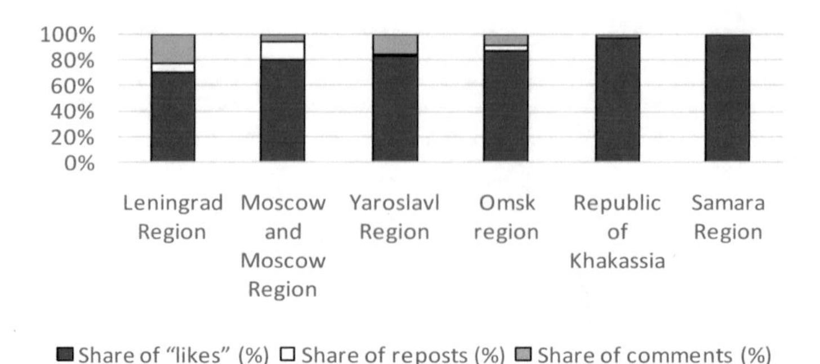

Fig. 3 Types of responses to publications in social networks of regional offices of ARPO DPMS (%) (*Source* author)

scores. Only organizations with social media accounts were included in the overall rating. The results of the score are shown in Table 1.

Organizations that demonstrate the most successful experience in building communications for people with disabilities took the first place in the ranking. The analysis of the content of their publications in social networks allowed us to determine the main content of the communication environment created by them. The main emphasis is shifted to the formation of a virtual educational space for people with disabilities, assistance in employment, socialization, and psychological support. The passive nature of responses to the publications of organizations in social networks is a consequence of the lack of skills among the activists of patient non-profit organizations to build a communication space where people with disabilities could realize their needs for self-realization and socialization.

Such a way of organizing communications is not able to meet the needs of target groups (people with disabilities) for full-fledged communications, to form confidence in information about existing possibilities for their professional and social self-determination.

The reasons for the low activity of non-profit organizations(that unite people with disabilities) in the Internet space lie in their attitude to the implementation of this function as a secondary one to other types of profile activities that are aimed at protecting the interests of people with limited health opportunities. This relates to the low level of skills formation for ensuring information support and promotion of their activities among permanent employees of limited health opportunities, the lack of competencies for purposeful involvement of people with disabilities in various social and professional activities.

In order to form necessary competences of employees of non-profit organizations to create and promote content in social networks, the author proposes a system of the activity indicators for administrator of non-profit organizations' groups in social networks representing the interests of patients.

When evaluating the organizational effectiveness of communication management in a social group, it is proposed to identify 3 groups of indicators: target indicators, economic efficiency indicators, customer focus indicators, and process indicators.

Table 1 Rating of regional offices of ARPO DPMS on communication activity in social networks (points)

Organization	Number of social media accounts	Number of participants in social groups	Number of publications	Quality of publications	The nature of the responses	Total points
Yaroslavl Region	1	3	3	3	3	13
Moscow Region	2	3	2	3	2	12
Omsk Region	2	2	3	3	2	12
Leningrad Region	1	3	1	3	3	11
Samara Region	2	1	3	3	1	10
Republic of Khakassia	2	1	2	2	1	8
Novgorod Region	1	1	1	3	0	6
Nizhny Novgorod Region	1	1	1	2	0	5
Sankt Petersburg	1	3	0	0	0	4
Chelyabinsk Region	1	2	0	0	0	3
Kirov Region	1	2	0	0	0	3
Novosibirsk Region	1	1	0	0	0	2
Voronezh Region	1	1	0	0	0	2

Source author

Target indicators determine the effectiveness of all communications in a social group, which is determined by the main goal—to transform responses of communication participants into ideas and projects for socialization and professional self-realization of people with disabilities. An indicator of the target effectiveness of a social group is the ratio of the number of interaction situations created within a social group and the strength of responses. When calculating the effectiveness of a social group, the number of situations created for the organization of interaction over a certain period and the level of responses received directly during the interaction are indicated. This level was determined by the scale of impressions developed by the authors of the "economy of impressions" [7].

This scale distributes impressions received depending on the activity degree of the communication participant in their formation (passive active communications) and on the transformation degree of the participant in the interaction situation (immersion—absorption). The strongest impressions (active immersion) form a response in the form of new ideas, materials that saturate the proposed content. The weakest impressions (passive absorption) are reflected in the viewing statistics. Impressions "passive immersion" are shown in giving "likes", impressions" active absorption"—in the number of reposts.

Each situation generates responses of different levels. To calculate the effectiveness of the interaction situation, each response level (Xi) is assigned a weight coefficient in the interval ($\alpha = 1; 0.4; 0.2; 0.1$). The effectiveness of each interaction situation is determined by the number of responses and their strength. And is calculated as follows (formula 1):

$$Ec = \sum_{i=1}^{n} xi \times ai \tag{1}$$

The indicators of the second group (process indicators) allow you to evaluate the administrator's activity in the scenario of situations in which there is interaction between members of the social group. These indicators are determined by the factors of activation of the communication process in each specific situation, which include:

- number of communication participants in the situation,
- the number of communications between participants in the social group built in each situation,
- number of participants' communications with the administrator,
- the time it takes the administrator to build a communication.

The effectiveness of the process is determined by the ability of social group administrators to maintain the maximum number of information channels, taking into account the time spent on organizing communication, and can be calculated using the formula (2):

$$\boldsymbol{Er} = \frac{Ca + Cv + Cp}{T} \tag{2}$$

Where *Ca* is the activity coefficient of communication participants in the social group, which is calculated as the ratio of the number of communication participants who responded to the publication to the total number of views. The interaction coefficient (*Cv*) reflects the possibility of the situation being played out to start the process of horizontal communications between the participants of the social group. It is calculated as the ratio of the number of communications between participants in the social group and the number of communications supported by the administrator. The planned interaction coefficient (*Cp*) shows the ratio of the number of administrator's posts in each interaction situation planned within the scenario of the situation and posts placed in the communication process. Time (T)—administrator's labor costs for maintaining interaction processes.

4 Discussion

The role of virtual communication platforms in the socialization and professional self-determination of people with disabilities is actively discussed in modern research. The global network is considered as a tool for overcoming the barrier between people with disabilities and professional, educational, and social space [1].

Many authors identify possibilities of social networks in the formation of the concept of intellectual education and offer models of unlimited access to this education for those users who choose a strategy of continuous education [5, 6].

The description of social practices of using social networks by people with disabilities allows researchers to conclude that there is a high and yet unrealized need for specialized content for this category of users, taking into account their relative closeness in building communications [3].

The content of publications that is posted in social networks and its impact on the role of these communication forms in socialization is the subject of deep discussions and is considered as a fundamentally different and largely innovative way of building personal communications with significant communities [2]. *In particular, researchers attach great importance to active content, which implies an active response and ways to promote it. Such content may include, for example, information about various events and invitations for users to participate in them. Therefore, the placement and promotion of such events are subject to a separate study* [10].

In general, the choice of content for placement in social networks and its promotion is, according to modern researchers, the most important factor for successful communication and creating a communication environment that increases the social and professional activity of users not only in the virtual, but also in the real space. Therefore, works that consider models for promoting various content and interacting with users are of particular value for developing indicators for social network administrators and moderators [4, 8].

5 Conclusion

Currently, the Internet resources created by patient organizations do not carry the potential of an active communication environment that can generate the interest of the target audience in the experience of successful professional and social adaptation of people with disabilities, and include them in the process of interaction to jointly solve socialization problems.

Using the proposed model for evaluating the effectiveness of building communications in social networks allows determining the main content of the activities of social group organizers to manage the processes of interaction with participants and inclusion of consumers in the process of creating joint products that reflect their unique ideas about the content and quality of services.

Focusing on communication performance indicators will allow the organizer of a social group to develop and implement an interaction scenario that takes into account all the interactivity parameters and will turn social networks into communication platforms for people with disabilities to access unlimited opportunities for self-realization and employment.

References

1. Alexandrova M, Zhivaykina AA (2017) The role of the Internet in the socialization of people with disabilities. Bull Med Internet Conf 7(1):48–49
2. Al-Qurishi M, Al-Rakhami M, AlRubaian M, Alamri A, Al-Hougbany M (2015) Online social network management systems: state of the art. Procedia Comput Sci 73:474–481
3. Efimov EG, Horoshunova EE (2018) Social networks as factor of formation of the educational circle of people with disabilities. Sign Probl Field Media Educ 3(29):127–132
4. Klepek M (2017) Systematic analysis of the current academic research on social media marketing. Sci Ann Econ Bus 64:15–27
5. Kravtsova O, Dubik A, Kotelevtsev N (2019) Using social networking opportunities in implementing elements of distance learning for people with disabilities. In: Guskova EA, Bok OV, Min LI (eds) Federal state educational service of Ukraine: experience, problems, prospects: materials of the All-Russian correspondence scientific and practical conference. OGAOU DPO "Belgorod Institute for the Development of Education", Belgorod, pp 174–177
6. Lytras MD, Visvizi A, Daniela L, Sarirete A, De Pablos PO (2018) Social networks research for sustainable smart education. Sustainability 10(9):29–74
7. Pine BJ, Gilmore JH (1999) The experience economy: work is theater & every business a stage. Harvard Business School Press, Brighton
8. Scuotto V, Del Giudice M, Carayannis EG (2017) The effect of social networking sites and absorptive capacity on SMES' innovation performance. J Technol Transfer 42(2):409–424
9. Sovietov B, Dubenetsky B, Tsekhanovsky B (2010) Theory of information processes and systems. Publishing Center "Academy", Moscow
10. Yoon S-W, Chung S (2018) Promoting a world heritage site through social media: Suwon City's Facebook promotion strategy on Hwaseong Fortress (in South Korea). Sustainability 10(7):21–89

Digital Autism as a Factor of Reduced Physical Activity of the Population

S. I. Zizikova, I. V. Nikolaeva, and O. B. Paramonova

Abstract This article examined digital autism from the perspective of current issues in the modern world. Studies of world psychologists and psychotherapists were studied, and a statistical summary of the development of this disease in children was also considered. The brain default system was considered for a better understanding of the problems. The causes of the active development of the disease in the population were identified. The relationship of information technology with the transformation of human thinking is shown. It is shown how digital autism reduces the physical activity of the population. Cases of aggressive behavior at the extreme stage of digital autism have been identified. The possibilities of addressing and preventing autism in the early stages of manifestation have been described. A methodology for parental behavior was identified. Distal thinking was considered, which orientates a person to future planning and noticeably weakens with digital autism.

Keywords Autism · Digitalization · "Brain default system" · Psychology · Information technology

1 Introduction

Today, humanity is developing rapidly with the advent of information technology and inventions. We are entering the digital age or the economy of things, absorbing the human brain and translating the civilization of text into the civilization of visual images that shift systemic thinking. In our country, the Internet appeared a little

---

S. I. Zizikova
Volga State Academy of Physical Culture, Sports and Tourism, Kazan, Russia
e-mail: zizikova@yandex.ru

I. V. Nikolaeva (✉)
Samara State University of Economics, Samara, Russia
e-mail: niv2017@bk.ru

O. B. Paramonova
Samara State Social and Pedagogical University, Samara, Russia
e-mail: paramonova71@mail.ru

S. I. Ashmarina and V. V. Mantulenko (eds.), *Current Achievements, Challenges and Digital Chances of Knowledge Based Economy*, Lecture Notes in Networks and Systems 133, https://doi.org/10.1007/978-3-030-47458-4_73

more than twenty years ago, but the IT sphere has very firmly entered everyday life and continues to radically change it daily. People today do not imagine their life without a computer, telephone, Internet. These inventions simplify our lives and save time so much that sometimes they take it and make children, and they are more vulnerable than digital adults with digital autism. These are people with developmental disabilities who minimize social life and live in their own world, where even close and dear people are not allowed. In fact, physical activity is reduced, because in the electronic world there is absolutely everything that a person needs. And his desire for physiological needs disappears. So what is digital autism and why is it a fact of a decrease in the physical activity of the population [2].

2 Methodology

The authors have considered digital autism from the perspective of the most pressing problem of our time and identify the causes of its appearance and formation. The units of analysis are: find out causes of digital autism; and consider research data on the current state of digital autism in the population. To achieve the goal and solve research problems, the research data of famous scientists, psychologists and psychotherapists were examined. The brain default system was studied to better understand the appearance of the problem. A biological relationship and comparison of humans and animals was carried out, as well as a historical parallel to the human evolution of the brain.

3 Results

The first unpleasant thing that young people face today is the suppression of the brain's default system. Researchers showed that when you activate the central performing network of neurons (and with constant consumption of information, it is always active), this means that energy does not enter the areas of the brain that are responsible for thinking. Essentially, your brain is hibernating [5]. About 40% of children in the USA and Russia under 10 years of age are almost constantly online, that is, they constantly consume information. European parents are more conscious in this regard. But by the age of 14–18, the situation is compared, and almost all teenagers in the world spend 60–70% of their time online. That is, they send their thinking server into hibernation; they simply do not form it. But this is not the only problem. As Gloria Mark told us, for the default system to turn on and start thinking about a serious problem, it needs time—about 23 min. If these 23 min are gone, you will continue to be in a different space. Look at the data from the 2016 study. Already then the average person conducted about 80 telephone Internet sessions per day. This means that it was interrupted every 15 min. In fact, he did not have time to launch his default system and load it with the necessary intellectual objects.

The default brain system was created so that we maintain relationships with other people. This is the most social system. But already in 1997, the amount of our screen time was equal to the amount of time we spend on live communication with people. At the time of the appearance of the iPhone in 2007, a person's screen time was more than 8 h, while face-to-face communication was less than two. And as you know, this trend only further intensified. What we have now is essentially an epidemic of digital autism. Digital autism is a condition in which young people cannot maintain long-term psychological contact with each other. They are not interested in the inner world of another person. People for them have actually become replaceable, because they do not see the values of each person individually [3]. Someone may say that this is a new civilization, perhaps we should just get used to it. For those who think so, I want to offer the data of a large study of 2018, made on very large samples. She testifies: if you spend more than 2.5–3 h a day on the phone, your indicators of depressive thoughts and suicidal tendencies sharply increase. Social networks in this sense also act very painfully, causing people to feel alienated.

The following graph: we have the opportunity to compare people who use a mobile phone and do without it. An Israeli study showed the presence of attention deficit disorder and hyperactivity all people who use cellular phones. Then they asked people who didn't use smartphones to do it. As a result, they increased social vigilance, aggression, internal stress and compromise [4]. In addition, if a person is not running the default system, so it lacks the distal thinking, he is unable to look ahead and plan for the future.

A famous study in 2017, in which three groups of students performed creative and mental tests. The first group had to leave the phone outside the room where they were tested. The second group was able to keep the phones near you—in your pocket or bag. The representatives of the third group had to put the phone in front of you face down and continue the tests. It turned out that the physical position of the phone affects the volume of operative memory and fluid intelligence. Small column—the phone is on the table, large—is in the other room. You have a better working memory and intelligence, when there is no phone. And you becomes stupid when he is there.

The difference between the classical and the virtual autism is that in the first case we are talking about the destruction and neurological underdevelopment, caused by biological factors. "In General, degradation is irreversible, in these areas no longer formed synapses, in our case, we are talking about restimulation, the development of appropriate neural areas, so the child quickly recovering," said Zamfir.

Statistics from the Center for Disease Control show that:

- in 1975, 1 out of 5,000 children was diagnosed with autism,
- in 2005, 1 out of 500 children was diagnosed with autism,
- in 2014, 1 out of 68 children was diagnosed with autism.

A study conducted by the Center for Autistic Children of Romania showed that over 90% of children aged 2–3 years had triggered (provoking, triggered—ed.). Development of autism spectrum disorders (ASD) over 4–5 h a day television programs or interaction with other forms and types of virtual reality. For adults, even 5 h of watching television and the Internet per day does not become a damaging factor, but for very young children (0–2 years old) electronic gadgets have a devastating effect. In most cases, babies who were left for more than 5 h in front of screens with virtual reality revealed delays in psychomotor development and speech functions, behavior disorders up to very serious ones—ADHD or even autism. This was confirmed in 90% of cases with new appeals of parents with children under 2 years old. Until a certain point, it was believed that adults are more resistant to the influence of the digital world. But this is not so. Andrei Kurpatov, author of scientific works on the methodology of thinking, devotes many of his speeches to this topic. In the book "Halls of the Mind. Kill the idiot in yourself", one of the sections is devoted to the problem of digital autism in adults.

According to his research, human thinking is deeply social, since evolutionarily people united in a group to exchange knowledge. Today everyone can google and find the answer to any question. Communication for knowledge is no longer necessary. Why put up with another person when all the knowledge you can get online. This leads to the alienation of people and a decrease in the density of communication, and as a result, the appearance of autistic symptoms.

"Gadget autism" occupies brain processes, a person ceases to control the time spent on the phone and physical activity is rapidly falling. According to recent studies, 38% of Russians are systematically involved in various forms of physical activity, of which: children 6–12 years old—67%, adolescents and youth—41%, adult population 30–59 years old—36%. It is worth noting that the trend towards a healthy lifestyle is actively approved by society today, but in fact people spend a day sitting behind a digital camera and often in a sitting position.

4 Discussion

Virtual autism is a modern disease affecting tech children. It was formerly called "thumb disease," but the disease mutated into the plane of an even more serious disease. New studies have concluded that most children spending too much time in front of TVs, laptops, and tablets have autism-specific symptoms. The phenomenon is described as "virtual autism" or autism caused by electronic screens, and was called so by the psychologist Marius Zamfir. The brain of modern children is not much different from the brain of our ancestors during the agrarian revolution, when 10 thousand years ago, communities of hunters and gatherers in the Middle East began to domesticate animals and grow cereals. Moreover, human physiology itself has remained the same, despite the phenomenal development of cognitive abilities: the human genome is not much more complex than the genome of a field mouse—2.9 billion nucleotides at the crown of nature, 2.4 billion nucleotides at the stocky rodent.

On the tablet screen, it is technically possible to cut, sculpt, draw and even build a family. But if the child does not do this with his own hands, with different parts of the palm of his hand, with all fingers, then the consequences will be catastrophic—not only the quality of life will suffer, where the townspeople still have to tie shoelaces and apply transport cards to the validators, but also the basis for implementation in society is speech and language skills.

Noam Chomsky included the study of his native language as part of the human genetic program. The child automatically distinguishes and assimilates grammatical constructions from oral speech. The extra hour with a tablet instead of a lively conversation with parents in early childhood is the danger of losing important syntactic structures. He identifies ten dangers that gadgets and modern technologies carry for a child: The physiological development of the brain. The human brain is tripled in the first two years of life. At this time, neural networks are developing that will ensure the operation of complex cognitive processes in the future. The environment sets the tone in the initial development, and the more multifaceted and wider it affects all the senses, the better. Early attachment to a smartphone, tablet and TV will lead to attention deficit disorder and reduce the possibility of self-control.

Lag in development. When a child looks at the screen, his position in space is strictly fixed—on the sofa, in bed, lying on the floor. The less the child moves, the higher the likelihood that he will have problems with basic literacy and mastering the school curriculum—typical problems of modern youth in developed countries. Direct correlation with the lack of physical activity that activates the blood circulation and brain activity of people.

Overweight. TV and video games—on any gadget—have a direct effect on childhood obesity, because it causes a decrease in activity.

Lack of sleep. Modern children are practically unlimited in the use of gadgets in their rooms. Hollywood cemented the image of a gifted child who reads books with a flashlight under the covers while his parents make him go to bed. With smartphones, night gatherings do not look so romantic. Children who spend a few extra hours after midnight with a phone in their hand are the first candidates for declining school performance. Without a healthy sleep, memory and other cognitive abilities will not work at full strength.

Mental disorders. Today, children make their own diagnoses. The era of classical subcultures has passed, and in its place came publics and communities where teenagers with an eating disorder, agoraphobia and imaginary or real depression try to help each other. On the one hand, it may seem that this is a fashion and another attempt to attract attention, but on the other, scientists note the growth of mental disorders in the younger generation and often associate them with dependence on gadgets. Already about 10% of children today take psychotropic drugs.

Aggressive behavior. Often, digital autism manifests itself in aggressive behavior, and there have been cases of death. When "his world" is taken away from the child, the person stands behind him with a mountain. And the children begin to raise their hands at their parents and try to return their things by various power methods.

Digital dementia. For children, this is very dangerous, since it is impossible to build complex neural networks in the brain with attention deficit, poor concentration and memory. An adult can return to the training of their cognitive abilities, when the child must build them literally from scratch.

Dependence. There is a painful attachment to technology—it's hard to just determine the threshold where it sets in. Human time is a valuable resource, and not only your loved ones are fighting for it, but also Internet giants and video game manufacturers.

Radiation. Canadian pediatrics still classifies modern gadgets as possibly dangerous for children. The harm of mobile phones for children today is more social, but the technological side is also under the close supervision of doctors.

Unsustainable consumption pattern. No one knows the norm. There are no strict guidelines according to which one application can be classified as developing or only mimicking a specific children's market. Nutritionists were able to offer basic numbers that people can focus on when choosing products. But the body responds quickly to an unbalanced diet. Psychiatrists do not have such luxury—if the child's brain began to function worse, memory and concentration worsened, then this often becomes noticeable only at the stages when it is time to prescribe medication. Children and gadgets today are developing together, so it is very difficult to isolate the decisive variable in this complex multi-factor model called life in a modern city [6]. The problem is becoming more urgent from year to year—in 90% of cases, parents of children with mental disabilities admitted that their babies spent 4–5 h a day in front of a TV, mobile phone or tablet. And this means that information technology.

5 Conclusion

Digital technology is a great tool for work but for everything else, should be cautious spending time with a gadget, and remember that there are people around with whom communication socializing we and does not turn into a "robot" covered by the digitalization. Communication is a great trainer for our thinking and the brain [1]. There is a memo about "digital hygiene". It was developed in order to avoid autism.

1. Disable your smartphone's sound, when possible.
2. Disable pop-up notifications on the smartphone. They are specially made to attract attention, but save it for more important things.
3. Tell yourself—you don't have to always be connected.
4. Not to carry a smartphone around the apartment. Determine the place where he always is, and keep him there. If you need a reason, you know where to find it.
5. Use the corresponding programs and watch your screen time. If you are not able to control yourself, use a program that normalized your screen time.
6. Remove from your smartphone especially "toxic" apps that kill your time.
7. Learn how to start your morning without a smartphone. First make what you need, and then you can check the phone. Remember: in the morning you must come in yourself, not in a smartphone.

8. Do not consume information from screens an hour before sleep and do not place the phone next to the bed. As an alarm clock use the alarm clock…. The book before going to sleep—much better than any information from the screen (it's a scientifically proven fact).
9. Remember that in addition to the alarm clocks in this world still have pens, pencils, notebooks, diaries, etc. do Not try to write to telephone: normal writing by hand has a positive effect on your psyche and increases the efficiency.

Digital autism today is growing rapidly, and an increasing number of people are unable to build communication outside gadgetboy system. They turn in on themselves and difficult susceptible to external changes. Children of the digital age are faster to adapt to innovations, but it is important not to forget about the world around us and observe not like about the rules to protect yourself from digital autism. Add physical activity, walking whether running normalizes blood circulation and does not saturate the cells with oxygen. It promotes analytical thinking and creativity. To evolve in the digital age need is also wise to not become a hostage to the "information miracle".

References

1. Aagaard T, Lund A (2019) Digital agency in higher education. Routledge, London. https://doi.org/10.4324/9780429020629-3
2. Aleksina AO, Chernova DV, Ivanova LA, Aleksin AY, Piskaykina MN (2019) The main directions in informatization of the sphere of physical culture and sports services. In: Popkova E, Ostrovskaya V (eds) Perspectives on the use of new information and communication technology (ICT) in the modern economy. Advances in intelligent systems and computing, vol 726. Springer, Cham, pp 473–479. https://doi.org/10.1007/978-3-319-90835-9_56
3. Cendrine M (2019) Measuring cognitive availability of young people with autism using digital agendas. Psychol Psychol Res Int J 4(6):000231
4. Drakopoulou E (2018) Autism & digital tablets. MOJ Addict Med Therapy 5(4):192–193. https://doi.org/10.15406/mojamt.2018.05.00117
5. Hurley-Hanson AE, Giannantonio CA, Griffiths AJ (2020) Autism in the workplace. Palgrave Macmillan, Cham. https://doi.org/10.1007/978-3-030-29049-8_10
6. Khaustov AV (2017) Autism & developmental disorders. Autism Dev Disord 15(1):3. https://doi.org/10.17759/autdd.2017150109

Socio-Cultural Consequences of Digitalization

Problems of Digital Processing of Documents in Interaction Between Tax Relations Parties

A. V. Azarkhin

Abstract The problem of using electronic document management in tax relations is quite relevant nowadays. Its relevance is expressed as follows. Now docflow between tax relations parties are provided by a variety of programs and services. The emergence and development of these services reflected positively on tax relations. However, these programs and services have both pros and cons in use and application. In this paper, the author attempts to analyze existing programs and services that make docflow possible for tax relations parties. In this paper, using the analysis method, we analyzed the existing programs and systems with which the relations between the parties are implemented. The results of this study are suggestions for resolving existing problems.

Keywords Document management · Electronic document management · Taxes · Tax relations · Parties

1 Introduction

It is necessary to mention that document management these days is an integral part of the work of individuals and legal entities, individual entrepreneurs, and government bodies. And tax authorities are sure no exception. However, one of the main tasks, the solution of which will improve the reliability and efficiency of the economic entity, is the effective organization of docflow. Considering the digitalization of the economy, it is important to improve mechanisms of fixation, storage and transmission of information, which can also eliminate the shortcomings of traditional ways of managing document processing [4]. All stages of the development of the state assumed the solution of problems consisting in ensuring the timely receipt of tax revenues in budgets, balancing the budget system, and reducing the deficit of external and internal debt. An important issue is also the increase in tax collection in all budgets

A. V. Azarkhin (✉)
Samara State University of Economics, Samara, Russia
e-mail: aazarkhin@mail.ru

Samara Law Institute of Federal Penitentiary Service of Russia, Samara, Russia

S. I. Ashmarina and V. V. Mantulenko (eds.), *Current Achievements, Challenges and Digital Chances of Knowledge Based Economy*, Lecture Notes in Networks and Systems 133, https://doi.org/10.1007/978-3-030-47458-4_74

of the budget system of the Russian Federation. This issue is one of the prior. In turn, since information and digital technologies are being introduced in all spheres of life, there is a clear need for automation of activities, including those of tax authorities.

2 Methodology

In this research, the author used the analytical method: available programs and services, and programs data were analyzed. The positive and negative features of the considered program documents were also studied. The author determined possible ways to solve the identified problems.

3 Results

The process of applying information technologies, including the creation of automated systems and databases for the effective functioning of tax authorities, effective interaction between tax relations parties is one of the priority tasks of the Federal Tax Service of Russia.

Let us characterize the tax authorities programs:

1. The program "Taxpayer". Using this program it is possible to draw up:
   - tax and accounting reporting,
   - calculation of insurance premiums,
   - tax statements,
   - statements related to taxpayers,
   - notifications on controlled transactions,
   - requests for information services, etc.
2. The program "Declaration". The specified makes it possible to automatically generate tax statements and returns using specified forms. In addition, it can be used to verify the correctness of the written declaration and reduce the likelihood of errors.
3. Service "Personal account of a taxpayer for individuals". This service allows the taxpayer:
   - monitor the status of settlements with the budget,
   - receive tax notices and receipts,
   - pay arrears and payments,
   - download programs to fill out a statement,
   - submit appeals to the tax authorities, without making a personal visit to the tax office [10].

4. Service "Preparation of a package of electronic documents for state registration". This service helps to prepare a complete set of documents that are necessary in order to create an LLC. At the same time, the service independently forms all the documents that are necessary for state registration (decision, charter, application, payment), which only need to be signed with an electronic signature and can be sent electronically to the registration authority, or in any other way. The result of work with the service is sent electronically to the email address, and if desired, these documents can also be obtained on paper. In the same way, you can submit documents to register as an individual entrepreneur, as well as to make changes to information about a legal entity and an individual, including those on termination of activity.
5. Service for obtaining information. Using this service, you can get extracts from the Unified State Register of Legal Entities and Individual entrepreneurs in electronic form; one can also obtain data from the register of disqualified persons; check arbitration manager, etc.

All of the above programs and services make it possible to receive information in a more efficient way, send applications by means of electronic methods of interaction, carry out activities on registration, change and terminate organization's and individual entrepreneur's data [1].

The listed positive aspects do not exclude the presence of negative aspects, which also need to be characterized.

Typical disadvantages include the following:

- most often an electronic document is tied to a specific information system and their migration to another information system is quite complicated or completely impossible,
- in relation to the programs used, there is a problem of updating these programs. The update implies the need for support and maintenance, which directly affects the cost of these programs and their maintenance,
- lack of control by the tax authority of those operations that are performed in the systems,
- complex software interface of electronic systems,
- the need for an electronic digital signature for legal entities and individuals to submit electronic applications and appeals.

Improving the tax system in the context of economic development requires the further development of a mechanism for recording, storing and transmitting information, which will increase the efficiency of tax administration, reduce the cost of tax measures, which will result in the ability to prevent tax violations, reduce the level of tax crime, etc.

4 Discussion

We consider it necessary to analyze possible solutions to the mentioned issues.

1. One of the proposals is the creation of so-called automatic technologies that allow timely processing of documents and make formal tax administration procedures more automated. This technology will make the process of calculating taxes, making payments, collecting arrears, etc. more automated. Thanks to the above, it will be possible to transfer several functions that are routine and bring formalized decision-making from inspections to information systems to a higher level.
2. Further development of the personal account service. Its further development will give taxpayers the opportunity not only to use certain information systems, but to use all the services provided by tax authorities.
3. A separate proposal can highlight a more progressive system of performance evaluation, which will be possible because the processes will be as automated as possible. The above, in turn, will help to consolidate internal and external information on the activities of tax authorities, which in turn will attract people who want to contact the tax authority using electronic document management.
4. The development of legislative acts regulating the procedure by which electronic document circulation between participants in tax relations will be carried out.
5. The implementation of electronic digital signatures. In order to implement this method of solving existing problems, it is necessary to identify trusted certification authorities that issue signature key certificates in order to use them in information interaction, and also to fix the order and sequence of how these keys will be issued.
6. To carry out training for personnel in tax authorities, as well as for taxpayers, on how to work with the tools and services that are available and are created, through which the docflow is implemented.
7. It is also important to legally regulate obligations of such tax services related to fixing the date when the taxpayer submits a declaration (calculation) and the specified documents are sent to all participants of electronic document management.
8. The next way to solve existing problems is to put a new way of organizing docflow into practice, which I would like to dwell on in more detail.

The basis of this method of organizing docflow is the technology of Blockchain—digital docflow. This method involves the organization of work with documents, including the creation, processing and storage of information by using a combination of automated processes. The use of smart contracts involves a single information environment in which the document is present throughout its entire life cycle, which makes it possible to track all operations performed [3, 7].

In a general sense, it can be said that a smart contract is a computer program through which obligations are monitored and ensured. The parties register all conditions, obligations and responsibilities. Also, it determines whether everything is executed and a decision is made: to complete the transaction, issue the required, impose a fine and penalty, close access to assets.

It is quite possible to use the example of smart contracts in electronic document flow between participants in tax relations. For example, to track the direction of documents, as well as all attached documents, take into account the date of receipt, in order to take into account the deadlines, if the documents are not submitted, are not presented in full, mark the error as indicated and notify the person who filed the application or appeal etc. Of course, as in any process, the process of applying smart contracts has both pros and cons. The cons include:

- lack of legal regulation. In fact, the use of smart contracts in the Russian Federation is not regulated, which can lead to a variety of problems,
- mandatory involvement of IT-specialists in order to create contracts in case of disputes over such,
- high costs for creating a smart contract,
- a rather high degree of vulnerability, since the user's device and the key record can be lost or hacked.

Despite the large number of negative aspects, we can also highlight the positive aspects of using smart contract technology between participants in tax relations:

- protection against unauthorized changes,
- system transparency (you can track all the stages and stages of the passage of electronic document management),
- confidentiality,
- high speed,
- self-fulfillment.

Smart contracts have the same advantages as electronic document management systems. However, this system will significantly reduce the burden on economic entities, while eliminating the formation and submission of tax reports, as well as timely tracking changes in tax legislation. In this case, the work of the tax authorities will become more effective because it will be possible to generate reports based on the information of smart contracts that are stored on the Blockchain and are confirmed. When organizing electronic document management according to the specified scheme, information will be transferred from the sender system to the recipient system via a secure communication channel using an electronic digital signature. The intermediary transfer will be the operator of the electronic document management system [8].

At the same time, the information transfer intermediary will not be able to influence the process of creating, processing and storing the document. In this case, the workflow will be the most effective and more organized. In foreign countries, Blockchain technology is used and developed quite widely.

For example, in the USA, Blockchain was developed in order to assist in the use of federal executive bodies (agencies) [6]. This technology is useful to document management professionals, since it discusses the possible consequences for federal electronic document management programs. The main attention is paid to the fundamental problems of document management, namely, issues of the existence of documents in the Blockchain, the integrity of documents, the establishment and monitoring of the shelf life of documents, as well as the transfer of documents for storage. The specified system is used in the US National Archives.

Foreign researchers also note that Blockchain technology can significantly improve existing technological applications, as well as apply new ones that have never been used in practice [5]. Blockchain, also known as distributed ledger technology, is capable of revolutionizing global economic changes, since it is unchangeable, transparent and provides trust, security, speed, reliability and transparency of decisions, both public and private [9].

The Office of Financial Regulation and Supervision of Great Britain sees the potential in the application of Blockchain technology, for example, a system is being developed to automate the reporting to the financial regulator based on the Blockchain (Project Maison) [2].

5 Conclusion

Electronic services related to tax relations are expected to undergo a quite serious development, since it is facilitated by an increase in Internet users, support for the need to develop electronic document management. Despite all the advantages and benefits, there is a number of problems in this matter that also need to be regulated and addressed. This will allow businesses and citizens to switch to electronic document management in tax relations, as well as to solve one of the most important and priority tasks for tax authorities—to create for taxpayers all the necessary conditions so that they can fully exercise their rights and fulfill all tax obligations. To summarize everything that was said above, it can be noted that the process of introducing new programs and services into the work of the tax authorities, as well as developing and improving existing ones, will enable all parties to have undeniable advantages, which subsequently will have a positive effect on the overall quality tax relations.

References

1. Barinov AYa, Mayburov IA, Ivanov YuB (2017) Prospects for improving the tax system. Innov Econ Dev 4(40):7–16 (in Russian)
2. Financial Conduct Authority (2017) Our work programme. https://www.fca.org.uk/firms/regtech/our-work-programme. Accessed 27 Jan 2020
3. Kardonov AV (2018) Scopes of application of smart contracts and risks at working with them. Bus Educ Econ Knowl 1:44–47 (in Russian)
4. Kosarin SP, Lebedeva YuA, Milkina IV (2019) Modernization of the information system of the tax authorities of the Russian Federation. E-Management 2(1):42–51 (in Russian)
5. Locurcio L (2017) Comment la Blockchain révolutionnera l'entreprise. Gérer, Prevoir, Optimiser: Le Magazine Des Dirigeants D'entreprises 98:42–44
6. National Archives and Records Administration (2019) Blockchain white paper. https://www.archives.gov/files/records-mgmt/policy/nara-blockchain-whitepaper.pdf. Accessed 05 Feb 2020

7. Nazarov MA, Mikhaleva OL, Chernousova KS (2020) Digital transformation of tax administration. In: Ashmarina S, Vochozka M, Mantulenko V (eds) Digital age: chances, challenges and future, ISCDTE 2019. Lecture notes in networks and systems, vol 84. Springer, Cham, pp 144–149. https://doi.org/10.1007/978-3-030-27015-5_18
8. Savelyev AI (2017) Some legal aspects of the use of smart contracts and blockchain technologies in Russian law. Law 5:94–118 (in Russian)
9. Stančić H (2018) New technologies applicable to document and records management: blockchain. Revista Catalana D'arxivística 41:56–72
10. Yunusova AK, Kashirina MV (2017) On the issue of the functioning of the personal account of a taxpayer in modern tax administration. Symb Sci 1(3):158–165

Technology Transformation in Education: Consequences of Digitalization

T. G. Bondarenko, T. P. Maksimova, and O. A. Zhdanova

Abstract In the article, the authors highlight current trends that are changing the landscape of the educational process and teaching technologies, as well as related methodological tasks and challenges that the teaching community faces. The active pedagogical and digital technologies considered in this contribution are presented as a joint tool of the educational environment, which makes it possible to implement principles of continuous education, build individual educational routes, and organize mixed learning. The issue of building an infrastructure related to the informatization at the same time opens up opportunities for the transition to a new level of education, and the application of active methods is aimed at training professionals who are guaranteed demand in the labor market in terms of their professional competences, and who easily use mobile and Internet technology that simultaneously develops their cultural competences.

Keywords Active methods · Learning process · Competencies · Best practices · Educational platforms

1 Introduction

The modern world, namely the integration processes that occur in all areas of life, shows that the development directions of scientific and cultural space are determined by the conditions of both globalization and glocalization, which, in turn, requires constant attention to the issues of adaptability of education (in a broad sense) to changing circumstances. Accordingly, the intensification of global integration processes and Russia's growing immersion in them cause the need for training a new generation of

T. G. Bondarenko · T. P. Maksimova (✉) · O. A. Zhdanova
Plekhanov Russian University of Economics, Moscow, Russia
e-mail: maksimova.tp@rea.ru

T. G. Bondarenko
e-mail: bondarenko.tg@rea.ru

O. A. Zhdanova
e-mail: zhdanova.oa@rea.ru

S. I. Ashmarina and V. V. Mantulenko (eds.), *Current Achievements, Challenges and Digital Chances of Knowledge Based Economy*, Lecture Notes in Networks and Systems 133, https://doi.org/10.1007/978-3-030-47458-4_75

research analysts and managers with the mentality of professionals who are able to think globally, to identify and effectively use opportunities of the rapidly changing business environment and are ready to respond adequately and in a timely manner to its challenges.

Effective education is an education that provides a solid basis (the basis of knowledge and search methods for obtaining and expanding it) and can be used in the future when solving tasks through the development of meta-subject relations in new, completely unrelated situations. Integration processes allow us to revise the concept of "learning" not only as a theory whose boundaries are defined by academic frameworks, where students' achievements are evaluated and their ability to recall previously received information, transmitted details, and the sequence of performing any traditional steps to solve conditional tasks.

In modern conditions, the transition *from skills and abilities* to remember, act and perform steps set by the standards *to competencies* requires the development of intersubject and metasubject relations, the ability to jump between related scientific subjects and use algorithmic knowledge and skills in solving problems in intersubject areas. This strategy of teaching and learning development should help to improve the competences of students, which is impossible without highlighting the specific attributes of the modern system of methods and techniques of teaching and learning, which are accompanied by a revolutionary development round characterized by artificial intelligence and digital economy.

2 Methodology

In the research process, general scientific methods were used: observation, description and comparison, conversations, and questionnaires (empirical methods); analysis of psychological and pedagogical literature, abstraction, analysis, logical and historical methods (theoretical methods). Private empirical methods were supplemented by general methods—pedagogical experiment, including diagnostics, and experimental training.

As a part of the literature analysis, the works of Russian and foreign scientists were studied, reflecting the main theoretical, methodological and practical approaches to the problems under consideration:

- research of modern approaches to the organization of professional education [7],
- works on the problem of designing modern pedagogical technologies [9],
- study on the problem of creating a digital educational space in the education system organization [8],
- research on the problem of organizational forms and methods that activate cognitive activity in modern conditions [10],
- materials of international scientific and practical electronic resources, including educational electronic resources, which also reflect the issues of interaction between educational institutions, science and practice (business community) in order to build effective models of professional education [5].

3 Results

From the analysis of the research results on the problems of transformation of modern pedagogy and methodology of the educational process, it can be concluded that the most popular active methods (in addition to the traditional model of training students through the organization of independent work) are the use of an advisory form by teachers, holding group discussions and organizing the discussion of essays. At the same time, it is necessary to focus on those forms that are stated only in a few work programs of the taught disciplines, annotations and educational programs, in order to enable them to form a situation of success in the learning process as a mechanism for restoring motivation to education and to form a need for self-education:

- interactive lectures and laboratory work,
- designing different models,
- projects and project-based learning,
- round tables,
- working with presentations, used in reality as a group reporting work, presented in many work programs,
- the case method,
- conducting scientific and practical research,
- discussion of reports.

Separately, it is necessary to emphasize the wider use of the following active teaching methods in the financial block disciplines, which allow promoting the development of masters' ability to argue their beliefs:

- scientific discussion,
- calculation and analytical tasks.

The most important competencies of graduates of educational programs will include the following:

- activity to plan (scientific aspect),
- skills to conduct an analysis (research aspect),
- ability to use information correctly (scientific aspect),
- ability to work in a team (professional aspect),
- skill to manage relationships (professional aspect),
- focus on results (scientific and professional aspects),
- ability to manage solutions (scientific and professional aspects).

Thus, the modernization of approaches to learning through the concept of effective interaction is a closed socio-economic structure with an active (key) role of the listener, which supports the possibility of rapid application of any corrective elements according to changing market conditions.

The study revealed the following fact: often students of any educational programs note their lack of motivation for the implementation of scientific and creative professional activities, both because of the lack of understanding of its essence and purpose in the framework of training for an educational program, and unwillingness to be engaged in it in the future.

The idea of switching to the maximum use of active methods in education is to modernize the structure of educational programs based on market requirements on the one hand, as well as personal interests of students, on the other hand, in order to understand the approach to developing a model of competencies that can be applied to a specific professional industry (economy) and provide an opportunity to build individual educational trajectories. Speaking about the applicability of the use of digital technologies in the educational process in the industrial aspect, the project approach can be used when preparing tasks that correlate with the implementation of national projects and priority state programs, for example, such a business case is described by Larina et al. [4].

Ways of effective use of active methods in the implementation of the educational process allow organizing an interactive interaction between the teacher and students, establishing a friendly, stimulating to active independent cognitive activity atmosphere in the classroom. Among the actual aspects of the modern problem of interaction between the individual and society, one of the most important is the question of timely formation of adaptive abilities in a person.

Pedagogical activity involves the presence of at least two sides: objective—a set of methods and techniques that the teacher traditionally uses, and personal—how the teacher uses these methods and techniques depending on his personal qualities and abilities. Today, we can say that VUCA, as a new environment for managing knowledge, talents, and personnel, is the most appropriate tool for integrating the educational process into the integration processes of our society (Fig. 1).

Education in the world of FINTECH innovation can be described as:

- high-tech, design,
- continuous, life long learning,
- with the educational biography and a unique educational route,
- digital transformation and the role of educational platforms shapes "education" as a life style,
- we are all students and teachers of the future skills,
- the development of educational startups requires increased investments in education.

It is particularly important to note the role of smartphones and tablets, which have significantly strengthened and accelerated the integration processes of the modern society [3]. Thanks to these gadgets, the emphasis in the formats of the educational process has shifted to such as:

- BYOD (Bring Your Own Device—"Bring your own device"). Students are encouraged to use their gadget in the classroom to search for information, watch video-materials,
- The format of the "flipped classroom". Learning the material and completing tasks takes place in the opposite sequence compared with the standard lesson.
- E-learning. These are no longer separate technologies, but the creation of a holistic environment that includes learning using mobile devices (mobile learning, or m-learning) and learning based on Internet technologies (web-based training, WBT),

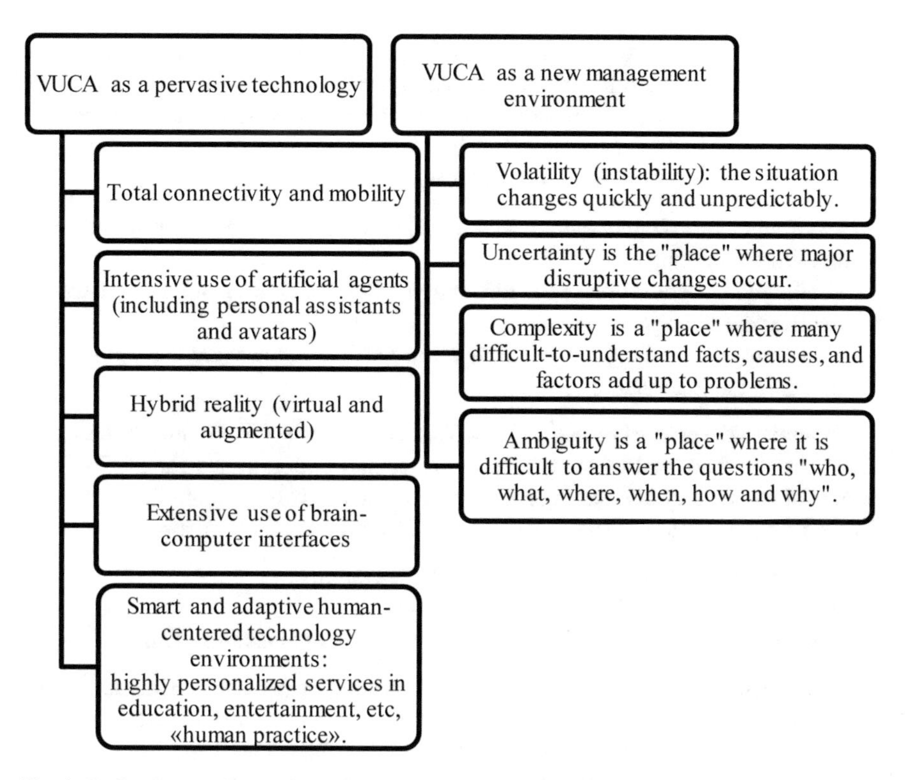

Fig. 1 Technology and education: VUCA as a norm (*Source* authors)

- MOOC platforms (Massive open online courses, MOOC) are a training courses with mass interactive participation. MOOCS help to create and support communities of students, teachers, and assistants.

An example of the impact of integration processes in the financial sector on education is the emergence of required future competencies within alternative currency systems (for example, cryptocurrencies such as bitcoin), the reputation (merit) economy (non-monetary commodity-money exchange, where a person's reputation in the community, including their experience, relations, achievements, etc.) becomes a currency, and direct investment in talented people (investing in someone else's education and projects on a paid basis).

Based on a detailed analysis of the Atlas of Emerging Jobs, we can say that the skills of the 21st century are: ability to learn, unlearn, and retrain during the whole lifetime, self-organization, concentration and attention management, collaboration (as a critical skill that should be embedded in different aspects of work and training), thinking: critical, problem-oriented, systemic, flexibility and adaptability.

For their formation in 2018, 17 educational services were available for students of different ages, which made the education process mass and accessible. Let's look at some educational projects that were able to integrate education and FINTECH innovations. Fintech Observatory Project. This project launched in Spain in 2012 to monitor and disseminate information about significant innovations in the banking and financial systems. Its main objective is to raise awareness of the opportunities currently offered by financial innovators by identifying needs, offering solutions and best practices wherever they are offered. Education and Fintech today are:

- development and implementation of training courses on banking, insurance and investment products,
- gamification of education in the financial field on examples from the real sector,
- involvement in projects,
- conducting and participating in simulation-business games and championships, including international ones [6].

Fundraising school (within the framework of the project V. Potanin "Charitable Foundation"). The goal of the project is to expand opportunities for professional growth of students by conducting educational events on ways to publicly raise funds for the implementation of student initiatives. The project objective are:

- achieving goals of the university student associations,
- implementation of student association programs,
- strengthening relations between university student organizations [2].

School games in education. The game technologies project is a model of the technology market. It clearly shows how technologies move from an idea to a finished product and what difficulties the project team may face. The game has investors, accelerators, employees, technologies at different stages of their readiness, labor market, auction, technology markets, and teams. The economy and balance of the game copy the real picture in the technology market [1].

Thus, the general patterns of influence of integration processes, including global ones, on modern learning processes are:

- technologies—a tool that provides many people with equal access to obtaining skills,
- mass knowledge and skills will be transferred primarily through automated solutions,
- the learning process is becoming more mass and more total,
- learning becomes truly individualized,
- the principle of "I can—I know" works.

4 Discussion

Mandatory attributes of the modern education in the context of the integration process are: 1. Understanding the process of how students learn (providing timely, specific feedback to the learner, tracking and recording the learning progress, starting with the learner's self-assessment); 2. Compliance with the principles of critical thinking in training and teaching (support for the acquisition of knowledge, skills of professional behavior in the learning process, providing timely, specific feedback to students); 3. Facilitating rather than controlling learning (individualization of the learning trajectory, expected and timely increase in complexity as the student progresses throughout the program along the "proposed" trajectory of the educational process), 4. Effective promotion of the competence approach through the formation of competence-based educational programs that meet the requirements of normative and instructional documents, including those implemented in various formats, for example, through open educational resources, using video materials, recorded lectures, and project training. Many foreign programs are already designed for students who have chosen an individual educational trajectory, implementing it at their own pace, rather than at the pace dictated by semesters or credit hours. This approach means that motivated and effective students can complete a dedicated educational level in less time, and vice versa.

The process of studying disciplines of the chosen educational program is aimed at forming a set of competencies, which allows us to talk about such final holistic results as the ability to use the information (knowledge, skills) for decision-making, which is achieved by using the following forms and methods of training that contribute to the formation and development of competencies, such as a lecture, essay, abstract, presentation, seminar, case study, test task, didactic game, control work, analytical reports, information reviews, practices, research work, as well as various types of independent work of students on the instructions of the teacher. That is, both active and interactive forms of work are used nowadays.

Integration processes occurring in the society and affecting scientific and cultural space—reduction of the autonomy of universities, a new world view, refusing the industrial model of the society and the transition to new models of production and logistic systems, the emergence of a promising professions card—form identical requirements for educational programs, but taking into account the fact that individual educational program is the institutional core of training and education.

5 Conclusion

Education and educational processes are under enormous pressure as a result of technological innovations such as radio lessons, remote lectures, distance learning, automatic machines (robots), and developing automation and autonomization technologies (robotics, the Internet of things, and artificial intelligence). Their influence on educational activities allows us to speak about the emergence of the VUCA world as an educational norm in the world of the future.

Modern technologies in the context of globalization and glocalization change education on a daily basis, which is reflected in the applicability and need for constant monitoring and adaptation of new technologies in the process of training, education, as well as the subsequent assessment of consequences of their implementation in educational activities. The skills of the future that are formed on modern educational platforms and meet the requirements of reality allow being throughout the life in the online and offline educational environment, while simultaneously plotting individual educational trajectories.

References

1. Barr M (2017) Video games can develop graduate skills in higher education students. Comput Educ 113:86–97. https://doi.org/10.1016/j.compedu.2017.05.016
2. Clark DB, Tanner-Smith EE, Killingsworth SS (2016) Digital games, design, and learning: a systematic review and meta-analysis. Rev Educ Res 86(1):79–122
3. Ernst & Young (2017) Penetration of financial and technological services in megacities of Russia and in the world. http://www.ey.com/Publication/vwLUAssets/EY-fintech-index-russia-rus-2017/%24FILE/EY-fintech-index-russia-rus-2017.pdf. Accessed 01 Feb 2020
4. Larina TN, Zavodchikov ND, Larin PE (2019) Agro-economic potential and prospects of implementation of precision farming technology in Russian regions. In: Soliman KS (ed) Proceedings of the 33rd international business information management association conference: education excellence and innovation management through vision 2020. IBIMA, Spain, pp 3437–3445
5. Lytras MD, Daniela L, Visvizi A (2018) Enhancing knowledge discovery and innovation in the digital era. IGI Global, Hershey
6. McKinsey & Company (2018) Fintech decoded: capturing the opportunity in capital markets infrastructure. https://www.mckinsey.com/industries/financial-services/our-insights/fintech-decoded-the-capital-markets-infrastructure-opportunity. Accessed 01 Feb 2020
7. Milovanov KYu, Nikitina EE, Sokolova NL, Sergeeva MG (2017) The creative potential of museum pedagogy within the modern society. Espacios 38(40):27
8. Santo R, Ching D, Peppler K, Hoadley C (2016) Working in the open: lessons from open source on building innovation networks in education. Horizon 24(3):280–295. https://doi.org/10.1108/OTH-05-2016-0025
9. Scharon CJ (2016) Long-term impacts of a museum school experience on science identity. PhD thesis. University of Washington, Washington
10. Swanger D (2016) Innovation in higher education: can colleges really change? www.fmcc.edu/about/files/2016/06/Innovation-in-Higher-Education.pdf. Accessed 01 Feb 2020

Using Innovative Technologies During Sports Training in the Additional Education

A. M. Danilova and A. D. Voronin

Abstract The article discusses the importance of using innovative technologies in the training process to achieve high results by athletes. The authors give examples of a number of mobile applications designed both for professional and non-professional sports, contributing to the development of physiological, cognitive and technical qualities in athletes. The authors also developed a methodology for the development and improvement of technical and tactical abilities in athletes using multimedia support in the process of sports training and subsequent reflection of the results of this process. In addition, that's proof of effectiveness using of mobile applications and other multimedia support tools to make the success of athletes in the educational process.

Keywords Sport · Health · Innovative technologies · Athletes · Education · Mobile applications

1 Introduction

At the moment, it is no secret to anyone there is a rapid development of innovative technologies in the modern world which affects all spheres of life, in particular sports and health. According to the fact that the standard of living of people is very actively developing, physical activity, and sports are becoming indispensable elements in the life of each of us. Thanks to the rapid development of world sports of the highest achievements, science in the field of sports began to develop very actively [1].

Nowadays, new pedagogical technologies are being developed and introduced to build multi-year training for athletes using the latest multimedia and communication tools. Improving the structure of the process of sports training is one of the main

A. M. Danilova (✉)
Samara State University of Economics, Samara, Russia
e-mail: vitr@list.ru

A. D. Voronin
Samara State Technical University, Samara, Russia
e-mail: rusasha-voronin-1994@mail.ru

S. I. Ashmarina and V. V. Mantulenko (eds.), *Current Achievements, Challenges and Digital Chances of Knowledge Based Economy*, Lecture Notes in Networks and Systems 133, https://doi.org/10.1007/978-3-030-47458-4_76

tasks of the concept of the Federal Target Program for the Development of Physical Culture and Sports. It is possible to increase the effectiveness of the training process of sports training and achieve a higher level of improvement of the mental, technical, tactical and physical potential of athletes through the development and subsequent introduction of new educational methods. However, even "introducing new innovative technologies, one can hardly surprise anyone with various programs that are directly designed to control both functional changes in the human body" [4, p. 20], and his mental abilities or psychological qualities.

2 Methodology

Over the past decade, the use of multimedia equipment has become a prerequisite for the process of sports training. Often the use of multimedia equipment came down to simply watching a video on the monitor, which shows successful technical actions, techniques of professional athletes, a student filmed on camera from certain competitions. Unfortunately, this condition can't always guarantee the development and improvement of technical and tactical abilities in adolescents in the process of playing sports. According to Makarov and Sysoev "this is due to the fact that in adolescence, athletes during the training process can't immediately see, hear and perform the action" [3, p. 113]. In addition, it is "difficult to evaluate and correlate their technical and tactical actions with the actions of professional athletes. So it is very important to carry out the reflection of technical and motor actions after studying and mastering the technical device" [3, p. 113].

Today, with the advent of the era of modern technology, the need for acquiring bulky and inconvenient technical equipment has disappeared. At the moment, such technical devices as computers, laptops and even tablets are becoming less popular in the field of professional sports. Smartphones, smart watches, fitness bracelets and smart clothes are already coming to the first place. In order to shoot and watch a video with the athlete's performance, an ordinary camera phone will suffice. And smartphones of the latest generation have built-in cameras that can record and play video in 4K format with a frequency of 30–60 frames per second. That is, in order to study in detail the technical effect and whether the reception performed by the athlete, the video has already been captured and recorded on the phone with high definition resolution and it can be stopped, slowed down, scaled up with the touch of a hand.

Thanks to such multimedia support, the trainer can individually indicate to everyone his mistakes and explain in detail what is required of him at each stage of the technical action. If the situation requires it, repeatedly repeat and work out the technique he needs. Thus, the principles of differentiability, individualization and educational reflection of the process of sports training are implemented. In addition, the trainer-teacher can send the video to the student through instant messengers for the self-sustained analysis by the athlete of the actions he performed, both verbally and in writing forms. For more successful reflective analysis, an athlete can be helped by certain mobile applications installed on gadgets, which can also be attributed to multimedia support. For example, using the Hudl Technique or Coach's eye—Video-analysis applications, a teenager will be able to compare his techniques with the same techniques performed by professional athletes by uploading two videos (one with his own technique and one with a professional technician) and simultaneously viewing with stops and highlighting your shortcomings. To properly perform exercises that develop physical strength, you can use the Keelo Lift application. Using this application, you can instantly view and study how correctly an athlete performs one or another exercise during the educational process. In addition to mobile applications that help to improve their skills in professional sports, there are a number of those aimed at promoting a healthy lifestyle and developing physical qualities during non-professional sports. Basically, these applications focus on the person involved in sports on the correction of physical activity, optimizing sleep and wakefulness and diet tracking. In other words, "the application intended to process the set of mobile phone owners—such as weight reduction, increased muscle mass, control motion activity,—to flow more effectively and accurately and provide desired result" [6, p. 112]. "Data are processed on kilometers traveled, weight, calories burned, total weight raised, number of approaches, exercise duration. Monitoring your own progress is a serious motivator for maintaining physical activity, which helps to accustom users to the training regimen" [2, p. 37].

Some time ago, in the sports, there were no such mobile applications. All the necessary information about maintaining a healthy lifestyle, the proper construction of the training process, as well as adhering to dietary regimes and the day in general, our citizens could receive in the framework of the state system of physical education (comprehensive schools, vocational education institutions, additional education institutions, sports schools, voluntary sports societies) or in commercial gyms and fitness clubs. Most of the population didn't seek help from such organizations due to various factors: lack of literacy, sociability, low motivation for playing sports or financial difficulties. In addition each time increasing the number of false information provided in the Internet (which has become one of the most important sources of information for the modern man). And due to the incorrect construction of the training process, independent exercise in physical exercises can be not only non-effective enough, but also can lead to negative consequences—overwork, injuries.

3 Results

Science in modern time is working hard on the problem of creating a model of an effective training process, since without a precisely set goal many of the tasks will simply be impossible to solve. One of these tasks is to prepare the physical, intellectual, and technical potential of professional athletes.

According to our opinion, for the quality of technical preparation for and athletes at every stage of their sports development should be carried out as a self-control, so the collection and analysis of data and discussion training-teaching staff.

To this end, we have developed a methodology for the development and improvement of technical and tactical abilities in athletes using multimedia support tools in the process of sports training and subsequent reflection of the results of this process (Fig. 1).

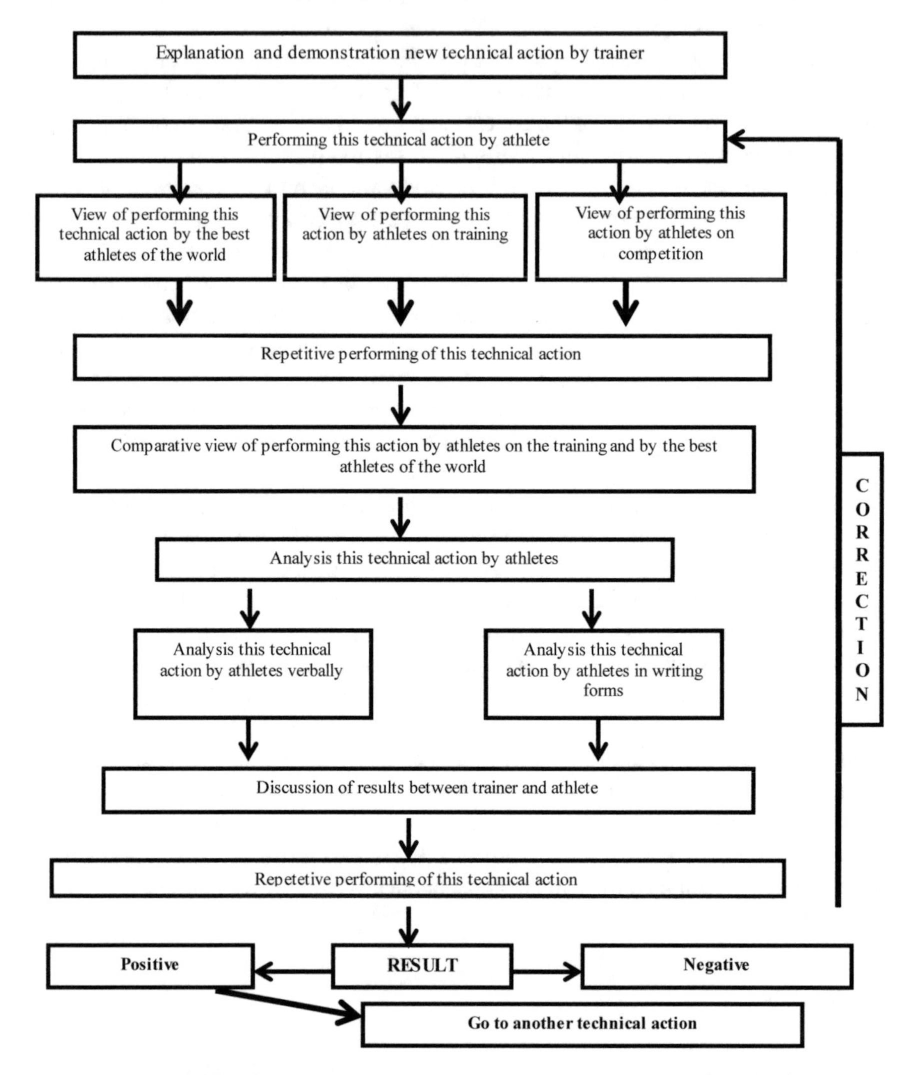

Fig. 1 An algorithm for the development and improvement of technical and tactical preparedness of teenage schoolchildren in the process of sports training using methods of reflective analysis and multimedia accompaniment (*Source* authors)

This algorithm assumes compliance with a certain step-by-step structure:

- the study by athletes of technical action using video materials,
- awareness by athletes of the technical actions shown in the video, the actions of other athletes, their actions using reflective analysis. The reflection is provided in oral and written forms with the obligatory further discussion of its results with the trainer-teacher.

4 Discussion

A trainer needs to maintain an interest in training and in mastering the techniques and technical actions of athletes. In addition, taking into account their individual capabilities and abilities. Athletes, in turn, improve independence in making decisions during the development of technical techniques; then, together with the coach, they monitor the results of the work done [5]. Due to these conditions, athletes will successfully formulate technical and tactical training and the level of its development will increase significantly. The lack of conditions for the development and improvement of technical and technical indicators of athletes in the process of sports training using the reflection method forms an incomplete algorithm of actions for them that doesn't consider the specifics of competition. As a result, during performances at competitions, the actions of the competing athletes, the rivals quickly adapt and significantly reduce the effectiveness of their technical and motor actions by using countermeasures.

There are a number of reasons why the use of mobile applications in the field of sports, both professional and non-professional, will be useful for solving its tasks: interactive learning by transferring the basic knowledge of various training programs, storage of training programs in electronic form; used as automation control, correction of the results of motor activity; these application allow optimizing the processes of testing the physical condition of the practitioner and technical readiness; they carry out operational collection and processing of information about private motor activity, including visualization of the current state of the athlete's body and the dynamics of his data.

5 Conclusion

Based on all provided information, we can conclude that the use of multimedia equipment in conjunction with the reflection method in the process of sports training will positively affect the improvement of technical and tactical indicators of the student. Therefore, this will positively affect the quality of not only their technical actions, but also their performances in competitions and the formation of their success.

References

1. Gorin KYu, Netbay SG, Gvozdkov PYu, Karimulin AA (2019) Digital technologies in the practice of physical training of specialists of the Ministry of internal affairs of the Russian Federation. Sci Notes Univ P.F. Lesgaft 6(172):44–48
2. Grishaeva OA, Shirshova EO (2017) Market research of mobile applications for assessing human motor activity. Tauride Sci Obs 12(29):34–40
3. Makarov YuM, Sysoev AV (2012) The development of cognitive activity using three factorial support of the educational process of basketball players aged 13–15. Sci Notes Univ P.F. Lesgaft 3(85):111–116
4. Ponomareva KV (2018) Trends in the market of sports accessories as a mirror of social trends of society. Econ Res Dev 1:17–24
5. Voronin AD, Danilova AM (2018) Cognitive activity of an athlete – a fighter in the educational process. OlymPlus 2(7):9–13
6. Zhigareva OG (2018) Mobile applications as a means of promoting a healthy lifestyle among students. Econ Socio-Humanit Stud 4(20):111–115

Features of Digital Transformation of Modern Banking Transactions

A. M. Mikhaylov and N. A. Petrov

Abstract Financial transactions by credit and institutions under current conditions are associated with the transformation of money functions as a means of circulation and payment and the emergence of digital platforms that mediate financial transactions of business participants and compete with the traditional banking system. As a result, classic credit institutions that have received a license to carry out depositary credit and cash operations. And intermediaries appearing in their place in making financial payments, for example, such as microfinance organizations, crowdfunding platforms and accumulating investments increase the risks of the non-return of the capital invested in them due to the specific nature of their activities and the relative underdevelopment of the institutional field in the domestic credit. The relevance of this problem determined the purpose of the study—the authors are to analyze transformation of current banking operations, their digitalization level and prospects for their further implementation. To achieve this purpose analysis and synthesis methods are used, and the descriptive statistics approach is also used. As the result of the study the authors justify the importance of traditional credit and financial institutions in mediating financial transactions between business participants and highlight their strengthening in the implementation of digital payments between counterparties.

Keywords Bank · Crowdfunding · Investments · Payment systems · Transactions

1 Introduction

With the development of the digital economy, new so-called "end-to-end" technologies appear that help a person carry out computational and communication economic operations, which significantly accelerates the technological development of

A. M. Mikhaylov · N. A. Petrov (✉)
Samara State University of Economics, Samara, Russia
e-mail: petrovnkt@gmail.com

A. M. Mikhaylov
e-mail: 2427994@yandex.ru

S. I. Ashmarina and V. V. Mantulenko (eds.), *Current Achievements, Challenges and Digital Chances of Knowledge Based Economy*, Lecture Notes in Networks and Systems 133, https://doi.org/10.1007/978-3-030-47458-4_77

states and the economy as a whole [10]. These technologies include online lending, crowdfunding and digital retail transactions.

Functions of the nowadays money determine the need for banking operations. Money used both in cash and non-cash, is also a standard of value of goods or services. The transfer of money from one owner to another is associated with the risk of loss of evidence that confirms that the money belonged to this particular owner. The safety of funds is the main task in performing banking operations. Due to banking operations, credit organizations can reliably attract, accumulate and place customers' funds through transactions.

According to the Federal Law FL No. 395-1 "On Banks and Banking Activities" (hereinafter—Federal Law No. 395-1), a bank is a credit institution that has an exclusive right to collectively carry out the following banking operations: attract funds from individuals and legal entities, place these funds on its own behalf and at its own expense on terms of repayment, payment, urgency, opening and maintaining bank accounts of individuals and legal entities [5]. A banking operation is a combination of actual and legal actions that are carried out exclusively by a credit institution and only on the basis of a license issued by the Central Bank of Russia.

The concept of banking operations can be and is applied in legislation in various meanings. Such an application depends on the context and purpose of the legislator. In the law, banking operations are used in the public law sense. Other sources apply the concept of banking operations in a broad sense and consider it as a completed series of actions or actions in technical activities of a credit institution [2].

Such operations are carried out by credit organizations due to their exclusive legal capacity. The part 1 Article 5 Federal Law No. 395-1 contains a list of classic banking operations, such as:

- raising funds of legal entities and individuals in deposits,
- opening and maintaining bank accounts,
- cash settlement,
- purchase and sale of foreign currency,
- placement of precious metals,
- collection of cash, settlement documents,
- cash services for individuals and legal entities,
- issuing bank guarantees,
- money transfer,
- lending [5].

It is worth noting that banking operations can be carried out not only by a credit institution, which has the right to do it, such as the Central Bank of Russia, the Deposit Insurance Agency. Particular attention should be paid to the norms of the Federal Law No. 395-1 and the Civil Code (hereinafter referred to as the Civil Code of the Russian Federation [6]) as part of regulation of banking operations [5]. For example, the Civil Code of the Russian Federation sets forth chapters that regulate banking. These chapters contain both civil law and banking law. Such norms have a dual nature—they regulate both transactions and banking operations. For example, the Civil Code of the Russian Federation contains a ban for legal entities to conduct settlement operations on a deposit bank account. Legal entities are required to use only a checking account. On the one hand, this norm is a norm of civil law, and on the other, banking law. Another example is that only the Civil Code of the Russian Federation contains a rule which indicates that the legal entity should have a license when providing a bank loan.

When attracting and placing funds, the bank must comply with established economic standards. Such standards are also set in accordance with legislation of the Russian Federation, and the numerical value is determined by the Central Bank of the Russian Federation. Compliance with such standards is an integral part of banking operations. For example, the maximum risk standard per borrower must be observed for loans. It is also worth noting that when conducting banking operations, special requirements are imposed on the subjects of such relations. The relationship between customers and the bank is established in the agreement that contains necessary conditions for banking relations.

Banking operations that relate to individuals are protected by the Law of the Russian Federation of 07.02.1992 No. 2300-1 "On the Protection of Consumer Rights" [7]. As an illustrative example, we can consider a situation where a customer, an individual, filed a lawsuit against a bank related to unauthorized debiting of his funds from the account via the electronic banking system. Such a citizen will rely on his requirements on provisions of the law on consumer protection. In court, he will prove that he stored the login, password and EDS keys in accordance with the terms of the banking service agreement and the bank's instructions. He claim that, despite this, an order (warrant) was given to write off and transfer a certain amount of money from his account from another city, from a computer and IP address that doesn't belong to him. And the bank, having verified the authenticity of EDS, wrote off these funds. According to the norms of the Law of the Russian Federation of 07.02.1992 No. 2300-1 "On the Protection of Consumer Rights", we can conclude that the client, losing money due to his own negligence, blames the bank. However, the bank was obliged to provide information security facilities during this operation [7]. If such conditions are met, then the client will not be able to count on a positive outcome.

2 Methodology

The authors used a systematic approach in order to group the main factors affecting digital payments by market participants and their choice of a platform for financial transactions for these purposes. An analysis and synthesis method were also used. And the authors used expert assessments of leading scientists of the world.

3 Results

Currently, to ensure the transparency of banking operations, the issue of developing repositories' activity in Russia is relevant. The following issues need to be addressed here:

- it is required to ensure that the parties agree on which repository all transaction reports will be submitted to,
- there is no clear definition of other transactions with securities and foreign currency concluded on the basis of a general agreement,
- the responsibility of the repository in case of insufficient protection of confidential information, as well as in case of incorrect entry of data into the registry, is not defined,
- there are no requirements for the capital of repositories, licensing procedures, no clear procedure for admitting organizations to perform corresponding functions; incomplete compliance with the classification of derivative financial instruments used in regulation, business practice, standard documentation and classifications of international organizations, such as ISDA and BIS.

The current practice of working with a credit bureau (hereinafter, CB) is criticized, where there is a formal disclosure of information on interaction with legal entities. The high cost of CB services remains, first of all, due to the lack of competition in the market. It seems that when improving the legislative framework on credit records, the possibility of a potential borrower to refuse to process personal data, as well as their transfer to the CB, should be excluded. This position should be an obligatory clause of the loan agreement that does not require the consent of the borrower.

It is advisable to provide banks and the credit bureau with quick access to information from government agencies in order to verify the reliability of data on the financial condition and discipline of existing and potential customers.

The definition of a consumer loan means such funds that are provided by the lender (bank or other financial institution) to the borrower under a loan agreement for various purposes with a fixed interest. If earlier these operations were carried out only in the bank's offices, then, online lending became more popular in Russia in 2012. This is due to the fact that an online loan is a service provided by a financial institution or organization using remote operations via the Internet.

We can divide existing financial products, such as "online loans", into:

- a microloan (a small loan that is characterized by high interest rates and a short repayment period; for example, the maximum rate on these loans was set at 1.5% per day in the Russian Federation in January 2019, and from July 1, 2019 it has been shrinking by 1% per day; the maximum amount of a microloan to an individual in a microfinance organization (Bystrodengi, MigCredit, Center of Loans)—1 million, in a microcredit company ("Good Money", "Jet Money Microfinance", "Agency for Express Lending")—500 thousand rubles),

- an express loan (a loan, up to 1 million rubles, for a period of several days to several months, with an interest rate of 30–60%; you can obtain it at the bank's office or on its website, for example, Tinkoff, Home Credit Bank, UralSib),
- a long-term loan (a loan that must be repaid over a long period of time (from 3 years), with an interest rate of 8–13%, you can obtain it at the bank's office and on the bank's website, for example, Sberbank, "Gazprombank", "VTB", and "Alfa-Bank").

The convenience of this type of lending is that you can receive a cash amount, as well as obtain a credit card through courier delivery or by direct transfer to an existing card [13]. This method is necessary for young and busy people, since time is not spent on obtaining a loan with a personal presence, but it is allocated for work and additional income. Therefore, the opportunity cost of visiting the bank's office is reduced and the productivity of the employee does not decrease.

Online lending is available only at banks that have their own website where this service is offered. For the convenience of site visitors there is a loan calculator that helps to calculate the cost of a loan [1]. For example, according to the website of PJSC Sberbank (based on calculations for November 2019), if a young person aged 18 to 21 wants to take a loan from Sberbank for the amount of 500,000 rubles for a period of 2 years, then the rate on the proposed loan will be 12.9% (provided that he receives a salary at this bank; without these conditions, the rate is 13.9%), the overpayment will be in the amount of 69,938 rubles, the monthly payment will be 23,747 rubles. An online banking resource offers loan conditions specified in documents and agreements that are in electronic format. After connecting to Internet banking (banking technology via the Internet), a loan application is drawn up: the electronic form of the borrower must indicate the personal data of the borrower, his income level, as well as passport data. Then the application form goes directly to the bank base, which considers it from a few minutes to 2–3 h.

There are certain advantages of online lending [12]:

- convenience (you can apply for a loan anywhere via the Internet),
- high speed of consideration of the application, in contrast to direct contact with the bank,
- saving time for filing an application, as the operation is performed online,
- ability to receive a certain amount of money immediately after considering the application,
- you can quickly find the loan that you need on the official website.

Along with the advantages, there are some disadvantages of this system:

- interest is higher on loans obtained through online appeals to microfinance organizations than at the bank's office, since the application is quickly reviewed and accepted by the lender,
- payment system commissions appear due to money transfers,

- online service increases the risk of fraud (attackers can fake the site of the online resource for providing a loan and get client data, his passport data, information from a credit or debit card).

Online lending is becoming increasingly popular in the credit market of Russia. Now it is a thriving segment of the financial market of the Russian Federation. So, according to individual experts, in 2017 online banks provided 67% more loans than in 2016. Among the cities with over one million people, Perm and Volgograd occupy high positions in this type of lending (in terms of the volume of loans), followed by Samara and Novosibirsk. The leaders in the regions are the Far Eastern and Siberian Federal Districts, and the North Western and North Caucasian Federal Districts are in third and fourth place in terms of the number of online loans.

Currently, digital payments are increasingly playing a role in household retail operations, gradually replacing cash. But it is worth noting that cash still continues to play a significant role in the modern global economy (Fig. 1).

Comparing the figures from 2013 to 2019, you can see that in the period from 2013 to 2014, the share of cash in circulation increased by 574.1 billion rubles (7.8%), and then decreased by 232, 3 billion rubles (2.9%) from 2014 to 2015. From 2015 to 2019 it steadily increased and increased by 2830.2 billion rubles (38.3%) in 2019 compared with 2013.

Analysis of the dynamics suggests that the proportion of cash in circulation continues to increase for the period from 2013-2019. This indicates the confidence of the majority of the population in cash, which makes it the main tool in the field of retail payments. However, the share of non-cash funds continues to increase (Fig. 2).

Based on the graph, we can conclude that the number of digital cash payments increased by 28.7% by the first half of 2019 compared to 2013.

Today, the limited use of digital payments in retail operations of households is associated with several factors [8]:

- imperfect current legal regulation,
- frequent cases of fraud,

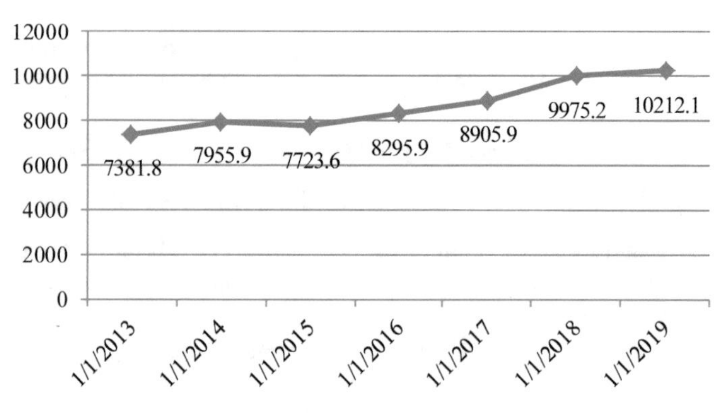

Fig. 1 Dynamics of changes in the amount of cash (M1 in trillion rubles) in circulation for the period 2013–2019 (*Source* authors)

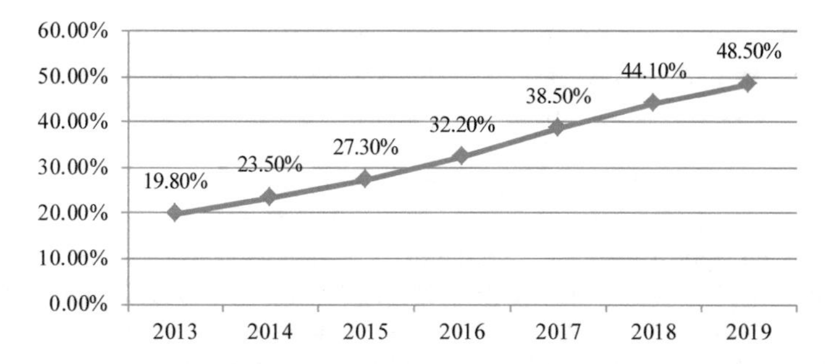

Fig. 2 Dynamics of digital payments in Russia for the period 2013–2019 (*Source* authors)

- limited direct transfer of funds from one person to another,
- need to connect to the Internet to make payment transactions,
- charging a bank commission for the transfer to the account of the counterparty via Internet banking and mobile transactions.

4 Discussion

With the development of Internet technologies and the spread of social networks, new opportunities are opening up for the economy. The structural changes that took place in the field of financing of economic entities provided access to online tools that made it possible to raise funds much easier, faster and more transparently in comparison with traditional lending [3]. One of them is crowdfunding, which is a new type of fundraising, due to which people interested in any project contribute to its financing. Crowdfunding acts as a source of capital, which is provided by a virtual community that wants to support a project. Initially, crowdfunding served as a way of financing creative projects (books, games, films), but in the 2000s, with the advent of crowdfunding platforms, this type of financing was used in the economy.

Crowdfunding specializes in financing specific projects, rather than the enterprise as a whole. It is its main feature. This form of financing is especially suitable for small or newly created enterprises that cannot use traditional sources of raising capital. Crowdfunding attracts external investment from a large number of people, in exchange for a certain fee or product. Crowdfunding solves specific tasks of developing, evaluating and selling a product or service through an open advertisement on the Internet. Unlike other financial institutions, crowdfunding involves attracting small financial contributions from numerous investors [11]. Thus, crowdfunding is a collaboration of entrepreneurs who need to attract the capital of investors, which are its source, and intermediary crowdfunding platforms that provide the legal basis and preliminary selection of viable projects.

Any crowdfunding model consists of three participants: the author of the project, investors and the crowdfunding platform. The authors of the project want direct access to the market and financial support from interested parties. Investors expect a certain reward. Therefore, they bear the risk by investing financial resources in proposed projects. And crowdfunding platforms act as an intermediary, bringing together project authors and investors for their interaction [4].

All crowdfunding platforms work according to a similar principle: the authors of the idea present their projects on the crowdfunding platform that selects viable projects. After approval of the application, the authors create a web page on the crowdfunding platform where the project and the reasons for its financing will be presented. For each project, a financial goal is established—this is the amount and term for collecting financial funds. After the launch of the project, information on the remaining amount and period, which is in the public domain, is published. Further, the author should arouse and maintain interest in the project [9]. Project information, as a rule, usually consists of a description of the project, a video call, a product sketch, and more. The main task is to create a compelling message that will arouse interest and encourage investing in the project. The advantage of crowdfunding over other investment methods is that it collects funds from the general public [14]. Usually the most successful projects receive 25–40% of their income from people with whom the author is somehow connected (family, relatives, friends, acquaintances, colleagues). Then the campaign receives support from strangers.

5 Conclusion

The types of investing financial resources that we examined and the ways of making digital transactions can be called the most promising, and they can become the fundamental basis for youth entrepreneurship in the future. Despite the preservation and strengthening of intermediary functions of classical credit and financial institutions in making payments, including using digital platforms (Internet banking, mobile banking, etc.), serious competitors appear, for example, such as Amazon and online retailers, like AliBaba, creating their own banks to expedite payments with customers. In our opinion, the transition to the full digitalization of financial transactions will occur in the near future, which is caused by the speed and relative cheapness of such methods of calculation.

References

1. Agarwal S, Chomsisengphet S, Liu C, Song C, Souleles N (2018) Benefits of relationship banking: evidence from consumer credit markets. J Monet Econ 96:16–32. https://doi.org/10.1016/j.jmoneco.2018.02.005
2. Baig AS, Blau BM, Whitby RJ (2019) Price clustering and economic freedom: the case of cross-listed securities. J Multinatl Financ Manag 50:1–12. https://doi.org/10.1016/j.mulfin.2019.04.002

3. Beck T, Pamuk H, Ramrattan R, Uras RB (2018) Payment instruments, finance and development. J Dev Econ 133:162–186. https://doi.org/10.1016/j.jdeveco.2018.01.005
4. Cash A, Tsai H-J (2018) Readability of the credit card agreements and financial charges. Finance Res Lett 24:145–150. https://doi.org/10.1016/j.frl.2017.08.003
5. Federal Law dated 02.12.1990 No. 395-1 "On Banks and Banking Activities". https://www.wto.org/english/thewto_e/acc_e/rus_e/WTACCRUS58_LEG_243.pdf. Accessed 01 Feb 2020
6. Federal Law dated 26.01.1996 No. 14 "Civil Code of the Russian Federation (part two)". http://www.consultant.ru/document/cons_doc_LAW_9027/d73c103b664fa1b30b63f8fa944a2e0f33cd67f0/. Accessed 01 Feb 2020
7. Federal Law dated 07.02.1992 No. 2300-1 "On the Protection of Consumer Rights". http://www.consultant.ru/document/cons_doc_LAW_305/. Accessed 01 Feb 2020
8. Immordino G, Russo FF (2018) Cashless payments and tax evasion. Eur J Polit Econ 55:36–43. https://doi.org/10.1016/j.ejpoleco.2017.11.001
9. Javadi M, Amraee T, Capitanescu F (2019) Look ahead dynamic security-constrained economic dispatch considering frequency stability and smart loads. Int J Electr Power Energy Syst 108:240–251. https://doi.org/10.1016/j.ijepes.2019.01.013
10. Konovalova M, Mikhailov A, Persteneva N (2015) Crisis phenomenon as a form of permission structural imbalances. Econ Law Issues 5:66–71
11. Presthus W, O'Malley NO (2017) Motivations and barriers for end-user adoption of bitcoin as digital currency. Procedia Comput Sci 121:89–97. https://doi.org/10.1016/j.procs.2017.11.013
12. Sá A, Pereira A, Pappa G (2018) A customized classification algorithm for credit card fraud detection. Eng Appl Artif Intell 72:21–29. https://doi.org/10.1016/j.engappai.2018.03.011
13. Song J, Yang F, Wang L (2017) Secure authentication in motion: a novel online payment framework for drive-thru internet. Future Gener Comput Syst 76:146–158. https://doi.org/10.1016/j.future.2016.06.011
14. Vishnever V, Konovalova M, Mikhailov A (2015) Institutional regulation as a form of institutional interests in the banking sector. Int J Econ Perspect 9(2):23–27

The Use of Digitalization in the Characterization of Illegal Banking Activity

I. E. Milova, A. S. Knyaz'kina, and A. M. Mozgunova

Abstract The relevance of the study is determined by the fact that economic activity is an integral part of any civilized society. In the context of Russia's transition to a regulated market, it became obvious that the organization and management of economic processes require the use of information and digital technologies. Analysis of statistical indicators reveals an increase in violations in the banking sphere. The purpose of this article is to study theoretical and applied issues related to the legality of banking operations based on the periodization of criminal legislation in combination with criminological approaches. The authors consider the problems of countering criminal manifestations in the banking sector, focusing on the issues of historical reconstruction, using elements of digitalization. The problem lies in the conflict between the qualitative legal construction of the criminal law in this area and its weak empirical application. An additional factor that reduces the effectiveness of the implementation of regulations is the lack of methodological and tactical developments for the disclosure and investigation of the crimes under consideration. For the purpose to optimize the practice of countering illegal banking operations, the authors make a number of specific proposals based on the study of the genesis of the development of criminal legislation.

Keywords Credit institutions · Criminal cashing · One-day firms · Financial fraud · Operational experiment · Economic extremism

I. E. Milova (✉) · A. S. Knyaz'kina · A. M. Mozgunova
Samara State University of Economics, Samara, Russia
e-mail: irina.milova@ro.ru

A. S. Knyaz'kina
e-mail: Angelina.knyazkina@yandex.ru

A. M. Mozgunova
e-mail: nastiniam@gmail.com

S. I. Ashmarina and V. V. Mantulenko (eds.), *Current Achievements, Challenges and Digital Chances of Knowledge Based Economy*, Lecture Notes in Networks and Systems 133, https://doi.org/10.1007/978-3-030-47458-4_78

1 Introduction

The market is characterized by a high degree of freedom of subjects, which led to the formulation of the problem of ensuring the proper legal status of participants in financial procedures carried out under the control of the state before the domestic legislator. The competent authorities control the dynamics of the banking system and those elements of it that are of a public nature. However, their intervention in the activities of credit institutions is quite limited [8].

An appeal to official statistics shows that the number of crimes in the field of banking violations is constantly growing [15]. This trend is not exclusively national, but rather belongs to the category of modern global challenges. At the same time, Russia is characterized by a large number of criminal prosecutions in cases of this type, with subsequent termination of proceedings in pre-trial procedures, and most often on non-rehabilitative grounds [14]. The reasons for this situation are the complexity of proof.

Analysis of the current criminal law, namely article 172 of the Criminal Code of the Russian Federation shows that the problem of formal application of this rule is its blank nature, since there is no general approach to the definition and interpretation of basic concepts [4]. At the same time, there are problems in its application, since credit institutions often refuse to provide information to the investigative authorities and the court, citing the presence of bank secrecy [1].

During the preparation of the article, the authors conducted a survey of investigators who specialize in solving economic crimes. It is significant that more than 80% of respondents stated that they have difficulties investigating the actions of this group, including qualification issues. As a result, a qualitatively formulated legal structure is almost not used in practice, turning into an exotic crime and legal fiction.

The authors believe that in order to change the current situation, it is useful to study the history of the development of legislation in this area; features of modern approaches; signs of the crime of illegal banking; ways to optimize the practice of its application, as well as to consider some criminological characteristics.

The article articulates the idea that there can be no formal rules in criminal law, all of them must have practical application, provided they are empirically relevant. Otherwise, the relevant provisions are subject to decriminalization. Analysis of publications in the media, materials of investigative and judicial practice shows that illegal banking activities have become a widespread manifestation. The criminal nature of this phenomenon is obvious, which requires the organization of a competent legal response.

On the basis of the Samara State University of Economics, during the training bachelors, specialists and masters, future lawyers develop skills of a competent approach to the investigation of crimes in the field of economics, and study methods and tactics for detecting illegal banking activities [11]. The materials of this article are of practical importance for law enforcement agencies-investigative and operational units, whose competence includes the fight against such negative manifestations, including the specifics of the disclosure and investigation of these crimes.

2 Methodology

The methodological basis of the introduced research is formed by the basic dialectical approach, according to which illegal banking is considered as a special object of knowledge. All financial and credit mechanisms are examined from the point of view of their epistemology, with an emphasis on relative characteristics. Private scientific methods are used quite actively, including historical-reconstructive, formal-logical, system-structural, analytical-synthetic, comparative, statistical, mathematical, expert-assessment, survey-sociological, questioning, interviewing, legal modeling, experiment, as well as the method of field research, implemented through a random sample of criminal cases on crimes in the field of illegal banking (the analysis of materials was carried out on the basis of a specially developed program, with the subsequent generalization and systematization of the obtained data).

3 Results

Chronologically, the institutions that were the forerunners of modern banks belong to the VIII century BC. The period of feudal fragmentation, the conditions of serfdom and landowner land ownership, the weak development of industrial production, were the reasons why banking structures in Russia appeared quite late, approximately in the XVIII century. Thus, the processes of formation of commodity-money relations in our country can be estimated as somewhat slow.

Historically, the emergence of credit relations dates back to the period of the collapse of Kievan Rus. We find information about this in economic and monastic books, in various charters, both of a secular and spiritual nature. These factors are mentioned in the Russian Pravda, Pskov court and Novgorod court certificates, which provide information about persons who, by their characteristics, resemble subjects of credit relations.

An important step was the development of the 1497 judicial code, since it unified approaches to regulating a number of controversial issues, including the loan agreement. This regulation contained a separate article that provided for liability for a merchant who had embezzled borrowed property. The responsibility of such an offender varied depending on whether he acted with or without malice.

These provisions were further developed in the 1550 law book, where the term "swindlers" was first used, which was equated with "Tat", that is, a thief. In the Cathedral Code of 1649 [2], there was a tendency to soften the position of debtors-nobles. They were forbidden to collect borrowed interest, because it was widely believed that this kind of relationship is gratuitous. The legislator described in detail the structure of property crimes, which mentioned the responsibility for violation of obligations on loans. For the first time, such acts were singled out as criminal offences. The debtor continued to bear personal responsibility for the unpaid debt, both in the material sense and in the physical sense [10].

However, this punitive mechanism was applied only if the loan was taken for commercial purposes. Sanctions for violations were very severe. However, along with restricting the freedom of the perpetrator, it was important to compensate for the losses incurred by the creditor. The amount of debt had no legal significance. Statistical indicators have not been preserved, so it is difficult to assess the effectiveness of the practical application of the above norm. However, an analysis of the relevant provisions leads to the conclusion that creditors' interests were guaranteed and protected. Thus, the legislator of that time demonstrated a creditor approach in this sense.

The Criminal Code of 1903 [3] regulates responsibility for economic offenses. Emerging commercial banks were engaged in promising financing of developing industrial enterprises. According to the established practice, they were included in the number of their shareholders. A little later, the code of penalties introduced a group of rules that establish liability for owners of banking institutions of money changers' shops for carrying out illegal operations, making prohibited currency transactions (for gold reserves and similar values).

At the beginning of the revolutionary events of 1917, the Russian credit system was very extensive. The period from 1909 to 1913 was marked by an increase in abuse of credit and promissory notes, which is directly related to violations in the activities of commercial institutions and banks.

The first world war enriched many domestic private banks, while the position of the State Bank weakened. The banks were essentially engaged in ordinary speculation. For the purpose of resale, credit institutions rented warehouses and engaged in repurchases. A high percentage of participation belonged to foreign capital (34%) of bank shareholders, which allowed them to form the policy of credit institutions. The October period post in this area was related to the CEC Decree "On the nationalization of banks" [5], which stated that financial activities in this area belong to the state, and all private banks are part of a single State Bank, being under its control. In just a few months, both private and foreign banks were liquidated in Soviet Russia.

The year 1922 was marked by the adoption of the first Criminal Code. There were crimes that are called economic in the modern interpretation, but there they were called against the order of management. At the end of the XX century, laws were adopted to provide regulatory support for the transformation of the economic structure and procedures for carrying out economic activities in Russia. The first stage was associated with the transformation of the banking system of the Soviet Union (late 80's), it was accompanied by the emergence of cooperative banks. The next period was associated with the adoption of the law "On banks and banking activities". Simultaneously with these regulatory innovations, the State Bank of the USSR, Republican central banks, cooperative and commercial credit institutions were created. At the last stage, independent banking systems were established in the former soviet republics. In addition, an inter-republican credit institution was established that operated throughout the CIS. Of great importance for the development of legislation in this area was the liberalization of the currency, which was officially used as a means of payment and as an equivalent in credit transactions.

The current criminal code of the Russian Federation has a separate rule—article 172 of the Criminal Code of the Russian Federation [4], which regulates liability for illegal banking activities. This legal structure allows us to objectively assess the content of banking activity for its legality [12]. Meanwhile, there are serious difficulties in implementing the above legislative model [9].

According to the studied criminal cases, it was revealed that a new type of criminal activity was created in our country, consisting in the registration of organizations that do not conduct economic activities, consist of a single founder, do not have a staff, do not enter into any transactions, but are engaged exclusively in cash-out activities, with the receipt of a percentage for such operations [6]. To solve such crimes, digital technologies can be used in the form of software products that systematically summarize tax audit materials, operational information, and monitoring data for media publications, including information posted on electronic platforms. Digital approaches could become part of research when conducting financial and analytical expertise in such cases. From the point of view of disclosure, it would be useful to compile a digitized psychological portrait of a subject prone to this type of illegal activity, highlighting its typical criminological characteristics. Such a digital model could be used for operational purposes.

4 Discussion

Criminal liability for such violations in the credit sphere is also known to foreign legislation. The authors studied the regulatory approaches that exist in other countries, justifying the idea that some provisions would be appropriate to implement in Russian criminal law. A distinctive feature of the legal regulation of liability for violations of banking operations in foreign countries is a higher degree of specificity. Of course, the rapid development of financial mechanisms in developed civilized countries is a big plus, although there are some criminal manifestations. In the vast majority of countries, violations of banking practices are traditionally referred to as fraud. This is the approach in the criminal law of England, Germany, India, the United States, and France. Russian criminal law is aimed at individuals, and in some foreign countries, legal entities are also punished for such violations (India, China, the United States, France).

Analysis of statistics shows that the most common crimes involving bank cards. Criminal elements make extensive use of social networks and software. In the United States, there has been an increase in this type of crime on the Internet, the so-called cybercrime. Credit institutions are focused on strengthening bank protection mechanisms.

In accordance with the decision of the International monetary Fund, the Chinese Yuan is included in the number of reserve currencies, so studying the experience of the PRC is very interesting. In Russia, there is a two-level banking system, in China it is three-level. The first level is represented by the People's bank as a central credit institution. Along with it, other banks operate in this segment, where the state owns a controlling stake. They are focused on specialized areas. Several large commercial banks also have a basic degree of state participation, and they have a large number of branches, both in the country and abroad.

In China, commercial banks (second level) can provide insurance services for money investments. Persons who have previously been convicted of corruption are not eligible to hold senior positions in the banking system. This option is not available to entities that previously held key positions in credit institutions, whose license was revoked for violations or declared bankrupt. This approach would also be of interest to the domestic banking system, since it strengthens the public's confidence in the credit sector. The third level is formed by agricultural and urban credit cooperatives. This structure provides free access to financial products for the population.

Russia could follow this path, but with the clarification that such activities will be financed from regional and municipal budgets, with the involvement of foreign participation. In this case, lending would become more affordable for small and medium-sized businesses.

Criminal liability under the current criminal code of the Russian Federation is established only for illegal banking activities, as a result of which a particularly large amount of income was extracted. At the same time, the legislator does not give the definition of "income" itself, which causes discrepancies in practice. Some countries, including former Soviet republics, have eliminated this regulatory gap. As a confirmation, we will refer to the criminal legislation of Tajikistan, where income refers to the difference between invested and received funds. The corresponding rule of the criminal code of the Russian Federation contains an indication that when calculating damage, not only the actual damage, but also the lost profit must be taken into account. This approach corresponds to the provisions of the Model Criminal Code, where this article is included in the section "Crimes in the sphere of economic activity".

As a qualifying feature, the domestic criminal law refers to an organized group that has a high degree of stability in engaging in such criminal activities. At the same time, there is no mention of preliminary collusion by a group of individuals, which allows a number of subjects to avoid responsibility for violations committed in this area. It is obvious that the degree of public danger of such a group is no lower than that of an organized group. A more correct approach in this direction is demonstrated by the criminal code of Uzbekistan, where the norm regulating liability for illegal banking activities specifies preliminary collusion by a group of persons as a qualifying feature.

There is no qualifying feature in the Russian legislation for committing this act using one's official position, although practice shows that this is actually the case. Meanwhile, it is obvious that the bank's management has access to all its products, determines the direction of development and is the guarantor of its effective activity. It is obvious that top managers should be held more responsible for performing these actions. It would be advisable for the courts to additionally apply a sanction to them in the form of deprivation of their right to hold senior positions in credit institutions, either for life or for a certain period.

As an illustration of this thesis, we will name the Spanish Criminal Code, which provides for such deprivation for a period of seven to ten years. It is noteworthy that such acts are considered to be committed against the state administration, since their harmfulness is directly linked to the undermining of the authority of banking institutions among the population. In the Criminal Code of Armenia, such activities are classified as violations in the sphere of entrepreneurship, and they are punishable by a fine or imprisonment. We find similar approaches in the criminal law of the Republic of Belarus.

In the United States, there is a tightening of sanctions against entities that commit such violations [13]. The literature provides a criminal scheme identified during the investigation of criminal cases of this type. Fraudsters used personal data of customers who took in the service system enterprises (restaurants, shops) on the territory of not only other countries, but even continents. After that, the criminals produced fake cards, which were then used to pay for purchases. The most active criminals purchased Apple products as extremely popular. After this purchase, the goods were resold outside the United States. Models for fake cards were made in Russia, Libya, and China. The material for copying media was obtained when the credit card holder tried to use it in a fake ATM specially installed for reading data. The equipment copied the pin code. Then a duplicate of the card was made, followed by the withdrawal of funds from it. Similar crimes are committed in Russia, so the experience accumulated in other countries is subject to careful study and reflection [7].

5 Conclusion

According to the results of the study, the authors note that the analysis of the dynamics of the development of domestic criminal legislation on illegal banking activity confirms the validity of the introduction of this crime. This kind of criminal business is very common and is a side-negative manifestation of the market economy [16]. The danger of it is that it undermines the financial basis of the state, which does not receive huge amounts of taxes. In addition, credit institutions cannot develop effectively in such conditions, because instead of banks that are legally engaged in these operations, underground entities that do not have a license begin to do so. In modern Russia, criminal business has become widespread, consisting in the creation of firms that do not conduct economic activities, but specialize in cashing money, with the payment of a certain percentage of the amount by the interested person for such a "service". The authors believe that it is appropriate to tighten the sanctions 2 articles 172 of the Criminal Code of the Russian Federation regarding the imposition of additional punishment in the form of a fine [4]. Taking into account the fact that such acts cause major material damage to the state, the imposition of a fine levied on the budget should not be an alternative, but mandatory in this case. With fairly severe sanctions, statistics show that these crimes tend to grow steadily. The authors believe this is due to the loyalty of the court in imposing punishment under article 172 of the Criminal Code. Analysis of criminal cases shows that most of the defendants are sentenced to probation, without being sent to correctional institutions. Due to the excessive leniency of the punishment for this category of crimes, there is a relapse. The solution to the problem could be the introduction of a qualifying attribute—"repeatability" into the norm under consideration. In the course of the study, the characteristic criminological features of persons who commit such acts were identified. The vast majority of them are men, in the age group of 30–40 years. They have higher education, most often economic (in some cases, the second legal), have knowledge in the field of digital and computer technologies. Most of those convicted under article 172 of the Criminal Code of the Russian Federation have a family, they have dependent minor children, and their parents are pensioners [4]. The motive for committing the crime in question is self-serving, the defendants seek to obtain income through illegal actions. The authors distinguish several types of economic criminals: (a) casual (they start to engage in such activities in connection with a certain combination of life circumstances); (b) malicious (these individuals have formed a stable anti-social attitude); (c) intellectual (plan such actions, carefully consider them; develop complex criminal schemes). In the studied criminal cases, most often there is a mixture of features of different types. At this stage, there is a need to establish criminal liability for illegal banking activities for legal entities. Proposals for the creation of a digital criminological portrait of the subject of illegal business activities are of an innovative nature. If implemented, it would be possible to create an electronic database for such individuals, with an analysis of their inherent determinants. Digital technologies would also contribute to the detection of these crimes and make operational work in this area more effective.

References

1. Alexandrova I (2019) Improvement of criminal legislation on liability for illegal banking activities. Bull Nizhny Novgorod Acad Ministry Intern Aff Russ 2(46):61–66 (in Russian)
2. Cathedral Code of 1649. http://www.hist.msu.ru/ER/Etext/1649.htm. Accessed 02 Feb 2020
3. Criminal Code of 1903. https://www.booksite.ru/fulltext/1/001/008/113/530.htm. Accessed 02 Feb 2020
4. Criminal Code of the Russian Federation from 13.06.1996 N 63-FZ. http://www.consultant.ru/document/cons_doc_LAW_10699/. Accessed 02 Feb 2020
5. Decree "On the nationalization of banks" dated 27.12.1917. http://www.hist.msu.ru/ER/Etext/DEKRET/banks.htm. Accessed 02 Feb 2020
6. Fetisenkova T (2016) Illegal caching funds as an element of actus reus. Bus Law 3:142–145 (in Russian)
7. Galanina E (2017) Questions for discussion the involvement of credit institutions accountable for violations of banking legislation. Young Sci 7:379–382 (in Russian)
8. Guryanova A, Smotrova I (2020) Transformation of worldview orientations in the digital era: humanism vs. anti-, post- and trans-humanism. In: Ashmarina SI, Vochozka M, Mantulenko VV (eds) Digital age: chances, challenges and future. Springer, Cham, pp 47–53
9. Khisamova A, Kamalieva L (2019) Criminal liability for illegal banking activities. Sci Electron J Meridian 7(25):39–41 (in Russian)
10. Kuznetsov S, Olimpiev A, Mikhailenko N (2018) Responsibility for crimes in the credit and banking sector in Russia: pre-revolutionary period. Bull Moscow Univ Ministry Intern Aff Russ 3:127–135 (in Russian)

11. Serper E, Khvostenko O, Pershin M (2020) Harmonization of financial and credit resources of commercial organizations in the digital economy. In: Ashmarina SI, Vochozka M, Mantulenko VV (eds) Digital age: chances, challenges and future. Springer, Cham, pp 335–341
12. Shahsuvarian K (2019) Illegal banking activities. Issues of science and education: theoretical and practical aspects. In: Vostretsov AI (ed) Proceedings of the international (correspondence) scientific and practical conference, pp 215–220 (in Russian)
13. Teti E, Dell'acqua A, Etro L, Volpe M (2017) The impact of board independence, CEO duality and CEO fixed compensation on M&A performance. Corp Gov (Bingley) 5:947–971
14. Titov YuP (1997) Textbook on the history of the state and law of Russia. Prospect, Moscow (in Russian)
15. Volkodavova E, Zhabin A, Yakovlev G, Khansevyarov R (2020) Priorities of business activities under economy digitalization. In: Ashmarina SI, Vochozka M, Mantulenko VV (eds) Digital age: chances, challenges and future. Springer, Cham, pp 71–79
16. Yavorsky MA, Milova IE (2019) On the issue of countering economic extremism. In: Bortnikov SP (ed) Actual problems of legal system development in the digital age. Samara State University of Economics, Samara, pp 138–143 (in Russian)

Civil Litigation Principles Transformation in Sight of Digitalization Impact on Legal Culture

E. N. Churakova

Abstract The introduction of information technology in the legal activities of state bodies of the Russian Federation reflects most in administration of justice. Use of information and technological tools in the activities of justice bodies inevitably raises the question of how the so-called "electronic justice" relates to the model of legal proceedings defined in the law, and what is the relationship between the legal principles of the administration of justice and the above-mentioned set of tools. Of greatest interest is the consideration of this issue within the framework of the economic justice model, since it is crucial in ensuring the activity of arbitration courts that the most noticeable results of introducing modern technologies have been achieved there.

Keywords Principles of legal proceedings · E-justice · Equality of parties · Transparency of justice · Principle of immediacy

1 Introduction

The influence of information technology on the development of the legal system of the Russian Federation requires an understanding of the essence of its legal model and determination of the laws governing changes in legal relations under the influence of the information environment. The legal model of the information society, on the one hand, should correspond to the rapid development of social and technological processes, and on the other hand, it should preserve and strengthen guarantees for the rights and freedoms of citizens, guarantees for the implementation of the principles of justice, the protection mechanism developed over the years by legal institutions. Currently, despite the high level of information technology influence on the legal culture of society, it cannot be concluded that the legal system of the Russian Federation can be described as an information one. However, there is a tendency to develop a real mechanism that has a tangible effect on procedural relations.

E. N. Churakova (✉)
Samara State University of Economics, Samara, Russia
e-mail: churakovaen@gmail.com

S. I. Ashmarina and V. V. Mantulenko (eds.), *Current Achievements, Challenges and Digital Chances of Knowledge Based Economy*, Lecture Notes in Networks and Systems 133, https://doi.org/10.1007/978-3-030-47458-4_79

The most significant issue in the implementation of e-justice in the judicial system should probably be its conjunction with the legal foundations of legal proceedings as a whole: the principles of equality of parties, openness of court proceedings, discretion, the combination of oral and written forms of communication within the proceedings, and the immediacy of the administration of justice.

It is important to note that the interaction of e-justice with the principles of the arbitration process is poorly analyzed and is characterized only from the point of the openness and accessibility of judicial protection for citizens arising from constitutional provisions [7]. Of course, the prerequisites and factors contributing to the active use of modern technology in judicial proceedings are quite obvious precisely in consideration of openness and accessibility of justice. And, of course, the implementation of the functionality of e-justice primarily serves the purpose of ensuring access to justice for everyone. However, evaluating the model of e-justice in isolation from all other principles, not considering the relationship with them and the degree of impact is fundamentally wrong [2].

2 Methodology

In this study such methods as dialectical method; comparative legal method; formal legal method were used. The study is conducted using functional, systematic and structural approaches. The systematic nature of the study allowed to observe aspects of e-justice from a legal and ethical perspective.

3 Results

An analysis of methods for implementing e-justice processes in Russia allows us to talk only about the active use of technological models, for example, such as an electronic document management system and, in some cases, the use of video conferencing and the Internet. Thus, the use of information and communication technologies in the Russian Federation in litigation is of an auxiliary nature. While in other countries, the model of conducting litigation in the electronic environment is already actively implemented, as evidenced by individual authors [3]. The use of information technology needs an additional settlement. In general, modern technical and organizational capabilities can improve the quality of life, increase the efficiency of law enforcement, modernize the judicial process, and promote the development of various elements of a legal culture. Informatization of justice does not cancel and does not change the traditional principles of the civil process, their meaningful content. Its resources are designed to create the conditions for their implementation in law enforcement practice and legalization for the benefit of existing principles. Any changes external to the law including informational progress must be subordinate to its principles, and not vice versa when they are reflected in law.

4 Discussion

The principle of equal rights of the parties in its essence is closely connected with ideas about ensuring accessibility and openness to judicial protection. In order to ensure equal rights to the parties it is necessary, among other things, to provide them with equal opportunities in the exercise of their procedural rights and obligations regardless of the territorial context [8]. Of course, the essence of equality is not limited to this component alone, but e-justice is determined by the scope of its functional capabilities, and its impact on the judicial system is revealed from the perspective of providing parties with equal opportunities to participate in the case. In the absence of such opportunities, the inequality of the subjects of procedural relations acquires not only legal, but also technological character.

Another component of the implementation of this principle through e-justice is the access of all persons involved in the case to the information array of legislation and the practice of court cases—in other words, legal information. This aspect in its implementation is also a factor ensuring the transparency of justice, which is also of great importance for the qualitative characterization of the justice model [5].

The openness, accessibility and transparency of the administration of justice for subjects of public relations are provided not only by electronic document management tools, but also by other technological means that are components of the electronic justice model: by holding court hearings using video conferencing and recording in audio format. It is important to note that the open publication of judicial acts on the network significantly contributes to the transparency of justice administration and is a factor contributing to the formation of a uniform practice for courts in similar cases, since the presence of a single information system that includes standardized documents and has search algorithms provides the ability to accumulate the experience of the courts, highlight patterns and predict the favour of ruling.

Another consequence of the existence of such a system may be a reduction in the number of abuses in the framework of litigation, because their presence is clearly reflected in the published materials, and the fact of the apparent contradiction of a decision to the current practice is not difficult to establish by analyzing it in conjunction with other rulings on this issue. At the same time, however, there is the other extreme of the desire for a uniform judicial practice and it is the deliberate or unconscious refusal of judges when considering cases to fully analyze all circumstances and search for new approaches to resolving a dispute in order to comply with the general direction of legal practice. It is impossible to deny that such manifestations exist at the present moment and one can find a clear contradiction to the principle of immediacy here because when considering certain circumstances the court will focus on the interpretation of the rule of law in a similar situation, set out in a decision by another court—in other words, it shall resolve a specific situation through the prism of another, previously considered, in connection with which it will not be able to directly evaluate and take into account all relevant circumstances of the case.

Another important aspect in the light of the issue of transparency of justice is the provision of the functioning of e-justice, which includes an open information base of judicial practice, the implementation of legal regimes for the protection of information applicable to information that should be restricted by law [6]. The implementation of these requirements of the law is possible when implemented by providing limited access to the direct participants in the case under consideration, and requires a reasonable balance while adhering to the principles. The existence of such a problem in relation to the protection of personal data is indicated, in particular, in the work of Gubulov [6].

It is noted in the literature that videoconferencing as an element of the current model of electronic justice significantly affects the implementation of the principle of immediacy (Article 10 of the Arbitration Procedure Code of the Russian Federation) [9]—in particular, the nature of such an impact is defined as facts, as a result of which this principle is transformed. Indeed, during a court session in the indicated format, the assessment of the circumstances of the case is carried out indirectly by the judge, using technical means and copies of evidence collected in writing—that is, there is no personal perception, or there is obedience but only partially [4]. At the same time, the use of video conferencing serves as an auxiliary factor, changes the conditions in which the principle of immediacy is implemented, but does not abolish it.

Regarding the impact of e-justice on the principle of combining oral and written proceedings in legal literature, it is suggested that the digitalization of legal proceedings will sooner or later lead to a change in the role of oral or its total exclusion from litigation and as a result the principle of orality will be eliminated. It is impossible not to admit that the introduction of information technology has a significant impact on the principle of oral and written proceedings. At the same time, it is difficult to agree with this opinion, since the oral beginning of the proceedings, characterized primarily by the oral form of conducting a court hearing, the examination of evidence in oral form, the record reflecting procedural actions performed orally, and evidence information obtained orally cannot be completely ruled out.

5 Conclusion

Based on the foregoing, we can draw the following conclusions. E-justice as a system has an auxiliary role in the activities of the courts as it reflects in the fact that the implementation of a number of key principles is ensured through technical means and in particular [1], the principles of the arbitration process, if we consider this issue in the context of economic justice.

Moreover, quite often the assessment of the impact of e-justice on the implementation of procedural principles is carried out quite narrowly, from the point of openness and accessibility of judicial protection for each subject of public relations, which is fundamentally wrong. Electronic proceedings as an auxiliary tool of the judicial system cannot but comply with all principles of the administration of justice formulated in the law. Although, of course, it is quite obvious that the perks for expanding the use of technical means in judicial activity are precisely considerations on the implementation of the ideas of openness and accessibility formulated in context of the principle of equality of parties involved in the case.

At the same time, in the context of several principles, ambiguous conclusions arise about the impact of e-justice on the implementation of these principles. First of all, this is due to the fact that the use of technological tools and, primarily, the use of the Internet creates certain risks of evidence reliability, as well as on protection of information, access to which is limited by law. In these circumstances, it is important to maintain a reasonable balance in the implementation of the principles of openness, transparency and at the same time ensure the implementation of the provisions of the law on legal regimes of information protection [10].

In the nature of the influence of e-justice on the principles of the arbitration process, there is such a thing as the transformation of principles as a result of their implementation through technical means. This is most clearly manifested in the case of the principle of immediacy during a court hearing using video conferencing. At the same time, one can judge the presence of a similar effect on the principle of combining oral and written proceedings. There are suggestions in the literature that the implementation of e-justice is a factor in which the combination of oral and written language is violated in favor of the latter, but given the use of video conferencing and recording in audio recording format, these concerns of researchers should be considered excessive.

References

1. Balogun AV, Zuva T (2018) Towards the adoption of software engineering principles for assessing and ensuring the reliability of digital forensic tools. In: Silhavy R, Silhavy P, Prokopova Z (eds) Cybernetics approaches in intelligent systems: computational methods in systems and software, vol 661. Advances in intelligent systems and computing. Springer, Cham, pp 271–282
2. Caianiello M (2019) Criminal process faced with the challenges of scientific and technological development. Eur J Crime Crim Law Crim Justice 27(4):267–291
3. Cashman P, Ginnivan E (2019) Digital justice: online resolution of minor civil disputes and the use of digital technology in complex litigation and class actions. Macquarie Law J 19:39–79
4. Denisov IS (2018) The development of e-justice in Russia. Bull St Petersburg Univ Ministry Intern Aff Russ 1(77):101–104 (in Russian)
5. Duguzheva MK, Simaeva EP (2019) Transformation of the legislation on culture in the context of digitalization. Vestnik Perm Univ Law Sci 2:190–208
6. Gubulov AK (2019) Problems of protecting personal data in the electronic justice System. Law Justice 8:148–149 (in Russian)
7. Kasper A, Laurits E (2016) Challenges in collecting digital evidence: a legal perspective. In: Kerikmäe T, Rull A (eds) The future of law and eTechnologies. Springer, Cham, pp 195–233
8. Kondyurina YA (2013) Implementation of the principles of the arbitration process in the electronic justice system. Bull Omsk Univ 1(34):157–161 (in Russian)
9. The Arbitration Procedure Code of the Russian Federation dated 07.24.2002 № 95-FZ. http://www.consultant.ru/document/cons_doc_LAW_37800/. Accessed 02 Feb 2020 (in Russian)
10. Valeev DK, Nuriev AG (2019) Electronic document flow in the field of justice in the digital economy. Vestnik Perm Univ Law Sci 3:467–489

Digital Technologies of Tax Administration in Resolving Tax Disputes

O. S. Skachkova

Abstract Reasonable taxation is the main condition for sustainable business development as it determines the trustful attitude of civil society to the state. This article discusses the issues of the advantages and disadvantages of tax administration informatization, including the purpose of the study in the framework of this article is to consider the concept of automating tax planning and control in Russia as an institution that ensures the trust of civil society to the state, as well as to clarify the significance of these automated systems when appealing acts of tax authorities. The study of practical world experience and the identification of difficulties in the process of informatization of the domestic tax system bases the conclusion about the use of these automated systems as a means of proof in the framework of the arbitration process. The study also aims to formulate general conclusions on the development of the information component of the tax system.

Keywords The tax system of the ASC on VAT-2 · Digital economy · Legal consequance · Litigation on VAT correction

1 Introduction

Electronic interaction of business in the tax system as a reflection of the digitalization of the economy is a modern direction of the economy that opens up not only new VAT opportunities, but also reveals additional tax risks for taxpayers during their interaction with the tax authorities. This issue is relevant not only for the Russian state, but is the subject of discussions at international conferences. So, Rodrigues, Azavedo, Reis assess the increase in the efficiency of tax administrations as a result of the introduction of e-government [11]. The role of information technologies use in financial management and taxation of startups in Indonesia [15, 16], the issues of improving the efficiency of tax administration are the subject of interest of scientists in Kazakhstan [19]. Hodzic emphasizes the need for digitalization of the Croatian

O. S. Skachkova (✉)
Samara State University of Economics, Samara, Russia
e-mail: yarmoluik@mail.ru

S. I. Ashmarina and V. V. Mantulenko (eds.), *Current Achievements, Challenges and Digital Chances of Knowledge Based Economy*, Lecture Notes in Networks and Systems 133, https://doi.org/10.1007/978-3-030-47458-4_80

tax authorities in order to organize coordinated work with other member states of the European Union, which is in the process of transition to digital supply [7].

One of the elements of digital economy on tax verification when reporting and checking the tax returns is a system of ASC VAT-2. Questions about the advantages and disadvantages of informatization of various spheres of public relations attract the attention of both legal practitioners and IT specialists. So, questions about the legal consequences of the introduction of ASC on VAT-2 into the tax system as one of the new elements of the digital economy is most relevant. But it is worth paying attention to the use of information technologies in tax administration and resolving tax disputes, for example, when acts of tax authorities on recover of fines and penalties for VAT are appealed.

The current tax legislation and the existing practice of its application do not allow the authorities to make unambiguous conclusions that information from ASC on VAT-2 can be used by both the tax legal authority and civil taxpayers in court as evidence the reliability/fallacy of the conclusions of the tax control staff as part of the arbitration process.

2 Methodology

The research methodology is based on applying general theoretical methods of cognition and special ones. The systematic method makes it possible to present the "tax administration system" as a set of certain elements. Comparative legal, logical methods are indispensable in the analysis of legislation and practical material of the judiciary and cash management, control systems of foreign states. Analysis of the case law of some constituent entities of the Russian Federation allows us to evaluate the role of automated tax control systems in the implementation of justice.

3 Results

1. The information of tax administration regarding the transition from the declaration system to tax monitoring should be accompanied by a parallel improvement in the legal regulation of taxation (in particular, the size of tax rates, the procedure for calculating taxes), as well as the development of the institution of preliminary tax regulation, the positive experience of which has been demonstrated by Western European countries.
2. Both foreign communities and Russian civil society remain urgent—this is a problem of information security and the protection of personal data in the context of comprehensive digitalization. The Russian state allocates considerable funds for the development of information security technologies, for the implementation of the "Information Security" project of the Digital Economy program, the task

of which is to create a safe and sustainable information infrastructure for citizens and representatives of business and the state in the digital space.
3. Evaluating the advantages and disadvantages of world and domestic experience in using technical innovations in the field of tax administration, one should pay attention to the studies of countries such as China and Korea, which offer the use of Blockchain and web scanning systems (crawler technologies) in the organization of tax management.
4. The active development of e-justice in Russia necessitates paying attention to the problem of using these automated systems as part of the resolution of arbitration disputes. So, now practice offers a rather vague position on whether information from ASC on VAT-2 can be used in court as evidence.

On the one hand, we can see rulings where information from ASC on VAT-2 is used in court as evidence in coherence with all other evidence submitted to a case. Taxpayers claims in such cases are rejected.

For example, the Ruling of the Ninth Arbitration appeal court from 01.29.2018 on the case № A40-94,741/17 on the appeal of the act of the tax authority has left the claim of appellant rejected [12]. The basis was the recognition by the court of evidence proving VTA tax fraud obtained during tax audit. The fraud was in creating and dealing with bad-faith suppliers in order to accomplish a non-commodity supple scheme to obtain profit through tax return in the absence of any real supply by creating formal document flow and conducting formal payments.

A similar position is reflected in the Ruling of the Eleventh Arbitration appeal court dated 02.27.2018 on the case № A65-26711/2017 [13]. The tax authority, having analyzed the chain of the counterparty's suppliers of the audited organization by the means of VAT-2 software complex did not reveal the manufacturer.

Organization's appeal on the ruling and proving the court ordered tax liability unlawful was rejected.

There is a different litigation practice that can be found where the information obtained with ASC on VAT-2 is not accepted as proof of taxpayers' bad faith and data is interpreted in favor of VAT taxpayers.

In the Ruling of the Ninth Arbitration appeal court of 01.29.2018 on the case № A40-94741/17 [12], satisfying the company's claim, the court took into account the argument about the reality of the counterparty as the latter has filed tax returns that were validated ASC on VAT-2.

The courts often account the fact that violations on filling out tax returns by counterparties cannot serve as proof of tax fraud on the side of the audited party. For example, in the Ruling of the Fifteenth Arbitration appeal court dated February 13, 2017 on the case № A32-26347/2016, the court stated that the audited organization, cannot be liable for the actions of third parties (suppliers of the 2nd, 3rd level), with which it has no economic relations, that apparently are outside of its reach and cannot be in any way dictated by the defendant [14].

Thus, it is obvious that the use of ASC on VAT-2 software in tax budget system as one of the elements of digital economy is effective with tax audit and detecting tax fraud. However, the use of data from it as sufficient evidence is controversial.

At present, it can be stated that the data obtained through the electronic control method is used by the budget only in conjunction with other evidence in court disputes.

5. Currently, the world community can observe different approaches to taxation of electronic commerce, including in determination of a permanent place of business and determining the nature of income, which causes difficulties for both tax authorities and taxpayers. The main problems regarding the administration of VAT in electronic commerce are related to the taxation of supplies from the organization to an individual, in which suppliers must know the status and location of the client because billing requirements are governed by the VAT legislation of the client's country.

4 Discussion

As part of the development strategy of the information society in the Russian Federation, measures are being taken to improve the institution of tax control. The introduction of modern information technologies contributes to the implementation of the digital economy reform in the activities of state bodies, accelerates the process of making management decisions, simplifies procedures, and reduces budget expenditures of the state. Tax authorities have a special place in the system of federal bodies and are leaders among other state bodies in the application of technical innovations. So, it is noted that the Federal Tax Service of Russia has recently become the largest and one of the most effective IT companies [5]. At present, the bases of tax and customs services are being combined, which will help stop most illegal tax schemes. The Federal Tax Service is changing the policy of mass audits of organizations for a preventive approach based on digital tax control. Nevertheless, the essence of domestic tax policy is to increase tax collection without changing tax rules [1]. If you turn to foreign experience it can be noted that many countries are strengthening control by improving the relationship between the taxpayer and tax authorities, and improving tax legislation. For example, the Swedish tax authorities contribute to the correct calculation and payment of taxes with frequent amendments to the laws governing taxation [4]. The most interesting experience is Belgium, where there is a system of preliminary agreements with tax authorities, or Canada, where there is a voluntary disclosure program, within which it is possible to provide for a reduction in the amount of penalties and sanctions [2]. States often, in addition to or instead of the computerized optimization of tax collection control make significant efforts to simplify taxation conditions and develop the institution of preliminary tax regulation.

With the digitalization of any area of public relations, public administration, whether it is e-justice, e-notary [12], e-government, the problem of information security and non-interference in the private lives of citizens is most acute. Automation of tax planning and control systems as an integral part of the digital economy is no exception. Maphumula and Njenga draw attention to the fact that trust and information security risks are the main factors that determine the full use of the potential of electronic document management systems [9].

Attention is also drawn to the work of U.S. scientists on identifying risk and protection factors associated with the victimization of identity theft. Moreover, persons with a higher socio-economic status, as the authors note, are more often victims of identity theft through credit cards/debit accounts [3].

The assessment of world experience in the use of technical innovations in the field of tax administration is directly related to information security. Many international conferences and seminars are devoted to the problems of developing reliable, safe and effective systems of interaction between subjects. So, Chinese scientists propose, for example, a new BIS system, a system of interaction between banks and tax authorities based on Blockchain [8]. The use of Blockchain is discussed more and more often: the results of research by Myeong and Jang suggest that it is possible to use Blockchain administration in the field of electronic tax collection, to increase the level of security and transparency of taxation [10]. The technology of web scanning (crawler technology) in the organization for effective tax management is also studied [6].

Questions about the advantages and disadvantages of informatization of various spheres of public relations attract the attention of both legal practitioners and IT specialists. In this respect, the informatization of tax control is no exception. So, issues of legal consequences from introduction of tax system ASC VAT-2 as one of the new elements of the digital economy are most relevant. For example, Wolf and Dale note that the introduction of new technologies by the tax authorities to protect and prevent tax violations in France helped to stop fraud with circulatory VAT refunds [20].

The VAT system-ASC VAT-2 is an information resource that provides the following information on tax payers of VAT and their contractors to tax inspectors:

- VAT tax returns, purchases, sales, tax journals in view of the accounting of invoices issued and received,
- information on differences of taxpayers' explanations,
- data on prior tax violations.

A distinctive feature of the latest version of the program complex is the industry approach when conducting tax control during which whole industry sectors are validated and verified as opposed to several companies.

Taxpayers bear the risk of being included in the chain of bad-faith taxpayers if their contracting parties do not fulfill their tax obligations.

The main benefit of the ASC VAT-2 program is the ability to establish inconsistency of information on VAT tax return automatically. It acts as follows:

- the taxpayer submits a tax return for VAT to the tax authority,
- the information on the submitted return form then transfers and accumulates in a Big Data system,

- the ASC VAT-2 program compares information on operations of organizations in tax returns (invoices, sales reports and purchases), for the sole purpose to identify non-compliance of data. The program reveals so-called VAT gaps in a chain of legal operations. For example, a taxpayer can claim a tax return and reflect an invoice when the seller did not pay VAT to the budget and did not reflect invoice in the sales book.

The current tax legislation and the existing practice of its application do not allow to make solid conclusions on whether the information from ASC VAT-2 can be used by both the tax authority and taxpayers as evidence in court for proof of the reliability/fallacy of the conclusions of the tax control staff in arbitration process.

At the same time, for additional charge of VAT tax payments to taxpayers a tax audit can be organized to collect evidence of tax fraud through violation of paragraph 1 of Art. 54.1 of the Tax Code of the Russian Federation, "distortion of information on the facts of economic life, on taxable objects, which are subject to reflection is in the tax data and (or) accounting systems" [18, p. 54].

Information obtained as a result of applying the program complex ASC VAT-2 can be included in materials transferred to law enforcement agencies, and, as a consequence serve as basis for criminal prosecution. It is necessary to pay attention that in 2014 amendments to the Code of Criminal Procedure of the Russian Federation came into force and to open a criminal case against a taxpayer the presence of a tax authorities' act on making a taxpayer accountable for tax fraud is not necessary.

The use of ASC VAT-2 can also result in serious legal consequences for the taxpayer in the event that the program establishes signs of schemes of tax fraud. In such a case, the system automatically generates a report that is further used in initiation of a criminal case, if the amount of the unpaid tax exceeds the one established by the Criminal Code. It is noteworthy that tax crimes can also be qualified under Article 159 of the Criminal Code of the Russian Federation "Fraud", and not by articles related to tax crimes such as Article 198 and Article 199 of the Criminal Code of the Russian Federation [17]. In addition, at the present time, the scheme as mass signer established it is a scheme where one person that has an electronic signature submits to the tax authorities digital tax returns on VAT for several taxpayers.

5 Conclusion

The scientific significance of the work lies in the conclusions based on an analysis of Russian and foreign experience of the role of the institute of tax administration as a necessary element in the development of digital economic relations, of the importance of electronic evidence in the exercise of the right to fair justice in tax disputes. The practical significance of the study is to identify the most acute problems, (including the legal problem of an ambiguous position of the information obtained from the system ASC on VAT-2 as evidence of the tax violation) and composing general directions of development of the legal regulation of informatization of tax administration in Russia.

References

1. Bolgova VV (2019) Some problems of digitalization of taxation in Russia. In: Bortnikov SP (ed) Actual problems of the development of the legal system in the digital era. Samara State University of Economics, Samara, pp 11–13 (in Russian)
2. Boyko NN (2015) The experience of foreign countries in the field of tax administration. Polit State Law 4. http://politika.snauka.ru/2015/04/2862. Accessed 02 Feb 2020
3. Burnes D, DeLiema M, Langton L (2020) Risk and protective factors of identity theft victimization in the United States. Prev Med Rep 17. https://www.ncbi.nlm.nih.gov/pubmed/32071847. Accessed 02 Feb 2020
4. Farikova EA, Vorontsov AD (2019) Analysis of the experience of tax control in foreign countries. Young Sci 47(285). https://moluch.ru/archive/285/64188/. Accessed 02 Feb 2020 (in Russian)
5. Gagarin P (2018) How automation changes tax control. CFO Café. https://cfocafe.co/new-tax-control/. Accessed 02 Feb 2020
6. Han J, Zhou Q, Han H (2019) Research on application of reptilian technology in tax management achievements. IOP Conf Ser: Earth Environ Sci 267(4). https://iopscience.iop.org/article/10.1088/1755-1315/267/4/042021. Accessed 02 Feb 2020
7. Hodžić S (2019) Tax administrative challenges of the digital economy: the croatian experience. EJournal Tax Res 16(3):762–779
8. Lu Z, Wan X, Yang J, Wu J, Zhang C, Hung PCK, Huang S-C (2019) Bis: a novel blockchain based bank-tax interaction system in smart city. In: O'Conner L (ed) Proceedings of the 17th international conference on dependable, autonomic and secure computing. IEEE, New Jersey, pp 1008–1014
9. Maphumula F, Njenga K (2019) Innovation in tax administration: digitizing tax payments, trust and information security risk. In: Ochara NM, Odhiambo JN (eds) Open innovations conference, OI 2019. University of Johannesburg, Johannesburg, pp 304–311
10. Myeong S, Jung Y (2019) Administrative reforms in the fourth industrial revolution: the case of blockchain use. Sustainability 11(14):3971
11. Rodrigues MHP, Azevedo PA, Reis JL (2019) Business intelligence, e-government marketing, and information technology to support tax planning decision making. RISTI – Revista Iberica Sistemas Tecnologias Informacao 2019(E24):198–207
12. Ruling of the Ninth Arbitration appeal court from 01.29.2018 on the case № A40-94741/17. https://sudact.ru/arbitral/court/reshenya-9-arbitrazhnyi-apelliatsionnyi-sud/. Accessed 02 Feb 2020
13. Ruling of the Eleventh Arbitration appeal court dated 02.27.2018 on the case № A65-26711/2017. https://sudact.ru/arbitral/court/reshenya-11-arbitrazhnyi-apelliatsionnyi-sud/. Accessed 02 Feb 2020
14. Ruling of the Fifteenth Arbitration appeal court dated February 13, 2017 on the case № A32-26347/2016. https://sudact.ru/arbitral/court/reshenya-15-arbitrazhnyi-apelliatsionnyi-sud/. Accessed 02 Feb 2020
15. Skachkova OS, Chugurova TV, Gubaydullina EKh (2020) Digital notary as a necessary element of digital economy: international experience. In: Ashmarina SI, Mantulenko VV (eds) Proceedings of the II international scientific conference "global challenges and prospects of the modern economic development". The European proceedings of social & behavioural sciences. Future Conference, London (in press). https://www.europeanproceedings.com/proceedings/EpSBS/volumes/gcpmed-2019-publication-agreed-future-conference. Accessed 02 Feb 2020

16. Supardianto FR, Sulistyo S (2019) The role of information technology usage on startup financial management and taxation. Procedia Comput Sci 161:1308–1315
17. The Criminal Code of the Russian Federation of June 13, 1996 № 63-FZ (as amended on August 2, 2019). http://www.consultant.ru/document/cons_doc_LAW_10699/. Accessed 02 Feb 2020
18. The Tax Code of the Russian Federation of July 31, 1998 № 146-FZ. http://www.consultant.ru/document/cons_doc_LAW_19671/. Accessed 02 Feb 2020
19. Turuntayeva A, Tlegenova F, Kassiyenova K, Abrakhmatova G, Radzhapov A, Alshymbek D (2019) Improving the effectiveness of tax administration through the example of the republic of Kazakhstan. J Legal Ethical Regul Issues 22(2):16
20. Wolf M, Dale S (2019) Tax administrations' adoption of new technologies to protect and ensure tax revenues. ERA Forum 19:457–464

Digitalization of the Openness Principle of Civil Proceedings: Enunciation Issues

G. E. Ageeva

Abstract Digital transformation of all social relations is a sound vector of their modern development. Civil litigation also does not evade this trend: e-justice, audio recording, video conferencing and other ways of digital optimization of justice administration are introduced and actively used. The principle of openness has also undergone digital transformation. In modern conditions, it is possible to broadcast the course of any open trial on the information and telecommunication network "Internet" in case the judge grants such permission. However, in this context, the problem of correct observation of all participants' legitimate interests balance arises as compliance with the principle of openness intersects with the rights of the subjects. This issue is primarily about personal data and a simple reluctance to appear on the Internet in what we might call "a bad light". In addition, the subjects of certain types of relations even though there is neither any secret protected by law nor various aspects of private life that may be affected might not want to provide information about themselves and their specific legal conflicts to an unlimited number of people. However, the digital broadcast of the trial is actively used in domestic civil litigation, and therefore mechanisms to maintain balance of interests of all litigation participants needs to be studied and developed.

Keywords Principle of openness · Trial broadcast · Civil procedure · Digitalization of personal data protection

1 Introduction

Openness and publicity of the proceedings foresee for timely, qualified informing of the public about the activities of the courts which helps to increase the level of legal awareness on the judicial system, and they remain the guarantee of fair justice, and also perform the function of public control of the courts represented by judges. Therefore the principle of openness represents the initial beginning of

G. E. Ageeva (✉)
Samara State University of Economics, Samara, Russia
e-mail: galinaageevva@mail.ru

S. I. Ashmarina and V. V. Mantulenko (eds.), *Current Achievements, Challenges and Digital Chances of Knowledge Based Economy*, Lecture Notes in Networks and Systems 133, https://doi.org/10.1007/978-3-030-47458-4_81

civil proceedings, guaranteeing, in the opinion of the Plenum of the Supreme Court of the Russian Federation [7], fairness, publicity and the impartiality of the trial. Openness is ensured by the access of all visitors to the trial or to information about it, including through the use of the Internet telecommunications network. Any open trial should take place in a room that would accommodate all persons wishing to take part in it. Journalists, along with other entities, have the right to be present during the proceedings [8]. Judges are prohibited from obstructing them for reasons of professional affiliation or other grounds not provided by law.

The rules on the openness of proceedings are not subject to application if there are grounds for conducting a closed trial. These include: proceedings in cases containing information constituting a state secret, the secret of adoption of a child, as well as in other cases if required by federal law [11]. In addition, if there is a motion from an interested person, the judge has the right to consider the case in closed order on the grounds of securing the right to privacy or preservation of information constituting a commercial or other secret protected by law, as well as information whose open discussion can interfere with justice or may lead to adverse consequences for the subject [5, 10]. In this context, it is important to emphasize that such motions will not constitute an unconditional basis for proceeding to a closed hearing.

Digitalization of the openness principle in civil litigation forms over the past ten years. First, it is worth mentioning the federal law of 2008 "On providing access to information on the activities of courts in the Russian Federation" [3], which fixed the obligation of the court to publish information on the activities of courts and post the texts of adopted judicial acts in the information and telecommunication network "Internet". In addition, the same federal law established the possibility of broadcasting court hearings on radio, television and the Internet. It can be stated that modern digital technologies are firmly set into the modern civil process.

At present, during an open trial, any party even the one that has nothing to do with the case under consideration but is actually present in court has the right, upon obtaining the prior permission of the judge, to broadcast the course of the trial on the Internet [12]. The decision on the admissibility of broadcasting the court session is determined by the court that must account for the interests of justice, ensuring the safety of participants of legal proceedings, preventing the disclosure of information referred in the manner established by federal law to information constituting a state or other secret protected by law. It is worth mentioning that the court does not have the right to refuse the request for broadcast only because of the subjective reluctance of the participants in the process. And such reluctance can be quite logical: we are talking primarily about personal data and a simple reluctance to appear on the Internet in perhaps not the "best light". In addition, the subjects of certain types of relations in which neither any secret protected by law nor various aspects of private life may be affected may not want to provide information about themselves and their specific legal conflicts to an unlimited number of people. The procedure for publishing judicial acts and rules on the exclusion of certain information from them is regulated in sufficient detail. With the peculiarities of the legal protection of the subjects of legal proceedings during the broadcast of the proceedings, the situation is more complicated. In view of this, the study of certain aspects of digitalization of the openness principle and the development of proposals to ensure the rights and legitimate interests of private parts are currently relevant and significant. This draws significance precisely to the sphere of civil litigation, since the main participants are most often individuals in cases arising from personal property and non-property relations.

2 Methodology

The main methodological approaches used in this study, besides general philosophical methods, were systemic and integrated approaches. Legal phenomena of modern digital society are considered as elements of a system that are interconnected and interdependent. Therefore, when discussing public law principles, it is necessary to pay attention to their impact on and relationship with private law.

In addition, when studying the elements of the system, it is important to use paired methods of scientific knowledge: analysis and synthesis; induction and deduction. Moving first from the particular to the general, then from general to particular so it is possible to reach more substantiated scientific results that, when tested, will have a positive impact on law enforcement practice.

In consideration of legal phenomena it is also necessary to actively use the observation method. First, attention should be paid to the genesis of the regulatory basis and emerging law enforcement practice, the advantages and disadvantages of which should be analyzed in detail. And when trying to build your own conclusions, it is also important to draw on existing scientific publications on the topic of research, using the method of theoretical experiment, you can come to more reasoned and therefore reasonable conclusions.

3 Results

Nowadays, one of the most striking manifestations of digitalization of openness principle in civil proceedings is that full, partial, live and delayed broadcasts of hearings can be transmitted on the radio, television and on the Internet [4]. Currently, broadcasting of any processes through modern digital technologies, not just legal ones grows in popularity [9]. Therefore, we believe that the digital transformation of the openness principle will very soon function actively in the direction of litigation broadcasting. However, there are concerns that a significant violation of the ratio of private and public interests balance will thereby occur. The fact is that modern, "digital" understanding of the principle of publicity, information on legal proceedings and the texts of judicial acts adopted are subject to publication in the public domain on the Internet. Exceptions are cases considered in closed meetings. In addition, when placing judicial acts, the legislator ordered the courts to exclude personal data from their text [6]. However, the legislation does not contain such a ban, restriction or other means to ensure the rights of subjects to the inviolability of personal data in case of broadcasting the justice process. Although at the same time it regulates the broadcasting rules on the Internet in broad detail. Thus, a slight advantage is already noticeable in observing the public interest of the judiciary—the principle of publicity overweighs and causes ignorance on the part of private interests of the parties. There is an urgent issue with the development of a mechanism to ensure the rights and legitimate interests of all participants in the process; the creation of basis for the expansion of existing grounds for refusing to broadcast due to objections from individuals, possibly using the constructions of the evaluation concept. It seems important to highlight the issue of protecting personal data in the context of digitalization of the openness principle. The Federal Law "On Personal Data" states that in case of a person's participation in constitutional, civil, administrative or criminal proceedings, court proceedings in arbitration courts, personal data processing is mandatory, and the consent of the subject is not required [4]. Moreover, for the purposes of justice, processing of special categories of personal data and even biometrics is allowed free of any restrictions. However, to comply with the subject's right to restrict any further distribution of personal data, as was mentioned before, when publishing the texts of judicial acts on the Internet, personal data is removed from the contents of the document. When broadcasting personal data can become vulnerable to any third parties without the consent of the subject and that not quite for the purpose of justice, which is a violation of the private interests of the subject. In this regard, it is worth mentioning that the attention to the proper protection of personal data in a digital society is very high. Some authors indicate that "the improvement of information technology requires the need of reflection of a citizen's right to protect personal data in the definition of legal personality" [13, p. 128].

4 Discussion

Digital transformation of the openness principle considering possible broadcast of a hearing on the information and telecommunication network "Internet" leads to the need to analyze the advantages and disadvantages of such an innovation. Democratic society requires maximum transparency of public authorities. This ensures the possibility of performing a very effective function of public control over the legality, objectivity and impartiality of legal proceedings, reducing the possibility of lawlessness, and maximizing the observance of the rights, freedoms and legitimate interests of entities not endowed with authority. Some scholars emphasize that "publicity is aimed at protecting the parties from secret justice, which is not subject to public control, and it is publicity that is one of the means of maintaining confidence in courts of all levels" [1, p. 47]. As for the shortcomings, it is possible to mention the significant costs for the federal budget that are so only if the broadcast is mandatory and recordings of hearings are to be posted on courts' websites [2]. However, the main drawback, in our opinion, is the insufficient legislative elaboration of the observance of personal interests on privacy and the right to protection of personal data. Talking about maintaining a balance of private and public interests, it would be possible to fix a rule on the mandatory broadcasting of the trial and its subsequent fixation on the court website in constitutional and administrative proceedings. Thus, the observance of the public interest would be ensured and access to justice will be guaranteed. In civil litigation, where an individual is always present, it is necessary to expand the grounds for broadcast restriction if, for example, there is a motion of any party on that.

5 Conclusion

Based on the results of this study, the author would like to mark the following in conclusion:

1. The fact that the digital era has long become a reality of our society does not mean at all that it is possible to introduce any modern technologies into law enforcement practice without their detailed scientific and legislative elaboration.
2. Digitalization of the principle of openness in civil proceedings is an inevitable and positive phenomenon of modern law enforcement. However, it is necessary to protect the subjective rights of participants in legal proceedings by expanding the grounds for refusal of hearing broadcast.
3. In order to ensure the protection of personal data and restrict access of an unlimited number of persons to it without permission, in our opinion, legislative regulation is required to prohibit continuous broadcast and to enforce withdraw from the record of those time intervals when a person discloses any personal data or other personal circumstances that it would like to leave out from the public eye.

References

1. Abrosimova EB (2009) Essays on the Russian judicial system: reform and results. Institute of Law and Public Policy, Moscow
2. Ageeva GE, Lang PP, Loshkarev AV, Chugurova TV, Churakova EN (2018) Peculiarities of protecting the rights of participants of financial markets in court. In: Popkova EG (ed) The future of the global financial system: down-fall or harmony. Lecture notes in networks and systems, vol 57. Springer, Cham, pp 545–552
3. Federal Law of December 22, 2008 № 262-FZ (as amended on December 28, 2017) "On Providing Access to Information on the Activities of Courts in the Russian Federation". http://www.consultant.ru/document/cons_doc_LAW_82839/. Accessed 02 Feb 2020
4. Federal Law of July 27, 2006 № 152-FZ (as amended on December 31, 2017) "On Personal Data". http://www.consultant.ru/cons/cgi/online.cgi?req=doc&base=LAW&n=286959&fld=134&dst=100257,0&rnd=0.9278161440984669#06065238188676227. Accessed 02 Feb 2020
5. Latypova EYu, Nechaeva EV, Gilmanov EM, Aleksandrova NV (2019) Infringements on digital information: modern state of the problem. In: Mantulenko V (ed) Problems of enterprise development: theory and practice. SHS web of conferences, vol 62, no 10004. EDP Science, Les Ulis

6. Nenkov N, Petrova M, Dyachenko Y (2016) Intelligence technologies in management and administration of justice. In: Cristea L, Antonietti A, Parga SL, Holman R (eds) Proceedings of the 3rd international multidisciplinary scientific conference on social sciences and arts. STEF92 Technology Ltd, Albena, pp 385–392
7. Plenum of the Supreme Court of the Russian Federation Resolution of December 13, 2012 N 35 "On the openness and transparency of legal proceedings and on access to information on the activities of the courts". http://www.consultant.ru/document/consdocLAW139119/. Accessed 02 Feb 2020
8. Proskuryakova MI (2018) E-Justice in Germany: current status and development prospects. Bull Saint-Peterburg Univ Law 9(3):433–477
9. Roscini M (2016) Digital evidence as a means of proof before the international court of justice. J Confl Secur Law 21(3):541–554
10. Taran KK (2020) Approaches determining the applicable law using internet technologies in the digital economy. In: Ashmarina SI, Vochozka M, Mantulenko VV (eds) Digital age: chances, challenges and future. Lecture notes in networks and systems, vol 84. Springer, Cham, pp 622–629
11. The Code of Civil Procedure of the Russian Federation of November 14 2002 №. 138-FZ (as amended on December 2, 2019). http://www.consultant.ru/document/cons_doc_LAW_39570/. Accessed 02 Feb 2020
12. Vaníčková R (2019) New functionality, security and protection of CCTV systems: technological progress and digital society development. In: Mantulenko V (ed) Proceedings of the conference "eurasia: sustainable development, security, cooperation". SHS web of conferences, vol 71, no 03003. EDP Science, Les Ulis
13. Zaman SK-U (2009) Civil status of individuals in Germany, Italy, France and Russia: PhD thesis. Peoples' Friendship University of Russia, Moscow

Recognition of Cryptocurrency as an Object of Civil Rights by Russian Courts

E. K. Gubaydullina

Abstract The article covers one of the most pressing and discussed topics in business—cryptocurrency. A narrow understanding of this category in our country and a cautious, rather negative even, attitude towards it is associated with a certain lack of understanding of the "Blockchain" technology in society as well as the lack of legal regulation by the legislator. The latter results in significant disagreement between the positions of state bodies regarding the technology. The relevance of the studied topic is that both the cryptocurrency itself and the legal basis for its functioning in Russia are in their infancy, however, its adoption and development can lead to significant economic benefits. At present, the need to develop the legal framework for the use of cryptocurrency and designate the further vector of the generation of the legislative framework aimed at legal regulation of this industry has become quite acute. In the article also the analysis of the emerging judicial practice of Russian arbitration courts and courts of general jurisdiction related to the recognition of cryptocurrency as an object of civil rights is carried out.

Keywords Cryptocurrency · Digital rights · Bitcoin · Objects of civil rights · Judicial acts

1 Introduction

The recent active development of cryptocurrency turnover in Russia in the absence of legal norms defining the concept of "Cryptocurrency" and the procedure for conducting transactions with it leads to various and often directly opposite application of the current legislation of the Russian Federation in Russian courts' rulings on the matter. As a rule, the study of cryptocurrencies starts with an analysis of bitcoin. Among other types of cryptocurrencies, such as Ethereum and litecoin, is bitcoin,

E. K. Gubaydullina (✉)
Samara State University of Economics, Samara, Russia
e-mail: elmira_zaripova@mail.ru

S. I. Ashmarina and V. V. Mantulenko (eds.), *Current Achievements, Challenges and Digital Chances of Knowledge Based Economy*, Lecture Notes in Networks and Systems 133, https://doi.org/10.1007/978-3-030-47458-4_82

which ranks first in importance [3]. Bitcoin is a digital currency that uses protocols and cryptographic algorithms to determine the security of transactions and create new ones [8].

2 Methodology

Various methods of scientific knowledge were used in this paper. The main ones are materialist dialectics, as a universal approach to objective knowledge of reality and a method of system knowledge. Also such general scientific methods as analysis and synthesis; induction and deduction were used. Private-scientific methods of cognition were applied too, such as formal legal, comparative legal and logical ones. For example, the formal legal method proved useful in the interpretation and study of regulatory documents.

3 Results

The legislator chose to enact legislation on transactions of digital objects progressively, and extensive interpretation of enacted amendments allows to accept cryptocurrency as an object of digital rights. In addition, on May 22, 2018, the State Duma adopted the Draft Federal Law № 419059-7 "On Digital Financial Assets" in the first reading (as amended by the State Duma of the Federal Assembly of the Russian Federation in first reading on 05.22.2018) (Draft Federal Law № 419059-7 "On Digital Financial Assets", 2018) which introduces the concept of "cryptocurrency" [7]. This bill directly establishes that cryptocurrency is property but is not a legal means of payment in the Russian Federation. However, to date, further consideration of this bill has been postponed, which makes its fate not quite certain, and the status of the cryptocurrency and operations with it are still unresolved by Russian legislation [2].

It is needless to say that the amendments to the Civil Code of the Russian Federation [10] (hereinafter—the Civil Code of the Russian Federation) on digital rights have fundamentally changed the situation with the unresolved issues of cryptocurrency transactions in Russian legislation for the better. Perhaps, it is just the beginning of changes on the issue. Accordingly, Russian courts will still have to make decisions in this area based on an extensive or restrictive interpretation of Art. 128 of the Civil Code of the Russian Federation [1, 10].

4 Discussion

According to Art. 128 of the Civil Code of the Russian Federation, that fell out of force on 01 October 2019, objects of civil rights included things, e.g. cash and documentary securities, other property, including non-cash funds, book entry securities, property rights, work results and the provision of services, the protected results of intellectual activity and equivalent means of individualization (intellectual property), intangible goods [10].

As a result of the adoption by the State Duma on March 12, 2019 of the amendments to the Civil Code of the Russian Federation, which entered into force on October 1, 2019, the concept of "digital rights" was enshrined at the legislative level in the new article 141.1 of the Civil Code of the Russian Federation. In addition, digital rights simultaneously became the subject of civil rights as a result of their inclusion in Art. 128 of the Civil Code of the Russian Federation. In accordance with the provisions of the new Article 141.1 of the Civil Code of the Russian Federation, obligations and other rights specified as such in the law are recognized as digital rights, the content and conditions for the implementation of which are determined in accordance with the rules of the information system that meets the criteria established by law [10]. Implementation, disposal, including transfer, pledge of digital law and any action that limits the use of digital law is possible only in the information system without contacting a third party [9]. Unless law stats otherwise the owner of digital law is a person who, in accordance with the rules of the information system, can dispose of this right. In cases and on the grounds stipulated by law, that person is recognized as the owner of digital right.

The first judicial act to recognize cryptocurrency as an object of civil rights was the decision of 05 of September 2018 № 09AP-17044/2019 of the Ninth Arbitration Court of Appeal on the appeal of the financial manager regarding the ruling of the Moscow Arbitration Court of 03.03.2018 on refusal in satisfying the requirement to resolve the disagreement between the financial manager and the citizen—the debtor regarding the inclusion of the contents of the cryptocurrency wallet located on the Internet telecommunication network in the bankruptcy estate of the insolvent (bankrupt) citizen. In this case, the court of appeal did not agree with the arguments of the arbitration court of the first instance, proceeding from the fact that cryptocurrency does not apply to civil rights objects on the territory of the Russian Federation, though transactions with cryptocurrency are executed, those transactions are not ensured by legislation, absence of the controlling center in the cryptocurrency system and the anonymity of cryptocurrency users does not allow to determine with certainty who cryptocurrency belongs to. According to the court of appeal, due to the optionality of civil law, the Civil Code does not have a closed list of civil rights [5]. Since the current civil law does not contain the concept of "other property" referred to in Art. 128 of the Civil Code of the Russian Federation, taking into account modern economic realities and the level of development of information technologies, its broadest interpretation is permissible. In addition, the court of appeal applied Article 6 of the Civil Code of the Russian Federation, according to which, if it is impossible to use an analogy of the law, the rights and obligations of the parties are determined on the basis of the general principles and the meaning of civil law (analogy of law) at the requirements of good faith, reasonableness and justice [10]. Accordingly, the appellate court ruled that cryptocurrency cannot be regarded as "other property" and should be included in the bankruptcy estate of a citizen declared insolvent (bankrupt).

Today we can say that the Ninth Arbitration Court of Appeal has set a precedent of recognizing cryptocurrency as an object of civil rights and its inclusion in the bankruptcy estate. The same composition of the court, which passed the first judicial act in Russia on the inclusion of cryptocurrency in the bankruptcy estate in connection with its classification as objects of civil rights, makes similar decisions on an ongoing basis. An example is the resolution of 18.04.2019 № 09AP-17044/2019, by which the court ordered the debtor, declared bankrupt, to provide the financial manager with access to his Bitcoin wallet in order to include its contents in the bankruptcy estate [5].

At the same time, there is a directly opposite judicial practice of considering similar cases. So, in a ruling from 18.10.2018 № F06-38270/2018, the Volga District Arbitration Court refused to satisfy the cassation appeal of the bankruptcy creditor against the determination of the Saratov Region Arbitration Court and the decision of the Twelfth Arbitration Court of Appeal, that is, it refused to consider the neglect to identify electronic wallets and transactions with cryptocurrency during formation the bankruptcy estate of the debtor by the financial manager illegal [4]. In this case, the arbitration court of cassation agreed with the arguments of lower courts that, from a direct interpretation of the rule of law, cryptocurrency does not apply to civil rights objects, as it is not legally regulated in the Russian Federation, transactions with cryptocurrency, are not overseen by law enforcement, the absence of a controlling center in the cryptocurrency system and the anonymity of cryptocurrency users does not allow to establish with certainty who cryptocurrency belongs to. Accordingly, in the court's opinion, the financial manager should not have taken measures to identify electronic wallets and transactions with the debtor's cryptocurrency [4].

In many cases, courts of general jurisdiction hold a similar opinion regarding cryptocurrency. In particular, the Volgodonsky District Court of the Rostov Region, when considering a civil case on the collection of unjust enrichment, refused to recover from the defendant money in rubles received as a result of cryptocurrency sale. According to the court, the plaintiff's arguments about receiving cryptocurrency income, which was later exchanged for rubles and transferred to the defendant's bank card account, are not based on the law, since operations with cryptocurrencies are performed outside the legal field of the Russian Federation, cryptocurrencies are not guaranteed and are not provided by the Bank of Russia. When making a decision of March 18, 2019 in this case, the court drew attention to the Information published by the Bank of Russia on the use of "virtual currencies", in particular, Bitcoins in transactions, which indicates that they have no collateral and entities legally bound by them. The status of cryptocurrencies in the Russian Federation is monetary surrogates, the issue of which, is prohibited throughout the Russian Federation. The court also indicated that cryptocurrency is not provided for by Art. 128 of the Civil Code of the Russian Federation as an object of civil rights, including, is not classified as non-cash money [6]. Bitcoins are not recognized as electronic money, since there is no operator that can control and maintain the type of currency in question. Cryptocurrencies are not regulated by payment system operators, thereby eliminating the totality of organizations necessary for the recognition of bitcoin by the payment system. At the same time, Bitcoin is not a foreign currency either, since it is not a monetary unit of foreign states and an international monetary or settlement unit located in a bank account.

The current practice of considering criminal cases on the legalization (laundering) of money or other property acquired by a person as a result of a crime when cryptocurrency acts as such property is also very interesting. In such cases, the courts of general jurisdiction show marked unanimity, recognizing cryptocurrency as not just property, but in some cases as cash. So, some courts indicate indicates that the money received as a reward for the work on equipping tabs came in the form of a Bitcoin cryptocurrency from depersonalized bitcoin wallets for the Bitcoin wallet in use by the accused in an Internet application designed to store and transfer electronic money—Bitcoin cryptocurrencies. Then the accused, realizing the illegality and social danger of his actions, by making repeated financial transactions with cryptocurrency received as a reward for committing a crime, in order to give it a legal form of possession, use and disposal, used the services of an online Internet exchanger to transfer it to rubles. Thus, the court concluded that the accused was involved in the scheme of money laundering.

Another example is when the defendants legalized (laundered) money received from criminal activity by acquiring digital cryptocurrencies and further using Internet resources to simulate the origin of criminal proceeds from legal sources. Namely, through the Internet resource "bestchange.ru" and the Internet application "Jaxx", they made exchanges for transactions with digital cryptocurrency, systematically bought and sold digital cryptocurrency "Lightcoin", while simultaneously receiving cash income from the difference in exchange rates. According to the court, the use by the defendants of a scheme for acquiring digital cryptocurrencies, which are not regulated by the financial system of the Russian Federation, and the use of which is not approved by the legislation of the Russian Federation, provided them with the opportunity to own, use and dispose of funds obtained by criminal means by imitating the origin of criminal proceeds from legal sources. Thus, in this case, there is a recognition by the court of the Litecoin cryptocurrency as property acquired as a result of the legalization (laundering) of money received by the accused as a result of a crime.

5 Conclusion

Analyzing judicial acts of Russian arbitration courts, we can conclude that the practice of recognizing cryptocurrencies as objects of civil rights is developing gradually. This trend was set in mid-2018 when the first such judicial act was passed. Since the beginning of 2019, more and more arbitration courts have been adhering to this practice, providing an opportunity for financial managers of individuals declared insolvent (bankrupt) to gain access to their cryptocurrency wallets in order to fill the bankruptcy estate. Apparently, now, after the amendments to the first, second and fourth parts of the Civil Code of the Russian Federation have come into force, which have defined the basic concept of "Digital Law" in the system of civil rights, this trend will only intensify.

As it is noted above, the courts of general jurisdiction in their rulings on criminal cases related to the legalization (laundering) of money or other property acquired by a person as a result of a crime when cryptocurrency acts as such property, unanimously recognize it as property, and in some cases, as money. Apparently, this circumstance is largely related to the general features of the criminal law practice of the Russian courts, aimed at convictions. As for civil rulings by the courts of general jurisdiction court as in each case it is a matter of dispute whether to determine the cryptocurrency's belonging to the objects of civil rights or not it is not yet possible to talk about any stable position on this issue.

References

1. Ageeva GE, Lang PP, Loshkarev AV, Chugurova TV, Churakova EN (2019) Peculiarities of protecting the rights of participants of financial markets in court. In: Popkova EG (ed) The future of the global financial system: downfall or harmony, vol 57. Springer, Cham, pp 545–552
2. Bortnikov SP (2020) The state sovereignty in questions of issue of cryptocurrency. In: Ashmarina S, Vochozka M, Mantulenko V (eds) Digital age: chances, challenges and future. Lecture notes in networks and systems, vol 84. Springer, Cham, pp 564–573
3. Chatterjee G, Edla DR, Kuppili V (2020) Cryptocurrency: a comprehensive analysis. Smart Trends Comput Commun 65:365–374
4. Decision the Volga District Arbitration Court refused of 18.10.2018 N F06-38270/2018 N A57-21957/2017. http://www.consultant.ru/cons/cgi/online.cgi?req=doc&base=APV&n=169565#05966734340590909/. Accessed 24 Feb 2020
5. Decision of 05 of September 2018 N 09AP-17044/2019 of the Ninth Arbitration Court. http://www.consultant.ru/cons/cgi/online.cgi?req=doc&base=MARB&n=1444056#002353643983540454З/. Accessed 24 Feb 2020
6. Decision of the Volgodonsk District Court of the Rostov Region N 2-4140/2018 2-505/2019 2-505/2019(2-4140/2018) M-3745/2018 M-3745/2018 of 18.03.2019 N 2-4140/2018. https://sudact.ru/regular/doc/CQABiQ3tU55V/. Accessed 24 Feb 2020
7. Draft Federal Law N 419059-7 "On Digital Financial Assets" in the first reading (as amended by the State Duma of the Federal Assembly of the Russian Federation in first reading on 05.22.2018. http://www.consultant.ru/document/cons_doc_LAW_216363/942772dce30cfa36b671bcf19ca928e4d698a928/. Accessed 24 Feb 2020
8. Faghih MJM, Heidari H (2020) Predicting changes in bitcoin price using grey system theory. Financ Innov 6:13
9. Korobeynikova EV, Yermoshkina SN, Kosilova AF, Sheptukhina II, Gromova TV (2019) Digital transformation of the Russian economy: challenges, threats, and prospects. In: Mantulenko V (ed) Global challenges and prospects for modern economic development. The European proceedings of social & behavioural sciences, vol 57. Future Academy, London, pp 1418–1428
10. The Civil code of the Russian Federation (part one) from 30.11.1994 № 51-FZ (as amended on 03.07.2016). http://www.consultant.ru/document/cons_doc_LAW_5142/. Accessed 24 Feb 2020

Social and Legal Aspects of Remote Employment

M. K. Kot

Abstract The study considers the legal, social aspects of the introduction of remote employment in Russia, to which we include distance work and telecommuters. The author reveals their distinguishing features, shows the connection between the use of remote employment and social and economic aspects of the employee and the employer, reveals the problems of legal qualifications of remote work that impede their use in the digital economy. As a result of the study, the author came to the conclusion that there is only one legal mechanism for ensuring remote employment as a condition for the development of a new type of economic relationship—this is distance work.

Keywords Remote employment · Distance work · Home work · Innovative economy · Digital economy

1 Introduction

The history of Russian labor law dates back more than a century: the period of the origin and formation of its main institutions took place during the industrial type of the country's economy. This circumstance has formed a stable connection of many labor law norms and institutions with the characteristics of the factory type of production. To the greatest extent, this applies to such sections of labor legislation as working hours and rest periods, labor protection, change and termination of labor relations. Actually, the classical type of the labor relationship includes the interaction of following features: personal, property and organizational, providing for the subordination of the employee to the power of the employer, primarily in terms of the organization and management of labor.

With the transition of the world economy from the industrial type to the post-industrial one, it became necessary to rework the classical concept of the labor relationship, since it was possible to replace the employee's direct attachment to the

M. K. Kot (✉)
Samara State University of Economics, Samara, Russia
e-mail: mkroz@mail.ru

S. I. Ashmarina and V. V. Mantulenko (eds.), *Current Achievements, Challenges and Digital Chances of Knowledge Based Economy*, Lecture Notes in Networks and Systems 133, https://doi.org/10.1007/978-3-030-47458-4_83

employer's organization with a new type of employment that does not require the employee to stay in employer's premises. In this case, the traditional element of the employee's legal status—his accountability and subordination is transformed into "final" control—control not of the labor process, but of its result. Thus, the institute of remote employment is formed, which is represented in the world economy in several forms—this is homework and telework. By virtue of existing international conventions [1], the differences in these types of employment emphasize the place of work (at home), and information technology when doing work outside employer's premises during telework [15]. In Russia and the former Soviet Union countries, the introduction of digital products and information methods of interaction in all spheres of the population's life is recognized as a priority area of economic development, especially when government agencies and recipients of public services interact. In the private sector, the idea of innovative development of enterprises and the use of electronic technologies are supported.

The innovative way of economic development involves the expansion of labor potential, involvement in production of the most active part of the population—young educated people who expect to get the opportunity to study, to advance in service, while effectively combining work and other social responsibilities. Only the flexible employment option, which provides for a deviation from traditional models of labor relations that developed in the Soviet era and are maintained inertia until today, fully corresponds to such expectations of workers. The traditional concept of labor relations is not only legally enshrined in Art. 15 of the Labor Code of the Russian Federation [12], it is also ensured by established personnel and judicial practices. In the framework of this research, we conducted a study aimed at assessing legal security of the idea of maximizing the use of flexible employment in Russia as a condition for the country's innovative development, and social effects of its application.

2 Methodology

The author used classical methods of interpretative jurisprudence—the analysis of regulatory legal acts and materials of judicial practice. In this case, the techniques of formal logic were used. To assess the economic and social effect of distance employment, the author analyzed sociological studies in this area conducted in Russia. The author did not carry out independent polls and other techniques within the framework of sociology. In addition, the official statistics on employment was used.

3 Results

As a result of our study, we identified the following legal models of flexible employment in Russia.

1. Remote work is a form of employment, providing for the possibility of an employee to be in official labor relations, but without going to a stationary workplace, to the employer's office, to perform a certain labor function using digital technologies, primarily the Internet. The positive side of this employment model is the provision of the employee with all social guarantees provided for by the labor legislation of the Russian Federation, with independent planning of his working time and the ability to combine labor and social, family responsibilities. A feature of the regulation of this type of employment is the need for a remote employee to have a condition on the remote nature of work in the employment contract, as well as the use of electronic document management, including signing of documents by electronic digital signature.

The legal fixing of this type of employment in Russia occurred in 2013 through the introduction of appropriate amendments to the country's Labor Code. In the same period, sociological studies of prospects for the development of this model were carried out, which confirmed the existence of expectations of the widespread use of remote work with the inclusion of more than a third of the country's population in this work by 2020 [13]. However, in general, the employment structure has not undergone significant changes in this direction. The main part of large and medium-sized businesses retains the model of interaction with employees that has developed over the decades, leaving only a small part to the share of remote employees. The analysis of Internet job sites showed a noticeable increase in the share of small businesses and microenterprises, individual entrepreneurs who allow remote employees to be part of their team. This is, first of all, the sphere of online stores, computer services, accountants and call managers. However, the frightening fact is that only a few employers offer formalization of relations as labor relations. The rest prefer to attract workers either without execution, or on the terms of civil contracts with remuneration for each task performed or for the day worked. For Russian law, such methods of organizing labor are illegal, and employers can be held administratively liable in the form of a large fine. The social consequence of this type of labor is the absence of pension savings and payments for temporary disability, pregnancy, childbirth [16].

Analysis of official databases of legal information showed us that the practice of applying the norms on distance work is extremely insignificant and concerns only two cases of contesting the termination of an employment contract with a remote employee. In both cases, the court supported the employer, recognizing his right in dismissal for failure to submit reports on the work done by a remote employee, as well as in dismissal as a result of a change in the organization's development concept [1, 2].

The author believes that the opportunity to establish conditions in the employment contract with a remote employee on additional grounds for terminating it in the law does not generally undermine the existing principles of international and Russian labor law. However, the classification of these conditions should be carried out taking into account the general position of the law regarding the classification of grounds for dismissal as guilty and innocent. Guilty grounds, which include failure to submit reports, absenteeism, require disciplinary offenses taking into account the evidence of employee guilt and prohibitions of dismissal during temporary disability, vacation.

Innocent grounds, as in the above case, a change in the concept of the organization's work, the closure of certain areas of business, should not be attributed to grounds for termination of relations, but to such institutions of the labor legislation of the Russian Federation as deadtime, reduction in the number of employees, or a change in the employment contract. The court in this case did not apply the principles of labor law. It applied the formal legalization of conditions on additional grounds for terminating the contract with a remote employee. In our opinion, the prevalence of the formal approach of the court to understanding the basic principles of labor legislation significantly reduces the level of protection of the rights of remote workers.

The social effects of the widespread adoption of distance work are mixed. Obviously, with this type of employment, a person has more opportunities to carry out his family functions [9]. However, the polls conducted among women working remotely showed their dissatisfaction with the lack of communication in the team, which indicates a low self-realization of such employees, the emergence of the effect of "disconnection" of the employee from society. In addition, remote employees are often deprived of the opportunity to improve their skills. They perform, as a rule, tasks of the same type, which ultimately leads to their degradation and reduced efficiency of their work.

There are social effects associated with the employer's point of view about distance work in organizations. For many managers, loss of control is a strong deterrent to hiring such an employee. In addition, for many managers, savings in creating jobs is not obvious in view of the need to purchase expensive equipment for organizing remote communications and the introduction of electronic document management [11]. Thus, it can be stated that the institution of distance employment is not in great demand due to strict formalization of labor relations (this obstacle is important, first of all for small businesses), and also due to the unwillingness of medium and large businesses to abandon traditional employment models in favor of remote ones.

2. Telecommuters—this type of employment involves the fulfillment of a labor function by an employee at home with or without employer materials. This type of labor relations was known even to the Soviet rule of law, since universal employment as a principle of legal regulation of labor in Soviet times could not be achieved only by creating jobs by the employer—people living in remote territories (Far North) or raising children could fulfill their job responsibilities only at home.

Thus, home-based labor is not a product of the post-Soviet economy. On the contrary, it is more consistent with the factory type of production only in small sizes, since telecommuters, unlike remote workers, perform work without special information equipment—the presence of information systems is not a defining sign of home-based labor. This type of employment is similar to handicraft work and labor under civil law contracts, since the telecommuter is obliged to transfer the result of the work to his employer. The condition that combines the two types of remote work—telecommuting and remote—is the condition of the place of work—outside the office owned by the employer. In our opinion, the preservation of the two indicated forms in the current legal system is unnecessary, since the actual place of work—at home or in any other premises not owned by the employer—does not have legal significance. In addition, the use of the Internet and other means of remote communication are necessary for the performance of almost any work, therefore this feature is currently inherent in all types of remote employment. Due to the characteristics described above, telecommuting is not widespread and competes with other types of employment, primarily self-employment.

4 Discussion

Remote employment issues make up a large part of the current legal debate both in Russia and abroad [6, 8, 13, 14, 16]. Most often, the problem of remote employment is considered in Russia in the aspect of introducing mechanisms of the digital economy [3, 5, 10]. In addition, the private issues of the legal classification of distance relations, telecommuting are also being developed in Russia. There are also economic assessments of distance employment as a condition for the country's innovative development. Conceptually, the problems of distance labor are solved in the works of Vasilieva et al. [17]. The doctrinal basis of these works is to study the signs of a classic labor relationship and identify signs that are transforming under the influence of new economic conditions (atypical employment) [4].

In our opinion, the progressive direction of scientific research in this area should be the work carried out at the intersection of economics, law and sociology, thus it will be possible to conduct a comprehensive assessment of remote labor in the aspect of the employee as a person, the employee as human capital in the economy, and remote labor as mechanism for improving the efficiency of the economy [7].

5 Conclusion

As a result of the study, the author came to the conclusion that there is only one legal mechanism for ensuring remote employment as a condition for the development of a new type of economic relationship—this is distance work. In general, one can state the sufficiency of this legal institution to meet the needs of employers and workers in organizing jobs outside the employer's office. The slowdown in the spread of this type of employment is due to a number of economic and social reasons, as well as the technological inability of many employers to switch to online cooperation with staff in view of the existence of territories that are not fully covered by the Internet.

References

1. Appeal ruling of the Moscow City Court of 12.04.2017 № 33-4427/2017. https://epam.ru/rus/media/view/kak-oformit-sotrudnika-na-udalennuyu-rabotu-kommentarii-anny-ivanovoi-dlya-portala. Accessed 28 Dec 2019
2. Appeal ruling of the Sverdlovsk Regional Court dated 05/11/2017 in case No. 33-7310/2017. http://www.ekboblsud.ru/sudpr_det.php?srazd=6&id=136&page=. Accessed 28 Dec 2019
3. Belitskaya IYa, Kuznetsov DL, Orlovsky YuP (2018) Features of the regulation of labor relations in the digital economy: monograph. Contract, Moscow
4. Chernykh N (2019) The influence of atypical forms of employment on the theoretical concepts of labor relations (on the example of norms on distance labor). Actual Probl Russ Law 8:108–117. https://doi.org/10.17803/1994-1471.2019.105.8.108-117
5. Davletgildeev R, Klimovskaya L (2019) Legal status of platform workers in Russia: right on unemployment and social assistance. Hum Soc Sci Rev 7(6):639–643
6. Ding N, Bagchi-Sen S (2019) An analysis of commuting distance and job accessibility for residents in a U.S. legacy city. Ann Am Assoc Geogr 109(5):1560–1582
7. Dorofeeva A, Nyurenberger L (2019) Trends in digitalization of education and training for industry 4.0 in the Russian Federation. IOP Conf Ser Mater Sci Eng 537(4):042070
8. ILO Convention No. 177 "On Home Work" (concluded in Geneva on June 20, 1996). The EAEU Member States have not ratified this Convention. https://www.ilo.org/dyn/normlex/en/f?p=NORMLEXPUB:12100:0::NO::P12100_INSTRUMENT_ID:312322. Accessed 28 Dec 2019
9. Karlik A, Platonov V, Yakovleva E (2019) System analysis of the implementation of the virtual forms human-computer interaction in the higher education sector results of the review of the experience of implementation of the virtual forms. In: Nordmann A, Moccozet L, Volkova V, Shipunova OA (eds) Proceedings of the scientific and theoretical conference communicative strategies of information society. ACM international conference proceeding series (157195). ACM, New York
10. Kot M, Shpanagel F, Beloserova O (2020) Problems of digital technologies using in employment and employment relations. In: Ashmarina SI, Vochozka M, Mantulenko VV (eds) Digital age: chances, challenges and future. Lecture notes in networks and systems, vol 84. Springer, Cham, pp 548–556
11. Kramer B, Engler S, Bischofberger I (2019) Distance caregiving-empirical insights from an employer perspective. Z Gerontol Geriatr 52(6):546–551
12. Labor Code of the Russian Federation № 197-FZ (30.12.2001). http://www.consultant.ru/document/cons_doc_LAW_34683/. Accessed 28 Dec 2019
13. Meske C, Kissmer T, Stieglitz S (2020) Bridging formal barriers in digital work environments – investigating technology-enabled interactions across organizational hierarchies. Telematics Inform 48:1–14
14. RBC (2015) A fifth of Russians will work remotely by 2020. https://www.rbc.ru/technology_and_media/17/06/2015/5580515f9a7947e7bf4bfc99. Accessed 28 Dec 2019
15. Saurin SA (2019) Features of the regulation of Russian norms on distance work for a circle of persons and their relationship with international standards. Actual Probl Russ Law 10:86–92. https://doi.org/10.17803/1994-1471.2019.107.10.086-092
16. Thomas M (2019) Employment, education, and family: revealing the motives behind internal migration in Great Britain. Popul Space Place 25(4). https://onlinelibrary.wiley.com/doi/full/10.1002/psp.2233. Accessed 28 Dec 2019
17. Vasilieva YuV, Shuraleva SV, Brown EA (2016) Legal regulation of distance work: problems of theory and practice: monograph. PSNIU, Perm

Shaping the Model of Digital Economy in Russia and Its Regions

V. I. Menshchikova, E. Yu. Merkulova, N. V. Molotkova, and E. P. Pecherskaya

Abstract The article reveals the essence of the digital economy resulting from the transformation of technological processes in the field of information and communication. It was revealed that digital technologies enhance education of the population, help to reduce costs in the exchange of knowledge, and are a basic resource in the development and implementation of innovative projects. The paper considers various approaches to assessing the implementation of digital technologies, presents the analysis of Russia's place in the field of digital technology development in the world. Russia is compared with other countries of the world by several indicators—the ICI Development Index, the E-government Development Index, the Global Cybersecurity Index, the international Digital Economy and Society Index, the global connectivity index. In accordance with the obtained values, the countries are divided into three clusters: developing, mid-developed and developed economies. The main indicators of the use of information technologies and information and telecommunication networks in the organizations of the Russian Federation are presented and the corresponding classification of the regions is constructed. Three main prerequisites for the formation of a digital economy model in modern Russia are identified.

Keywords Digital economy · Digital technologies · Economic growth · Region · Quality of life

V. I. Menshchikova · E. Yu. Merkulova · N. V. Molotkova (✉)
Tambov State Technical University, Tambov, Russia
e-mail: nmolotkova@list.ru

V. I. Menshchikova
e-mail: menshikova.vi@mail.tstu.ru

E. Yu. Merkulova
e-mail: merkatmb@mail.ru

E. P. Pecherskaya
Samara State University of Economics, Samara, Russia
e-mail: pecherskaya@sseu.ru

S. I. Ashmarina and V. V. Mantulenko (eds.), *Current Achievements, Challenges and Digital Chances of Knowledge Based Economy*, Lecture Notes in Networks and Systems 133, https://doi.org/10.1007/978-3-030-47458-4_84

1 Introduction

In modern conditions, the competitiveness of any state depends on the pace of development of the digital sector of the economy. Moreover, the economic development of modern countries is increasingly influenced by digital technology. It should be noted that the technological lag does not allow moving to a new level, since progress in the field of digitalization is taking place at a high speed, and the replacement of software and equipment used takes an average of three years. Therefore, the creation and development of the areas of information and telecommunications infrastructure, as well as the introduction of their products in various sectors of the economy, becomes extremely important. As world experience shows, digitalization can improve the quality of services provided to the population, the efficiency and effectiveness of management procedures, as well as the stability and security of the national economy.

Today, for the socio-economic development of Russia, digitalization processes in all areas of the national economy are becoming increasingly important. In the near future, Russia will face the challenge of ensuring the economic growth through the introduction of breakthrough innovations. However, the structural flaws in the digital transformation ecosystem, the lack of digital skills, limited access to capital markets and the lack of an open innovation culture limit Russia's ability to achieve fundamental technological goals. Therefore, it becomes extremely necessary to study the possible ways of forming the digital economy model in Russia, as well as in a regional context. To achieve the goal of the study, the article discusses the problems of: determining the essence of the digital economy and its main features; identifying the role of digital technology for economic growth; disclosing various approaches to assessing the implementation of digital technologies; analysing Russia's place in the field of digital technology development in the world; creating a classification of Russian regions by indicators of the use of information technologies and information and telecommunication networks in organizations; determining the basic prerequisites for the formation of a model of the digital economy in modern Russia, influencing further development of digitalization. The aim of the article is to identify key trends in the implementation of the digital economy model in Russia through its comparison with other countries by a number of indicators, as well as to determine the degree of readiness for digitalization in the context of regions.

2 Methodology

The research methodology is based on the methods of comparative analysis (comparing the values of digitalization indicators of the socio-economic system of Russia with those of other countries) and trend analysis (to determine the growth rates of these indicators in Russia and its regions). The methods for collecting primary economic information, including analysis of legislative and regulatory acts of the

Russian Federation, official statistics, and analysis of other open sources of information were used. Also, we used analytical introspection to determine the estimated phenomena of the studied object using practical materials analysis, summarized the scientific research of foreign and Russian scientists in the field of digital economy, and assessed the existing trends in the development of the digital economy. In the study, we used the latest theoretical and applied research on the formation of the digital economy in modern Russia. These included the works of Vertakova and Plotnikov [13], Banche et al. [4], Dobrynin et al. [6], Efimushkin et al. [7], Ivanova and Nikitin [9], Kolmykova et al. [10], Sudarushkina and Stefanova [12]. This study summarized some empirical data on the priorities for implementing the digital economy model in Russia. We relied on the hypothesis that the pace of economic growth and improvement of the living standards of the population of Russia depend on the pace of development of the digital economy.

3 Results

The digital economy is rapidly gaining ground. In Russia, in 2019, the transition from analogue to digital television was completed; this allowed the population to receive a video signal with high image quality. Domestic costs for the development of the digital economy amounted to 3.3 trillion Russian rubles, which was 3.6% of GDP. Costs in the household sector amounted to 1.2 trillion Russian rubles or 1.3% of GDP. In recent years, Russia has significantly improved its position in international ratings of the development of the digital economy. Thus, according to the ICI Development Index in 2018, Russia ranks 45th out of 176 countries in the world, and if we look at the dynamics, since 2008 it has improved by 4 positions. This index includes three subindexes: access to ICT (Access sub-index), use of ICT (Use sub-index), and practical skills in the use of ICT (Skills sub-index). Iceland occupied the first position in 2017 on this index, in comparison with the previous year, it was ahead of the Republic of Korea and Switzerland is in third place. Russia in terms of the development of the ICT index is between Portugal and Slovakia [1].

According to the E-government Development Index, in 2018, Russia ranked 32th out of 193 countries of the world, and had the most significant breakthrough in this area; compared to 2008, the improvement was by 28 positions. This index includes three key positions: the development of online government services (Online Service Index), the telecommunication infrastructure of ICT (Telecommunication Infrastructure Index), the development of human capital (Human Capital Component). Denmark was the leader in the ranking of this index, Australia was second and Republic of Korea was third. Russia was between Israel and Poland. Considering the components of the e-government development index, it should be noted that Russia has significant potential, since it occupied the 25th position in the development of online public services, and the 28th position in the human capital development index, which proves a positive trend. A restraining factor was the development of telecommunications infrastructure, where Russia occupied 45th position; compared to the previous year Russia went down by 7 positions [1].

According to the Global Cybersecurity Index, Russia took 26th place in 2018, which was by 16 positions lower than the previous year. This index includes an assessment on the following components: legal aspects of cybersecurity (Legal), technical aspects of cybersecurity (Technical), organizational aspects of cybersecurity (Organizational), the country's skills in building a cybersecurity system (Capacity Building), international cooperation in the field of cybersecurity (Cooperation). The first three leaders are the UK, the USA, and France. Russia ranked between Italy and China [1].

The International Digital Economy and Society Index was calculated by the Director General of the European Commission's communication networks, content and technologies for non-EU countries in accordance with the methodology of the European Digital Economy and Society Index, DESI. In 2017, Denmark had the best value of this indicator—0.76. The calculation of this index includes: connectivity, human capital, use of the Internet, integration of digital technologies, digital public services. Russia occupied the 39th position between Greece and Chile; its level of development against the leaders was 63% [2].

Currently, the Global Connectivity Index (GCI) is used to assess the level of informatization of the economy in different countries of the world; it includes 40 indicators and is calculated for 50 countries of the world (Global Connectivity Index, 2017). The top thirty countries in terms of GCI are presented in Fig. 1.

According to the data obtained, all countries can be divided into three clusters: developing, mid-developed and developed economies. Developing economies are at

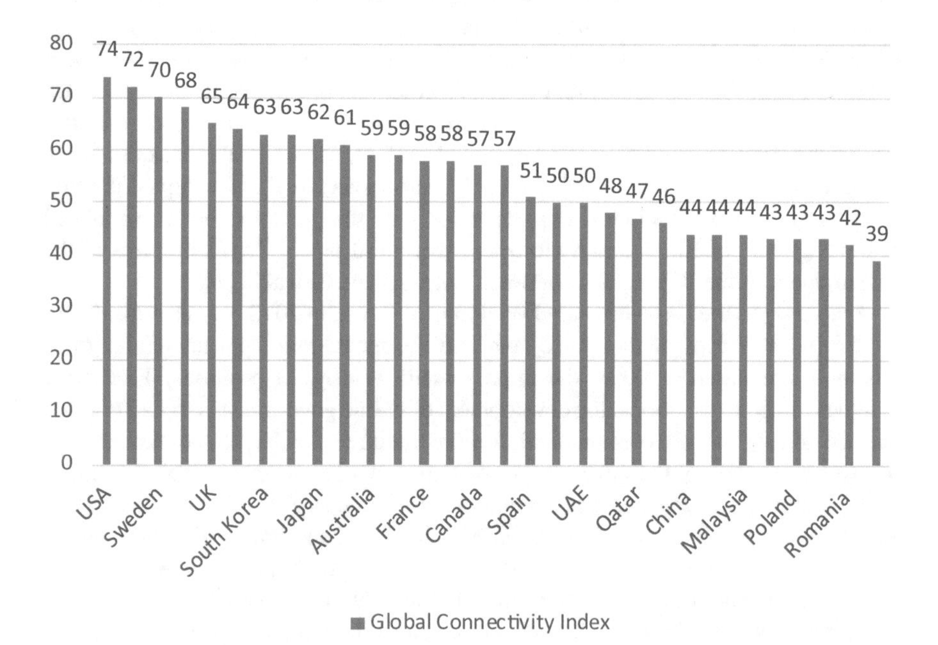

Fig. 1 Top 30 countries in terms of GCI (Global Connectivity Index) (*Source* authors based on [8])

an early stage of development of ICT infrastructure; their main priority is to attract as many people as possible to the development of the digital industry. The mid-developed economies ("middle cluster"), which includes Russia, currently have the most significant GDP growth rates due to the expansion of the information technology market. In particular, Russia has good prospects for the development of mobile broadband. There are prospects for the introduction of 5G networks, which will allow more active use of cloud technologies and ultimately change the business landscape. The first 16 countries in the ranking are leaders in the use of ICTs. Internet commerce is the most widely developed in them, and technologies to improve the quality of life of the population are being introduced.

Considering the structure of the Global Innovation Index for 2018, it should be noted that Russia took the 46th place in the world. The main components of this index are human capital and science (22nd place), in the development of technology and the knowledge economy, Russia was on 47th place and in business development—on the 33rd place, in import of services—on the 28th place, in graduates of scientific and engineering specialties—on the 15th a place. Factors holding back the innovation index are infrastructure development (63 positions), development of creative activities (72 positions) and the level of development of institutions (74 positions) [2].

One of the indicators of the quality of life of the population is access to information, including Internet sources. Between 2010 and 2018, Internet access in households increased from 48.4% to 76.6%, a positive trend is also the narrowing of the gap in broadband Internet access. In 2018, household security was 73.2%. The increasing availability of Internet resources contributes to its use. For example, in 2010 only 49.3% of the population had access to the Internet, but in 2018 the proportion of the population using the Internet grew to 87.3%, with 68.8% of the population using it daily [2]. In a regional context, the readiness for digitalization, estimated on the basis of indicators of the use of information technologies and information and telecommunication networks in organizations, is presented in Table 1.

Considering the degree of mastering the digital skills of the population, it can be noted that more than 40% of the population can work with text editors, use e-mail, manage files by copying, moving, transferring between different devices. Moreover, more than 20% of the population use photo-, video- and audio-editing programs s, and can also work with spreadsheets. The proportion of the population who can independently change the parameters and settings of the software configuration is about 3%; and 1.5% of the population can create software tools independently. Thus, over the past 10 years, the list of population's skills in working with digital devices has expanded significantly and education in this area is constantly growing. The most popular areas of Internet use by the population are: participation in social networks (77.8%); search for information on goods and services (54.1%), negotiating via Skype and similar Internet resources (52.6%); downloading movies, music, images, information (42%); playing computer games (29.8%); job search (7.8%); distance learning (3.1%) and other uses [1]. Thus, three main prerequisites for the formation of a model of the digital economy in modern Russia can be distinguished. The first one is related to the necessary regulatory support for the information economy,

Table 1 Grouping of Russian regions by the main indicators of the use of information technology and information and telecommunication networks in organizations

Group	Ranking	Regions
I—high	**1–10**	Moscow, St. Petersburg, Leningrad Region
II—sufficiently high	**11–20**	Stavropol Territory, Nizhny Novgorod Region, Yaroslavl Region, Republic of Ingushetia, Republic of Tatarstan, Sverdlovsk Region, Khabarovsk Territory, Moscow Region, Sakhalin Region, Belgorod Region, Republic of Crimea, Republic of Karelia, Novgorod Region, Republic of Bashkortostan, Voronezh Region, Tambov Region, Nenets Autonomous Region Okrug, Kaliningrad Oblast, Khanty-Mansiysk Autonomous Okrug Ugra, Kamchatka Krai
III—above average	**21–30**	Ivanovo Region, Astrakhan Region, Orenburg Region, Chelyabinsk Region, Kaluga Region, Lipetsk Region, Chuvash Republic, Vologda Region, Republic of Adygea, Vladimir Region, Murmansk Region, Tyumen Region, Ryazan Region, Perm Krai, Tula Region, Yamalo-Nenets Autonomous Okrug, Irkutsk Region, Penza Region
IV—average	**31–40**	Smolensk Region, Altai Republic, Kemerovo Region, Chukotka Autonomous Region, Primorsky Territory, Arkhangelsk Region, Pskov Region, Krasnodar Region, Karachay-Cherkess Republic, Bryansk Region, Komi Republic, Oryol Region, Krasnoyarsk Territory, Ulyanovsk Region, Khakassia Republic, Udmurt Republic, Magadan region, Kursk region, Kirov region, Trans-Baikal Territory, Tomsk Region, Republic of Mari El
V—below average	**41–50**	Rostov Region, Chechen Republic, Amur Region, Republic of Sakha (Yakutia), Kostroma Region, Altai Territory, Sevastopol, Omsk Region, Samara Region, Novosibirsk Region, Tver Region, Kabardino-Balkarian Republic, Republic of North Ossetia-Alania, Republic of Tyva
VI—low	**51–60**	Volgograd Region, Saratov Region, Jewish Autonomous Region, Republic of Kalmykia, Republic of Buryatia, Republic of Mordovia, Kurgan Region, Republic of Dagestan

Source authors

which determines the procedure for its formation. The second one is related to the readiness of infrastructure (mobile communications, the Internet, an e-commerce system and online payments) for its implementation. The third one is related to significant progress in the digitalization of the socio-economic system, as a result of which Russia exceeds the world average.

4 Discussion

The concept of the digital economy goes back to the previous concepts of economic development and emphasizes the importance of technological and human capital, applied information and communication technologies and the achievements related to the highly efficient creation, storage, dissemination and use of information. They were critically reassessed and filled with a new meaning, proclaiming information as the most valuable resource, and ensuring free and continuous interaction of economic entities is the highest priority for the development of socio-economic systems.

Digital economy can be understood a system of public relations, which is being developed in order to optimize the processes of socio-economic development using electronic technologies, electronic infrastructure and services, technologies for analyzing large volumes of data and forecasting [3]. The researchers [5, 11, 13] state that the world is currently in the process of building an information economy, the main features of which are the following:

- availability of the information resource, which, along with traditional factors of production (labor, land and capital) provides the process of reproduction of capital,
- increased importance of creative potential from the use of human capital,
- changes in the direction of product flows during the formation of the volume and range of products; while in the industrial period the demand was influenced by manufacturers, now production depends on the needs of customers,
- increased pace of information flows, causing more competition among companies and putting those that keep up with the changes ahead of competition,
- the development of information technology allows you to create different types of business on the Internet trading platforms, which requires less financial cost, but a large amount of "knowledge",
- virtualization of capital leads to an increase in the role of intangible assets and a decrease in tangible assets.

The scale and pace of development of the digital economy is evidenced by the fact that Internet technologies are being shifted from dealing with the inquiries from individual consumers and manufacturers to the ones made by various institutions, including government bodies. Many forms and types of paper workflows have lost their relevance and are being replaced by electronic document management systems in education, healthcare, banking, government bodies (tax services, pension funds, social insurance funds, law enforcement agencies, etc.). The development of the digital economy is currently actively carried out in the following areas: improving the concept of "electronic government"; development and implementation of technological projects of "smart city" and "smart village"; development and promotion of unmanned vehicles; expanding the scope of application of 3D printer technologies (in architecture, medicine, industrial production); cheaper technologies for building a "smart home"; intensification of biotechnological research aimed at increasing the duration and quality of human life; expanding the scope of electronic activities in the field of economics and management (automation of the accounting and analytical complex, financial and credit operations, the creation of electronic services in the tourism, transport, hotel industry, etc.); ICT improvement in the field of Internet commerce, agriculture, various mediation operations; digital lifecycle management of products based on a model. All of the listed types of economic activity can significantly reduce production and circulation costs, accelerate capital turnover by reducing the time of information processing and decision-making.

5 Conclusion

Thus, increasing the transparency of the business environment, investing in digital technology, introducing new technologies in key areas that provide competitive advantages, fostering links between all key players in the digital ecosystem, including the public sector, the private sector and the academic and educational community are crucial for the successful development of Russia in the medium term. At the same time, the main directions in the formation of the digital economy model in Russia will include:

- integrating technologies with business models, and focusing main innovations on business models rather than technologies,
- creating products and services by borrowing all the necessary resources for a required period of time on a project basis,
- combining all the necessary resources (human, computing and communication, technological, digital procurement, online production, finance, business models) for the design, production, operation and organization of a business in a single information space (using cloud technologies),
- creating the operator of the global platform of industrial ecosystems, i.e. production environment in the new conditions (production of capital goods),
- arraigning sales through industrial ecosystems, which will lead to personalization (lack of warehouses and unsold goods),
- introducing universal standards, regulations, methodologies and implementing global brokerage of products and services (without borders and distances), which will keep the existing customers,
- developing a program-controlled life cycle of products and services (design, production, maintenance) and supply chains,
- creating a software-controlled infrastructure (computers, memory, networks), which allows the economy to function at a new production and technological level, which in turn will significantly affect the change the borders between industries.

The above directions of the digital economy necessitate further development of algorithms and mechanisms for their implementation. Ahead is the direction of efforts to accelerate the pace of transformation of the private and public sectors; raising public awareness of the use of digital technology; strengthening the interaction of the scientific and educational community with the private and public sectors; as well as creating a business environment conducive to the implementation of innovations, the development of R&D and entrepreneurship—all the main elements of the digital economy culture that are currently lacking in Russia. Improving legislation and the tax environment, stimulating investment in innovation and developing entrepreneurship should be priority areas of the government policy.

References

1. Abrakhmanova G (2019) Digital economy 2019: national statistical compilation. HSE, Moscow
2. Abrakhmanova G (2019) Indicators of the digital economy 2019: statistical compilation. HSE, Moscow
3. Babkin A (ed) (2017) Trends in the development of the economy and industry in the context of digitalization. St. Petersburg Polytechnic University Publisher, St. Petersburg
4. Banche B, Boutenko V, Kotov I, Rubin G, Tuschen S, Sycheva E (2016) Russia online? Catching up or falling behind. The Boston Consulting Group, Boston
5. Bodrunov S, Plotnikov V, Vertakova Y (2017) Technological development as a factor of ensuring the national security. In: Soliman KS (ed) Proceedings of the 30th international business information management association conference, IBIMA 2017 – vision 2020: sustainable economic development, innovation management, and global growth. IBIMA, Madrid, pp 2666–2674
6. Dobrynin A, Chernykh K, Kupriyanovsky V, Kupriyanovsky P, Sinyagov S (2016) The digital economy – the various ways to the effective use of technology (BIM, PLM, CAD, IOT, Smart City, BIG DATA, and others). Int J Open Inf Technol 4(1):2–10
7. Efimushkin V, Ledovskikh T, Scherbakova E (2017) Infocommunication technology space of the digital economy. T-Comm Telecommun Transp 11(5):15–20
8. Huawei (2019) Global connectivity index. http://www.huawei.com/minisite/gci/en/. Accessed 04 Mar 2020
9. Ivanova E, Nikitin A (2018) Cluster-cooperative project of innovative development of agriculture. Qual-Access Success 19(S2):8–14
10. Kolmykova T, Nesenyuk E, Halameeva K (2019) The development of the digital economy in the transition to the sixth technological order. News Southwest State Univ Ser Econ Sociol Manag 9(1(30)):57–64
11. Polyanin A, Golovina T, Avdeeva I, Dokukina I, Vertakova Y (2017) Digital strategy of telecommunications development: concept and implementation phases. In: Soliman KS (ed) Proceedings of the 30th international business information management association conference, IBIMA 2017 – vision 2020: sustainable economic development, innovation management, and global growth. IBIMA, Madrid, pp 1792–1803
12. Sudarushkina I, Stefanova N (2017) Digital economy. ANI Econ Manag 18(6.1):182–184
13. Vertakova Y, Plotnikov V (2016) Innovative and industrial development: specifics of interrelation. Econ Chronicle-XXI 156(1–2):37–40

Social and Ethical Problems of Digital Technologies Application in Human Resource Management

V. G. Konovalova and A. E. Mitrofanova

Abstract The purpose of this contribution is to study social and ethical problems and risks caused by the application of digital technologies and their manifestations in the field of human resource management. Summarizing data from international and national studies, the authors identify the following as the main problems: structural changes in the labor market and associated risks of reducing the professionalism level and the educational system; "digital inequality", the emergence of new types of discrimination, including when working with Big Data and making decisions based on a proactive approach; changing the boundaries of confidentiality in the professional and private spheres; problems related to the development of the practice of tracking and monitoring the effectiveness of the personnel activities, including their impact on labor relations; problems of responsibility and ethical boundaries of the use of autonomous and intelligent technologies.

Keywords Digital technologies · Structural changes in the labor market · Digital inequality · Big Data · Discrimination · Privacy

1 Introduction

The unprecedented application of digital technologies in various fields of activity requires paying attention not only to new opportunities that are opening up, but also to potential or already manifested threats, problems and risks. Among the obvious challenges and specific threats of digitalization, it is possible to allocate the following issues: ensuring human rights in the digital world, preservation of digital user data, ensuring confidence in the digital environment, as well as the threat for the individual, business and government, capacity building of the external information technology impact on the information infrastructure, the growing scale of computer crimes,

V. G. Konovalova (✉) · A. E. Mitrofanova
State University of Management, Moscow, Russia
e-mail: vg_konovalova@guu.ru

A. E. Mitrofanova
e-mail: ae_mitrofanova@guu.ru

S. I. Ashmarina and V. V. Mantulenko (eds.), *Current Achievements, Challenges and Digital Chances of Knowledge Based Economy*, Lecture Notes in Networks and Systems 133, https://doi.org/10.1007/978-3-030-47458-4_85

including international ones, the gap with the leading foreign countries in the development of competitive information technology, the dependence of socio-economic development on the export policy of foreign countries, the lack of efficiency of scientific research related to the creation of advanced information technologies, the low level of implementation of domestic developments, and the insufficient level of personnel support in the field of information security. The consequences of management digitalization for human resource management are also ambiguous.

2 Methodology

In order to analyze the digitalization consequences for human resource management, the authors turned to the data of international and national studies in this field at the macro and micro levels and experts' opinions. According to an international study by Aruba Networks (an HPE company), "The Right Technologies Unlock the Potential of the Digital Workplace" [2], employees working in the digital workplace increase not only their productivity, but also motivation, job satisfaction, and an overall assessment of their own well-being.

Employees who are fully employed in digital workplaces are 51% more likely to be satisfied with their work and 43% more likely to have a positive view of the work-life balance than those who have limited access to digital technologies in the workplace ("outsiders"). 60% of people working in digital workplaces are also more likely to report a high level of motivation at work, 91% of them have a positive opinion about their company's strategy, 73% report a positive impact of digitalization on the productivity, and 70% connect their improved interaction with each other with the impact of digital technologies.

The digital workplace also contributes to professional development: 65% of those working in digital workplaces report that the use of digital technologies has had a positive impact on their professional growth, 72% confirmed that their ability to acquire new working skills has increased due to digital workplaces, and among "outsiders" only 31% of respondents noted a professional growth and 58% of respondents—the acquisition of new working skills.

Almost all respondents (93%) said that their work space will improve due to the wider use of technology, and 64% of respondents are confident that their company will lag behind competitors if it does not apply innovations. The same number of respondents (64%) believes that the traditional office will become obsolete as a result of ongoing technological progress. The majority of respondents expressed the opinion that digital technologies as a result form a more efficient (56%) and attractive (47%) working environment with extensive opportunities for collaboration (52%). The advantages of the digital workspace also include opportunities for collaboration, mobility, reduction of stress and overloading, and increased employee engagement [2].

At the same time, according to the study "The Future of Jobs" [5], in the coming years, along with the creation of 2 million new jobs in intellectual and high-tech areas where data analysis and management of complex technological processes are required, about 7 million jobs will be cut in the real sector and in the field of administrative work, where the share of routine, unskilled labor is quite high. Experts suggest that by 2030, the unemployment rate will reach 7.5% or higher, presumably due to robotics and the displacement of people from the production.

One of the significant consequences of technological trends in the labor market will be the spread of the phenomenon of so-called "extra people"—those whose basic level of skills and willingness to change will not allow them to keep employment in competition with robots or algorithms. "Extra people" in the modern economy can join the ranks of informal employment, become a source of social tension.

McKinsey experts estimate that by 2030, between 75 and 375 million workers (3 to 14% of the world's workforce) will have to retrain, and the rest will have to adapt as their professions evolve together with the automation development [11]. At the same time, retraining older people in new professions (due to age specifics) will lead them to less qualified and, therefore, to less paid positions. Highly qualified professionals may find themselves in a worse position than unskilled employees, since they will be involved in the transition from traditional to digital technologies until the last moment and will be "thrown out" on the labor market when the positions corresponding to their status are already occupied, which may also contribute to increasing social tension and rejecting the idea of the economy digitalization.

"Digital inequality" can manifest itself at many levels (from the individual, society, corporations and regions to the global level) and provoke an increase in social inequality in the society, as it affects the social status, increases the prestige of those who can use new communication technologies, acquire new specializations, or skills and increase intellectual abilities in the digital world in order to create new opportunities for achieving their goals in the social, political or economic spheres. However, equality of access to technology is not a sufficient condition for achieving digital equality, since factors such as the availability of skills and opportunities, social and cultural attitudes towards technology, the institutional environment, and social transformations will greatly influence whether technologies are used effectively and correctly. The use of digital technologies changes the boundaries of privacy in the professional and private spheres, even to the point where privacy disappears.

3 Results

At the micro level, the application of digital technologies in companies, on the one hand, creates a new, more "transparent" model of human resource management, provides an individual approach to evaluating candidates and finding the right people, optimal training and development, identifying fraud attempts, factors and manifestations of employee's stress and other problems in the organization, and, on the other hand, creates conditions for the emergence of new social and ethical problems and risks that affect human resources in one way or another and are manifested in such areas such as discrimination, privacy, security, information overload, digital divide, artificial "intermediaries", virtual reality, reliable and up-to-date information, etc. [4, 6, 9, 14]. One of the factors of "digital inequality" can be the use of Big Data. Currently, only large companies have access to really large amounts of social data, and many of them (IBM, Microsoft, etc.) are not ready to share information with everyone. Researchers often become employees of such companies only for this reason, or send their students and postgraduates to these companies to get a full access to their databases. The very use of Big Data generates and will generate inequalities in access to it.

Making decisions in the field of human resource management based on Big Data analysis can also lead to new forms of discrimination—when making decisions about hiring, evaluating the potential of employees, etc. If earlier a person was evaluated by his/her actions, then in the era of Big Data, they expect to identify personal inclinations and evaluate the candidate based on the forecast of what he/she will do. A proactive approach is common for the entire range of Big Data applications (human resources management uses both external data that is outside the company's information space—candidate profiles in social networks, job applicants' resumes and job descriptions on specialized job search sites, and internal data—information about employees, their responsibilities and performance indicators from corporate systems, Excel spreadsheets and regulatory documents, as well as informal feedback from colleagues and management). Nowadays, a data bank is obtained and collected from various sources before determining the full range of their current and potential uses, mobilized algorithms and analytics not only to understand the past sequence of events, but also to predict them and intervene *before* actions, events, and processes [1, 10, 12, 15]. A candidate or employee is evaluated using a large amount of information about people and their behavior that has nothing to do with the tasks of a real work environment. Indirect evaluation leads to erroneous rejections (the employer rejects potentially good candidates) and erroneous admission of candidates (unsuitable people are hired for the wrong reasons).

As data sets get larger and more detailed, it becomes easier to identify an individual using them. Individuals are often not even aware that certain aspects of their lives are converted into data, social media users do not have access to data control tools, and often are not even aware that this data is being used by someone. Many do not realize the multiplicity of agents and algorithms that are currently collecting and storing data for future use.

Social networks and other large-scale digital platforms are a key tool for spreading false information (fake news). Research by Edelman Trust Barometer shows that 63% of respondents cannot distinguish verified information from false information when it is received via the Internet [7]. Often, information functions as a tool for fine manipulation, based (among other things) on the analytics of raw Big Data. Spreading false information affects not only political and ethical issues, but also directly influences the economy, creating a basis for making false investment decisions.

An important issue directly related to privacy is tracking practices, that is, any systematic, routine, and focused attention to personal details for a given purpose (such as management, influence, or empowerment) that are implemented on the basis of Big Data and web tracking. More and more companies are thinking about employee performance monitoring systems that help to solve a set of tasks to prevent information leaks, employee fraud, as well as control business processes and identify inappropriate use of the working time, which entails very specific economic losses for the organization.

Uncontrolled use of the Internet (social networks, messengers, personal mail) can lead to leaks of confidential information, which in turn will cause significant harm to business. According to the study "The Right Technologies Unlock the Potential of the Digital Workplace" [2], despite the fact that employees reported high awareness of cybersecurity issues (52% of respondents think about security "often or daily"), they recognized that they allow more and more risk in actions with the company's data and devices (70% of respondents admitted to risky behavior, such as passing passwords and devices to others). A quarter (25%) of employees have connected to potentially insecure open Wi-Fi networks over the past twelve months, 20% reported using the same password in multiple apps and accounts, and 17% confirmed that they write down passwords to remember them.

For these reasons, many companies are already currently monitoring staff, do not limit themselves to the usual passes and turnstiles: employee performance monitoring systems are currently used to solve the following tasks: accounting for employees' working hours, evaluating and monitoring staff performance, identifying disloyal employees and fraudulent schemes within the company, searching for possible information leaks and protection from insiders, investigating information security incidents, operational control of employees and identifying risk groups.

Depending on the solution concept, systems can be differentiated as follows:

- the first group of systems allows you to collect a large amount of data and provide it in the form of analytical reports, as well as keep statistics of productive and unproductive working hours,
- the second group provides the manager with the ability to monitor employees by viewing online broadcasts or video recordings of their actions, as well as monitor violations,
- the third group of systems has a fairly narrow set of functions that are limited only to accounting for staff working hours.

In addition to the task of monitoring the effectiveness of users, monitoring systems solve a number of problems related to monitoring data channels and preventing leaks of confidential information—systems can track the leakage of confidential information by monitoring various data channels and working with external devices.

The technology development allows you to easily monitor employees in other areas. For example, the Boston startup Humanyze produces identification badges for employees with built-in biometric sensors. These devices track movements around the office, the duration of conversations, processing audio information, can determine whether a person dominates the conversation, and record the tone, volume, and speed of speech. According to the company's representatives, these "people meters" can help to assess any parameters that affect the employee's productivity, including the effectiveness of programs to attract people of different nationalities and sexual orientations, as well as people with various disabilities.

If the job involves driving a car, GPS and sensors on the phones can be used to track the employees' location, and sometimes their driving habits. Companies claim that this helps to improve results: 45% of respondents of a survey conducted by technology and service company Aberdeen Group said that tracking technologies were a key factor in improving the quality of their work.

A lot of companies consider the access to employees' email and phone records as a mandatory security measure necessary to protect the organization against conflicts of interest, insider trading, public sharing of confidential information, and other reputation threats. Tracking employees' behavior is designed to help not only to maintain the company's reputation, but also make sure that employees are working on their tasks, i.e., manage their performance.

4 Discussion

Gartner's research has shown that companies can generate useful information by processing Big Data from multiple sources, but experts estimate that individual members of our society, organizations, and governments are increasingly concerned that personal data may be used for malicious invasion of privacy, for creating control systems, or for commercial or political manipulation [8].

New rules, institutions, and experts will be needed to solve the problems that arise, in order to interpret the complex algorithms that generate conclusions from Big Data, and to protect the interests of those who may suffer from these conclusions. The tightening of regulatory requirements in the field of information security and responsibility for compliance with them, especially in the field of personal data protection, has become one of the global trends: currently, many countries are actively developing draft laws on regulating the distribution of false information, while the regulation concerns both the activities of online platforms (social networks) and the activities of individuals who directly distribute false information.

The regulation development is determined by the search for answers to the following questions: (1) whether new rules are needed to combat false information distributed on the Internet, or whether existing mechanisms are sufficient; (2) under what conditions false information should be subject to regulation; (3) who should be authorized to evaluate information in order to classify it as false information; (4) who should be responsible for removing/limiting false information; (5) what areas should new rules apply to: politics, economy, public security, etc.

Three strategies are proposed for managing information fairly and effectively in the new era of Big Data: shifting privacy protection from individual consent to accountability for data users, prioritizing the human factor over forecasts, and creating a new class of Big Data auditors.

Improving employees' productivity by monitoring all kinds of data requires a thoughtful and flexible approach. Surveillance may force you to follow the rules in the short term, but it doesn't form habits. The employer's refusal to consider the negative aspects of monitoring not only complicates or even destroys working relationships, but can also violate the law and be a criminal offense. Even if monitoring is justified, it is necessary to find a balance between the legal right to conduct business at your own discretion and respect for the personal information of your employees. Constant monitoring can lead to social conformity, reduced authenticity, and self-censorship [13].

Systems that use artificial intelligence technologies are becoming increasingly autonomous in terms of the complexity of tasks they can perform, their potential impact on decisions, and the diminishing ability of humans to understand, predict, and control their functioning. Not only the collection of information goes from a person to a machine, but also its analysis and conclusions. Such systems can learn from their own experience and perform actions that were not intended by their creators. These characteristics raise questions related, first, to predictability and, second, to the ability of systems to act independently, but not to bear legal responsibility. Moreover, independent learning of algorithms calls into question the subject of actions [3, 13].

Ethical problems also arise in relations between people when using "technical intermediaries" (for example, when providing remote services—medical, educational, legal assistance, etc., transferring a part of professional responsibilities to automated self-learning systems).

5 Conclusion

Summing up, it should be emphasized that technological changes are the most important factor of modern economic growth, stimulating an increase in the labor productivity, but also a source of new problems and risks, new trends in the labor market. Issues related to the use of digital technologies often go beyond the established ethical norms and require their revision not only in terms of changes in public values, but also in terms of the legislation development.

References

1. Andersen M (2017) Human capital analytics: the winding road. J Organ Eff People Perform 4(2):133–136. https://doi.org/10.1108/JOEPP-03-2017-0024
2. Aruba (2018) The right technologies unlock the potential of the digital workplace. https://www.arubanetworks.com/assets/eo/Aruba_DigitalWorkplace_Report.pdf. Accessed 12 Feb 2020
3. Asaro P (2019) AI ethics in predictive policing: from models of threat to an ethics of care. IEEE Technol Soc Mag 38(2):40–53. https://doi.org/10.1109/MTS.2019.2915154
4. Braganza A, Brooks L, Nepelski D, Ali M, Moro R (2017) Resource management in big data initiatives: processes and dynamic capabilities. J Bus Res 70:328–337. https://doi.org/10.1016/j.jbusres.2016.08.006
5. Committed to Improving the State of the World (2016) The future of jobs. Employment, skills and workforce strategy for the fourth industrial revolution. http://www3.weforum.org/docs/WEF_Future_of_Jobs.pdf. Accessed 12 Feb 2020

6. Dery K, Sebastian IM, Meulen NV (2017) The digital workplace is key to digital innovation. MIS Q Exec 16:135–152
7. Edelman (2017) Edelman trust barometer special flash poll. Research. https://www.edelman.com/trust2017/trust-barometer-media-fake-news-flash-poll. Accessed 12 Feb 2020
8. Gartner (2019) Gartner top 10 strategic technology trends for 2019. https://www.gartner.com/smarterwithgartner/gartner-top-10-strategic-technology-trends-for-2019. Accessed 12 Feb 2020
9. Hildebrandt M (2016) Law as information in the era of data-driven agency. Mod Law Rev 79(1):1–30
10. Levenson A, Fink A (2017) Human capital analytics: too much data and analysis, not enough models and business insights. J Organ Eff People Perform 4(2):145–156. https://doi.org/10.1108/JOEPP-03-2017-0029
11. McKinsey & Company (2017) What the future of work will mean for jobs, skills, and wages. https://www.mckinsey.com/global-themes/future-of-organizations-and-work/what-the-future-of-work-will-mean-for-jobs-skills-and-wages. Accessed 12 Feb 2020
12. O'Brolchain F, Jacquemard T, Monaghan D, O'Connor N, Novitzky P, Gordijn B (2016) The convergence of virtual reality and social networks: threats to privacy and autonomy. Sci Eng Ethics 22(1):1–29
13. Reyt J, Wiesenfeld B (2015) Seeing the forest for the trees: exploratory learning, mobile technology, and knowledge workers' role integration behaviors. Acad Manag J 58(3):739–762. https://doi.org/10.5465/amj.2013.0991
14. Royakkers L, Timmer J, Kool L, Est R (2018) Societal and ethical issues of digitization. Ethics Inf Technol 20:127–142. https://doi.org/10.1007/s10676-018-9452-x
15. Stuart M, Angrave D, Charlwood A, Kirkpatrick I, Lawrence M (2016) HR and analytics: why HR is set to fail the big data challenge. Hum Resour Manag J 26(1):1–11. https://doi.org/10.1111/1748-8583.12090/

Digital Economy and Intelligent Supply Chain Management: International Experience

K. Doan, S. Carrino, N. V. Ivanova, and T. E. Evtodieva

Abstract Digital economy formation is the main driver of the modern society development where added value is no longer a physical product and the distinction between the physical and virtual worlds is erased. Technological and intellectual innovations require a revision of the system of business relations between entities involved in the creation and distribution of tangible and intangible benefits. Speed becomes a key factor in the development of the company, including the rate of change in the nature of the business, the speed of business processes managing, the lifestyle of consumers and their requests changing dynamics, influenced by the increasing availability of information. The aim of the study is to conduct a comparative analysis of the digitalization level of individual countries, determine the capabilities of digital technologies in intelligent supply chain management at the level of a business entity and determine the positive effect of their use. The main results of the study include the allocation of the main directions of the use of modern technologies in the logistics supply chains and consideration of the experience of using digital technologies in the logistics activities of companies.

Keywords Digital economy · Intelligent supply chain management · Logistics process mining · Supply chains

K. Doan · S. Carrino
HE Arc, Neuchatel, Switzerland
e-mail: karine.doan@he-arc.ch

S. Carrino
e-mail: stefano.carrino@he-arc.ch

N. V. Ivanova (✉)
Samara State University of Economics, Samara, Russia
e-mail: nataliaivanova86@yandex.ru

T. E. Evtodieva
Rostov State University of Economics, Rostov, Russia
e-mail: evtodieva.t@yandex.ru

S. I. Ashmarina and V. V. Mantulenko (eds.), *Current Achievements, Challenges and Digital Chances of Knowledge Based Economy*, Lecture Notes in Networks and Systems 133, https://doi.org/10.1007/978-3-030-47458-4_86

1 Introduction

During the ongoing economic transformations associated with fundamental changes in the economy logistics has become a key area of business and a driver for improving the efficiency and development of business entities. Today, no one doubts the need to invest in the logistics components of a business and actively apply logistic tools to solve multi-criteria tasks of managing the material flow through the supply chain. However, the transformation of the logistics implementation conditions dictates the need to track the direction of development of the macro environment to project the necessary actions to adapt logistics technologies to the new environment of interaction between producers, consumers and infrastructure elements.

As part of the research on the problems of economic relations transformation, it is noted that digitalization of the economy is becoming a source of long-term economic growth in modern conditions [4]. The digital economy is a new paradigm of accelerated economic development. The digital economy transforms the business model, goods and services, forms new criteria and parameters for evaluating the companies efficiency. In a digital economy, a necessary requirement for interaction is the tight integration of business processes and the active use of end-to-end technologies that ensure transparency in business, equal participation in making management decisions and the ability to track the flow of streaming processes in real time.

It should be noted that today the actions direction is set by various countries at the macro level. Specialists of the McKinsey Institute, determined that by 2025 GDP growth due to digitalization in a number of leading economies of the world will increase dramatically. According to McKinsey assessments GDP growth in China could increase up to 22%, in the USA about 1.6–2.2 trillion US dollars, in Russia from 19 to 34% of the total expected GDP growth [1]. International consulting company Accenture predicts that by 2020 the digital technologies use will add 1.36 trillion U.S. dollars or 2.3% of GDP in the total GDP of dozens of the world's leading economies. Digitalization will affect other countries. The study noted that the GDP of developed countries will increase due to the "digital economy" by 1.8%, and GDP of developing countries—by 3.4% [10]. As a result, it becomes actual to analyze the level of digitalization of the economies of different countries as well as generalization of experience in the use of digital technologies in the formation of supply chains and determine the degree of innovative technologies influence on supply chain management. Thus the importance of the study is to determine the possibility of technological innovations of the digital economy applying in the practice of supply chain management and the theory of logistics development in modern conditions.

2 Methodology

Theoretical and methodological basis of the research are the works of the Russian [2, 5, 12, 13], who studied the paradigms of the formation of a new post-industrial society and determine technological and organizational innovations in logistics. General scientific and private methods of cognition as the main methods for studying the problem of digitalization of the economy and intelligent supply chain management in the digital environment were used. The application of general scientific methods of cognition allowed us to objectively and comprehensively explore the conceptual framework of logistics and supply chain management in the modern economy, systematize and summarize the existing experience of business structures on the application of digital technologies in logistics. Private research methods were used at the stage of processing, analyzing and summarizing the obtained scientific information.

3 Results

The economic system transition to the digital economy requires a rethinking of approaches and methods of logistics management. In the works of both foreign and Russian scientists, the influencing factors of the world's economies digitalization are identified and ranked. It is determined that the activity of countries and their impact on the formation of the digital economic depends on the overall level of development of the macroeconomic system and their integration degree into the global economy. In almost all economies of the world, the main object of digitalization is the global supply chains that ensure the movement of material flows at the intercountry level. However, at present the scientific basis for purposeful supply chain management in the conditions of digitalization is not clearly formed. The development of scientific thought today is based on the principles of pioneer projects analysis in the field of supply chain management and their adaptation. In this regard, it becomes important to systematize and summarize practical knowledge and experience in order to rethink approaches to supply chain management in the modern world and determine the productive technologies and methods of their digitalization, which served as the basis for this research and determined its subject, relevance and focus. The aim of the study is to consider the experience of the modern digital technologies use in the practice of supply chain management and the definition of the main functional areas of logistics, priority in terms of digitalization, which contributes to the development of the logistics organization theory in modern conditions.

4 Discussion

The development of the global economy is based on the information technologies application, active automation and robotization, the exchange of information in real time, additive manufacturing and technologies that provide cybersecurity and augmented reality [5]. It should be noted that the application and penetration level of the mentioned innovations has differences and peculiarities in different countries. Despite the interest of different countries in the early transition to a digital economy there is a varying degree of digitization of business processes and an active use of innovative tools in the practice of business entities. In particular, China's digital market, despite its capacity, exists in isolation from the global market due to the small number of international players. Nevertheless in 2016 China came in second place in the world in terms of the level and scale of development of the digital economy [12]. India has a large market potential due to the infrastructure development problems that are actively solved by the government can also be a full participant in the digital market [3]. Even the European Union's market is fragmented despite the European Union's Europe 2020 development strategy. There are seven priority areas of digital development are defined and phased objectives are formulated. It's implementation will allow reaching a certain level of digitalization specifically the European Union's GDP growth by 5% over the next eight years [15].

At present time the level of digitalization of the economy of a particular country can be analyzed on the basis of the international Digital Evolution Index rating, conducted by the Fletcher School of Law and Diplomacy at Tufts University, USA. The level of digital economy development in different countries determine is based on the following factors: the supply level (availability of Internet access and infrastructure development degree); consumer demand for digital technology; institutional environment (state policy, legislation, resources); innovative climate (investment in R&D and digital start-ups) [9]. According to the cumulative total four groups of countries were ranked according to the economy digitalization rate and level. In 2017 a high level of digital development is noted in the UK, Singapore, New Zealand, Estonia, Japan, Hong Kong and the United Arab Emirates. China, Russia, Brazil, Mexico, Philippines, Chile, and India were included in the category of countries with sustainable growth rates. For countries that have been leading in the digitalization of the economy for a long time but have significantly slowed growth, researchers attributed South Korea, Australia, Canada, the United States, Scandinavian countries and Western European countries. Among European countries, leaders are Iceland, Switzerland, Denmark and the United Kingdom. Particular attention should be paid to Switzerland, which despite crises characteristic for all economies of the world provides high-quality, sustainable development (2.4% at the end of 2018) through the use of the fifth-generation ultrafast Internet in the industrial sector and active development of robotics [14]. Digitalization outsiders are identified: Greece, Pakistan, Egypt and South Africa. Development rate shows that while the transition from the traditional economy to the digital one incentive measure are required to introduce and maintain innovative technologies in all countries regardless of the level of digitalization achieved at the present time. The digital market is dynamic and it is impossible to be a leader and hold positions without constant improvement and the development of technologies that ensure the possibility of stable interaction in a turbulent market environment.

Let us dwell on the defining of the recent news that brings digitalization to the economy of the any country. Theoretical views on the digital economy analysis [2] allows to determine the following features that have a direct impact on all areas of business activity:

1. There is a mechanisms of self-organization and harmonization of the economic system transformation based on the all spheres of society informatization.
2. Rapidness becomes a key factor in the companies' development. Today under the influence of the general availability of information for any participant in a business process, whether it is a manufacturer or a consumer, the needs and requirements are changing responsively, which requires an instant reaction of the business to their satisfaction and increasing the efficiency of managing business processes.
3. Under the scientific and technological progress influence there are shifts towards innovative-active spheres characterized by a short product life cycle.
4. In the modern product offer material values are replaced by the information which contributes to ensuring the intensive growth of the intellectual component of the products produced.
5. The sphere of the labor force knowledge and skills application is changing.

The digital economy development leads to a radical review of the nature of the relationship between market participants, implementing various functions and activities in the business including logistics. This is facilitated by digital technologies that form the basis for the development of modern society and provide:

- combination of virtual and real world,
- automating most of the same constantly repetitive operations,
- making management decisions at any level of the hierarchy operatively on the basis of processing large amounts of data and taking into account the explicit and implicit factors influencing this decision.

In the conditions of digitalization of the economy and changing priorities in the organization of the movement of economic flows in supply chains they actively apply innovative technologies that ensure the improvement of quality and speed of customer service and contribute to the minimum gap between production and service level. The emphasis on the high-speed rhythm of organizing material flows within the supply chain is an essential condition for maintaining the competitive position of a modern enterprise. In this regard, there is an active reorientation of participants in the supply chains to digitalize the main operations and business processes, allowing to qualitatively increase the efficiency of interaction and performance results.

A special place in the digitalization of supply chains and logistics at the international level is occupied by infrastructure projects for the digital development of transport. As mentioned in the issue [7]. The digital railway will be the digital transport basis, which will allow a 50% increase in the capacity of the railway infrastructure and reduce the cost of transportation by the same amount. Vision of the railway sector and technical development strategy "Europe's railway industry of the future". The result of the stimulation and development of railway transport is the emergence of a digital railway in the UK, Germany, USA, Switzerland, France, Austria. Russia also developed a comprehensive program for the innovative development of the holding company Russian Railways for the period 2016–2020, one of the priorities of this program is the Digital Railroad project implementation.

Being a promising country in terms of the digital development dynamics, in Russia there are strategic initiatives that contribute to improving the efficiency of supply chains by improving transport operations parameters. Thus, on the basis of the Platon system, the global navigation system ERA-GLONASS and the infrastructure of the Russian railways, which have proved their effectiveness, it provides development and launch of a single digital platform of the transport complex in the period 2019–2022, which will track the goods, create legal documentation [11]. It is expected that this platform will have a positive impact such as transportation costs optimizing and transport and logistics solutions unification.

In 2018 the Ministry of Transport of Russia adopted an initiative and signed an agreement on the establishment of the Digital Transport and Logistics Association based entirely on Russian solutions and software and has the main goal of a single multimodal digital transport and logistics space formation, which will allow getting current and reliable information about the transport services market based on digital technologies and services and the integration of decisions into a single digital platform.

Today we can say confidently that a certain countries specialization in the digital sphere is being formed. For example, Germany is the clear leader in "smart plants" and "smart manufacturing systems". The United States has leadership in advanced analysis and intelligent data processing. Japan can become an advanced country for the production of robotics for the Internet of things. Already today, the Japanese government has identified opportunities for sales of robotics from 600 billion ¥ per year to ¥ 2.4 trillion per year (about £ 19 billion per year) by 2020 [7]. The UK plays a leading role in the field of engineering solutions for the digital industry integration and logistics, which allows transforming traditional supply chains into digital ones.

Innovative digital technologies with potential in logistics and supply chain management are:

- cognitive technologies such as technologies for big data of unstructured (usually textual) information processing,
- cloud technologies as technologies that provide fast simple network access to a large array of resources generated by user request,
- the Internet of things as a technology that allows for the collection of a variety of data through remote access and to manage them in real time,
- machine learning, as a set of mathematical methods of analysis and statistical processing of information, optimization and programming methods, allowing to

reveal hidden patterns and take them into account when developing management decisions,
- Blockchain technology as a set of encryption algorithms that allow to protect databases in which information about all transactions within distribution systems with a time stamp and a link to previous information blocks, which eliminates the falsification of the accumulated data array. Moreover, today only 1% of organizations currently use Blockchain in their supply chain operations, and 35% only study the possibilities of its use, in the future about 1/3 see the potential of Blockchain technology to create a competitive advantage of their company over the next 10 years. About 10% of respondents believe that during this period it will change the industry [8].

It is important to understand that the rapid development of digital technologies leads to significant changes in logistics and supply chain management, changing the operating model of their organization, identifying new market opportunities and increasing cost-effectiveness. If the first wave of digital innovations in logistics came down to automating existing technologies and business processes, today the key to success and viability of any supply chain are intelligent technologies based on machine learning and artificial intelligence, which provide additional competitive advantages and sustainable development. The survey of supply chains focal companies in 2017, showed that "47% of the big business supply chains leaders believe that artificial intelligence is a breakthrough and an important technology in relation to the development strategy of the supply chain" [13, p. 2].

According to analysts, by 2030, about 70% of companies around the world will use at least one type of artificial intelligence technology in their activities. The active use of artificial intelligence will provide an increase in aggregate GDP by 16%, which will amount to 1.2% of the annual additional increase in GDP [6]. The main goal of the application of machine learning and artificial intelligence is currently costs reducing and the quality of the proposed product improving. All innovative technologies of the digital economy allow communicating to all participants of logistics activities without the influence of the human factor, thereby ensuring the objectivity of the information presented and processed and structured, which is the basis for making decisions on major issues of logistics cooperation. Modern digital society technologies make it possible to determine the optimal vector of spatial movement of products in the supply chain in such a way as to ensure the possibility of maximum total profit of the chain, taking into account the rapid response to changes.

5 Conclusion

The digital direction of the society and economic structure development determines the need to adapt the activities of all supply chain actors to the changes in the macro environment. Efficient operation of supply chains today is possible only if there is a constant current action plan focused on maximizing the current and expected in the future situation, which is impossible without technological innovation. The basic innovative technologies that determine the success of supply chains are such as self-driving vehicles and intelligent transport systems, "smart" warehouses; automation and robotics; augmented reality, cloud computing; smart sensors, as well as full transparency of data sets and their clear structuring. The degree of activity of technology application at the business level largely depends on the interest and the degree of the government participation in innovative activities support. It is determined that the level of development and implementation of digital technologies is directly dependent on the territorial location of the business. In particular, while in the CIS countries the main emphasis is still on the automation of logistics business processes, in the European Community priority is given to the formation of integration interactions with customers and suppliers on the basis of advanced information technologies, in China and the United States are beginning to work on the global digitalization of logistics processes using Blockchain technology. Thus, the considered experience of digital technologies use in the companies' logistics activities and support for digital development in different countries allowed us to identify promising areas of supply chains and their priority links digitalization.

Acknowledgements This paper goes within the framework of scientific and academic cooperation on the project "Smart Supply Chain" between the HE ARC, Neuchatel (Switzerland) and Samara state University of Economics, Samara (Russia). The authors thank administration of organizations and all the members participated in the project.

References

1. Aptekman A, Kalabin V, Klintsov VI (2017) Digital Russia: a new reality. Report McKinsey. http://www.tadviser.ru/images/c/c2/Digital-Russia-report.pdf. Accessed 20 Feb 2020
2. Borisova VV (2018) Formation of a digital logistics ecosystem. In: Albekov AU (ed) The digital revolution in logistics: effects, conglomerates and growth points. Rostov State University of Economics, Rostov-on-Don, pp 47–51
3. Chakravorti B, Bhalla A, Chaturvedi RS (2017) The most digital countries in the world. https://hbr-russia.ru/innovatsii/trendy/p23271. Accessed 20 Feb 2020
4. Dybskaya VV, Sergeev VI (2018) Digital logistics and supply chain management: development prospects. Logistics: current trends. Publishing House of GUMRF, St. Petersburg
5. Evtodieva TE, Chernova DV, Ivanova NV, Wirth J (2019) The internet of things: possibilities of application in intelligent supply chain management. In: Ashmarina S, Mesquita A, Vochozka M (eds) Digital transformation of the economy: challenges, trends and new opportunities. Advances in intelligent systems and computing, vol 908. Springer, Cham, pp 395–403
6. Freedom Finance (2019) The impact of artificial intelligence on the global economy. https://24.kz/ru/news/delovye-novosti/item/284820-mckinsey-predstavil-model-potentsialnogo-vliyaniya-tsifrovykh-tekhnologij-na-mirovuyu-ekonomiku. Accessed 20 Feb 2020
7. Kupriianovskii VP, Evtushenko SN, Dunaev ON, Bubnova GV, Drozhzhinov VI, Namiot DE, Sinyakov SA (2017) Cognitive-information technology in the digital economy. Mod Inf Technol Educ 1(13):74–96
8. Lagoon D (2018) Supply Chain. How artificial intelligence and blockchain will change logistics. https://www.forbes.ru/biznes/357749-cepi-postavok-kak-blokcheyn-i-iskusstvennyy-intellekt-izmenyat-logistiku. Accessed 20 Feb 2020

9. Mastercard (2017) TOP 10 most advanced digital economy countries. http://web-payment.ru/article/250/top-10-cifrovaya. Accessed 20 Feb 2020
10. Rostec (2016) At the forefront of the digital economy. Annual report of ROSTEC Corporation for 2016. http://ar2016.rostec.ru/digital-g20/. Accessed 20 Feb 2020
11. Rudycheva N (2017) IT market in transport: the end of stagnation. http://www.cnews.ru/reviews/transport2018/articles/rossijskij_transport_obedinit_edinaya_tsifrovaya_platforma. Accessed 20 Feb 2020
12. Skrug VS (2018) Digital economy and logistics. Bull BSTU named V.G. Shukhov 5:138–143
13. Supply Chains (2018) Chat bots in purchases: what will help artificial intelligence? http://supplychains.ru/2018/10/30/chat-bot-zakupki. Accessed 20 Feb 2020
14. SWI Swissinfo (2018) Switzerland's Economy in excellent condition. https://www.swissinfo.ch/rus/%D1%8D%D0%BA%D0%BE%D0%BD%D0%BE%D0%BC%D0%B8%D0%BA%D0%B0-%D0%B8-%D0%B6%D0%B8%D0%B7%D0%BD%D1%8C_%D1%8D%D0%BA%D0%BE%D0%BD%D0%BE%D0%BC%D0%B8%D0%BA%D0%B0-%D1%88%D0%B2%D0%B5%D0%B9%D1%86%D0%B0%D1%80%D0%B8%D0%B8-%D0%B2-%D0%BF%D1%80%D0%B5%D0%BA%D1%80%D0%B0%D1%81%D0%BD%D0%BE%D0%BC-%D1%81%D0%BE%D1%81%D1%82%D0%BE%D1%8F%D0%BD%D0%B8%D0%B8/44386278. Accessed 20 Feb 2020
15. Zubakov GV (2017) Digital transformation in the logistics outsourcing. Hum Cap Vocat Educ 1(21):63

Innovative Solutions for Ensuring Information Security of Modern Enterprises

A. V. Balanovskaya, A. V. Volkodaeva, and A. Yu. Smol'kova

Abstract One of the key factors determining success of modern enterprises in the field of information security is their openness to innovations, work in promising development areas and their innovative activity. Enterprises of the information security market work exclusively in the aspect of innovative activities that ensure their sustainable economic development and strengthening of competitive positions in the market. The growing use of information technology leads to increased digital security risks. Cyber incidents lead to various types of losses for companies. To solve these problems, there are many technical and software products for ensuring information security of enterprises. Acquisition and development of innovative products in the field of information security will allow companies to build an effective protection system against threats and ensure successful operation and long-term development.

Keywords Innovative activity · Information security · Innovative activity · Digital risks · Cyber incidents · Information technologies

1 Introduction

Information security of the Russian Federation is a protection of the individual, the whole society and the state from internal and external information threats, which ensures the implementation of the constitutional rights and freedoms of humans and citizens, decent quality and standard of living, sovereignty, territorial integrity and sustainable socio-economic development of the Russian Federation, defense and

A. V. Balanovskaya
Samara State University of Economics, Samara, Russia
e-mail: balanovskay@mail.ru

A. V. Volkodaeva (✉) · A. Yu. Smol'kova
Moscow Federal Government Funded Educational Institution "Moscow City University", Samara, Russia
e-mail: arina-21@mail.ru

A. Yu. Smol'kova
e-mail: smolkovaanna1986@mail.ru

S. I. Ashmarina and V. V. Mantulenko (eds.), *Current Achievements, Challenges and Digital Chances of Knowledge Based Economy*, Lecture Notes in Networks and Systems 133, https://doi.org/10.1007/978-3-030-47458-4_87

security of the state [5]. In the Russian Federation, information security issues are regulated at the legislative level by the Federal law "On information, information technologies and information protection" of July 27, 2006, No. 149-FZ (last.ed. 02.12.2019 No. 427-FZ.), Decree of the President of the Russian Federation of 17.03.2008 No. 351 "On measures to ensure information security of the Russian Federation when using information and telecommunication networks of international information exchange" (last.ed. from 22.05.2015) and Decree of the President of the Russian Federation from 05.12.2016 No. 646 "On approval of the information security Doctrine of the Russian Federation" [5–7].

Innovative activity of an enterprise in the field of information security is an activity aimed at using research and development in the field of information technologies to expand the range and improve the quality of information security products, improve their production technology, implementation and effective use in the activities of enterprises in various industries. Innovative activity involves a complex of scientific, technological, organizational, legal, financial and commercial measures that together lead to innovations [2]. Legal aspects play a significant role in the development and improvement of the information security products. At the legislation level, there is a number of requirements for data protection and the use of information technologies that correspond to the current stage of development and technological equipment of enterprises.

2 Methodology

The problem of using innovative solutions to ensure the information security of enterprises was studied using scientific research methods:

- theoretical, including the study and analysis of foreign and domestic experience in the field of innovative solutions for ensuring information security of modern enterprises; analysis and classification of losses from cyber incidents; a comparative method for comparing indicators of official statistical data of the Russian Federation for the period 2015–2018; a systematization method when building the correspondence of innovative solutions for ensuring the information security with the functionality of these systems,
- statistical methods of data processing, such as calculating the cost structure by the type of innovations and economic activity; calculating growth rates and growth rates of indicators: investments of Russian enterprises in fixed capital for equipment with information and communication technologies; indicators of innovative activity of enterprises engaged in the development of computer software; quantitative indicators of organizations that use digital information protection tools.

3 Results

Due to the need to meet modern information security requirements, more and more companies are investing in fixed capital for equipment with information and communication technologies. In 2018, compared to the previous period, this indicator increased by 24.21% and amounted to 484034.6 million rubles (Fig. 1).

The increase in investment in information and communication technology equipment, including the information security, determines the growth trend in the number of organizations that have used means to protect information transmitted over global networks. In particular, in 2018, the share of organizations that used encryption tools increased by 3.39%, while the share of organizations that used electronic digital signature tools increased by 2.41% (Fig. 2).

The presented statistics do not determine the growth of the share of internal expenditures for research and development in the information and communication

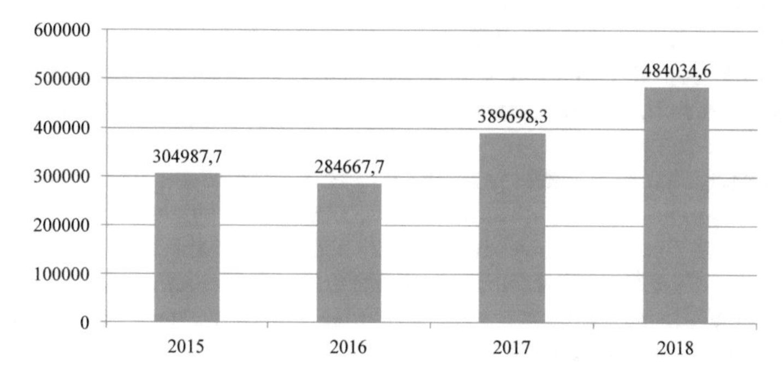

Fig. 1 Volume of investments in fixed capital for equipment with information and communication technologies in the Russian Federation, in actual prices (million rubles) (*Source* authors based on [8])

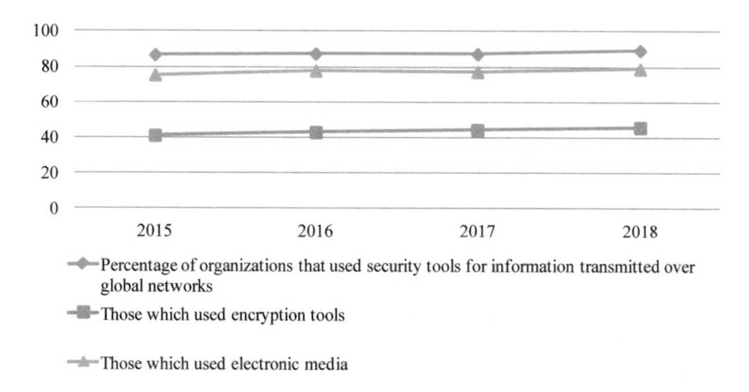

Fig. 2 Share of Russian organizations that used means of protecting information transmitted over global networks in the total number of surveyed organizations, % (*Source* authors based on [8])

technologies sector in the total volume of internal R&D expenditures in the Russian Federation. However, due to the relevance of innovative activities in the field of information technologies and the need to improve them in accordance with the environment requirements, this indicator should have higher development rates. Thus, according to the monitoring of the information society development in the Russian Federation, it decreased from 2.5% to 2.4%.

Innovative activity of information technology organizations (when compared by the type of economic activity) has a relatively small weight in the total number of organizations. Thus, the innovative activity of enterprises engaged in the development of computer software, providing consulting services in this area and other related services in 2018 was 6.6% in technological innovations, 2.8% in process innovations and 3.9% in product innovations. Innovation activity of enterprises engaged in information technology activities was lower for technological innovations (4.5%) and product innovations (1.7%), but it was higher for process innovations and amounted to 3.1%.

The cost of technological innovations in organizations that develop computer software, consulting services in this area and other related services amounted to 14.8 billion rubles from the total cost of enterprises for all types of economic activities. For organizations that operate in the field of information technology, this indicator was 3.5 billion rubles (Table 1).

The largest amount of funds is spent on research and development, but there are no costs for purchasing patents and licenses, as well as engineering and marketing research. Thus, like any innovative project, a project in the field of information technology and information security includes a set of research, design, production, organizational, financial, commercial and other activities, the implementation of which requires appropriate costs.

Table 1 Expenses for technological innovations of organizations, by types of innovative and economic activities in the field of information technologies in the Russian Federation, billion rubles

Cost items	Development of computer software, consulting services in this area	Activities in the information technology field
Research and development	7.6	0.1
Purchase of machinery and equipment	1.8	2.3
Acquisition of new technologies	0.2	0.0
Purchase of software tools	0.4	0.9
Design	2.0	0.0
Training of personnel	0.0	0.1
Other expenses	3.8	0.1

Source authors based on [9]

Table 2 Potential losses from common types of cyber incidents

Types of cyber-incidents	An example of a cyber-incident	Potential losses
Denial of service	A distributed denial-of-service attack on a server that causes the company's website to be unavailable	- A break in the business - Costs of responding to the incident - Reputational damage - Responsibility of directors and officers - Loss of data and software
Privacy violation	Unauthorized disclosure of personal data of third parties	- Costs of responding to the incident - Privacy compensation - Protection costs - Fines and penalties - Reputational damage - Loss of data and software - A break in the business
Cyber-extortion	Ransomware that prevents access to data or the network until the ransom is paid	- Cyber-extortion and extortion - A break in the business - Costs of responding to the incident - Reputational damage - Responsibility of the director and officials - Loss of data and software
Cyber fraud	Illegal financial transfers as a result of social engineering	- Financial theft and/or fraud - Costs of responding to the incident - Responsibility of the director and officials

Source authors

The increasing use and dependence on information technology in economic activities creates significant benefits in terms of productivity and efficiency, but also leads to significant risks. Among them there are "digital security risks" that, when materialized, can disrupt the achievement of economic and social goals by violating the confidentiality, integrity, and availability of information and information systems. It is widely believed that most companies will not know that they have been affected by any cyber incidents. Although, quantitative measurements are still emerging and causing significant problems. Reports on the frequency and scale of (reported) cyber incidents regularly show a significant increase in both the number of incidents and the percentage of companies affected by them. This led to the situation when the cyber risk was identified as a risk of the highest (or second highest) interest in doing business in the five G7 countries in 2017 at the world economic forum.

Cyber incidents, such as privacy violations, denial-of-service attacks, cyber fraud, and cyber extortion, can lead to a number of different types of losses for companies (Table 2). In the world practice, there are examples of physical damage and destruction caused by cyber attacks, including damage to a steel plant in Germany in 2014 and large-scale power outages in Ukraine in 2015.

Innovative activity of enterprises in the field of information security determines strategic prospects and expands the scope of their activities, ensuring a high level of competitiveness of the company and its products. There are many technical and software solutions to ensure the enterprise information security, but a number of them are more in demand due to the compliance with modern market requirements and legislation. There are many ways to create a reliable security system for wireless networks. Among the existing systems, we can single out the WPA2 encryption algorithm, since the WEP algorithm (Virtual Private Network application method) is considered as obsolete. It is also possible to use the 802.11i standard as a whole, aimed at increasing the security of wireless networks: it involves the use of AES encryption [13]. «Security information» and «event management» (47% of responses), «security awareness software» (46% of responses), and «web application firewall» (33% of responses) are among the TOP 3 areas of information security that customers are interested in [10]. «Security information» and «event management» is a real-time analysis technology that allows you to respond to threats before significant damage occurs. «Security awareness software» is an innovative solution in the field of cybersecurity, combining the latest educational technologies and ensuring maximum awareness of employees and the formation of new models of cybersecurity behavior (Table 3).

The Threat Stack API allows suppressing and rejecting alerts from existing tools, optimizing workflows for the response to incidents, and reducing the average response time. Coherence Anti-Ransomware serves as the main protection against ransomware, provides anomaly detection, and the ability to find and delete infected files throughout their global data space. Sentinel SCA Essentials Edition identifies third-party and open source components, and Sentinel Source Essentials Edition prioritizes vulnerabilities based on their severity [11].

In the advanced version of bizhub SECURE Platinum, SSL/TLS and automatic logging are enabled, as well as user authentication and automatic logout from the administrator and user accounts. PayControl and SafeTech provide adaptive user authentication, confirmation of transactions with an electronic signature, and scoring of clients devices. Titus Accelerator for Privacy helps to ensure the data privacy and compliance with cybersecurity policies and information security standards. HITRUST MyCSF includes a redesigned user interface and the ability to create your own system for evaluating certain regulatory or management requirements. ComplianceAlpha 2.0 is built on a new API-based infrastructure that allows platform components to interact with each other and with third-party systems. Solar Dozor 7 allows you to use automated analysis to detect early signs of violations by employees of the company, and Solar appScreener 3.3 is a version of the application security analyzer [12].

Table 3 Innovative solutions for ensuring the information security in accordance with the functional opportunities

	The information security solution												
	Security information and event management	Security awareness software	Web application firewall	Threat stack API	Cohesity anti-ransomware	Sentinel source essentials edition и sentinel SCA essentials edition	SIOS protection suite	Bizhub SECURE	PayControl и SafeTech	Titus accelerator for privacy	HITRUST MyCSF	ComplianceAlpha 2.0	Solar appScreener 3.3
Cloud infrastructure protection				+									
Protection against ransomware attacks and financial fraud					+				+				
Security testing, application monitoring			+			+	+						+
Encryption and development of strong passwords								+					
User identification and personal data protection		+	+						+	+			+
Security risks management	+	+	+								+	+	

Source authors

Every year, leading companies in the field of information security develop new solutions and make changes to existing ones in accordance with the market requirements for protecting information from viruses and spam, internal fraud and cybercrime, repelling DDoS attacks, protecting cloud systems and various threats of a different spectrum.

Governments play an important role in countering cyber threats by recommending security standards in key industries (for example, the NIST Cybersecurity Framework), as well as strengthening law enforcement prosecutions of cyber criminals and imposing harsher penalties. The European Union introduced the General Data Protection Regulation (GDPR) on 25 March 2018, the legal basis of which sets out guidelines for the collection and processing of personal information of individuals within the European Union. Government funding is needed to support law enforcement agencies and government agencies in the fight against cybercrime. In the US, for example, a federal budget of $15 billion was proposed for the fiscal year 2019, which is 4.1% more than in the previous year, which does not include spending on cybersecurity in 50 states and municipalities. Using the U.S. share of the global GDP, we can estimate global public sector spending at about $60 billion in 2017. A part of the public sector's spending is directed to building national security capabilities, such as cybersecurity, defense, and cyberwarfare.

4 Discussion

In the private sector, global spending on cyber security products and services reached more than $114 billion in 2018, up 12.4% from $101 billion in 2017, according to Gartner [1]. Firms invest in cybersecurity to protect their own computer systems, including prevention measures (installing firewalls, encryption, and access control), detection measures (such as intrusion detection systems), and incident response measures (computer forensics, reverse engineering malware).

Despite the increase in spending on cybersecurity from year to year in both the public and private sectors, the number of cybercrime cases continues to grow, and the growth of cybercrime does not seem to be restrainy. There are problems in the fight against cybercrime: international law enforcement agencies worry about the rampant spread of cybercrime and the rapid growth of dark web and underground cybercrime platforms [3]. The level of coordination observed between law enforcement agencies around the world appears to be lacking, and the prevailing atmosphere is that there is no effective exchange of information about cyber incidents between organizations because of non-disclosure, antitrust, or privacy laws [4]. Private sector firms spend money almost exclusively on protecting their own 'walls', which do not directly affect external participants in cybercrime. There is a need to involve the private sector and business solutions to counter cyber threats [14].

5 Conclusion

Developing innovations in the field of enterprise information security involves a high risk of investing resources in research and production of new solutions and not getting the expected commercial success. This is determined by both the usual updating of the existing security system at the enterprise as opposed to the application of a new solution and the lack of training of the personnel of consumer enterprises for the introduction and use of innovative information security products. Thus, it is necessary not only to apply innovative information security products, but also to ensure the activity of enterprises to meet the new conditions of its functioning. The development of innovative products in the field of information security allows companies that use them to build an effective system of protection against threats. High-quality security of the company's information resources contributes to its confidence in protection from the encroachments of competitors, its successful functioning and long-term development by directing the main forces to ensure effective activity, and not to counteract cyber threats.

References

1. Aitken R (2018) Global information security spending to exceed $124B in 2019, privacy concerns driving demand. Forbes. https://www.forbes.com/sites/rogeraitken/2018/08/19/global-information-security-spending-to-exceed-124b-in-2019-privacy-concerns-driving-demand/#5cb66de47112. Accessed 12 Feb 2020
2. Amirova ES, Agalakova AV (2012) Features of the organization of innovative activity of enterprises in modern conditions. In: Dmitrieva NV (ed) Proceedings of the XIV international scientific conference "economics and modern management: theory and practice. Sibak, Novosibirsk, pp 36–40
3. Anderson R, Moore T (2006) The economics of information security. Science 314(5799):610–613
4. Biener C, Eling M, Wirfs JH (2015) Insurability of cyber risk: an empirical analysis. Geneva Pap Risk Insur – Issues Pract 40(1):131–158
5. Decree of the President of the Russian Federation from 05.12.2016 No. 646 "On approval of the information security Doctrine of the Russian Federation". http://base.garant.ru/71556224/. Accessed 12 Feb 2020
6. Decree of the President of the Russian Federation of 17.03.2008 No. 351 "On measures to ensure information security of the Russian Federation when using information and telecommunication networks of international information exchange" (last.ed. from 22.05.2015). http://www.consultant.ru/document/cons_doc_LAW_75586/. Accessed 12 Feb 2020
7. Federal law "On information, information technologies and information protection" of July 27, 2006, No. 149-FZ (last.ed. 02.12.2019 No. 427-FZ). http://www.consultant.ru/document/cons_doc_LAW_61798/. Accessed 12 Feb 2020
8. Rosstat (2019) Monitoring the development of the information society in the Russian Federation. https://www.gks.ru/folder/14478. Accessed 12 Feb 2020
9. Rosstat (2019) Russia in numbers 2019. Rosstat, Moscow
10. Security Code (2019) Information security in practice. Results of 2018, prospects for 2019. https://www.securitycode.ru/upload/iblock/119/Info_Security_in_practice_2019.pdf. Accessed 12 Feb 2020

11. Security Lab (2019) New IS solutions of the week: January 15, 2019. https://www.securitylab.ru/news/497459.php. Accessed 12 Feb 2020
12. Security Lab (2019) New IS solutions of the week: October 11, 2019. https://www.securitylab.ru/news/501711.php. Accessed 12 Feb 2020
13. Skrypnikov AV, Denisenko VV, Evteeva KS (2019) Protection of data when transmitting wireless communication channels. Int J Humanit Nat Sci 8(2):35–38. https://doi.org/10.24411/2500-1000-2019-11485
14. World Economic Forum (2017) Global Risks Report 2017. www.weforum.org/reports/the-global-risks-report-2017. Accessed 12 Feb 2020

System Approach to the Control Organization of Management Decisions

S. V. Sharokhina and T. A. Shevchenko

Abstract The use of new technologies aimed at the reconstruction of business processes, such as business process redesign (BPR) and total quality management (TQM), as well as the distribution of responsibilities, significantly improves the quality of the management process. The implementation of modern technologies should encourage employees who are at the lowest levels of the organizational structure to be responsible for solving problems, and in many cases to make their own decisions. The purpose of the article is to form theoretical positions on the organization of control in the process of forming management decisions in the enterprise information system. Based on the analysis of scientific works of domestic and foreign researchers, using a systematic approach, the authors argue that beyond the control process, as a condition for its implementation, there are organizational, methodological and information support and, as a manifestation of the latter, setting for condition parameters of controlling objects. The authors revealed the interaction process between administrative and economic control in the enterprise information system.

Keywords Control · Administrative control · Economic control · Enterprise information system · Management decisions

1 Introduction

Fundamental changes in management practices under economic conditions, which are characterized by a sharp activation of new and supernew technologies, have significantly affected the style of management work at enterprises in various industries. If they are not effective, it is only because the process of granting the right

S. V. Sharokhina (✉)
Syzran Branch of Samara State University of Economics, Syzran, Russia
e-mail: sharokhinatv@gmail.com

T. A. Shevchenko
Branch of the Military Training and Research Center of the Air Force "Air Force Academy named after Professor N.E. Zhukovsky and Yu.A. Gagarin", Syzran, Russia
e-mail: privet7770@rambler.ru

S. I. Ashmarina and V. V. Mantulenko (eds.), *Current Achievements, Challenges and Digital Chances of Knowledge Based Economy*, Lecture Notes in Networks and Systems 133, https://doi.org/10.1007/978-3-030-47458-4_88

to solve problems has not reached the lower levels of the organizational structure of enterprises. A large number of managers at all levels blocked change processes, which means the loss of additional advantages [10]. Problems of development of management decisions are in the focus of theorists' attention, as it is evidenced by publications of scientific papers on the technology of decision-making [6, 8, 13].

However, the main research method for the system approach to the control function is the work of V.A. Shevchuk, where the emphasis is placed on the fact that from the managerial (cybernetic) view, control can be interpreted as a function, subsystem, or management process. According to this scientist, the study of the managerial essence of control using the graphoanalytic method allows abstracting it from other management functions and considering it separately as a subsystem (system), which, in turn, consists of a set of interacting elements that form a certain integrity [14].

The problem of a systematic approach to the organization of control was studied in detail by Kocherin [7]. In his opinion, the control system is a set of subject, object and control tools that interact as a whole in the process of establishing results of activities, as well as measuring the condition of the control object, analyzing and evaluating measurement data, and developing corrective actions. These research works have a little affect on the above mentioned problems, do not give detail characteristics of the monitoring values that would be adequate to needs and requirements of the development process of management decisions the accounting function is identified with the analytical one.

2 Methodology

The methods of the theoretical level used in the study include: abstraction, formalization, analysis and synthesis, induction and deduction, generalization. The research methodology is based on a systematic approach, as well as on the analysis of scientific works of domestic and foreign authors on management, control and analysis issues.

Generalization of scientific ideas in the field of management allows us to state that managing or managing an enterprise means constantly making decisions, the implementation of which will allow the company to maintain its business activity [5]. At the same time, there is an identification of such concepts as "manager's decision" and "management decision". The first concept characterizes decision-making by the manager (it is about any decisions that are a manifestation of the nature of his activities: administrative, economic, industrial) [6]. The concept of "management decision" involves making decisions in the context of the implementation of management functions.

This differentiation of concepts gives grounds to assert that the value of a management decision is determined by the ability to link all management functions. This indicates that the decision-making function plays a special role in the management process—it is necessary for the implementation of other functions, including the control one [11]. You can embed all management functions in each management decision-making function.

Therefore, the rationale for a systematic approach to the organization of control in management decision-making should be carried out taking into account the fact that the control function is a part of the management process; the control function cannot be studied as an analytical component, the purpose of which is to assess the level of performance of planned tasks. The main element of the control system—the control process should be considered as a set of actions that the controlling entities perform in relation to the controlled objects in order to achieve the control goal.

3 Results

Beyond the control process, there are: organizational, methodological and information support and setting parameters for the controlling objects as necessary conditions for its implementation. These aspects of the control essence are closely related to its organizational support, which implies achieving a certain state of ordering the system elements, maintaining its qualitative certainty [12].

The main elements of any system are its mechanism, statics and dynamics. The mechanism is characterized by the theory (defining a goal, subject, and method) and the practice. A system in statics is characterized by its structure, elements, and relations between them, as well as its technique. The system in dynamics is characterized by technical processes that are reflected in it.

Ordering all elements of the control system is inherent in its organization. Such an organization is in purposefully improving the mechanism, structure and process of control and can mean either organizing control as a system, or organizing its functioning in time and space. The first interpretation corresponds to the concept in a broad sense. The second interpretation corresponds to the concept in a narrow sense. So, analyzing the control of financial and economic activities of the enterprise, we come to the conclusion that it is subject-separated in the management system as a management function (type of the human activity), is a component of the named system. Besides, the control of the financial and economic activity belongs to the system of the highest symbolic level—the system of economic information [4].

The basis of the mathematical description of control is the balance method of managing the economy. The essence of it is to achieve:

- inter-related coordination of needs and resources within the relevant economic system of an enterprise, association, organization, institution, corporation, etc.,
- mutual coordination of actions of various economic formations,
- ensuring a balance in the economy and proportionality in accordance with economic laws [9].

A systematic approach to the control organization involves:

- development of a general logic for building a functional system of preliminary, current and subsequent control,
- definition of specific principles on which the implementation of control actions of personnel should be based,
- determination of specific features of the control organization in the conditions of automated processing of economic information [3].

Therefore, the organization of the main elements of the control system is a set of individual cycles, the main of which are:

- defining a purpose, tasks, and control objects,
- collecting information about controlled objects,
- identification of deviations and violations in the implementation of control actions,
- adoption of corrective measures aimed at elimination and prevention of violations.

The first cycle is related to the mechanism, and the last three are related to the system in dynamics, that is, to technological processes that occur within this system. Statics as a system element (the structure, personnel, and equipment) has remained out of our attention, which leads to the conclusion that it is necessary to develop an organizational structure of control, regulatory and technical support for control actions.

It is the regulatory support of control actions that should consist of internal control standards, control regulations and instructions of its subjects [2, 7, 15]. The subjects of control goal-setting within the enterprise are the owners of capital. In the public sector, the owner is the state.

Therefore, control in the management system (that is, in the management decision-making system) can be defined as a combination of administrative and economic (financial and economic) control. In turn, administrative control ensures overall functioning of the economic system in accordance with the socio-economic goals for which it functions.

Administrative and economic control can be considered as certain parts of the internal control, the quality of which determines the effectiveness of management decisions [1]. But these types of controls cannot be represented as the ratio of parts to the whole. The scheme of interaction between administrative and economic control in the enterprise information system is shown in Fig. 1. The formalized representation of the nature of interaction between administrative and economic control confirms the existence of relatively independent parts of both types of control.

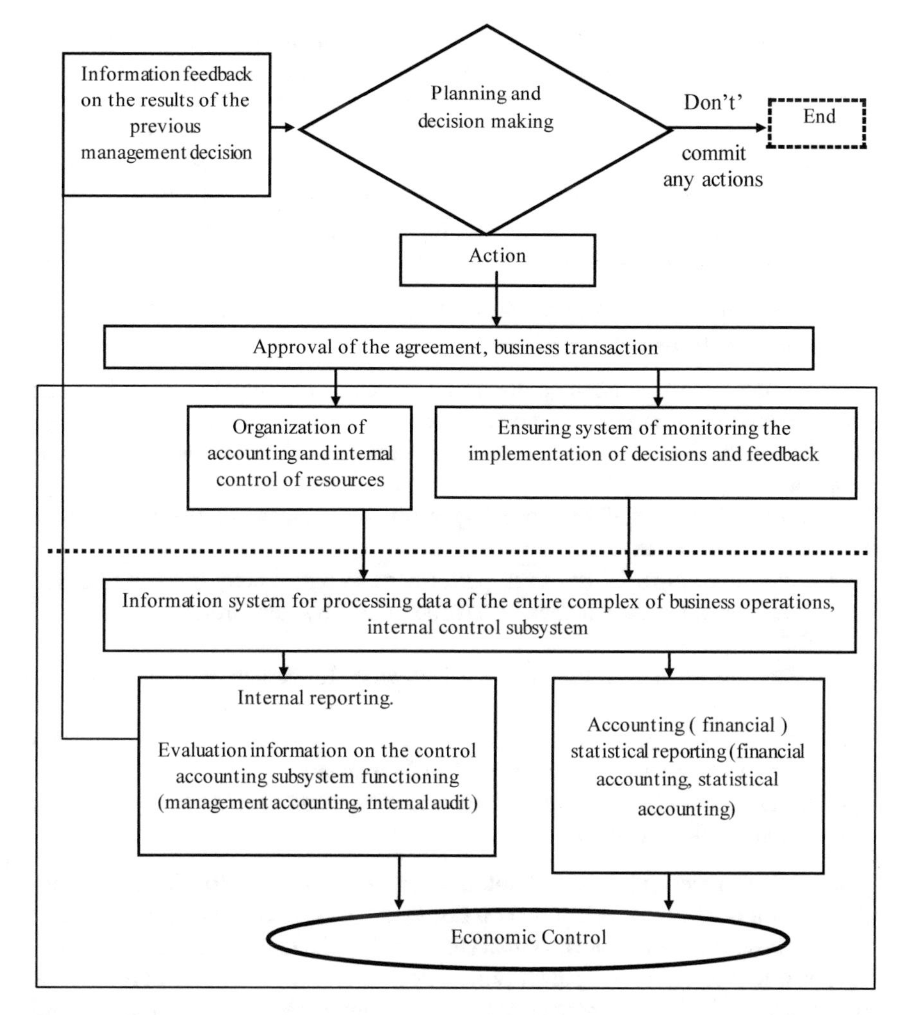

Fig. 1 Diagram of interaction between administrative and economic control in the enterprise information system (*Source* authors)

4 Discussion

According to the above, the purpose of control in the management decision-making system is precisely to promote the rational and efficient use of labor, material and financial resources of any enterprise in any form of ownership. The goal defines the main control tasks that can be related to:

- prevention of illegal, economically impractical steps in the economic activity that contradict the interests of the company and its business partners at the stage of developing management decisions (proactive control),

- elimination of deviations, violations at the stage of making and implementing management decisions (current control),
- checking the legality, economic feasibility of already completed operations, implemented management decisions during the evaluation of: execution of the estimate; achieving the goal of the development strategy; fulfilling obligations to partners and the state; management condition; quality of accounting and economic information as a basis for developing, making and implementing management follow-up decisions (final control).

Analysis of the control essence creates the necessary conditions for determining the principles of its operation, such as:

- comprehensiveness—dissemination at all stages of development, adoption and implementation of management decisions,
- combining the interests of the owner of the capital, each worker and all business partners,
- delegation of responsibilities (powers), rights and responsibilities to the lower levels of management,
- objectivity and impartiality—each performer of control actions should be free from bias and external influence, should be guided by standards,
- expediency, economy,
- specialization—concentration of control activities for certain objects, each stage of the development process and management decision-making,
- effectiveness—the ability of the control system to prevent crisis situations and ineffective management decisions in a timely manner,
- professionalism—the ability of specialists to competently perform control actions in relation to management decisions.

The control process, like any technological process, consists of operations. A control operation is an action performed on economic data (information) and aimed at obtaining various intermediate and final (final) indicators that are suitable for evaluating performance and making management decisions. The set of control operations forms a procedure. Several procedures form a technological phase of control, and phases respectively form a stage of the control process.

In the organization of the control process, there are three stages: the first is the preparation of data for control; the second is the verification of data by various methods of control; and the third is the generalization of control results. The objects of the control process organization should be the control information carriers; the control process technology (the movement of carriers during the implementation of the control method).

5 Conclusion

A systematic approach to the control organization (in the broad sense of this phenomenon) is characterized as:

- definition of a range of subjects, their structuring, and systematization of control objects,
- justification of the purpose and objectives of the activity,
- definition of legal and regulatory support for activities,
- systematization of sources of information support for the control process,
- justification of the main principles of the control process organization,
- highlighting the main stages of the control process,
- organization of the control process technology,
- determination of technical support for control actions.

The determination of organization objects is based on the content characteristics of the control process stages and the nature of the interaction between the administrative and economic control in the enterprise information system. Not only on the effectiveness, but also on the efficiency of control largely depends on the degree of organizational support for the implementation of the control function in the management process, as a process of making management decisions.

References

1. Briciu S, Dănescu AC, Dănescu T, Prozan MA (2014) Comparative study of well-established internal control models. Procedia Econ Financ 15:1015–1020
2. Chang YT, Chen H, Cheng RK, Chi W (2019) The impact of internal audit attributes on the effectiveness of internal control over operations and compliance. J Contemp Account Econ 15(1):1–19
3. Dreving SR, Khrustova LY (2016) Modern understanding of internal financial control: problems and perspectives of investigation. Manag Sci 3:30–44
4. Garifova LF (2015) Infonomics and the value of information in the digital economy. Procedia Econ Financ 23:738–743
5. Gilboa I (2017) Making better decisions: decision theory in practice. Wiley-Blackwell, Chechister
6. Ji X, Lu W, Qu W (2018) Internal control risk and audit fees: evidence from China. J Contemp Account Econ 14(3):266–287
7. Kocherin EA (2000) Fundamentals of state and management control. Filin, Moscow
8. Kupriyanov YuV (2015) Balance method of strategic planning in industrial policy. World New Econ 3:16–20
9. Kuznetsova NV (2015) Management decision - making methods. Infra-M, Moscow
10. Santoro G, Quaglia R, Pellicelli AC, Bernardi P (2020) The interplay among entrepreneur, employees, and firm level factors in explaining SMEs openness: a qualitative micro-foundational approach. Technol Forecast Soc Chang 151:119820
11. Swami S (2013) Executive functions and decision making: a managerial review. IIMB Manag Rev 25(4):203–212
12. Tereshchenko NG (2017) The term “management” in a series of similar categories. Concept 10:75–83

13. Timchenko TN (2017) System analysis in management. RIOR, Moscow
14. Tokareva AA (2016) Features of monitoring and controlling the financial results of the organization. Symb Sci 11(1):192–196
15. Wilford AL (2016) Internal control reporting and accounting standards: a cross-country comparison. J Account Public Policy 35(3):276–302

Digitization in the Development of Program Budgeting in Russia

L. N. Mulendeeva

Abstract The article traces the development of program budgeting under the digital economy in Russia. Information society shaping and digital economy formation are the priorities for strategic development in our country, which makes the research ever more relevant. Currently various digital tools are being used extensively in program budgeting. This study aims at the investigation of program budgeting digitization in the Russian Federation. The methodology used in this study combines the method of generalization, system and comparative analyses, factor analysis, and graphical method. The findings demonstrate that digitization of program budgeting might harmonize the efforts of different ministries, which develop and implement government programs. Moreover, the automatization of program budgeting could allow increasing the efficiency of budget spending.

Keywords Digital economy · Information society · Program budgeting · Budget planning · E-Government

1 Introduction

Today, digital economy is an integral part of shaping the information society, which, in its turn, is considered to be the strategic priority of the Russian Federation development. This goal involves providing citizens with high-quality reliable information through the development of information and communication infrastructure, as well as the implementation of competitive information and communication technologies. The economic and socio-cultural life of citizens in the information society is characterized by the enormous influence of information and the degree of its application. The development and application of the latest information technologies contributes to the development of a new stage of economy—the digital economy. Its efficiency is due to the possibility of using technology for processing large amounts of data, which helps to reduce costs for the production of goods and services. The digital

L. N. Mulendeeva (✉)
Samara State University of Economics, Samara, Russia
e-mail: muln@mail.ru

S. I. Ashmarina and V. V. Mantulenko (eds.), *Current Achievements, Challenges and Digital Chances of Knowledge Based Economy*, Lecture Notes in Networks and Systems 133, https://doi.org/10.1007/978-3-030-47458-4_89

economy is spread throughout all spheres of economic life in Russian society, and the budget planning process is not an exception. Digitalization of the budget process is aimed at improving the efficiency of using budget funds and improving the quality of services provided by the government.

2 Methodology

The application of general scientific methods, comparative and the generalization methods among them, allowed determining the scope of digital economy and its role in program budgeting as well as in the socio-economic development of Russia. Methods of economic, statistical, factor and comparative analysis were used to determine the international ratings of the Russian Federation according to the degree of e-government development, the amount of costs for foreign communications and telecommunications development.

3 Results

Studies over the past decades have provided many definitions of digital economy. For example, the World Bank defines the digital economy as a system of economic, social and cultural relations based on the use of digital information and communication technologies (ICT).

Digital data are considered by the Russian legislation as the core factor of production in digital economy. The efficiency of various industries, technologies, equipment, storage, sales and delivery of goods and service could be sufficiently increased with big data processing [2].

In general, the digital economy is a segment of economic relations driven by technological advances, a global network, and information systems [11]. The tools of the digital economy in the world are represented by cloud computing, big data processing technologies, mobile technologies, the Internet of things, geolocation, etc. [9]. In addition to increasing production, the digital economy is aimed at improving the efficiency of public administration. In this regard, Russia is actively working on the introduction and implementation of e-government infrastructure that provides state and municipal services to citizens and organizations through the use of information and communication technologies.

The purpose of e-government is to increase the transparency of state authorities' activities and involve the public in solving national issues. As a result, the quality of services provided by the state increases, and the needs of the citizens are taken into account. Belokurova, Pizikov, Petrenko, and Koshebayeva carried out an investigations into institutional model of digital economy development. The economists have found that the formation of the digital economy is preceded by the stages of formation of e-government and information society. It is argued in their study that the digital economy is created by the government of the country artificially [3].

This idea is developed in Hanna [8] and Osnovin [13]. The position according to which the state has a leading role in the development of the digital economy is fully shared throughout this paper. At the same time, the state's task is to ensure that the actions of the ministries both responsible for economic development and finance of the country and directly involved in the development and implementation of information and digital technologies are coordinated.

It is stated that one of the restraining factors for the development of the digital economy in Russia is the inefficient distribution of powers among the subjects of information society. The state should coordinate the actions of all participants in the process of digitalization of the economy and the relationship of material, labor and financial resources [15].

According to official data from the UN Department of Economic and Social Development, Russia moved from the 60th position in 2008 to the 32nd position in 2018 in the world ranking of the. The value of this index increased from 0.5120 to 0.7969 (Fig. 1) [1].

According to the index of citizens involvement in using e-government services, Russia ranked 23rd in the international ranking in 2018. Next, the dynamics of research and development costs in the domain of «Information and Telecommunication Systems» in the Russian Federation was studied (Fig. 2) [1].

As it can be seen from the diagram during 2010–2017, there was an increase in research and development funding of information and telecommunications technologies. In 2017, this indicator reached the value of 81390,7 million rubles, which is more than 2 times higher than the amount of these costs in 2010. The federal budget of the Russian Federation has the biggest share in the financing for the development of the information and telecommunications system (in 2017, 49284,2 million rubles, or 60.55% of the total funding).

Thus, the development of the digital economy in our country is initiated by the government and receives enormous financial support from the budget.

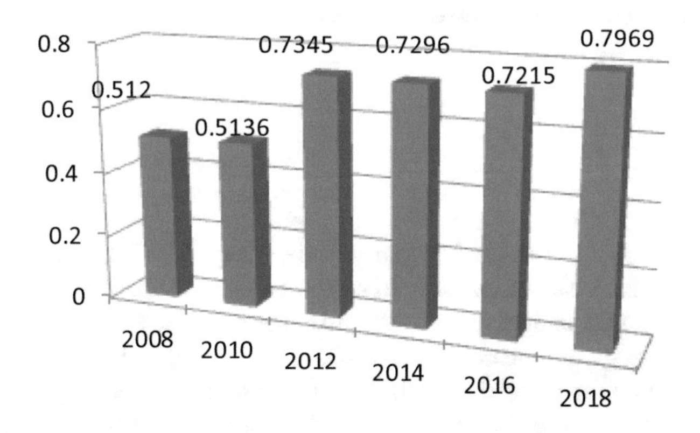

Fig. 1 Dynamics of e-government development index in Russia (*Source* author based on [1])

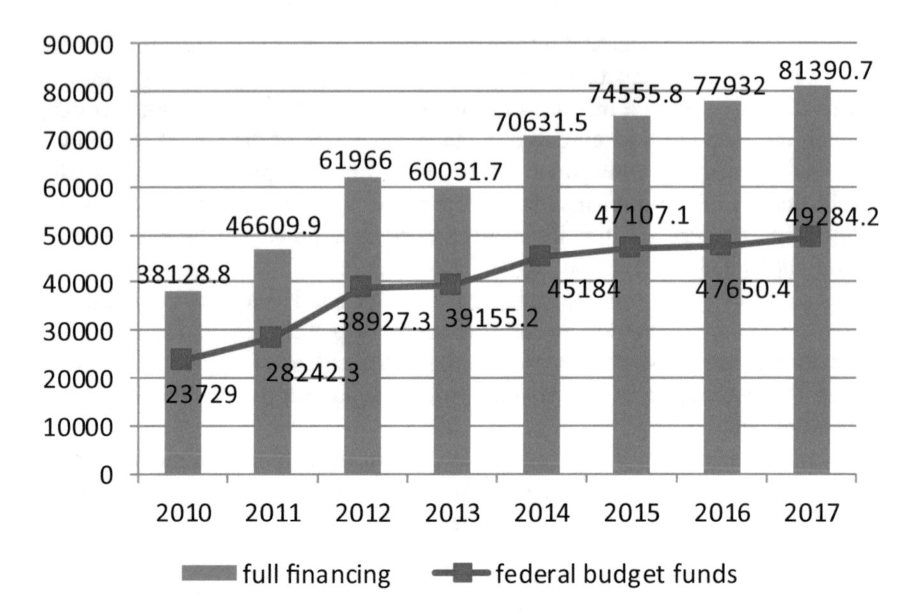

Fig. 2 Dynamics of research and development funding in the domain of «Information and Telecommunication Systems» in Russia in 2010–2017, mln. rub. (*Source* author based on [1])

Bogoviz, Alekseev, Ragulina argue that the existing deficit of the Federal budget of the Russian Federation hinders the formation of the digital economy [5]. This situation is compounded by insufficient funding for ICT from private investors (their share in 2017 was 11.7%).

In 2018, 77% of Russian citizens had Internet access from various devices. At the same time, only 54.5% of the population interacted with state and municipal authorities via the Internet, using official websites and portals [1]. Despite the increase in the use of e-government by three times compared to 2015, the activity of Russian citizens in this area is not active enough. The most popular types of electronic services are health care and medicine, taxes and fees, state road safety inspection, housing and utility services.

4 Discussion

The population of the Russian Federation shows very little interest in the service "Electronic Budget". The «Electronic Budget» is an information portal that provides information about the budget process and the situation in the Russian budget system [10]. Program budgeting method underlies budget planning in Russia and has transformed it into program budgeting [12]. The use of program budgeting in the Russian Federation must correspond with the task of developing the digital economy, which is characterized by the use of data in digital form in all areas of social and economic activity. Information infrastructure, which is among the five main areas of development of the digital economy specified in the program «Digital Economy of the Russian Federation» [14], is of particular importance for improving program budgeting. This is due to the fact that program budgeting is, at its core, budget planning, which is not limited only to the formation of budgets, but also involves the analysis and control of the planned and actual financial indicators compliance.

In practice, the implementation of this control is facilitated by the placement of up-to-date information on the implementation and evaluation of the effectiveness of state programs on the websites of relevant ministries. Thus, any user of a personal computer has open access to information about spending budget funds, applying the infrastructure of the digital economy. Program budgeting expands the functionality of monitoring the implementation of programs. It is performed with the use of modern information technologies thereby it increases the efficiency of government and municipal finance management. However, the development of program budgeting in the Russian Federation is constrained by a number of problems that can be partially solved through the use of information and telecommunications technologies. For example, many experts mention the lack of interagency coordination in the development and implementation of state programs. The presence of a unified digital platform would help to coordinate the goals, objectives, targets, and activities of government programs implemented jointly by several ministries.

Moreover, budget expenditure planning in our country is often carried out using the indexation method, i.e. reporting data on the same expenditures for the previous period are indexed by the inflation rate (or another coefficient). As a result, the efficiency of budget spending decreases. It is evident, that the application of ICT to the budget process, the use of a large amount of internal and external information about the state of the economy and social sphere of the state might allow switching to a fully automated process of program budgeting in the future. In this case, various forecast scenarios of economic development might be taken into account. The target indicators of state programs can be identified, and adjustments to state programs in accordance with the real socio-economic situation of the country can be timely made. It will result in the efficiency of budget planning in Russia. Program budgeting under budget process digitization was thoroughly examined by Bogoslavtseva et al. [4].

5 Conclusion

Thus, it could be concluded that today the development of any country in the world is accompanied by the introduction of digital technologies in almost all economic processes. Currently, the idea of creating the digital economy is being identified in the strategic documents of many countries, including Russia, although in our country, the role of the government is reduced mainly to creating conditions for the development of the information society and the digital economy, rather than to the development of advanced information technologies [7].

Nevertheless, much has been done to ensure that information is accessible and open to the public via the Internet and nowadays the government provides many services to citizens through e-government portals.

It is important to note the essential role of information and telecommunications technologies in the implementation of the budget process in the Russian Federation. In practice, the electronic budget makes it possible to implement the principle of transparency of the budget system [6]. In addition it allows determining the efficiency of spending budget funds.

Information technologies provide citizens with up-to-date information about the implementation and evaluation of the effectiveness of state programs. That increases the effectiveness of program budgeting, which involves constant monitoring of the use of budget funds. Automation of program budgeting in the future will bring budget planning in Russia to a fundamentally new level and ensure sustainable and rational use of the public financial resources.

References

1. Abdrahmanova GI, Vishnevskiy KO, Gohberg LM, Demyanova AV, Kevesh MA, Kovaleva GG, Fursov KS (2019) Digital economy indicators: 2019: statistical digest. https://www.hse.ru/data/2019/06/25/1490054019/ice2019.pdf. Accessed 20 Jan 2020
2. About the Strategy for the Development of the Information Society in the Russian Federation for 2017–2030. Decree of the President of the Russian Federation of 09 May 2017 N 203. http://www.consultant.ru/document/cons_doc_LAW_216363/. Accessed 20 Jan 2020
3. Belokurova EV, Pizikov SV, Petrenko ES, Koshebayeva GK (2020) The institutional model of building the digital economy in modern Russia. In: Popkova E, Sergi B (eds) Digital economy: complexity and variety vs. rationality. ISC 2019. Lecture notes in networks and systems, vol 87. Springer, Cham, pp 64–70
4. Bogoslavtseva LV, Karepina OI, Bogdanova OY, Takmazyan AS, Terentieva VV (2020) Development of the program and project budgeting in the conditions of digitization of the budget process. In: Popkova E, Sergi B (eds) Digital economy: complexity and variety vs. rationality. ISC 2019. Lecture notes in networks and systems, vol 87. Springer, Cham, pp 950–959
5. Bogoviz AV, Alekseev AN, Ragulina JV (2019) Budget limitations in the process of formation of the digital economy. In: Popkova E (ed) The future of the global financial system: downfall or harmony. ISC 2018. Lecture notes in networks and systems, vol 57. Springer, Cham, pp 578–585
6. Budget code of the Russian Federation. Accepted 17 July 1998. http://www.consultant.ru/document/cons_doc_LAW_19702/ac7eb6cf1d15c2a4225e64864e980f435d2142bc/. Accessed 20 Jan 2020
7. Dyachenko OV, Istomina EA (2018) Theoretical foundations of the digital economy in strategic planning documents. Bull Chelyabinsk State Univ 8(418):90–102
8. Hanna N (2018) A role for the state in the digital age. J Innov Entrep 7:5
9. Ivanenko LV, Karaseva EA, Solodova EP (2020) Clusters, digital economy and smart city. In: Ashmarina S, Mesquita A, Vochozka M (eds) Digital transformation of the economy: challenges, trends and new opportunities. Advances in intelligent systems and computing, vol 908. Springer, Cham, pp 291–295

10. Kovaleva TM, Valieva EN, Popova EV (2020) Characteristics of Russian government financial resources: historical overview and the situation under digital economy. In: Ashmarina S, Mesquita A, Vochozka M (eds) Digital transformation of the economy: challenges, trends and new opportunities. Advances in intelligent systems and computing, vol 908. Springer, Cham, pp 635–645
11. Krukova AA, Mihalenko JA (2017) Tools of the digital economy. Karelian Sci J 6(3(20)):108–111
12. Mulendeeva LN, Glukhova AG, Glukhov GV (2019) Development model for program-target budgeting in constituent entities of the Russian Federation. In: Mantulenko V (ed) Eurasia: sustainable development, security, cooperation – 2019. SHS web of conferences, vol 71, no 02003. EDP Sciences, Les Ulis
13. Osovin MN (2018) Features of the construction of the Russian model of the digital economy: problems and solutions. Probl Mod Econ 3(67):27–31
14. Program "Digital Economy of the Russian Federation". Decree of the Government of the Russian Federation of 28 July 2017 N 1632. http://static.government.ru/media/files/9gFM4FHj4PsB79I5v7yLVuPgu4bvR7M0.pdf. Accessed 20 Jan 2020
15. Vovchenko NG, Ivanova OB, Kostoglodova ED, Nerovnya YV, Rykina SN (2020) Digital transformation of the system of public finances management. In: Popkova E, Sergi B (eds) Digital economy: complexity and variety vs. rationality. ISC 2019. Lecture notes in networks and systems, vol 87. Springer, Cham, pp 940–949

Procurement Analysis by Regions of Russia: Data from the Unified Information System

I. S. Pinkovetskaia and I. N. Nikitina

Abstract One of the greatest achievements in the development of the information society is the transition of all purchases for state and municipal needs in Russia to a digital platform. This platform is a Unified Information System in the field of procurement, which, in accordance with the law, should be used for competitive procurement in Russia starting from 2019. The purpose of the article is to assess specific indicators that describe regional features of the procurement of goods and services. As these indicators, we considered the average cost of one contract for each of the regions of Russia, the level of involvement of small businesses in the implementation of contracts, the share of savings in the total cost of the contract, the share and value of the contracts with small businesses participation in the total number and value of all concluded contracts. The study used data from the Unified Information System for 82 regions of Russia. The calculations of the values of indicators describing regional aspects of the procurement were made; average values and intervals of change for most regions of five indicators under consideration were estimated; regions with high and low values of indicators were identified.

Keywords Contract system · Public procurement · Regions · Unified information system · Digital platform

1 Introduction

One of the main directions of the Strategy for the Development of the Information Society in Russia for 2017–2030 [3] is the implementation of the tasks of the digital economy, that is, the processing of large amounts of information in the interests

I. S. Pinkovetskaia
Ulyanovsk State University, Ulyanovsk, Russia
e-mail: judy54@yandex.ru

I. N. Nikitina (✉)
Samara State University of Economics, Samara, Russia
e-mail: i.n.nikitina@gmail.com

S. I. Ashmarina and V. V. Mantulenko (eds.), *Current Achievements, Challenges and Digital Chances of Knowledge Based Economy*, Lecture Notes in Networks and Systems 133, https://doi.org/10.1007/978-3-030-47458-4_90

of enterprises and organizations, as well as federal, regional and municipal governments. One of the greatest achievements in the development of the information society is the transition of all purchases for state and municipal needs in Russia to a digital platform. The Unified Information System in the field of procurement [13] acts as such a platform. In Russia, in accordance with the law, starting from 2019, only electronic procedures are applied when conducting competitive public procurement, including the signing of contacts. The Unified Information System generates real-time data arrays, including procurement plans and information on their implementation, data on procurements and contracts concluded; registry of unscrupulous suppliers; catalogs of goods, services and works for the provision of municipal and public services. The presentation of data in a structured form provides wide opportunities to find all the necessary information on procurement, both for customers and contractors. It is essential that all the information is processed automatically. Through integration with regional information systems and electronic platforms, a unified information service is provided.

Let's consider some performance indicators of the Russian competitive procurement system in recent years. The total value of contracts concluded in the Russian Federation in 2019, according to the Unified Information System in the Field of Procurement, reached 9.6 trillion rubles. Compared to 2018, the growth in the total value of contracts amounted to 15%. In total, federal, regional and municipal customers signed 3.4 million contracts in 2019. In the process of concluding these contracts, customers obtained total savings in the sum of 361 billion rubles.

Digitalization of purchases, as shown by the accumulated experience and conducted research [4, 8, 14], can provide an increase in the efficiency of spending budget funds at all levels, reduce corruption, simplify procurement procedures, reduce their time and cost. All this can improve the quality of state and municipal management based on the openness and transparency of purchasing the necessary products of appropriate quality. In addition, bureaucratic procedures and administrative barriers are reduced, and personal interaction with contractors and customers is excluded. The article [1] emphasizes that the electronic contract system requires government support, the adoption of appropriate legislation, the development of digital infrastructure, and the security of online payments and transactions. Regional characteristics have a great influence on the volume and structure of public procurement. This is shown in particular by the example of the Italian electronic market in [7].

Much attention in the formation of the contract system should be given to ensuring unhindered access of small businesses to tenders. Some aspects of this problem are reflected in scientific research. Thus, article [12] considers the disproportionate share of government contracts with small businesses in the European Union, despite the fact that the relevant law recognizes the important role of such enterprises in procurement. Using the data from 271 customers in Ireland, the article [5] proves that there is a significant difference between the declared state policy of the need to assist small businesses in concluding contracts and the real situation. Analysis of public procurement in European countries has shown [11] a significant difference in cost savings for these countries and their areas, due to objective and subjective reasons.

The availability of the Unified Information System in the Field of Procurement makes it a pressing issue to study the indicators that characterize state and municipal purchases, as well as to obtain new knowledge on public procurement immediately after the end of each time-period (month, quarter and year).

In the course of the study, we ask the following three questions:

1. What are the general characteristics of contracts for the considered regions of Russia?
2. What are the characteristics of contracts involving small businesses in each region of Russia?
3. What cost savings does the contractual procedure using the Unified Information System in the Field of Procurement provide for each region of Russia?

At the same time, we respond to recent calls in the literature for a more systematic study of procurement for state needs [6].

In the course of the study, the normal distribution density functions were evaluated, which characterize the relative indicators of tenders held in 2019.

2 Methodology

The procurement analysis for state and municipal needs, including the work of the Unified Information System in Russia, is given some attention in the studies of Russian scientists [2, 9]. At the same time, the main attention is paid directly to the general issues of the procurement system, the regional features of the concluded contracts, their indicators and interregional comparative analysis have not yet been properly reflected in conducted studies. At the same time, analysis of the structure of tenders in Russia demonstrates the absolute predominance of customers at municipal (71.1%) and regional levels (19.4%). They account for more than nine out of every ten contracts concluded. Federal authorities account for only 9.5% of all customers [13]. Taking this into account, the study of the problem of regional features of the contract system functioning in Russia is of the greatest relevance at present.

It should be noted that a comparative analysis of the absolute values of regional indicators of state and municipal procurements does not seem appropriate, since the regions of Russia differ significantly in terms of population, socio-economic situation, climatic conditions, geographical location, which has a significant impact on the volume and structure of such purchases. Therefore, a comparative analysis of procurement for state and municipal needs on the basis of relative indicators seems logical.

The purpose of the article is to assess specific indicators that describe regional features of the procurement of goods and services for state and municipal needs based on the data for 2019. We consider the following indicators arising from the results of the analysis of the literary sources given above. The first indicator describes the average cost of one contract for each of the regions. The second indicator characterizes the involvement of small enterprises (SEs) in the implementation of contracts as main performers or subcontractors (co-executors) and reflects the average cost per contract in the region. The third indicator is the savings obtained in the process of concluding contracts. The fourth indicator is the share of contracts with SEs participation in the total number of contracts concluded. The fifth indicator is the share of the value of contracts with SEs participation in the total value of all concluded contracts. The use of specific indicators allows conducting a comparative analysis by region, which is relevant for studying the problem of improving the contract system in Russia.

During the study, three hypotheses were tested:

- hypothesis 1: currently there are significant differences in the values of each of the five considered indicators by the regions of Russia,
- hypothesis 2: the values of indicators are not determined by the geographical location of the regions,
- hypothesis 3: the values of the indicators do not depend on the level of economic development of the regions.

The hypotheses testing was based on modeling empirical data using normal distribution density functions. The development of these functions, as shown by the authors' earlier work, allows one to obtain unbiased characteristics of the studied economic processes [10]. Official data from the Unified Procurement Information System for 82 regions of Russia were used as initial data. A fragment of this data for six regions is shown in Table 1.

Table 1 A fragment of data on state and municipal procurement in Russia's regions

Regions	Number of concluded contracts	Value of contracts, billion rubles	Relative savings, %	Number of contracts with SEs participation	Value of contracts with SEs participation, billion rubles
Altai Krai	64,618	73.54	6.15	44,063	33.5
Amur Region	17,957	26.81	3.8	12,325	16.34
Arkhangelsk Region	26,149	50.53	6.5	16,283	23.96
Astrakhan Region	23,042	56.85	6.57	15,876	25.31
Belgorod Region	48,237	75.25	3.06	29,182	39.19
Bryansk Region	27,194	34.07	3.77	15,825	15.44
…	…	…	…	…	…

Source authors based on [13]

3 Results

This article presents the models developed by the authors that describe the distribution of the values of the five indicators by region. The development of these models was based on relative indicators calculated by the authors using the data for 2019, listed on the website of the Unified Information System in the field of procurement. Models are, as indicated earlier, density functions of the normal distribution. Such functions (y) that describe the distribution of the values of procurement indicators for state and municipal needs by region (x) are given below:

- the cost of one contract, x_1, million rubles

$$y_1(x_1) = \frac{41.07}{0.69 \times \sqrt{2\pi}} \cdot e^{\frac{-(x_1-1.69)^2}{2\times 0.69\times 0.69}}, \quad (1)$$

- the cost of one contract with SEs participation, x_2, million rubles

$$y_2(x_2) = \frac{38.51}{0.54 \times \sqrt{2\pi}} \cdot e^{\frac{-(x_2-1.27)^2}{2\times 0.54\times 0.54}}, \quad (2)$$

- the resulting savings, x_3, %

$$y_3(x_3) = \frac{77.20}{1.62 \times \sqrt{2\pi}} \cdot e^{\frac{-(x_3-5.35)^2}{2\times 1.62\times 1.62}}, \quad (3)$$

- the share of contracts with SEs participation in the total number of contracts, x_4, %

$$y_4(x_4) = \frac{410.03}{5.42 \times \sqrt{2\pi}} \cdot e^{\frac{-(x_4-64.77)^2}{2\times 5.42\times 5.42}}, \quad (4)$$

- the share of the value of contracts with SEs participation in the total value of all contracts, x_5, %

$$y_5(x_5) = \frac{1288.57}{13.88 \times \sqrt{2\pi}} \cdot e^{-\frac{(x_5-50.46)^2}{2\times 13.88\times 13.88}}. \quad (5)$$

The quality of the developed models was evaluated using three tests: Kolmogorov-Smirnov, Pearson and Shapiro-Wilk tests. Testing has shown that the developed models well approximate the original data over the entire range of their changes and are of high quality.

4 Discussion

Density function of the normal distribution (1)–(5) allow us to characterize the considered five indicators of procurement for state and municipal needs in the regions of Russia. The regional average values of indicators based on functions (1)–(5) are shown in column 2 of Table 2. Column 3 of this table shows the change intervals in the values of indicators for the majority (68%) of regions. The boundaries of these intervals are calculated as follows: the average square deviations are respectively added and subtracted from the values given in column 2.

As can be seen from the data shown in column 2, in 2019 the average cost per contract in the regions of Russia was 1.69 million rubles. In most regions, this indicator ranged from 1.0 to 2.4. The higher level of this change interval was noted in the city of Moscow, Murmansk, Sakhalin, Penza, Tyumen and Magadan regions, the republics of Ingushetia, Buryatia, Dagestan and Chechen. The lower level of this change interval was observed in Ryazan, Vladimir, Voronezh, Smolensk, Tomsk, Lipetsk and Omsk regions, the republics of Khakassia, Udmurtia and Mari El.

The average cost of one contract with SEs participation in the regions under review was 1.27 million rubles. In most regions, the value of this indicator was in the range from 0.73 to 1.87 million rubles. The higher level of this change interval given in column 3 of Table 2 took place in Novgorod, Moscow, Penza, Kaluga, Sakhalin and Tyumen regions, the city of Moscow, the republics of Sakha (Yakutia), North Ossetia-Alania, Dagestan, Buryatia. Values of contracts with SEs participation less than 0.7 million rubles were noted in Oryol, Voronezh, Omsk, Lipetsk, Tula, Ryazan regions, the republics of Udmurtia, Mari El and Komi.

The average resulting savings in 2019 amounted to 5.35%. At the same time, in most regions this indicator ranged from 3.73% to 6.97%. Savings above the upper limit of this range were noted in Tomsk, Ryazan, Kurgan, Vladimir, Lipetsk and Omsk regions, the republics of Tuva, Chuvashia, Bashkortostan, Mari El, and Udmurtia.

Table 2 Indicators characterizing regional procurement in Russia in 2019

Indicators	Average values	Values typical for most regions
1	2	3
Cost of one contract, million rubles	1.69	1.00–2.38
Cost of one contract with SEs participation, million rubles	1.27	0.73–1.87
Resulting savings, %	5.35	3.73–6.97
Share of contracts with SEs participation in the total number of contracts, %	64.77	59.35–70.19
Share of the value of contracts with SEs participation in the total value of all contracts, %	50.46	36.58–64.34

Source authors

Values less than the lower limit of the range occurred in Murmansk, Tambov, Penza, Belgorod, Orel, Tyumen and Vologda regions, the republics of North Ossetia-Alania, Tatarstan, Adygea and Kabardino-Balkaria.

The average share of contracts with SEs participation in the total number of contracts concluded in 2019 was 64.77%. That is, SEs attended more than two-thirds of the concluded contracts. For most regions, this indicator ranged from 59.35% to 70.19%. More than 71% of purchases with the participation of SEs took place in Kaliningrad, Tula, Voronezh, Oryol and Tyumen regions, the Khabarovsk Krai, the city of St. Petersburg, the republics of Dagestan and Udmurtia. The lowest levels of the indicator from 5% to 59% were observed in Vladimir, Volgograd, Nizhny Novgorod, Kostroma, Tver, Moscow and Bryansk regions, the republics of Kalmykia, Chechen, Kabardino-Balkaria, as well as the Stavropol Krai.

The average share of the value of contracts with SEs participation in the total value of all contracts amounted to 50.46%. In most regions, this indicator ranged from 36.58% to 64.34%. Values greater than the upper limit of the range were in Kaluga and Sakhalin regions, the republics of Crimea, Khakassia, Ingushetia and Altai Krai. The lowest values of the indicator were in 2019 in Murmansk, Vologda, Magadan and Tambov regions, the Republic of Karelia and Tatarstan, the city of Moscow.

The data presented in Table 2, as well as their analysis presented above, demonstrates the presence of a significant differentiation of the values of each of the five indicators considered in our study by region. Thus, we can conclude that hypothesis 1 has found its confirmation. The analysis of the lists of regions, which are characterized by high and low values of each of the five indicators, shows that both high and low values occur in the regions located in the center of the country, its north and south, as well as in the west and east. This allowed us to conclude that hypothesis 2 has been confirmed. The situation is similar with hypothesis 3, since the regions with both high and low indicators values include regions with different levels of economic development.

5 Conclusion

The findings of the study, which contain scientific novelty and originality, are as follows:

- it was proposed to use the information provided by the Unified Information System in the field of procurement for evaluating the indicators of state and municipal procurement in Russia,
- calculations of the values of five indicators that characterize the regional aspects of procurement for state and municipal needs were carried out,
- the possibility of using normal distribution functions to describe the differentiation of relative indicators characterizing procurement for state and municipal needs by regions of Russia was shown,

- the average values and change intervals of the five indicators under consideration for most regions were estimated,
- regions with high and low values of indicators were identified,
- it was shown that the average value of cost savings when concluding contracts in 2019 was 5.35%,
- it was proved that the values of each of the indicators were significantly differentiated by 82 regions of Russia,
- the absence of dependencies between the values of each of the indicators and such factors as the level of economic development of the regions and their location was confirmed.

The obtained research results have a certain theoretical and applied value. The results of the study can be used to develop technology for processing data on the activities of the contract system, using the proposed indicators. The proposed indicators can be used in subsequent research on competitive procurement. The new knowledge obtained can be used in the educational activities of universities. The government, regional and municipal authorities can apply the results of the study to implementation of projects and programs for the development of elements of the contract system. In addition, the research results are of interest to all public procurement participants as well as potential customers and potential contractors.

References

1. Altayyar A, Beaumont-Kerridge J (2016) An investigation into barriers to the adoption of e-procurement within selected SMEs in Saudi Arabia. J Bus Econ 7(3):451–466. https://doi.org/10.15341/jbe(2155-7950)/03.07.2016/009
2. Barkatunov VF, Larina OG (2019) Elements of digital technologies, legal innovations and efficiency of financial resources management in the procurement process for state and municipal needs. Proc Southwest State Univ Ser Hist Law 9(3):53–73
3. Decree of the President of the Russian Federation of 09.05.2017 No. 203 "On the strategy for the development of the information society in the Russian Federation for 2017–2030". http://publication.pravo.gov.ru/Document/View/0001201705100002. Accessed 27 Feb 2020
4. Fernandes T, Vieira V (2015) Public e-procurement impacts in small- and medium-enterprises. Int J Procure Manag 8(5):587–607. https://doi.org/10.1504/IJPM.2015.070904
5. Flynn A (2018) Investigating the Implementation of SME-friendly policy in public procurement. Policy Stud 39(4):422–443
6. Koala K, Steinfeld J (2018) Theory building in public procurement. J Public Procure 18(4):282–305. https://doi.org/10.1108/JOPP-11-2018-017
7. Lbano GL, Antellini RF, Castaldi G, Zampino R (2015) Evaluating small businesses' performance in public e-procurement: evidence from the Italian government's e-marketplace. J Small Bus Manag 53:229–250. https://doi.org/10.1111/jsbm.12190
8. Lewis-Faupel S, Neggers Y, Olken B, Pande R (2016) Can electronic procurement improve infrastructure provision? Evidence from public works in India and Indonesia. Am Econ J Econ Policy 8(3):258–283. https://doi.org/10.1257/pol.20140258
9. Meretukova SK, Meretukov ST, Guseva SS, Shishova SK (2019) Standardization and interaction of software systems as tools for optimizing the public procurement contract system. Bull Adygeya State Univ Ser 4 Nat-Math Tech Sci 1(236):103–110

10. Pinkovetskaia IS, Nikitina IN, Gromova TV (2018) The role of small and medium entrepreneurship in the economy of Russia. Monten J Econ 14(3):177–188. https://doi.org/10.14254/1800-5845/2018.14-3.13
11. Popa M (2019) Uncovering the structure of public procurement transactions. Bus Polit 21(3):351–384. https://doi.org/10.1017/bap.2019.1
12. Stake J (2017) Evaluating quality or lowest price: consequences for small and medium-sized enterprises in public procurement. J Technol Transf 42(5):1143–1169. https://doi.org/10.1007/s10961-016-9477-4
13. Unified information system in the field of procurement (2020) Statistics on Federal Law № 44. https://zakupki.gov.ru/epz/main/public/home.html. Accessed 27 Feb 2020
14. Vaidya K, Campbell J (2016) Multidisciplinary approach to defining public e-procurement and evaluating its impact on procurement efficiency. Inf Syst Front 18(2):333–348. https://doi.org/10.1007/s10796-014-9536-z

Prospects for the Development of the Russian Telemedicine Market: Legal Aspect

A. Sidorova, A. Bezverkhov, and A. Yudin

Abstract The goal of this contribution is to make an overview of the proposed directions of telemedicine technologies application in the Russian Federation, to generalize factors that negatively affect the development of information technologies in Russian medicine. This study was conducted using general and specific research methods: synthesis, modeling, deduction, generalization, concretization; observation method, modeling, and description method. In his research, the author grouped factors that prevent the widespread application of telemedicine technologies in the Russian Federation, analyzed various points of view of representatives of the medical community on this issue. The article investigates problematic aspects of the Russian market for telemedicine technologies and outlines some prospects for the development of these technologies in the near future. This should have a positive impact not only on the quality and availability of medical services, but also on their cost.

Keywords Telemedicine · Healthcare · Patient · Doctor · Monitoring

1 Introduction

The mass market development of telemedicine services in Russia became possible on January 1, 2018, when the Federal law No. 242-FZ "On amendments to certain legislative acts of the Russian Federation on the use of information technologies in the field of health protection" [3] legalized a range of services that can be provided by telemedicine. In 2018, the Russian budget provided 740 million rubles for the development of telemedicine, a similar amount was planned for 2019. For the creation

A. Sidorova (✉)
Samara State University of Economics, Samara, Russia
e-mail: an.sido@bk.ru

A. Bezverkhov · A. Yudin
Samara National Research University, Samara, Russia
e-mail: bezverkhov-artur@yandex.ru

A. Yudin
e-mail: yudin77@ssau.ru

S. I. Ashmarina and V. V. Mantulenko (eds.), *Current Achievements, Challenges and Digital Chances of Knowledge Based Economy*, Lecture Notes in Networks and Systems 133, https://doi.org/10.1007/978-3-030-47458-4_91

of a single digital circuit in healthcare from 2019 to 2024, it is expected to spend 177.6 billion rubles [13]. According to the National Agency for Financial Studies (NAFI), Russians are wary of innovations in the field of telemedicine: 56% of Russians are not ready to get consulting services from a doctor via video link (56%), 42% of respondents are ready for remote consultations (2% found it difficult to answer). According to the results of previous NAFI research, 61% of respondents would like to receive medical care remotely; the most popular telemedicine services are registration of certificates (44%), emergency consultations (39%), issuing prescriptions (38%) and scheduled consultations (22%) [6]. As the main prerequisites for the development of telemedicine can be considered the advantages that it provides to all interested participants in the process, first of all, health care institutions, patients and medical professionals. The prospects for the development of the global and domestic telemedicine services markets are evaluated differently. According to experts, the global mHealth market was expected to grow by 33.5% to $14 billion by 2020 [16], and the market for medical devices, services and applications, which relates to the Internet of things, will reach $136.8 billion by 2021 with an average annual growth rate of 12.5% in the period from 2015 to 2021 [2]. Currently, the main share of the national and international market is occupied by blood pressure monitoring systems, followed by blood glucose meters and heart monitors. According to some experts, the global telemedicine market will reach $44 billion by 2021 [11]. According to former British health Minister Andrew Lansley, approximately 80% of all calls to the UK national health service can be transferred to a remote service. Changing the format of consultations (full-time to part-time) will reduce budget expenditures. If only 1% of face-to-face medical consultations are carried out remotely, the annual budget will be reduced by about 250 million pounds [12]. The U.S. experience shows that only preliminary telemedicine approval of the treatment plan has reduced the cost of transportation in case of emergency hospitalization from $2.2 million up to $1.4 million, which provided annual savings of $500 million [10]. According to specialists of the Russian state corporation “Rostec”, the prospects for the development of the Russian telemedicine market by 2020 looked very optimistic—the market volume was estimated at 300 billion rubles, and the number of consultations was expected to reach more than 1 million. At the beginning of the year 2021, it will be still too early to talk about the implementation or non-implementation of these prospects, but it is already possible to sum up some intermediate results.

2 Methodology

This study was conducted using general and specific methods of cognition. To display the objective reality, the author used such research methods as synthesis, modeling, deduction, generalizations, concretization, and some special methods: observation method, modeling method, and description method. Using the above methods together allowed achieving the set goals.

3 Results

A brief PEST-analysis allows determining prospects for the development of telemedicine, taking into account political, economic, social and technological trends. Today, the problems of telemedicine in the Russian Federation can be divided into two groups: implementation problems and development problems. Problems of telemedicine implementation and development have a national scale. Not enough funds are allocated from the federal and regional budgets to develop up-to-date training, retraining and professional development programs that will ensure the smooth operation of telemedicine projects.

In order to implement telemedicine, every state budgetary health institution should be equipped with everything necessary, including telecommunications, which will ensure uninterrupted communication with patients who need help, and the timely exchange of knowledge and experience between medical professionals. In addition, it is impossible to ignore the fact that the level of availability of modern digital technologies to the Russian population is very different. Not all regions have gas and electricity, much less—high-speed Internet. There is also a problem of purchasing and maintaining high-quality audio and video equipment.

Providing central and district health care institutions with a reliable Internet connection and reducing telemedicine capabilities to the "specialist-specialist" format will not provide a real improvement in quality and efficiency, while budgets at all levels reduce the cost of medical care for the population. Subjective factors that prevent the application of telemedicine should also be taken into account. The degree of readiness of medical professionals and patients for changes of this kind can significantly affect the implementation speed of telemedicine services. There is no certainty that medical professionals and especially potential patients will pass identification and authentication procedures, learn Skype or other remote communication technologies.

4 Discussion

In Russia, telemedicine was initially developed mainly by private providers, but by 2023, according to the national health project [14], it will also be available within the framework of public health. Weaknesses and related threats lead to the fact that today the practice of using telemedicine technologies in health care in the Russian Federation is not too extensive. Problems with the application of telemedicine technologies lie in different areas:

- *a limited range of medical services provided by the nomenclature of medical services is used* [9].

Mikhail Men attributed this use of technologies to the peculiarities of the Russian telemedicine market, stating that from 10 medical services specified in the nomenclature, only seven were provided. At the same time, the main part of them relates to only two services: decoding of descriptions, interpretation of electrocardiographic data (80.4%) and decoding of x-ray studies (16%)" [1]. Most of the services using telemedicine technologies are provided in the Siberian and Ural federal districts. This is determined, among other things, by a lack of funds in the budgets. Today, the Samara region is recognized as one of the rapidly developing regions in this area. In 2019, a unified medical information system has been created and is being implemented in the region. In 2019, more than 2000 workplaces of doctors and paramedics, and more than 150 emergency medical teams were equipped with computers connected to it. In the regional districts, 96 paramedic and midwifery centers were connected to the Internet last year, and all 496 are planned to be connected by the end of 2021. The services of the regional contact center of the Ministry of Healthcare, which opened at the end of 2019, have already been used by about 100 thousand people. In 2019, Samara hospitals conducted more than 2000 telemedicine consultations, including with national medical research centers. In 2020, it is planned to increase this number at least four times [8].

- *insufficient development of the regulatory framework.*

As mandatory requirements for the development of telemedicine in the Russian Federation should be considered the following aspects:

- the presence of standards for remote telemedicine services (telemedicine consultation),
- a list of criteria that will ensure an informed decision and appointments in the format of teleconsultation "specialist-patient",
- normative documentation containing criteria for assessing the quality of telemedicine services,
- procedures for mandatory recording teleconsultations of any format,
- technologies for recording and preserving the entire virtual consultation procedure, available to heads of medical institutions, practitioners and patients, in order to determine the legal responsibility of the doctor-teleconsultant,
- a system of standards for information exchange between personal devices (mobile applications, gadgets, etc.) and archival repositories of electronic patient records,
- unified analytical algorithms for systematization, processing, storage and transmission of multi-format information.

Among the problems of telemedicine development, there are also the lack of nomenclature of medical services with the use of telemedicine technologies, the lack of telemedicine technologies in clinical recommendations (treatment protocols), restrictions on the use of telemedicine for diagnosis and remote correction of treatment. There are no uniform requirements for working spaces intended for such consultations, and there are no requirements for equipment used outside of medical organizations.

- *lack of interaction between federal and regional ministries on the implementation, regulation and development of telemedicine.*

Organizational and technical issues related to the application of telemedicine services should be addressed by the federal and regional ministries of health. The standard for conducting telemedicine consultations (providing remote telemedicine services) will enable the collection and exchange of data, the necessary degree of formalization, and the implementation of a systematic approach within the framework of teleconsultation of any format, including preventive services for the population and dispensary observation. The introduction of a single list of criteria for making informed decisions and appointments will help to improve the quality and safety of medical services for the population. Normative documentation containing criteria for evaluating the quality of telemedicine services will provide a unified approach to providing telemedicine care to the patient.

The procedure for mandatory recording of a teleconsultation of any format will allow implementing a systematic approach to conducting teleconsultations. The use of a single technology for recording and preserving the virtual consultation procedure will not only ensure that information is available to a wide range of interested parties—heads of medical institutions, doctors and patients, but also the ability to determine the legal responsibility of the doctor-teleconsultant in each case. In addition, the database of teleconsultations will eliminate the isolation that exists between medical institutions of our country, domestic and foreign specialists, will allow implementing current methods of diagnosis, treatment and prevention, using unique personal experience of various specialists, and can become a training material for medical students and novice doctors who will be able to master the standard procedures for collecting anamnesis, diagnosis, treatment and communication features with different categories of patients.

The introduction of a unified system of standards for information exchange between archived data—medical electronic patient records and personal devices (mobile applications, gadgets, etc.), as well as unified analytical algorithms for systematization, processing, storage and transmission of multi-format information will automatically form a single database (big data) and process huge amounts of statistical data, determining significant cause-and-effect relations and patterns. Another free niche in the telemedicine market is a system of compulsory medical insurance (CMI). If telemedicine services are included in the compensation for CVI expenses, they may become interesting for major players [5]. Today, payment for telemedicine services in the public sector is made at the expense of the CMI on the basis of point tariff agreements formed at the regional level by the territorial fund of compulsory medical insurance (TFCMI), regional authorities and insurance organizations. English specialists conducted a study in 2008 and found out that the use of medical technologies reduces tariff costs by 8%, the level of planned hospitalizations by 14%, the number of beds by 14%, and calls for accidents and emergency assistance by 15% [1].

- *the lack of a model of payment for telemedicine services.*

This problem also occurs in other countries. In Russia, there are no federal recommendations for such payments. The regions themselves set the optimal cost experimentally. In 2018, according to the accounting chamber, its cost differed by almost

25 times in the regions. Many experts associate the new breakthrough in the development of telemedicine in Russia with large investments in teleradiology. This segment is the most prepared in the domestic healthcare sector. Similar telemedicine projects already exist. For example, the national teleradiology network forms an extensive network of remote consultations of radiologists and offers advice to doctors, patients and medical institutions [7]. Others believe that the prospects for the development of telemedicine are in the B2B sector, in particular, in increasing sales of medical services to large companies, in case they are included in corporate insurance packages [4].

Boris Zingerman, who heads the group of the expert council "Electronic medical card" under the Ministry of Healthcare of the Russian Federation and the direction of digital medicine of INVITRO LLC, believes that telecommunications and high-tech methods should become everyday and familiar tools for all practitioners. He considers the integration of classical and innovative medicine to be the result of the application of telemedicine in Russian healthcare [1]. In his opinion, telemedicine has a future in Russia. The market for telemedicine gadgets and home appliances is already actively developing, the telecommunications segment has formed, any information can be easily delivered to the cloud, and the market for integration platforms is actively developing, where this information can be collected, aggregated, analyzed and used. And the problem is that modern medicine is not used to operational interaction with the patient.

Another problem was announced on 13.02.2019 at the symposium "Digital medicine: From e-health to artificial intelligence": the convenience of such communication between the patient and the doctor will lead to the assessment of doctors not based on their medical qualities, but on their communication abilities. Co-founder of the Doctor Smart Service Pavel Roitberg recommends entering the telemedicine market only to those players who have expertise in the field of healthcare, information technology and Internet business [1]. Russian healthcare Minister Mikhail Murashko has identified telemedicine as a key area for the whole healthcare system and, in particular, for the field of oncology.

5 Conclusion

For our country, the development of telemedicine technologies is not only of social importance, but also economic. Telemedicine in Russia is developing at an accelerated pace, but even today there are factors that negatively affect its widespread implementation. These include the lack of qualified personnel capable of efficiently and quickly interacting with telemedicine systems; the problem of ensuring compatibility and standardization of devices and technologies used in the field of telemedicine; poor regulatory framework and lack of international standards and, as a result, a large number of poor-quality and unreliable solutions; patients are not ready to use a new type of medical care; data protection and confidentiality issues; telemedicine services are often not covered by insurance. The state pays great attention to domestic projects in the field of telemedicine and is ready to fully support them. It is proposed to fix the priority of domestic technologies in the field of healthcare, and to ensure the storage of telemedicine data using Russian information systems. Of great interest is the digitalization of medical data, which significantly facilitates the access of doctors to information about the health of Russian citizens. Now, for example, not all doctors from clinics can quickly get data from the hospital where their patient was treated. The development of distance medicine makes it possible to form a single set of data on diagnoses, treatment methods and recommendations given to patients by various specialists [15].

References

1. Beskaravaynaya T (2019) The first official data on the telemedicine technology market in Russia are presented. Medvednik. https://medvestnik.ru/content/news/Predstavleny-pervye-oficialnye-dannye-o-rynke-telemedicinskih-tehnologii-v-Rossii.html. Accessed 17 Feb 2020
2. EverrCare (2016) The medical Internet of things market will reach $136.8 billion by 2021. https://evercare.ru/allied-market-research-2016. Accessed 19 Feb 2020
3. Federal Law of July 29, 2017 N 242-FZ. On amendments to certain legislative acts of the Russian Federation on the use of information technologies in the field of health protection. http://www.consultant.ru/cons/cgi/online.cgi?req=doc&base=LAW&n=221184&fld=134&dst=1000000001,0&rnd=0.6012367395368097#026070283170878383. Accessed 19 Feb 2020
4. Gorodnova NV, Klevtsov VV, Ovchinnikov EN (2019) Prospects of telemedicine development in the context of digitalization of the Russian economy. Rus J Innov Econ 9(3):1049–1066. https://doi.org/10.18334/vinec.9.3.41173
5. Melik-Huseynov DV, Khodyreva LA, Turzin PS, Kondratenko DV, Gozulov AS, Emanuel A (2019) Telemedicine: regulatory support, realities and prospects of application in domestic health care. Exp Clin Urol 1:4–10. https://doi.org/10.29188/2222-8543-2019-11-1-4-10
6. NAFI (2017). Russians in good health are interested in telemedicine. https://nafi.ru/analytics/rossiyane-s-krepkim-zdorovem-zainteresovany-v-telemeditsine/. Accessed 19 Feb 2020
7. National Teleradiological Network (2020) Remote consultation service for radiologists X-ray, CT, MRI, PET. https://teleradiologia.ru/%D0%B4%D0%BB%D1%8F-%D0%BA%D0%BB%D0%B8%D0%BD%D0%B8%D0%BA/. Accessed 19 Feb 2020
8. NIASAM (2020) Andrey Turchak together with Dmitry Azarov held a meeting on the development of digital medicine at Samara State Medical University. https://www.niasam.ru/Zdravoohranenie/Andrej-Turchak-vmeste-s-Dmitriem-Azarovym-provel-v-SamGMU-soveschanie-po-razvitiyu-tsifrovoj-meditsiny146108.html?utm_source=yxnews&utm_medium=desktop&utm_referrer=https%3A%2F%2Fyandex.ru%2Fnews. Accessed 19 Feb 2020
9. Order of the Ministry of Health of Russia of October 13, 2017 N 804н (as amended on April 16, 2019). On approval of the nomenclature of medical services (Registered in the Ministry of Justice of Russia on 07/07/2017 N 48808). http://www.consultant.ru/cons/cgi/online.cgi?req=doc&base=LAW&n=327766&fld=134&dst=100015,0&rnd=0.276862310037445#06422156420045091. Accessed 19 Feb 2020
10. Popova M, Ragimova S (2017) Tele acceleration: healthcare. Komersant Inf Technol 95:17. https://www.kommersant.ru/doc/3311052. Accessed 19 Feb 2020
11. Rambler Finance (2017) Experts: the global telemedicine market will grow by 33.5% by 2020. https://finance.rambler.ru/economics/37000761-eksperty-mirovoy-rynok-telemeditsiny-k-2020-godu-vyrastet-na-33-5/?updated. Accessed 19 Feb 2020

12. Statsenko N (2017) The Russian telemedicine market and its problems. Rusbase. https://rb.ru/analytics/kto-krainii-k-ortopedu/. Accessed 19 Feb 2020
13. Strategy 24 (2018) National project "Health Care" (2028-2024). https://strategy24.ru/rf/health/projects/natsionalnyy-proekt-zdravookhranenie. Accessed 19 Feb 2020
14. The Ministry of Health of the Russian Federation (2019) National projects "Healthcare" and "Demography". https://www.rosminzdrav.ru/poleznye-resursy/natsproektzdravoohranenie. Accessed 17 Feb 2020
15. United Russia (2020) Digital Board has developed proposals for the introduction of digital technology in telemedicine. https://er.ru/news/190670/. Accessed 17 Feb 2020
16. Zdrav.Expert (2016) MHealth applications (global market). http://zdrav.expert/index.php/Статья:Приложения_mHealth_%28мировой_рынок%29. Accessed 19 Feb 2020

Information Technologies as a Tool for Preventing Corruption in Organizations in Russia

R. R. Khasnutdinov and N. S. Maloletkina

Abstract The article deals with certain aspects of prevention of corruption offenses committed in enterprises, organizations and corporations. Corporate corruption is one of the many types and manifestations of corruption that causes especially great damage to the country's economy. That's why, anti-corruption activities at enterprises should be considered as one of the main and priority areas of the Russian state anti-corruption policy. Prevention of corruption is a priority in anti-corruption activities in enterprises, where measures are taken to prevent corruption, including identifying and subsequently eliminating the causes of corporate corruption. An important role in prevention is assigned to enterprises and organizations themselves, which, in accordance with the law, are obliged to create conditions and take measures to prevent corruption. The effectiveness of the implementation of anti-corruption measures in the organization is significantly increased when using information technologies. The study analyzes the possibilities of information technologies on the example of carrying out certain anti-corruption measures recommended to organizations, shows their impact on reducing corruption risks, increasing the transparency of business processes, automating anti-corruption audit, and discusses debatable issues regarding their implementation.

Keywords Corruption · Prevention · Organization · Information technology

1 Introduction

Corruption is a complex socio-cultural phenomenon that has evolved in parallel with the development of society. It poses a serious threat to society and the state in all spheres of life. Being multi-faceted, corruption creates obstacles not only for state and

R. R. Khasnutdinov (✉)
Samara State University of Economics, Samara, Russia
e-mail: Khasnutdinoff@mail.ru

N. S. Maloletkina
Samara Law Institute of the Federal Penitentiary Service of Russia, Samara, Russia
e-mail: levkovka707@mail.ru

S. I. Ashmarina and V. V. Mantulenko (eds.), *Current Achievements, Challenges and Digital Chances of Knowledge Based Economy*, Lecture Notes in Networks and Systems 133, https://doi.org/10.1007/978-3-030-47458-4_92

municipal services, but also for business activities, hinders the solution of political and economic problems, and complicates the development of trade and economic relations with other countries. According to the report of the Prosecutor General Yu. Chaika at the meeting of the Federation Council of the Russian Federation in 2018, the damage from corruption crimes amounted to 65.7 billion rubles, which is 66% more than in 2017 (39.6 billion) [11].

The state of the economy of the country as a whole and of the regions is of fundamental importance for the development of society. The objective result of the process of transformation of the Russian economy should be the acquisition of competitive positions in the world market, and, consequently, the role of the business sector should increase. The development of competitiveness is hindered by corruption in enterprises and organizations, the so-called corporate corruption. Corporate corruption, unlike other types of corruption, is particularly damaging not only to organizations themselves, but also to the country's economy as a whole. Anti-corruption activities at enterprises should be considered as one of the major and priority areas of the Russian state's anti-corruption policy.

In modern conditions of widespread digitalization and development of the information society, information technologies are becoming increasingly important, which are already partially used and must be actively used in preventing corruption in enterprises and organizations.

2 Methodology

This study was conducted on the basis of both general scientific methods of analysis, synthesis, deduction, generalization, and using private scientific methods: comparative legal, system-structural. Comparative legal analysis was used to identify the features and capabilities of certain information technologies in the anti-corruption policy of Russian organizations. The system-structural method made it possible to generalize the main activities of organizations for preventing corruption through information technologies, as well as to determine the vector of their further development. The research was based on works, scientific research, articles and publications of Russian and foreign legal scholars, practitioners who consider various aspects of preventing and countering the spread of corruption in organizations, and the use of information technologies in this work.

3 Results

Corruption in enterprises and organizations has a huge number of manifestations. These are situations when a management decision is made not for the benefit of the enterprise, organization, or company, but for the personal enrichment of the persons involved in making this decision, and situations involving illegal remuneration on behalf of and in the interests of the organization, and commercial bribery, and corporate fraud, and much more. As a result of such various manifestations of corruption, there are negative consequences in the form of increased expenses of the enterprise, an increase in the cost of purchased goods and services by the end user, shadow cash flows, and damage to the reputation of the enterprise.

No matter how effective the means of fighting corruption are, it is almost impossible to identify and bring to criminal and administrative responsibility the majority of corrupt officials. The latency of this type of crime is extremely high. Therefore, in anti-corruption activities in enterprises, the priority should be its prevention, that is, the prevention of corruption, including the identification and subsequent elimination of the causes of corruption. Enterprises and organizations themselves should play an important role here for a number of reasons.

First, Article 13.3 of the Federal law "On combating corruption" [5] stipulates the obligation to develop and take measures to prevent corruption. Supervision of the implementation of this legislative order is actively carried out by the Prosecutor's office [7]. Secondly, very high fines are imposed on legal entities for committing corruption offences. For example, the upper threshold of the sanction of article 19.28 of the Code of administrative offences of the Russian Federation [2] for illegal remuneration on behalf of a legal entity is up to 100 times the amount of illegal remuneration, but not less than 100 million rubles with mandatory confiscation of the subject of the offense (money, securities). Of course, such high fines greatly affect the financial condition of enterprises and can even lead to the bankruptcy of small companies. This is why it is extremely unprofitable for businesses. Furthermore, additional negative consequence to the legal entities involved according to article 19.28 of the Code of administrative offences of Russian Federation, is a two-year ban on participation in procurement for state and municipal needs established by the requirements of section 7.1 of part 1 of article 31 of the Federal law dated 5 April 2013 No. 44-FZ "On contract system in procurement of goods, works, services for state and municipal needs" [6]. In order to implement this provision of the law for interested persons, a Register of legal entities brought to administrative responsibility for corruption is posted on the website of the General Prosecutor's office of the Russian Federation. And third, it is beneficial for organizations themselves to take anti-corruption measures and implement anti-corruption activities, since corruption has the above-mentioned negative consequences for them, which adversely affect their sustainable development.

The Ministry of labor and social protection of the Russian Federation published the guidelines "Measures to prevent corruption in organizations" in the 2019 edition, which proposed key tools that organizations should implement in order to effectively prevent corruption [8]. Many enterprises, fulfilling the legal obligation, actively develop and take the measures recommended by them to prevent corruption.

Information technologies play a huge role in implementing anti-corruption measures in organizations. Currently, an increasing number of organizations are implementing information technologies to solve economic and other problems, which optimize business processes, increase competitiveness, automate many internal processes, provide a competitive advantage, and more. A lot has been said about the role and opportunities of information technologies in the sustainable development of enterprises and organizations [10], including in foreign literature [4, 9, 13–16]. Bizhoev described information technologies as technical methods to combat abuses and corruption in the procurement sphere in detail [1]. Information technologies should certainly be considered as a serious tool for preventing corruption in enterprises. Consider their impact on improving the effectiveness of certain anti-corruption measures recommended for implementation in enterprises.

The fundamental anti-corruption measure is the development and adoption of the organization's anti-corruption policy, which is determined by the specifics of its activities. In addition to the adoption of this policy, it is important to inform employees about the established approaches in the fight against corruption in the organization, about legal acts, and anti-corruption standards. It is advisable to place such information on the organization's website in the Internet in a specially provided section, which should be clear and easy to use. Access to it should be as simple as possible from the main page of the site (in 1–2 clicks). It would be reasonable to provide subsections in which the legal regulation, the main anti-corruption measures in the organization, explanations of certain anti-corruption provisions, examples of corruption manifestations and judicial (law enforcement) practice on them, etc. should be placed separately. Information about changes in anti-corruption regulations, anti-corruption rules, and standards can be periodically updated by means of information technologies, for example, by sending them automatically to e-mail or to office computers.

In addition to informing the organization, it is necessary to provide periodic advice and training on issues related to anti-corruption. And if the organization carries out these activities independently, the use of information technologies for distance learning will greatly facilitate and accelerate these processes. E-courses are easy to learn at any time from your desktop or home computer, including completing the tasks provided in the training program and passing control tests. The use of information technologies in training saves human and other resources as much as possible, and also allows you to organize a periodic check of employees' knowledge about the main provisions of anti-corruption legislation, to assess the knowledge of employees as objectively as possible, to analyze and see a picture of the level of education of employees on the issue of anti-corruption legislation for the enterprise as a whole.

Step-by-step development of anti-corruption work at the enterprise should be preceded by an assessment of corruption risks, in order to eventually develop a plan of measures to minimize them. Among the main principles that should be followed when assessing corruption risks, we highlight the following: analysis of business processes, not personal qualities, checking the presence of corruption risks in all business processes, rational allocation of resources, maximum specification of the description of corruption risks, regularity in assessing corruption risks. Solving problems to minimize corruption risks is greatly facilitated by the use of information technologies. The use of such systems as CRM (Customer Relationship Management system customer relationship management), SCM (Supply Chain Management system supply chain management), ERP (Enterprise Resource Planning system enterprise resource management) enable: to make transparent the relationship of the various divisions with clients (CRM); to automate the processes associated with the planning, execution and control of all costs of raw materials, resources from the moment of order receipt to the time of exit of the finished product (SCM); maximize automation of processes related to planning, accounting, control, analysis of business processes, and solving business problems throughout the organization (ERP). Through the introduction of these information technologies, the company effectively operates on the basis of a process approach [10], when management in the organization occurs through processes, rather than the functions of individual employees or divisions. As a result, significantly decrease corruption risks: maximum removed the human factor in making any decisions in the processes that become automatic, eliminates barriers and levels the hierarchy solves the problem of ineffective interaction between departments and services, transparency of processes, decreases the need for intermediate structures of the enterprise. Gazprom has implemented «The antirutina corporation» artificial intelligence system to analyze purchases, manage costs, minimize the risks of forecasting errors, and ensure transparency of processes. Many trading companies have actively implemented and successfully operate the Forecast NOW intelligent system [12].

Speaking about measures to reduce corruption risks in the organization, we cannot say about such a software tool as smart contracts (smart contracts), which are actively gaining popularity and have already had a positive reputation. A smart contract is a computer program that tracks and ensures the fulfillment of the parties' obligations under the contract. Previously, the parties prescribe all the terms of the transaction in the smart contract, sanctions for their violation and non-fulfillment, and seal them with electronic signatures. After that, the contract execution process is automated and becomes controlled by a self-executing computer code recorded in the Blockchain. A smart contract can independently determine whether everything is fulfilled, and then make a decision about completing the transaction, issuing money, goods, imposing prescribed fines and penalties, blocking access to assets, etc. Smart contracts can be reasonably used for various transactions, such as license payments, delivery, lease, purchase of equipment, etc. Smart contracts are easy to track, they are characterized by automation, maximum transparency, and accuracy of execution. Clear mathematical algorithms of computer code exclude various intermediaries, the creation of any additional barriers, the possibility of bribery, other human factors, which provides a high minimization of corruption risks.

The above guidelines also refer to the anti-corruption audit of individual transactions as one of the measures required for implementation in organizations. Organizations, if they have the capacity and resources, should provide additional analysis of transactions and transactions with high corruption risks. These may include various large purchases, sales or leases of property, transactions related to investments or loans. To do this, it is necessary to develop so-called "corruption indicators", that is, a list of quantitative and qualitative indicators that indicate possible corruption offenses when concluding transactions, performing operations, and making decisions, as well as algorithms for their detection. It is advisable to automate the anti-corruption audit using various information technologies. For example, using data mining technology that contains various mathematical and statistical algorithms, you can not only display and predict business-critical processes, but also identify transactions, operations, and business processes that are subject to risks of corruption, conflicts of interest, fraud, and other offenses. As an anti-corruption software tool, the technology (Data Mining) should be developed to detect corruption and corporate fraud in order to quickly respond to them [8]. We should also mention such analytical tools as Self-Monitoring, Analysis and Reporting Technology (SMART), which, in accordance with the prescribed algorithms for analysis and risk assessment, check transactions and business processes both online and offline, identify and detect various violations, stop payments for incorrectly formed or questionable transactions when making purchases, such information technology capabilities for checking internal processes prevent criminal corruption and fraudulent practices, identify and prevent illegal financial flows [8]. For example, since 2017, the accounting Chamber of the Russian Federation has been using information technologies to conduct remote audits, which allow detecting more than 70 types of violations automatically without involving an inspector [3].

Corruption risks can arise not only within the organization, but also when working with external partners, customers, and suppliers who may be involved in corruption situations. In this regard, in order to exclude unscrupulous business partners, companies are recommended to conduct a procedure for evaluating the integrity of partners. This procedure, called Due diligence, allows you to collect the most objective information about the counterparty for honest business conduct and intolerance to various types of corruption offenses. Such information can be collected in open sources (media, Internet), as well as in various databases of a free and paid nature. And, of course, in such work it is necessary to resort to the help of information technologies, which not only facilitates the search for information in terms of time and volume, but also can create its own internal database of partners of the organization. This database can be checked periodically in automatic mode for possible corruption risks of working with contractors.

In implementing anti-corruption measures in organizations, it is quite logical to use a variety of other information technologies that, in addition to solving other tasks, will also provide significant assistance in preventing corruption. For example, SIEM-class information technologies, as part of information security tasks, will allow you to protect restricted access information. And the use of DLP systems (Data Leak Prevention), in addition to preventing leaks of confidential data, allows you to control the use of working time and resources by employees of the enterprise, monitor communication between them in order to disclose fraudulent and corrupt schemes at the planning stage, and monitor the legality of employees' actions.

4 Discussion

Systematic implementation of anti-corruption measures in the organization, as well as the use of information technologies in this work, is associated with certain expenses.

This is partly why there are still quite a large number of organizations where anti-corruption work is not carried out or is carried out formally. Iliy wrote about the detection of a large number of violations of the requirements of the law on internal prevention of corruption by the Prosecutor's office [7]. The introduction of information technologies in the implementation of anti-corruption measures by Russian organizations is at the very beginning of the road. This is a fairly time-consuming process, and you don't have to wait for quick results. However, despite this, this work should be carried out, and in the medium and long term can bring the organization a number of significant advantages. The organization's commitment to the law and high ethical standards in business relations help to strengthen its reputation among other companies and clients. At the same time, the reputation of an organization can serve to some extent as a protection against corruption attacks by unscrupulous representatives of other companies, state bodies and local governments. In addition, the implementation of measures to prevent corruption significantly reduces the risks of applying measures of responsibility for bribing officials to the organization.

We believe that since the state is directly interested in reducing the level of corruption, including through the performance by organizations of their duty to implement anti-corruption measures, it is for real action and implementation of article 13.3. The anti-corruption law should provide for separate funding items, including those aimed at developing and further providing information technologies that reduce corruption risks and effectively prevent various types of corruption in organizations. Kuznetsov and Kuznetsov are rightly noted that the introduction of information technologies will not quickly solve the problem of corruption. This process depends on financial and technical investments at all levels of its implementation, and for success it is necessary that the public sector actively cooperates with the private sector in this direction [8].

5 Conclusion

Prevention of corruption offenses in organizations is a priority in the system of combating corporate corruption. The state has assigned a responsible role in this work to the organizations and enterprises themselves, which, according to the law, are obliged to create conditions and take measures to prevent corruption. The effectiveness of the implementation of anti-corruption measures in the organization is significantly increased when using information technologies. They are important because they perform not only economic tasks to optimize and automate internal processes, but also reduce corruption risks, increase transparency of business processes, increase the quality of anti-corruption audit, remove various barriers, intermediary structures and human factors that can potentially lead to corruption situations. Currently, the introduction of information technologies in the implementation of anti-corruption measures by Russian organizations is at the very beginning of the road. This is a time-consuming process that requires significant financial investments. However, in spite of this, this work should be carried out not only by the forces and means of private organizations, but also with significant financial, organizational and other support from the state.

References

1. Bizhoev BM (2018) Fundamentals of intellectual contract system in the sphere of public procurement. J Econ Regul (Issues Econ Regul) 9(1):110–122
2. Code of administrative offences of Russian Federation dated 30.12.2001 N 195-FZ (as amended on 12.27.2019). http://www.consultant.ru/document/cons_doc_LAW_34661/. Accessed 30 Jan 2020
3. D-Russia (2017) The accounting chamber's it system for remote state audit detects more than 70 types of violations. http://d-russia.ru/it-sistema-schetnoj-palaty-dlya-udalyonnogo-gosaudita-vyyavlyaet-uzhe-bolee-70-vidov-narushenij.html. Accessed 16 Feb 2020
4. Farooq R, Ganaie GH, Ahirwar GS (2020) Role of information and communication technology in small and medium sized enterprises in J & K. Smart Innov Syst Technol 160:355–360
5. Federal law "On combating corruption" of 25.12.2008 No. 273-FZ. http://www.consultant.ru/document/cons_doc_LAW_82959/. Accessed 16 Feb 2020
6. Federal law "On contract system in procurement of goods, works, services for state and municipal needs" of 05.04.2013 No. 44-FZ. http://www.consultant.ru/document/cons_doc_LAW_144624/. Accessed 16 Feb 2020
7. Iliy SK (2018) Prevention of corruption in organizations and institutions by prosecution bodies. Bull All-Rus Adv Train Inst Inter Minist 2(46):90–95
8. Kuznetsov AA, Kuznetsov PA (2019) Implementation of modern information technologies in the fight against corruption in the Russian Federation. Legal Fact 61:37–40
9. Munz J, Gindele N, Doluschitz R (2020) Exploring the characteristics and utilisation of farm management information systems (FMIS) in Germany. Comput Electron Agric 170:105246
10. Popova SL (2014) Influence of information technologies on the formation of sustainable development of the enterprise. Bull Saratov Socio-Econ Univ 1:73–77
11. RIA News (2019) The Prosecutor General's office called the damage from corruption crimes for 2018. https://ria.ru/20190409/1552499622.html. Accessed 16 Feb 2020
12. The Ministry of labor and social protection of the Russian Federation (2019). Measures to prevent corruption in organizations. https://rosmintrud.ru/uploads/magic/ru-RU/Ministry-0-106-src-1568817692.8748.pdf. Accessed 16 Feb 2020
13. Ustinova A (2017) Artificial intelligence will think about Gazprom's purchases. ComNews. http://www.comnews.ru/content/110511/2017-11-16/ii-podumaet-nad-zakupkami-gazproma. Accessed 16 Feb 2020
14. Voytovych N, Smolynets I, Hirniak K (2020) The role of technology innovation in food systems transformation. Qual Access Success 21(174):128–134
15. Williams O, Olajide F, Al-Hadhrami T, Lotfi A (2020) Exploring process of information systems and information technology for enterprise agility. In: Saeed F, Mohammed F, Gazem N (eds) Proceedings of the 4th international conference of reliable information and communication technology. Advances in intelligent systems and computing, vol 1073. Springer, Cham, pp 1042–1051
16. Zhangmin (2020) Construction of enterprise accounting risk control based on information technology. J Phys Conf Ser 1437(1):012126

Regional Labor Market: Supply and Demand in the Context of Digitalization

M. Simonova, Y. Lyachenkov, and E. Kostikova

Abstract The study analyzes changes in the structure of supply and demand on the labor market, which are determined by changing consumer preferences and technological innovations. The study of the labor market was conducted based on the example of the Samara region to identify the impact of information technology on changes in supply and demand. A sharp increase in the number of resumes was noted with significantly lower growth rates in the number of vacancies with a slight increase in proposals in the field of IT technologies. The most significant increase in the share of resumes is observed in the sales sector, which is caused by reorientation of enterprises to production of goods to order, high competition of goods and services. Orientation to the quality and new consumer properties of products leads to an increase in requirements for the quality of the workforce, as a key factor determining economic growth. It was determined that an increase in the number of resumes compared to the number of vacancies with a low level of registered unemployment shows an increase in hidden unemployment and informal employment. The labor market is saturated with qualified specialists with competencies that do not correspond to modern conditions.

Keywords Professions · Digitalization of professions · Labor market · Popular professions · Qualifications · Job sites

---

M. Simonova (✉) · Y. Lyachenkov
Samara State University of Economics, Samara, Russia
e-mail: m.simonova@mail.ru

Y. Lyachenkov
e-mail: vsg63@hotmail.com

E. Kostikova
Samara State Technical University, Samara, Russia
e-mail: elenatopo@mail.ru

S. I. Ashmarina and V. V. Mantulenko (eds.), *Current Achievements, Challenges and Digital Chances of Knowledge Based Economy*, Lecture Notes in Networks and Systems 133, https://doi.org/10.1007/978-3-030-47458-4_93

1 Introduction

The large-scale use of digital technologies in all sectors of the economy entails a change not only in the structure of employment in the labor market, which has already become a noticeable phenomenon, but also in the content of labor functions across almost the entire spectrum of professions and specialties [13]. Changes in the content of labor presuppose the availability of new competencies and qualifications. Requirements for the quality of the workforce are increasing significantly and rapidly, which is reflected in requirements for job applicants and for busy personnel [3]. Low-efficient employees are gradually forced out organizations through restructuring, staff optimization, and replacing routine processes with automated systems. At the same time, vacancies are being created with a new set of functions and qualifications corresponding to them, which require professional and super-professional skills. However, they are still not widely distributed on the labor market [4, 8]. Both the number of vacancies and the number of non-employed applicants who, for one reason or another, most often qualifying, cannot find a job with satisfactory working conditions, are growing [9]. Specialists with a fairly large experience of work, but who have not made a professional or service career in their field, are not in demand, despite the presence of education and qualifications. Often these are specialists of the middle or older age group who are still fully functional, but not inclined to learn and get a new education, specialty. The measures taken by the Government of the Russian Federation regarding people of pre-retirement age do not cover this segment of the population and do not take into account the mental characteristics of such a group. Thus, the structural imbalance of the labor market is gradually increasing, which requires additional research and development of new models for regulating social and labor relations.

This is most pronounced on regional labor markets, because, unlike the capital regions, they are attended by less qualified job seekers who do not have the necessary experience in the relevant field [1]. There is practically no competition in high-tech fields of activity, which leads either to vacancy or to hiring specialists without the necessary knowledge and skills. At the same time, specialists with high but somewhat outdated qualifications who do not have modern over professional or cross-cutting skills cannot find a job. Such new cross-cutting skills include, first of all, the possession of new digital technologies in relation to the field of activity [14] where the candidate wants to find a job and communicative, business qualities that would allow the employer to reach a new level of business [6]. An extremely high level of competition in all types and fields of activity makes the competence and qualifications of personnel as a key factor in the economic growth of companies.

2 Methodology

Analysis of the balance of the regional vacancy market and resume will reveal the most popular and attractive areas of activity for applicants, employment opportunities and the choice of the most suitable candidate for employers [15]. As an example, the Samara region was chosen as a developed industrial, but not metropolitan region, having a dynamically developing business environment with a fairly high level of education of the employed population (Fig. 1).

With a stable share of employees with higher education, it is necessary to note the growing share of specialists with secondary vocational education and, accordingly, a decrease in the share of persons with basic general education, which shows the growing demand for skilled labor of workers and secondary technical personnel. At the same time, it is of interest to employ specialists in various professional groups in the labor market [2].

In the selection of personnel, information technologies are becoming more widespread, allowing both applicants and employers to find the best option [12]. According to the data of the hh.ru job site, more than 40% of employers use automated tools provided on job sites as the most affordable financially for staff recruitment. Many employers, about 25%, use the services of E-Staff, which in turn create adapted software for personnel services of companies and recruiting agencies. Programs such as MS Office (Excel, Word, etc.), their own system, 1C, SAP are less often used by employers. The development of technology allows for more precise selection of personnel according to certain criteria, but at the same time, it does not significantly reduce the imbalance of supply and demand in the labor market. It requires a new approach to capabilities of artificial intelligence. The following possibilities of using information technologies in staff recruitment can be noted:

- speed of processing resumes and candidate databases,
- initial interview and screening of inappropriate candidates,

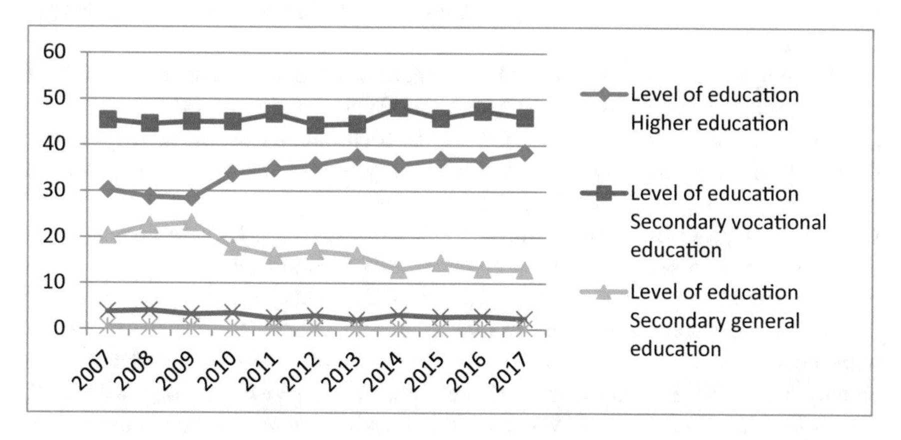

Fig. 1 A level of education of the employed population (*Source* authors based on [11])

- processing large databases with information about applicants,
- automated initial assessment of candidates using combined assessment tools,
- verification of the accuracy of information provided by applicants.

At the same time, the number of barriers for jobseekers in job search and job placement is increasing, which are becoming increasingly difficult to overcome precisely because of the automated approach. That is, there is a paradoxical situation when the more the selection process is automated, the greater the gap between the requirements of employers and the capabilities of applicants. It is necessary to consider the specifics of automated selection, when the more requirements are laid down in the selection criteria, the less likely it becomes to find such a candidate. Potentially strong candidates from related fields of knowledge are cut off from participation in the selection, they are not able to go on a full-time tour and find a compromise employment option.

Let us analyze demand and supply trends in the Samara region using the most common job sites. According to the aggregator of job sites GorodRabot.ru, one can distinguish the most common areas—sources of vacancies in the Samara region labor market: Industrial worker jobs—20.7%, Production—10.7%, Sales—10.6%, Transport, logistics—9.4%, Medicine, pharmaceuticals—6.9%, Information technology, Internet, telecom—4.9%, Installation and service—3.9%, Construction, real estate—3.9%, Top management—3%, Others—25.9% [7]. The number of vacancies for the year increased almost 2 times across the entire spectrum. Wages are growing very weakly. However, it is necessary to note the absence of sharp jumps and downturns, a weak stable growth is observed over a long period. Moreover, the wages of highly skilled workers are correlated and, in many cases, exceed the wages of senior management.

Comparison of the number of vacancies posted on different job sites makes it possible to assess the structure of vacancies and preferences, both for applicants and employers for choosing tools to fill vacancies and job search. Based on the data available for analysis, we will analyze the number of vacancies posted on the most well-known job sites that are used by both job seekers and employers to fill vacancies (Table 1).

The analysis allows us to conclude that there is a different specialization of job sites for groups of professions and specialties. With the obvious priority of HeadHunter.ru and Superjob.ru sites in terms of the total number of vacancies posted in the group of professional fields in demand, it is possible to single out priorities of employers for the placement of individual professional groups. HeadHunter.ru provides more employment opportunities for sales, medical and IT professionals, while Superjob.ru and Rabota.ru specialize in Industrial worker jobs, Production and Transport. We should note a more uniform distribution of vacancies on the HeadHunter.ru website, which makes it possible to maneuver for applicants in related industries when posting resumes. A free information resource Jobs in Russia.ru related to the employment service of the Russian Federation, developed by analogy with well-known job sites, has several advantages, the most important of which is the lack of

Table 1 The number of vacancies of the most popular economic areas of labor in the Samara region (January 2019)

Economic sphere of labor	Number of vacancies, job sites							
	HeadHanter		Superjob		Rabota		Jobs in Russia	
	Units	%	Units	%	Units	%	Units	%
Industrial worker jobs	1 395	13,7	5 578	47,7	1 761	42,4	22	0,4
Sales	3 575	35,2	1 627	14	950	22,9	525	10,2
Production	1 464	14,4	1 987	17	1 095	26,3	2 752	53,5
Transport	1 047	10,3	1 992	17	122	2,9	349	6,7
Medicine	957	9,4	157	1,3	16	0,4	1 345	26,2
IT	1 264	12,4	77	0,7	159	3,8	126	2,5
Installation and service	469	4,6	280	2,4	53	1,3	23	0,4
Total (in popular professional areas)	**10 171**	**100**	**11 698**	**100**	**4 156**	**100**	**5 142**	**100**

Source authors

payment for both sides of the employment process, but the scale of audience coverage does not allow us to talk about the effectiveness and complexity of activities. The prevailing approach to employment as a formal process, not aimed at satisfying the needs of the applicant and employer, is still affecting. This also affects the site's orientation towards user convenience, when the complexity and duration of registration procedures can alienate potential customers. The employed population in the Samara region is distributed by professional groups in accordance with the regional structure of the economy and has a pronounced production character, which can be estimated from the data of Samarastat [11], the calculations for which are presented in the Table 2.

There is a decrease in employed in the Samara region by 3.9 thousand people or 0.3%, due to a wave of decline in the birth rate of the nineties of the twentieth century and an echo of the 2nd World War. Such a structure is sufficiently comparable with the structure of vacancies, and activities of OKVED can be grouped according to the same criteria as demanded vacancy groups for analyzing supply and demand on the labor market. Let us analyze the resumes for popular professional groups posted on the most popular job sites (Table 3).

A significant difference in the number of resumes posted on two popular job sites shows differences in marketing and pricing policies of sites, focuses on different job and qualification categories. However, despite this, it is necessary to note general trends in the preferred professional areas of applicants. Applicants place the largest share of resumes in the professional Sales sector, which is almost 2 times more than the number of resumes in Production and Transport sectors. Industrial worker jobs do not have quantitative advantages, although their share is quite high, but 4 times less than in Sales. It should be noted that job sites are not an effective channel to attract job seekers and do not reflect the true number of vacancies or job seekers.

Table 2 The average annual employed population of the Samara region by type of professional activity (thousand people)

Type of professional activity	2017	2018	Percentage of total	
			2017	2018
Agriculture, forestry, hunting, fishing and fish farming	88,9	86,2	5,4	5,2
Mining	18,3	19,4	1,1	1,2
Production	318,5	318,9	19,2	19,3
Providing electric energy, gas and steam; air conditioning	42,3	41,0	2,6	2,5
Water supply; water disposal, waste collection and disposal	20,2	18,6	1,2	1,1
Construction	139,7	133,3	8,4	8,1
Wholesale and retail trade; repair of motor vehicles and motorcycles	277,1	283,7	16,7	17,2
Transportation and storage	136,4	139,5	8,2	8,4
Hotels and catering	42,9	43,7	2,6	2,6
Information and communication	32,4	34,6	2,0	2,1
Finance and insurance	34,9	33,1	2,1	2,1
Real estate	45,6	45,3	2,8	2,7
Professional, scientific and technical activities	67,8	63,5	4,1	3,8
Administration and related additional services	49,2	49,2	3,0	3,0
Public administration and military security; social security	70,1	70,1	4,2	4,2
Education	113,4	112,3	6,8	6,8
Health and social work	98,8	100,0	6,0	6,1
Culture, sports, leisure and entertainment	23,1	22,8	1,4	1,4
Other types of services	36,1	35,8	2,2	2,2
Total	**1656,8**	**1652,9**	**100**	**100**

Source authors

There is a low percentage of applicants for IT-technologies, which shows the lack of development of this area due to the relatively recent increase in the number of jobs and related training areas.

3 Results

A balance of labor market and labor resources is necessary to meet the needs of the economy and provide the population with opportunities for reproduction of labor. Qualitative and quantitative parameters of the distribution of labor resources by priority areas of activity for a given region show the state of the labor market and the success of the training system in the region. The balance of supply and demand shows

Table 3 Analysis of the candidate's resume by popular types of professional activity in the Samara region based on data from popular job sites (January 2019)

Type of professional activity	HeadHunter		Superjob	
	Units	%	Units	%
1. Industrial worker jobs	9 017	8,1	31 032	13,8
2. Sales	39 463	35,5	90 950	40,4
3. Production	22 937	20,6	30 251	13,4
4. Transport	21 106	19	43 663	19,4
5. Medicine	4 365	3,9	6 658	3
6. IT	9 344	8,4	10 910	4,8
7. Tourism, service	4 972	4,5	11 749	5,2
Total	**111 204**	**100**	**225 213**	**100**

Source authors

the labor shortage or labor redundancy of a particular situation, which determines the state policy in the social and labor sphere. Consider the balance of supply and demand in the Samara region in the most popular professional spheres of labor (Table 4).

Calculation in relative indicators in fractions of the whole allows us to evaluate the structure of vacancies and resumes by professional areas, which characterizes the distribution within the selected parameters and the change over time. With a consistently high share of vacancies and resumes in Sales and Medicine, there was an increase in the share of Production, Transport and services. Decrease in shares is

Table 4 Dynamics of the balance of labor resources in demand types of professional activity using the example of the Samara region for the period from 2017 to 2019

Type of professional activity	2017		2018		2019	
	Number of vacancies, %	Number of resumes, %	Number of vacancies, %	Number of resumes, %	Number of vacancies, %	Number of resumes, %
1. Industrial worker jobs	28,3	18,7	25,3	16,5	23,6	12,9
2. Sales	33,6	33,2	36,1	39,0	32,7	35,6
3. Production	10,2	13,5	9,9	12,0	12,7	14,9
4. Transport	6,4	8,7	7,4	10,4	14,8	18,9
5. Medicine	5,1	7,3	5,0	7,1	5,8	7,1
6. IT	13,1	14,2	12,6	9,9	5,7	6,2
7. Tourism, service	3,3	4,4	3,7	5,2	4,8	4,4
Total	**100**	**100**	**100**	**100**	**100**	**100**

Source authors

observed in the IT sphere and Industrial worker jobs. The result should be considered in conjunction with absolute indicators, which we will give later, but it is necessary to emphasize them, since they allow you to identify a trend in the structure of the economy, which is characterized by an increasing need to find consumers of goods and services. Indicative is the decrease in the shares of the IT sector and Industrial worker jobs, both in vacancies and in resumes. In our opinion, IT is increasingly becoming a multifunctional sphere when IT management skills become a necessary component of other specialties and professions that are not directly related to programming, but which can no longer be successfully implemented without them [10]. A similar trend can be observed in Industrial worker jobs, they increasingly require additional skills, higher qualifications. Let us analyze the balance of supply and demand in the labor market of the Samara region in absolute terms (Fig. 2).

In the study period, the total number of vacancies grew unevenly and with a slight decrease in 2018, in 2019 there was an increase of 20% compared to 2017. The number of resumes for the same period increased by 44%, that is, the growth rate of the number of resumes was practically 2 times higher than the number of vacancies. The ratio of aggregate supply and demand over the period increased from 3 to 5 times, the labor market is becoming increasingly labor-surplus. For applicants, the situation is not favorable, the length of the job search period is increasing, many are forced to agree to work with a decrease in position and income, while losing their qualifications. The discrepancy between estimates and the level of registered unemployment, which is about 3.9% in the region, raises in large cities of the region within 1–1.5%,

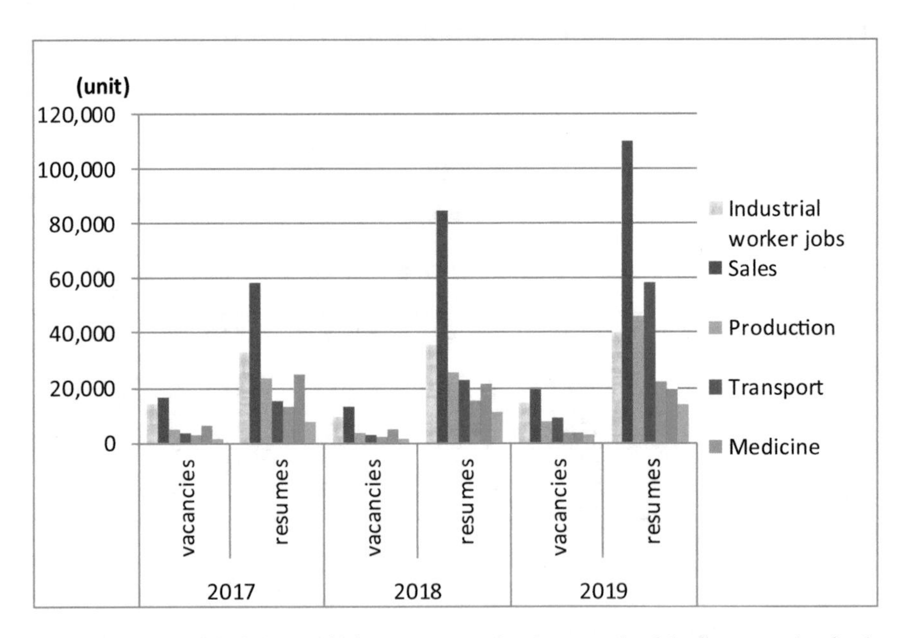

Fig. 2 Dynamics of the balance of labor resources using the example of the Samara region for the period from 2017 to 2019, units (*Source* authors)

which indicates an increase in informal employment. In certain professional areas, the ratio of supply and demand varies 4–8 times. The largest difference in the number of vacancies and resumes is in Sales and Transport sector. There is a simultaneous decrease in the number of resumes and vacancies in the IT sphere. New IT specialties at universities and colleges in the region cannot reverse this unfavorable trend. Education is a fairly inertial field, the results of which have long-term consequences, which should lead to certain results in the medium term.

4 Discussion

The widening of the quantitative gap in supply and demand on the labor market shows unsatisfied demand from applicants. We are to identify qualification and age parameters of unmet demand, which are supposedly of average values, and it is a rather serious potential threat to the balance of the labor market. Population aging occurs in many civilized countries, which leads to the development of national educational programs and employment promotion programs [5]. At present, the professional structure of the regional labor market has not undergone significant changes due to the preservation of priority economic sectors for the region, which the structure of jobs mainly depends on. The massive digitalization of economic processes has not significantly affected the professional structure of the Russian regional labor markets yet, but it has significantly affected the increase in quality requirements of the workforce. At the same time, employers attribute the economic growth of the business to qualitative changes in their products, which entails a shift in emphasis in labor productivity at enterprises, work mainly on order, and, consequently, a sharp increase in sales vacancies. Customer focus has made this area of activity one of the key one to the competitiveness of the enterprise. In our opinion, this type of activity is not in the focus of the education system. The training of sales managers is carried out mainly through additional education, which reduces the quality of specialist training and leads to unmet demand from employers.

5 Conclusion

The transformation of the economy and the social and labor sphere are interrelated processes, any change in which causes significant consequences, both for each person and for society. Our studies have shown significant changes in the structure of supply and demand in the labor market, which are determined by changing consumer preferences and technological innovations. At the same time, there is no significant increase in the share of vacancies related directly to IT technologies on the regional labor market in the developed technological region of Russia with a rather high educational level of the population. A sharp increase in the number of resumes was noted, significantly lower than the growth rate of the number of vacancies. The most significant increase in the share of resumes is observed in the sales sector, which is caused by reorientation of enterprises to production of goods to order, high competition of goods and services. Orientation to the quality and new consumer properties of products leads to an increase in requirements for the quality of the workforce, as a key factor determining economic growth. It was determined that an increase in the number of resumes compared to the number of vacancies with a low level of registered unemployment shows an increase in hidden unemployment and informal employment. In the course of the research, we proposed that, with increasing requirements of employers for the quality of the workforce, middle-aged or older workers who do not meet increased requirements are forced out workplaces. The labor market is saturated with qualified specialists with competencies that do not correspond to modern conditions. As a result, the search term for a job is lengthened under conditions most often with a decrease in the level of position and income, which requires additional research.

References

1. Alpek L, Tésits R (2019) Measuring regional differences in employability in Hungary. Appl Spat Anal Policy. https://doi.org/10.1007/s12061-019-09306-6. Accessed 11 Feb 2020
2. Behaneck M (2019) Employment websites: personnel by mouse click. [Online-Jobborsen: Personal per Mausklick]. Betonwerk Und Fertigteil-Technik/Concr Plant Precast Technol 84(1):42–51
3. Bogoviz AV, Gulyaeva TI, Semenova EI, Lobova SV (2019) Transformation changes in the system of professional competences of modern specialists in the conditions of knowledge economy's formation and the innovational approach to training. In: Popkova E, Ragulina Y, Bogoviz A (eds) Industry 4.0: industrial revolution of the 21st century. Studies in systems, decision and control, vol 169 (18373566). Springer, Cham. https://doi.org/10.1007/978-3-319-94310-7_19
4. Civilcharran S, Maharaj MS (2019) A framework to determine the digital skills preparedness of graduates for industry. In: Soyjaudah KMS (ed) International conference on intelligent and innovative computing applications. IEEE Inc., London, pp 1–6. https://doi.org/10.1109/iconic.2018.8601250
5. Earl C, Taylor P, Roberts C, Huynh P, Davis S (2017) The workforce demographic shift and the changing nature of work: implications for policy, productivity, and participation. In Profili S, Sammarra A, Innocenti L (eds) Age diversity in the workplace. Advanced series in management, vol 17. Emerald Publishing Limited, Bingley, pp 3–34. https://doi.org/10.1108/s1877-636120170000017002
6. Engelhardt C (2015) The DigCurV review of training needs in the field of digital preservation and curation. An overview of the main findings. In: Cirinnà Ch, Fernie K, Lunghi M, Casarosa V (eds) Proceedings of the framing the digital curation curriculum conference. CEUR workshop proceedings, vol 1016. CEUR-WS, Aachen. http://ceur-ws.org/Vol-1016/. Accessed 12 Feb 2020
7. City of Works (2020) Salary statistics—Samara oblast. https://gorodrabot.ru/salary?l=%D1%81%D0%B0%D0%BC%D0%B0%D1%80%D1%81%D0%BA%D0%B0%D1%8F+%D0%BE%D0%B1%D0%BB%D0%B0%D1%81%D1%82%D1%8C. Accessed 12 Feb 2020
8. Howard J (2019) Artificial intelligence: implications for the future of work. Am J Ind Med 62(11):917–926. https://doi.org/10.1002/ajim.23037

9. Kruger TM, Clark-Shirley LJ, Guest MA (2019) Careers in aging: how job-seekers search for and how employers advertise positions in aging-related markets. Educ Gerontol 45(7):483–494. https://doi.org/10.1080/03601277.2019.1658853
10. Lovcheva MV, Konovalova VG, Simonova MV (2020) Development of corporate digital training. In: Ashmarina S, Vochozka M, Mantulenko V (eds) Digital age: chances, challenges and future, vol 84. Lecture notes in networks and systems. Springer, Cham, pp 473–479. https://doi.org/10.1007/978-3-030-27015-5_57
11. Ramkumar A, Rajini G (2019) Innovative way of using social media networks for e-recruitment and selection. Int J Sci Technol Res 8(10):175–181
12. Rosstat (2019) Regions of Russia. Socio-economic indicators. https://gks.ru/storage/mediabank/Region_Pokaz_2019.pdf. Accessed 22 Jan 2020
13. Shelomentseva VP, Ifutina YA, Frezorger LA (2017) Structural changes in economics and their impact on the labor market. J Adv Res Law Econ 8(4):1315–1321. https://doi.org/10.14505/jarle.v8.4(26).30
14. Szabó I, Ternai K (2018) Process-based analysis of digitally transforming skills. In: Tjoa A, Zheng LR, Zou Z, Raffai M, Xu L, Novak N (eds) Research and practical issues of enterprise information systems, vol 310. Lecture notes in business information processing. Springer, Cham, pp 104–115
15. Theys T, Deschacht N, Adriaenssens S, Verhaest D (2019) The evolution of inter-regional spatial mismatch in the USA: the role of skills and spatial structure. Urban Stud 56(13):2654–2669. https://doi.org/10.1177/0042098018803017

Internet-Banking as a Part of Russian Digitalization: The Key Trends

S. Y. Salomatina

Abstract The growth in a number and a range of the distant banking services is the most evident sequence of digitalization in Russian economics in general, and in the market of financial services in particular. Modern technologies in the Internet-banking increase the efficiency of the banks resources, minimize the costs and improve the client service. Moreover, the larger number and the higher quality of the distant banking services are the obvious benefit on the competitive market. However, in Russia, Internet-banking is a comparatively new system: potential clients will have to identify advantages and disadvantages of it, in order to use the system completely. The aim of the article is to reveal the trends in development of the Internet-banking in Russia and to evaluate its' role in the path of the Russian economics' digitalization. This aim is achieved using the systematic approach, which possessed the issue in an integrated manner.

Keywords Internet-banking · Distant banking services · Mobile banking · Economics' digitalization · Digital transformation

1 Introduction

Nowadays, competitiveness, along with economical growth, depends dramatically on the usage of modern technologies and innovations. This is the reason why involving distant banking services in the banking strategies is crucially important. The rapid development of the Internet technologies has started the numerous changes in each area of life, including economics and banking in particular. The Internet-banking was created due to that digital transformation. The Internet-banking increased the efficiency of relationship between a bank and a client, as well as the speed of the data analysis and operations management.

The Internet-banking technologies develop non-stop; this is what attracts new clients and users to the system. Interaction between a bank and a client become

S. Y. Salomatina (✉)
Samara State University of Economics, Samara, Russia
e-mail: salom771@rambler.ru

S. I. Ashmarina and V. V. Mantulenko (eds.), *Current Achievements, Challenges and Digital Chances of Knowledge Based Economy*, Lecture Notes in Networks and Systems 133, https://doi.org/10.1007/978-3-030-47458-4_94

easier, faster and more comfortable when using distant banking services. More of that, it helps to reduce losses in such resources as time and information [2].

Commercial banks put an effort in increasing the quality of their distant services in order to reduces their costs and to attract more clients. Obviously, this fact is a sign that integrated banking services, along with the usage of telecommunication systems, become especially vital.

The modern trends in the banking involve the combination of traditional banking services and the newest technological and scientific innovations; these factors maximise the efficiency of the financial market. The integration of the newest technologies leads to the usage of the distant banking services, which clearly are the element of the economics' digitalization. The distant access to the services improves the targeted communication with a client, helps in analysing their needs and wishes. It cuts the gap between a bank and a client and increases the competition in the banking market, which is one of the key aspects in the Internet-banking development.

Therefore, the Internet-banking development and the complete usage of the distant banking services, as the modern trends in the banking, increase the quality and the efficiency of the services, minimise the costs, which leads to the growth in competitiveness and stimulates the digital transformation of Russian economics.

2 Methodology

The toolkit of this article is based on the principles of the functional and systematic methods, along with general academic methods of knowledge (analysis and synthesis, induction and deduction, abstraction) and the instruments of the statistical analysis. This toolkit is claimed to enhance the knowledge on the role of the Internet-banking development in the overall digital transformation of the banking services.

3 Results

The Internet-banking development comes up the overall digital transformation of the financial services. In 2019, there was a number of the key trends of this transformation [5]:

- the mobile banking (m-banking) development. The modern client searches for the reliable, secure, comfortable and simple app with 24-h access to the banking services. Such app would attract more clients than comfortable and luxious client relations office. Obviously, the m-banking will adapt its' services to a client's wishes and needs,
- the "aside" mobile apps development. The modern bank not only competes with other banks, but also with the providers of alternative financial services (micro

loans, investment management, budgeting etc.). To avoid the excessive competition, the bank has to integrate such services into its overall strategy, deciding whether to be the information provider only, or to deal with the "aside" financial service providers, or to become a provider itself,
- the Blockchain development in payments, depositing services, lending and other financial spheres with high demand for the security [10],
- the usage of the Big Data as the tool of the machine learning in order to fulfil effectively a client's specific wishes and needs.

The mainstreams of the banking industry in the digital economics, in turn, are: converting the information into the knowledge; the processes automatization; the decline in the cost and speed of renovating a range of services. The information in the digital economics has a huge impact: information is the main and priceless intangible asset [9]. When choosing a bank, a client pays attention to the services and time management, which is the reason why distant banking services are actively involved in the financial sphere. The Internet-banking operations are significantly cheaper and definitely more comfortable than that in the offices.

The usage of the banking innovations is designed for:

1. Acceleration of the transactions,
2. Simplified identification of a client using modern technologies,
3. The best possible security of the information about transactions and of the personal data,
4. Provision of the wider range of affordable and comfortable services,
5. Adoption of the paperless workflow [8].

For clients, the Internet-banking is a tool for the comfortable access and usage of their financial resources [3]. Currently, to open a bank account, a client only have to submit the personal data and the application form on a bank's website, and the account will be opened in the next few days.

Investment services have also become more accessible. Clients may easily open brokerage accounts and start trading and investing on stocks. Along with that, they don't need to communicate with a broker personally: clients can buy and sell securities, derivatives and any other financial asset within the specific application.

More of that, clients may proceed payments between their own accounts immediately, along with payments for the goods and money withdrawal. For example, VTB, one of the largest versatile commercial banks of Russia with the Public participation, offers its' clients the completely functional Internet-banking services, which diminish the need in visiting the office. There are numerous distant operation services:

1. Provision of the accounts information: loans, deposits, incomes and spendings, penalties and percentage incomes, balance etc. Moreover, it provides the information about the new services and personal offers.
2. Transactions: between personal accounts; to the other banks; using the phone number or the card number; loan repayments; investment.

3. Payments: patternal and automatical payments; commodities, telecommunication and tax payments; charity payments to Russian and foreign funds; transactions to the electronic wallets (Yandex.Money, QIWI, WebMoney, ONpay.ru etc.); payments to the pension funds: VTB Future, Gazfond etc.; private and public insurance: SOGAZ, Alliance, VTB-Insurance etc.; payment to mutual funds.
4. Currency exchange using the actual daily information of VTB; Exchanging currencies: USD, EUR, GBP, CHF, JPY and SEK.
5. Investment: open of the brokerage account, investment in treasuries, shares and futures; access to the MOEX [1].

Such distant services are classical for the Russian Internet-banking. The comparison of the m-banking provided by different Russian commercial companies reveals that the personal account app is a useful tool for every client: they can use every service fully and manage operations with personal accounts 24/7.

The consulting company «Markswebb» publishes the annual rating of the best Internet-banks for private clients. «Markswebb» evaluates the banks differently within different segments: in the Daily Banking segment, «Markswebb» evaluates how fully and comfortably a client can manage his assets on the debet account; in the Digital Office segment they evaluate the possibility and efficiency of the solutions, which traditionally needed a visit to the bank office or contact center.

In 2019, the best banks in the Daily Banking segment, according to the «Markswebb» are Tinkoff Bank, «Levoberezhnyi», «AkBars», «UralSib», Reiffeisen and PochtaBank. For the same year, the top-5 in the Digital Office segment are Tinkoff Bank, «Levoberezhnyi», PromSvyazBank, «AkBars», Sberbank [4].

The analysts of Markswebb state that the clients' everyday tasks on asset management are fulfilled in the most of the banks, however, these banks still limit the development of the Internet-banking and m-banking services. Nevertheless, the market is not stagnating: it is developing the payment proceeds services and fighting for the payment loyalty of the clients. The more significant development of the Internet-banking comes up with the digital office concept—the complete distant facilities. It includes distant sign up and log in, as well as distant usage of the financial services and products—as a mandatory minimum. Banks start to transit the documents' submission and assignation, along with the service or product cancellation, to the Interner-banking; As for the loans, opening a pre-confirmed credit account is typical today. To consult the clients, banks use online chats. Moreover, the online bank should soon become a path for investment: nowadays, it is possible to invest in simple financial products (mutual funds, equity units etc.), however, harder tasks have to be solved otherwise. Another developing concept is the «Internet-banking for paperflow»: online profile of a client can be used as storage for his personal documents and data. This information may be used when creating new services or developing the existing ones.

Experts say that further development of the Internet-banking depends on the number of clients that will continue to use the system: despite the fact that the Internet-banking already has a loyal market group, experts doubt whether it is appropriate to maintain the system for them only. On the other hand, the mobile app has its' own advantages even for the conservative clients: for example, they can scan QR-code instead of filling the payment list, or scan the passport to submit a request or to send it to the chat. Thereby, the development of the Internet-banking depend crucially on the data about the target group: if the share of users will continue to decline, the investment into the system must be limited and more actions must be undertaken to maintain the current level of the payments. But if the target group remains stable and loyal, the investment must be restructured basing on the needs of users in order to increase their payment proceeds and to simplify the user experience in the m-banking—recommendation of the rating's authors.

4 Discussion

Economic progress connected inseparably with the newest technologies' invention. The development of innovations and modern solutions demands a large quantity of working labor, time and financial resources; however, the result nearly always worth it. The question is whether the development of the online services is profitable for the banks; and the answer is yes—the Internet services expand the target market, sequentially attracting new clients with new needs for various banking products.

It is vital to mention that the gradual implementation of the innovations is enshrined in the «Concept of the long-term social and economic development of Russian Federation». The Concept states that the global economical growth, as well as the development in the electronic payment systems, is the promising direction for Russia to strengthen its' positions on the global market, to increase the import of the innovations and capital and to increase the ratio of the economical integration.

We expect the enhancement of the impact of innovations on the Russian economics: it is planned that in the period from 2020 to 2030 will be created and integrated the technological base of economic systems, built with the use of innovations in nanotechnologies, IT etc. Innovations and effective technological solutions will become the comparative advantages in the competition between the countries. The banks, in turn, will become the drivers of the progress; they will predetermine the path of the development of innovative systems.

The specific feature of Russian Federation, in comparison to other markets, is high level and rapid ratio of the indirect contribution of the Internet technologies to the national economics. According to preliminary estimates, in 2021 the contribution must reach 35% (for example, in the USA and China this indicator must be 25% and 31% respectively [6]). Investigations reveal that the share of the Internet economics in GDP is expected to increase from 3,8% in 2016 to 4,7% in 2021.

In the period from 2017 to 2021, more than 7,5% of the GDP growth will be caused by the Internet economics exclusively. According to IMF, the GDP growth in Russia in the period from 2017 to 2021 is expected to reach 6,0% growth, along with 10,7% growth in the Internet economics [7]. The impact of the Internet technologies on GDP will only increase, as the number of the mobile phones users grows, and the mobile networks spreads. It is expected that the further growth will depend on the balance changes: people will spend more on modern mobile technologies than on the traditional ones. Also we expect the income increase in the content-making segment, especially in mobile commerce, advertisement and app development.

Every organization involved in the digital economics creates the market for numerous distributors, providers and service companies. The positive impact of the Internet economics nowadays spreads far beyond the areas of functioning of the organizations, which were the kick-starters of the mobile technologies. This impact was evaluated as 1,53% of GDP in 2017—or 1303 billion of rubles. Banks started to cross the borders of their traditional functions: the Internet-banking affords the clients to do their financial tasks in the faster, comfortable and secure manner; banks reduce costs on the personal services by using their own Internet-platforms. Due to the widely spread network and huge growth of technological innovations, the Internet-banking must maintain and increase its' impact on the industry, which were evaluated as at least 15% of all banking incomes in 2017, which is 114,0 billion of rubles, and must reach up to 40% of all banking incomes in 2021 which is expected to be 304,0 billion of rubles [7]. In conjunction, Russian mobile technologies sector may enter the top-80 countries' GDP global rating, surpassing the economics of Belorussia, Lithuania and Azerbaijan.

5 Conclusion

The Internet-banking in Russia is already the 11th largest economic activities; it competes with the agricultural activity. The modern digital economics is not only the sphere of business activity, but also a main feature of the development in modern society. The world leader in the share of the digital economics in GDP among the G-20 countries is Great Britain. Today, Russian Federation only takes 16th place in this rating. The analysts of the Boston Consulting Group counted that Russian backlog, in comparison to the flagship countries, is 5–8 years in average. If no activities on the digitalization are undertaken, the backlog would only increase—in the next 3–5 years it may reach 15–20 years of backlog. Meanwhile, this indicator is one of the key ones within the frames of the «Industry 4.0» concept, which rates the competitiveness of the national economics, basing on the investment climate of the country [7].

Purposeful and intelligent steps in that direction will help to avoid the dependence of Russian operational and technological activity on the foreign digital platforms, technologies and standards. We need the clarity of vision on the problems and step-by-step strategy in order to use the appearing possibilities in-time, and to keep the digital and physical independence. Therefore, the Internet-banking, as the element of the digital transformation of the economics, has a huge impact nowadays. The digital technologies spread in Russian Federation have already reached significant level and currently continues to grow due to spread of mobile technologies, increase in the usage of the mobile Internet and services, along with massive transition of the everyday tasks to the online world.

References

1. Banki.ru (2019) Markswebb named the best online banking for individuals in Russia. https://www.banki.ru/news/lenta/?id=10904889. Accessed 17 Feb 2020. (in Russian)
2. Chaouali W, Yahia IB, Souiden N (2016) The interplay of counter-conformity motivation, social influence, and trust in customers' intention to adopt Internet banking services: the case of an emerging country. J Retail Consum Serv 28:209–218. https://doi.org/10.1016/j.jretconser.2015.10.007
3. Konovalova ME, Kuzmina OY, Salomatina SY (2020) Transformation of the institution of money in the digital epoch. In: Ashmarina S, Mesquita A, Vochozka M (eds) Digital transformation of the economy: challenges, trends and new opportunities, vol 908. Advances in intelligent systems and computing. Springer, Cham, pp 315–328. https://doi.org/10.1007/978-3-030-11367-4_31
4. Markswebb (2019) Business internet banking rank 2019. https://markswebb.ru/report/business-internet-banking-rank-2019/. Accessed 17 Feb 2020. (in Russian)
5. Newman D (2019) Top 7 digital transformation trends in financial services for 2019. Forbes. https://www.forbes.com/sites/danielnewman/2019/01/16/top-7-digital-transformation-trends-in-financial-services-for-2019/#107fd0055310. Accessed 17 Feb 2020
6. Pramanik HS, Kirtania M, Pani AK (2019) Essence of digital transformation—manifestations at large financial institutions from North America. Future Gen Comput Syst 95:323–343. https://doi.org/10.1016/j.future.2018.12.003
7. RAEC, OC&C, & Google (2017) Mobile internet economy in Russia 2017. http://mobile2017.raec.ru/assets/mobile-internet-economy-in-russia-eng.pdf. Accessed 17 Feb 2020
8. Salihu A, Metin H, Hajrizi E, Ahmeti M (2019) The effect of security and ease of use on reducing the problems/deficiencies of electronic banking services. IFAC-Papers OnLine 52(25):159–163. https://doi.org/10.1016/j.ifacol.2019.12.465
9. Shaikh AA, Karjaluoto H (2015) Mobile banking adoption: a literature review. Telemat Inform 32:129–142. https://doi.org/10.1016/j.tele.2014.05.003
10. Thakor AV (2020) Fintech and banking: what do we know? J Financ Intermed 41:100833. https://doi.org/10.1016/j.jfi.2019.100833

Reforming Issues of Entrepreneurship in the Digital Economy

G. I. Yakovlev, A. V. Streltsov, and E. Yu. Nikulina

Abstract In the context of the digital economy, effective industrial platforms are being formed, where a special place belongs to the development of production capacities of manufacturing enterprises with high scientific and technical potential. Purpose: consideration of innovative business models that are deeply specific in relation to resources aimed at increasing the operational efficiency of enterprises to create unique successful technologies and products. Methodology: to analyze digital modernization of enterprises, groups of objective laws for updating fixed assets have been identified: development, synergy, composition and decomposition, proportionality and time saving. When considering the problems of developing the technical base of mechanical engineering, it turned out to be expedient to systematize according to certain criteria, including the interaction between government bodies and business, industrial community, characteristics of industry demand, product quality, production and financing efficiency, innovation, resource component, structural characteristics of the industry. The results of the study are the main solutions of the problems of the Russian engineering development in the digital economy with new business models.

Keywords Updating · Modernization · Business models · Efficiency · Digital platforms · Fixed assets

1 Introduction

Currently, profound transformational shifts are taking place in the scale and models of reproductive activities of business entities at the level of national economies and

G. I. Yakovlev (✉) · A. V. Streltsov · E. Yu. Nikulina
Samara State University of Economics, Samara, Russia
e-mail: dmms7@rambler.ru

A. V. Streltsov
e-mail: oisrpp@mail.ru

E. Yu. Nikulina
e-mail: katerina_nikulina@list.ru

S. I. Ashmarina and V. V. Mantulenko (eds.), *Current Achievements, Challenges and Digital Chances of Knowledge Based Economy*, Lecture Notes in Networks and Systems 133, https://doi.org/10.1007/978-3-030-47458-4_95

individual firms. At the same time, representatives of the manufacturing business are required to understand the essence of ongoing changes in order to develop new lines of development, competencies and methods of organizing production. Digital technologies form new rules of the game in the business space and pose new challenges to the industry. If you do not solve them, you can slip into the number of hopelessly lagging business actors. A significant number of special publications deal with the formation of a new model of digital industrial platforms that affect conceptual foundations of traditional business models. They are deeply specific with respect to resources, network and social interaction and open new opportunities for creating unique technologies and products that can successfully compete in global markets. A special role in the formation of industrial platforms belongs to manufacturing sectors of the economy as carriers of high technologies. Characterizing the development of the machine-building complex, it should be noted that its functioning in any developed country is the basis for the development of productive forces, technological, economic security due to objective belonging to the advanced scientific and technical level. This complex has the main role both in increasing the efficiency of the innovation activity of productive forces, and in ensuring the national security of the country. Due to the low entry threshold, digitalization in our country is mainly prevalent in the tertiary sector of the economy—commerce and services. At the same time, there is a big gap between the leading companies that are at the forefront of the digital transformation, and the gray mass of companies that have not even reached the basic level of automation. Digital transformation is a completely new phenomenon in social development, as long as there is no universal effective roadmap for firms to pass this newly emerging powerful socio-economic phenomenon, which allows them to have unprecedented competitive advantages.

2 Methodology

In the digital economy, the requirements for ensuring the efficiency of modern production are based on a group of objective laws for updating fixed assets: development, synergy, composition and decomposition, proportionality, time saving, which ultimately determine the competitiveness of enterprises. The leading element of modern production modernization is the intensive renewal of fixed assets, giving a new quality to even significantly age-old equipment of industrial enterprises and entrepreneurial structures to support digital transformation while creating an effective stream of consumer values. It is possible to provide a competitive advantage on the basis of digitalization of existing fixed assets of enterprises by introducing the industrial Internet of things and integrated information systems, which was noted by Volkodavova et al. [15]. The socio-economic importance of Industry 4.0 is great. It is a cardinal breakdown of the old business model and the formation of a new, socially-oriented and innovative one, through the creation of electronic industrial platforms, with a large number of scientific and client services. To build a digital transformation management system, it is advisable to take a closer look at the relationship between updating production and other stages of modernization (Fig. 1).

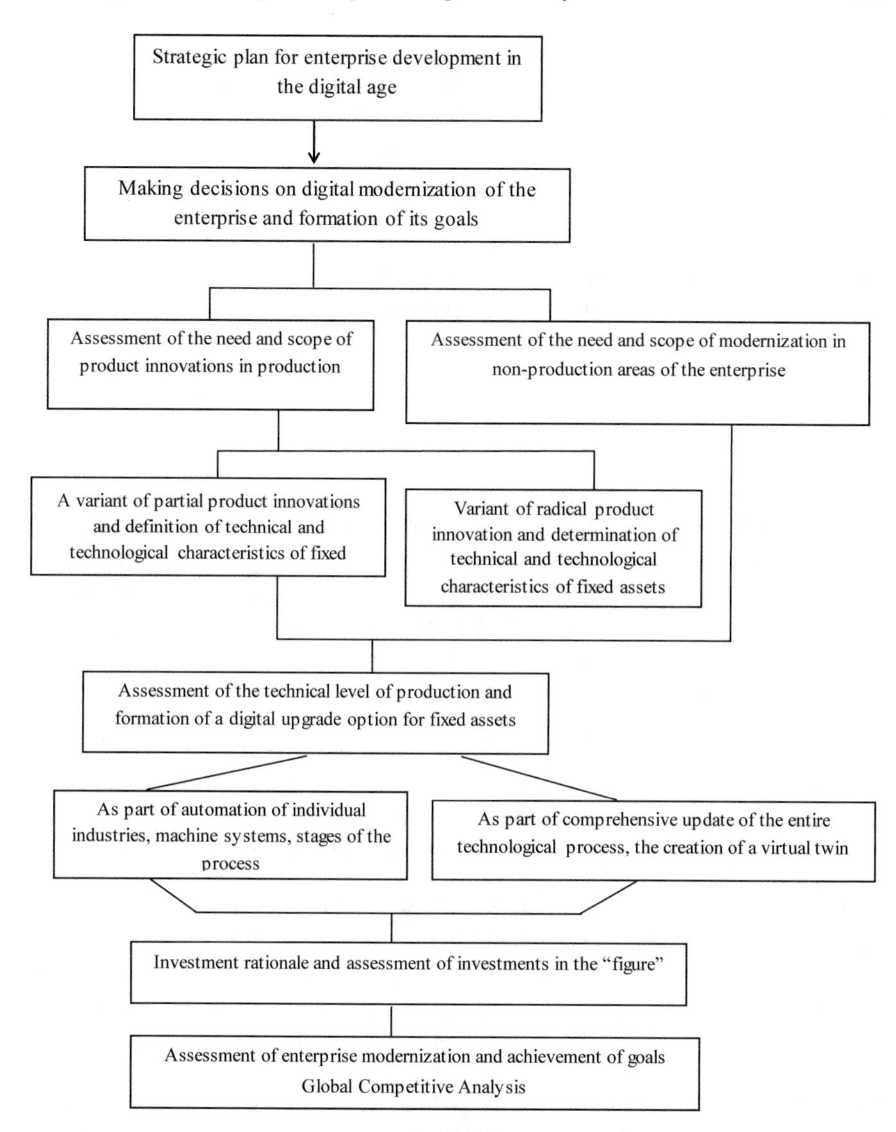

Fig. 1 Updating fixed assets in the enterprise modernization management system (*Source* authors)

In the digital economy, consumer needs not only the basic service or product, but the whole range of services for his consumption and disposal, including training, insurance, lending, after-sales service, up-grade, and all this should be provided on a single platform, research and production service. The industrial Internet of things, even on traditional equipment, machines equipped with special sensors, based on data analysis, allows you to quickly change and optimize production processes by modeling real processes on “digital twins”. Both digital modernization of the enterprise as a whole and the local renewal of fixed assets, as its component predetermines the development of the enterprise for the long term [10], taking into account the laws of development, synergy, composition and decomposition, proportionality and time saving. To improve the efficiency of updating fixed assets in the process of digital modernization, it is necessary to identify and use these laws. We can also identify certain properties of the renewal of fixed assets as an essential element of enterprise modernization. They are cyclicity, long-term, staged, risky, cost-intensive, which require separately detailed analysis in further studies.

3 Results

Products of the machine-building complex currently account for about 35% of the value of world industrial products. The share of engineering products in world exports is 6 trillion dollars or 42% [11]. Russian enterprises are lagging behind the world level in terms of digitalization in key industries—industry, transport, agriculture, as in the leading countries of the world the share of the digital economy in GDP is calculated in twin digits, instead of our unit percentage points.

In addition, the current macroeconomic situation, the constant pressure of sectoral and other non-economic sanctions and political restrictions negatively affect the development of the engineering complex. As a result, if in industrialized countries mechanical engineering accounts for 30 to 50% or more of the total industrial production, for example, in Germany 53.6%, in the UK—39.6%, in Japan—51.5%, in China—35.2%, then in Russia it is about 20% [1]. In the late Soviet era, this indicator reached about 40% [10].

There are a large number of publications that address the problems of domestic engineering, both in the general and industrial context. The range of these problems is wide enough. Therefore, for their detailed consideration and solution, it is advisable to systematize the problems of the Russian mechanical engineering according to certain criteria. It seems that these signs should include: interaction between government bodies and industrial community, characteristics of industry demand, product quality, production and financing efficiency, innovation, resource component, and structural characteristics of the industry. The systematization of problems according to these characteristics is presented in Table 1.

Describing the problems of domestic mechanical engineering by the first sign—by the interaction between government bodies and industrial community, we can note, first of all, the lack of a full-fledged market. There is no former, administrative regulation of the industry. At the same time, the market mechanism for regulating production and financial relations has not been developed yet, since there is no effective competition. Relatively successful engineering enterprises are oriented toward fulfilling a state order, and they are precisely regulated by non-market methods.

In many ways, the above problems are a consequence of the low efficiency of production in engineering and the associated chronic underfunding. A large group

Table 1 Problems of the Russian engineering development

Feature description	Problems
Interaction between public authorities and industrial community	• The market is not fully formed • Monopoly of the resource producer • There is no systematic approach to modernization • There is no coordination in the development and implementation of federal and regional target programs • Industrial Policy Act does not work
Demand characteristics	• Over the years of market transformations, demand for domestic innovative engineering products has been lost, and the image of high-quality civilian industrial products has been lost • Supply chain system (contractual activity) to meet municipal and state needs for engineering products is inefficient • High loan rates
Product quality	• Share of defective products is increasing • There are no incentives for the full implementation of quality standards • Quality certificates do not provide benefits and they are formal in nature • Updating technical regulations and standards is slow • In domestic engineering products, as a rule, there are no complex digital embedded programs and subsystems
Efficiency of production and financing	• Low production efficiency. More profitable (cost-effective) in many cases, the phaseout and turning into dealerships for the sale of foreign mechanical engineering products • Profitability of industry products is lower than profitability in raw materials industries • There is no system of innovative financing • A single system of innovation and investment activity has not been formed
Innovations	• There is no mechanism for transferring developments in the field of R&D and development work from research (research institutes) and design centers (design bureaus) to production, including developments corresponding to the world level • A significant number of closed research institutes and design bureaus • Intellectual property of developers is not adequately protected by Russian law • Prevalence of the evolutionary scientific and technological process over the revolutionary one • Low demand for new technologies from enterprises • There is lack or insufficiently effective programs for the integrated technological development of enterprises
Resource component	• Low share of the active part of fixed assets of enterprises • Strong deterioration of fixed assets and, especially, machinery and equipment • Low rates of renewal of fixed assets • Low level of development of domestic machine tool industry • Lack of qualified personnel • Ineffective system of professional training • Low remuneration and prestige of work at machine-building enterprises and associated with R&D services

(continued)

Table 1 (continued)

Feature description	Problems
Structural characteristics of the industry	• Incomplete structuring of the industry • Narrow industry orientation • Continuing deep specialization of enterprises • Lack of synergy and cooperation between enterprises with domestic and foreign owners

Source authors based on [1, 7, 10, 13–15]

of problems in the Russian machine-building are related to innovation. Here, first of all, it is necessary to note the absence of a mechanism for transferring developments from R&D to the sphere of industrial production. The absence of this mechanism leads to a whole range of negative consequences. Without the sale of their own developments in mechanical engineering, research institutes and design bureaus do not receive necessary financing and close. World-class developments are transferred to foreign competitors. Without the introduction of new developments, enterprises are not stimulated by revolutionary changes in technology.

To solve the above problems, a whole range of systemic solutions is needed, which should be formed and implemented with the participation of the state. They can be systematized according to a number of signs (Table 2).

Describing these areas, it should be noted that the current economic situation in Russia does not allow simultaneously solving the problem of complex modernization of the production potential of machine-building enterprises. But the priority approach and its phased implementation based on the developed strategy for modernization of domestic engineering can mitigate the above problems. The proposed main solutions of the problems of Russian engineering suggest the multifaceted participation of the

Table 2 Main solutions the problems of the Russian engineering development in the digital economy

Sign of sys-thematization	Solutions
Improvement of management efficiency within the framework of industrial policy	(1) Make up inventory of machine-building enterprises by machine-building sub-sectors (2) Restructure the engineering complex (3) Improve the efficiency of engineering management based on the study of innovation and investment cycles of its development (4) Study technological stages of mechanical engineering (5) Form a mechanism for the interaction of leading sub-sectors of engineering (6) Develop a coordination mechanism for civil engineering and the military-industrial complex (7) Enter international science-intensive engineering projects, industrial platforms on the basis of strategic partnership and alliance (8) Form state orders for breakthrough engineering technologies

(continued)

Table 2 (continued)

Sign of sys-thematization	Solutions
Opportunities for international cooperation	(1) Organize joint ventures in high technology sectors (2) Organize intermediate stages of the production cycle for high-tech products abroad with the mandatory provision of final assembly in Russia (3) Stimulate the influx of foreign direct investment (4) Promote export of products and participate in export production chains
State financing	(1) Develop federal targeted programs for machine building, especially for technological and innovative development (2) Provide targeted state funding for the development of production and technological potential of enterprises and sub-sectors (3) Provide targeted financing for the development of agricultural and road construction machinery in conjunction with the above complexes (4) Organize innovative and risky lending
Technology	(1) Form and support integrated innovative and technological solutions for the development of enterprises and sub-sectors (2) Identify and stimulate technological breakthroughs
Coordination	(1) Develop and provide comprehensive organizational, technical, marketing and investment support of import substitution (2) Form a system of cooperative ties in mechanical engineering, including with foreign companies (3) Develop an investment support mechanism for the development of Russian engineering based on the analysis and consideration of the interests of all its participants (4) Study technological patterns and stimulate the transition from existing technological patterns to promising ones (5) Reconstruct the R&D chain—production—sales, develop the system of research institutes and design bureaus (6) Identify current and future tasks for the development of machine building and provide conditions for the implementation of the latter
Resource component	(1) Form a personnel training system for mechanical engineering, increase wages and the prestige of working in Russian mechanical engineering (2) Provide comprehensive modernization of the engineering production infrastructure (3) Provide mass renewal of fixed assets in engineering, coupled with technological innovations, mainly of their active part, with digitalization and the creation of a "twin"

Source authors

state in their implementation, including measures of direct financing, as well as organizational and coordination functions, and especially the development of the infrastructure for platform business solutions.

4 Discussion

Vivek Ghosal and Usha Nair-Reichert found a significant effect of enterprise investments in modernization and incremental innovation, as a factor in productivity growth, especially in the competitive analysis in the long term [5]. The development strategy of a modern enterprise provides an assessment of the need for periodic modernization and updating of fixed assets when implementing programs for the reconstruction and technical re-equipment of enterprises on a digital basis [14]. Kukartsev et al. simulated the reproduction of fixed assets in the development strategy system of the machine-building enterprise [7]. Anisimova and Zagoretska offered to consider new types of reproduction of fixed assets: narrowed, innovative and stop production [2].

The importance of digitizing manufacturing activities of companies was shown by Coreynen et al., including the possibility of dynamic reconfiguration of resources, while reducing the level of input barriers [3]. Research by Nambisan showed that new digital technologies have changed the nature of uncertainty and risk of entrepreneurial processes and results, while enriching the theory of success of digital entrepreneurship [8]. Subsequently, Nambisan et al., also revealed a change in the nature and direction of entrepreneurial activity in various industries, their implemented business models and production apparatus, carried out as a result of the platforming phenomenon, which provides for the transition to more open and distributed innovation models, as well as the growing importance of digital industrial platforms as a platform for creating and acquiring consumer values [9].

Digital platforms today radically transform almost every industry, according to de Reuver et al. [4]. At the same time, digital industrial platforms are a complex subject of research due to their distributed nature and interweaving with institutions, markets and technologies. The scale and complexity of platform innovations are growing, and the diffusion of methods of organizing production based on digital platforms to many industries is accelerating. Hsieh and Wu showed how entrepreneurs can take advantage of the innovation ecosystem based on the industrial platform in the main stages of the innovation process (i.e., inventions, production and commercialization) [6].

The effect of digitalization is achieved not only in achieving a high level of competitiveness and efficiency, but also in reducing the cost of the production process by the enterprise. The priority in digital modernization is to assess the need and scope of product innovations in production [14]. Skuras et al., marked off the need to upgrade production assets with innovative products in favor of [12] new construction on digital principles (the Greenfield principle).

5 Conclusion

An important condition for advancing Russia along the path of economic progress and strengthening the competitiveness of manufacturing sectors is, first of all, the digital development strategy of the basic sectors of the economy, the concentration of efforts and resources on the formation of the effective, technologically developed and market economy, a class of initiative entrepreneurs who build business on a solid digital industrial base. New business concepts provide for an intensive update of fixed assets, development of the ability to support digital transformation from the technical and managerial side while creating an effective stream of consumer values.

Despite the absence of a well-established "road map", different initial levels of enterprises, the processes of modernization and reproduction of fixed assets in most cases are satisfactorily resolved through digital technologies. Taking into account the laws, principles and properties of updating fixed assets formulated above will allow for the most rational "embed" of this updating in enterprise modernization based on digital platform solutions, and increase its efficiency. However, in order to achieve this in relation to Russian conditions, it is necessary to implement a set of fundamentally new decisions in the management mechanism, as well as legislative and normative measures to improve accounting, tax accounting, to overcome the complex of problems revealed in this study.

References

1. Abrahamyan SI, Fedotov AA (2015) Problems of modern engineering in Russia and approaches to their solution. Manag Econ Syst Electron Sci J 8(80):1999–4516
2. Anisimova MV, Zagoretska OY (2014) New methods of fixed assets reproduction at enterprises. Actual Probl Econ 156(6):190–196
3. Coreynen W, Matthyssens P, Bockhaven W (2017) Boosting servitization through digitization: pathways and dynamic resource configurations for manufacturers. Ind Mark Manag 60:42–53
4. De Reuver M, Sorensen C, Basole RC (2018) The digital platform: a research agenda. J Inf Technol 33(2):124–135
5. Ghosal V, Nair-Reichert U (2009) Investments in modernization, innovation and gains in productivity: evidence from firms in the global paper industry. Res Policy 38(3):536–547
6. Hsieh YJ, Wu YJ (2019) Entrepreneurship through the platform strategy in the digital era: insights and research opportunities. Comput Hum Behav 95:315–323
7. Kukartsev VV, Boyko AA, Antamoshkin OA (2018) The simulation model of fixed assets reproduction of mechanical engineering enterprises. In: Nagorny A, Nagorny V, Tisenko V (eds) International Russian automation conference, RusAutoCon 2018 (85017769). IEEE, Piscataway. https://doi.org/10.1109/rusautocon.2018.8501776
8. Nambisan S (2017) Digital entrepreneurship: toward a digital technology perspective of entrepreneurship. Entrep Theory Pract 41(6):1029–1055
9. Nambisan S, Siegel D, Kenney M (2018) On open innovation, platforms, and entrepreneurship. Strateg Entrep J 12(3):354–368
10. Nikolaeva VI (2017) The goals of modernization of fixed assets of the enterprise. In: Merzlikina GS (ed) Collection of materials of the XVII annual open competition of research works of students and young scientists in the field of economics and management "Green Sprout". Volgograd State Technical University, Volgograd, pp 70–72
11. Savinov Y (2017) International trade in machinery and equipment: dynamics and structure. Rus Foreign Econ Bull 1:32–47
12. Skuras D, Tsegenidi K, Tsekouras K (2008) Product innovation and the decision to invest in fixed capital assets: evidence from an SME survey in six European Union member states. Res Policy 37(10):1778–1789

13. Stolyarova MA, Shulgaty OL, Dzagoeva MR, Bestaeva LI, Kaitmazov VA (2017) Generalization of foreign experience in the reproduction and recording of fixed assets. Int J Appl Bus Econ Res 15(12):241–250
14. Stremmel SA (2017) The influence of the innovation process on the need to modernize the basic production assets of the enterprise. In: Merzlikina GS (ed) Proceedings of the XVII annual open competition of research works of students and young scientists in the field of economics and management "Green Sprout". VolgGTU, Volgograd, pp 82–83
15. Volkodavova EV, Zhabin AP, Yakovlev GI, Khansevyarov RI (2020) Key priorities of business activities under economy digitalization. In: Ashmarina S, Vochozka M, Mantulenko V (eds) Digital age: chances, challenges and future, vol 84. Lecture notes in networks and systems. Springer, Cham, pp 71–79. https://doi.org/10.1007/978-3-030-27015-5_9

Statistical Analysis of Scientific and Technological Potential of the Russian Federation

N. Kulikova and N. Persteneva

Abstract This paper discusses the development of digital economy in the Russian Federation. The purpose of the study was a statistical assessment of the scientific and technological potential of the regions of the Russian Federation during the formation of the digital economy. Scientific and technological potential manifests itself as the main link that ensures competitiveness, reveals the potential and independence of the state. The importance of this issue is confirmed by the fact that our country has adopted the national program "Digitalization of the Economy of the Russian Federation". In our study, special attention was paid to the formation of a system of statistical indicators of scientific and technological progress. During research, non-parametric statistical multivariate estimation methods, cluster analysis, differentiation estimation methods were used. Based on these methods, the regions of the Russian Federation were ranked according to the indicators of scientific and technological progress, and territorial regularity of their distribution was revealed. The results obtained allow us to characterize the regions of the country in terms of scientific and technological potential, as well as to develop strategic programs for the transition to digitalization of the economy.

Keywords Statistics · Digital economy · Scientific and technological potential · Cluster · Region

1 Introduction

Today, one of the global trends to which Russia has a direct relationship is the digitalization of the economy. The directions of the digital economy are progressively developing by the government of the Russian Federation, marking 2018 as the starting point for the implementation of the state program "Digitalization of the

N. Kulikova · N. Persteneva (✉)
Samara State University of Economics, Samara, Russia
e-mail: persteneva_np@mail.ru

N. Kulikova
e-mail: Nataliya.kulikova@gmail.com

S. I. Ashmarina and V. V. Mantulenko (eds.), *Current Achievements, Challenges and Digital Chances of Knowledge Based Economy*, Lecture Notes in Networks and Systems 133, https://doi.org/10.1007/978-3-030-47458-4_96

Economy of the Russian Federation" [8]. Obviously, such policy affects all spheres of life, which causes a 3-fold increase in the costs of developing the information and communication technology sector by 2024 compared to 2017 [7].

Assessment of scientific and technological potential is relevant in the framework of the digitalization of the economy. Firstly, due to the fact that the digitalization of the Russian economy can become a source of long-term economic growth. In terms of price level of 2015, by 2025, the increase in the share of GDP from increasing the efficiency of R&D and product development is expected to be in the range of 0.2–0.5%, and in general, the contribution of digitalization to the growth of the total expected GDP is estimated at 19–34% [6, 8]. Secondly, rapid advance of the world's leading economies in the digital economy, among all other factors, is directly related to the infrastructure of the scientific sector, the stimulation of research, the ecosystem of startups, favorable tax policy in relation to organizations characterized by high science-intensive production, etc. [1].

Scientific and technological potential is presented as a complex phenomenon that contains material and technical, personnel, organizational, information and financial components. World experience shows that in the era of digitalization, the creation of new scientific and technological infrastructure from scratch is more effective than the transformation of the existing ones [5, 9]. Therefore, the creation of such systems and structures requires high accuracy and productivity of research and development. Russia will need to put great effort in achieving the global level of development of both the scientific sector and the digitalization of the economy.

The purpose of our study is statistical assessment of the scientific and technological potential of the regions of the Russian Federation during the formation of the digital economy. The subject of the research is scientific and technological potential in the framework of digitalization. The object of the research is 80 subjects of the Russian Federation.

2 Methodology

The authors used the following statistical methods to achieve the goals of the research.

1. PATTERN (Planning Assistance Through Technical Evaluation Relevance) method.

This method belongs to the class of nonparametric methods of comparative analysis. Such methods have several advantages over traditional parametric ones. In particular, they can be used in relatively small samples; by standardizing the values of the initial indicators, the necessary level of information compression is achieved. In addition, they are simple enough to interpret and are not sensitive to measurement errors. The method is used for multidimensional grouping of Russian regions by the indicators of scientific and technological potential. The result of the calculation is the average integral estimate of scientific and technological potential—in the range from 0 (worst value) to 1 (best value).

2. Cluster analysis method.

Cluster analysis is the common name for the set of computational procedures used to create a classification, as a result of which clusters, that is, groups of similar objects, are formed. Clusters are created on the basis of the criterion of minimum distance between individual units in the space of several variables—the characteristics of these units.

For determining the optimal number of clusters, hierarchical clustering method in combination with Ward clusterization method was used. The algorithm for clustering is the divisible method of K-means.

3. Estimation of the level of differentiation between the regions of the RF.

For the assessment, both the indicators calculated by Rosstat [2] and the author's indicator were used. Rosstat indicators include the traditional ones—the quartile differentiation coefficient and the Gini coefficient.

We suggest an indicator, calculated according to the following formula:

$$STP = \frac{Iu}{Il} \tag{1}$$

where:

- Iu is the mean value of the aggregate indicator in the upper 25% group;
- Il is the mean value of the aggregate indicator in the lower 25% group.

The indicators were standardized as the relation of the absolute value of the indicator in a particular region of the RF to the average value of the indicator in the RF.

3 Results

Scientific and technological potential is an integral indicator, so it can be represented as a system. Given the research topic, it makes sense to include indicators that combine both the component of the scientific and that of technological potential and the digital economy, but within the framework of science.

As a basis for assessing the scientific and technological potential of the regions of the Russian Federation using the PATTERN method, 29 indicators characterizing the scientific and technological potential were used. The source of data for the assessment is the Rosstat database [2]. Thus, 80 subjects of the Russian Federation were evaluated, regardless of the presence of overshoot. An equal-interval grouping into 4 groups was made according to the criterion of the average integral assessment of scientific and technological potential.

Group 1 includes the following regions of the Russian Federation: Moscow, St. Petersburg and the Moscow Region. The city of Moscow stands out among the subjects of the aggregate, revealing a large gap between itself and both the city of

St. Petersburg and the Moscow region. The largest group was Group 3, occupying the medium position in terms of the integral indicator. The worst results are shown by such regions as the Republic of Ingushetia, the Sakhalin Region, Zabaikalsky Krai, the Orenburg Region, the Magadan Region, the Leningrad Region, etc.

For a more accurate presentation of the state of regions of the Russian Federation in the sphere of the scientific and technological potential, cluster analysis based on the same system of indicators was carried out. During the analysis, 5 factors were excluded as insignificant, namely: "the number of advanced production technologies", "the share of researchers in the number of employees performing research and development", "the share of the cost of the overall development of science in the internal cost of research and development", "the share of investment in reconstruction and modernization in the total volume of investment in fixed assets" and "the proportion of new technologies, in the total number of advanced production technologies developed."

In addition, Moscow, St. Petersburg, and Sevastopol were excluded from cluster analysis as having abnormal values. As a result of the analysis procedures, 5 clusters were formed. For a more accurate assessment of cluster rating, it seemed appropriate to rank the mean values of the clusters for each indicator, taking into account the nature of the factor. After this procedure, the ranks were summed up, and the final rank was awarded to the final values. The result was an expert definition of Cluster 1 as the best and Cluster 5 as the worst.

According to the results of the analysis, it is clear that the most numerous cluster is Cluster 3, which occupies the medium position in relation to the other clusters. Cluster 1 is the leader by a large margin. It includes the Vladimir Region, the Kaluga Region, the Moscow Region, the Tula Region, the Yaroslavl Region, the Rostov Region, the Republic of Tatarstan, the Perm Territory, the Nizhny Novgorod Region, the Samara Region, the Ulyanovsk Region, the Sverdlovsk Region, the Chelyabinsk Region, the Novosibirsk Region, the Tomsk Region, and the Khabarovsk Territory. This cluster is ahead of the previous ones on 11 of the available factors. Thus, for example, the regions included in this cluster are far ahead in terms of volume and share of internal expenditure on research and development, the number of patents granted for inventions, the share of innovative goods in the volume of exports and investment in fixed assets in the field of information and communication. The only indicator by which this cluster shows the worst result is the share of the costs of the overall development of science in the internal costs of research and development.

This kind of phenomenon is explained by the specifics of these regions: they show priority of expenditure on research and development in such areas as industrial production, agriculture and rational use of energy in comparison with the development of science in general. These regions have a substantial number of qualified scientific personnel, solid financial resources and high research effectiveness. Thus, they are able to redirect their funding onto development of high-tech approach in other types of activities. Cluster 1 also occupies the third and fourth positions in terms of the following indicators: "the share of researchers under the age of 39," "the share of researchers with a scientific degree" and "the share of completely worn-out fixed assets for information and communication activities". This situation indicates a low interest in the work in the research direction and the threatening state of fixed assets, especially given the declared goal of the transition of the economy to digitalization.

The smallest and the most underperforming cluster among the subjects of the Russian Federation in terms of scientific and technological potential is Cluster 5, accounting for slightly more than 11.6% of the total population. The cluster consists mainly of the regions of the North Caucasus Federal District, with the exception of the Republic of Khakassia and the Amur Region. Despite the fact that the scientific and technological potential of the cluster can be described as very low, it has several leading characteristics in the Russian Federation, namely: the predominance of the share of people with a scientific degree among researchers, the high level of expenditure on research and development aimed at developing the economy and a large share of expenditure of general development of science. Also, the share of completely worn-out basic funds in the field of information and communication is one of the smallest among the available clusters. You can notice that the parameters of Cluster 5 are opposite to those in Cluster 1.

To check the adequacy of the assessment of the scientific and technological potential of the constituent entities of the Russian Federation, it is necessary to compare the results of cluster analysis with the results obtained using the PATTERN method (Table 1).

According to the results of the comparison, it is clear that the subjects that fell into Clusters 2 and 5 are close to each other according to one or another criteria, namely, they are in the category of regions with low scientific and technological potential, therefore their rank fluctuates.

To present the state of the scientific sector and its weaknesses, it is necessary to pay attention to some indicators characterizing it. More than 20 factors under consideration are characterized by a large coefficient of variation, which indicates the presence of not only overshoots, but also inequality among the regions.

Thus, the difference between the smallest investment in fixed assets in professional, scientific and technological activities of 10% of the most developed regions exceeds the maximum value of investments of 10% in the less developed regions by almost 155 times. The same picture is observed in terms of investment in fixed assets by type of activity in the field of informatization and communications, showing an excess of 13 times. Such a discrepancy in many indicators may be due to the specialization of the regions. So, for example, in the Samara region, money will

Table 1 Composite rating of the clusters

Index	Cluster number				
	1	2	3	4	5
Mean aggregate estimate (Pattern)	0.116	0.077	0.094	0.109	0.079
Cluster rank according to the aggregate estimation results (Pattern)	1	5	3	2	4
Cluster rank according to the cluster analysis results	1	4	3	2	5

Source authors

be directed to space engineering and the automotive industry [3, 4, 9], and in the Krasnodar Territory, it will be given to financing the development of life sciences, rational environmental management, and the development of transportation systems.

Based on the methodology of Rosstat for assessing the degree of differentiation between the regions of the Russian Federation in terms of information development, an assessment was made of the degree of differentiation of subjects by scientific and technological potential [2, 6].

The indicators were standardized as the ratio of the absolute value of the indicator for a particular region of the Russian Federation to the average value of the same indicator in the RF. As a result, the calculations show a difference in the scientific and technological potential of the subjects of the Russian Federation by 2.3 times. However, if we calculate the degree of differentiation through the quartile differentiation coefficient, the difference is 1.6 times. The distribution of regions by the integral indicator of scientific and technological potential is close to uniform—the Gini coefficient is 0.187, although there is a slight discrepancy, as evidenced by the Lorentz curve.

The development of the subjects of Russia in the framework of scientific and technological potential is highly differentiated, which may be related both to the specialization of the regions and the length of the transition to the digital economy in connection with various initial conditions.

4 Discussion

The development of the scientific sector depends, to a large extent, on the starting conditions, such as geographical location, socio-economic policy, and living standards of the population.

There is a noticeable discrepancy between the regions of Russia in terms of the development of scientific and technological potential, which requires a special approach to the digitalization of the regions, taking into account their local features. A large role is played by statistical studies, which are designed to quantify the phenomenon. However, at the stage of developing the legislative framework and official methods for calculating indicators for monitoring the digitalization of both the economy in general and science in particular, assessing the scientific and technological potential presents certain difficulties.

In February 2019, Rosstat approved the methodology for calculating indicators within the framework of subprograms of national projects, but in reality the methods have not yet been developed for all indicators. The used system of indicators is not complete enough; it contains indicators developed by Rosstat and used to characterize the STP of the constituent entities of the Russian Federation.

Based on official data, we can conclude that the digitalization of the economy in foreign countries takes place more intensively and from an earlier date, so Russia, with its ambitious goal of becoming one of the 5 leading global economies, will have to make a lot of effort to implement the relevant federal programs.

5 Conclusion

Scientific and technological potential can be considered a synthesis of the resource potential and the result of the process of using these resources. Thus, it means combining the science personnel, the material and technological base and the financing of R&D in one group, and the effectiveness of R&D in another. Based on this, a system of statistical indicators is formed to describe the scientific and technological potential.

At the stage of the digital economy formation, the Russian Federation and its subjects have the most worn-out fixed assets in the field of informatization and communication in comparison with other types of activities, which can explain the high costs of digitalizing the economy.

The specialization of the subject of the Russian Federation, its primacy in a particular type of activity, significantly determine the direction of development of science and the priority of costs for research and development. Only 35% of the regions of the Russian Federation possess sufficient scientific and technological potential for the successful development of state policy.

References

1. Alcacer J, Cantwell J, Piscitello L (2016) Internationalization in the information age: a new era for places, firms, and international business networks? J Int Bus Stud 47(5):499–512
2. Federal State Statistics Service (2019) State statistical data on science and innovation in the RF. https://gks.ru/folder/14477. Accessed 10 Feb 2020
3. Firulev OV (2016) The role of innovation potential in the enterprises of space and rocket industry. New Sci Chall Oppor 1:181–187
4. Firulev OV, Erygin Y (2016) The role of innovative projects of advanced development in sustaining the innovation potential in the space and rocket industry. Bull Transbaikal State Univ 10(22):126–133
5. Massini S, Caspin-Wagner K, Chilimoniuk-Przezdziecka E (2016) Global sourcing and the unbundling of innovation: challenges and opportunities for emerging countries. In: Lewin AY, Kenney M, Murmann JP (eds) Building China's innovation capacity: overcoming the middle-income trap. Cambridge University Press, Cambridge, pp 267–297
6. Report on National Project "Digital Economy". Official website of the Ministry of digital development, communications and mass communications of the Russian Federation. https://digital.ac.gov.ru/upload/iblock/f4b/%D0%BE%D1%82%D0%BA%D1%80%D1%8B%D1%82%D1%8B%D0%B5%20%D0%B8%D0%BD%D0%BD%D0%BE%D0%B2%D0%B0%D1%86%D0%B8%D0%B8.pdf. Accessed 10 Feb 2020
7. Strategy of development of information society in the Russian Federation for 2017–2030, approved by the decree of the President of the Russian Federation from 09.05.2017 № 203. http://publication.pravo.gov.ru/Document/View/0001201705100002. Accessed 10 Feb 2020
8. The state program "Digitalization of the Economy of the Russian Federation", approved by order No. 1632-p of July 28, 2017. http://tadviser.com/index.php/Article:National_program_Digital_economy_of_the_Russian_Federation. Accessed 10 Feb 2020
9. Zysman J, Kenney M (2018) The next phase in the digital revolution: abundant computing, platforms, growth, and employment. Commun Assoc Comput Mach 61(2):54–63

Social and Cultural Consequences of Digitalization in Higher Humanitarian Education

A. Klyushina, O. Shalifova, L. Stoykovich, and G. Stoykovich

Abstract Digitalization covers all spheres of life in modern society, this is our objective reality. In our opinion, in terms of importance and influence on the development of society, digitalization can be compared with the scientific and technological revolution of the twentieth century. In our article, we summarize the results of studies on the impact of digital technologies on the process and the results of training students who major in humanities, conducted on the basis of SSUSSE from 2012 to 2019.

Keywords Digitalization · Digital technologies · Society · Digital culture · Higher humanitarian education · Electronic environment

1 Introduction

Nowadays, digitalization covers all spheres of life in modern society, this is our objective reality, the significance of which is still not fully understood by most people. Among the broad cross-sections of the population, the understanding of digitalization is not clearly formed, it is very vague or absent at all. As the survey showed, even students of higher educational institutions experience difficulties in determining the term digitalization. So, some people call digitalization a quick access to information resources (62%), others, believe, it's the science related to calculations (31%), and others think, that digitalization is the introduction of digital technologies, i.e. "Digitalization is digitalization…" (7%). Experts use this term in a more functional way,

A. Klyushina · O. Shalifova · G. Stoykovich
Samara State University of Social Sciences and Education, Samara, Russia
e-mail: klyushina@pgsga.ru

O. Shalifova
e-mail: shalifova@pgsga.ru

G. Stoykovich
e-mail: yugast@mail.ru

L. Stoykovich (✉)
Samara State University of Economics, Samara, Russia
e-mail: liliyastoikovich@gmail.com

S. I. Ashmarina and V. V. Mantulenko (eds.), *Current Achievements, Challenges and Digital Chances of Knowledge Based Economy*, Lecture Notes in Networks and Systems 133, https://doi.org/10.1007/978-3-030-47458-4_97

as complete automation of processes and stages of production, from the design of the product to its delivery to the consumer, as well as subsequent after sales service [8]. According to Katz, ITU expert, digitization refers to the transformations triggered by the massive adoption of digital technologies that generate, process, share and transfer information [7]. In the future, we will adhere to this definition, as part of the article.

In our opinion, in terms of importance and influence on the development of society, digitalization can be compared with the scientific and technological revolution of the twentieth century, however, it is only gaining speed both in technology development and the scope of production, communications, the economy, and in involving all the sections, ages and categories of the population. Only by 2025 it is expected that 40% of the Russian population will have digital skills [9]. Digitalization is in the center of attention of scientists of both technical and humanitarian specialties. A large number of works on this subject have been written [1, 3, 4, 6, 7, 9], the vast majority of them are written in the field of technical sciences, IT, computer technology. Foreign researchers are still leaders in this sphere. In our opinion, there are still not enough Russian studies on the topic, and in the available ones, the humanitarian aspect of this global phenomenon is not adequately described in comparison with the technical and hard sciences.

Digitalization in education can be considered as one of the most important areas of digitalization. Some specialists believe that digitalization of education is the availability of a sufficient number of computers, computer labs in educational institutions, and equipping the educational process with programs that provide access to various educational resources, electronic libraries to get acquainted with the results of modern scientific research [2]. Although, in our opinion, like many others, this definition is limited by the material side of the problem, while mastering, implementation and being fluent and feeling comfortable in this sphere by a person is not less important. In our article, we summarize the results of studies on the impact of digital technologies on the process and the results of training pedagogical student conducted on the basis of Samara State University of Social Sciences and Education (SSUSSE) from 2012 to 2019. Studies were carried out simultaneously in three interrelated areas:

1. Self-assessment of the degree of digital skills in students of humanitarian specialties.
2. Assessment of the formation of oral and written communication skills in the conditionally formal “pedagogical student” situation.
3. A comparative analysis of the results of text tasks using electronic and printed media in the process of learning a foreign language.

2 Methodology

In modern realities, studying is impossible without the use of mixed technologies, as well as all kinds of digital teaching aids, which implies a high level of digital skills for both students and teachers. For the first direction of the study, we made regular surveys of students of 1–4 courses of the humanitarian profile of SSUSSE in the period from 2012 to 2019. For the second direction of the study, we used the method of questioning students of 1–4 courses in the period from 2012 to 2019.

The third direction of the study we carried out as an experiment. In the period from 2012 to 2019, 4 groups of first-year students of the humanitarian profile of the "Foreign Language" direction were selected by the method of targeted sampling. The level of proficiency in the studied language ranged from B2 to C1 according to CEFR (Common European Framework, the official European scale for assessing level of foreign language skills), while the students themselves were unaware of participation in the experiment. The students received tasks for reading and understanding the text and grammatical transformations. Two groups worked with print media, the other two groups worked with electronic devices. The experiment included the timing of fulfilling tasks by every student. At the analysis stage, the following characteristics were taken into account: the current level of language proficiency, current academic performance, the information medium that the students used, the speed of completing tasks and the degree to which tasks were completed.

3 Results

Our study showed that students are fluent in the electronic environment, feel confident, highly value their capabilities in this sphere, the average rating of respondents is 8 on a 10-point scale. This is not only because of an exaggerated self-assessment, but also the result of educational activity in subjects teaching electronic technologies. In schools and universities, serious attention is paid to the formation of competencies in the field of digital technologies and the electronic environment, which is expressed both in special disciplines and in the steady and high-quality implementation of electronic technologies in all taught disciplines. At the same time, students openly acknowledge their dependence on electronic means, gadgets, and the Internet: 92% of respondents can do without a phone no more than 2–3 h (not to mention night time), 6%—a day, 2%—a few days. There was only one answer: I can do as long as I wish. The absence of a telephone or the Internet, according to the majority of respondents, causes feelings of panic, anxiety, isolation, loneliness.

According to the results of the survey, 95% of the students prefer live communication. However, the question of what is the real share of live communication in their life against the virtual, showed a ratio of 25%–75%, respectively. We would like to notice that despite all the unfavorable trends some traditional habits and preferences among students are still kept: 97% of respondents prefer printed—books to electronic ones, and they read e-books only when they have no opportunity to read printed books and under certain circumstances. It is generally recognized that computer games cause serious harm to social adequacy because the most part of them are cruel and contain elements of violence, they have a negative emotional background, players transfer the game to reality, connect it with the outside world and people, and the psyche. The tested group of students did not show new trends in this direction.

98% of respondents play computer games, 30% of them do that every day, the rest play 1–2 times a week. The decline in the cultural level of the younger generation is manifested in the inability to communicate correctly in the electronic environment. The communication of young people in various social networks, Viber, Whatsup, via SMS, etc., its cultural, social, linguistic aspects haven't been studied yet. As our personal experience shows, the "student-teacher" electronic communication, that takes place almost daily, a significant part of students do not follow the rules of ethics and use a rather primitive or even vulgar language. The absence of greetings, formal apply, simple etiquette forms in emails, the use of straightforward expressions, often containing vernacular, slang and even familiar expressions (I dropped my work to you, you are my scientific advisor, term paper, "kursach", but what do you mean? etc.), a direct imperative and questions (check my work, are you already through my work? check my work, give me a task, When will you be at work? Have you already read my work? etc.). In our opinion, can be explained by various reasons such as lack of upbringing and the sense of the electronic environment as a space without "extra" words and language and the feeling that it saves time, which seems the norm for many users, but it is not. The distant way of this communication, when the face and eyes of the partner are not visible, there is no mutual reaction, and communication has a surrogate way also contributes to this situation. There is an obvious shift in the value-semantic guidelines of functioning in the digital environment. Some scientists call this "digital multiculturalism" [4, p. 220]. As the researchers note, and we adhere to this position, digitalization and the establishment of these guidelines should become the most urgent social problem of the current period.

Analysis of the studies showed that students, who have excellent marks, completed their tasks approximately equally correctly, however, when they used paper textbooks the tasks were completed a little faster. The students with a medium and low level of knowledge showed different results: those who worked with paper textbooks completed tasks with higher results than those who worked with electronic devices. They have approximately the same run time. Of course, we do not pretend to be unconditional and finite, the lag of the second group may be explained by poor skills in using electronic means, in addition, the number of studied people is limited. To make the results of the investigation more valid we need a wider scale and more time. This is just the initial attempt of a serious research. But even this makes us think over whether it is worth to plunge ourselves in the electronic environment or whether to leave the place to materiality, reality. It seems promising to continue experiments in this direction.

4 Discussion

Rapid and overwhelming development of digital technology transforms the reality and some cultural norms are changing. Scientists note that this generation perceives the format of traditional culture with the established system of values through the prism of digitalization and that leads to such consequences as clip culture, TV culture, computer games and digital culture [5]. At the same time the change in an individual person, personality (a large or even the most part of the population), his adaptation to this new trend is much slower. That leads to psychological and social problems, to a human-computer conflict. G.L. Tulchinsky draws attention to this problem. He introduces the concept of sociocultural engineering as a separate type of integrated, interdisciplinary scientific approach aimed at scientific and expert reflection, forecasting and creating new relevant educational programs on this basis. New requirements for the basic skills of specialists in the labor-market are coursed by redistributing labor in the society. "These changes lead the society to a change in the standards of behavior and communication, the ways of perceiving and thinking, mastering the profession and requirements for educational technologies, transforming the worldview and generational habits" [13, p. 3]. Thus, we are talking about a change in the whole value-semantic paradigm of modern society, a change in the language of culture and ways of producing cultural meanings and images. The role of education cannot be overestimated because of new requirements to the skills of a modern man, as long as education is the key instrument to their formation.

Wide access to information is one of the main and indisputable advantages of digitalization in general and in the field of education. This promotes to accumulation and growth of knowledge and, accordingly, an increase in the intellectual level of the individual and society as a whole. Online learning is becoming more and more popular. It includes watching a video of a lecture course online, participation in seminars, webinars, and online courses. According to political scientists, sociologists, specialists in the field of education and information technology, these forms will gradually replace traditional methods of education [5]. Society and education are facing a serious question of whether we should abandon traditional forms at all.

The fact that in the near future Russian schools are planning to replace printed textbooks by electronic ones embedded in special certified devices makes us worry. In our opinion, this fact, will lead to the aggravation of influence of negative factors of digitalization on the child. The textbook is the most important, crucial and central element of the school space, a unifying factor, an element of culture, in which educational role plays both the content and the formal features. Color design, pictures, structure, print quality, and even paper quality can influence on the understanding of educational material. One of the significant advantages of paper textbooks is that the student cannot be limited in time during the day, while the use of electronic textbooks must be limited. Foreign researchers also say that printed books have advantages and that it is more efficient to read printed text compared to reading an electronic device. The results of a large-scale study conducted by Delgado et al. point out that there is huge necessity of using paper books and limit "electronic" reading due to its negative impact on the psyche and health [3]. At the same time, the authors acknowledge the inevitability of using electronic means and call for a reasonable balance. So, one of the problems, connected with digitalization is the widespread use of electronic means and its effect on the student's ability to work with texts. The experience of working with students of various specialties (economic, linguistic, pedagogical, legal science, etc.) shows, that students face difficulties in mastering large texts, use fast and superficial reading, do not try to understand the content, do not accept long sentences [6, 10, 12, 13].

Another serious problem in the scientific society is the so-called digital inequality. Some sections of the population have poor availability of digital resources [3, 14] due to geographical factors (rural areas, suburbs), social factors (age groups, some professional groups) and property factors (families with low-income, asocial individuals). People who do not know how to use a computer, the Internet, begin to feel themselves excluded from the general course of life, handicapped, marginal. In future this process may become stronger.

The universal computerization of education solves the problems of digitalization at all its stages—from elementary to higher education. Computer skills are a development factor. As a result, at the same time, social groups that quickly and easily adapt to computers and digital technologies, and have programming knowledge, such as young people, pupils, students gave the advantage. Undoubtedly, the future belongs to those who are fluent in digital technologies. In this regard, social groups of poorly educated people, elderly people, who has no opportunity to increase their skills in technology, turn out to be problematic, and all their involvement in digitalization is watching TV. This is a serious problem for society. The study of these problems, the identification of its various aspects, the prevention of undesirable consequences have the crucial importance for scientists. This problem must be solved also by means of education.

5 Conclusion

The importance of implementing of digitalization is confirmed by the state policy, according to which the digital economy is an economic activity, the key production factor of which is digital data, which contributes to the formation of information space, taking into account the needs of citizens and society in obtaining high-quality and true information, the development of the information infrastructure of the Russian Federation, the creation and application of Russian information and telecommunication technologies, as well as the formation of a new technological basis for the social and economic sphere [11, 12].

There is no doubt that digitalization is an essential part of our life, a factor of progress and the community development, it has a lot of perspectives and gives many advantages in different areas of life. More and more people and types of activities are involved in it. Like any phenomenon, it has its pros and cons. As we know, the scientific and technological revolution has allowed mankind to develop rapidly and improve their lives, but the rapid development of industry has led to a violation of the ecological condition of the planet and the need to limit industrialization. The consequences of digitalization have not been studied yet and we only can predict them. But it is necessary to think about it in order to be ready to solve, prevent or minimize possible problems.

The convenience and speed of performing a variety of actions and tasks at all levels: at the level of government, industry, institutions, at the level of an individual—household level, personal space, level of professional activity are advantages of digitalization. Of course, it is mostly used in technical fields, in the economy, in the media, in managing the country at all levels, in trade, medicine, and banking operations. The advantage of digitalization in education is the use of computer technology that makes the process of education more effective both in studying different disciplines, especially hard sciences and natural sciences, and in controlling the level of knowledge.

At the moment, many of the consequences of digitalization can only be predicted. Our goal and prospect of research is to expand the experimental base of research, to study it deeper and identify trends of the impact of digitalization on society, on students in order to create the basis for developing special methods of using digital technologies in education, taking into account all the risks of digitalization. Awareness of the inevitability of digitalization makes us look for ways of preventing its negative effects and developing immunity in the younger generation and society. In our opinion, we cannot allow a complete rejection of the traditional ways of educational and cultural values, and the main task is to find a reasonable balance between digitalization and tradition. Critical outweighing of digitalization can lead to the degradation of human activity, and our task is to prevent this.

References

1. Afanasyev A (2018) Digitalization of education: all the disadvantages of the electronic school. VC. https://vc.ru/flood/43800.pdf. Accessed 29 Jan 2020
2. Aleksankov AM (2017) The fourth industrial revolution and the modernization of education: international experience. Strateg Prior 1:53–69
3. Delgado P, Vargas C, Ackerman C, Salmerón L (2018) Don't throw away your printed books: a meta-analysis on the effects of reading media on reading comprehension. Educ Res Rev 25:23–38
4. Gegenfurtner A, Ebner C (2019) Webinars in higher education and professional training: a meta-analysis and systematic review of randomized controlled trials. Educ Res Rev 28:100293
5. Gnatyshina EV, Salamatov AA (2017) Digitalization and formation of digital culture: social and educational aspects. Bull Chelyabinsk State Pedagog Univ 8:19–24
6. Jeong H, Hmelo-Silver CE, Joc K (2019) Ten years of computer-supported collaborative learning: a meta-analysis of CSCL in STEM education during 2005–2014. Educ Res Rev 28:100284
7. Katz R (2017) Social and economic impact of digital transformation of the economy. GSR-17 Discussion Paper. https://www.itu.int/en/ITU-D/Conferences/GSR/Documents/GSR2017/Soc_Eco_impact_Digital_transformation_finalGSR.pdf. Accessed 02 Feb 2020
8. Kudlaev MS (2018) The process of digitalization of education in Russia. Young Sci 31:3–7
9. Lisenkova AA (2018) Challenges and opportunities of the digital age: the sociocultural aspect. Lib Arts Russ. 7(3):217–222
10. Oganov SR, Kornev AN (2017) Oculomotor characteristics as an indicator of the formation of written text analysis skills in children 9–11 and 12–14 years old. Spec Educ Sci Methodol J 3(47):112–122

11. Shlee IP, Volgin Y (2016) The formation of professional and communicative competence among employees of internal affairs bodies as a factor in the success of their professional activities. Vocat Educ Russ Abroad 2(22):144–149
12. Strategy of development of information society in the Russian Federation in 2017–2030, approved by the decree of the President of the Russian Federation № 203 dated 09.05.2017. http://static.government.ru/media/files/9gFM4FHj4PsB79I5v7yLVuPgu4bvR7M0.pdf. Accessed 24 Jan 2020. (in Russian)
13. Tulchinsky GL (2017) Report "Socio-cultural engineering and digitalization" at a meeting of the Academic Council of St. Petersburg State University on December 25, 2017. https://spbu.ru/openuniversity/documents/itogi-zasedaniya-uchenogo-soveta-spbgu-69. Accessed 05 Feb 2020. (in Russian)
14. Volkova T, Pavlov G, Schlee IP (2018) Ecological and legal regime of the subsoil use (by the example of Kuzbass coal industry). In: Tyulenev M, Zhironkin S, Khoreshok A, Vöth S, Cehlár M, Nuray D, Abay A et al (eds) Proceedings of the IIIrd international innovative mining symposium. E3S web of conferences, vol 41. EDP Sciences, Les Ulis, p 02027. https://doi.org/10.1051/e3sconf/20184102027

The Digital Economy, Cyber Security and Russian Criminal Law

S. P. Bortnikov and A. V. Denisova

Abstract All branches of law are constantly under the influence of the social environment, especially national politics and economy, and therefore the author has set the goal of studying the impact of the development processes of the digital economy of the Russian Federation on the Russian criminal law as an important component of cyber security. The authors describe the problems hindering the development of the Russian digital economy, which can be resolved using criminal law instruments. The study revealed the gaps in the Russian criminal law regarding the socially dangerous acts committed with the use of modern information technologies. The paper shows the importance of intersystem relations to the national digital economy and Russian criminal law for lawmaking work and their impact on the effectiveness of the branch. The authors proposed to criminalize some socially dangerous acts committed with the use of modern information technologies, considering the traditions of national lawmaking and the successful experience of the international community and the foreign countries in countering of cybercrimes.

Keywords The digital economy · Cyber security · Russian criminal law · Cybercrimes · Informational economic crimes

1 Introduction

National politics is the dominant, defining phenomenon in the processes of lawmaking and law enforcement, it entangles the law system with its strong and rigid links. Therefore, it is so important today to take into account its priority areas related to the increasingly active implementation of information and communication technologies, the development of the information society in the Russian Federation and

S. P. Bortnikov (✉)
Samara State University of Economics, Samara, Russia
e-mail: serg-bortnikov@yandex.ru

A. V. Denisova
University of Prosecutor's Office of the Russian Federation, Moscow, Russia
e-mail: anden2012@yandex.ru

S. I. Ashmarina and V. V. Mantulenko (eds.), *Current Achievements, Challenges and Digital Chances of Knowledge Based Economy*, Lecture Notes in Networks and Systems 133, https://doi.org/10.1007/978-3-030-47458-4_98

the formation of a national digital economy [3]. Considering the fact that law regulation is a part of state administration, and resources of the branch are the instruments of state management, with can solve complex tactical and strategic tasks of the state, there is no doubt that the law will be demanded in solving problems of ensuring cyber security in the territory of the Russian Federation.

The modernization of the national economic system raised the question of improving measures to ensure information security in all sectors of the economy. According to a survey of two-thirds of Russian companies, the number of crimes in the digital environment has increased by 75% over the past 3 years, and it will only increase in connection with the ongoing digitalization of the country's economy [1].

In this regard, the National Program "Digital Economy of the Russian Federation" [6] focuses on the issues of criminal law ensuring the protection of the rights and interests of the individuals, business and the state against threats to information security in the digital economy. The article 17.17 of this document indicates the need to criminalize new types of socially dangerous acts committed using modern information technologies. The main research question is what should be these new crimes. Purpose of the research is to give a solution to the cyber security problems in the digital economy that will be resolved using criminal law instruments. The study should reveal the gaps in the Russian criminal law regarding the socially dangerous acts committed with the use of modern information technologies. The research will show the significance of intersystem relations to the national digital economy and Russian criminal law for lawmaking work and their impact on the effectiveness of the branch.

2 Methodology

The methodological framework of this study constitutes a set of methods of scientific knowledge, among which the main place is occupied by the methods of systematic, analysis and comparative law. The research is based on national and foreign legislation, international legal acts and opinions of the competent scientific community on the interaction of economic and legal systems, processes and the results of their mutual influence. The author conducted comparative research to know whether and how the cybercrimes are regulated differently in Russia and other countries.

3 Results

3.1 Cybercrime Law of Foreign Countries

The Ministry of the Interior of the Russian Federation is obliged to develop draft of new crimes committed using information technology, by June 30, 2020. For the successful realization of this order it should analyze the experience of foreign countries, which have in the legislation the so-called "Informational economic crimes". The articles about this crimes appeared several decades ago (e.g., in the USA the Computer Fraud and Abuse Act was adopted in 1984, in the UK the Computer Misuse Act—in 1990).

Most of the foreign countries that have the articles about "information economic crimes" in the legislation criminalized the following types of socially dangerous acts: illegal access to a computer system; illegal actions related to entering into a part of a computer system or into a computer system without permission or legal justification; hacking; illegal interference with a computer system; illegal actions that interfere with the operation of a computer system; denial of service attack, damage to a computer system; illegal interference with computer data (actions related to damage, deletion, deterioration, alteration or blocking of computer data without permission or justification; deleting files of computer systems without permission); illegal interception or access to computer data (illegal actions related to gaining access to computer data without permission or justification, including receiving data during the transfer process, which is not designed to be public, as well as receiving computer data (such like copying data) without permission, recording the transfer of data without the right to do so in a wireless network, copying computer files without permission, and others [11].

Thus, according to article 202 of the German Criminal Code, the person's unlawful receipt of computer data that were not intended for him and were protected from unauthorized access, for the purpose of extracting benefits for himself or for a third person, is punishable by imprisonment for the term up to three years. Erasing, destroying, disabling, changing computer data or attempting to commit such actions is punishable by a fine or imprisonment for the term up to two years. Paragraph b of article 303 of the German Criminal Code criminalizes "DNS attacks" (computer sabotage) and the creation of malware. Computer sabotage means interference with data processing that could cause significant harm to an enterprise, government agency, or business. An appropriate act may be carried out by destroying, damaging, disabling, modifying a computer system, or interfering with data transmission [9].

The Dutch Government criminalized the intentional, to benefit for itself or for a third party, use of technical devices to intercept or record data from telecommunication systems or connected equipment, if the data is not intended only for the relevant person (Article 139c of the Criminal Code of the Netherlands). The contributors to this crime are criminally liable for supplying funds for the illegal interception and recording of data transmitted via telecommunication or automated systems (Article 139d). Persons involved in the crime who possess data and know or should know that ones were obtained as a result of illegal listening, recording or interception of data from automated systems or telecommunication systems, are also criminally liable (Article 139e). Also, according to Dutch law, computer crimes include: unauthorized access to computer networks; unauthorized copying of data; computer sabotage; spread of viruses; computer spying. Some articles of the Criminal Code of the Netherlands, providing for responsibility for committing "common" crimes (extortion, fraud, forgery, etc.), were added with explanations allowing the use of them to combat cybercrimes [8].

3.2 *Russian New Cyber Security Law*

The Russian legal system is the main reference point for the criminalization of new offenses committing with the use of modern information technologies, for the needs of national law and enforcement. First of all, the legislator should take into account the peculiarities of national legal technology, the existing criminal law norms and the practice of their application. Secondly, the legislator should consider the successful experience of the international community and some foreign countries in countering attacks on information security. These government measures will improve the protection of systems, networks and software applications from digital attacks (national cybersecurity) and will contribute the fight against transnational cybercrimes.

In most cases both in Russia and abroad, criminals use "cyber weapons" in the economic activity of business entities to extract material benefits or to cause property damage to victims—bona fide users of information and communication technologies. Considering that these crimes are dangerous not only for citizens, society, business and the state, but also threaten the security infrastructure of the digital economy of Russia, it is advisable to recognize them as informational economic crimes [10]. This group of crimes, locating at the junction of the institutes of economic crimes and cybercrimes, is undoubtedly related to the "young" sub-sector of "economic criminal law" in the Russian criminal law system, which protects investments, credit relations, the consumer market and reduces economic costs, making the market more efficient [4, pp. 961–969; 5, pp. 48–58].

Philosophers define formation of something as a directed realization of an internal goal [7, p. 636]. From our point of view, the formation of the "economic criminal law" sub-sector in Russian criminal law is determined by the need to realize the following goals—the beneficial effect of the criminal law on the state's economic development, including the development of the national digital economy, creating favorable conditions for honest market participants and minimizing risks and threats to their information security.

Currently, the state quite seriously controls the modern economy through legal facilities. Consequently, the law also has certain opportunities for influencing economic activity, the impact on it. Moreover, an efficiently functioning national economy requires effective criminal law protection of its key systems including national cybersecurity.

The Doctrine of Information Security of the Russian Federation [2] states that now the number of cybercrimes is increasing on the territory of Russia, especially in the credit and financial sphere, the number of crimes related to the violation of the constitutional rights and freedoms is increasing too, especially violation of privacy, personal and family secrets in the processing of personal data using information technology. Moreover, the methods of committing such crimes are becoming more sophisticated. In addition, experts note an increase in coordination of computer attacks on objects of critical information infrastructure of the Russian Federation, increased threats of using information technologies to damage the sovereignty, territorial integrity, political and social stability of the Russian Federation. Therefore, one of the strategic goals of ensuring the information security of the Russian Federation is to increase the effectiveness of prevention of offenses committed using information technology, and combating such offenses. However, concrete ways to realize this goal are not indicated anywhere, how they will be developed and implemented by the responsible public authority (in this case, the Ministry of Internal Affairs of the Russian Federation), and how this will affect on the criminal law—we can only guess.

The state maintains the stability of various dynamic systems for the sake of public interests, makes responsible decisions about the objects managed through its policy. Legal regulation is the part of public administration, but at the same time it is under the state control. Ideally, this impact on the criminal law should develop it, support and optimize the system characteristics of the branch, and produce structure effects on it.

However, in practice, sometimes the controlling effect of the state on the criminal law can "undermine" characteristics of the branch and reduce their "efficiency". When external influence is constantly carried out on the criminal law and its elements, "external" information is introduced into the branch, pushing it to certain changes. In such cases, the concrete plan should be available it should provide for the possibility of reorganizing the branch to make internal changes tailored to the needs of social environment. Therefore, it is so important for law-enforcement agencies not only to describe in general terms the directions of the criminal policy for ensuring information security, and to indicate the need of development some abstract measures to improve the criminal law, but also to provide specific guidelines for these activities.

4 Discussion

Cyber security problems impede the development of the digital economy of Russia and include the problems of ensuring human rights in the digital world, the identification (correlation of a person with his digital image), the safety of the user's digital data, the problems of ensuring citizens' trust in the digital environment; the growth of cybercrimes, including international crimes; new threats to individuals, businesses and states associated with trends towards the construction of complex hierarchical information and telecommunication systems that widely use virtualization, remote (cloud) data storages and heterogeneous communication technologies and terminal devices; enhancing the capabilities of external information and technical impact on the information infrastructure, including critical information infrastructure. Some of these problems clearly relate to the jurisdiction of Russian criminal law and should be resolved using the arsenal of appropriate resources of the branch.

5 Conclusion

Finally, we conclude that Russian criminal law is closely linked by intersystem interactions with other social phenomena and processes (first of all with politics and economy). These relations are even more important than the relations of criminal law with other legal phenomena, as the branch is permanently updated because of social pressure. Secondly, the quality of relations of Russian criminal law with the national economy and their consideration in the process of legal reform are the significant conditions for ensuring the effectiveness of the criminal law in the economic sphere. Thirdly, Russian criminal law should be recognized as an important component of cyber security which one is the state of security of systems, networks and software applications from digital attacks aimed at gaining access to confidential information, changing and destroying it, at extorting money from users, or at disrupting the normal activity of companies. Due to the future amendments Russian criminal law will be more successful in protection of the individuals, society and the state from internal and external information threats, ensuring sustainable socio-economic development of the Russian Federation in the digital space.

References

1. Decree of the Government of the Russian Federation of July 28, 2017 № 1632-r "On Approving the Program "Digital Economy of the Russian Federation" (expired). https://www.garant.ru/products/ipo/prime/doc/71634878/. Accessed 08 Feb 2020. (in Russian)
2. Decree of the President of the Russian Federation of December 5, 2016 № 646 "On approval of the Doctrine of information security of the Russian Federation". http://base.garant.ru/71556224/. Accessed 08 Feb 2020. (in Russian)
3. Decree of the President of the Russian Federation of May 09, 2017 № 203 "On the Strategy for the Development of the Information Society in the Russian Federation for 2017–2030". http://kremlin.ru/acts/bank/41919. Accessed 08 Feb 2020. (in Russian)
4. Esakov GA (2013) Economic criminal law: concept, content and prospects. Lex Russica 9:961–969 (in Russian)
5. Klepitsky IA (2013) Economics and criminal law. Law 8:48–58 (in Russian)
6. National Program "Digital Economy of the Russian Federation" (approved by the Presidium of the Presidential Council for Strategic Development and National Projects, 06/04/2019). https://digital.gov.ru/uploaded/files/natsionalnaya-programma-tsifrovaya-ekonomika-rossijskoj-federatsii_NcN2nOO.pdf. Accessed 08 Feb 2020
7. Prokhorov AM (ed) (1983) Philosophical encyclopedic dictionary. Soviet Encyclopedia, Moscow (in Russian)
8. The Criminal Code of the Netherlands of 3 March 1881. https://www.legislationline.org/documents/section/criminal-codes/country/12/Netherlands/show. Accessed 08 Feb 2020
9. The German Criminal Code of November 13, 1998. https://ec.europa.eu/anti-trafficking/sites/antitrafficking/files/criminal_code_germany_en_1.pdf. Accessed 08 Feb 2020
10. Turyshev AA (2006) Information as a sign of crimes in the sphere of economic activity. Academy of Economic Security of the Ministry of Internal Affairs of Russia, Omsk. (in Russian)
11. UNODC (2015) International classification of crimes for statistical purposes: Option 1.0. United Nations Office on Drugs and Crime, Vienna

Cryptocurrencies in the Regulatory Field of International Organizations

Ilya Lifshits

Abstract The widespread use of virtual currencies has raised a question of the perimeter and ways for regulating transactions with these assets. First of all, there was a problem of determining the subject of regulated transactions, as well as a problem of minimizing risks of laundering criminal proceeds, consumer fraud, and negative impact on the financial stability. The relevant agenda has been mastered by international financial organizations, which have developed a number of recommendations, reports and analytical references using a soft legal method. These materials can be used by states to develop a model for regulating transactions with crypto assets. The recipients of these approaches are not only states, but also integration associations that have, like the EU, or seek to have, like the Eurasian Economic Union, the appropriate competence. It is obvious that the application of legal regulation of operations with virtual assets under the influence of international organizations will lead to the loss of advantages of such assets that cause their popularity at present.

Keywords Crypto currency · Virtual assets · Tokens · Taxation · EU · EAEU

1 Introduction

The global financial architecture formed after the 2008–2010 crisis can be represented as a combination of two components. The first of them, the regulatory one, covers international legal regulations, as well as soft-legal recommendations, called "international financial standards". The second, institutional component, includes a network of bodies (*standard setting bodies—SSBs*) that exist in the form of international intergovernmental or non-governmental organizations and forums that develop and monitor the implementation of international financial standards in national legal systems. A certain challenge for the global financial architecture was the emergence and rapid development of operations with cryptocurrencies based on the Blockchain

I. Lifshits (✉)
Russian Foreign Trade Academy, Moscow, Russia
e-mail: lifshitsilya@yandex.ru

MGIMO MFA of Russia, Moscow, Russia

S. I. Ashmarina and V. V. Mantulenko (eds.), *Current Achievements, Challenges and Digital Chances of Knowledge Based Economy*, Lecture Notes in Networks and Systems 133, https://doi.org/10.1007/978-3-030-47458-4_99

technology. On the one hand, it is obvious that cryptocurrencies have the potential to have a beneficial impact on the economy, a new digital technology, a tempting opportunity to attract investment in the financial sector, and a way to get rid of unnecessary intermediaries in payments. On the other hand, a newcomer is fraught with many dangers, the most important of which are risks of using crypto assets in criminal activities [17], in particular, money laundering and corruption, the creation of financial pyramids, as well as the obvious weakening of state influence on the financial system. The report, prepared under the auspices of the OECD, indicates that corrupt behavior includes fraudulent acts, market abuse, theft, hostage-taking, black market operations, terrorist financing, and tax evasion [20]. Despite the obvious dangers, international institutions have been trying for the past few years to integrate crypto currencies into the global financial architecture, "digest" them, and adapt them to the existing regulation.

2 Methodology

This article investigates the current position of crypto currencies in the regulatory field of international bodies and some integration associations (the EU and the EAEU) on the basis of the historical, comparative legal method, the method of dialectical connection between logical and historical ways of knowledge, the concept of a systematic and comprehensive research. The used research methods are a system of philosophical, general scientific and special legal means and methods of knowledge that provide objectivity, historicism and comparativism of the study of international law and the law of integration associations.

3 Results

Regulation of operations with a certain asset involves its *definition*, which in the case of crypto currencies, due to the novelty of the phenomenon, presents a certain problem. This is why the publication of a glossary of terms related to this phenomenon was the first task that was implemented by the Financial Action Task Force (FATF) in 2014 [12]. It is worth noting that the definition of concepts related to crypto assets at the level of international institutions was first given by a body whose goal is to counter money laundering. A virtual currency (this generic term is used) is defined as a digital representation of value that can be digitally traded and functions as (1) a medium of exchange and/or (2) a unit of account and/or (3) a store of value, but does not have a legal tender status… in any jurisdiction. Thus, the definition indicates only digital characteristic of the object, and then lists the usual properties of money, indicating that the virtual currency is not a legal, i.e., state-approved means of payment. The FATF defines crypto currency as a math-based decentralized, convertible virtual currency that is protected by cryptography. In 2019, the Eurasian Economic Union prepared a report on the development of crypto currencies and the Blockchain technology, with a glossary attached [8].

A digital asset is understood as "property" in the electronic form, the creation and use of which is carried out using digital technologies, a digital sign (token) is a type of digital asset, a tool to legitimize an obligation and other rights, and a cryptocurrency is a type of digital sign (token), which is an entry in the register of transaction blocks (Blockchain), other distributed database and is accepted as a means of exchange and (or) a unit of accounting and (or) a means of storing (accumulating) value [8]. The authors of the study, carried out by the Cambridge Centre for Alternative Finance, note that there is no uniform terminology, and regulators use different terms, but the "virtual currency" has been the most popular term overall, although the terms "crypto currency" and "digital currency" have often been used interchangeably [1].

In June 2019, FATF issued the Guidance for a risk-based approach to virtual assets (VA) and service providers associated with virtual assets (VASP) [13], and even earlier, in October 2018, the main document of this body—Recommendations on combating money laundering and the financing of terrorism—was also changed. In recommendation No. 15, there was an indication that countries should ensure that countering money laundering and terrorist financing (CML/FT) regulation, licensing or registration are applied for providers of services related to virtual assets in order to comply with measures taken by the FATF [14]. The corresponding change was enshrined in a resolution of the UN Security Council, which "urged" all states to comply with the comprehensive international standards [24]. In the communiqué of the G20 meeting of finance ministers and Central Bank governors in Fukuoka the intention was also highlighted to apply the FATF recommendations on virtual assets and related providers [22]. As it is indicated in the explanatory notes, virtual assets should be considered as "property", "proceeds", "funds", "funds or other assets" or other "corresponding value" [14], i.e., for purposes of combating money laundering, virtual assets are given a broadest meaning. In addition, the recommendations require service providers to deanonymize participants in payments in virtual currency [14], which, to a large extent, deprives crypto currency operations of their main advantage.

The Financial Stability Board (FSB) issued a document in 2018 examining the impact of crypto assets on the financial stability in the world [15]. The Board notes that due to the small size of the market, crypto assets do not pose a threat to the global financial stability, but require certain measures from regulators. The main conclusion of the report is that such assets, devoid of the main attributes of sovereign currencies, are not a common means of payment, a stable store of value and a mainstream unit of account. In another document, the FSB reviews approaches to regulating crypto assets by 25 states and 7 international financial organizations [16]. For example, the International Organization of Securities Commissions (IOSCO) provides the development of methodological materials related to platforms for trading crypto assets [19].

The first EU institution that addressed the issue of virtual currencies at the official level was the European Central Bank, which in a 2012 report pointed out the lack of legal regulation as a distinctive feature of the circulation of such currencies [9]. The report was released 3 years after the appearance of Bitcoin, the most popular and largest crypto currency in the world by capitalization (for example, the capitalization of Bitcoin is more than 165 billion dollars at the beginning of March 2020, or more than 63% of the total capitalization of crypto currencies). In 2014, the European Banking Authority has published its opinion, proposing a series of measures designed to prevent the use of virtual currencies for illegal activities, deanonymization of operations, as well as regulating providers of crypto currency transactions similar to financial services, i.e. prohibiting market manipulation, introducing minimum capital requirements, etc. [5].

In the 2019 report, the European Banking Authority defines crypto currencies as a private asset that depends primarily on cryptography and distributed ledger technology, as a part of their perceived or inherent value, and crypto assets include payment/exchange tokens, investment-type tokens, and tokens that provide access to a good or service [6]. The definition of a crypto asset, which is given in the European Central Bank report (ECB's 2019), emphasizes that such an asset is not and does not represent a financial claim or obligation of any particular entity [10].

The European Banking Authority emphasizes that, as a rule, crypto assets are outside the scope of financial services regulation in the EU, which leads to the emergence of different approaches to the legal regime for transactions with such assets in the member states. Such differences may be an obstacle to functioning of the EU single market, in which each participant is granted equal rights [6]. The ECB further points out that under the existing regulatory infrastructure, crypto assets cannot be included in the settlement system, since they are not qualified as securities, in addition, the existing system of capital requirements for banks and investment companies is not adapted for crypto assets; however, the risks to the financial stability are still insignificant due to the small spread of relevant transactions [25].

Another regulatory challenge is the issue of taxation of transactions with crypto currencies. The Court of Justice of the European Union in the *Hedqvist* case in 2015 considered the tax regime of the most popular crypto currency Bitcoin in relation to exchange operations [7]. The essence of this case was as follows: Mr. Hedqvist, intending through his firm to provide services for the exchange of traditional currency to the virtual currency "Bitcoin" and back, applied for clarification to the Swedish Revenue Law Commission, and received a preliminary decision to exempt such operations from VAT. The Swedish tax authority (Skatteverket) challenged this preliminary decision in the Supreme Administrative Court of Sweden (Högsta förvaltningsdomstolen), which referred to the EU Court the request for interpretation of the EU Directive on the common value added tax system [2]. The referring court gave a legal description of the virtual currency, defining it with reference to the ECB report as a type of unregulated digital currency that is issued and controlled by its developers and accepted by members of a specific virtual community. Virtual currencies are similar to other convertible currencies in terms of their use in the real world, but they differ from electronic money in the sense of the relevant EU Directive, since virtual currencies are not expressed in traditional monetary units [7].

At the same time, the virtual currency Bitcoin is not a material object, therefore, operations with it are not subject to the regime of delivery of goods for VAT purposes, but to the regime of delivery of services [7]. Thus, operations for exchanging crypto currencies for fiat money for remuneration are the delivery of services for the purposes of the VAT Directive, but such services are exempt from VAT as operations involving currency, banknotes and coins used as a legal means of payment (article 135(1)(e) of the Directive) [7]. The latter conclusion was made based on the repeated thesis that the virtual currency has no other purpose than to be a means of payment [7]. At the same time, the court quite clearly refused to exempt transactions with Bitcoin from VAT on other grounds: as transactions with deposit and current accounts, payments, transfers, debts, checks and other negotiable instruments (article 135(1)(d) of the Directive), as well as transactions with shares, shares in companies and associations, debt obligations and other securities (article 135(1)(f) of the Directive) [7]. It is obvious that taking into account the value of the court's decision in this case, it resolves a rather narrow issue of taxation for exchange operations with a monetary token—Bitcoin. The solution left out the fundamental problems of the procedure for taxing mining operations, payments in crypto currency for taxable goods and services, and issues of taxing transactions with other types of tokens, such as property and investment ones. As the researchers point out, this decision only opens up a discussion about taxation of transactions with crypto assets [21].

To date, the European Union has established a legal regime for virtual currencies only for the purpose of countering money laundering. Thus, with reference to the FATF in the Directive of the European Parliament and the Council on countering money laundering in the financial system in May 2018, changes were made to define a virtual currency and an e-wallet storage service provider [4], to establish obligations of member states to ensure the registration of such providers, as well as virtual and fiat currency exchange enterprises, and the identification of their managers and beneficial owners [4]. It seems that it is the FATF standards that will force all countries (and integration associations, if they have the appropriate competence) to introduce rules on virtual assets into national legislations. And the change in legal acts on anti-money laundering will necessarily entail the establishment of a regime for regulating such assets in other legislation areas. Within the framework of the Eurasian Economic Union, as it is indicated by the EEC, to date, no agreed vision has been developed for understanding the essence and regulation of crypto currencies [8]. At the same time, in the Republic of Belarus, a member of the EAEU, the Decree of the President of the Republic at the end of 2017 defined the key terms [3] and the main development institute in this area—a high-tech park.

4 Discussion

International financial organizations, in accordance with their specialization, have started developing approaches to regulating virtual assets over the past 5 years. First of all, the main problem points that require immediate intervention were identified: the use of crypto currencies for laundering criminal assets, high volatility of crypto assets, the possibility of building financial pyramids, taxation of exchange operations and mining, and risks to the financial stability. Obviously, states and integration associations are actively cooperating with international financial organizations to develop effective approaches to regulating crypto currencies, which will significantly reduce risks of using new technologies.

The Eurasian Economic Commission notes an important role of international organizations in the process of forming a system for regulating crypto currencies and the Blockchain technology. These organizations, as it is indicated by the EEC, are engaged in developing recommendations in the interests of common understanding of the essence of crypto currencies by all countries [8]. The expert committee of the European Commission, which developed at the end of 2019 recommendations for regulating FinTech (financial technologies) in the EU, also involves a close cooperation between European regulators (European Supervisory Authorities and the European System of Central Banks) and international bodies that develop standards in the financial sector [11]. Thus, two integration associations, faced with the appearance of an unknown phenomenon in the form of operations with virtual currencies, are moving step by step towards the goal of developing a legal regulation system or the relevant operations. At the same time, researchers note that direct centralized regulation of a decentralized asset, which is crypto currencies, meets with insurmountable challenges [23].

5 Conclusion

Summarizing positions of international financial rulemakers, it should be noted that they all profess a very cautious approach to operations with crypto currency. As Torkunov points out, such a general scientific method as an experiment is not available to the science of international relations [26]. It seems that the emergence of crypto currencies is seen as an inevitable evil, which is currently not very noticeable because of the lack of dissemination of relevant transactions. As experts note, expectations of benefits from operations based on the Blockchain technology may be overstated, and obsessive calls (hype) to use it will eventually subside, and real benefits will be clearer [27]. It is obvious that the main trend of regulating crypto assets today is to put crypto currency operations under the supervision of regulators, deanonymize operations with them, to introduce licensing, reporting, capital adequacy and liquidity requirements for professional market participants. For example, in January 2020, the district court of New York ordered the company "Telegram" to provide a report on financial transactions for the placement of tokens, and the company's founder Pavel Durov was required by the U.S. Securities and Exchange Commission to testify in this case [18]. This tendency will deprive the operations with virtual currency of the main advantage that ensures their attractiveness, namely, non-control and decentralization. Time will tell whether this trend will be implemented.

Acknowledgments The author thanks Vladislav Ponamorenko, Associate Professor of the Department of Public Law of the Russian Foreign Trade Academy for valuable recommendations on the development of the topic of this article.

References

1. Blandin A, Cloots AS, Hussain H, Rauchs M, Saleuddin R, Allen JG, Zhang BZ, Cloud K (2018) Global cryptoasset regulatory landscape study. University of Cambridge, Cambridge
2. Council Directive 2006/112/EC of 28 November 2006 on the common system of value added tax. OJ L 347. https://eur-lex.europa.eu/eli/dir/2006/112/oj. Accessed 29 Feb 2020
3. Decree of the President of the Republic Belarus №8 from December 21, 2017. On the digital economy development. http://president.gov.by/ru/official_documents_ru/view/dekret-8-ot-21-dekabrja-2017-g-17716/. Accessed 29 Feb 2020
4. Directive (EU) 2018/843 of the European Parliament and of the Council of 30 May 2018 amending Directive (EU) 2015/849 on the prevention of the use of the financial system for the purposes of money laundering or terrorist financing, and amending Directives 2009/138/EC and 2013/36/EU, OJ L 156. https://eur-lex.europa.eu/legal-content/EN/TXT/?uri=CELEX%3A32018L0843. Accessed 29 Feb 2020
5. EBA (2014) Opinion of virtual currencies, EBA/Op/2014/08, 4 July 2014. https://eba.europa.eu/sites/default/documents/files/documents/10180/657547/81409b94-4222-45d7-ba3b-7deb5863ab57/EBA-Op-2014-08%20Opinion%20on%20Virtual%20Currencies.pdf?retry=1. Accessed 29 Feb 2020
6. EBA (2019) Report, 9 January 2019. Report with advice for the European Commission on cryptoassets. https://eba.europa.eu/eba-reports-on-crypto-assets. Accessed 29 Feb 2020
7. ECJ (2015) Judgment of the Court of 22 October 2015 Skatteverket v David Hedqvist, Case C-264/14, Published in the electronic reports of cases, ECLI:EU:C:2015:718. http://curia.europa.eu/juris/document/document.jsf?docid=170305&doclang=EN. Accessed 29 Feb 2020
8. Eurasian Economic Commission (2019) Cryptocurrencies and blockchain as attributes of the new economy. http://www.eurasiancommission.org/ru/nae/news/Pages/22-07-2019-1.aspx. Accessed 29 Feb 2020
9. European Central Bank (2012) Virtual currency schemes. https://www.ecb.europa.eu/pub/pdf/other/virtualcurrencyschemes201210en.pdf. Accessed 29 Feb 2020
10. European Central Bank (2019) Crypto-Assets: Implications for financial stability, monetary policy, and payments and market infrastructures, No 223, May 2019. https://www.ecb.europa.eu/pub/pdf/scpops/ecb.op223~3ce14e986c.en.pdf. Accessed 29 Feb 2020
11. Expert Group (2019) 30 recommendations on regulation, innovation and finance. Final report to the European Commission. https://ec.europa.eu/info/sites/info/files/business_economy_euro/banking_and_finance/documents/191113-report-expert-group-regulatory-obstacles-financial-innovation_en.pdf. Accessed 29 Feb 2020
12. FATF (2014) Report virtual currencies key definitions and potential AML/CFT Risks, June 2014. https://www.fatf-gafi.org/media/fatf/documents/reports/Virtual-currency-key-definitions-and-potential-aml-cft-risks.pdf. Accessed 29 Feb 2020
13. FATF (2019) Guidance for a risk-based approach to virtual assets and virtual asset service providers. FATF, Paris
14. FATF (2019) International standards on combating money laundering and the financing of terrorism & proliferation. FATF, Paris
15. Financial Stability Board (2018) Crypto-asset markets potential channels for future financial stability implications. https://www.fsb.org/wp-content/uploads/P101018.pdf. Accessed 29 Feb 2020
16. Financial Stability Board (2019) Crypto-assets regulators directory. https://www.fsb.org/2019/04/fsb-publishes-directory-of-crypto-assets-regulators/. Accessed 29 Feb 2020
17. Foley S, Karlsen JR, Putnins TJ (2018) Sex, drugs, and bitcoin: how much illegal activity is financed through cryptocurrencies? Rev Financ Stud 32(5):1798–1853
18. Fomin D (2020) Telegram ICO. Date and place set. RBC. https://www.rbc.ru/crypto/news/5e12fea69a79477c542e6d40. Accessed 29 Feb 2020
19. IOSCO (2019) Issues, risks and regulatory considerations relating to crypto-asset trading platforms, consultation report, CR02/2019. https://www.iosco.org/library/pubdocs/pdf/IOSCOPD627.pdf. Accessed 29 Feb 2020

20. Katarzyna C (2019) Cryptocurrencies: opportunities, risks and challenges for anti-corruption compliance systems. Warsaw school of economics. OECD. http://www.oecd.org/corruption/integrity-forum/academic-papers/Ciupa-Katarzyna-cryptocurrencies.pdf. Accessed 29 Feb 2020
21. Kollmann J (2019) The VAT treatment of cryptocurrencies. EC Tax Rev 3:164–170
22. Ministry of Finance, Japan (2019) Communiqué G20 Finance Ministers and Central Bank governors Meeting Fukuoka. http://www.g20.utoronto.ca/2019/2019-g20-finance-fukuoka.html. Accessed 29 Feb 2020
23. Nabilou H (2019) How to regulate bitcoin? Decentralized regulation for a decentralized cryptocurrency. Int J Law Inf Technol 27:266–291
24. Security Council Resolution 2462 of 28 March 2019 S/RES/2462. https://undocs.org/ru/S/RES/2462(2019). Accessed 29 Feb 2020
25. Sukhodolov AP, Ivantsov SV, Sidorenko EL, Spasennikov BA (2018) Street-level corruption in Russia: basic criminological parameters. Russ J Criminol 12(5):634–640. https://doi.org/10.17150/2500-4255.2018.12(5) (in Russian)
26. Torkunov AV (2019) International studies: chaos or pluralism? Polis Polit Stud 5:7–18. https://doi.org/10.17976/jpps/2019.05.02
27. Tredinnick L (2019) Cryprocurrencies and the blockchain. Bus Inf Rev 36(1):39–44

Printed in the United States
by Baker & Taylor Publisher Services